AAPG Studies in Geology, No. 42
and SEG Geophysical Developments Series, No. 5

# Applications of 3-D Seismic Data to Exploration and Production

Edited by
**Paul Weimer**
and
**Thomas L. Davis**

AF180760

Published jointly by
The American Association of Petroleum Geologists
and
The Society of Exploration Geophysicists
Tulsa, Oklahoma, U.S.A.
Printed in the U.S.A.

## American Association of Petroleum Geologists

Association Editor: Kevin T. Biddle
Science Director: Richard Steinmetz
Publications Manager: Kenneth M. Wolgemuth
Special Projects Editor: Anne H. Thomas
Production: Custom Editorial Productions, Inc., Cincinnati, Ohio

**This and other AAPG publications are available from:**

**AAPG Bookstore**
P.O. Box 979
Tulsa, OK 74101-0979
USA
Tel (918) 584-2555; (800) 364-AAPG (USA—book orders only)
Fax (918) 560-2652; (800) 898-2274 (USA—book orders only)

**Geological Society Publishing House**
Unit 7, Brassmill
Enterprise Centre
Brassmill Lane
Bath BA1 3JN
United Kingdom
Tel 0225-445046
Fax 0225-442836

**Australian Mineral Foundation**
AMF Bookshop
63 Conyngham Street
Glenside
South Australia 5065
Australia
Tel. (08) 379-0444
Fax (08) 379-4634

**Canadian Society of Petroleum Geologists**
#505, 206 7th Avenue Southwest
Calgary, Alberta T2P 0W7
Canada
Tel (403) 264-5610

Copyright © 1996
By the American Association of Petroleum Geologists
All Rights Reserved
Published October 1996
ISBN: 0-89181-050-1

THE AMERICAN ASSOCIATION OF PETROLEUM GEOLOGISTS (AAPG) DOES NOT ENDORSE OR RECOMMEND ANY PRODUCTS OR SERVICES THAT MAY BE CITED, USED, OR DISCUSSED IN AAPG PUBLICATIONS OR IN PRESENTATIONS AT EVENTS ASSOCIATED WITH AAPG.

AAPG grants permission for a single photocopy of an item from this publication for personal use. Authorization for additional copies of items from this publication for personal or internal use is granted by AAPG provided that the base fee of $3.00 per copy is paid directly to the Copyright Clearance Center, 222 Rosewood Drive, Danvers, Massachusetts 01923. Fees are subject to change. Any form of electronic or digital scanning or other digital transformation of portions of this publication into computer-readable and/or transmittable form for personal or corporate use requires special permission from, and is subject to fee charges by, the AAPG.

## Society of Exploration Geophysicists

SEG Editor: Sven Treitel
Geophysical Developments Series Editor: Leon Thomsen

**This and other SEG publications are available from:**

**Society of Exploration Geophysicists**
P.O. Box 702740
Tulsa, OK 74170-2740
USA
Tel (918) 497-5500
Fax (918) 497-5558

AAPG and SEG
wish to thank the following
for their generous contributions
to

***Applications of 3-D Seismic Data to Exploration and Production***

**Major Corporate Contributors**

CGG, France

Landmark Graphics Corp.

Schlumberger Geco-Prakla

Western Geophysical

**Other Corporate Contributor**

Seismic MicroTechnology

Contributions are applied against the production costs of publication, thus directly reducing the book's purchase price and making the volume available to a greater audience.

# About the Editors

## Paul Weimer

Paul Weimer is an associate professor in the Department of Geological Sciences at the University of Colorado at Boulder and serves as the Director of the Energy and Minerals Applied Research Center. He received his B.A. from Pomona College (Claremont, CA) in 1978 with honors in geology. He received his M.S. in geology in 1980 from the University of Colorado at Boulder and his Ph.D. from The University of Texas at Austin in 1989. During the summer of 1978, he worked as a geologist with Amoco Production Company (Denver, CO). From 1980 to 1984, he worked as an exploration geologist with Sohio Petroleum Company (San Francisco, CA) doing regional exploration of the North Slope of Alaska. He worked with Mobil Oil Corporation (Dallas, TX) as a research and exploration geologist from 1988 to 1990. He co-edited four other volumes: *Alaskan North Slope Geology* (SEPM-Pacific Section), *Seismic Facies and Sedimentary Processes of Submarine Fan and Turbidite Systems* (Springer-Verlag), *Siliciclastic Sequence Stratigraphy* (AAPG Memoir 58), and *Submarine Fans and Turbidite Systems* (SEPM-Gulf Coast Section). He has authored more than 50 papers.

Paul's research interests are in applied sequence stratigraphy and basin analysis, petroleum geology of turbidite systems, 3-D seismic interpretation, and reservoir geology. He received the 1992 J. C. "Cam" Sproule Award from the AAPG. He is a member of AAPG, SEPM, GSA, EAPG, SEG, and several other professional societies.

## Thomas L. Davis

Tom Davis is currently Professor of Geophysics at the Colorado School of Mines (CSM) and Director of the Reservoir Characterization Project. He began his career in geophysics at the University of Saskatchewan in the late 1960s as a singular student in the geophysics option of the geological engineering program. Following graduation with a B.E. in 1969, he went on to complete an M.S. in geophysics at the University of Calgary (1971) and a Ph.D. at Colorado School of Mines (1974). His career in the oil patch began when he worked as a summer hire for Amoco (1968) and Chevron (1969–1970) and then as a permanent employee of Amoco (1971–1972).

Except for a 3 1/2 year sojourn to the University of Calgary (1977–1979), Tom has been a faculty member of the Department of Geophysics at the Colorado School of Mines since 1974. He is dedicated to his students and currently teaches courses on seismic interpretation and integrated exploration and development. Since 1980 he has team-taught integrated, multidisciplinary courses at Mines. Having dedicated his career to education and research, he still takes time to consult and lecture worldwide and to watch the wheat grow in Saskatchewan.

Tom joined SEG in 1971 and has taught extensively in SEG's continuing education program from 1980–1994. He served as Second Vice President during 1989–1990 and was SEG's Spring 1995 Distinguished Lecturer. He is a member of the European Association of Geoscientists and Engineers, AAPG, and a host of other professional societies.

# Contents

# Foreword

The impact of 3-D seismic reflection data in the petroleum industry has been truly significant during the past three decades. Theoretical studies of the late 1960s and early 1970s have evolved into a technology that has significantly changed the petroleum industry. 3-D seismic data are now an affordable technology that can be utilized by major companies and independent petroleum geologists. Nearly all issues of the *AAPG Explorer* or *The Leading Edge* feature some aspect of 3-D seismic data and their significance to the petroleum industry. But, in spite of this technology's great impact on the geosciences community, there are still relatively few published case studies clearly documenting its importance.

The purpose of this atlas is to present 30 case studies demonstrating 3-D seismic interpretation and its confirmation from drilling and production. The atlas includes examples from around the world, including both exploration and production applications of 3-D seismic data. Topics included are: (1) the 3-D seismic stratigraphic expression of features (such as carbonate buildups, fluvial sandstones, deltas, turbidite systems), (2) the 3-D seismic expression of different structural styles (such as growth faults, salt features, basement-involved faulting, reverse and thrust faulting, wrench faults), (3) case studies that illustrate how 3-D seismic data changed production strategies in field development, (4) how 3-D seismic data have impacted the exploration of individual prospects, (5) case studies emphasizing the integration of petrophysical data with 3-D seismic data, and (6) case studies involving the prediction of petrophysical properties from 3-D seismic data.

This atlas is unique for several reasons. First, it is the first joint publication of the AAPG and SEG since *AAPG Memoir 15/SEG Special Publication # 10, Stratigraphic Oil and Gas Fields* was published during 1972. Second, the atlas is designed to complement the highly successful *AAPG Memoir 42, Interpretation of Three-Dimensional Seismic Data,* written and edited by Alistair Brown (3 editions published), which contained several case studies. Third, because 3-D seismic is best done in color, this atlas is extremely expensive to print. AAPG and SEG have solicited funds from several companies to help defray the printing costs of this atlas, making it more affordable for their members. And fourth, this is the only publication of its kind.

The atlas developed from conversations in the AAPG Geophysics and Integration Committee during the spring of 1992, chaired by Ray Thomassen. This atlas exists because many people freely gave their time and ideas to making this project work. During the planning and compilation of this volume, several meetings were convened in various cities between the AAPG and SEG, SPWLA, and SPE. We thank the following for their help and input in making this atlas a viable project: AAPG representatives—Ray Thomassen, Gary Howell, Fred Dix, Ken Wolgemuth, Anne Thomas, Cathleen Williams, Susan Longacre, Kevin Biddle, Ken Frakes; SEG representatives—Ken Brenkman, William Brumbaugh, Pat Lindsey, John Hyden, Don Steeples; SPE representative—Craig Van Kirk; and SPWLA representatives—Eric Pasternack, Steve Prensky. We are grateful to the following individuals for helping us to solicit manuscripts: Ray Buchanan, Robert Graebner, H. P. Mohr, and Woody Nestvold. We especially extend our thanks to Bob Hardage for his help in editing the final manuscripts for the atlas. A special thanks to Mark Chapin, Shell Offshore, and Shell Development Company for permission to use the figures on the front cover.

In addition to the many planning meetings, advertisements soliciting papers were published in national and local society newsletters. We sent solicitation letters for contributions to the atlas to about 1000 people . We convened a poster session at the 1994 AAPG National Meeting in Denver with interactive displays. Still, we found publishing 3-D seismic data to be problematic. We concluded that everyone wants to see 3-D seismic studies, but most companies are reluctant to share their examples in print. We hope that in the future, as the technology matures, many more case studies will be published, especially integrating the reservoir engineering information that is so critical to the integrated studies. In the near future, CD-ROM and interactive displays are critical to properly see the information that we print here.

We thank the authors for their patience, perseverance, and dedication with their manuscript preparation. We are extremely grateful to Colleen Viele, Peter Varnai, and Jennifer Crews for their assistance during various stages of the atlas preparation. Finally, we thank Laurie, Lou, Rudy, Pat, Ryan, Kiefe, and Brett for their forbearance and patience during the preparation and assembling of this atlas.

We feel that this atlas is an important step in the continuing effort to document 3-D seismic interpretation and its efficacy in petroleum geology. We hope that this volume serves the petroleum geology community well and helps people find and develop reservoirs more effectively.

" so there ain't nothing more to write about, and I am rotten glad of it, because if I'd a knowed what a trouble it was to make a book I wouldn't a tackled it and I ain't a going to no more... "

Huck Finn

Paul Weimer
AAPG Representative
University of Colorado

Tom Davis
SEG Representative
Colorado School of Mines

Boulder-Golden, Colorado
December 1995

**OVERVIEW**

Nestvold, E. O., The impact of 3-D seismic data on exploration, field development, and production, *in* P. Weimer and T. L. Davis, eds., *AAPG Studies in Geology No. 42* and *SEG Geophysical Developments Series No. 5*, AAPG/SEG, Tulsa, p. 1–8.

# The Impact of 3-D Seismic Data on Exploration, Field Development, and Production

E. O. Nestvold*


## ABSTRACT

Three-dimensional (3-D) seismic surveys have proved to be powerful tools for imaging the subsurface since their introduction in the mid-1970s. Today, 3-D seismic surveys demonstrate high cost/benefit ratios by reducing dry-hole risk and by providing better well placement for flow rates and drainage. Additional benefits include improved reserves estimates and shorter cycle times for appraisal and development project planning. Also, old fields with declining production profiles are being rejuvenated. 3-D seismic surveys are revolutionizing the geophysical industry, with far-reaching effects on the exploration and production business worldwide. With improvements provided by substantial investments in technology, 3-D seismic surveys will play an even more significant role in field development and production strategies in the years to come.


## INTRODUCTION

The significance of 3-D seismic data in the petroleum industry during the past decade cannot be overstated. What started as a technology too expensive to be utilized except by major oil companies is now routinely used by independent operators in the United States and Canada. As with all emerging technologies, documentation of successes has been limited. There are some, however, which will be cited in this paper when appropriate. The purposes of this paper are (1) to review some significant trends in activity as 3-D acquisition has grown and come to dominate, (2) to review some examples of 3-D surveys, (3) to discuss developing technology trends, (4) to summarize industry experience with 3-D imaging—especially rejuvenation of mature fields with declining production profiles—and, (5) to give my view of the impact of 3-D technology. This paper is designed to complement and update previous work (Nestvold and Nelson, 1992a, b; Nestvold, 1992).

## WORLDWIDE MARINE GEOPHYSICAL EXPENDITURES

I will start by looking at the recent history of marine geophysical expenditure (Nestvold and Jack, 1994, 1995; Jack and Nestvold, 1994). The trend in global spending on marine seismic acquisition closely followed the oil price crash in the mid-1980s to a low of only $500 million in 1988. However, expenditures then rebounded strongly to a peak of $1.4 billion in 1992, nearly the same as the previous expenditure peak in the early 1980s. The late 1980s and early 1990s were the boom times for marine 3-D seismic; expenditures grew at a rate of 20% per annum (Farestveit, 1992; Morgan, 1992).

However, 3-D seismic acquisition expenditures have declined from 1993 to 1995. In fact, 1993 and 1994 were terrible years for the marine geophysical service industry, as can be appreciated, for example, from the demise of a major seismic contractor, Halliburton Geophysical Services.

Three-dimensional marine acquisition may have peaked in the Gulf of Mexico and the North Sea for the time being, because large proportions of prospective acreage have been surveyed. Acquisition capacity in terms of numbers of available streamers is increasing in these and other less mature offshore areas. Thus, the market for first-time surveys is expected to decrease in many areas.

On the other hand, surveys will be repeated in many basins when significant improvements in resolution can be demonstrated. The historical seismic survey repeat rate based on technology advances has been up to ten years.

But marine technology is dynamic. There is a change of focus from appraisal to more production-oriented activities. Thus, monitoring reservoir production will become increasingly important, with surveys being repeated at much shorter intervals. This activity is called 4-D seismic surveying. For example, fixed ocean-bottom sensors have been installed in BP's Foinaven field west of Shetlands (Shirley, 1995). Also, there is a trend into exploration 3-D activity. Large speculative surveys are now available in the Gulf of Mexico and the North Sea as geophysical companies are becoming more creative in marketing their services.

## GROWTH LAND SURVEYS

The history of 3-D activity on land is very different (Nestvold, 1994; Nestvold and Jack, 1994, 1995; Jack and Nestvold, 1994). Also, worldwide activity levels for land are much harder to come by. Estimates can be made, however, based on the activity of a major seismic service company—Geco-Prakla. That company began acquiring land 3-D seismic surveys in the mid-1970s, but activity remained at a relatively low level until the early 1980s, when activity started to accelerate. Although interrupted by the oil price crash, that company's land activity has grown steadily during the last ten years, at an average annual rate greater than 30%.

In total, Geco-Prakla acquired more than 51,000 $km^2$ (20,000 $mi^2$) of 3-D data through 1993. Based on historical annual proportional crew counts, we estimate that this represents between one-fourth and one-fifth of cumulative land 3-D acquisition world-wide, excluding the CIS, China, and India.

Compared to marine use, 3-D activity on land has been much less common, but the potential market is much larger. In North America, for example, activity throughout the latter part of the 1970s and the 1980s consisted mainly of small-field "postage-stamp" surveys of a few square miles in areal extent. It was not until the early 1990s that the tremendous power of 3-D seismic imaging as a tool to map the subsurface became known to many oil and gas operators and also to investors. As a result, there has been an onshore boom in its use, particularly in West Texas, Louisiana, and Alaska. Activity will grow when land 3-D seismic use gains the same wide acceptance in the oil companies as has marine 3-D seismic acquisition, and this depends on efficiency and unit-cost improvements.

## EXAMPLES

### COMPARISON 3-D VERSUS 2-D (ALASKA)

What is 3-D seismic imaging? It is a tool which allows us to correctly see into the earth—for the first time.

The image in Figure 1, presented by Speers and Dramgoole (1992) and Baird (1992), is that of the top reservoir of BP and Arco's Kuparuk field in Northern Alaska. The two-way traveltime surface for the top reservoir event is shown as color dipping down to the northeast from red to blue. The image is also artificially illuminated from the northeast and displayed as intensity which accentuates changes in dip and azimuth, highlighting the dense fault pattern within the reservoir.

Notice that a 3-D survey is slotted into the center of the image. Notice also that the image incorporates the widely spaced 2-D seismic data acquired in the areas surrounding the field. The difference between the level of detail in the 3-D dataset in the center of the image and in the surrounding 2-D data is dramatic. For example, one does not expect that the small-scale faulting would stop suddenly at the edge of the 3-D survey. To recognize the true degree of reservoir complexity from the 2-D seismic data alone is simply not possible.

This, to me, is a very clear illustration of the power of 3-D imaging relative to that of 2-D geometry!

Due to reservoir compartmentalization, a significant infill drilling program is needed to maximize recovery in Kuparuk Field. Speers and Dramgoole (1992) report that this 3-D survey has helped position wells at optimum locations by confirming the identity of small-scale faulting and by helping to resolve areas of questionable fault

*IBM Corp., Two River Way, Houston, TX 77056

continuity. The economic return from this 3-D survey for BP and Arco is immediate, because wells that are drilled tend to be successful and those that would have been marginal, at best, are not drilled.

### RESERVOIR QUALITY EXPRESSED BY SEISMIC AMPLITUDE (CANADA)

The second example is from a 3-D survey over a field in Saskatchewan. Figure 2 is a vertical section from this survey. The strong event in the lower third of the section is the Devonian horizon.

One can also see well tracks with annotated depth values and a variety of logs. Note that one well on this section penetrated the Devonian, but the others were targeted for the shallower Waseca objective. Two of the wells were drilled horizontally into the Waseca reservoir.

First, the reflection time of the Devonian was interpreted. Then amplitude extraction was started with the Devonian horizon. A data cube was formed with successive amplitude extractions shallower and parallel to the Devonian horizon.

Figure 3 shows the seismic amplitude of the reservoir horizon, the Waseca, which is about 100 ms shallower than the Devonian.

Red depicts the highest amplitude. The lower amplitudes, in shades of gray, reveal major channel features. Well tracks are also shown (in plan view), with the surface location depicted by a derrick symbol. One can see that wells have been drilled into the low amplitudes in the channels. Clearly, this reservoir quality indicator on the 3-D seismic amplitudes has been used to target these wells.

### POREFILL EXPRESSED BY SEISMIC AMPLITUDE (GULF OF MEXICO)

The vertical section in Figure 4 is from a field in offshore Louisiana, displayed four different ways. The upper left profile is variable density, the upper right is reflection strength, the lower left is the cosine of instantaneous phase, and the lower right is reflectivity displayed as wiggle traces.

The strong horizon reflection is the J sand. Bright spots and flat spots are clearly visible. In fact, the response on the left displays a classical vertical amplitude profile through a hydrocarbon accumulation. Since flat spots represent positive impedance contrasts for downgoing waves, they also allow one to establish the polarity of the seismic section. One can also establish the relative impedance of the sealing shale and the sand below it. In this example, the sand is lower in impedance than the sealing shale because the polarity of the top reservoir is the same down flank (below the water contact) as it is at the crestal bright spot.

Also, with representative petrophysical information, one can identify the porefill to a high degree of confidence.

## TECHNOLOGY DEVELOPMENTS

Four technological developments have contributed significantly to the wide acceptance of 3-D seismic data during the past decade. These are:

(1) multi-streamer, multi-source, multi-vessel 3-D marine technology;
(2) onboard processing of navigation and seismic data;
(3) depth migration—now coming into its own, principally in the Gulf of Mexico—and
(4) workstation technology.

### MULTI-ELEMENT MARINE ACQUISITION

During the last decade, large investments have been made specifically in new surveying equipment as a total of 20 new vessels have been launched or upgraded with new equipment by a number of seismic contractors, such as Western, PGS, Geco-Prakla and CGG. As a result of these advances in acquisition, the production rate for 3-D marine surveys has increased dramatically, with a corresponding decrease in unit cost (Farestveit, 1992). This investment contributed to the boom times for 3-D marine seismic surveying, particularly in the Gulf of Mexico and the North Sea.

The vessel shown in Figure 5 is a typical acquisition boat for 3-D surveys. The Geco Emerald represents a capital investment of more than $30 million.

The first 3-D marine surveys, in the late 1970s, were acquired by vessels equipped with a single streamer and a single source group. During the boom times of the late 1980s and early 1990s, technology developments by service companies, encouraged by client companies, resulted in a dramatic increase in the number of streamers deployed (Farestveit, 1992). For example, the Geco-Prakla quad-quad system uses two vessels, each with two streamers and two source groups, to acquire a wide swath of 16 lines during a single pass.

The number of streamers deployed continues to increase. In fact, by early 1995, three vessels acquired the capability to tow up to 12 streamers. These vessels have been used largely in the North Sea and the Gulf of Mexico.

The wider swaths of equipment have reduced unit costs significantly. Equally important is the fact that survey acquisition time has decreased dramatically, which contributes significantly to shorter project cycle times.

### SEISMIC PROCESSING ONBOARD MARINE VESSELS

Until the early 1990s, the increased complexity of recorded data, particularly navigation data, introduced delays in processing. These delays have been largely eliminated by onboard processing, as illustrated in Figure 6.

Onboard processing, as well as reducing time and cost, has actually increased quality in the acquisition. Practical onboard processing objectives are: quality control of acquisition and finalization of navigation; optimization of program; seismic and navigation data merging; trace editing; and initial seismic processing. Realization of these objectives eliminates many of these tasks from the onshore processing program.

However, there is an important additional advantage. Onboard processing delivers immediately useful products, such as a reduced fold 3-D volume and a fully processed 2-D subset of the 3-D volume. In the future, complete onboard 3-D processing will be practical, given hardware and software developments now underway. For some time these have included Unix-based systems with single processors from Sun and IBM. Currently, massively parallel systems (IBM SP2 scalable parallel Unix machines) are being installed on board a number of seismic vessels.

### 3-D DEPTH IMAGING

3-D depth imaging with massively parallel processors (MPP) is fundamental for the new pre-salt play in the Gulf of Mexico (Weber and Ratcliff, 1994). MPPs provide the capability to attack massively complex problems like 3-D depth migration.

These systems are now commercially available. Their importance has been highlighted by the 1993 sub-salt discovery (Mahogany) by Phillips, Amoco, and Anadarko in the Gulf of Mexico (George, 1994a, b), which kicked off a new play. In this play, pre-stack depth migration is necessary because the objective structures lie below the salt, where they cannot be imaged properly with conventional 3-D time migration.

Figure 7 shows a velocity model together with migrated data. The velocity model is depicted by the colored planes that intersect the velocity data cube. Blue means low- and red means high-velocity values. The white surfaces are isosurfaces enclosing salt velocity values; the horizontal plane cutting the far salt body shows high-velocity (red) values inside the body. The gray horizontal plane is another isosurface; it is perfectly flat because of the simple linear vertical gradient of the velocity field around the salt.

Figure 7 emphasizes the essential role played by the velocity model in 3-D depth migration. All seismic migration algorithms depend on the velocity model that is used, because it is the velocity model that determines the vertical and lateral positioning of reflectors and fault planes after depth migration. The skill of the interpreter/processor in using all clues from nearby well control and seismic diagnostic data to lock in to the proper velocity model (usually after several iterations) determines the success of the process.

### WORKSTATION TECHNOLOGY

The fourth technology advance that has contributed to the explosion of 3-D seismic applications is the use of workstation technology. In the early days of 3-D seismic development (until the early 1980s), interpreters were very inventive in the manipulation of paper sections. I remember one such display, called the "accordion display," which consisted of several sections from a 3-D survey glued together in such a way that they could be folded and unfolded to provide an overview of several lines at a time.

Today, workstations are ubiquitous. One cannot imagine interpreting 3-D seismic data without using them, as is clearly illustrated in the papers in this volume. In addition to straightforward structural interpretation, workstations are also increasingly used to compute and display seismic attributes, for both structural and stratigraphic features (Dalley et al., 1989; Stark, 1995; Bahorich and Farmer, 1995).

## INDUSTRY EXPERIENCE

### SHELL GROUP COMPANIES

The power of 3-D seismic technology is understood and appreciated at the highest levels in the Shell Group (see, for example, Moody-Stuart, 1993). In that article, which is a published version of the keynote address he gave at the SEG convention in Moscow the previous July, Mark Moody-Stuart said that he was convinced that it would be very difficult for him to overstate the importance of 3-D seismic imaging, which, in his opinion, could really be appreciated only by those who have worked with it.

Evidence of reserves additions achieved in several projects have been published (see, e.g., Nestvold, 1987, 1991, 1992; Nestvold et al., 1991; Nestvold and Nelson, 1992a, b; Pink, 1992). Increasingly, these additions result from large "fairway" 3-D surveys in highly prospective but largely undrilled areas, as well as surveys around producing fields.

In Nigeria, for example, the first four 3-D surveys (850 sq km) acquired in the swamps during 1985 through 1987 cost $28 million. However, added reserves (proved by 10 appraisal wells) were 400 million barrels. Thus, the 3-D seismic cost per added barrel was 7¢! See Nestvold (1992) and Nestvold and Nelson (1992b).

In the Netherlands, the Groningen field was discovered by drilling the Slochteren 1 well in 1959. During the 1980s, extensive (fairway) 3-D coverage was acquired onshore. This paid off with the discovery of two new fields, Grijpskerk and Munnekezijl (Kruijs and Donze, 1993). In fact, sub-salt bright spots on 3-D seismic data have proved to be indicators of gas trapped in Rotliegend sand reservoirs below the Zechstein salt.

The confidence Shell companies have in 3-D seismic data can be judged from the steady growth in the amount of 3-D seismic activity; Rijks (1994) reported that Shell companies' 3-D activity peaked at more than 20,000 km$^2$ in 1992.

Moreover, the swing from 2-D to 3-D data is amply illustrated by the fact that relative expenditure on 3-D seismic surveys had risen markedly since 1979. In 1992, some 90% of Shell Companies' seismic expenditure was invested in 3-D surveys (Rijks 1994).

## EXXON WORLDWIDE

Recently, Exxon recounted its experience with 3-D seismic data and presented its perspective on the technology (Johnson et al., 1992, 1993; Greenlee et al., 1994). From the late 1970s through 1993, Exxon had participated in some 400 3-D seismic surveys in 12 countries. Exxon's share of the expenditures for these surveys was $620 million. Performance highlights include a 70% exploration success rate using 3-D data in the Netherlands, based on 23 wells drilled between 1987 and 1992. Other successes include the UK offshore (47% exploration success rate, based on 38 wells 1990–1992), offshore Malaysia, and the Gulf of Mexico (70% exploration success rate for 44 wells located on basis of 3-D surveys from 1987 to 1992).

Exxon published thorough analyses of the economic results for two provinces. In the Gulf of Mexico they spent some $35 million (Exxon share) for 30 surveys. As a result, they added some 70 million barrels (Exxon share). In the Malay basin, 10 surveys cost $13 million (Exxon share). Here, the resulting added reserves totalled 100 million barrels (Exxon share).

The use of 3-D seismic surveys has led to addition of new reserves, drilling of fewer dry or marginal exploration wells, and optimization of the number and placement of delineating, development, and secondary recovery wells. These benefits, according to Exxon, are the result of superior structural definition, more detailed reservoir descriptions, reservoir fluid content characterizations, and quantitative interpretation methods.

The authors see 3-D seismic imaging as the technology offering the greatest possibility for improvement in Exxon's exploration and development performance. Furthermore, with 3-D the future is seen to be one of reduced exploration and development drilling risk coupled with more efficient recognition and recovery of reserves, in both new and mature ventures.

## SOME PRODUCING FIELDS REJUVENATED WITH 3-D SEISMIC TECHNOLOGY

In the early days of 3-D seismic technology use (1970s and early 1980s), surveys were commissioned largely to appraise a discovery, and then only over a small area surrounding the discovery well. Since then, the use of 3-D surveys has grown to include exploration surveys, on the one hand, and surveys over producing fields, on the other. Today, large exploration surveys are being commissioned in highly prospective areas which may have little or no well control. The rationale is that exploration targets in the survey area will be identified and mapped by the 3-D data, even though the location and size of these targets are not known beforehand.

Also in the early days, surveys were not commissioned over producing fields, on the grounds that an abundance of wells provided adequate subsurface control and 3-D seismic acquisition would constitute overkill. In recent years, that theory has been dispelled over and over again as operators have added significant value with 3-D surveys over producing fields. Several examples, listed in Table 1, are discussed below.

### STRATTON FIELD

The Stratton field, located in South Texas, is operated by UP Resources (Shirley, 1994). The field was discovered in 1937.

UP Resources and the Bureau of Economic Geology at The University of Texas at Austin acquired an 8-mi$^2$ (20-km$^2$) 3-D survey in fall 1992 (Levey et al., 1994; Hardage et al., 1994). In late 1993, a 55-mi$^2$ (140-km$^2$) 3-D survey was acquired covering almost all of the Stratton and Agua Dulce fields. The survey was shot in conjunction with Pennzoil.

For further details regarding the geological setting and the recent 3-D surveys, see the article by Hardage et al. in this volume.

The finding cost in the last set of wells drilled with 3-D data (including one mechanical failure) was $3 per BOE. As a result of the 3-D surveys and subsequent drilling, UP Resources estimated that there are some 180 wells yet to be drilled in the field.

### CHOCOLATE BAYOU FIELD

Chocolate Bayou Field is operated by Texas Meridian (Taylor, 1994; Pennington et al., 1994a, b). It was discovered in 1938. Located about 25 miles (40 km) southeast of Houston in Brazoria county, it has produced more than 2 tcf of gas and 65 million barrels of oil from 30 sands through 300 well bores. Most of the production has come from sands of the upper and middle Frio (Oligocene) section.

The license was offered by Phillips in 1991, together with a 3-D survey that was acquired in 1988. Texas Meridian noticed a separate, buried structure south of the main producing field.

The first two wells drilled by Texas Meridian were successful. The first found 65 ft (20 m) of gas pay, while the second located 115 ft (35 m). Production rates are in excess of 22 million cubic feet of gas and 300 barrels of oil per day.

**TABLE 1. Some Producing Fields Rejuvenated by 3-D Seismic Data Acquisition***

| Field | Disc | Wells pre-3D |
|---|---|---|
| Yibal | 1962 | 280 |
| Ship Shoal 113 | 1955 | 230 |
| Bay Marchand | 1949 | 800 |
| Grand Isle 16 | 1948 | 400 |
| Belle Isle | 1941 | >50 |
| Chocolate Bayou | 1938 | 300 |
| Stratton | 1937 | 400 |

*Associated with each field is the year of discovery and the number of wells drilled before the first 3-D survey was acquired.

### BELLE ISLE FIELD

Belle Isle field, located in St. Mary Parish, Louisiana, was discovered in 1941. It is operated jointly by Oryx and Texaco (Box, 1993; Shirley, 1994; LeLeux, 1994) and it has produced more than 1 tcf of gas from 54 reservoirs. More than 50 wells have been drilled.

In 1991, a 56-mi$^2$ (140-km$^2$) 3-D survey was acquired. When interpreted, the 3-D survey indicated several untested deep fault blocks and many additional opportunities in the shallow Miocene sands that were not apparent with 2-D seismic data.

The new 3-D data pointed to new drilling opportunities and workovers on existing wells. For example, in late 1993, a well on a newly identified structure was drilled to 17,500 ft (5330 m), and produced 2.5 MMCFGD. After including the wells that were drilled on the basis of the 3-D data, daily production increased from 2–3 MMCFGD to 14 MMCFGD (some 5–7 times).

### GRAND ISLE 16 FIELD

Operated by Exxon, Grand Isle 16 Field is located about 7 mi (11 km) off the Louisiana coast. The field was discovered during 1948 by Humble Oil and became Exxon's first offshore development. Grand Isle 16 is associated with a large salt dome and contains 12 active platforms with over 400 wells. The field has produced a total of about 270 million bbl of oil and 420 bcf of gas from Tertiary sandstones ranging in depth from 7,000 to 16,000 ft (2130 to 4880 m) (Johnson et al., 1992, 1993).

3-D seismic data were acquired despite the dense well control because of the geologic complexity associated with salt tectonism and the highly variable fluvial reservoir quality.

Exxon reports that the Grand Isle 16 Field has benefited in many ways from the 3-D survey, e.g., successful development wells were sited and Exxon continues to drill infill wells (Johnson et al., 1992, 1993; Shirley, 1993). Bypassed oil and gas pools, some stratigraphically trapped, were identified by the amplitude response of the porefill (bright spots) on the 3-D data.

### BAY MARCHAND FIELD

The Bay Marchand Field is located off Louisiana approximately 65–70 mi (105–110 km) south of New Orleans in state coastal waters. Water depths range from 10 to 50 ft (3 to 15 m). The first well was drilled in 1930 by Gulf. The Bay Marchand discovery well (well no. 13 drilled on the prospect) was drilled by Chevron in 1949 (Abriel et al., 1991).

The field covers some 45,000 acres. There are 50 productive Pliocene–Miocene sands in the interval between 1000 and 13,000 ft (300 and 3960 m). The field has produced some 580 million bbl, with an additional 580 million bbl of reserves. A total of 800 wells have been drilled; presently there are 165 active wells.

A 3-D survey was acquired during 1987; its objectives were to delineate new pools, to review drilled areas, and to assist in reservoir management. The result is that a number of new productive fault blocks were discovered. Since 1990, the 3-D survey has been used more and more to assist reservoir management by identifying bypassed oil.

Production steadily declined to a low of 18,000 bopd during 1985. By the early 1990s, production had increased to 32,000 bopd after drilling additional wells located by incorporating data from the 3-D survey.

## SHIP SHOAL 113 FIELD

Ship Shoal 113 Field was discovered during 1955. From then until 1988, some 230 wells were drilled, with 157 completions and 73 dry holes—a 68% success rate. Production from 1963 to 1988 was 80 million bbl and 305 bcf (Hurley et al., 1993; Bailey and Chassaniol, 1994). Oil and gas are produced from 70 sands from lower Pleistocene to upper Miocene in age in the depth range of 2500 to 14,000 ft (760 to 4270 m).

In 1987, an 80-mi$^2$ (205-km$^2$), or 2600 line mi (4160 line km), 3-D survey was acquired. In 1989, a second survey of 4500 line mi (7200 line km) was acquired to the north and east of the first.

Based on the two 3-D surveys, more than 30 wells have been drilled, with a 90% success rate.

Oil production rates that were declining have stabilized, with increasing contribution from the recent wells drilled on the basis of 3-D interpretation. Gas production rates have stabilized also. As a result of the 3-D surveys and subsequent drilling, estimated oil reserves have increased by 21% and gas reserves by 30%.

## YIBAL FIELD, OMAN

The Yibal Field in Oman was discovered during 1962 and has been on stream since 1969 (Nestvold, 1991, 1992; Holleman, 1992; Pink, 1992). Oil is trapped in a faulted, domal structure of 130-m (425-ft) vertical closure. The producing reservoir is the Shuaiba carbonates. A 3-D survey of nearly 160 km$^2$ was acquired during 1988 at a cost of some US $2.2 million. One of its objectives was to optimize Shuaiba infill locations by avoiding water-conducting faults in the shallower Natih formation. By the time the survey was available for interpretation, some 280 wells (producers and injectors) had been drilled.

As confidence in the amplitude interpretation grew within Petroleum Development Oman, the team became bolder in targeting wells based on seismic amplitudes. Over 100 wells were drilled in two years, and early water breakthrough in producers has been cut by two-thirds.

Recently, Shell (Rijks, 1994) reported that cumulative production prior to the 3-D survey was 700 million bbl. At that time the remaining reserves were estimated to be some 560 million bbl. After incorporating the 3-D interpretation and taking into account results from the subsequent wells, the estimated remaining reserves increased (due to improved recovery) to some 820 million bbl, an increase of some 260 million bbl.

Other benefits of the 3-D survey include fewer wells and smarter wells, as well as reduction of facilities costs and operating costs.

A second 3-D survey was acquired at the end of 1993 (Rijks, 1994).

## IMPACT OF 3-D SEISMIC TECHNOLOGY

3-D seismic data have had a profound effect on the E&P business worldwide. This is because 3-D seismic images provide a clear definition of the subsurface for the first time.

Oil companies are looking for low-risk ways to increase reserves. The impact of 3-D seismic technology has been to increase reserves and, in so doing, to reduce finding costs (for exploration) and development costs (for production) by improving the drilling success rate dramatically. In fact, the impact of 3-D seismic data extends beyond drilling into location and sizing of production platforms (offshore) and surface production facilities (onshore).

Regarding field rejuvenation in particular, I have emphasized in this paper the success that has been enjoyed by operators that commission 3-D surveys over mature fields, even those with hundreds of wells and declining production profiles. This capability of 3-D imaging to rejuvenate producing fields is one of the exciting applications of the technology, and one that will become even more important, particularly on land, in the future.

Thus its impact ranges from exploration, through delineation, to development planning, and field rejuvenation.

Also, 3-D seismic data have had an important impact in integrating technologies, because all disciplines, i.e., geophysics, geology, and reservoir engineering, are beginning to use 3-D surveys as a basis for reservoir modeling during the life-history of each field. In many cases, as fields are produced, reservoir models will be updated continuously based on all the field data, including repeat 3-D surveys.

## REFERENCES CITED


Abriel, W. L., P. S. Neale, J. S. Tissue, and R. M. Wright, 1991, Modern technology in an old area: Bay Marchand revisited: The Leading Edge, v. 10, n. 6, p. 21–35.

Bahorich, M., and S. Farmer, 1995, The coherence cube: The Leading Edge, v. 14, n. 10, p. 1053–1058.

Bailey, J. H., and R. J. Chassaniol, 1994, Ship Shoal 113/South Pelto 20 Fields: Detection of Subtle traps utilizing 3-D technology and detailed development geologic practices: AAPG Annual Convention, Denver, AAPG Bulletin, v. 78, p. 831.

Baird, E., 1992, New technology in support of improved oil recovery: Offshore Northern Seas 1992, Stavanger, Norway, August 25–28, 1992.

Box, J. W., 1993, Reorganization at Oryx Energy focuses on teamwork, technology: Oil and Gas Journal, November 8, 1993, p. 54–61.

Dalley, R. M., E. C. A. Gevers , G. M. Stampfli, D. J. Davies, C. N. Gastaldi, P. R. Ruijtenberg, and G. J. O. Vermeer, 1989, Dip and azimuth displays for 3-D seismic interpretation: First Break, v. 7, p. 86–95.

Farestveit, A., 1992, Changing seismic and drilling roles in exploration markets: Ocean Industry, October 1992, p. 44–56.

George, D., 1994a, Technology that led to Mahogany driving global exploration of salt bodies: World Oil, January 1994, p. 30–34.

George, D., 1994b, Reliable subsalt surveying requires special 3-D acquisition techniques: World Oil, January 1994, p. 35–36.

Greenlee, S. M., G. M. Gaskins and M. G. Johnson, 1994, 3-D seismic benefits from exploration through development: an Exxon perspective: The Leading Edge, v. 13, n. 7, p. 730–734.

Hardage, B. A., R. A. Levey, V. Pendleton, J. Simmons, and R. Edson, 1994, A 3-D seismic case history evaluating fluvially deposited thin-bed reservoirs in a gas-producing property: Geophysics, v. 59, p. 1650–1665.

Holleman, N., 1992, Flood front detection with 3-D seismic in Yibal Field, Oman, 62nd SEG Annual Meeting, New Orleans, Expanded Abstracts, p. 1382.

Hurley, S. C., J. R. Murphy, and S. S. Mastoris, 1993, 3-D, CAEX help independent add reserves inexpensively in Ship Shoal area of Gulf: Oil and Gas Journal, January 18, 1993, p. 65–69.

Jack, I. G., and E. O. Nestvold, 1994, 3-D Seismic—the next step: PETEX 94, London, Conference Proceedings, p. 1.1–1.10.

Johnson, M. G., G. M. Gaskins, and S. M. Greenlee, 1992, 3-D seismic benefits from exploration through development: an Exxon perspective: 62nd SEG Annual Meeting, New Orleans, Expanded Abstracts, p. 261–264.

Johnson, M. G., G. M. Gaskins, and S. M. Greenlee, 1993, 3-D seismic benefits from exploration through development: an Exxon perspective: AAPG Annual Convention, New Orleans, Program, p. 125.

Kruijs, E., and A. Donze, 1993, Revival of the Northeast Netherlands onshore Rotliegende play: AAPG International Conference and Exhibition, The Hague, October 17–20, 1993, AAPG Bulletin, v. 77, p. 1638–1639.

LeLeux, D. E., 1994, What's happening in exploration: World Oil, April 1994, p. 25.

Levey, R. A., B. A. Hardage, and R. J. Finley, 1994, An integrated approach to identifying natural gas reservoir compartments using 3-D seismic, depositional facies analyses, and production tests: examples from Middle Frio fluvial reservoirs, Stratton Field, South Texas: AAPG Annual Convention, Denver, AAPG Bulletin, v. 78, p. 831.

Moody-Stuart, M., 1993, Opportunities from partnership: The Leading Edge, v. 12, n. 4, p. 269–273.

Morgan, N., 1992, Better technology brings costs down: Petroleum Economist, October 1992, p. 35–36.

Nestvold, E. O., 1987, The use of 3-D seismic in exploration, appraisal and field development: Proceedings of the Twelfth World Petroleum Congress (Houston), v. 2, p. 125–132.

Nestvold, E. O., 1991, 3-D seismic: is the promise fulfilled?: 61st SEG Annual Meeting, Houston, Expanded Abstracts, v. 1, p. 717–720.

Nestvold, E. O., 1992, 3-D seismic: is the promise fulfilled?: The Leading Edge, v. 11, n. 6, p. 12–19.

Nestvold, E. O., 1994, The 3-D seismic revolution: impact on development: SEG International Exposition and 64th Annual Meeting, Los Angeles, Expanded Abstracts, p. 1621.

Nestvold, E. O., Y. H. Eskes, P. H. H. Nelson, and E. J. H. Rijks, 1991, 3-D seismic in exploration, Proceedings of the Thirteenth World Petroleum Congress (Buenos Aires), v. 2, p. 289–292.

Nestvold, E. O., and I. G. Jack, 1994, 3-D Seismic—The next step: 7th Geophysical Congress of the Venezuelan Society of Geophysical Engineers (SOVG), Caracas, September 4–8, 1994.

Nestvold, E. O., and I. G. Jack, 1995, Looking ahead in marine and land geophysics—a conversation with Woody Nestvold and Ian Jack: The Leading Edge, v. 14, n. 10, p. 1061–1067.

Nestvold, E. O., and P. H. H. Nelson, 1992a, Explorers still hope to improve on 3-D seismic's wealth of data: Oil and Gas Journal, March 16, 1992, p. 55–61.

Nestvold, E. O., and P. H. H. Nelson, 1992b, 3-D seismic: the promise fulfilled: PETEX 92, London, Conference Proceedings, p. 75–79.

Pennington, A., C. Plant, and C. Davis, 1994a, 3-D seismic results in the discovery of significant reserves bypassed for 55 years: AAPG Annual Convention, Denver, AAPG Bulletin, v. 78, p. 835.

Pennington, A., C. Plant, and C. Davis, 1994b, 3-D seismic results in the discovery of significant reserves bypassed for 55 years in the Chocolate Bayou Field: SEG International Exposition and 64th Annual Meeting, Los Angeles, Expanded Abstracts, p. 1622.

Pink, M., 1992, Exploration and appraisal technology—maximising rewards by integration: Offshore Northern Seas 1992, Stavanger, Norway, August 25–28, 1992.

Rijks, E. J. H., 1994, The use of and experiences with 3-D seismic in Shell, VII

Venezuelan Geophysical Congress, Caracas, September 4-8, 1994.
Shirley, K., 1993, Seismic boosts Exxon: 3-D value shows on bottom lines: AAPG Explorer, v. 14, n. 6, p. 18–19, 28.
Shirley, K., 1994, Independents find 3-D seismic improves Gulf Coast odds: The American Oil and Gas Reporter, v. 37, n. 4, p. 65–72.
Shirley, K., 1995, 4-D gets tough Shetlands test: AAPG Explorer, v. 16, n. 11, p. 8–9.
Speers, R., and P. Dramgoole, 1992, Managing uncertainty in oilfield reserves: Middle East Well Evaluation Review 12, p. 30–41.
Stark, T., 1995, Surface slices as a stratigraphic interpretation tool, 65th SEG Annual Meeting, Houston, Expanded Abstracts, p. 865–868.
Taylor, G., 1994, Taste for chocolate oil revived: AAPG Explorer, v. 15, n. 4, p. 17–19.
Weber, D. J., and D. W. Ratcliff, 1994, Seismic advances changing image of the subsalt: American Oil and Gas Reporter, v. 37, n. 2, p. 49–55.

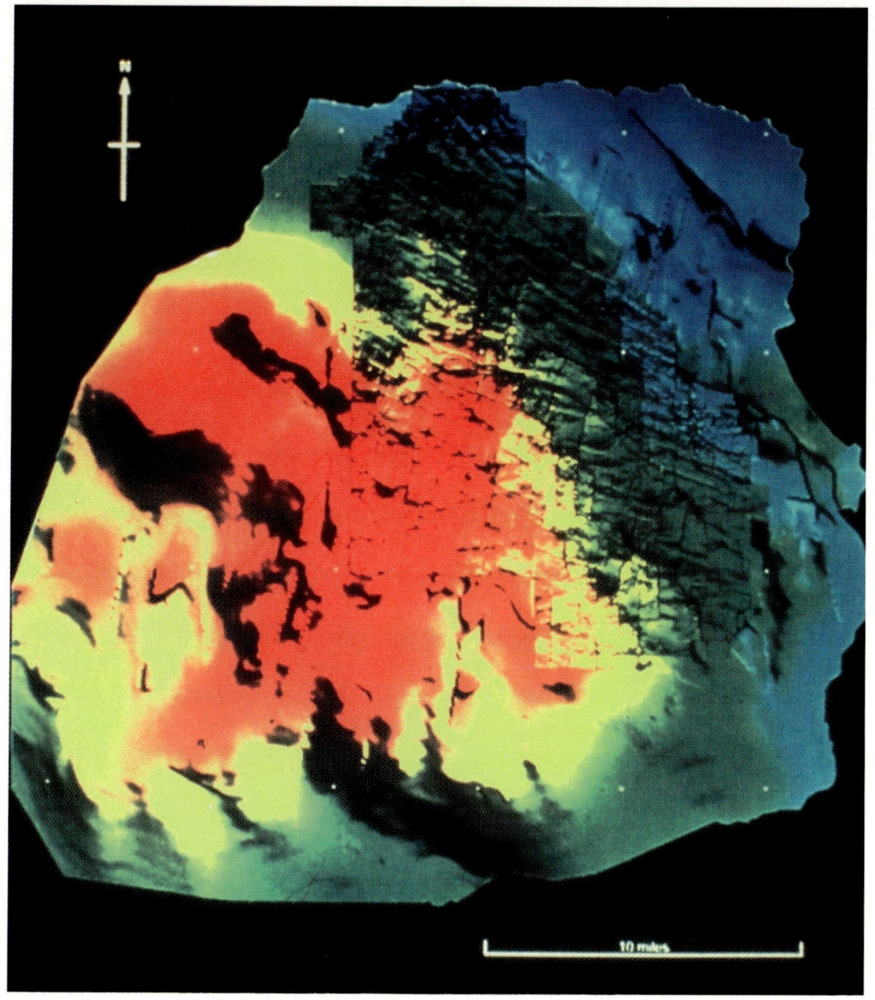


FIGURE 1. Two-way traveltime surface of top reservoir event, Kuparuk field, north Alaska. The two-way traveltime surface for this event is shown as color dipping down to the northeast from red to blue. The image is also artificially illuminated from the northeast and displayed as intensity which accentuates changes in dip and azimuth, highlighting the dense fault pattern within the reservoir. Courtesy BP Exploration (Alaska) and Arco. Figure reproduced from 35-mm slide.

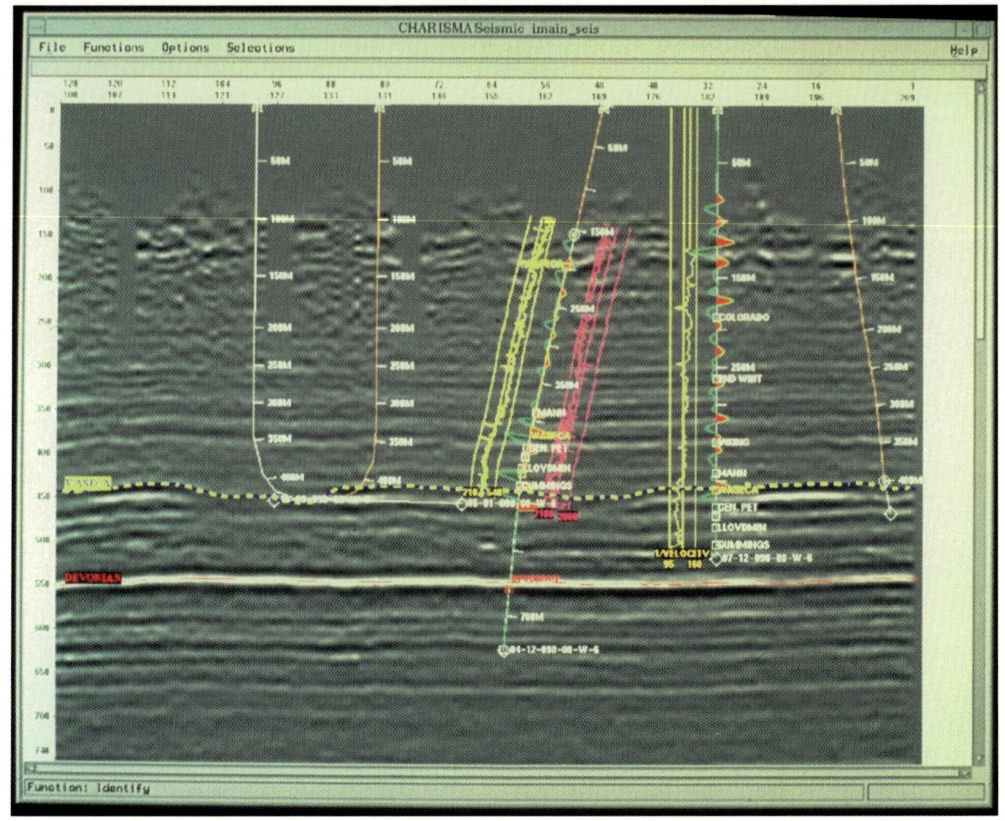


FIGURE 2. Vertical section from a 3-D survey in Saskatchewan. The strong seismic event in the lower third of the section is the Devonian horizon. Also, one can see well tracks with annotated depth values and a variety of logs. Courtesy Geoquest. Figure reproduced from 35-mm slide.

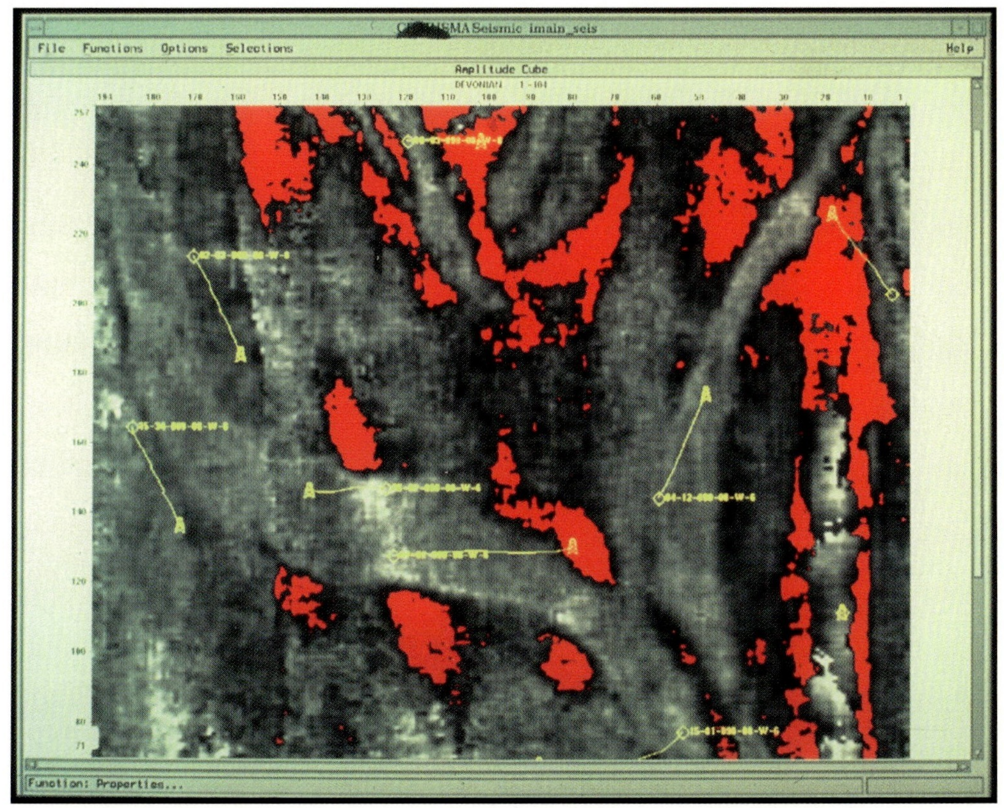


FIGURE 3. Amplitude map of the Waseca horizon from the same 3-D survey in Saskatchewan. The Waseca is about 100 ms shallower than the Devonian. Red depicts the highest amplitude. The lower amplitudes, in shades of gray, reveal major channel features. Well tracks are also shown (in plan view), with the surface location depicted by a derrick symbol. Courtesy Geoquest. Figure reproduced from 35-mm slide.

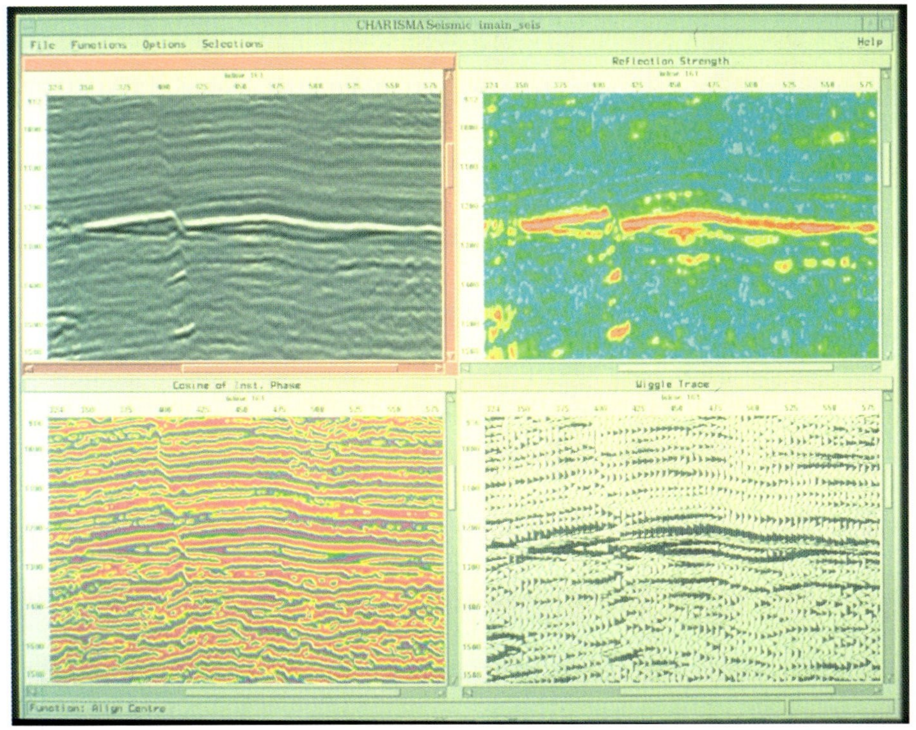

FIGURE 4. This vertical section is from a field in offshore Louisiana, displayed four different ways. The upper left profile is variable density, the upper right is reflection strength, the lower left is the cosine of instantaneous phase, and the lower right is reflectivity displayed as wiggle traces. Courtesy Geoquest. Figure reproduced from 35-mm slide.

FIGURE 5. Aerial photo of marine survey vessel. The vessel shown is a typical acquisition boat for 3-D surveys. The Geco Emerald represents a capital investment of more than $30 million. Courtesy Geco-Prakla. Figure reproduced from 35-mm slide.

FIGURE 6. Onboard processing improves data turnaround time and quality. This figure shows seismic processing staff working on their Unix workstations processing 3-D data onboard a marine vessel. Courtesy Geco-Prakla. Figure reproduced from 35-mm slide.

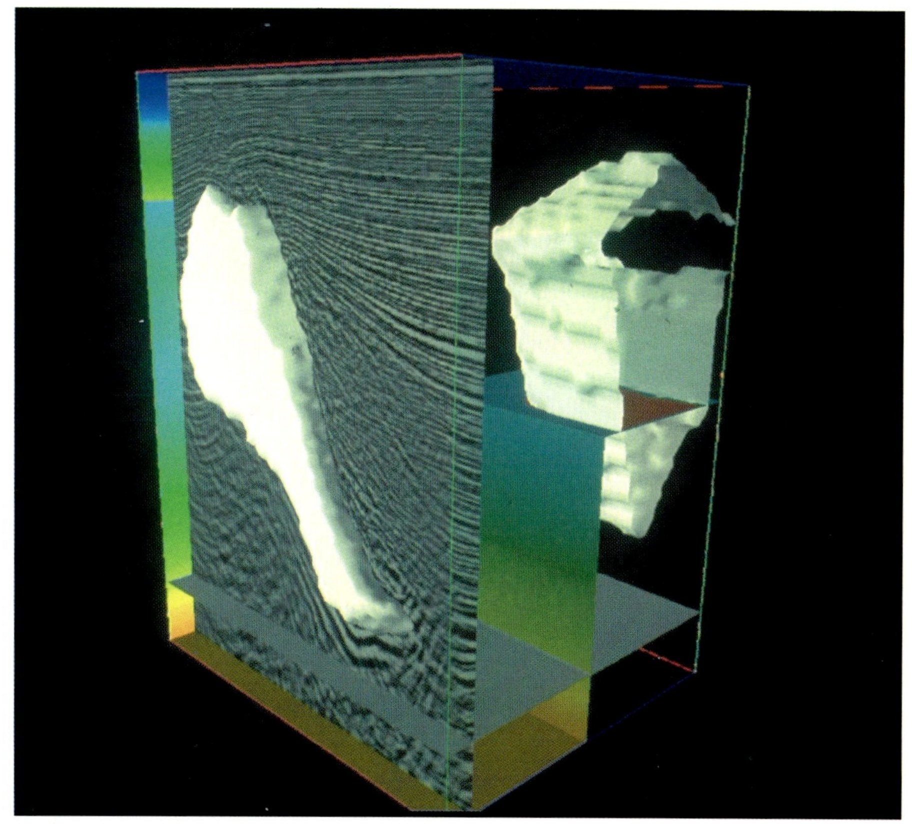

FIGURE 7. Velocity model combined with migrated data for 3-D depth migration work. The velocity model is depicted by the colored planes that intersect the velocity data cube. Blue means low and red means high-velocity values. The white surfaces are isosurfaces enclosing salt velocity values; the horizontal plane cutting the far salt body shows high-velocity (red) values inside the body. Courtesy Geco-Prakla. Figure reproduced from 35-mm slide.

CHAPTER 1

# 4-D Seismic Monitoring of Reservoir Production in the Eugene Island 330 Field, Gulf of Mexico

Anderson, R. N., A. Boulanger, Wei He, Y. F. Sun, Liqing Xu, and B. Hart, 4-D seismic monitoring of reservoir production in the Eugene Island 330 Field, Gulf of Mexico, *in* P. Weimer and T. L. Davis, eds., *AAPG Studies in Geology No. 42* and *SEG Geophysical Developments Series No. 5*, AAPG/SEG, Tulsa, p. 9–20.

**Roger N. Anderson*, Albert Boulanger*, Wei He*, Y. F. Sun*, Liqing Xu*, and Bruce Hart****

## ABSTRACT

We have begun the integration of rock physical properties, production data, reservoir modeling, and 4-D seismic monitoring from multiple generations of 3-D surveys to track changes of seismic attributes with pool drainage. Here we present the 4-D seismic monitoring technologies in order to (1) predict reservoir characteristics from seismic data, (2) locate bypassed pay, and (3) isolate drilling strategies that will maximize additional recovery for future fields. The test study is from the Eugene Island 330 Field of the offshore Gulf of Mexico. These results will have general application to other fields in the Gulf of Mexico, Nigeria, the North Sea, the Caspian Sea, and Indonesia—those with multiple generations of 3-D seismic coverage and seismically illuminated hydrocarbons.

## INTRODUCTION

3-D seismic technologies have progressed to the point that the data are being applied to reservoir production and engineering problems in addition to the original exploration uses. In addition, 4-D, or time-lapse, seismic monitoring is beginning to offer the possibility that control of field development may someday be possible through the merging of seismic reservoir simulations and field-monitored changes in impedance. Understanding the link between stratigraphy, fluid content, and changes in seismic response is a prerequisite to being able to confidently identify and quantify hydrocarbon drainage through the monitoring of 4-D seismic amplitude changes over time.

Changes in seismic amplitude within a given reservoir during its production history can be caused by changes in gas/oil/water ratios, in reservoir pressures, and/or in fluid contact levels, as well as by acquisition and geometry changes between surveys. We use wireline (sonic, density, porosity, water saturation [$S_W$] variations over time) and production (pressure, water cut, gas/oil ratio) data to produce forward models of acoustic impedance changes as reservoirs are produced. The model results allow us to predict the changes in seismic response that should be imaged by our 4-D seismic monitoring technologies. In order to extract the observed changes, we normalize different vintages of 3-D seismic surveys and examine resultant similarities and differences. The match between model and observed changes allows the prediction of location and composition of bypassed pay for future targeting.

*Lamont Doherty Earth Observatory, Columbia University, New York, NY 10964
**Pennsylvania State University, State College, Pennsylvania; now at New Mexico Bureau of Mines, Socorro, NM 87801

## EUGENE ISLAND (EI) 330 FIELD

To develop these seismic monitoring technologies, we analyzed two fault blocks of the LF reservoir along the EI 330/338 block boundary in the EI 330 field of offshore Louisiana (Oil and Gas Journal), Nov. 4, 1991; April 26, 1993; June 6, 1994; March 28/April 4, 1995, and May 18, 1996) (Figure 1). This field has produced more than 500 MMBOE since 1972, with oil and gas produced from various stacked sand reservoirs (Figure 2) in rollover anticlines and fault blocks associated with large regional (the Red) and counter-regional (the Blue) growth faults and smaller antithetic faults (e.g., the F fault) (Figure 3) (cf. Holland et al., 1990; Alexander and Flemings, 1995). The ongoing development of this field and the remarkable amount of seismic, production, geochemical, and wireline data available from the field make this an ideal location to research the effects of production and fluid migration on changing acoustic (seismic) signals over time.

In the EI 330 field, we have access to three vintages of 3-D seismic surveys (1985, 1988, and 1992; Figure 1). These surveys include the 1992 Shell/Exxon survey centered on Block 331 (Sh) (reflection and acoustic impedance versions), as well as the Texaco/Chevron 1988 survey over Blocks 338 and 339 (Tx) and the Pennzoil/Mobil/BP 1985 survey centered on Block 330 (Pz). We also have compiled >550 wireline logs from the field and have access to the production data from more than 300 wells in the field. We have conducted stratigraphic studies on multiple scales throughout the field (cf. Alexander and Flemings, 1995, and Hart et al., 1995). Combined stratigraphic, structural, and seismic attribute analysis have shown how depositional features and faults can compartmentalize reservoirs in this field. Hart et al., this volume, for example, imaged clinoforms controlling drainage in the GA reservoir of blocks 338/339. We have deepened a well into the Red fault zone, thought to be the major conduit for hydrocarbon migration in the area, collected oils from the fault zone; cored the fractured, geopressured shales adjacent to the fault; and conducted in situ stress and permeability measurements in the fault zone (purple line in Figure 3; Anderson et al., 1995a, 1995b.) (A CD-ROM of this field test is also available from the authors.) We have characterized the pressure field and examined the origin of geopressures (He and Anderson, in preparation; Hart et al., 1995).

## SEISMIC PREDICTION OF 4-D SEISMIC DRAINAGE EFFECTS

Laboratory measurements document the changes in acoustic reflection coefficients caused by changes in oil, gas, and water mixes, as well as effective pressure decreases. These results have been observed in the field in water and steam floods that have produced noticeable acoustic differences over time (e.g., Nur, 1982; Dunlop et al., 1988; Breitenbach et al., 1989).

In order to quantify the seismic effects expected from the production of hydrocarbons in the LF reservoirs, we conducted a seismic amplitude modeling study by varying the acoustic properties across a hypothetical constant-thickness sandstone reservoir undergoing pressure depletion (Figure 4). Gas/oil ratio (GOR), oil/water (O/W) contact, and gas/water (G/W) changes also affect seismic amplitudes through changes in acoustic impedance across the reservoir boundaries. Tuning was avoided in this modeling study.

Differential pressure (lithostatic minus reservoir fluid) increases as hydrocarbons are drained from a reservoir, reducing seismic amplitudes in both oil and gas sands (Domenico, 1976) (Figure 4). If, however, a GOR increase occurs during production, as with the formation of a secondary gas cap, seismic amplitudes are predicted to increase, or "brighten," over time, because the acoustic effects of the fluid change dominate over the pressure depletion effects. Similarly, seismic amplitudes can dramatically decrease, or "dim out," if in addition to pressure depletion of a high GOR oil, the O/W contact migrates in a reservoir. Here, the amplitude decrease is caused by pressure depletion and is further enhanced by the drop in impedance caused by the replacement of low-velocity oil and/or gas by relatively high-velocity water.

In addition, bypassed hydrocarbons can be identified by "near-zero" changes in high seismic amplitude regions over time, if there has been little increase in differential pressure. Also, areas of sustained high seismic amplitudes can be the highest permeability drainage pathways through which hydrocarbons are moving to get to well bores. Thus, the oil, water, and gas volumes remaining in a reservoir might be derived from observing changes in seismic amplitudes over time, after carefully calibrating these observations to changes measured within wells. The question is whether time-dependent 4-D seismic monitoring can detect such changes during natural field development.

## 4-D SEISMIC MONITORING TECHNIQUE

Our 4-D seismic monitoring technique consists of differencing regions of high amplitudes, grown from normalized 3-D seismic surveys acquired years apart over the same field, coupled iteratively with high-resolution seismic forward and inverse modeling used to interpret acoustic changes (Anderson, et al., 1995a, patent applied for, 1995). Pattern recognition, rather than interpolation and recomputation, is used to examine similarities and differences among the seismic amplitudes of the datasets. This technique minimizes the difficulties in matching acquisition parameters, variable geometries, and technology improvements that have accompanied the development of 3-D seismic methodologies over the years.

The Pz, Tx, and Sh 3-D seismic surveys were first converted into a common coordinate grid by interpolating the surrounding bins of the Tx and Sh surveys to produce bins that correspond to the coordinate grid of the Pz survey. The preparation of the multiple 3-D seismic surveys into a 4-D dataset ready for intercomparison then requires extraction of similar power spectra from the diverse datasets. However, the normalization technique cannot completely correct for differences in reflection strength caused by orientation changes relative to rock geometry in the subsurface. Consequently, the amplitude normalizations never result in exact spectral matches (Figure 5).

Two or more 3-D seismic surveys cannot be immediately used in 4-D seismic applications because acquisition and processing parameters applied to them by different geophysical service companies often vary markedly. Consequently, the seismic attributes, such as amplitude, phase, and the frequency bandwidth of different surveys, may not be similar enough to each other even though the surveys were acquired over the same geographic location. As part of our 4-D seismic technologies software, we apply several normalization techniques to extract similar power spectra from multiple surveys. First we use spectral matching, then amplitude normalization, and finally, phase correction to minimize the differences between seismic volumes so that the processed volumes can be directly input into our region-growing and differencing analysis. As an example, consider two seismic traces observed at the same location in the 1985 and 1992 surveys (Figure 6).

To match the frequency bandwidth of the different surveys, we first compute power spectra. The frequency bandwidth common to both surveys is then uniquely identified in the frequency domain. The elimination of frequency components outside of this common frequency bandwidth is accomplished by using a zero-phase bandpass filter. This filter is then applied to both volumes, resulting in similar power spectra for the two surveys (Figure 6). Matching of the magnitudes of seismic amplitudes of the different surveys is then implemented by using histogram matching techniques on reflection strength volumes (Figure 5).

Location problems must also be dealt with. In addition to navigation errors in older surveys, discussed further below, some 3-D surveys are processed with zero-phase seafloor reflectors whereas others used minimum phase. This produces a depth mismatch between surveys, which must be corrected. We use cross-correlation techniques between individual waveforms in the various datasets to correct for this phase mismatch. Then we have to seek out the changes in spectra and phases implemented over different time windows within each survey. (A routine processing practice is to increase amplitude and decrease frequency content at depth in most seismic volumes.) We carry out numerical experiments to determine the best filters to apply to these different time windows.

The comparisons are accomplished in attribute-derivative space because we have found that only the lowest frequency amplitude spectra preserve commonality. Vertical slices that run east-west along the EI 330/338 boundary through the reflection strength volumes of each of the three surveys demonstrate the improvements in frequency content and noise abatement that have historically developed in the 3-D seismic industry (Figure 7). The Pz survey has lower frequency content than the Tx survey, and both have considerably lower signal-to-noise ratios than the Sh survey. Specifically, the waveform envelope of the reflection strength (or second reflection strength) is calculated (He and Anderson, in preparation). Figure 8 shows one such extraction from real waveforms of the Pz and Tx surveys in which the LF reflector packet is marked (Figure 8).

## REGION GROWING

High-amplitude events (HAE) in any volumetric dataset are the accumulated, next-neighbor voxels of the "brightest" or most coherent energy (highest amplitudes) found within the volume. Feature-extraction operators are used to "grow" regions of HAE from seed points, which are the very highest amplitude voxels within the volume. Growth is controlled in 3-D away from the seed points by testing the gradient in amplitudes in each direction and continuing the growth only if the gradient is below a threshold. The feature extraction methodology is robust in the sense that the final derived geometry for each volume is relatively insensitive to signal quality and ensonification parameter variations because the operation is in "gradient space."

The data fields are then segmented into volumes of similar high amplitude regions (HARs) through the use of region-growing algorithms, and data from outside these HARs are excluded from future analysis (Figure 8). It is important to note that the HARs are not bounded by "isosurfaces" connecting equal seismic amplitudes. Instead, nonlinear, region-growing operators are applied to each dataset to isolate HARs, beginning from initial "seed" points. Compare the differences between the second reflection strength and region-grown amplitude extractions (Figure 8). Rough-cut connectivity between HARs within each dataset can be obtained at this stage by properly choosing the "manifolding" operator so that low-pass spatial filtering, dilation, and erosion "grows" the connections between segmented HAR seed points.

The 4-D seismic differences between the HARs are then computed, with voxels of "near-zero" change in high amplitudes delineated as green, "brightened" amplitudes over time shown by red, and "dimouts" shown by blue (Figure 8). Edge effects are created by location errors and by the velocity differences between the different fluid mixes in this time-domain example. For example, gas cap formation significantly slows the velocities within the top of the reservoir, producing a mismatched basal edge effect on the differences between HARs (Figure 8).

The task of determining growth and shrinkage from complexly changing images is analogous to the use of nuclear magnetic resonance to quantify the growth of a cancer. The similarities and differences from one observation period to the next are used to map growth of individual HAR "tumors" (red in Figure 8), shrinkage (blue), and locations where no change in size or density has occurred (green).

## THE EASTERN LF SAND

In order to test whether such theoretical seismic changes can be observed in a real reservoir, we examined the partially depleted Eastern LF oil reservoir that is located along the EI 330/338 boundary (Figure 3). The Eastern LF sand extends east-west over a 240-acre area at a depth of 6900–7000 ft (2100–2135 m) (Figure 9). A large, high-amplitude seismic anomaly (bright spot) is imaged from 1.98 to 2.06 seconds of two-way traveltime in both seismic surveys (see Figure 10). Average porosity is 27%, water saturation is 35%, and permeability is 500 md. There are proven reserves of up to 3.3 million bbl of oil in 6600 ac-ft, and probable reserves of 0.9 million bbl of oil in an additional 1800 ac-ft.

## SIMILARITIES AND DIFFERENCES IN THE EASTERN LF RESERVOIR

The HARs of the Eastern LF and surrounding reservoirs were then computed. The 1985 survey is shown in green in Figure 10A and the 1988 survey in red in Figure 10B. The similarities and differences in the Eastern LF region from 1985 to 1988 were visualized, and the locations were indicated where seismic amplitudes have decreased (blue), increased (red), or remained similar (green and yellow) (Figure 10C).

Another problem with the comparison of different vintages of 3-D seismic surveys is that the navigational accuracies of the surveys have greatly improved over the last ten years. We use a cross-correlation relocation technique whereby we select an unproduced amplitude anomaly region (salt forms can also be used). That amplitude should not have changed over the time interval under consideration. An example in our study area is the IC sand, imaged in detail in the insert in Figure 10C. There was no production from this sand until after both surveys were acquired. The green and yellow areas with similar amplitudes cover 80% of the volume of the reservoir, whereas red and blue areas, indicating increases and decreases, respectively, are found around the perimeter. There are "edge effects" present that should not be there, since no production occurred. These "edge effects" are most likely caused by x-y location errors between the 3-D surveys. In order to correct for this navigation error, we use a cross-correlation technique to find the optimal fit for the IC reservoir between the two surveys. We found that the Pz survey required a two-bin shift in the inline direction and a one-bin shift in the cross-line locations. (A bin is 41 ft [12.5 m].) Such a correction is then applied to the entire datasets, resulting in an improved correlation coefficient of >90% in non-productive intervals of the two surveys (Figure 11).

Changes in amplitude over time estimate the drainage of hydrocarbons from 1985 to 1988 and identify zones of possible bypassed hydrocarbons for future drilling. Near-zero seismic differences indicate locations where there were minimal change within the HARs of the surveys, which we interpret to be possible bypassed pay (green in Figure 12A). Decreases in amplitude could be caused by water encroachment or by pressure depletion between the times of the two surveys (blue in Figure 12A), and increases in amplitude could be from GOR increases and secondary gas dissolution (red in Figure 12A).

In the eastern fault block of the Eastern LF reservoir, 105,000 bbl of oil were produced from the A-5 well over the three years between the two seismic surveys. The 4-D seismic mapping indicated that significant dimout occurred over about 220 ac-ft of the reservoir, from the east and north (blue). Production efficiencies of 480 bbl/ac-ft would be required to produce this volume from this acreage, and the efficiency from some clean sands within the EI 330 field has exceeded

800 bbl/ac-ft (Anderson et al., 1995b). Further calibration of drainage volumes versus seismic amplitude changes will require study of more extensively drained reservoirs. See the analysis of the larger and thicker (up to 120 ft [37 m]) Western LF reservoir.

## VOLUMETRIC ANALYSIS OF DRAINAGE

The 4-D seismic differencing analysis produces a detailed, volumetric representation of the predicted drainage infrastructure within the bypassed pay interval, as well (yellow in Figure 12B). In fact, the 3-D pattern of bypassed hydrocarbons has surprising connectivity and logic to it. The amplitude pattern is aligned 30° to the major bounding faults. Possibly, these apparent lineations are small faults caused by a component of shear to the west accompanying extension to the south in the vicinity of the Eastern LF reservoir (thus the 30° rotation). This fault pattern, which might affect reservoir permeability, cannot be recognized in either of the primary 3-D seismic surveys but becomes a recognizable amplitude signal in the differenced, 4-D dataset.

## A-8ST HORIZONTAL WELL

During the spring of 1994, well A-8ST was drilled into the Eastern LF reservoir by Texaco, as operator for the partners (Chevron, Exxon, Penzoil, Mobil, and Cochrell) (Figure 13). A 1200-ft (356-m) horizontal sidetrack was then drilled along the property line between EI 330 and 338 and the well completed. Though located before the 4-D analysis was completed, the horizontal wellpath runs along one of the yellow "no high-amplitude change" lineations imaged by the 4-D analysis. Flow rates reached a maximum of more than 1500 bbl/day of low-GOR oil during production tests in August (Figure 14). By October 1995, the well had stabilized at approximately 900 bbl/day, and cumulative production had reached more than 420,000 bbl of oil. No water has been produced to date.

## SEISMIC MODEL OF DRAINAGE AT EASTERN LF TUNING THICKNESS

Quantification of the observed reservoir drainage effects is difficult for the Eastern LF reservoir because it is near the tuning thickness of the seismic sources used for the 3-D surveys. By calibrating the modeling results with the observed pressures and GORs, it was possible to compute the expected seismic amplitude changes, even at tuning thicknesses, using a 2-D finite-element forward model. First, we modeled the fluid content of a 35-ft (11-m) thick LF reservoir as 60% oil ($S_W$ of 40%) (Figure 15A) (Sun, 1994), then we inserted a small gas cap (100% gas) in the center and drained the oil ($S_W$ of 80%, oil saturation of 20%) to the west, replacing it with lower-pressured water, but leaving the eastern portion of the reservoir with bypassed oil (same $S_W$ of 40%) (Figure 15B). Drainage and replacement of pressured oil with lower-pressured water produced a pronounced dimout in the predicted seismic amplitudes (blue over red) in the seismic differences plot even at tuning thickness (Figure 15C). The formation of a gas cap, even at low pressure, produced a large amplitude increase (red over blue) in the difference plot (Figure 15C). Bypassed oil remained at similar amplitudes (yellow over light blue) in the difference plot (Figure 15C).

## SIMILARITIES AND DIFFERENCES IN THE WESTERN LF RESERVOIR

We then turn our attention to the Western LF reservoir, which extends along the boundary of EI 330/338 to the four corners (Figure 3). This reservoir has 3-D coverage from three surveys (Pz, Tx, Sh). Within the A fault block, the Western LF dips to the southwest from the crest of the rollover anticline in the central EI 330. It is part of a JD/KE/LF stack that is the main oil and gas producer in the field. Structural dips are on the order of 5–7 degrees, and the sand top deepens from about 6300 ft (1920 m) to over 7600 ft (2315 m) in this fault block. Figure 16 shows the locations of wells and the positions of fluid contacts, illustrating their movement with time as interpreted from the production data. Regions in red illustrate areas that have had oil replaced by gas due to expansion of a secondary gas cap up-dip. Regions in green are predicted to be oil-filled and are thought to have undergone no change in fluid composition. Water sweep is indicated in blue. This figure shows a large region that has been water-swept between 1972 and the present, whereas between the B-7 and B-6ST wells in the east, a secondary gas cap has grown between the time production started in 1972 and 1994 (Figure 17).

One of the interesting aspects of the oil-water contact evolution, even though derived from the sparse well coverage, is that the contact was not horizontal in 1994 but appears to cut across structural contours toward the northeast. This could indicate that the thicker sands that are located in this zone are more permeable, and fluids are being preferentially drawn from these higher-quality sands.

## HORIZON AMPLITUDE EXTRACTION

We then autotracked the major amplitude reflection that corresponds to the Western LF reservoir along the structural contour of the LF in each of the surveys (Figure 18A, B, C) and extracted the variation in amplitude for each survey. These "traditional" 2-D images clearly show dimouts that are approximately parallel to structural contours (compare with Figure 16), particularly down-dip, corresponding to the oil/water contact in the southwest, and up-dip, corresponding to the shale-out to the northeast. In general, the area of high reflection decreases with time, as might be expected for a reservoir with decreasing fluid pressures and oil saturations.

Because the three datasets have been normalized, we can compute differences between surveys along this Western LF trough (Figure 18D and E). The amplitude differences, Pz–Tx, show the brightened area marking gas cap formation (red in Figure 18D) and water sweep (blue in Figure 18D) similar to those observed from the well (Figure 17), but the boundaries are more complex and less tied to structure. Depletion of the gas cap from Tx–Sh time is imaged as an amplitude dimout up-dip in Figure 18E, with brightening indicated within the newly depleted oil zone, suggesting migration down-dip of the gas cap.

The production history of the B-7 well reflects this change in the gas cap (see Figure 17). The well was put on production in 1982, and prior to 1987, the GOR remained relatively constant at about 500. However, in 1987, between the Pz and Tx survey acquisition times, the GOR increased to >20,000. In 1989, the GOR fell and the well was shut in, prior to the acquisition of the Sh survey in 1992.

## VOLUMETRIC VARIATIONS OF SEISMIC DIFFERENCES

It is at this point that our 4-D seismic analysis technique departs markedly from traditional 2-D amplitude extractions on "wiggle" throughs. Region-growing within each double reflection strength dataset effectively expands the acoustic thickness of the Western LF reservoir so that volumetric variations within the HARs can be examined. We difference HARs to create volumetric representations of the seismic amplitude changes of Pz–Tx and Tx–Sh.

Because 3-D variations are difficult to convey on paper, extractions of the amplitude differences were then made into 2-D slices through the Pz–Tx and Tx–Sh HARs along planes that run parallel to the Western LF structural surface but cut through the top, middle, and bottom of the reservoir HARs. Because the differences carry memory of previous changes, we present both a fluid contact interpretation of the differences (Figure 19A) and the absolute difference image (Figure 19B). For example, a change near the B-7 well from oil at Pz time to gas at Tx time and to water at Sh time is indicated in Figure 18 first by an increase in amplitudes in the Pz–Tx difference (red), then a decrease in amplitudes in the Tx–Sh difference (blue).

The 4-D analysis indicates considerable change in the 3-D character of the fluid contacts over the time period examined. Specifically, the gas cap has indeed formed up-dip between 1985 and 1988, but has been depleted and moved down-dip by 1992, whereas the remaining bypassed oil is predicted to be located along a thin stringer running—perhaps not coincidentally—along the property line and within the hardline among the four block boundaries. The 4-D volumetric analysis predicts that the fluid contacts do not strictly follow either structure or sand quality but are strongly influenced by their proximity to producing wells (Figure 19A).

Compare the volumetric amplitude extractions from the region-grown HARs (Figure 19A) with the planar extractions from the 2-D surfaces (Figure 18, compare with Figure 19A, B) to get some feel for the three-dimensionality of the differenced HARs. Both show the gas cap formation from Pz to Tx time, but the HAR extraction shows the down-dip migration of the high-amplitude anomaly to have migrated deeper to the south by Sh time (Figure 19A). This information is not present in the 2-D surface extraction analysis.

## SUMMARY

In the future, sparsely distributed well information will be used to calibrate seismic changes observed during acoustic monitoring of entire fields, leading to reservoir characterization that can predict pressure distributions, interface changes such as O/W contact movements, porosity anomalies, permeability boundaries, GOR changes, and in particular, bypassed hydrocarbons—all in true 3-D and near real-time. However, the prediction of changes in reservoir physical parameters such as pressure and fluid content from the 4-D analysis of changes in seismic amplitudes over time is ultimately an ill-posed inverse problem. Therefore, in the future, refinements to the 4-D solution must be strongly coupled to the results of forward models of the different physical species (reservoir fluids, elastic, and acoustic).

*Acknowledgments*

An accompanying CD-ROM containing detailed accounts of this and related field studies of the Eugene Island 330 field is available for shipping and handling fees of $10 from David Roach, GBRN, Lamont-Doherty Earth Observatory, Palisades, NY 10964 (Internet address roach@ldeo.columbia.edu, phone 914-365-8330, fax 914-359-1631). We wish to thank the corporate affiliates of the GBRN for supporting this work: Amoco, Arco, BP, Chevron, Conoco, Elf, Exxon, Mobil, Pennzoil, Shell, Texaco, and Unocal. The work was cost-shared through DOE contract # DE-FC22-93BC14961.

## REFERENCES CITED

Alexander, L., and P. B. Flemings, 1995, Geologic evolution of a Plio-Pleistocene salt withdrawal mini-basin: Block 330, Eugene Island, South Addition, Offshore Louisiana: AAPG Bulletin, v. 79 n. 12, p. 1737–1756.

Anderson, R. N., and W. He, 1995, 4-D seismic interpretation techniques for the detection of drainage and migration in oil and gas reservoirs: U.S. Letters patent applied for.

Anderson, R. N., A. Boulanger, W. He, Y. F. Sun, L. Xu, 1995a, 4-D seismic imaging of drainage and pressure compartmentalization in the Gulf of Mexico: Oil and Gas Journal. Part 1, March 28; Part 2, April 4.

Anderson, R. N., P. B. Flemings, and S. Losh, 1995b, The GBRN/DOE Eugene Island Dynamic Enhanced Recovery Project: CD-ROM, Lamont Press.

Breitenbach, E. A., G. A. King, and K. N. B. Dunlop, 1989, The range of applications of reservoir monitoring: SPE Paper 19853.

Domenico, S. N., 1976, Effect of brine–gas mixture on velocity in an unconsolidated sand reservoir: Geophysics, v. 41, p. 882–894.

Dunlop, K. N. B., G.A. King, and E. A. Breitenbach, 1988, Monitoring of oil/water fronts by direct measurement; SPE Paper 18271.

Hart, B. S., P. B. Flemings, and A. Deshpande, 1995, Porosity and pressure: role of compaction disequilibrium in the development of geopressures in a Gulf Coast Pleistocene basin: Geology, v. 23, p. 45–48.

Hart, B. S., P. B. Flemings, D. M. Sibley, and B. Bishop, 1996, Facies Architecture of a shelf margin lowstand complex, Eugene Island Block 330 Field, Louisiana Offshore: AAPG Studies in Geology No. 42/SEG Geophysical Development Series No. 5, AAPG/SEG, Tulsa.

Holland, D. S., J. B. Leedy, and D. R. Lammlein, 1990, Eugene Island Block 330 Field-U.S.A. Offshore Louisiana, *in* E. A. Beaumont and N. H. Foster, compilers, Structural Traps III: Tectonic Fold and Fault Traps: Treatise of Petroleum Geology, Atlas of Oil and Gas Fields, AAPG, Tulsa, p. 103–143.

Nur, A. M., 1982, Seismic imaging in enhanced recovery; SPE Paper 10680.

Sun, Y. F., 1994, On the foundations of the dynamical theory of fractured porous media and the gravity variations caused by dilatancy, Ph.D. dissertation, Columbia University.

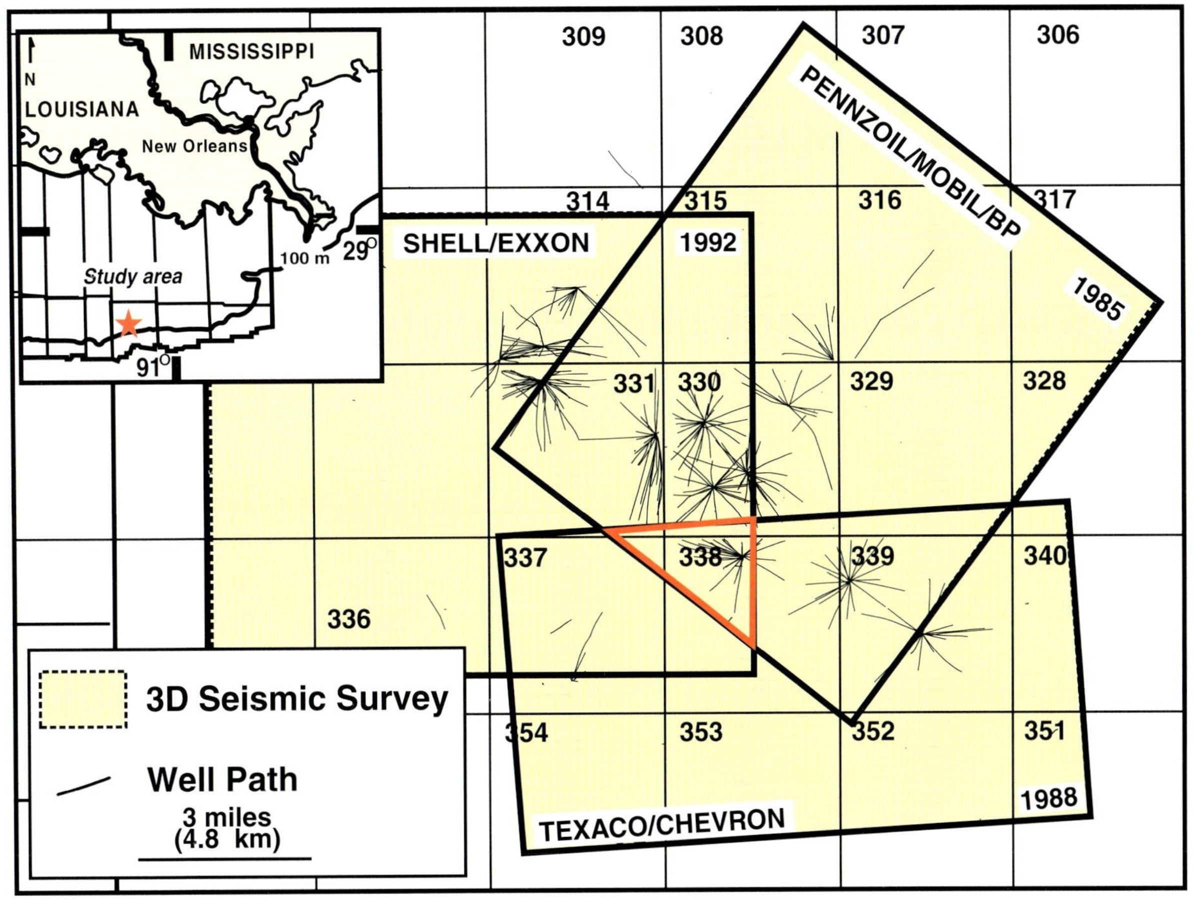


FIGURE 1. Location map of overlapping 3-D surveys. Red triangle is study area.

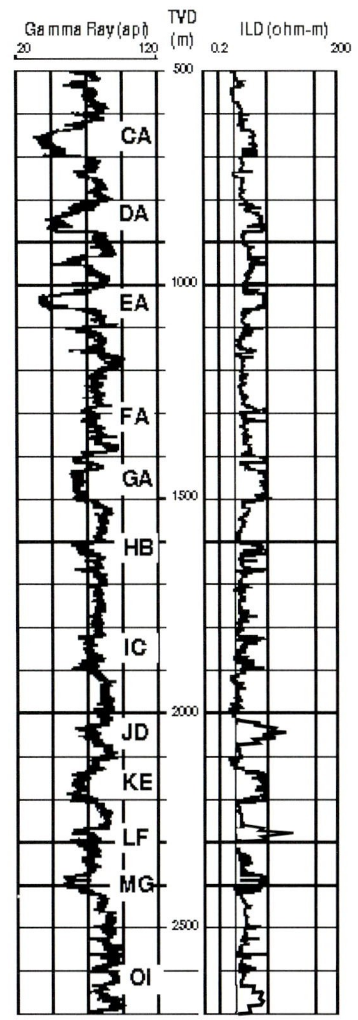


FIGURE 2. EI 330 type log, from Holland et al., 1990.

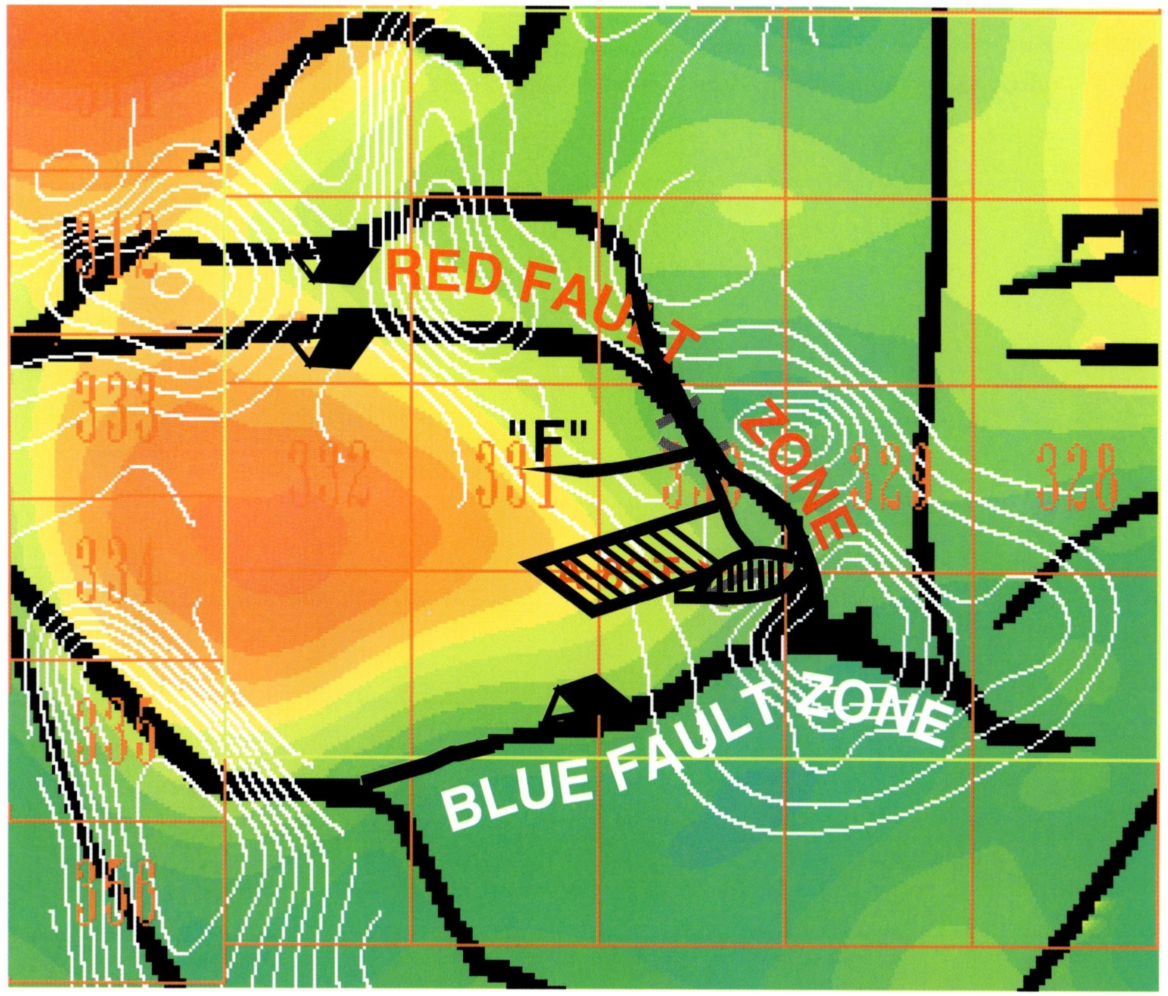


FIGURE 3. The Eugene Island 330 minibasin, shown by depth to the top-of-geopressure surface. Hot colors are deep; red = 15,000 ft. Green colors are shallow (darkest green = 5000 ft). Contour interval = 1000 ft (He and Anderson, in preparation). White contours are top-of-salt (top contour = 10,000 ft; interval = 2000 ft), and black lines are faults. Vertical bars are "Eastern LF," and diagonal bars show the distribution of the "Western LF" reservoirs.

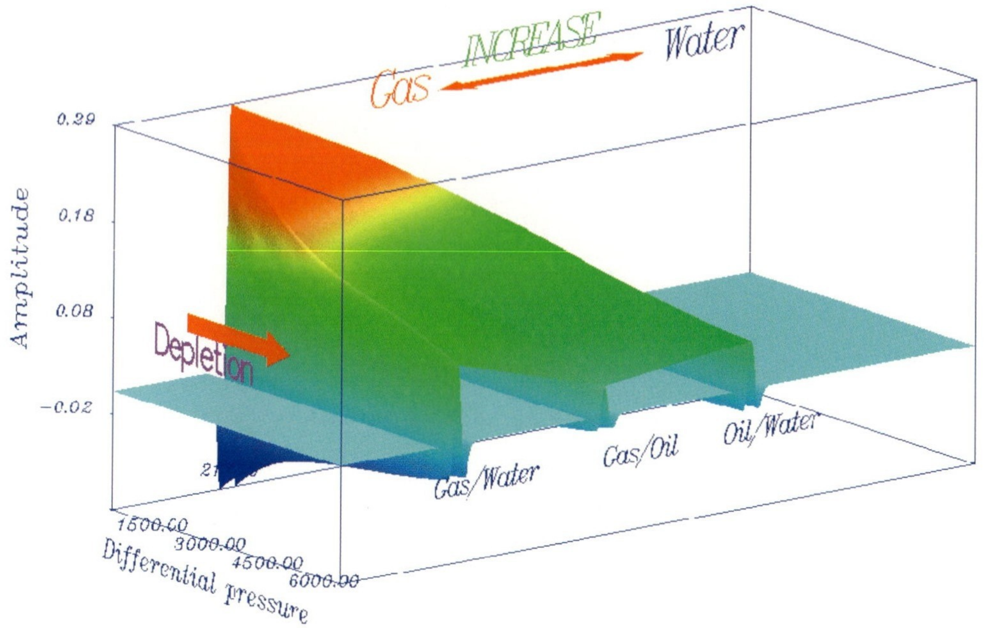


FIGURE 4. Model study of effects of fluid change on reflectivity in the EL 330 sands. A waveform modeling study was carried out to map the changes in seismic amplitude caused by the variation in oil/gas/water and formation pressure during depletion of a reservoir. The study is tracking the wavelet change across the upper interface of a thick sand with various fluid contrasts across the boundary.

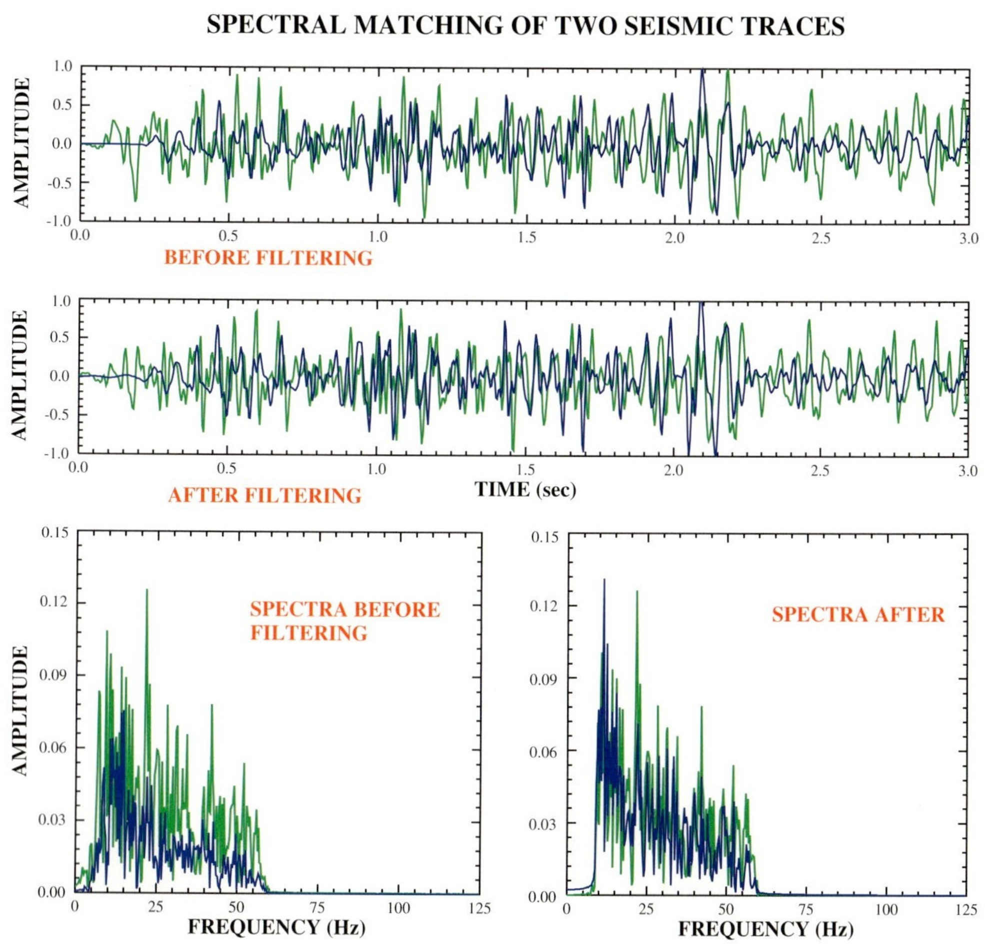


FIGURE 5. Spectral matching of two seismic traces.

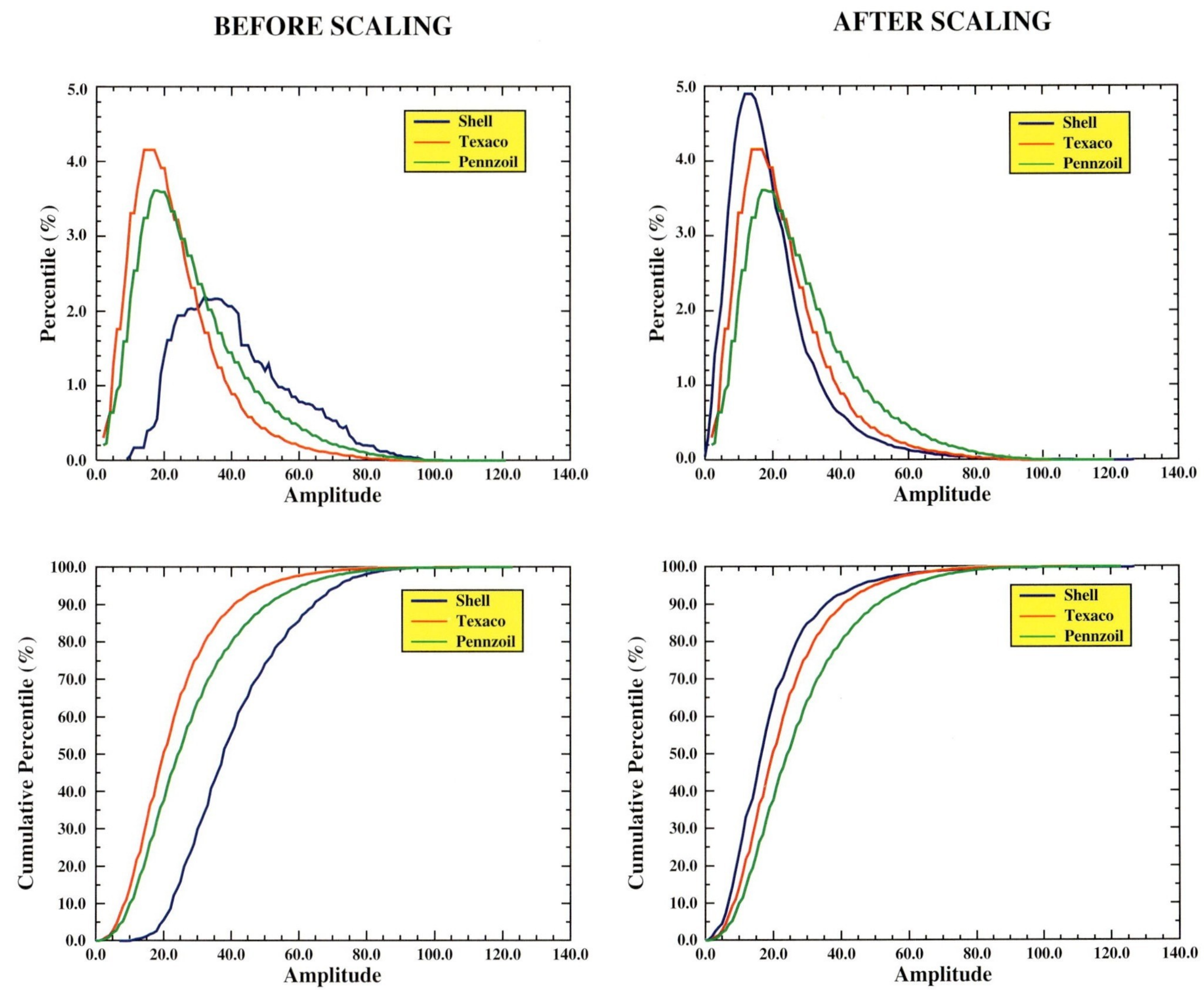


FIGURE 6. Spectral matching in 3-D seismic datasets. Histograms of amplitude content of reflection strength datasets before (left) and after (right) normalization.

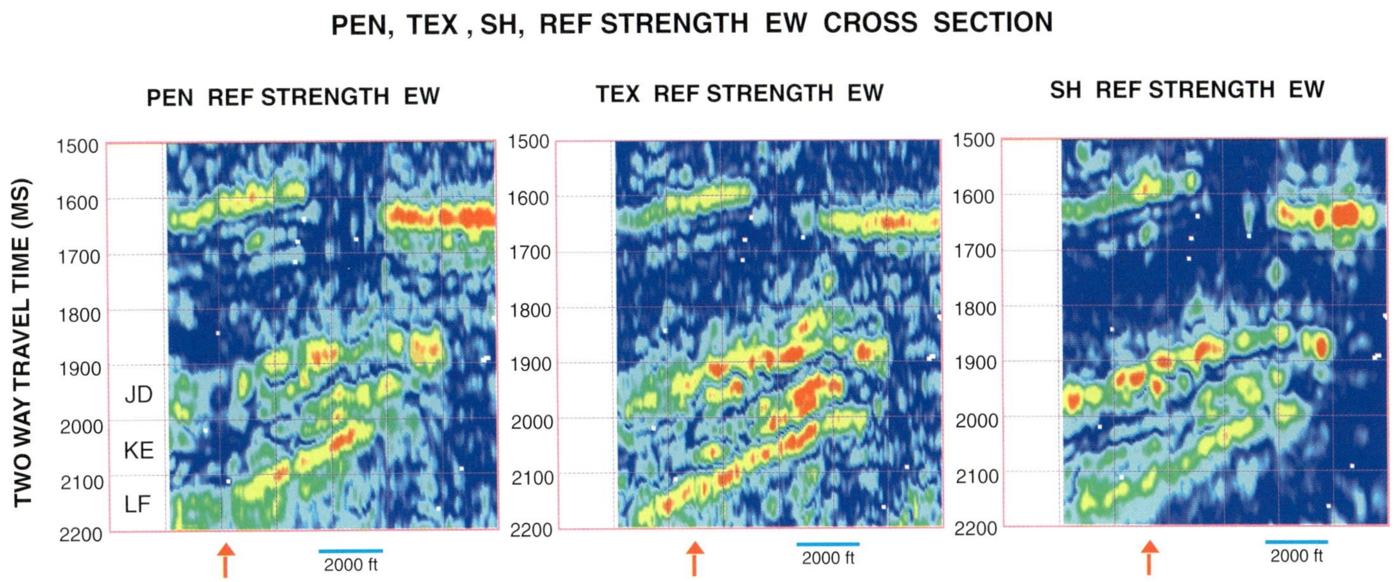


FIGURE 7. Seismic profile along the EI 330/338 property line. Amplitude attribute images of three surveys running along the boundary of EI blocks 330 and 338. Arrow is 4 corners.

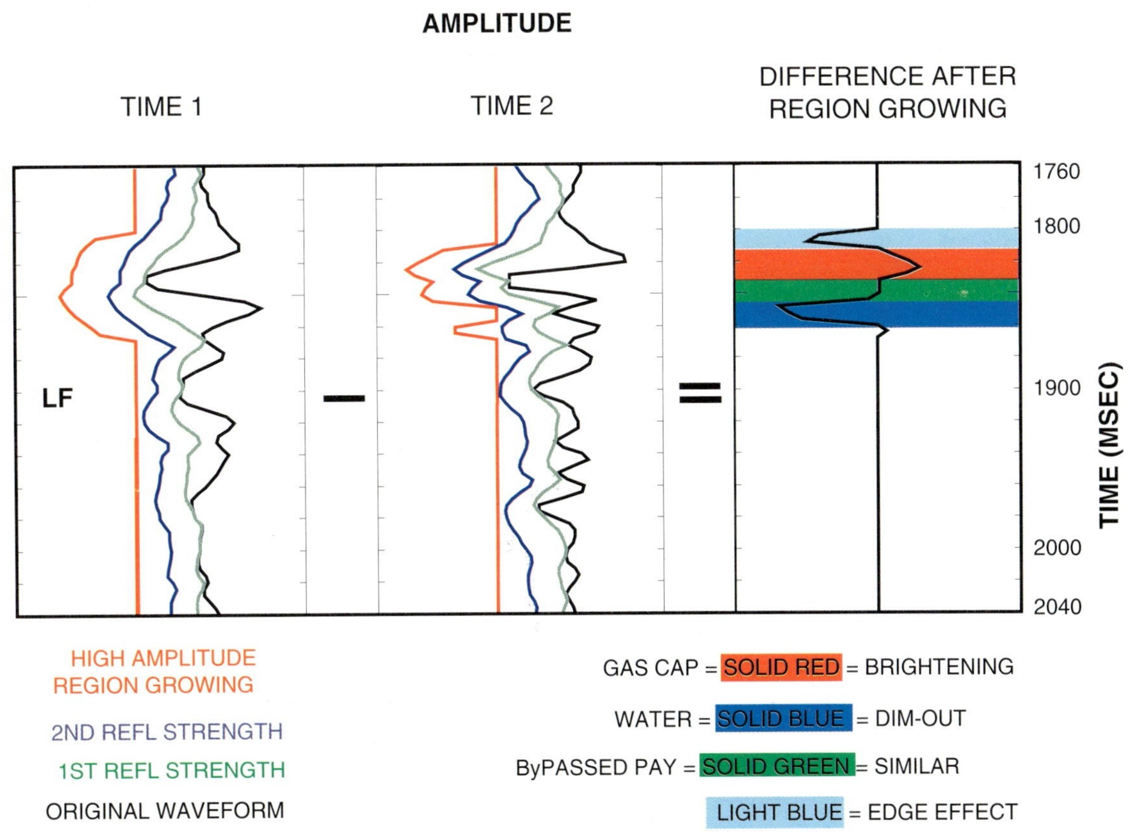


FIGURE 8. 4-D seismic differencing technique; example of growth of a high-amplitude region.

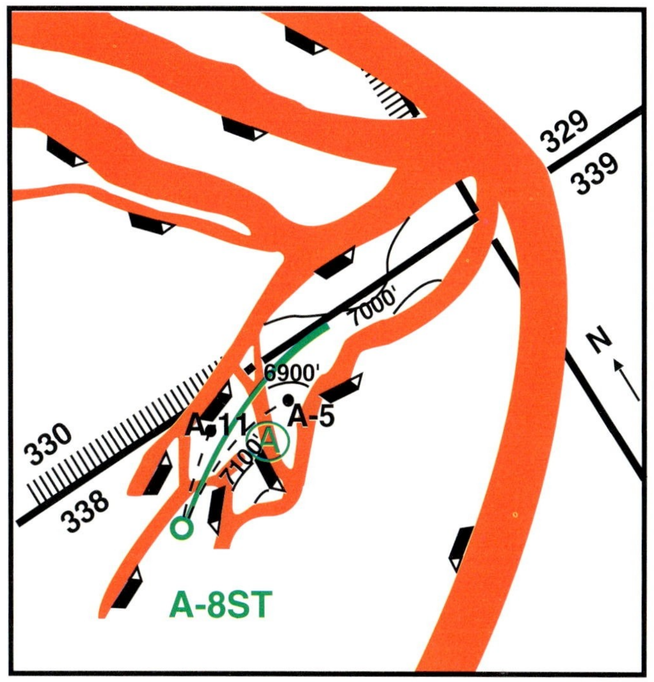


FIGURE 9. Location of Eastern LF reservoir. Structure of Eastern LF sand is indicated by contour. A-8ST is indicated in green. "A" fault is not distinct in the seismic data, but is clear in the 4-D seismic differences.

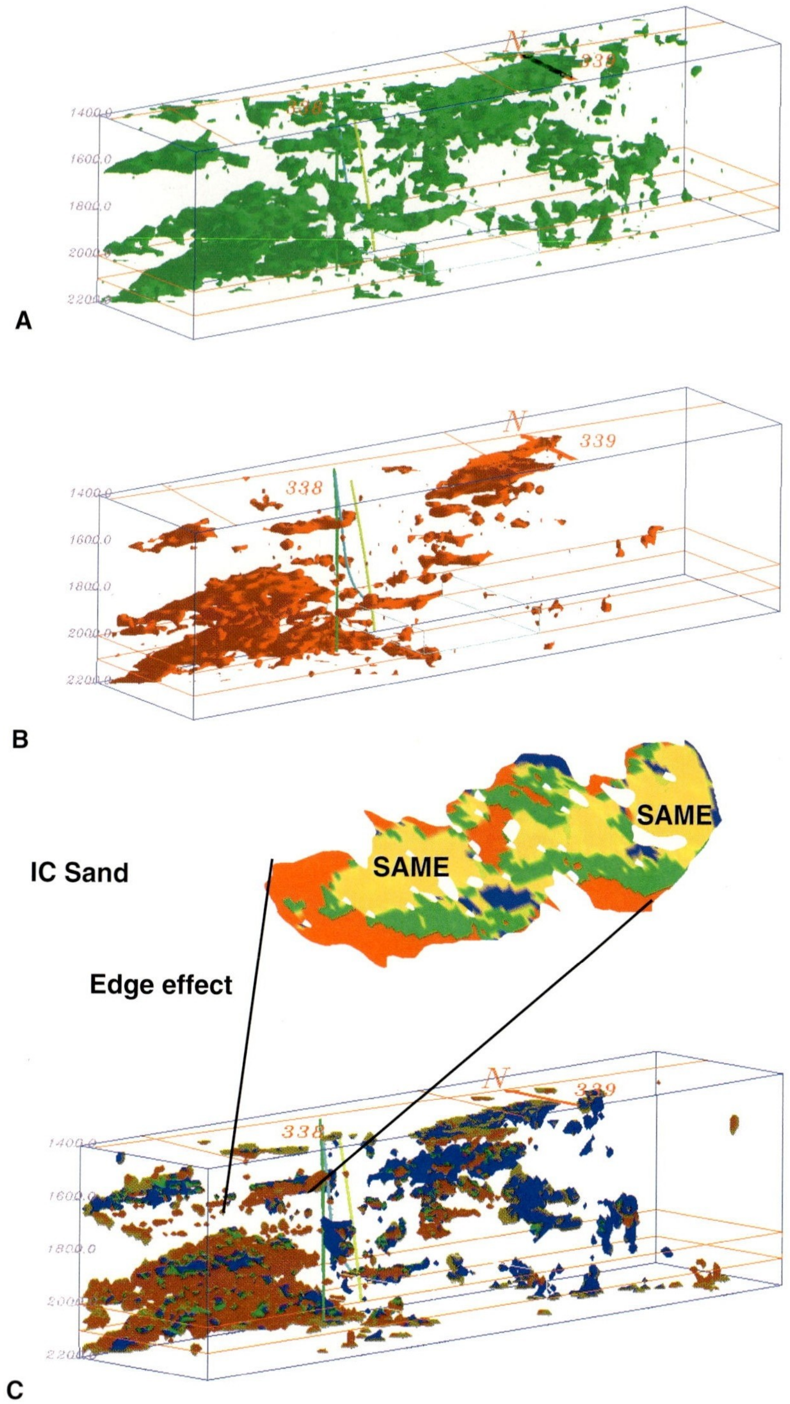


FIGURE 10. Seismic amplitude along the EI 330/338 boundary. (A) Amplitude envelopes around high seismic amplitudes from 1.4 to 2.2 seconds of traveltime for the Pennzoil (1985) survey, shown in green. The small box indicates the LF sand. (B) Texaco (1988) survey, shown in red. (C) The differences between Figure 10A and 10B in seismic amplitudes within the high-amplitude surfaces with dimouts from 1985 to 1988 indicated in blue and amplitude similarities shown in gold and green. Brightening is red.

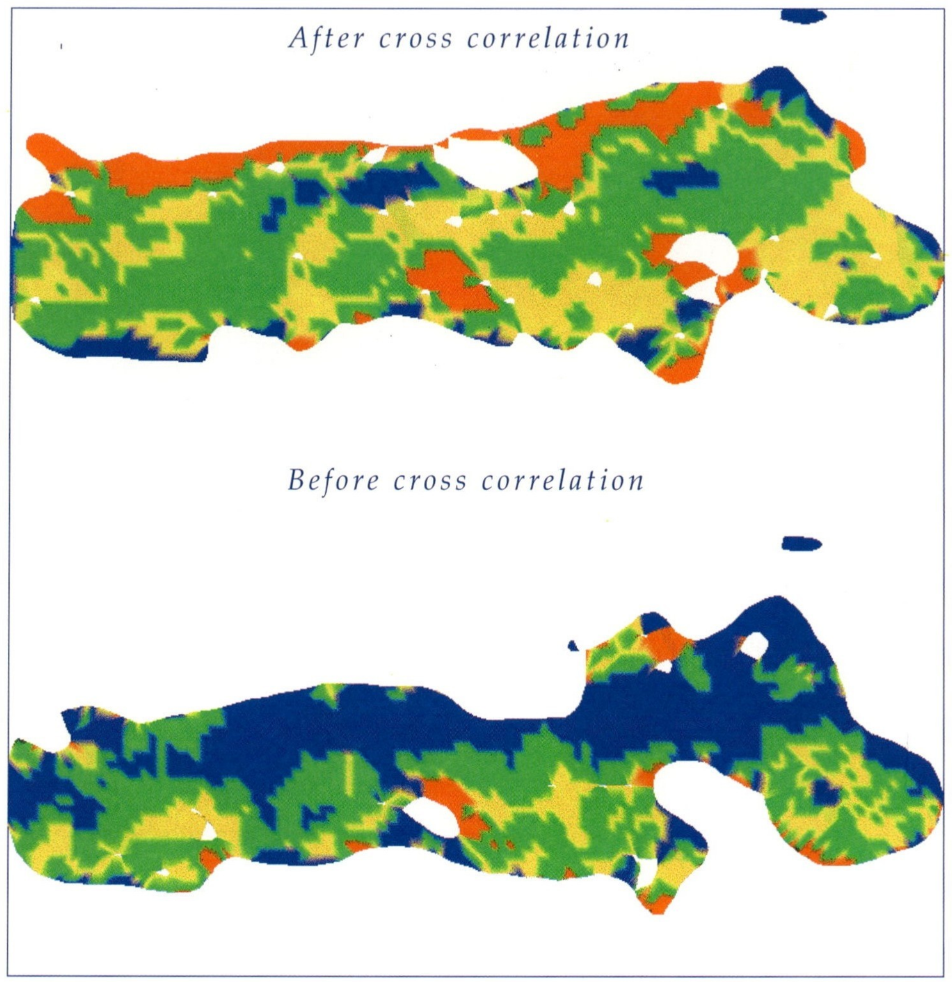


FIGURE 11. IC sand reservoir (preproduction). Cross correlation is used to correct navigation errors in older surveys. The fit between the two surveys has been improved from 80% (bottom) to 90% (top) in the IC reservoir. The amplitude differences between the two surveys are ≤5% (yellow), ≤10% (green), >10% old-new (blue) and >10% new-old (red).

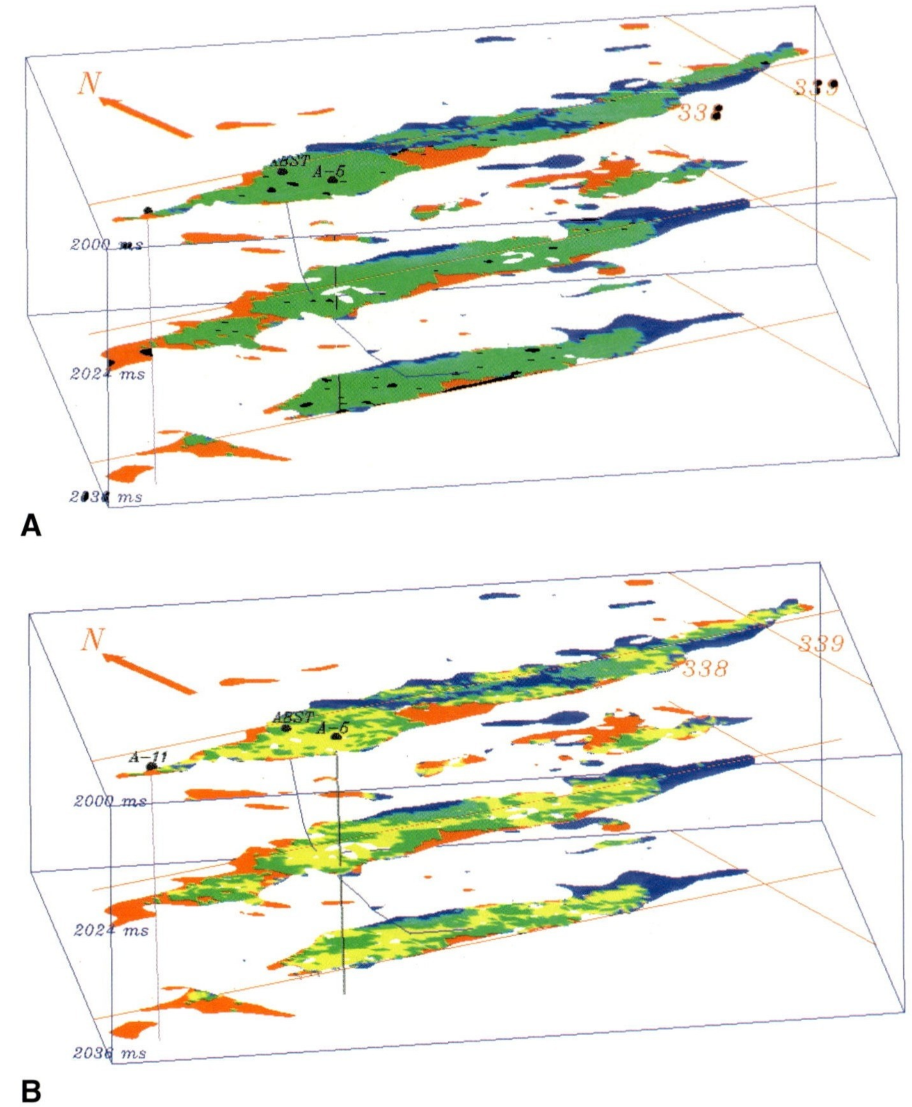


FIGURE 12. Eastern LF seismic amplitude differences. (A) Green = no change in high amplitudes, the most prospective region; blue = dimout from production; red = amplitude increase caused by gas/oil ratio increase with production. (B) Yellow = the region within the HAE with no measurable change (<10%) in seismic amplitude from 1985 to 1988. Green = the region that still has high amplitudes, but within which some change (10–30%) has been detected.

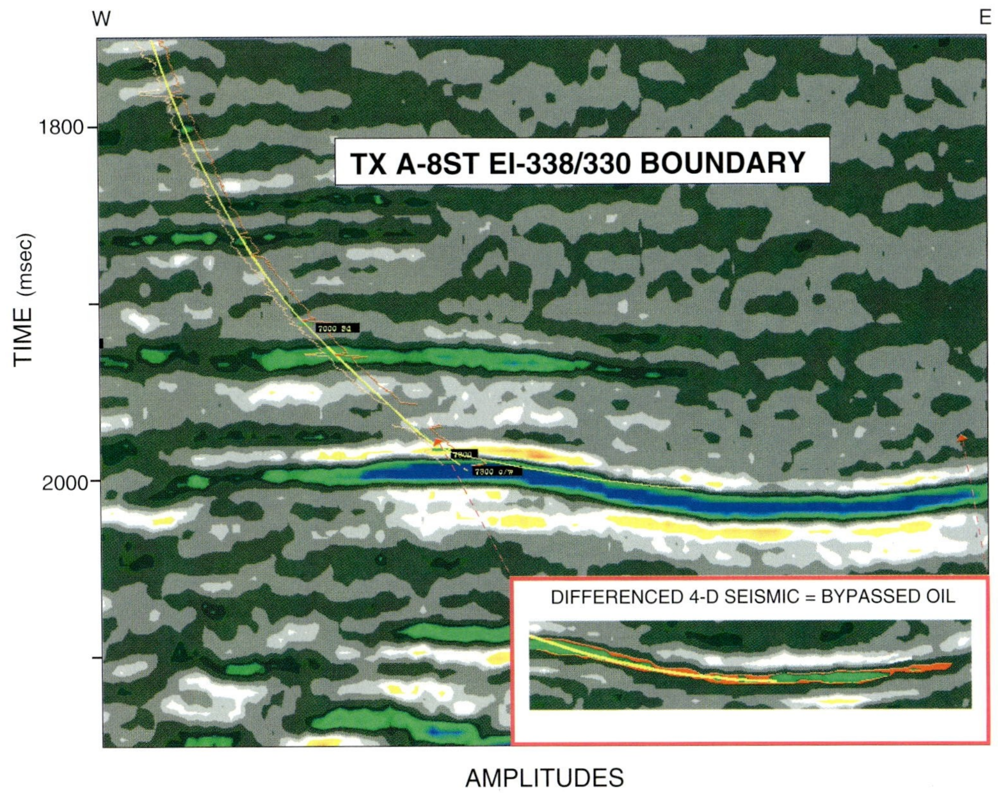


FIGURE 13. Horizontal well into Eastern LF reservoir in 1994. West-east seismic profile through the property boundary of EI 330/338, showing Eastern LF sand reservoir and blowup of the similarities and differences between seismic amplitudes from 1985 to 1988. The A-8ST horizontal well (yellow) targeted bypassed pay (red).

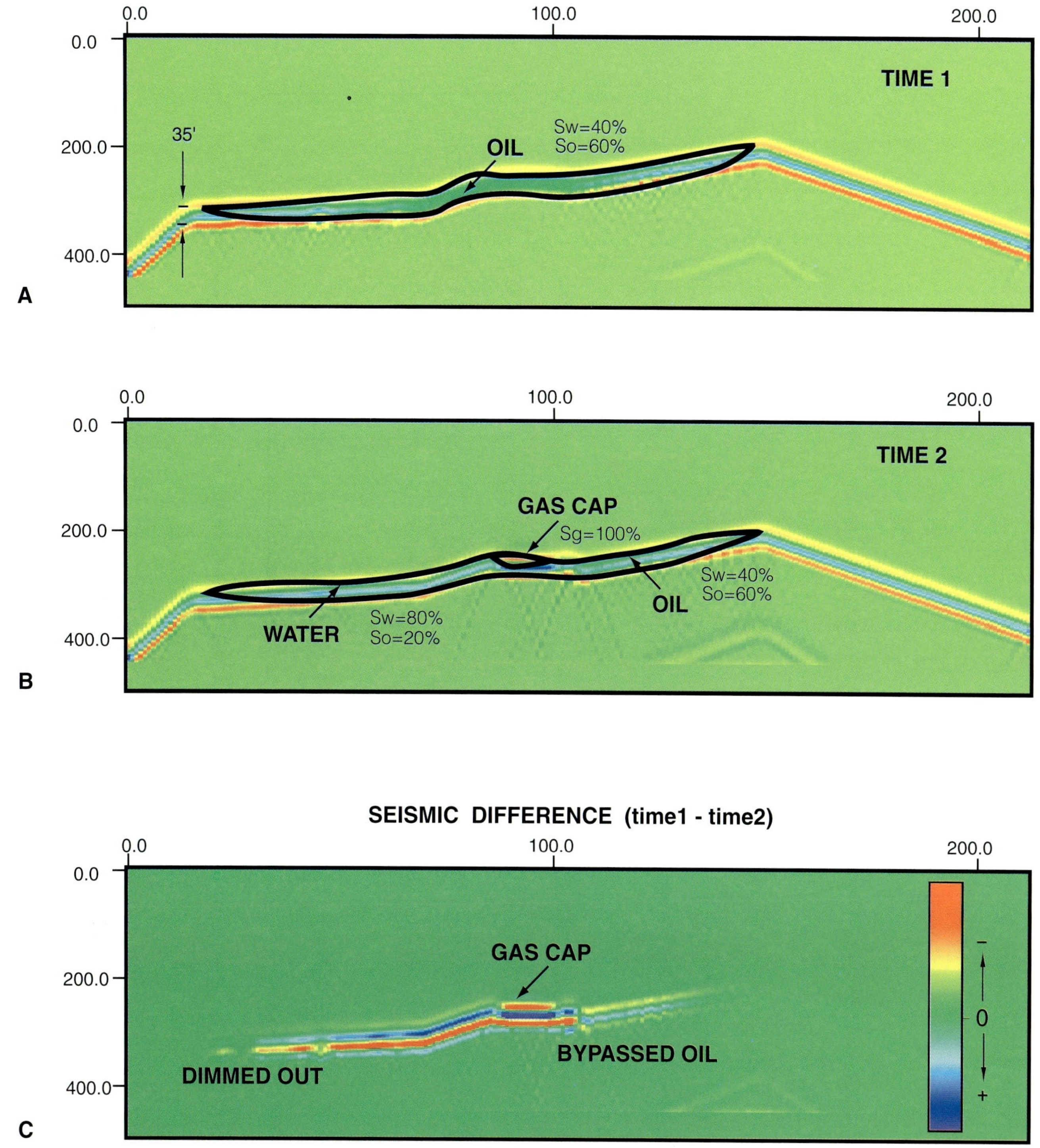


FIGURE 15. LF reservoir forward model. Seismic model of eastern LF reservoir, with oil/gas/water and pressure changes that were observed between 1985 (Time 1) and 1988 (Time 2), as indicated by production logs and pressure monitoring. Dimouts are predicted where the $S_W$ has increased, and brightening where the central gas cap formed.

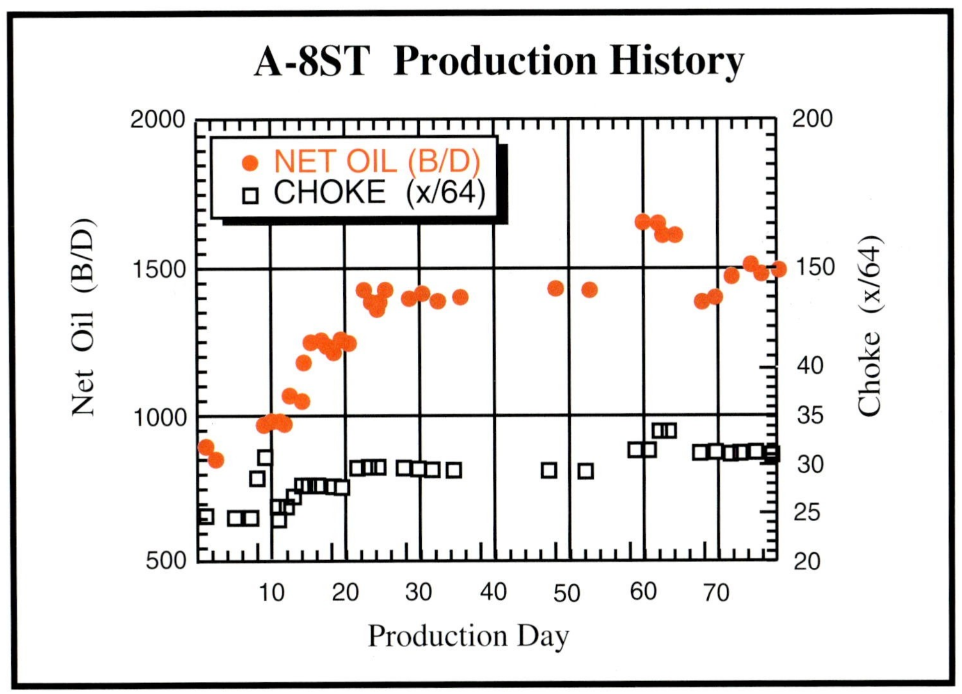


FIGURE 14. The production history of the A-8ST horizontal well for the first 80 days of production. More than a year into production, the well was producing about 1000 bbl/d.

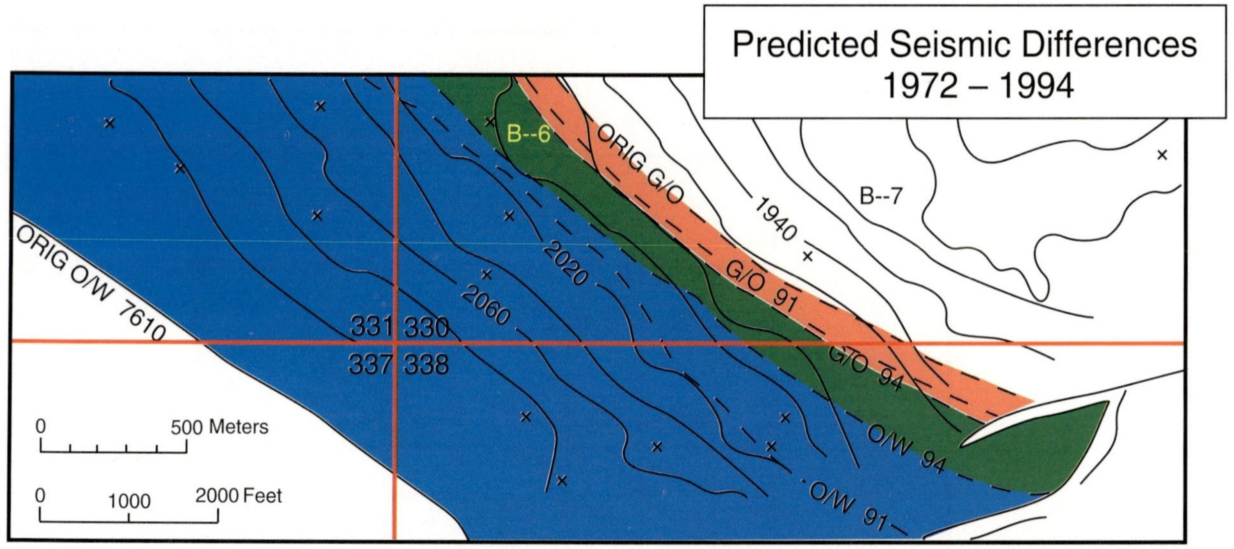


FIGURE 16. Map showing interpreted changes in fluid composition based on production data from wells only (locations indicated by x's). Compare this with observed seismic changes in Figure 18.

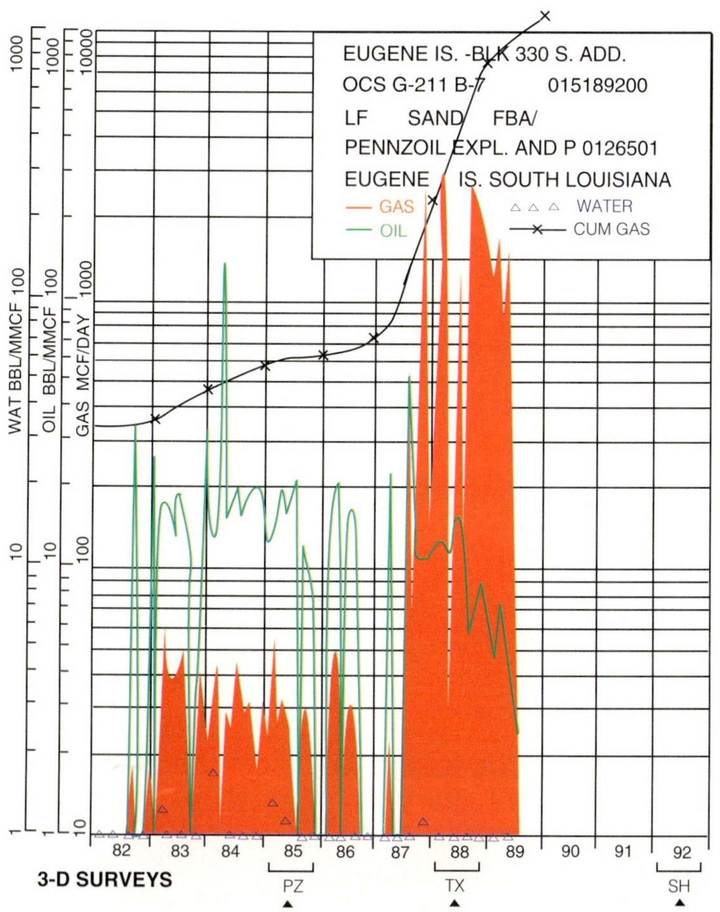


FIGURE 17. Production history of the B-7 well showing the increase in gas production between the Pz and Tx seismic surveys, then the watering-out of the well before the Sh survey.

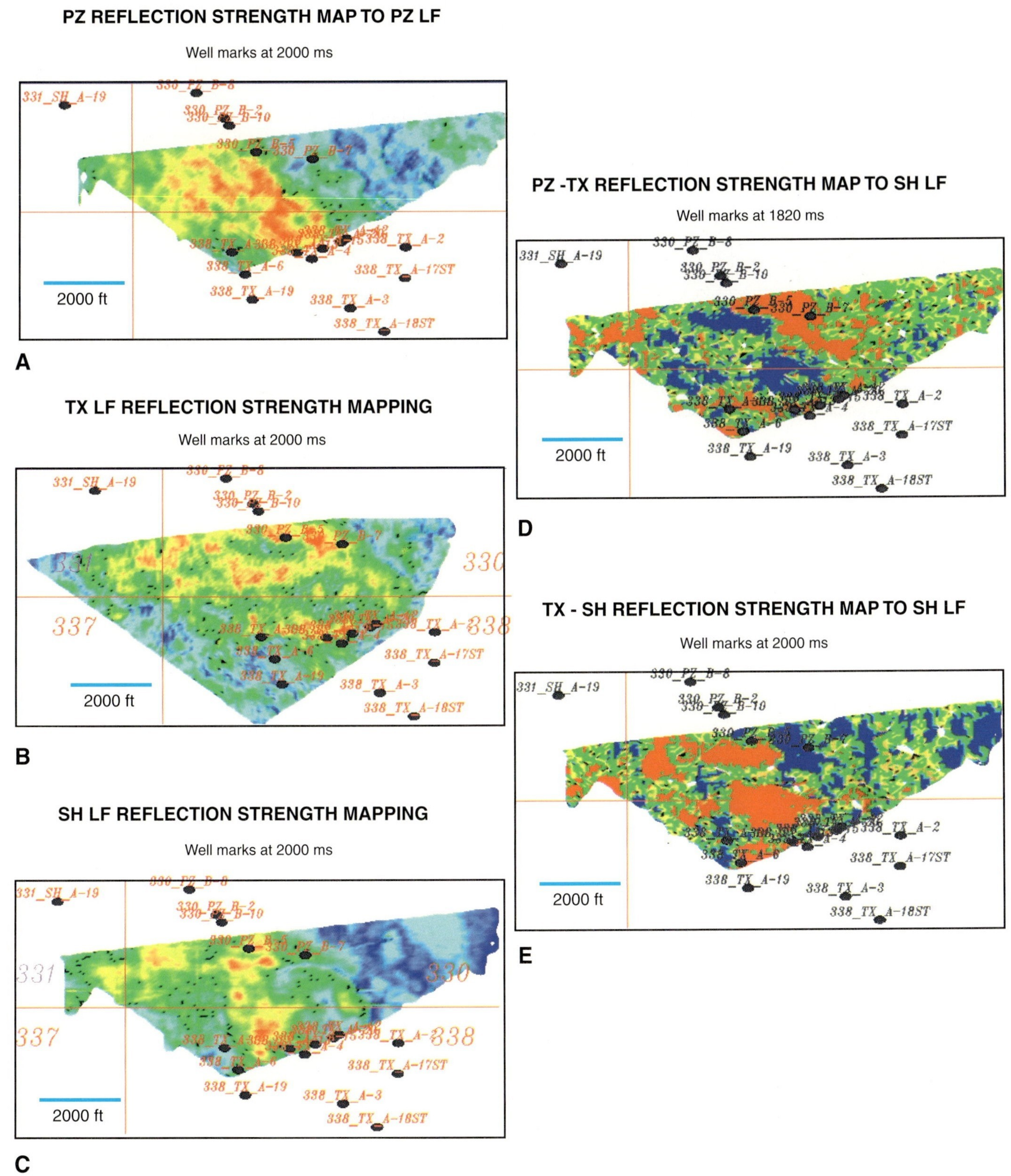


FIGURE 18. Traditional 2-D displays of amplitude variation along the "wiggle" trough of the main Western LF reflector.

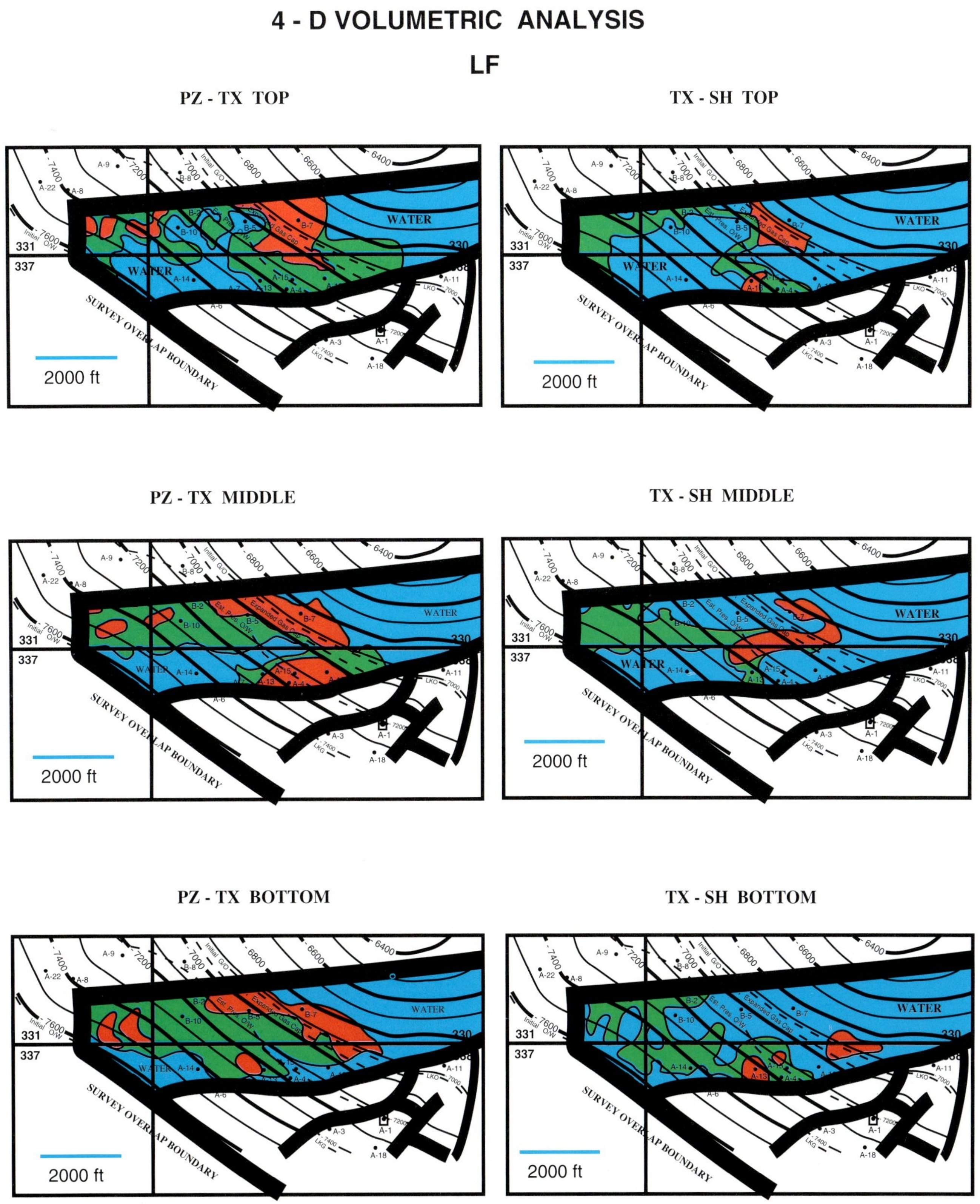


FIGURE 19A. Fluid contact interpretation of drainage in the Western LF reservoir. Blue is water, red is gas, and green is predicted location of bypassed oil.

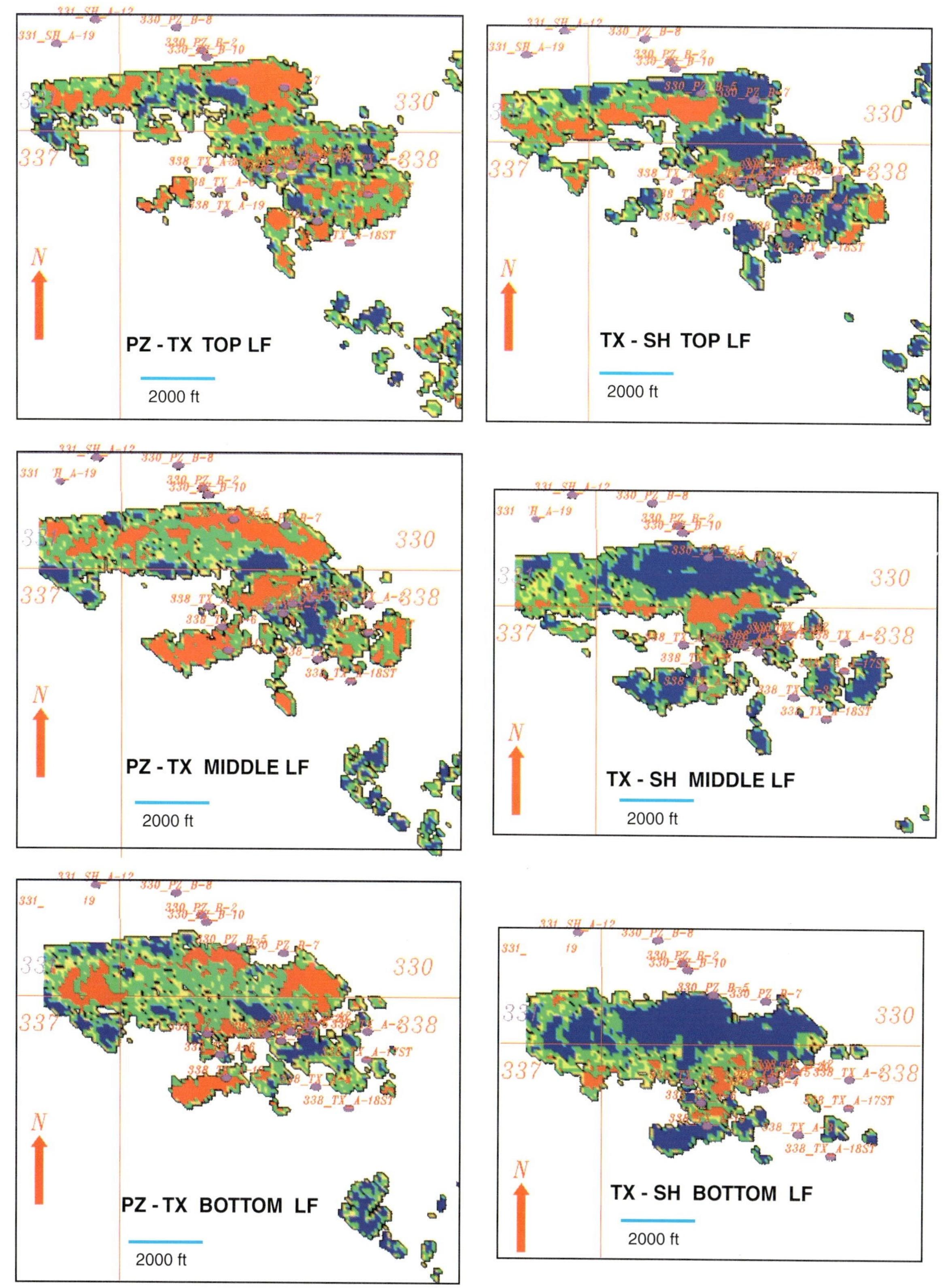


FIGURE 19B. Difference images of Western LF reservoir used to determine oil/gas/water boundaries in Figure 19A.

## CHAPTER 2

Hart, S., D. M. Sibley, and P. B. Flemings, Reservoir compartmentalization by depositional features in a Pleistocene shelf margin (lowstand) delta complex, Eugene Island Block 330 Field, Louisiana Offshore, *in* P. Weimer and T. L. Davis, eds., *AAPG Studies in Geology No. 42* and *SEG Geophysical Developments Series No. 5*, AAPG/SEG, Tulsa, p. 21–26.

# Reservoir Compartmentalization by Depositional Features in a Pleistocene Shelf Margin (Lowstand) Delta Complex, Eugene Island Block 330 Field, Louisiana Offshore

**Bruce S. Hart*†, David M. Sibley§, and Peter B. Flemings***

## ABSTRACT

We integrate 3-D seismic, wireline log, and production data to demonstrate that seismic amplitude maps of two pools in a Pleistocene lowstand delta complex of the offshore Gulf of Mexico image reservoir compartmentalization by depositional features and faults. The GA interval of the Eugene Island Block 330 Field is composed of stacked deltaic lobes, each of which consists of complex associations of distributary channels, clinoforms, and base-of-slope failure complexes. Production is from "updip" facies (including delta front and mouth bar sands) at the top of the interval. The highest amplitudes associated with the seismic horizon that marks the flooding surface at the top of the reservoir are associated with high-porosity (and high-permeability) clean, charged sands that depositional processes have not distributed uniformly. One of the two pools has a gas cap, and so the high amplitude response is expected (classic "bright spot"). The other pool has no gas cap, but the charged sands are also associated with high amplitudes because the in situ gas-oil-ratio of the oil is high (400 to 500 scf/bbl). In this second pool, there is a good qualitative relationship between production character and the combination of seismic amplitudes and structural position, with the best, most consistent production from high-amplitude areas on the crest of an anticline.

## INTRODUCTION

Mapping of amplitude anomalies associated with hydrocarbon accumulations ("bright spots") using 3-D seismic data from Plio-Pleistocene deposits in the Gulf of Mexico has proven to be extremely useful for tasks such as delineating pool size and locating bypassed pay. Uniform amplitude anomalies with boundaries that are parallel to structure contours are often interpreted as representing horizontal fluid contacts (e.g., Anderson et al., this volume). When the limits of the anomaly do not appear to be horizontal or the anomalies have a patchy distribution, some other phenomenon, such as patchy cementation, thickness variations, or porosity trends, is typically inferred.

There is a growing recognition of the role that depositional features can play in compartmentalizing reservoirs. In such cases, the compartmentalization may or may not be evident in pressure histories of specific wells in the field. This might arise, for example, if the lower, "wet" parts of sand bodies are in fluid communication but the upper hydrocarbon-charged parts of the sand are not. Wireline log, core, and production data can be used to examine reservoir compartmentalization when well spacing is sufficiently dense, but the integration of 3-D seismic data with these and other datasets is proving to be extremely helpful (e.g., Badgett et al., 1994; Sibley and Mastoris, 1994; Mills et al., 1995).

Shelf margin deltas are localized, wedge-shaped bodies that typically form during relative sea level lowstands on the upper continental slope seaward of fluvial channel networks. These deposits can be prolific reservoirs, but proper characterization of facies architecture (and hence reservoir heterogeneity) can be vital when exploiting hydrocarbon reserves located in them, especially during expensive offshore production operations. Here we present seismic transects, wireline log cross sections, extracted amplitude maps, and production data from one such deposit, the late Pleistocene (approximately 0.8 Ma) GA interval of the Eugene Island Block 330 Field (hereafter EI 330) of the Louisiana offshore (Figure 1A). Our goal is twofold: (1) to illustrate the stratigraphic character of the interval and (2) to show how seismic amplitude maps can be used to image reservoir compartmentalization by depositional features.

## SETTING

The EI/330 Field is the world's most prolific Pleistocene oil field. The general structural and stratigraphic setting, as well as the production history, of the field were described by Holland et al. (1990) and Alexander and Flemings (1995). Production from this field is primarily from several stacked sand bodies that have been folded into a rollover anticline southwest of a major basinal growth fault system (Figure 1B).

The GA interval is the youngest major reservoir in the field (Figures 1B,C), with production from the crest of the anticline in the central portion of EI Block 330. Original oil-in-place in this pool is estimated at 40 million bbl, of which 10 million bbl was originally considered recoverable. This reservoir has significantly overproduced, with cumulative production in excess of 18 million bbl by early 1995. Much of this overproduction is attributed to low resistivity pay reserves that were not originally recognized (cf. Holland et al., 1990). This interval is also productive from a separate pool, the Teal Field, that is located in a small graben in Blocks 338 and 339, to the southeast of the Block 330 reservoir (Figure 1D).

## STRATIGRAPHY

The GA interval is the product of shelf margin delta deposition during a period of eustatic sea level fall and lowstand at about 0.8 Ma. During deposition of this interval, relative sea level changes and local structural development (growth fault movement and continued development of the rollover anticline) strongly influenced depositional patterns. In these and other respects, the GA interval has many similarities with the shelf margin deltas described by Morton and Suter (in press) and the prograding complexes of Mitchum et al. (1993).

We recognize within the GA interval complex associations of erosional channels, channel fills, prograding clinoforms, and failure deposits. The main stratigraphic elements in this interval are: (1) a sequence boundary at the base of the interval; (2) the main deltaic complex of the lowstand prograding wedge, that can be further subdivided into at least two stacked to shingled deltaic lobes (channel fill, clinoform, and base-of-slope failure deposits are all present in this part of the GA); and (3) an upper, thinner sand body that pinches out over the crest of the anticline (the GA-1 sand, Figure 2) that was deposited at the onset of transgression. The principal reservoirs in the GA interval are located in the clinoform seismic facies; our main focus in this paper will be on these deposits.

The sequence boundary at the base of the GA interval is represented in well logs as an abrupt contact between shales (below) and sands (above). This surface is a fluvially incised unconformity on the crest of the anticline and the base of a prograding complex downdip (Figures 2, 3). Above the sequence boundary are complex vertical and lateral associations of channel fills, clinoforms, and failure deposits that are the product of deltaic progradation and lobe switching. Chaotic reflections, thought to be stacked channel-fill sands, are the main seismic facies of the updip areas of Blocks 330 and 331 (Figures 2, 3). Gamma ray logs through these deposits indicate that they are primarily composed of sand. Southward-dipping clinoforms are the

*Department of Geosciences, Pennsylvania State University, University Park, PA 16802
†Current Address: New Mexico Bureau of Mines and Mineral Resources, Socorro, NM 87801
§Chevron U.S.A. Inc., Lafayette, LA 70593. Current Address: Wapet, Perth, WA 6001 Australia

principal depositional features visible in seismic images from the GA interval in the downdip areas of the southern part of Block 330 and north of the counter-regional fault in Blocks 337 (Figure 3), 338, and 339. We believe that these clinoforms are the product of prograding deltaic lobes. Gamma ray logs show that the deposits imaged as clinoforms shale out towards the south.

At least two discrete prograding packages (delta lobes) are present in Blocks 330, 338, and 339, and they are separated by a surface that is labeled the "green surface" in Figures 2–5. This surface is the contact between two stacked progradational lobes, and, like the underlying sequence boundary, it changes from a channel base near the top of the interval in Block 330 to a downlap surface in the middle of the GA interval in Blocks 338 and 339. The green surface is identifiable in both the log and the seismic data in this area, but it becomes obscured in the seismic data where the overlying sands are charged in central Block 330 (Figures 2–4).

The fining-upwards trend seen at the top of the GA interval (above the green surface) in the gamma ray logs from the central part of Block 330 (Figures 2–4) resembles a log signature that has been associated with fluvial channel deposits in other areas (e.g., Galloway, 1981). A wireline stratigraphic section across the crest of the anticline illustrates toplap and downlap relationships, indicative of low-angle clinoforms, for the oil-charged, clean sands of this part of the GA (Figure 4A). The seismic data show that these sands are at the updip limit of clinoforms that downlap to the south, and recognition of this pattern has allowed us to make correlations that otherwise might be problematic. The wireline correlations are, however, below the scale of the seismic resolution (Figure 4B), especially because of the low-frequency, high-amplitude response of the oil-charged sands. We interpret these deposits to have been deposited in a river-mouth setting characterized by the interaction of fluvial and marine influences.

Clinoform geometries (without recognizable channel deposits) are also visible in well log correlations from the Teal Field (Figure 5A). We interpret these deposits to have formed by delta front progradation in an interdistributary area. As in the Block 330 example, the high-amplitude response of the charged sands obscures some of the details of the stratigraphy in the seismic data (Figure 5B). Note that this 3-D seismic dataset has a higher frequency content than the dataset shown in Figure 4B, so more details of the stratigraphy are visible in Figure 5B.

## RESERVOIR CHARACTER AND AMPLITUDE MAPS

The delta front and mouth bar deposits at the tops of the clinoforms form the principal reservoirs of the GA interval. These deposits lie directly below the flooding surface at the top of the interval. Base-of-delta-slope sandy failure deposits might form attractive reservoir units in other lowstand deltaic deposits but are unproductive here, probably because they sit too low on the flank of the anticline structure.

We use maps of seismic amplitudes from the top of the GA interval to help identify the location of hydrocarbons. Oil in Pleistocene sands, such as those discussed here, produces a high-amplitude response because of the low velocities that result from the high gas-oil-ratio (GOR) (e.g., Clark, 1992). The initial GOR for oils of the GA interval ranged from 400 to 500 scf/bbl (Holland et al., 1990; Sibley and Mastoris, 1994). A gas cap is present in the Teal Field (EI 338, 339), but is absent from the GA interval in central EI Block 330.

Figure 4C shows a map of the amplitudes associated with the GA top seismic pick in central EI Block 330. In a general way, the highest amplitudes are located at the crest of the anticline (as would be expected of hydrocarbons in a structural trap), but there is a "patchiness" to the distribution of the amplitudes that indicates that the structure is not the only control on seismic amplitude. Production data from this field show that wells that were producing in 1985 (the year of the seismic data collection) were generally from regions of high amplitudes in the seismic data and that there is a good qualitative relationship between amplitudes and production character (Figure 4C). Wells that have watered out are associated with the downdip limit of the high-amplitude regions (e.g., 330-A-13ST), whereas the strong amplitudes located higher on the structure are associated with the best-producing wells. The high amplitudes are associated with oil-charged, clean, porous, and permeable sand beds of the delta front facies (e.g., 330-C-2). Production is virtually water free on the crest of the anticline, but rates are lower because lower permeability facies are located there (e.g., 330-C-1). The amplitudes associated with the top of the GA interval are lower at this location. Together, the stratigraphic, seismic, and production data strongly suggest that the patchiness of the distribution of high seismic amplitudes (Figure 5C) is a result of reservoir compartmentalization by depositional features.

Figure 5C shows high-amplitude regions associated with the top of the GA interval in the Teal Field of Blocks 338 and 339. The trend of the linear anomalies is approximately E-W, roughly perpendicular to the direction of delta front progradation as interpreted from clinoform orientation in the seismic and wireline data. As in the EI Block 330 example, there is a complex relationship between the location of the amplitude anomalies and the structure contours. A well-log cross section oriented approximately perpendicular to the trend of the amplitude anomalies shows internal downlapping relationships and some toplap (Figure 5A) that are similar to the trends seen in the GA on the crest of the anticline in central EI Block 330 (Figure 4A). Based on the seismic data and well-log correlations, we interpret the curvilinear amplitude trends to represent the uneven distribution of hydrocarbons in shore-parallel sand bodies at the tops of clinoforms (delta front sands). Sibley and Mastoris (1994) describe how this conceptual model was integrated with production data to target the drilling of the 339-B-19ST horizontal well into the southernmost amplitude anomaly trend (Figure 5C) to access bypassed pay that was not being produced from existing wells. Another well in that drilling program found bypassed pay to the NW of the main reservoir, which was represented in the seismic data as an undrilled high-amplitude patch (Figure 5C).

## SUMMARY

Our mapping of the shelf margin deltaic deposits of the GA interval has revealed complex associations of depositional features such as channels and clinoforms that interfinger (laterally and vertically) over relatively short distances (few km). Hydrocarbon reserves in the GA interval are primarily associated with the updip facies at the tops of clinoforms, immediately below the flooding surface that caps the deltaic package. Identification of the clinoform geometries helped us to define reservoir heterogeneities with wireline log data and to interpret the amplitude maps associated with the top of the interval. The complex relationships between structure contours and amplitude anomalies in two fields that produce from this interval are a product of reservoir compartmentalization by depositional features in delta-front and delta-mouth bar deposits. This type of amplitude distribution contrasts quite clearly with that associated with a more continuous, block sand in this same field, where amplitude anomalies follow structure contours more closely (Anderson et al., this volume). Production data from the GA interval indicate that seismic amplitudes can be used to qualitatively predict reservoir character in similar exploration settings and to identify potential bypassed pay in development areas.

### *Acknowledgments*

Funded by Department of Energy grant DC-FC22-93BC14961 (R. N. Anderson, principal investigator), the Global Basins Research Network, and a Shell Faculty Fellowship to Flemings. Pennzoil Exploration and Production Company, Shell Offshore Inc., and Exxon donated wireline data to Penn State. We thank J. Austin of Pennzoil for his interest, insight, and encouragement.

## REFERENCES CITED

Alexander, L. L., and P. B. Flemings, 1995, Geologic evolution of a Plio-Pleistocene salt withdrawal mini-basin: Block 330, Eugene Island South Addition, Offshore Louisiana: AAPG Bulletin, v. 79, no. 12, p. 1737–1756.

Anderson, R. N., A. Boulanger, W. He, Y. F. Sun, L. Xu, and B. S. Hart, *this volume*, 4-D seismic monitoring of drainage in the Eugene Island 330 Field in the Offshore Gulf of Mexico: *in* P. Weimer and T. L. Davis, eds., AAPG Studies in Geology No. 42 and SEG Geophysical Development Series No. 5, AAPG/SEG, Tulsa, p. 9–20.

Badgett, K. L., P. L. Hill, W. H. Mills, S. P. Mitchell, G. S. Vinson III, and K. L. Wilkins, 1994, Team combines technologies to target horizontal wells in Gulf of Mexico oil field: Oil and Gas Journal, March 14, p. 44–49.

Clark, V. A., 1992, The effect of oil under in-situ conditions on the seismic properties of rocks: Geophysics, v. 57, p. 894–901.

Galloway, W. E., 1981, Depositional architecture of Cenozoic Gulf coastal plain fluvial systems, *in* F. G. Etheridge and R. M. Flores, eds., Recent and ancient nonmarine depositional environments: models for exploration: Society of Economic Paleontologists and Mineralogists Special Publication 31, p. 127–156.

Holland, D. S., J. B. Leedy, and D. R. Lammlein, 1990, Eugene Island Block 330 Field—U.S.A. Offshore Louisiana, *in* E. A. Beaumont and N. H. Foster, compilers, Structural Traps III: Tectonic fold and fault traps: Treatise of Petroleum Geology, Atlas of Oil and Gas Fields, AAPG, Tulsa, p. 103–143.

Mills, W. H., K. L. Badgett, S. P. Mitchell, and G. S. Vinson III, 1995, Using pressure transient analysis to optimize field development in compartmentalized shelf margin deltaic reservoirs: 1995 AAPG Annual Meeting, Houston, TX, Program with Abstracts, p. 66A.

Mitchum, R. M., J. B. Sangree, P. R. Vail, and W. W. Wornardt, 1993, Recognizing sequences and systems tracts from well logs, seismic data, and biostratigraphy: examples from the Late Cenozoic of the Gulf of Mexico, *in* P. Weimer and H. W. Posamentier, eds., Siliciclastic sequence stratigraphy: recent developments and applications: AAPG Memoir 58, p. 163–197.

Morton, R. A., and J. R. Suter, 1996, Sequence stratigraphy and composition of late Quaternary shelf-margin deltas, northern Gulf of Mexico: AAPG Bulletin, v. 80, no. 4, p. 505–530.

Sibley, D. M., and S. S. Mastoris, 1994, Rethinking conventional field development: 3-D shows direct detection of bypassed oil reserves, *in* Society of Petroleum Engineers International Petroleum Conference and Exhibition of Mexico Proceedings: SPE Paper 28719, p. 509–519.

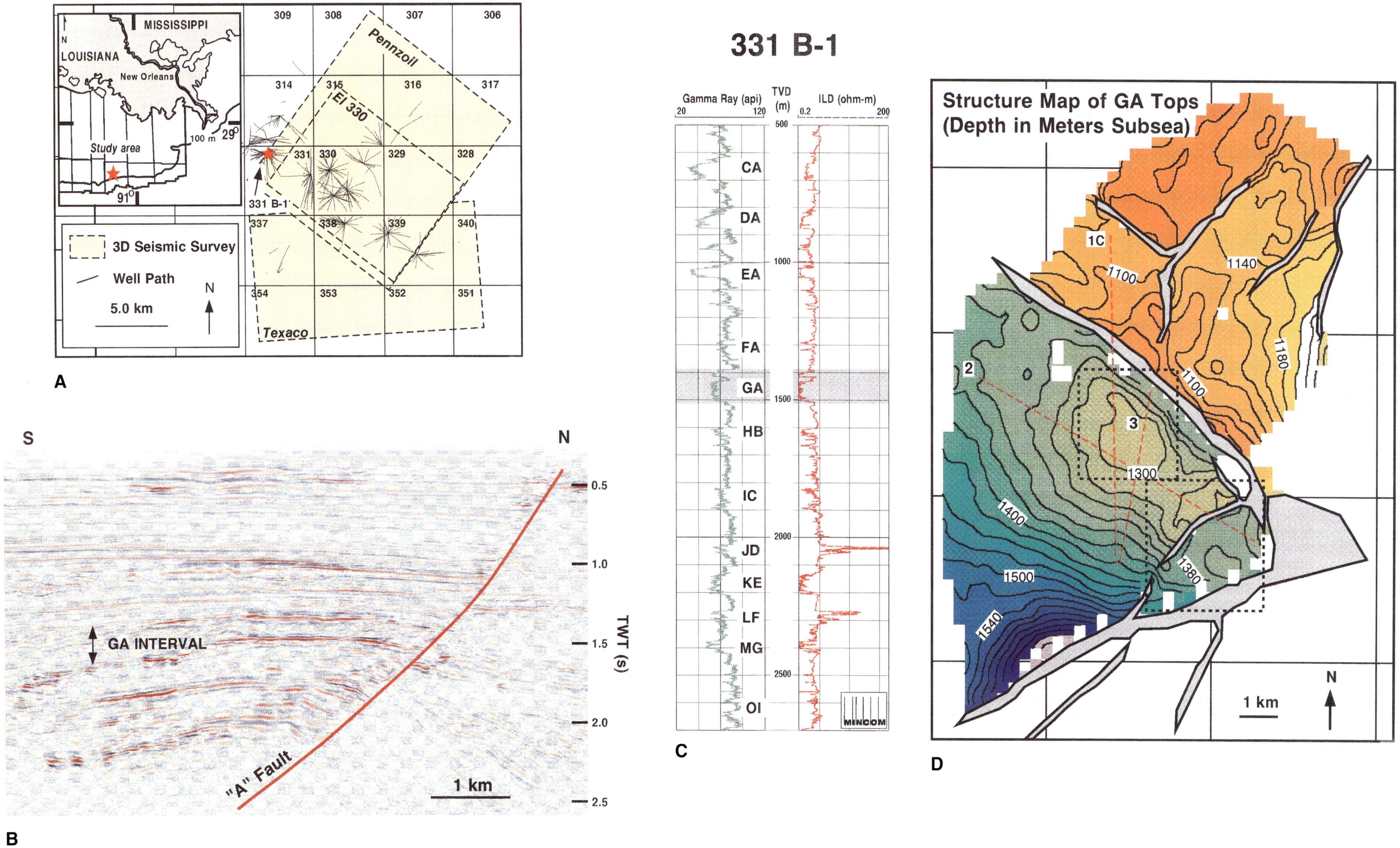


FIGURE 1. (A) Location of study area, showing well and 3-D seismic database. (B) Seismic transect through EI 330 Field showing location of GA interval and principal structural elements. Rollover anticline has formed basinward of principal regional growth fault. (C) Wireline log showing stratigraphic position of GA interval. Major sands labeled CA–OI following Holland et al. (1990). Well location shown in 1A. (D) Structure map of top of GA interval (depth in meters subsea) showing approximate locations of subsequent figures (labeled with figure numbers). Productive area of Block 330 in box labeled 4C, Teal Field in 5C.

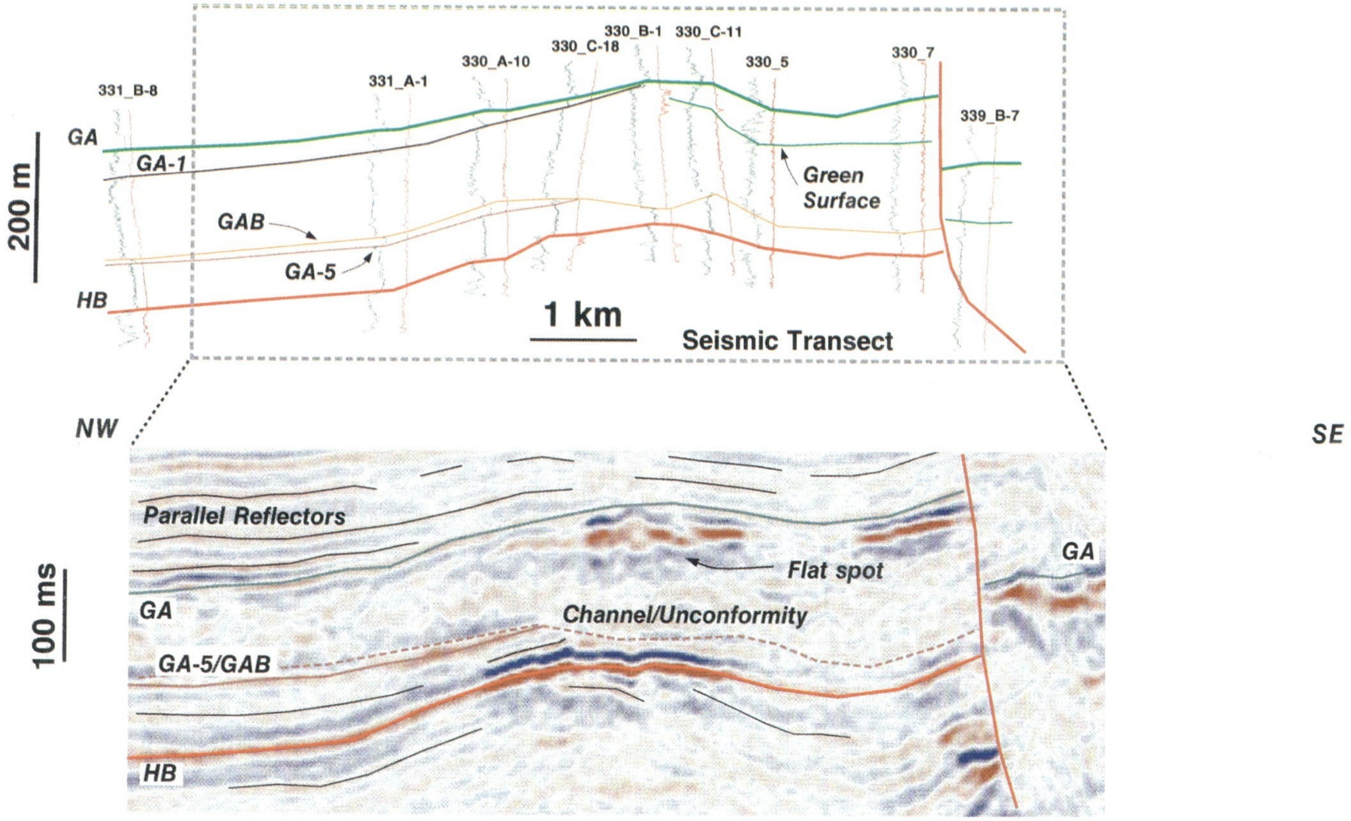


FIGURE 2. NW (left) to SE (right) wireline log cross section (above) and corresponding 3-D seismic transect (below) across EI Block 330, showing channelized nature of sequence boundary at the base of the GA interval. Stratigraphic surfaces shown are: (A) HB—flooding surface capping underlying lowstand deltaic complex; (B) GA-5—sand capping backstepping shaley parasequence; GAB—sequence boundary at base of GA interval; green surface—contact between two stacked progradational lobes (see Figure 3); GA-1—sand deposited at the onset of transgression and preserved in the low-lying area northwest of the anticline in Block 330. Seismic transect shows seismic facies, principal horizons, and extent of amplitude anomalies associated with hydrocarbons. Some reflections highlighted for clarity. Channel incision on the crest of the anticline during lowstand has removed the GA-5 sand that is present on the western flank. Although the GA-5 sand is not recognized in the 330 C-11 well on the eastern flank of the anticline, the gamma ray log in that well clearly shows the base of the blocky GA sand rising back up in that direction. Location of section shown in Figure 1D.

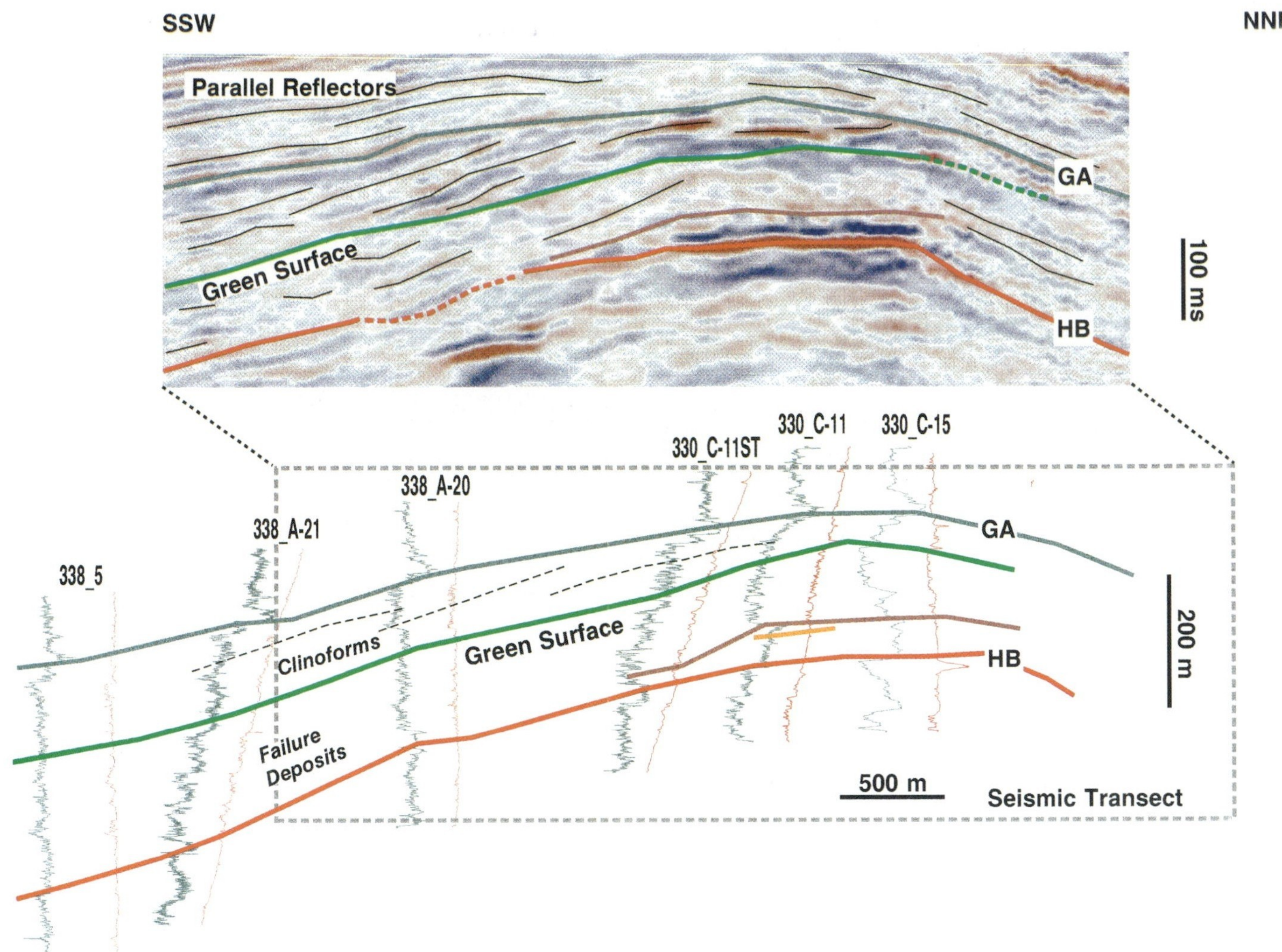


FIGURE 3. North (right)–South (left) 3-D seismic transect (above) and wireline log cross section (below) through central portion of EI Block 330 into northern part of Block 338 (see Figure 1D for location). Stratigraphic surfaces as in Figure 2. Section ties with Figure 2 at 330 C-11 well. The GA interval is divisible into two progradational units that are separated by a surface ("green surface") that shows a southward (offshore) transition from a channel base on the crest of the anticline to a downlap surface. Some clinoforms are visible below the green surface, but typically that interval is characterized by chaotic reflections that are failure deposits to the south (e.g., 338 A-21) or channel deposits to the north (e.g., 330 C-11).

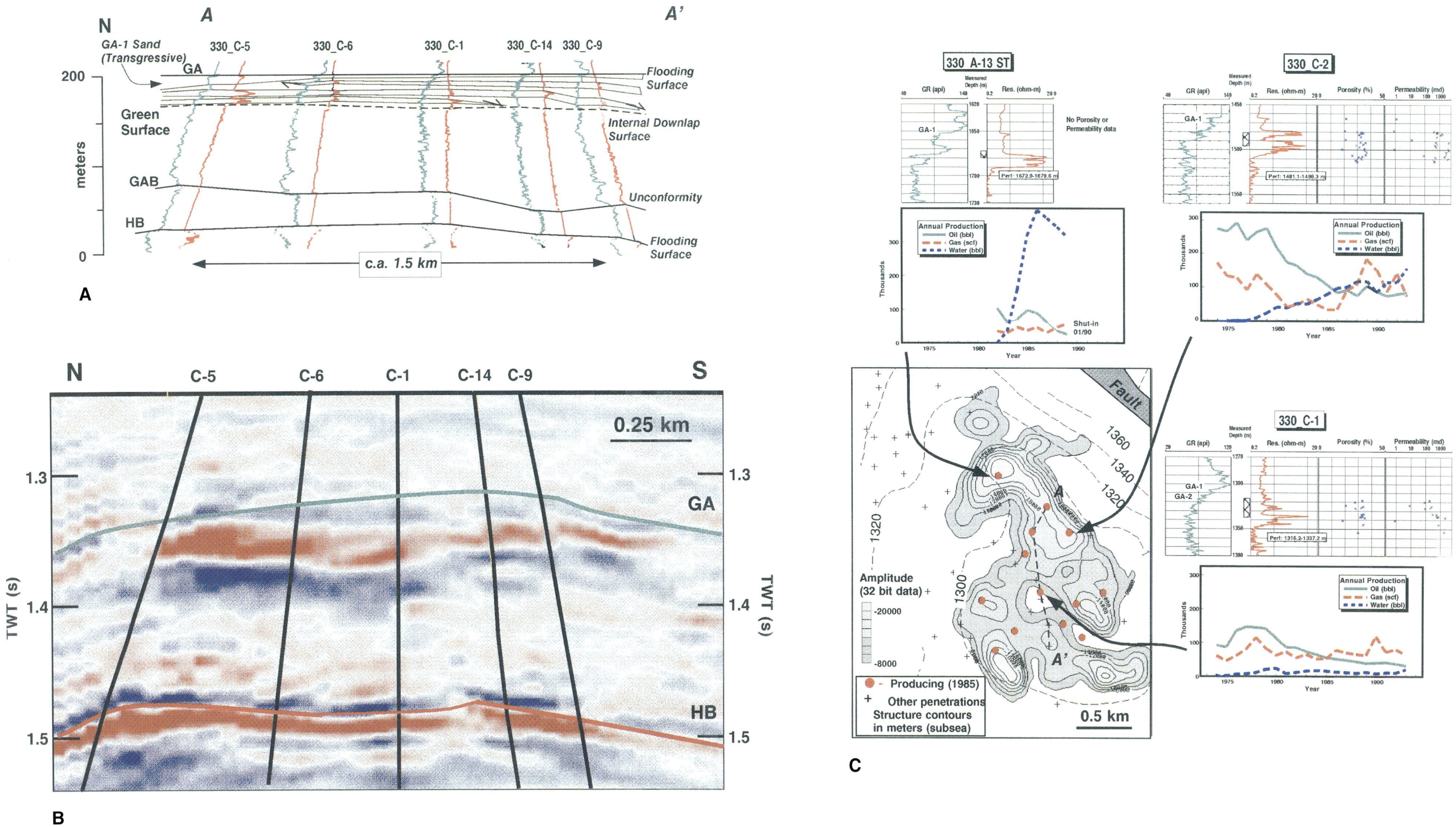


FIGURE 4. (A) Stratigraphic cross section, showing detailed correlation of oil-bearing sands in heterolithic deposits on crest of anticline (see Figure 1D for location). Section hung on GA pick (top of GA-1 sand and correlative siltstones to south). Note transition from gentle toplap in north to downlap of sands to south onto green surface. (B) Seismic transect that approximately corresponds to wireline log cross section of 4a. (C) High-amplitude regions below GA seismic pick (within 60 ms twt of pick), simplified structure contours, location of wells producing from GA sand in 1985, and production histories for some wells. Location of perforated zone shown on log display (depths in measured depth below kelly bushing). Porosity and permeability measurements from sidewall cores.

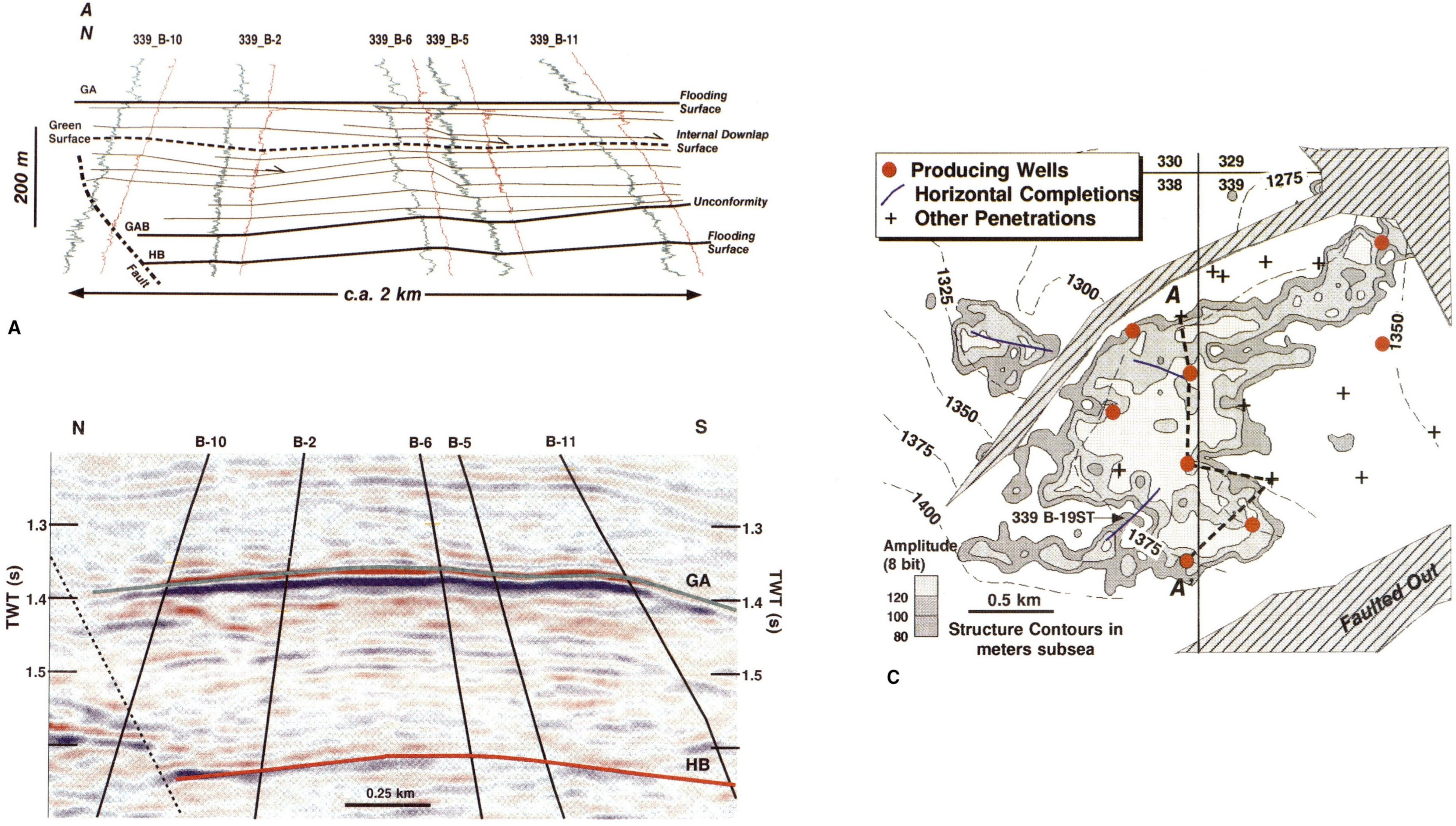


FIGURE 5. (A) North–south stratigraphic cross section through the GA interval in the Teal Field, showing detailed correlations of sands and shales and composite stratigraphy (stacking of deltaic lobes) of the interval (see Figure 1D for location). Section hung on GA top marker (siltstones correlative to GA-1 sand). Thickening to the north probably due to syn-depositional growth fault activity. (B) Seismic transect that approximately corresponds to wireline section. (C) High-amplitude regions associated with the top of the GA interval in Teal Field showing effects of reservoir compartmentalization by faults and clinoforms. Amplitudes thought to diminish to the right because of improper stacking during processing and to the north by shale-out (Sibley and Mastoris, 1994). Note location of 339 B-19ST well, discussed in text.

CHAPTER 3

# 3-D Seismic Imaging and Interpretation of Fluvially Deposited Thin-Bed Reservoirs

Hardage, B. A., R. A. Levey, V. Pendleton, J. Simmons, and R. Edson, 3-D seismic imaging and interpretation of fluvially deposited thin-bed reservoirs, *in* P. Weimer and T. L. Davis, eds., *AAPG Studies in Geology No. 42* and *SEG Geophysical Developments Series No. 5*, AAPG/SEG, Tulsa, p. 27–34.

**Bob A. Hardage*, Raymond A. Levey*, Virginia Pendleton†, James Simmons*, and Rick Edson***

## ABSTRACT

We investigated the following problem: "How do fluvial depositional processes create compartmented gas reservoirs?" Using vertical seismic profile (VSP) data to define where selected thin-bed gas reservoirs were positioned in a 3-D seismic data volume, we created horizon slices through this 3-D image that showed the reflection amplitude behavior across the depositional surfaces where targeted thin-bed reservoirs were located. We saw intriguing meander features on these 3-D amplitude displays, which appeared to be realistic depictions of intermeshed fluvial channels. We then overlaid well-log cross sections on these 3-D seismic images, which inferred the depositional environments that were found by wells that penetrated the reservoir system, and these geologic constraints confirmed that the imaged meander features were indeed channels.

The most important nonseismic data that we used to understand how 3-D seismic images can reveal reservoir compartment boundaries were various forms of reservoir engineering data that proved which wells shared a common pressure compartment and which wells did not. Using these engineering constraints, we showed that many of the seismically imaged channel features created reservoir compartment boundaries that impeded lateral fluid flow. Equally important, we showed that some seismically imaged channels had minimal effect on lateral flow and did not form compartment boundaries. We concluded that in any effort where 3-D seismic data are used to infer the internal compartmentalized architecture of a reservoir system, good quality reservoir engineering control, such as pressure interference tests and pressure decline curves, must be incorporated into the 3-D interpretation.

## INTRODUCTION

This case history summarizes the results of a field study that analyzed how fluvial deposition affects gas reservoir compartmentalization. We performed this study in a portion of Stratton Field in Kleberg and Nueces Counties of South Texas, and the stratigraphic interval we studied was the Oligocene Frio Formation, a thick, fluvially deposited sand-shale sequence that has been a prolific gas producer in Stratton Field and in several other fields along the FR-4 depositional trend (Figure 1). The regional structure and stratigraphy of the Frio system are well documented by Nanz (1954), Galloway (1977, 1982), Han and Scott (1981), and Galloway et al. (1982) and will not be repeated here. A more detailed explanation of this 3-D seismic investigation can be found in Hardage et al. (1994).

We acquired 3-D seismic data from a 7.6-sq-mi (19.7-sq-km) study area (Figure 2). A large number of wells, some as old as 40 years, existed inside this 3-D grid, and the logs recorded in these wells allowed us to make a reasonably thorough geologic analysis of the Frio reservoirs. As shown in Figure 2, we acquired additional data in several wells (the circled dots) to supplement the historic well log, production, and reservoir pressure data bases. These supplemental data consisted of modern well logs, cores, and various pressure tests. Vertical seismic profile (VSP) data were recorded in the well inside the triangle shown near the center of the 3-D grid.

*Bureau of Economic Geology, The University of Texas at Austin, University Station, Box X, Austin, TX, 78713-8924
†Formerly at Bureau of Economic Geology, The University of Texas at Austin, University Station, Box X, Austin, TX, 78713-8924; currently at Integrity Geophysics, 1503 Palma Plaza, Austin, TX 78703-3434.

## THIN-BED INTERPRETATION PROCEDURE

The emphasis in this case study was on demonstrating how geologic and engineering data are essential in interpreting depositionally generated reservoir compartment boundaries in 3-D seismic images. The seismic interpretation at Stratton Field was particularly challenging, because most of the Frio reservoirs were thin (<15 ft [<5 m]), and they were closely stacked, in some areas separated by only 10–15 ft (3–5 m) vertically. These conditions required precise calibration of stratigraphic depth versus seismic traveltime to extract a depositional stratal surface from the 3-D data volume that would reliably depict the areal distribution of a particular Frio thin-bed reservoir.

We used VSP data as the primary measurement to define where a specific thin-bed reservoir was positioned in the 3-D seismic data volume. The locations of the VSP calibration well we used is shown in Figure 2. The zero-offset VSP data recorded in this well were used to establish the precise depth-versus-time control needed for the thin-bed interpretation. These VSP data are shown in Figure 3, where the zero-offset image is spliced into a north–south vertical slice from the 3-D data volume passing through the VSP well. The VSP image has a wider range of frequency components (10–120 Hz) than does the 3-D seismic image (10–80 Hz), and we made no attempt to equalize the spectral bandwidths of these two images when constructing Figure 3. The figure also shows a graphic representation of the stratigraphic column penetrated by the VSP well. Only producing or potentially producing Frio reservoirs are shown in this diagram, and not all of the reservoirs are labeled by name. The top and base of each reservoir are accurately positioned in terms of two-way VSP traveltime, and because there is no difference between the VSP time datum and the 3-D time datum in this instance, the reservoirs are also correctly positioned vertically inside the 3-D seismic data volume at the VSP well.

Using these VSP traveltime control data, we knew exactly where each thin-bed reservoir belonged in the 3-D seismic reflection waveform at the VSP well. We then extended this thin-bed calibration away from the VSP well and across the entire 7.6-mi$^2$ (19.7-km$^2$) area imaged by the 3-D data.

We now show the results of this thin-bed interpretational procedure at Stratton Field and support the interpretations with geologic and engineering control.

## F37 RESERVOIRS

A horizon slice showing seismic reflection amplitude behavior across the F37 reservoir system is displayed in Figure 4. (Refer to Figure 3 to see where the F37 reservoir is located in the stratigraphic column.) This image reveals several intermeshed, meandering features that suggest a fluvial deposition is being portrayed. Log interpretation (Figure 5) implies the F37 reservoirs found in wells 189 and 185 were deposited as channel fill, and the seismic image shows these wells to be directly atop a meander feature. The initial depositional interpretation for the extremely thin F37 reservoir in well 211 was that this well bore could have penetrated either splay or channel fill. The splay option is indicated in Figure 5. Because the 211 wellhead is approximately 91 m (300 ft) north of the meander feature (Figure 4), the log-based interpretation of the F37 depositional environment at the 211 well is also supported by seismic evidence, because the bottom-hole location could be in either channel fill or splay.

We analyzed pressure histories recorded in several F37 reservoirs near these seismic meander features to determine whether reservoir compartmentalization existed. These pressure histories, summarized in Figure 6, show at least three, and perhaps four, individual F37 reservoir compartments in this area of the field. We relied heavily on these pressure data to guide the interpretation of the thin-bedded F37 reservoirs.

A reservoir model that honors all three data bases—the seismic, the geological, and the reservoir engineering—is proposed in Figure 7. This model assumes that the F37 reservoir in the southeast quadrant of the 3-D grid is composed of three intermeshed channels, labeled A, B, and C; a grid overlay of seismic inline and crossline coordinates is provided so these channels can be correlated with features in the 3-D seismic image. The location of the F37 stratigraphic cross section (Figure 5) is shown, but this geologic information defines channel locations along only a single 2-D profile of the model. The important information again is the reservoir pressure data, because without this engineering data there would be no reason to conclude that a three-channel model would be appropriate. The model thus places well 129 in channel A and well 185 in channel B, which allows these two wells to be in different F37 pressure regimes, that is, in different compartments (channels). Wells 127 and 161 are proposed to be in channel C, south of the 3-D seismic coverage. Only one meander loop of this hypothesized channel C extends into the 3-D seismic grid. The rapid F37 pressure decay observed in well 189 (Figure 6) implies that this well is not in pressure communication with well 185, even though both wells are in channel B. For this reason, we show a stratigraphic variation in channel B where there may be an intrachannel compartment boundary.

We concluded that the seismic image in Figure 4 revealed not just one meander channel system but at least three intermeshed thin-bed channels. By employing common existing reservoir engineering data (that is, pressure histories), we were able to use 3-D seismic images to define where compartment boundaries most likely existed in the interwell spaces.

## D11 RESERVOIRS

Our second example of a compartmented fluvial reservoir is the D11 system, which is one of the extremely thin reservoirs immediately above the D35 reservoir shown in Figure 3. Again, using our VSP calibration technique to define the appropriate flattened time slice that equated to the D11 stratigraphic level, we generated the reflection image in Figure 8.

The information in Figure 9 summarizes the D11 geologic and engineering data available in this portion of the 3-D seismic grid. The D11 reservoirs in these four study wells were all interpreted from log shape analysis to be channel fill; some log character indicated splay deposition in the younger channel system. The pressure information was again the key information that told us how many reservoir compartments should be seismically imaged. As noted in Figure 9, the oldest production well (77) was drilled in 1956 and encountered a reservoir pressure of 1740 psi. This well was abandoned in 1972 when the pressure declined to uneconomic levels. Wells 135, 150, and 156 were drilled in the 1980s, and all three wells encountered virgin D11 reservoir pressure, implying these wells were in a single D11 compartment that was depositionally segregated from well 77.

We thus interpreted our D11 seismic image with the knowledge that we were dealing with at least two intermeshed or juxtaposed thin-bed channel systems. The reservoir model we inferred from Figure 8 is shown in Figure 10. This model honored the geologic information by placing each of the four wells in a seismically defined D11 feature that appeared to be a channel and honored the pressure data by placing wells 135, 150, and 156 in the same channel complex and well 77 in a separate channel system.

## F21 RESERVOIRS

In studying the F21 reservoir system, we focused our attention in the area of wells 130, 164, 176, 195, and 201, where we had the geologic and engineering control shown in Figures 11 and 12. This control showed that we were dealing once again with a splay/channel-fill depositional regime (Figure 11) and that wells 176, 195, and 201 (completions A, B, and C) produced from the same F21 reservoir compartment (Figure 12). Transferring this information to the seismic image in Figure 13 showed that we had as much depositional heterogeneity in the interwell spaces at this F21 reservoir level as we had observed at the F37 (Figure 4) and D11 (Figure 8) levels; yet the reservoir pressure behavior implied the F21 reservoir system had no interwell compartment boundary, whereas the other reservoir levels (F37 and D11) did.

Our proposed model for the F21 reservoir system imaged in Figure 13 is shown in Figure 14. Although the meander channels in this model overlay and intersect each other like the channels shown in the models proposed in Figures 7 and 10, pressure data forced us to conclude that all of the intermeshed F21 channels between wells 176, 195, and 201 (Figure 13) comprise a single reservoir compartment, whereas the intermeshed channels in Figures 7 and 10 create distinct reservoir compartments.

## CONCLUSION

We show three examples of seismic thin-bed interpretation in a fluvially deposited gas reservoir and support these interpretations with geologic and reservoir engineering data. In two of these examples, the 3-D seismic data reveal stratigraphic variations where reservoir pressure information implies that a compartment boundary should exist. In the third example, the seismic data show a degree of depositional heterogeneity equivalent to that in the first two cases, yet reservoir pressure behavior indicates that none of these latter stratigraphic variations creates a compartment boundary. These examples illustrate that although fluvial deposition creates numerous compartment boundaries, determining which seismically imaged stratigraphic changes are compartment boundaries requires that geologic and reservoir engineering data (particularly reservoir pressure data) be incorporated into the seismic interpretation.

We found it was particularly important to have an accurate and reliable way to translate thin-bed stratigraphy (known in depth) into precisely defined seismic time windows. We are convinced that VSP data, when properly recorded and processed, are the best information to establish the detailed depth-versus-time calibration required to seismically distinguish closely spaced thin beds.


### *Acknowledgments*

Funding for this research was provided by the Gas Research Institute under Contract No. 5088-212-1718 and the U.S. Department of Energy under Contract No. DE-FG21-88MC25031. Landmark and IBM provided the workstation technology that was used in this investigation. Publication has been authorized by the Director, Bureau of Economic Geology.


## REFERENCES CITED


Galloway, W. E., 1977, Catahoula Formation of the Texas Coastal Plain—depositional systems, composition, structural development, groundwater flow history, and uranium distribution: The University of Texas at Austin, Bureau of Economic Geology Report of Investigations 87, 59 p.

Galloway, W. E., 1982, Depositional architecture of Cenozoic Gulf Coastal Plain fluvial systems, *in* F. G. Ethridge and R. M. Flores, eds., Recent and ancient nonmarine depositional environments: models for exploration: Society of Economic Paleontologists and Mineralogists Special Publication 31, p. 127–155.

Galloway, W. E., D. K. Hobday, and K. Magara, 1982, Frio Formation of the Texas Gulf Coast basin—depositional systems, structural framework, and hydrocarbon origin, migration, distribution, and exploration potential: The University of Texas at Austin, Bureau of Economic Geology Report of Investigations 122, 78 p.

Han, J. H., and A. J. Scott, 1981, Relationship of syndepositional structures and deltation, Vicksburg Formation (Oligocene), South Texas: Society of Economic Paleontologists and Mineralogists, Gulf Coast section, 2d Annual Research Conference, Program and Abstracts, p. 33–40.

Hardage, B. A., R. A. Levey, V. Pendleton, J. Simmons, and R. Edson, 1994, A 3-D seismic case history evaluating fluvially deposited thin-bed reservoirs in a gas-producing property: Geophysics, v. 59, p. 1650–1665.

Nanz, R. H., 1954, Genesis of Oligocene sandstone reservoir, Seeligson Field, Jim Wells and Kleberg Counties, Texas: AAPG Bulletin 38, no. 1, p. 96–117.

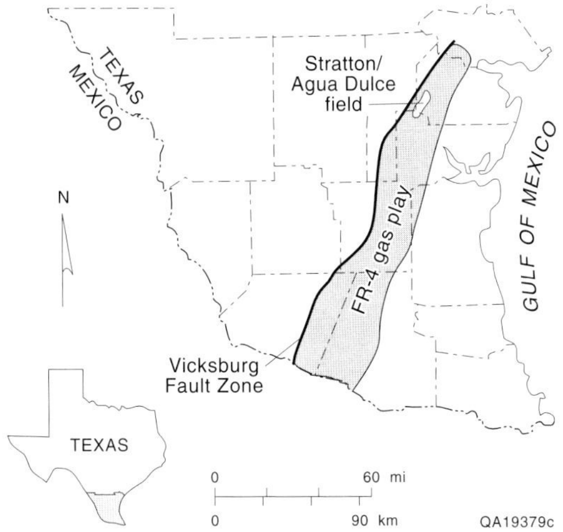


FIGURE 1. Map of the prolific Frio FR-4 gas trend in South Texas, showing the location of Stratton Field.

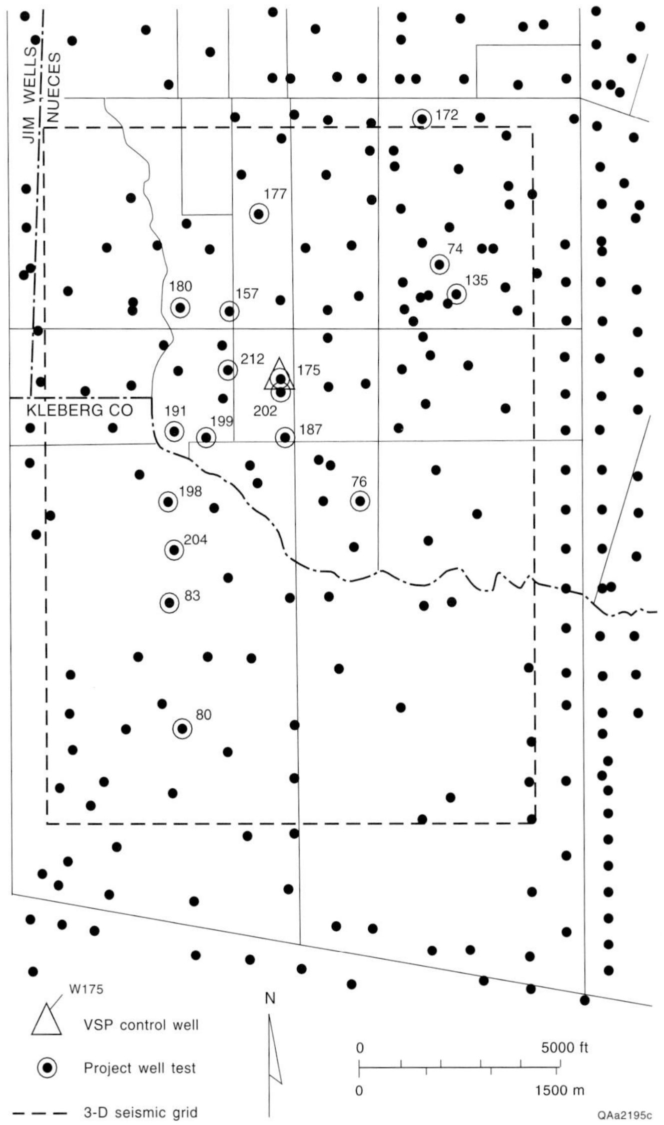


FIGURE 2. Generalized map of the study area. The solid dots show existing production wells, many drilled at 40-acre spacings. The circled dots indicate wells where additional geologic and engineering control was acquired in the form of modern logs, cores, or pressure tests. The 3-D seismic area is approximately 7.6 sq mi (19.7 sq km). VSP control data were recorded in the well marked by the triangle near the center of the 3-D seismic grid.

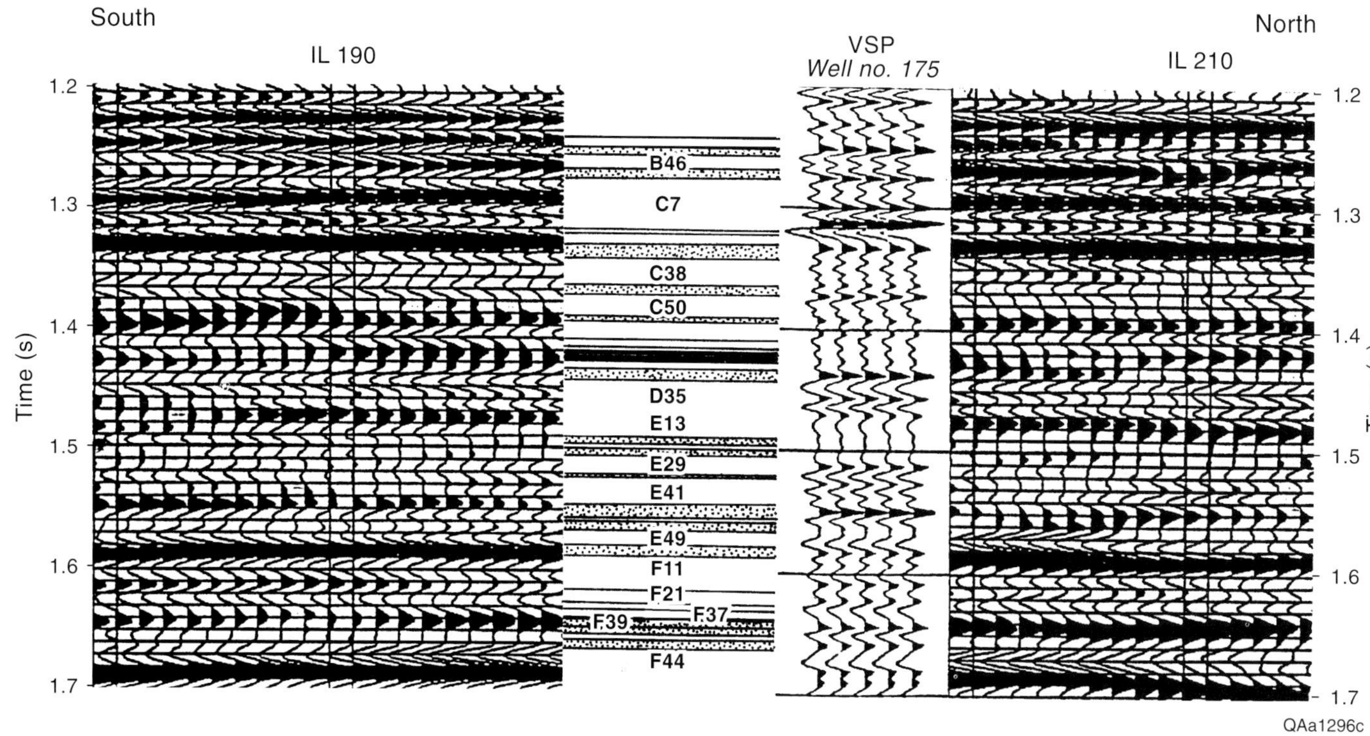


FIGURE 3. This display compares a north–south vertical section through the final migrated 3-D data volume with the zero-offset VSP image obtained at the VSP control well. The stratigraphic column shows most of the reservoirs penetrated by the VSP well. Note how many reservoirs there are, how thin they are, and how closely spaced they are vertically. The time positions of the top and base of each reservoir are defined via the VSP traveltime data. Precise positioning of the seismic time window for each thin-bed reservoir is the key to thin-bed seismic interpretation, and this VSP calibration technique was critical in defining where the depositional surface for each thin-bed reservoir was located in the 3-D seismic image. The VSP image has a much wider spectral bandwidth than does the 3-D seismic image.

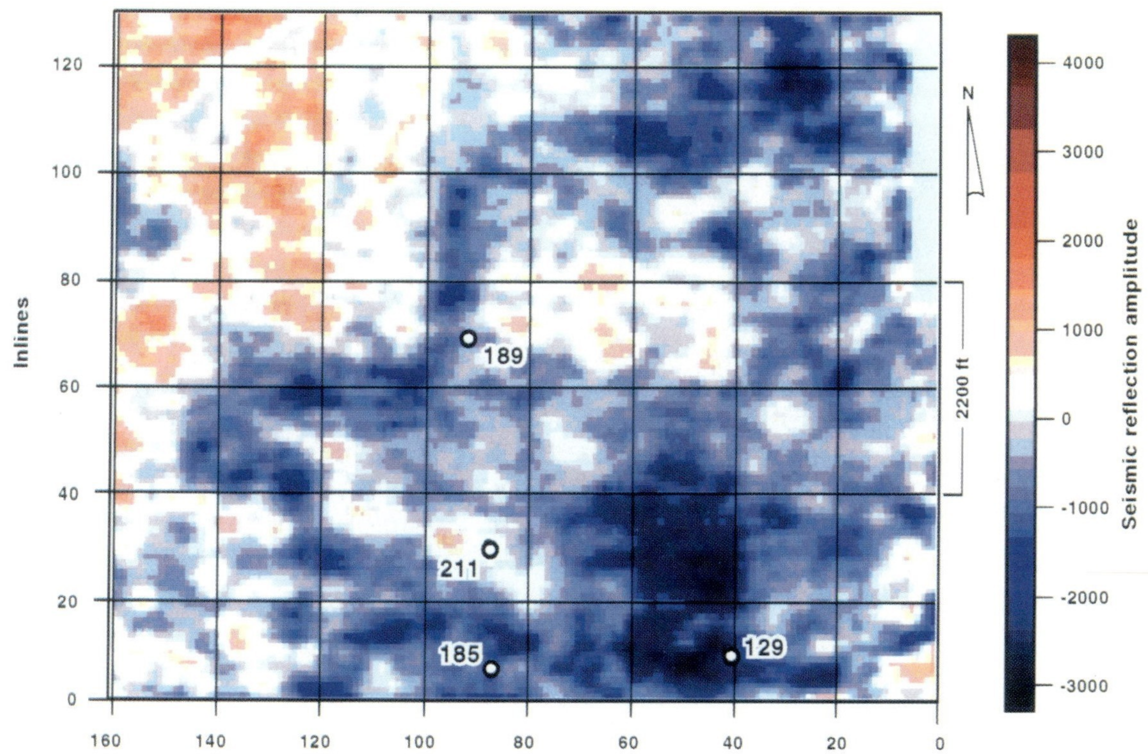

FIGURE 4. Seismic reflection amplitude behavior on the F37 stratal time surface in the vicinity of four critical information wells. An interpretation of the stacked thin-bed channel features revealed in this image follows as Figure 7.

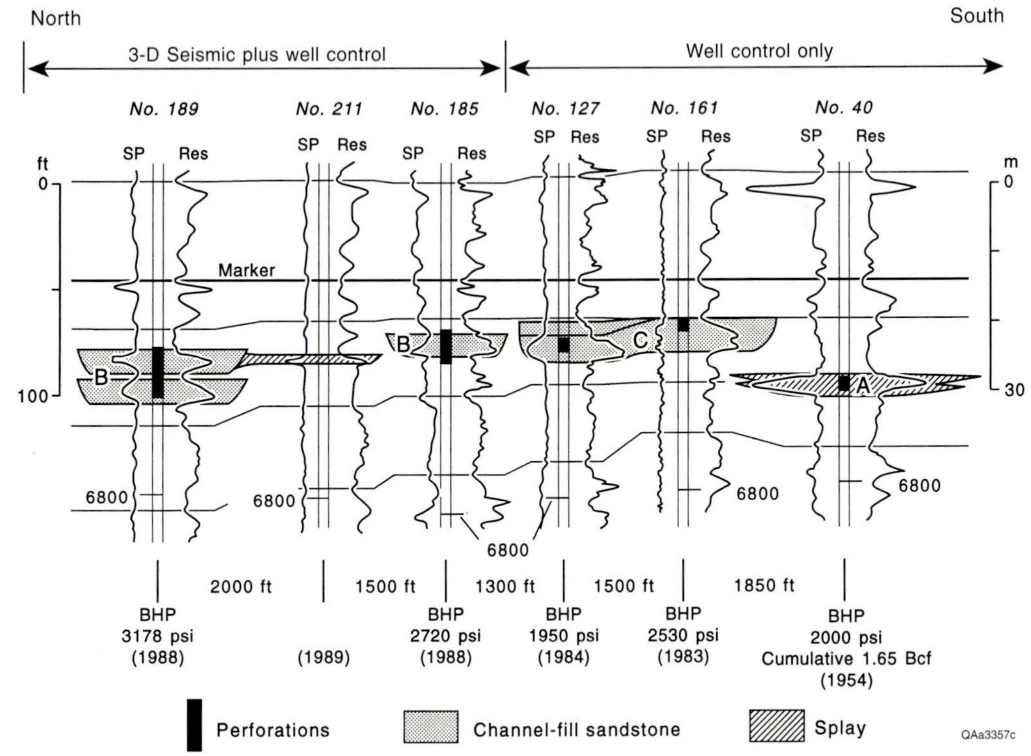

FIGURE 5. Stratigraphic cross section of the F37 reservoir, showing the depositional environments interpreted from log shapes and the initial bottom-hole pressure (BHP) observed in each well. All log curves are depth-shifted to a marker datum. The channels labeled A, B, and C refer to meander features shown in map view in Figure 7. The northern three wells are inside the 3-D seismic grid; the southern three wells are not.

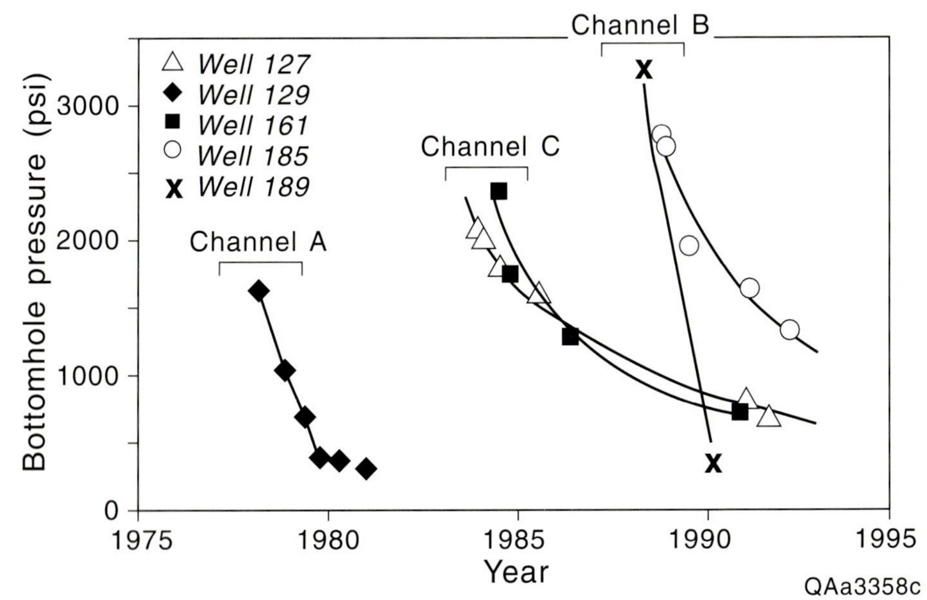

FIGURE 6. F37 reservoir pressure histories observed in wells near the seismically imaged meander features. These pressure decline curves indicate that these wells are positioned in at least three different F37 compartments, labeled as channels A, B, and C (same labeling notation as that used in Figures 5 and 7). There may be an additional, intrachannel, F37 compartment boundary in channel system B that segregates well 189 (the bold Xs) from well 185 (the open circles).

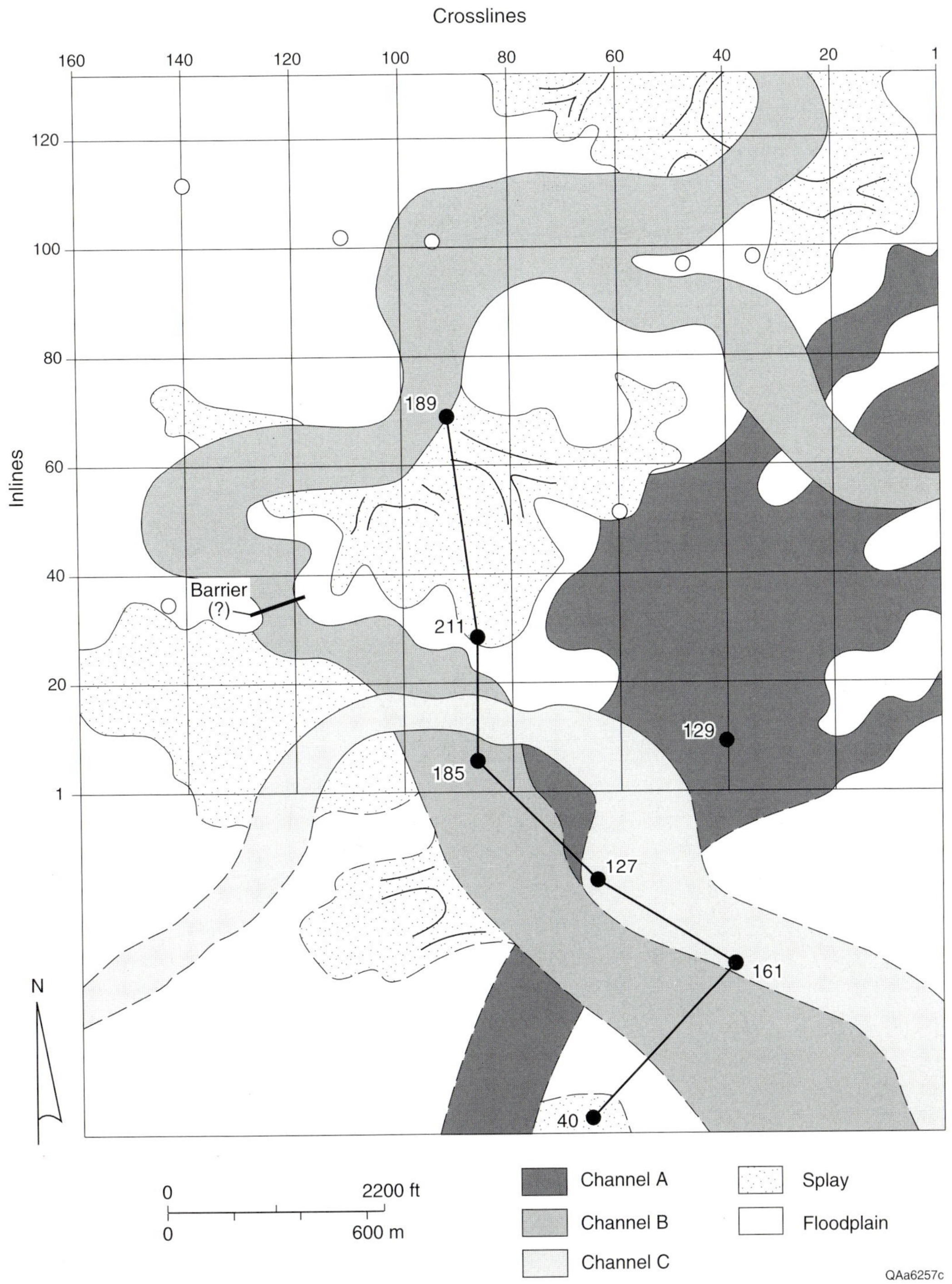


FIGURE 7. Proposed model for the F37 reservoir system imaged in Figure 4. This model honors the stratigraphic cross section shown in Figure 5 (the bold line connecting the six labeled wells) and the pressure history shown in Figure 6. Channel A is arbitrarily drawn as the deepest and channel C as the shallowest. An intrachannel barrier is proposed near crossline and inline coordinates (125 and 35, respectively) to explain the different pressure behaviors in wells 185 and 189 (see Figure 6). South of the 3-D seismic coverage area, the boundaries of channels A, B, and C are drawn as dashed lines to show that the channel shapes and positions are highly speculative.

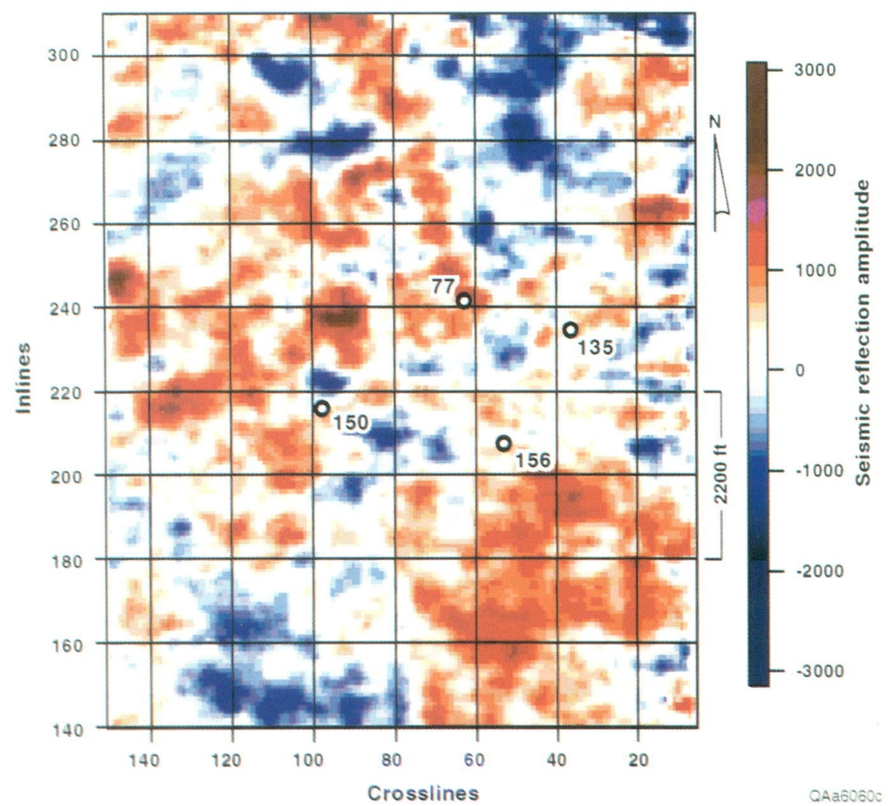


FIGURE 8. Seismic reflection amplitude behavior on a stratal time surface that passes through the D11 reservoir at the VSP calibration well.

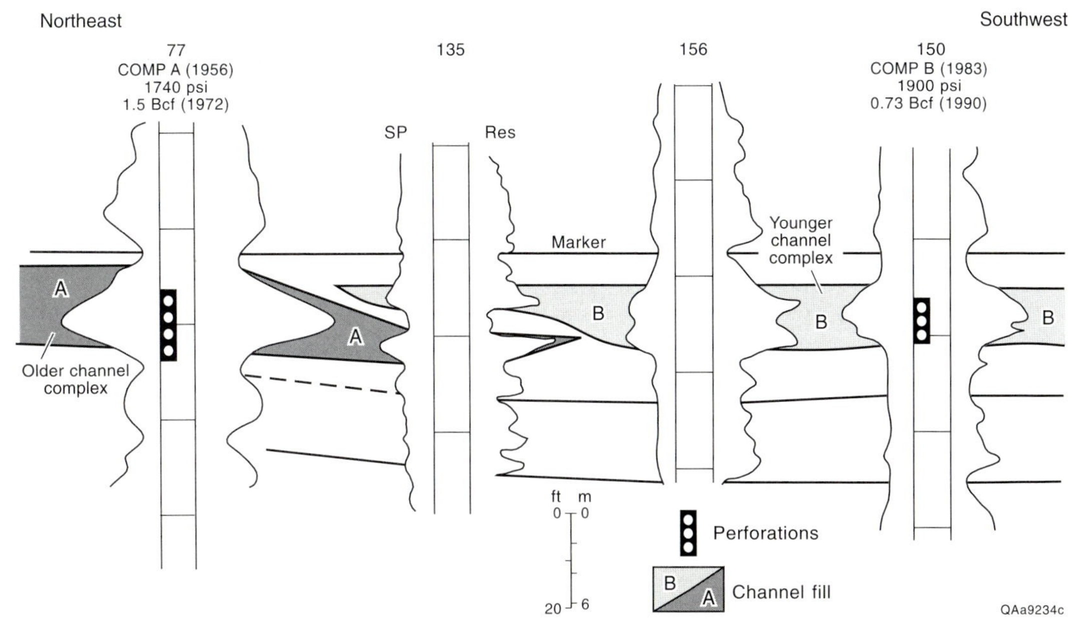


FIGURE 9. Stratigraphic cross section of D11 reservoirs showing the depositional environments interpreted from log shapes and the initial D11 reservoir pressures observed in two critical wells (77 and 150). These pressure data show that well 150, drilled in 1983, penetrated a D11 compartment with virgin reservoir pressure. This compartment evidently has no effective communication with the D11 compartment where well 77 was drilled, because well 77 was abandoned in 1972 because of depleted pressure. Consequently, an intermeshed two-channel reservoir system is proposed for the D11 reservoir level in this portion of Stratton Field.

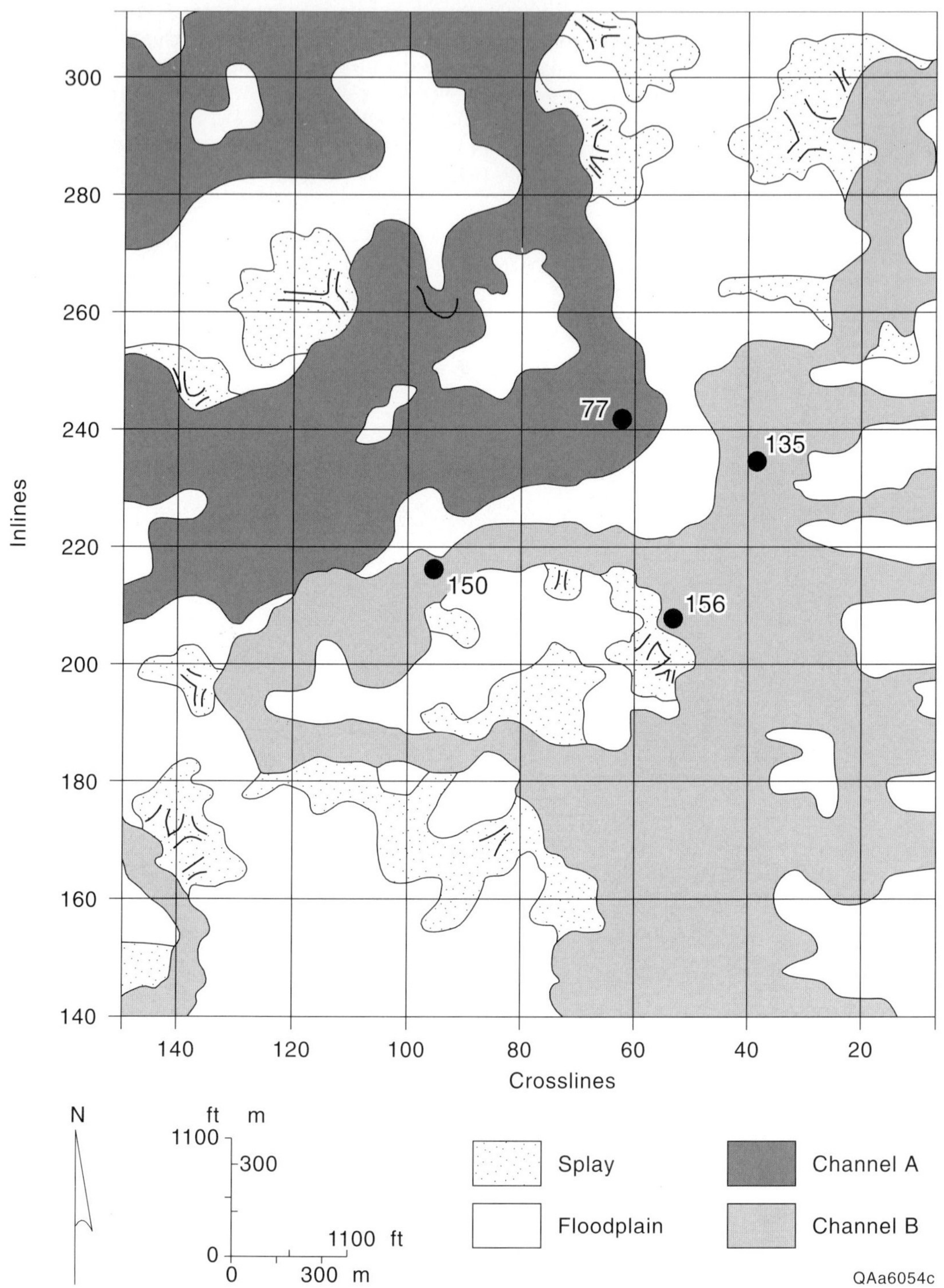


FIGURE 10. Proposed model for the D11 reservoir system imaged in Figure 8. This model honors the geologic and engineering data summarized in Figure 9 and mimics the meander pattern shown in Figure 8.

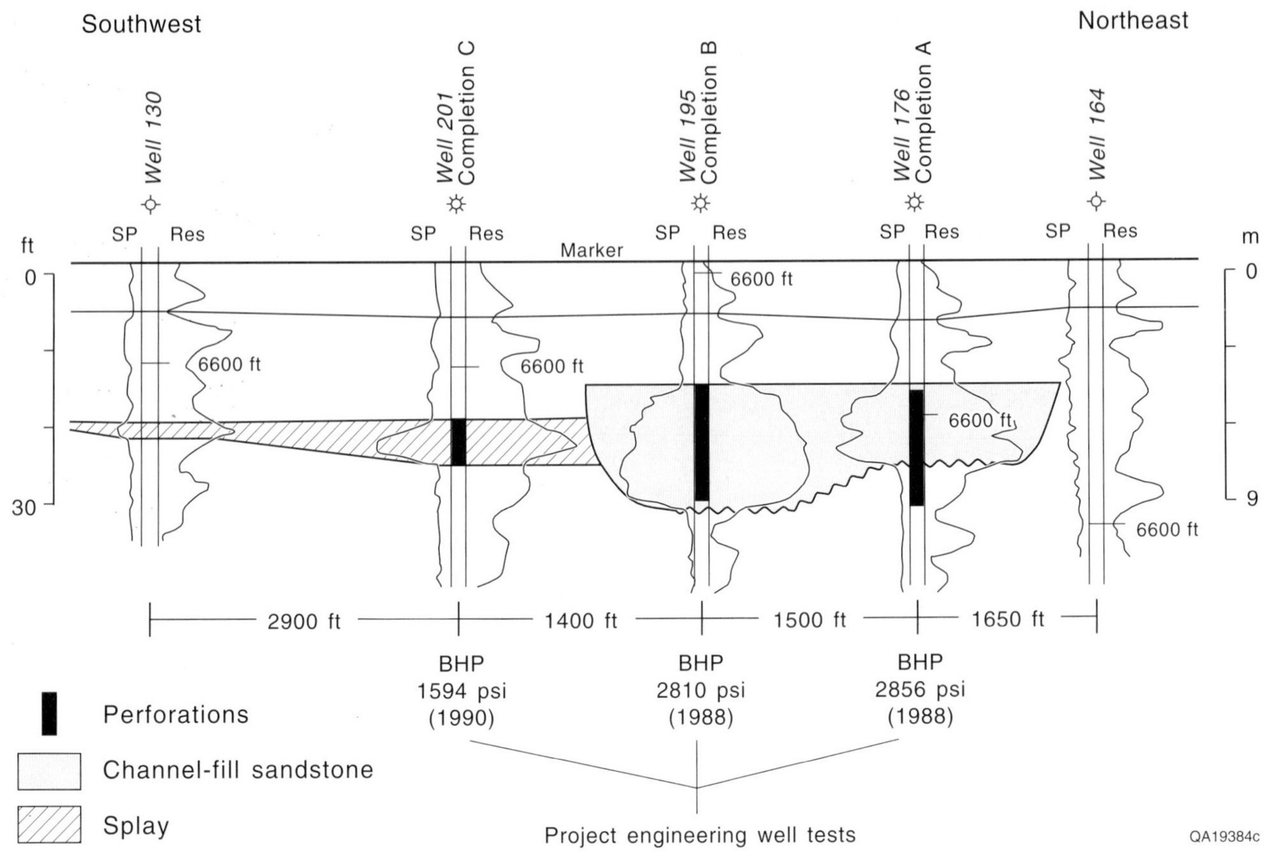


FIGURE 11. Stratigraphic cross section of the F21 reservoirs, showing the depositional environments interpreted from log shape analysis and the initial reservoir pressures (BHP).

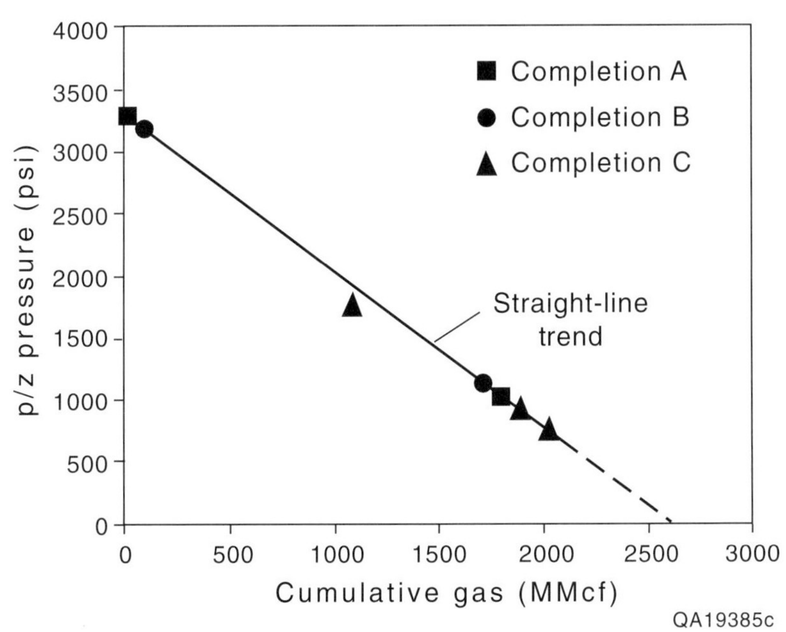


FIGURE 12. F21 reservoir pressure (normalized by the compressibility factor) versus cumulative gas production for completions A, B, and C, defined in Figure 11. Because the data points follow a single straight-line trend, all three completions are assumed to share the same reservoir compartment.

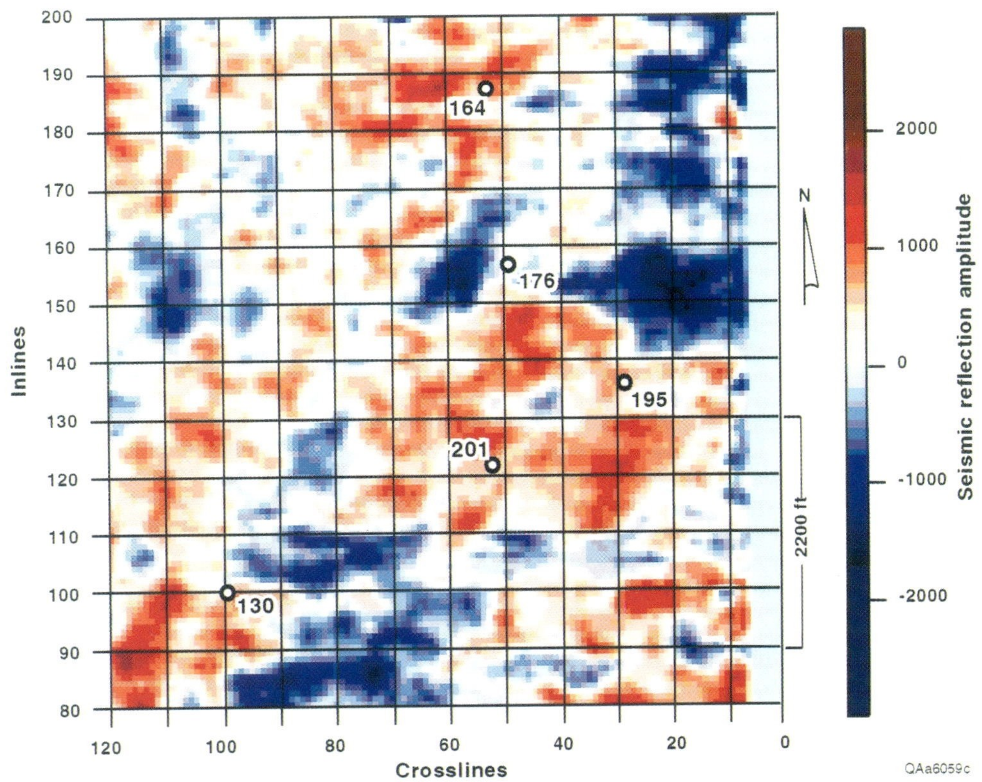


FIGURE 13. Seismic reflection behavior on a stratal time surface that passes through the F21 reservoir at the VSP calibration well. Note the stratigraphic resolution in this 3-D image. The intermeshed channels are as narrow as 200 ft (61 m), and log control indicates the channel fills are approximately 10 ft (3 m) thick (Figure 11). This subtle stratigraphy is imaged even though the channels are at a depth of 6600 ft (2012 m).

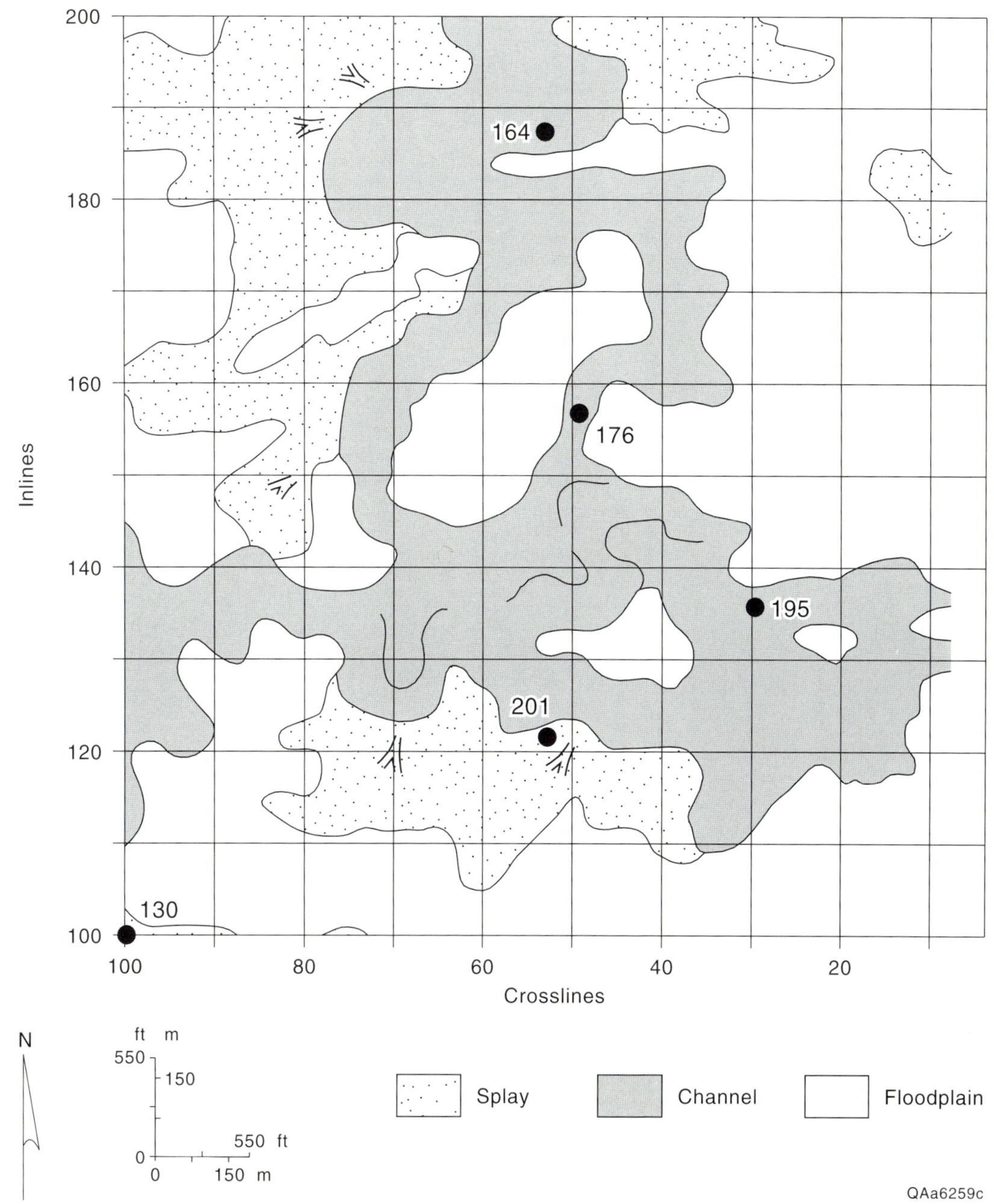


FIGURE 14. Proposed model for the F21 reservoir system imaged in Figure 13. This model honors the available geologic control (Figure 11) and engineering control (Figure 12), and it duplicates the meander pattern revealed in Figure 13.

# CHAPTER 4

# Seismic Delineation of Thin Sandstone Reservoirs in a Shale-Rich Sequence Using Instantaneous Frequency and Reflection Amplitude Attributes from 3-D Data, Texas Gulf Coast

Horkowitz, K. O., and D. R. Davis, Seismic delineation of thin sandstone reservoirs in a shale-rich sequence using instantaneous frequency and reflection amplitude attributes from 3-D data, Texas Gulf Coast, *in* P. Weimer and T. L. Davis, eds., *AAPG Studies in Geology No. 42* and *SEG Geophysical Developments Series No. 5*, AAPG/SEG, Tulsa, p. 35–44.

Kathleen O. Horkowitz* and David R. Davis**

## ABSTRACT

A combination of seismic trace attributes including instantaneous frequency and reflection amplitude was successfully used to seismically delineate gas-prone reservoir facies consisting of thin, discontinuous sandstone reservoirs in a shale-rich stratigraphic sequence. This type of reservoir section poses special problems for seismic interpretation due to a high degree of lateral and vertical variation in the ratio of net sandstone to gross interval thickness. Individual sandstone reservoirs are also below the limit of seismic bed resolution, further complicating the evaluation.

Instantaneous frequency maps were used directly to map the areal distribution of reservoir-prone intervals. The addition of calibrated average amplitude maps provided the ability to discriminate between homogenous shale-rich sections and relatively homogenous sandstone-rich sections characterized by amalgamated siltstones and sandstones. The lithofacies prediction model developed using this integrated geological, petrophysical, and seismic interpretation technique was confirmed by the results of two new wells which penetrated the target interval described in this study. Instantaneous frequency maps with overlays of amplitude and time structure contours successfully identified stratigraphic intervals favorable for gas accumulation and trapping.

## INTRODUCTION

Large volumes of natural gas are trapped in thin, discontinuous sandstone reservoirs in numerous South Texas Gulf Coast fields. Delineation and characterization of these reservoirs pose special problems to seismic interpreters involved in field development and exploration. Trapping mechanisms include both structural and stratigraphic components, making stratigraphic delineation an integral part of any analysis. Productive reservoirs in many fields are commonly less than 25 ft (8 m) thick, and therefore individual reservoir bed boundaries are not seismically resolvable. Additionally, interbedded sandstone and shale sections exhibit a high degree of vertical and lateral variability in the ratio of net sandstone thickness to gross interval thickness. This combination of geologic characteristics makes both delineation of individual reservoirs and estimates of pay thickness problematic.

This paper describes an integrated seismic interpretation technique which was successfully used to evaluate a mature gas field in the Rio Grande Embayment of South Texas (Figure 1). The field is associated with a large faulted anticline, downthrown to a major growth fault within the Frio Fault Zone, and is representative of numerous fields throughout this area.

A 3-D seismic survey over the field covering a migrated area of 18 $mi^2$ (46 $km^2$) was acquired in 1993 (Figure 2). The goal of the seismic interpretation was twofold: (1) to locate new development wells offsetting prior or existing production and (2) to propose locations for exploration wells in untested acreage. The structural framework for the study area was established using conventional seismic evaluation techniques. The more complex problem of identifying and mapping reservoir distribution was achieved through an integrated analysis using reflection amplitude, instantaneous frequency, and well log data.

## STRATIGRAPHIC SECTION

The Oligocene Frio Formation is the principal gas-producing formation of the Texas Gulf Coast (Kosters et al., 1989) (Figure 3). Gas is produced in the area from 12 Frio reservoirs ranging in depth from 5000 to 12,000 ft (1525 to 3660 m) subsea. Reservoirs in the specific interval of interest are in the proximal Frio Formation in a mixed aggradational/progradational section where thin, discontinuous sandstone reservoirs are interbedded with shales from fluvial, delta plain, and interdistributary-bay paleoenvironments (Galloway et al., 1982).

The target interval is a depositional package approximately 150–200 ft (46–61 m) thick composed of interbedded sandstones, siltstones and shales. This stratigraphic interval was evaluated over a 5-$mi^2$ (13-$km^2$) area using log data from 25 wells from a series of well and seismic transects (Figure 4A). The top of the stratigraphic interval is defined by a locally extensive, low-resistivity, low-velocity, silty shale (stippled zone in Figure 4B). Four potential sandstone reservoirs occur within this interval. Individual sandstone beds range from 12 to 25 ft (4 to 8 m) thick, and up to three of the four reservoirs may occur together in a given well.

Petrophysical and geological evaluation of well logs defined three log facies that characterize lithologic variation in the target interval from well to well (Figure 5):

Log Facies 1—dominantly shale to silty shale
Log Facies 2—thin, clean sandstones in a shale-rich sequence
Log Facies 3—amalgamated siltstone and sandstone

Gas reservoirs in the target interval typically are thin, clean sandstones in a shale-rich sequence and are characterized by Log Facies 2 .

Seismically, the target interval is defined by a basal erosional surface where truncated reflectors are overlain by a trough (Figure 6A). A small, discontinuous peak (shown tracked in yellow in Figures 6A and 6B) onlaps the basal event, and a strong continuous trough (shown tracked in dark blue in Figures 6A and 6B) defines the top of this sequence. This seismic package is generally wedge-shaped in cross section, both thinning and pinching out to the west and south and thickening to the east, where it is offset by faults. Syndepositional faulting strongly influenced the areal distribution of the depositional wedge as well as the facies distribution within it. Seismic and well-log traverses show both thickening of the target interval and an increase in the percentage of sandstone to the east .

A synthetic seismogram generated from a key well ties the low-velocity silty shale at the top of the stratigraphic section to the continuous trough bounding the top of the seismic package (Figure 7). This upper trough was tracked and used as a datum horizon for evaluating the mixed sandstone and shale section below it. The sandstone reservoirs tie to the small discontinuous peak below the trough tracked in yellow.

## METHODS

One of the primary goals of this study was to attempt to delineate the areal distribution of potential reservoir sandstones using seismic trace attributes. Once the distribution of the reservoir facies has been defined, potential targets can be evaluated within the overall structural framework. Since reservoir thickness is below the limit of seismic bed resolution, reservoir top and base cannot be mapped directly. Therefore, a technique to identify and map potential reservoir facies was required. A combination of seismic trace attributes including

*Resource Analysts, Houston, TX 77077
**Sanchez–O'Brien Oil and Gas Corporation, Houston, TX 77057

reflection amplitude, instantaneous frequency, and instantaneous phase was used to delineate potentially productive intervals. An integrated geological and petrophysical evaluation of log data from 25 wells within the study area provided a strong base for geologic calibration of the seismic trace attributes.

## INSTANTANEOUS FREQUENCY

Instantaneous frequency is defined as the temporal rate of change in the instantaneous phase function (Taner and Sheriff, 1977; Yilmaz, 1987) and is calculated as:

$$\omega(t) = \frac{d\varphi(t)}{dt}$$

where

$\omega(t)$ = instantaneous frequency,
$\varphi(t)$ = arctan $[y(t)/x(t)]$
$x(t)$ = seismic trace, and
$y(t)$ = Hilbert transform of $x(t)$.

Instantaneous frequency is a measure of the slope of the instantaneous phase function and is described by Robertson and Fisher (1989) as "a sample by sample estimate of the trace's dominant frequency—as distinct from the average frequency over a window." High instantaneous frequency values (i.e., steep slopes in the instantaneous phase function) will occur where closely spaced reflection coefficients of opposite sign create a rapid change in instantaneous phase from one time sample to another as a result of wavelet and sidelobe interference (Robertson and Fisher, 1989). Instantaneous frequency can be thought of, then, as representing the local reflection frequency of the seismic trace (Lindseth, 1982). Geologically, closely spaced reflection coefficients of opposite sign occur when lithologies with sufficient contrast in acoustic impedance are interbedded. This occurs, for example, along the margins of a thinning bed (Robertson and Nogami, 1984), or, as in this study, within a generally low-impedance unit punctuated by thin, higher-impedance sandstone beds.

Robertson and Nogami (1984) described complex trace attribute analysis of a synthetic model of a wedge. A principal result of their study was the recognition of "frequency tuning by beds that thin laterally." In the model, anomalously high instantaneous frequency values occur as bed thickness approaches one period (T) of the dominant frequency and again at bed thicknesses less than T/4. Their application of trace attribute analysis to a field case study demonstrated the occurrence of anomalously high instantaneous frequency values where sandstone beds thinned to less than T/4. In addition to showing the response of trace attributes to the synthetic model, their results illustrate the power of an integrated evaluation utilizing multiple trace attributes geologically calibrated by well data to define stratigraphic features not obvious on conventional seismic sections.

In the current study of the Frio Formation, composite values for amplitude and instantaneous frequency were calculated over a 40-ms window and geologically calibrated using 25 wells to map lateral changes in lithofacies. In the target interval, reservoirs are thin, high-impedance sandstones within a lower-impedance shale-rich interval (Table 1). Therefore, maps of instantaneous frequency should detect their presence as suggested by the wedge model studies and case study of Robertson and Nogami (1984).

**TABLE 1. Acoustic Properties of Sandstones and Shales in the Stratigraphic Section of Interest**

| Lithology | Velocity (ft/s) | (m/s) |
|---|---|---|
| Sandstone | 8300–8700 | 25/5–2636 |
| Shale | 6700 | 2030 |
| Silty-shale datum | 6400 | 1940 |

The following sections describe an integrated interpretation using instantaneous frequency, reflection amplitude, reflection time, and well-log data to:

(1) develop an understanding of the success/failure characteristics of existing wells,
(2) determine the seismic trace attribute characteristics of lithofacies in the study interval,
(3) map the areal distribution of the key lithofacies, and
(4) evaluate the distribution of the reservoir facies within the structural framework to identify development well locations and new exploration well locations.

## SEISMIC EVENT MAPPING

Two events were mapped to evaluate the target interval, the trough correlating with the locally extensive low-impedance silty shale at the top of the section and the small peak between the upper and basal troughs (Figures 6A, 6B, and 7). The trough (tracked in dark blue on the seismic lines in Figure 6 and in yellow in Figure 7) was used as a datum, and average amplitude and root-mean-square (RMS) instantaneous frequency were calculated for each trace over the 40-ms interval below the datum horizon and stored as a new horizon. The seismic attributes were calculated as follows:

Average amplitude = sum of amplitudes/number of samples

$$\text{RMS instantaneous frequency} = \sqrt{1/n\,\Sigma_n^1 (ifreq_n)^2}$$

A 40-ms window was initially chosen, to correlate with the depth range of interest in the well logs; however, attribute values were calculated for several different time windows to determine the optimal window for analysis. The 40-ms window provided maps which correlated best with the depth interval in the well logs and provided the best data for geologic calibration. Thicker intervals (50- and 60-ms intervals) showed similar gross patterns in attribute variation but lacked smaller-scale detail which could successfully distinguish important differences between wells. Thinner intervals (20- and 30-ms intervals) showed similar gross patterns in attribute variation but were also more noisy. Attribute maps in the following sections are all based on the 40-ms window.

The discontinuous peak between the upper and basal troughs was tracked to its visual extent, or to the edge of the survey, to provide a structural surface for the reservoir interval (Figure 8). The small peak pinches out to the west and south and terminates to the east into a fault. The character and extent of the event to the north is unknown, since it is beyond the area of 3-D coverage.

# RESULTS

## AVERAGE AMPLITUDE MAPS

Average amplitude values calculated for each trace over the 40-ms window are displayed as a horizon map in Figure 9. Calculated values range from –300 to +100, with the bulk of the data in the range from –120 to 0. Systematic evaluation of well-log data indicates that patterns of variation in the average amplitude map correlate with changes in percent sandstone in the study interval. Large negative values of average amplitude correlate with shaly lithologic sections, whereas positive values and small negative values (close to zero) correlate with sandstone-rich sections (Figure 9B). This relationship is also illustrated by a crossplot of average amplitude versus percent sandstone (Figure 10), which shows a positive, linear relationship between average amplitude and percent sandstone.

Average amplitude for a given percent sandstone ranges between 15 and –15 amplitude units. Comparison of representative well logs across the amplitude range indicates that this variation is related to lithologic bedding configuration. Wells with a lower amplitude (small negative values) for a given percent sandstone have a log signature similar to that of Log Facies 3, representing amalgamated siltstones and sandstones. In this case, the calculated average amplitude value is a good estimate of the compositionally homogenous 150-ft (46-m) interval. Wells with a higher (more negative) average amplitude for the same percent sandstone are more similar to Log Facies 2 and are characterized by thin, clean sandstones in a shale-rich sequence. In this case, the composite value is calculated from a bimodal population and appears to be dominated by the low(er) impedance shale-rich background.

The average amplitude map can be used directly to delineate the areal distribution of the shale-rich lithofacies (Log Facies 1), using an average amplitude cutoff value of –80 (Figures 9A and 9B: red, yellow, and white). These intervals are nonprospective due to lack of reservoir. Based on amplitude alone, however, it is not possible to map the distribution of intervals characterized by Log Facies 2 and Log Facies 3. This distinction is critical, since gas is produced from wells characterized by Log Facies 2, while structurally equivalent wells characterized by Log Facies 3 are nonproductive due to inadequate seal. Further lithofacies discrimination was required to distinguish between areas composed of thin, clean sandstone beds in a shale-rich sequence (Log Facies 2) and amalgamated siltstones and sandstones (Log Facies 3). By combining instantaneous frequency maps with average amplitude maps, it was possible to discriminate between all three lithofacies in the target interval.

## RMS INSTANTANEOUS FREQUENCY MAPS

Instantaneous frequency and instantaneous phase data volumes were generated for the entire 3-D seismic survey. As described previously, instantaneous frequency is a measure of the slope of the instantaneous phase function. Higher instantaneous frequency values will occur where closely spaced reflection coefficients of opposite sign create a rapid change in instantaneous phase from one time sample to another as a result of wavelet and sidelobe interference (Robertson and Fisher, 1989).

Figures 11A and 11B are variable-density displays of instantaneous frequency overlain by the real seismic trace $x(t)$ and the instantaneous phase trace $\varphi(t)$, respectively, where:

$$x(t) = A(t) \cos\varphi(t)$$

and

$$\varphi(t) = \arctan[y(t)/x(t)].$$

$$A(t) = \sqrt{x^2(t) + y^2(t)}$$

The color bar is scaled in Hertz with low instantaneous frequency values in shades of red and higher values in yellow and blue. The datum horizon is tracked in blue in Figure 11A, and the target interval is annotated on the time scale in both figures.

In Figure 11B, the instantaneous frequency display is overlain by the instantaneous phase trace. Note the thick red zone in the target interval where instantaneous frequency is low and the instantaneous phase trace has fairly broad, slowly changing character. In comparison, the instantaneous phase trace has a much steeper slope from sample to sample in the yellow and blue zone directly to the east. Instantaneous frequency is a measure of the rate of change in the instantaneous phase function and can be an indication of the frequency of occurrence of contrasting reflection coefficients within a given interval.

RMS instantaneous frequency was calculated for each trace over the 40-ms window and stored as a new horizon (Figure 12). The instantaneous frequency map reveals a prominent zone of low instantaneous frequency (reds) trending northwest to southeast. To the east, a zone of high instantaneous frequency values parallels the red zone, and the transition between the two zones is fairly abrupt. In contrast, the low-instantaneous-frequency zone grades smoothly to the west throughout the entire range. The abrupt transition on the east, from the low-instantaneous frequency zone to the higher-instantaneous-frequency zone, correlates with the stratigraphic pinchout of the depositional package as seen in Figures 11A and B in the vicinity of Line 1108 and Trace 65.

Correlation of well logs with variations in the instantaneous frequency map indicates that low-instantaneous-frequency intervals are characterized by a relatively homogenous section, either shale-rich or sand and silt-rich (Figure 5; Log Facies 1 and 3, respectively). Intervals with high instantaneous frequency (yellow to blue) exhibit a greater degree of vertical lithologic interbedding and are characterized by thin sandstone beds in a shale-rich sequence (Figure 5, Log Facies 2).

Wells drilled in the low instantaneous-frequency zones (red and orange) were unsuccessful in the target interval, even in structurally high positions, because they either were shale-rich and lacked sufficient reservoir or were composed of amalgamated sandstones and siltstones and lacked adequate seals. Wells drilled in zones exhibiting high instantaneous frequency penetrated good quality sandstones in a shale-rich sequence, providing both sufficient reservoir and adequate vertical and lateral seals for gas accumulation and trapping.

The areal extent of the potential reservoir facies (Log Facies 2) can be mapped directly from the instantaneous frequency maps using a cutoff value of 20 Hz (Figure 12, pale yellow color). However, low instantaneous-frequency zones on the map can be either shale-rich or sandstone-rich (Log Facies 1 or Log Facies 3, respectively). To fully map the three lithofacies in this depositional package and to differentiate between Log Facies 1 and Log Facies 3, it is necessary to evaluate average amplitude and instantaneous frequency together.

A crossplot of average amplitude and RMS instantaneous frequency over the interval of interest for 18 of the 25 wells identifies two distinct data clusters (Figure 13). Seven of the 25 wells do not have calculated attribute values due to gaps in the tracked datum horizon. Evaluation of this crossplot in conjunction with a plot illustrating representative well logs over the range of amplitude and instantaneous frequency (Figure 14) reveals the following relationships:

(1) One data cluster is characterized by high instantaneous frequency and low average amplitude (Figure 13). Five of the six wells in this cluster have gas reservoirs in the interval of interest; the sixth well has good-quality sandstones but is in a structurally low position (Well B). The interval of interest in each of these wells is characterized by Log Facies 2, thin clean sandstones in a shale-rich section.

(2) A second data cluster is characterized by lower instantaneous frequency and, in general, larger negative amplitude values. Wells with instantaneous frequency less than 20 Hz and low average amplitude (0 to –60) are characterized by amalgamated sandstone intervals which lack sufficient interbedded shales to trap gas (Well O). Wells with high average amplitude (–80 to –120) are shale-rich and lack sufficient reservoir rock (Figures 13 and 14, Well A).

(3) The intermediate amplitude range (–60 to –80) contains samples with the right mix of sandstone and shale, but the samples differ in terms of their bedding configuration. As instantaneous frequency increases at a given amplitude, the target interval becomes more distinctly bedded, more similar to Log Facies 2. In Figure 14, note the difference in bedding configuration between wells O and B, and again between wells X and F. Two of the three wells with instantaneous frequencies between 20 and 25 Hz on Figure 13 contain gas reservoirs; the third is structurally low.

These results clearly show that reservoir-prone intervals can be discriminated from reservoir intervals based on RMS instantaneous frequency and average amplitude. Once the areal distribution of potential reservoirs has been defined, the area can be evaluated within the overall structural framework for effective exploitation.

## SUMMARY AND CONCLUSIONS

The principal benefit derived from this evaluation was the ability to seismically delineate the areal distribution of reservoir facies using seismic trace attributes. Geologically calibrated instantaneous frequency maps were used to map the areal distribution of the potential reservoir facies: thin, clean sandstones in a shale-rich section. A combination of instantaneous frequency and average amplitude maps provided the ability to discriminate between homogenous shale-rich intervals, amalgamated sandstone, and siltstone intervals. Finally, instantaneous frequency maps with overlays of amplitude and time structure contours successfully identified areas favorable for gas accumulation and trapping.

The model developed using this technique was confirmed by the results of two new wells drilled to test different targets, but which penetrated the study interval described here. Figures 15A and 15B show the locations of these two new wells as vertical white lines in the 3-D instantaneous frequency data volume. The datum horizon is shown tracked in dark blue. The first well (Well A) penetrated a low-instantaneous frequency, moderately high-amplitude interval (Figure 15A). The target interval in this well is amalgamated silty sandstone and shale characterized by Log Facies 3. Figures 13 and 14 show where Well A plots on the attribute crossplots. The second well (Well F) penetrated a higher-instantaneous-frequency, lower-average-amplitude interval (Figure 15B). The target interval in this well contains thin, clean, gas-bearing reservoir sandstones and is classified as Log Facies 2 (attribute crossplots Figures 13 and 14, Well F). These well results illustrate the use of geologically calibrated seismic trace attributes as a tool for mapping areal reservoir distribution. A third well will test the model in a separate area of the field before extending the technique to surrounding untested areas.

This interpretation technique requires a thorough integration of seismic, well, and production data to effectively calibrate lateral, trace-to-trace variation in seismic trace attributes with variation in lithology and bedding configuration. A systematic approach to geologic calibration of each interval of interest should also include an assessment of other potential factors which might contribute to variation in instantaneous frequency within a given interval, such as the processing sequence applied to the seismic data, fault shadow effects, or thick gas sand effects. This technique was successfully applied in this study to the problem of reservoir detection and is being further applied in reservoir characterization and seal analysis.


### *Acknowledgments*

The authors express appreciation to Sanchez-O'Brien Oil and Gas for permission to publish these results and to Don Herron, Robert Sheriff, Bob Hardage, and other colleagues for critically reviewing this paper.


## REFERENCES CITED


Galloway, W. E., D. K. Hobday, and K. Magara, 1982: Frio Formation of the Texas Gulf Coast Basin—Depositional systems, structural framework, and hydrocarbon origin, migration, distribution and exploration potential: Report of Investigations, 122, Austin, Bureau of Economic Geology, The University of Texas at Austin, p. 78.

Kosters, E. C., D. G. Bebout, S. J. Seni, C. M. Garret, Jr., L. F. Brown, Jr., H. S. Hamlin, S. P. Dutton, S. C. Ruppel, R. J. Finely, and N. Tyler, 1989, Atlas of Major Texas Gas Reservoirs: Austin, Bureau of Economic Geology, The University of Texas at Austin.

Lindseth, R. O., 1982, Digital Processing of Geophysical Data—A Review: Calgary, Society of Exploration Geophysicists, p. 282.

Robertson, J. D., and H. H. Nogami, 1984, Complex seismic trace analysis of thin beds: Geophysics, v. 49, no. 4, p. 344–352.

Robertson, J. D., and D. A. Fisher, 1989, Complex seismic trace attributes, *in* Society of Exploration Geophysicists Seismic Interpretation Series, 1, p. 49–53.

Taner, M. T., and R. E. Sheriff, 1977, Applications of amplitude, frequency and other attributes to stratigraphic and hydrocarbon determination, C. E. Payton, Ed., *in* Seismic Stratigraphy—Applications to Hydrocarbon Exploration: Memoir 26, Tulsa, p. 301–327.

Yilmaz, O., 1987, Seismic data processing, S. M. Doherty, ed., *in* Society of Exploration Geophysicists Investigations in Geophysics, v. 2: Tulsa, p. 536.

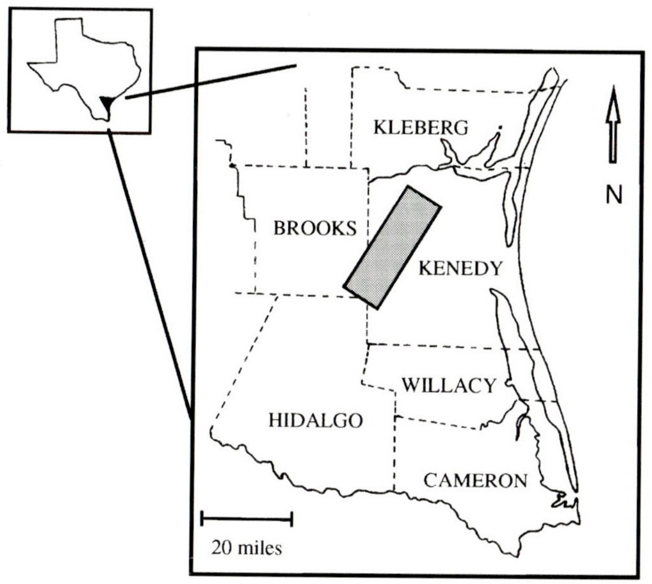


FIGURE 1. Location of study field in Kenedy County, Texas, within the Frio Fault Zone.

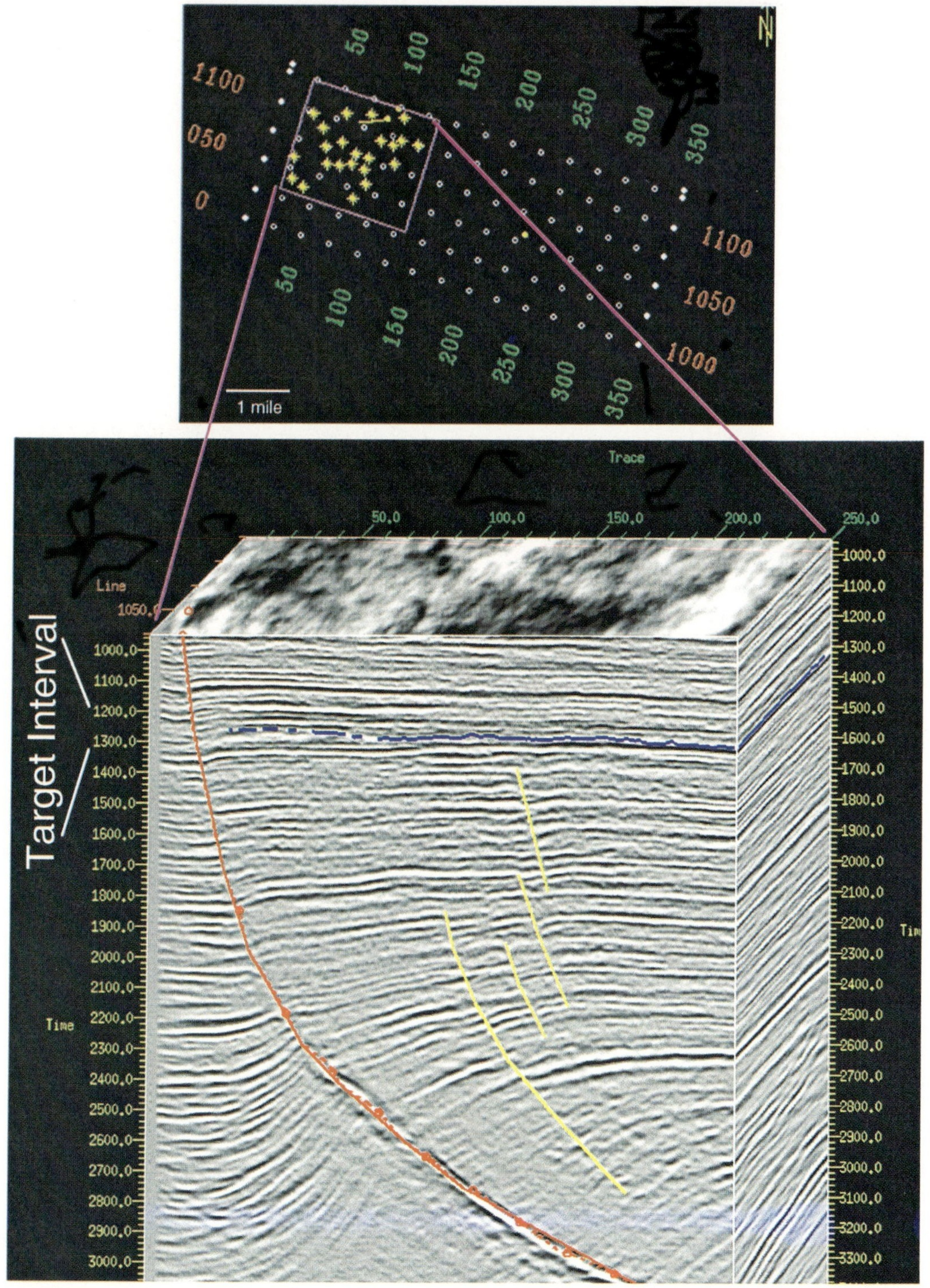


FIGURE 2. Base map from the 3-D seismic survey covering a migrated area of 18 $mi^2$. The 3-D survey covers two fields associated with the faulted, rollover anticline shown in the 3-D seismic data cube. The variable density amplitude display is normal polarity and the trace spacing is 110 ft (33 m). The seismic interval described in this paper is a 40 ms window below the trough tracked in dark blue.

| SYSTEM | SERIES | GROUP/FORMATION |
|---|---|---|
| Tertiary | Pliocene | Goliad |
| | Miocene | Fleming<br>Anahuac |
| | Oligocene | **Frio**<br>Vicksburg |
| | Eocene | Jackson<br>Claiborne<br>Wilcox<br>Midway |

FIGURE 3. Stratigraphic column for Tertiary of south Texas.

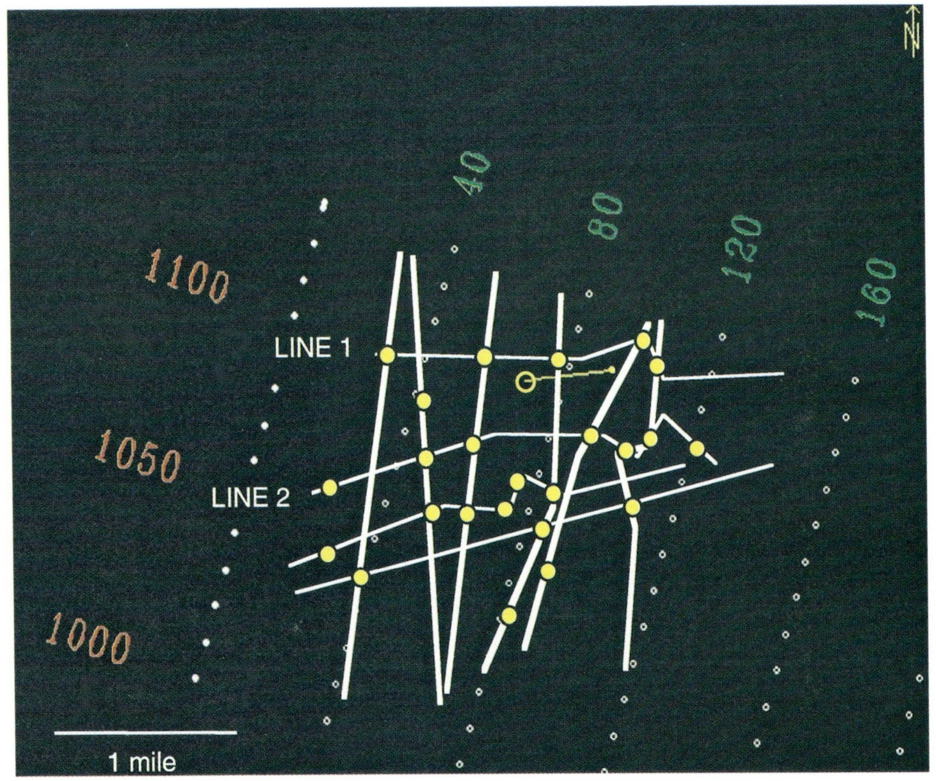


A

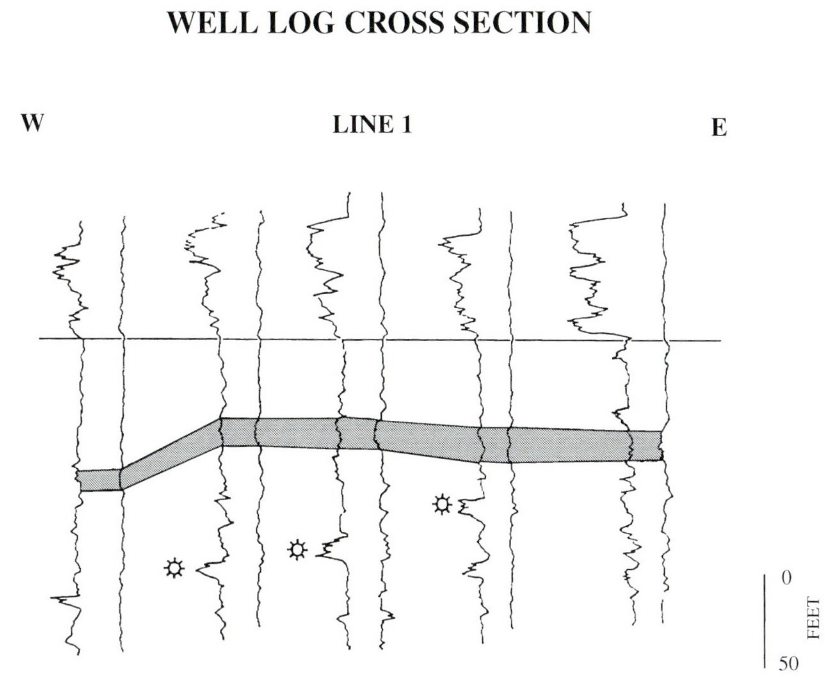


B

FIGURE 4. Well locations and well and seismic transects used to geologically calibrate seismic trace attribute maps of the target interval. Representative well-log cross section along Line 1 showing SP and resistivity curves. The top of the target interval is defined by a low resistivity, low velocity, silty-shale (stippled). Reservoirs are commonly 12 to 25 ft (4 to 8 m) thick and occur as thin, clean sandstones in a shale-rich sequence.

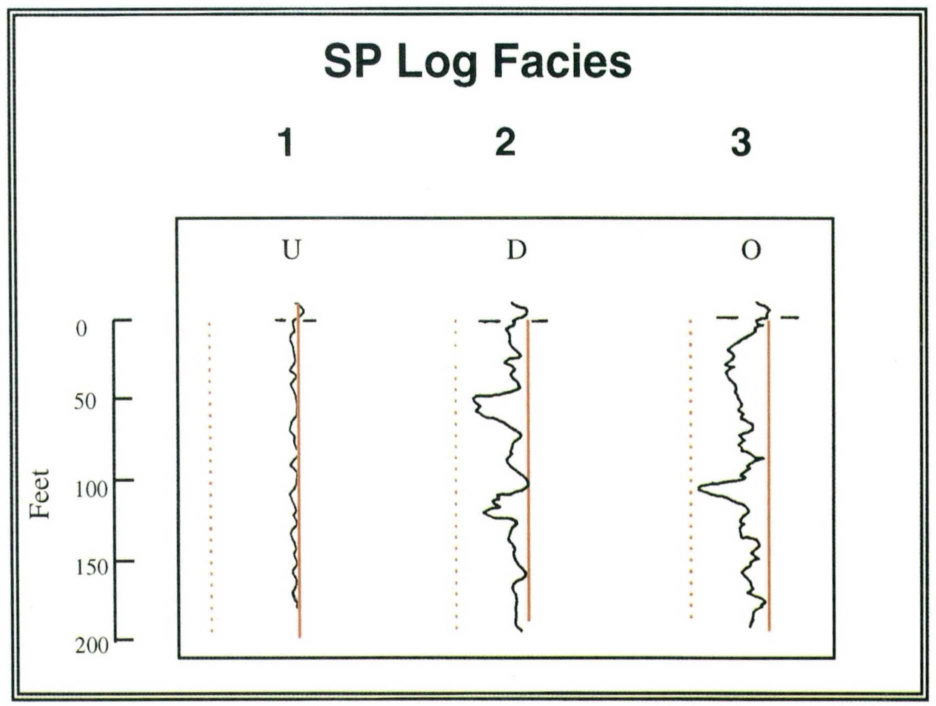


FIGURE 5. Three log facies characterize stratigraphic variability within the target interval: Log Facies 1—dominantly shale to silty shale; Log Facies 2—thin, clean sandstones in a shale-rich sequence; and Log Facies 3—amalgamated siltstone and sandstone. End member log facies are best represented by Well U, Well D, and Well O, respectively. The tick mark at the top of the log displayed here is the base of the low-velocity silty shale shown stippled in Figure 4B.

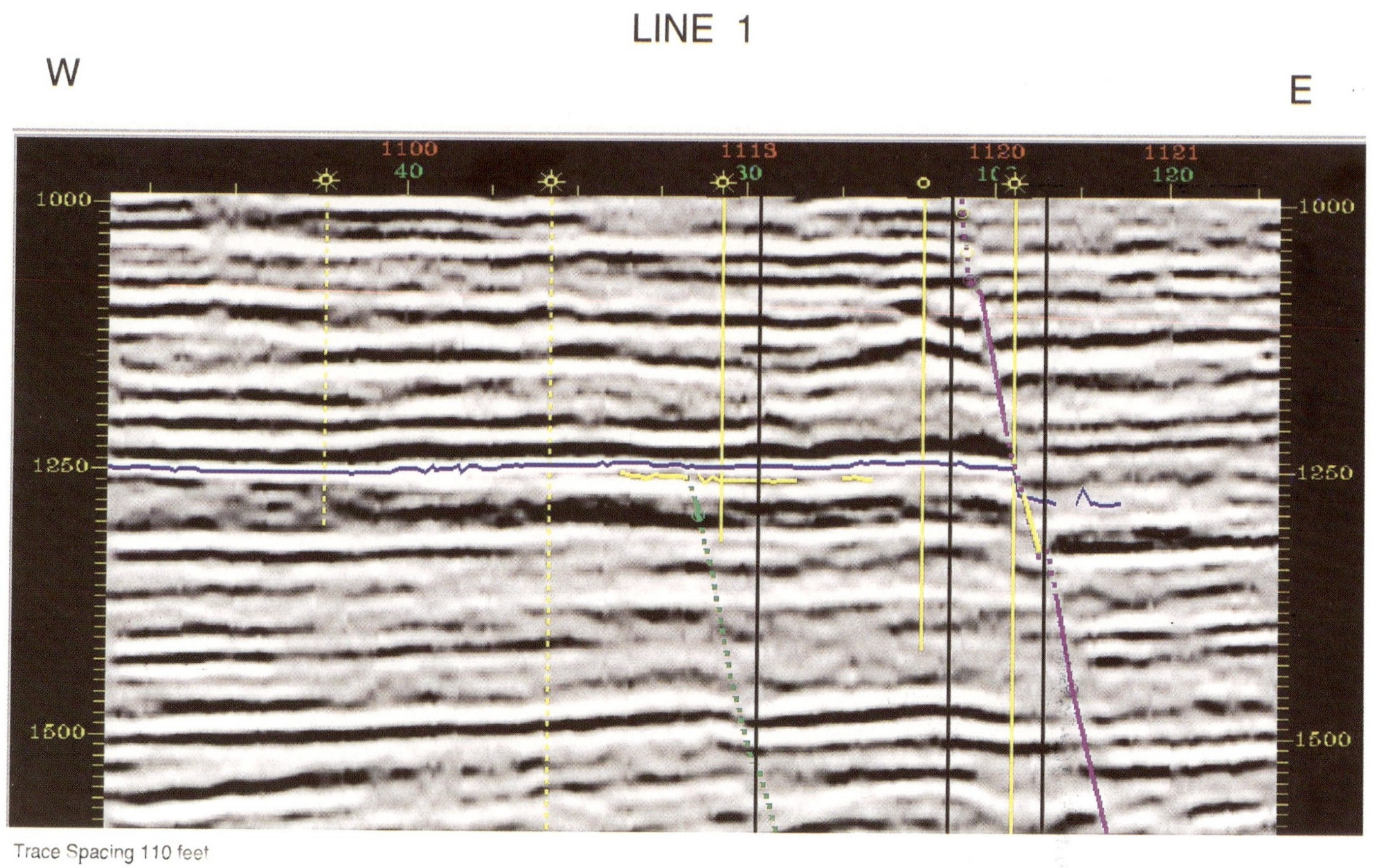


A

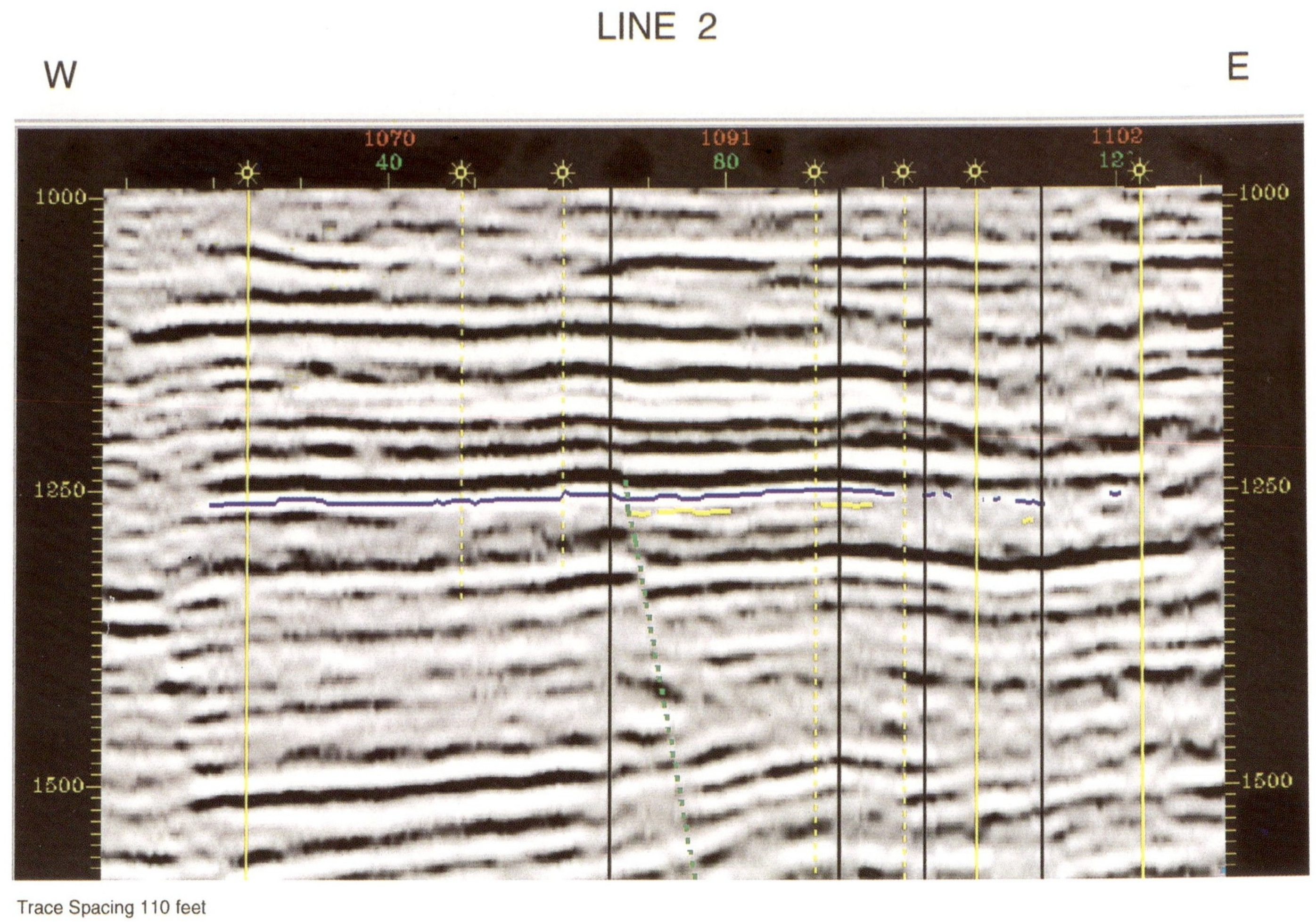


B

FIGURE 6. The locations of seismic transects Line 1 (A) and Line 2 (B) are shown on the map in Figure 4A. Seismically, the interval of interest is defined by a basal erosional surface where truncated reflections are overlain by a trough (the white trough below the small peak tracked in yellow). A discontinuous small peak, tracked in yellow, onlaps the basal trough. A bright, continuous trough, tracked in dark blue, defines the top of the seismic package.

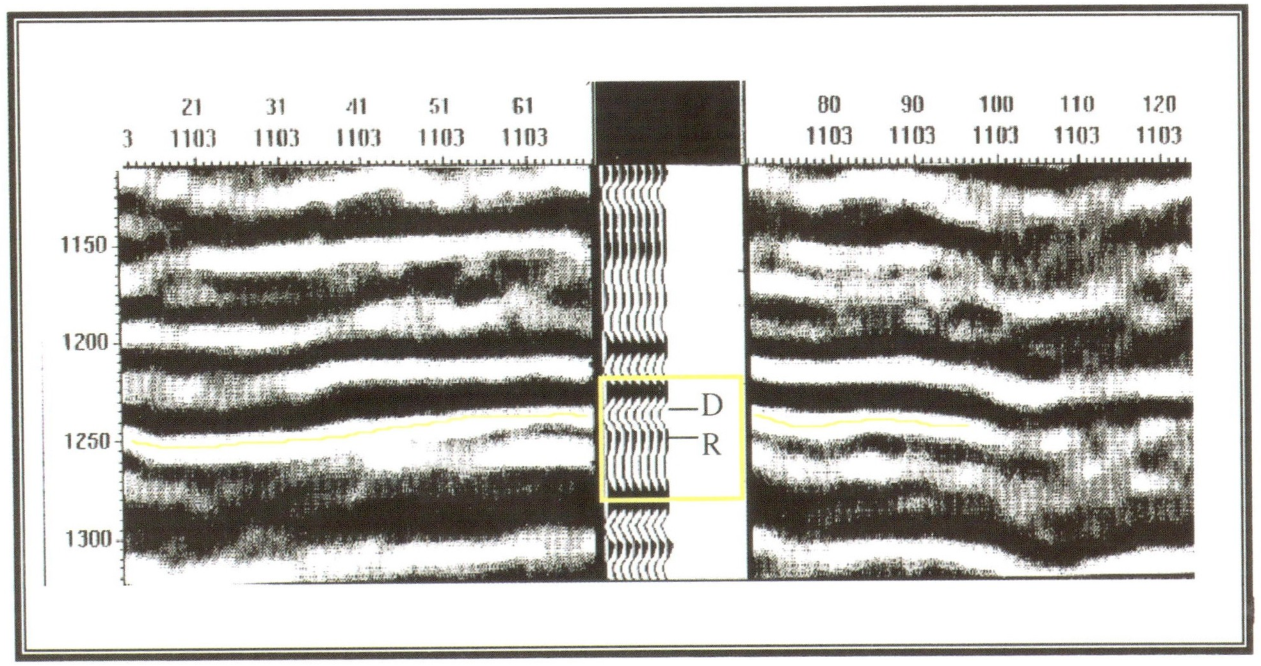


FIGURE 7. A synthetic seismogram (shown here as eight identical traces) generated from a key well is displayed at the well location on a segment of a seismic line from the 3-D survey. The top of the low-velocity silty shale at the top of the stratigrpaphic section is annotated on the synthetic traces as "D." This shale ties to the continuous trough tracked in yellow that bounds the top of the seismic package. The reservoirs, labeled "R," tie to the small discontinuous peak onlapping the basal trough. Trace spacing is 110 ft (33 m), and the display scale is 20 traces/inch.

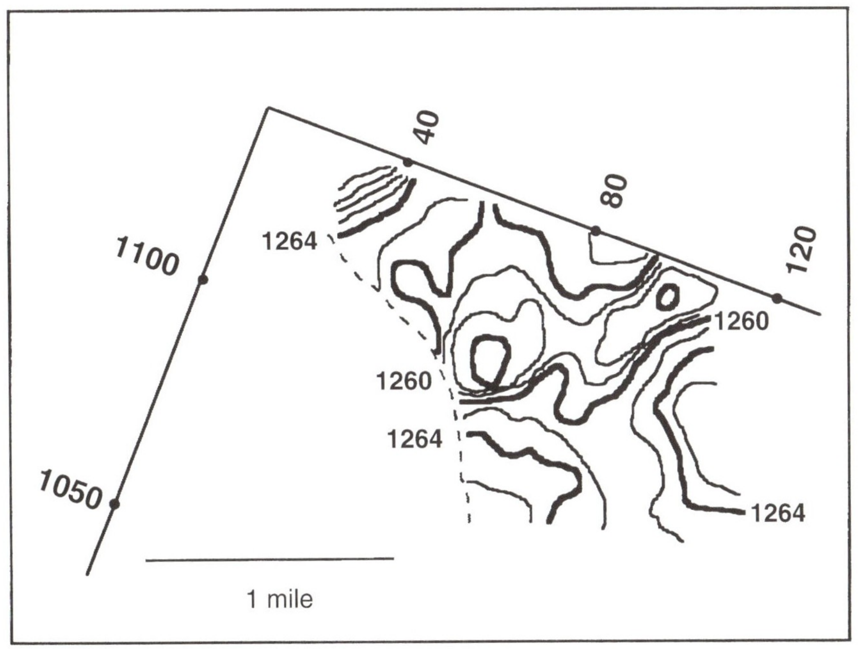


FIGURE 8. Two-way time structure map on the peak event which correlates with the reservoir sandstones. The dashed line follows the western termination of the peak where the stratigraphic package pinches out. Contour interval is 2 ms.

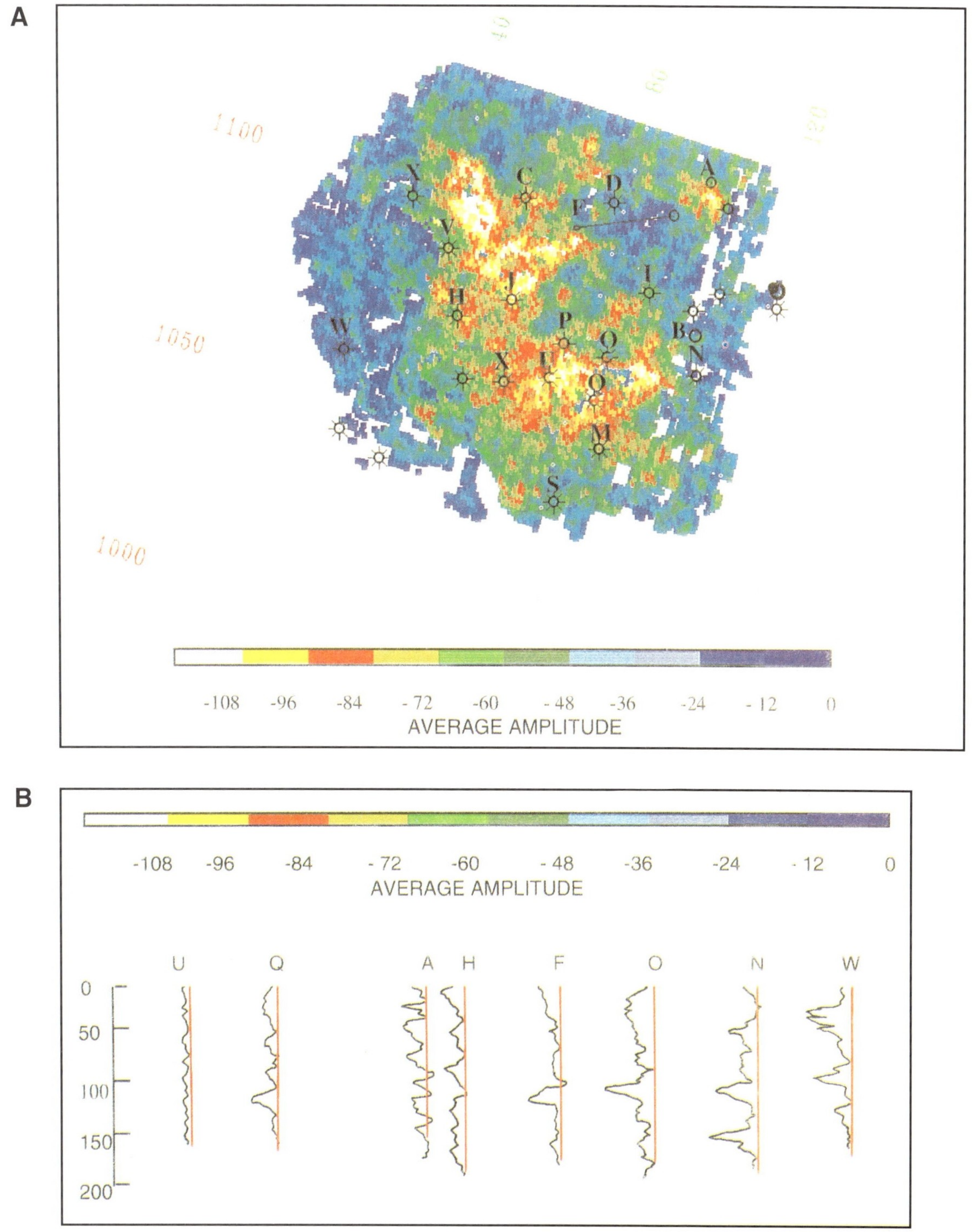


FIGURE 9. (A) A map of average amplitude values calculated over the 40-ms target interval illustrates the areal variation in the composite amplitude value. Wells used in the geologic calibration of attribute maps are identified by letter. (B) SP curves representative of lithofacies changes across the average amplitude range are ordered from shale-rich to sandstone-rich along the average amplitude scale. The color bar along the top of the figure is the same color bar shown in the average amplitude map in Figure 9A. Large negative values of average amplitude (yellow and red) correlate with shale-rich lithologic sections, whereas positive values and small negative values (green through blue) correlate with sandstone-rich sections.

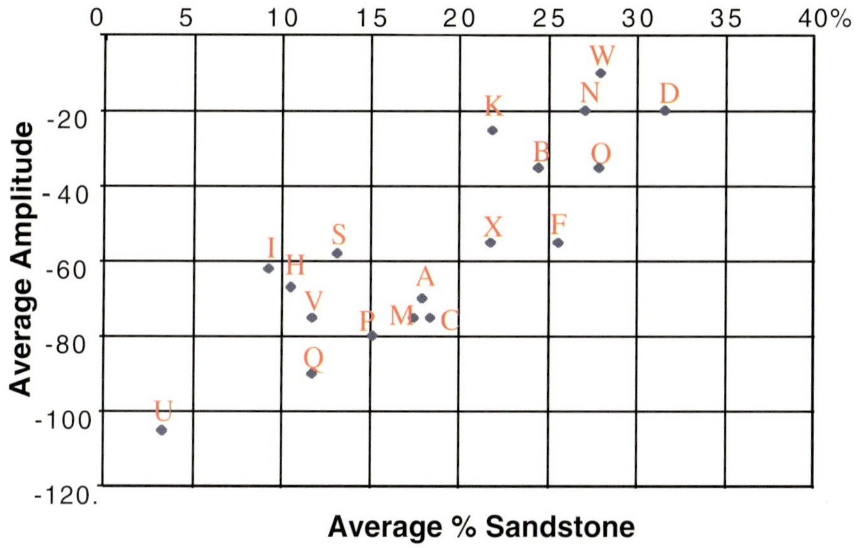


FIGURE 10. Crossplot of average amplitude from the 40-ms seismic interval and percent sandstone derived from the SP log over the target depth interval. The relationship indicates that large negative amplitude values are correlated with low percent sandstone and amplitudes closer to zero are correlated with higher percent sandstone.

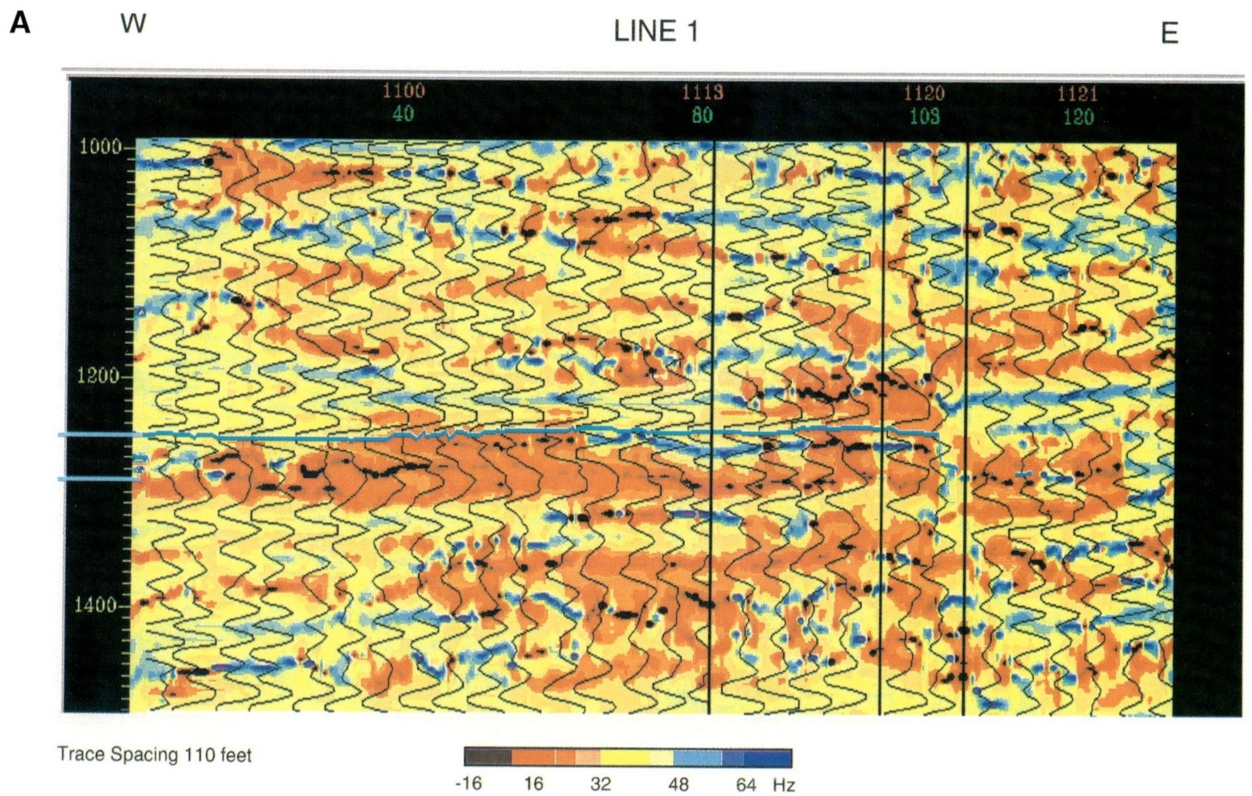


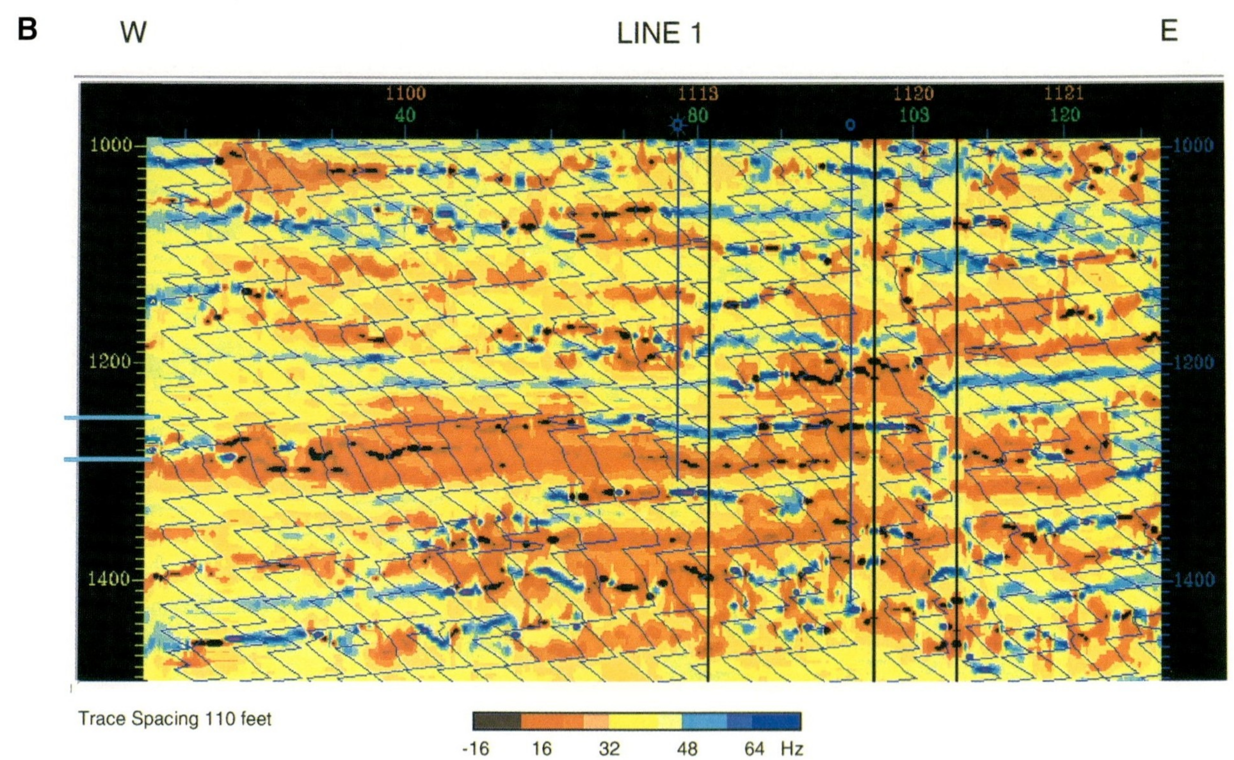


FIGURE 11. The color bars for A and B are scaled in hertz. (A) Variable-density instantaneous frequency display with an overlay of the real seismic trace (every fifth trace) for Line 1. The datum horizon is shown tracked in blue and the 40-ms target interval is annotated on the time axis for both plots. (B) Variable density instantaneous frequency display with an overlay of the instantaneous phase trace (every fifth trace) for Line 1. The 40-ms target interval is annotated on the time axis. Instantaneous frequency is a measure of the temporal rate of change in the instantaneous phase function.

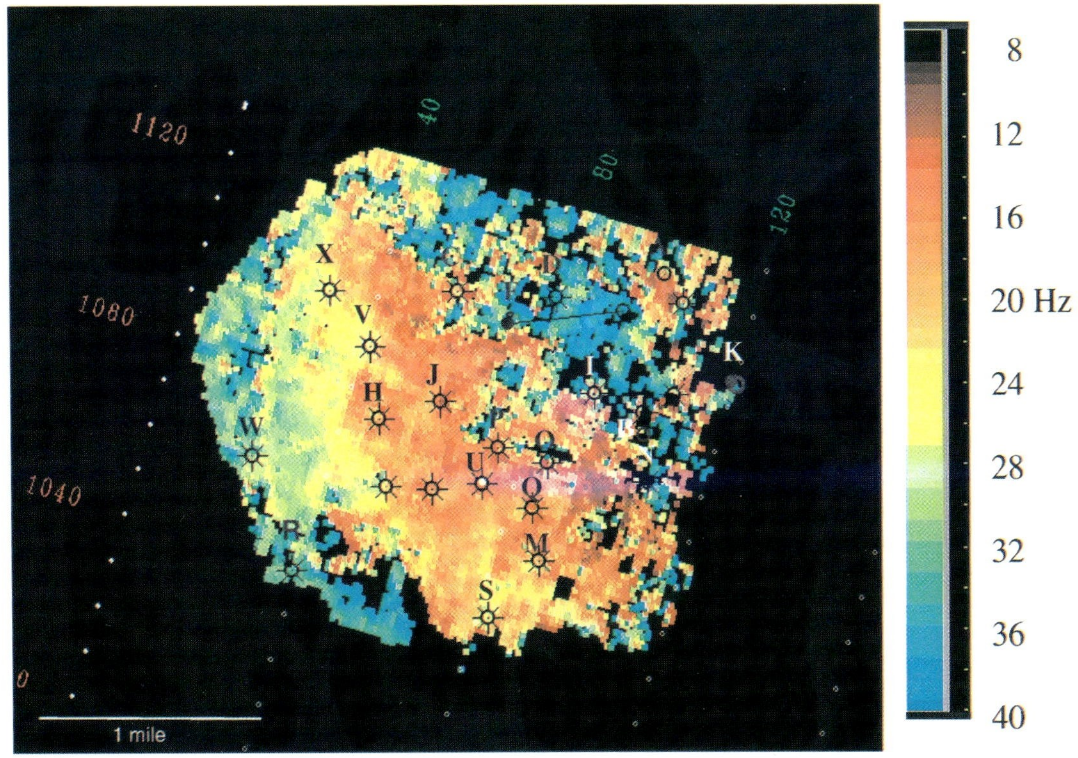


FIGURE 12. Map of RMS instantaneous frequency calculated over the 40-ms target interval. The map illustrates variability in the RMS value over the study area. Wells used to geologically calibrate the attribute map are identified by letter.

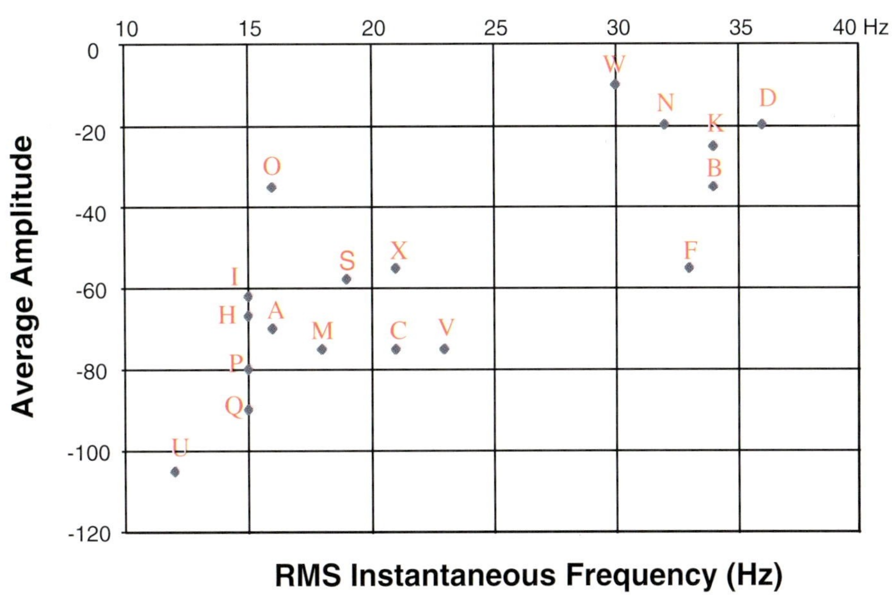


FIGURE 13. An attribute crossplot of RMS instantaneous frequency and average amplitude identifies two distinct data clusters. Five of the six wells in cluster 1, high instantaneous frequency and low amplitude, contain gas reservoirs. The sixth well has good-quality reservoir sandstones and adequate shale seals but is in a structurally low position. Wells with instantaneous frequency values less than 20 Hz do not contain gas reservoirs, either because they lack sufficient reservoirs (high amplitudes) or because they are too sandstone-rich and lack adequate seals (low amplitudes).

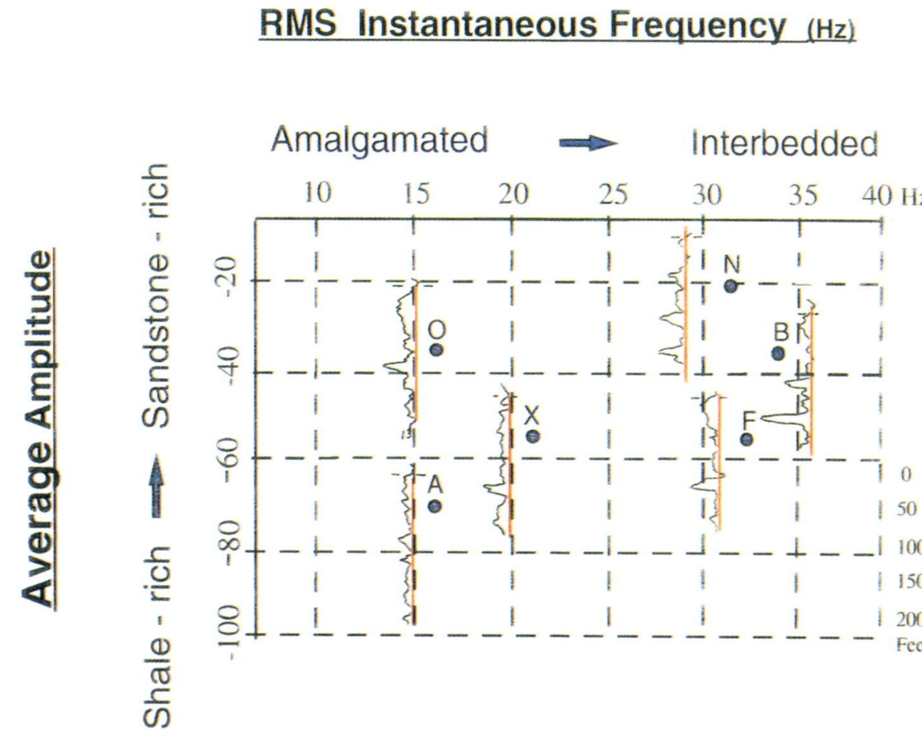


FIGURE 14. Combined analysis of instantaneous frequency and average amplitude can be used empirically to characterize variation in lithology and bedding configuration in this interval. Wells with equivalent average amplitude may have very different bedding configurations as illustrated by Well O and Well B. Wells with low instantaneous frequency (<20 Hz) are nonprospective due to either lack of sufficient reservoir (Well A), or lack of adequate lateral and/or vertical sealing shales (Well O). Well intervals with high instantaneous frequency (>20 Hz) have both sufficient reservoir and adequate shale to accumulate and trap gas.

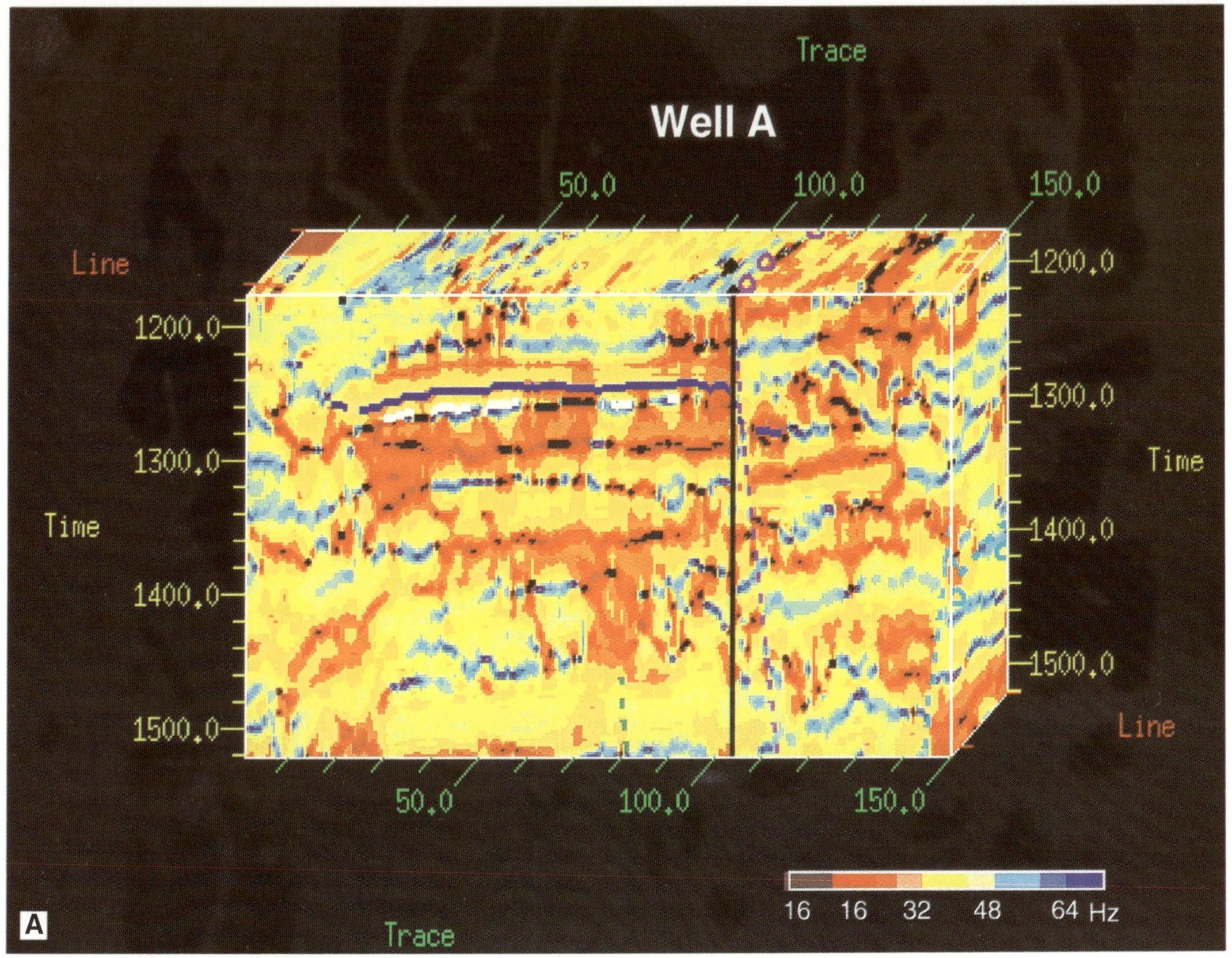


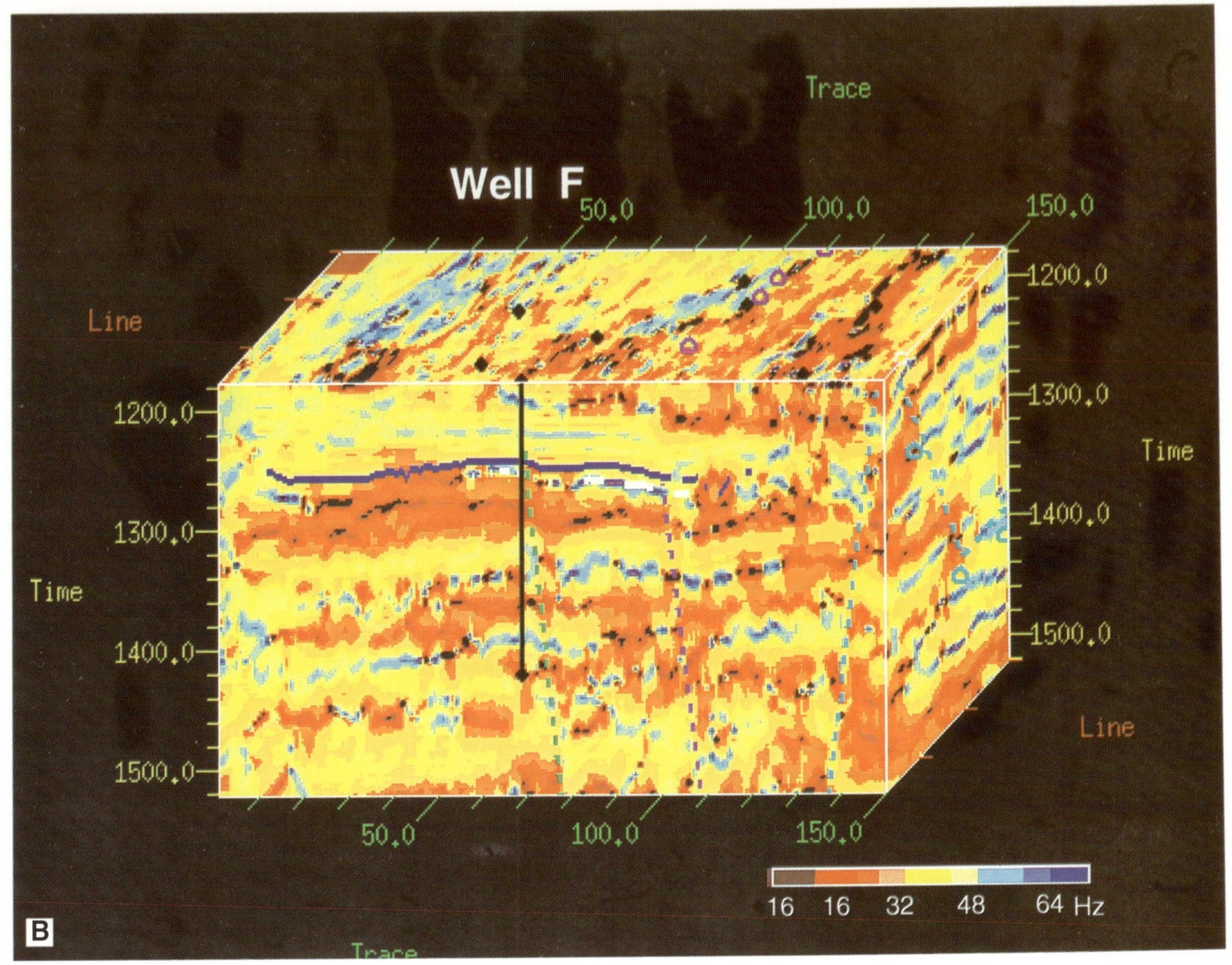


FIGURE 15. Two views of the seismic data cube volume display instantaneous frequency variation in the 3-D volume. The location of two new wells is shown as a vertical white line on the front face of the cube displays in the inline direction. (A) The first well (Well A) penetrated a low-instantaneous frequency interval (16 Hz, red), with an average amplitude of –70. The interval is composed of amalgamated siltstone and shale. The model indicates that low-instantaneous frequency intervals will be non-prospective. (B) The second well (Well F) penetrated an interval with high instantaneous frequency (33 Hz, yellow) and average amplitude of –58. The model indicates that high instantaneous frequency intervals in structurally high positions will be prospective. The interval is characterized by Log Facies 2, and contains thin, clean gas sandstone reservoirs in a shale-rich sequence.

Burnett, M., 3-D Seismic expression of a shallow fluvial system in West Central Texas, *in* P. Weimer and T. L. Davis, eds., *AAPG Studies in Geology No. 42* and *SEG 3-D Geophysical Developments Series No. 5*, AAPG/SEG, Tulsa, p. 45–56.

CHAPTER 5

# 3-D Seismic Expression of a Shallow Fluvial System in West Central Texas

Mike Burnett*


## ABSTRACT

3-D surveys have proven to be a powerful structural risk reducer for oilmen in West Central Texas. While the biggest structures were found using single-fold and 2-D common depth point (CDP) shooting, the success rate for stepout Mississippian and Canyon reefs drilled on 3-D surveys has increased substantially in this area: 3-D surveys have gained wide acceptance by the oil community in North and West Central Texas.

Whereas 2-D has been relatively ineffective in stratigraphic situations, this paper will showcase a 3-D survey which was used as a tool to pinpoint a shallow 1500-ft (460-m) meandering channel in the upper Pennsylvanian Cisco system.

Regional studies report evidence of a lowstand of sea-level for this formation, locally called the King, which occurred near the end of deposition (Brown et al., 1990). The 3-D data appear to demonstrate the presence of another relative lowstand of sea level, which occurred early during deposition of this formation.

Significant detail can apparently be seen in these data, including point bars, levees, and overbank deposits. The data appear to correlate with all logs that penetrate the 3-D volume: i.e., where a well penetrates the channel location depicted in the seismic image, the logs show evidence of a channel.

This 3-D survey was designed for objectives at 3000–6000 ft (425–1830 m). Because of this, the fold at the shallow fluvial unit is quite low (three- to four-fold). Additionally, the calculated bandwidth of the data is extremely high, in the 200 Hz and greater range. These data were acquired with an I/O II system and deep-hole dynamite, illustrating the increase in resolution available with modern 24-bit systems.

Because an oil show was reported in this zone during drilling for a deeper objective, a well was drilled to test a point bar identified in the 3-D data. Although no oil was found, an excellent point bar was encountered, which further authenticated the seismic survey's ability to provide stratigraphic detail about this shallow fluvial unit. This is significant, because most of the oil remaining to be found in this and other areas will come from stratigraphic traps.


## AREA OF STUDY

Baylor County, Texas, is located at the northeastern edge of the Eastern Shelf of the Midland Basin, on the southern edge of the Knox-Baylor Trough and off the northwestern tip of the Bend Arch (Figure 1). Post-Canyon (Late Pennsylvanian) through Early Permian deposition here consisted of a series of 16 limestone transgressional episodes interspersed with prograding fluvial–deltaic systems from the east and north (Brown et al., 1990). Production from the immediate area is from the Mississippian Lime at approximately 5300 ft (1615 m), the Caddo Lime at 5000 ft (1525 m), the Strawn sand at 4500 ft (1372 m), and the lower Canyon-age Palo Pinto reef at 3000 ft (915 m) (Figure 2). Two wells producing from the Mississippian strata within 2 mi (3.2 km) have a combined cumulative production of 650,000 BO, and the Caddo, Strawn, and Palo Pinto have added more than 5 MMBO from 20–30 more wells. The nearest shallow pay is encountered in Gunsight sands of the Cisco series approximately 7 mi (11.2 km) to the south.

A 14-mi$^2$ (36-km$^2$) 3-D program was acquired in eastern Baylor County, Texas, during the fall of 1992. This program was designed to image Canyon, Strawn, and Mississippian objectives. These horizons have all produced oil in the area, and the 3-D survey was designed to enhance further exploration potential in these horizons.

However, the focus of this paper is on a Pennsylvanian (Cisco age) unit much shallower in the section, this unit being responsible for significant production in Concho County, at the southern end of the Eastern Shelf. The location of the 3-D survey coincides with a Cisco fluvial system studied by Brown et al. (1990). The 3-D survey provides evidence of an additional relative lowstand in sea level that occurred early during deposition of this unit. A well was subsequently drilled expressly to test this newly identified unit. Excellent agreement was found between the resulting log and the geology indicated by the 3-D data. This survey appears to show details of a fluvial channel system including point bars, mid-channel bars, levees, and flood deposits.

## 3-D SURVEY PARAMETERS

In the late summer of 1992, a group of mineral owners and option holders agreed to a joint shoot of 8000 acres in eastern Baylor County. A program was designed for a high-resolution survey with the parameters listed in Table 1.

TABLE 1. 3-D Acquisition Parameters

| | |
|---|---|
| Recording System | I/O II |
| Seismic Source | Dynamite, 10 lb @ 80 ft |
| Geophone Group and Line Intervals | 220 ft and 660 ft |
| Shotpoint and Shotline Intervals | 220 ft and 1320 ft in a "brick" pattern |
| Recorded Patch | 8 lines × 60 stations/line |
| Resulting Bin Size | 110 ft × 110 ft |

## GEOLOGY OF STUDY INTERVAL

Work by Brown et al. (1990) details the depositional sequences in the Cisco and Wolfcamp series of the Eastern Shelf. They interpreted the Middle Cisco Ivan Limestone as having been deposited during a transgressive phase as far east as Jack and Eastland Counties, where it becomes lenticular and pinches out. It is this formation which the 3-D survey indicates has been incised. In the study area, the Ivan resides in the subsurface at an average depth of 1550 ft dipping gently towards the northwest at less than 1°. It is overlain by the King Formation, which consists of thin fluvial–deltaic strata sourced from the Ouachitas to the east and north. Most of this deposition resulted in net sand deposits in the 2- to 10-ft (1- to 3-m) range. Figure 3A shows the net sand isolith map for this interval from the study. Note that the 3-D survey is located where a relatively thick delta has prograded from the southeast. Figure 3B shows this area in detail, along with the location of the cross section in Figure 4. This cross section illustrates the varying net sand thicknesses and orientation of King sands in this area. Note also from Figure 3B that the inferred basinward limit of these highstand coastal deposits is only about 10 mi (16 km) to the west. An intermediate lowstand and deposition of a shelf-margin systems tract to the west is inferred from sandstone isolith map patterns shown in Figure 3A. The following relative rise in sea level marked the end of deposition of King strata and the start of deposition of the Blach Ranch Limestone. The Cisco and Wolfcamp series on the Eastern Shelf of the Midland Basin consist of 16 such transgressive–regressive events (Brown et al. 1990).

## INTERPRETATION

As interpretation progressed on the deeper horizons, an interesting anomaly was noted in the Ivan Limestone (Figure 5). While initial investigations of this marker revealed a possible channel, little effort was expended on this feature until the deeper work was completed, as no production above the Palo Pinto Formation had been established in the immediate area. The project was moved from the Landmark system to a PC-based interpretation workstation, where

*Integrated Seismology, Garland TX 75044

time slices quickly confirmed a well-developed fluvial system meandering from southeast to northwest across the study area (Figure 6). The gentle regional dip to the northwest makes interpretation of the entire channel difficult from time slices alone, because one cannot see the channel in its entirety from a single time slice. Since this fluvial system has incised into the Ivan limestone, it is reasonable to pick the limestone throughout the survey and flatten this horizon. Now the regional dip is removed from the data, allowing time slicing to show the complete channel. This is demonstrated in Figure 7. The zero crossing at the bottom of the peak has been picked and set to a time of 320 ms. The slice seen is 323 ms, or about in the center of the trough which should represent the base of the Ivan. The entire channel can now be seen, implying that whenever this stream was instigated, it was persistent enough to erode through the Ivan, or at least cut enough of this limestone so as to change its seismic signature.

The seismic event running south-to-north, then curving away to the west, starting from the southwest quarter of the map, is a subcropping unit shown in Figure 4. It is difficult to tell at this point where the shallowest expression of the fluvial system commences, but a flattened slice 10 ms above the reference 320 ms shows a part of the channel in the southeast. Since the section from Blach Ranch to Ivan is about 50 ms (100 ft) thick, this suggests evidence for another lowstand early during the King deposition, one which was active for at least the first 20 ft (6 m) of deposition. At this depth, the channel appears to range from 200 to 500 ft (60 to 150 m) in width. Since the bin size is 110 × 110 ft (33 × 33 m) for this survey, the data limit lateral resolution of stratigraphic features like this. If a project were being designed for imaging such a target, a 55-ft (17-m) bin size would be more appropriate. Note that at least four wells have penetrated the channel, reflecting the Ivan to be missing entirely by erosional cutout. Figure 8 shows such an example. The Chardonnay X2 well has penetrated the normal section of Ivan Limestone, whereas the X1 well penetrated nothing but fluvial sandstone where the Ivan should be and what is interpreted as limestone debris at the bottom of the channel, 14 ft (4 m) below the original base of the limestone.

Now that the exact location of the channel is known, one can pull arbitrary seismic lines as desired to see aspects of the unit. Figures 9A and B show two such lines. Figure 9A shows a seismic line oriented to show the axis of the channel. The eastern edge of the channel cut is clearly seen, as well as mid-channel sandstone bodies. On the western side of the channel, there is a cross section of a meander and apparently a levee as well as overbank deposits lying close to the top of the Ivan limestone. If this interpretation is correct, it would support the premise that this channel was the result of an intermediate lowstand which occurred early during King deposition. It is also possible that since we see evidence of this channel over a 20-ft (6-m) section of the base of the King Formation, the same stream bed may have been re-occupied several times by successive lowstand events during early King time.

Standard procedure in this area is to drill nearly to the Palo Pinto limestone with water only (no drilling mud). While increasing penetration rates dramatically, this practice produces no useable cuttings and suppresses any oil shows. When there is no shallow production, most operators will choose to save the cost of drilling mud and rig time, especially when there are a number of older penetrations in an area.

The Chardonnay X1 well was drilled as a Caddo test as a result of the 3-D survey and drilled as described above except through the King zone. Special care was taken to recover any samples, and a thick sandstone was encountered with the normal Ivan limestone missing, just as predicted by the 3-D seismic data. An oil show in the sandstone was reported in the form of fluorescence and odor. No drillstem test was run, and the well was completed in the Caddo Formation. This oil show in the King zone provided enough encouragement to test the interpreted point bar shown in Figure 9B. Figure 10 shows the resulting log in a cross section with nearby wells. This confirms the seismic interpretation of the point bar. As compared to the adjacent logs having 12–15 ft (4–5 m) of Ivan limestone, the No. 1 Jinks had only 5 ft (2 m) of limestone at the base of the Ivan. There is an Ivan limestone erosional remnant, indicating that the fluvial system did not cut through the limestone completely at this location. A comparison of the sand body in the new well with the X1, which penetrated mid-channel, shows that the Jinks indeed tested a point bar, as evidenced by the thick sandstone with a fining-upward sequence on top, whereas the X1 drilled several stacked mid-channel sandstone bodies.

Although the test of the point bar was not successful from a petroleum standpoint, it was from a geological perspective. Many shallow fluvial units in West Central Texas are quite prolific and very elusive. 2-D seismic techniques to date have had limited success at best in predicting fluvial orientation and detail. In contrast, 3-D seismic data have provided accurate pre-drill information about the location of the point bar and channel. This technique promises to revolutionize exploration for shallow fluvial targets on the Eastern Shelf.

Another interesting property of this 3-D survey is that even though there was 660 ft (200 m) between receiver lines, the fold at 1300–1500 ft (400–460 m) is fairly low. Table 2 summarizes the fold vs. depth relationship for this survey.

Note that the fold at the channel (1500 ft [460 m]) is 3 to 4. This gives testimony to the accuracy of 3-D migration, although amplitude artifacts can be seen in the timeslices in Figure 6. This is probably due to the low and variable fold at this depth, as these artifacts are not visible deeper in the dataset.

The other aspect of the seismic data which allows such detail to be resolved is the bandwidth of this survey. Figures 11 and 12 illustrate the unusual bandwidth recovered in this seismic survey. This resolution is due to several factors. The source was dynamite (see Table 1); standard geophysical vibrators often cannot generate much more than 110 Hz of usable bandwidth. We used a 24-bit seismic recording

**TABLE 2. Depth vs. Nominal Fold**

| Reflector Depth (ft) | Nominal Fold |
|---|---|
| 1300 | 1 |
| 1500 | 3 |
| 3000 | 7 |
| 5000 | 14 |
| 7000+ | 20 |

system, and careful attention was paid during fieldwork and processing so as to retain all available bandwidth while rejecting noise. To illustrate the high-frequency content of the data, the time separation between the two yellow dots at the point bar in Figure 9B is between 4 and 5 ms (200–250 Hz). As a further demonstration, Figure 13 shows seismic section D-D′ from Figure 9B next to a bandpassed version. The details of the point bar have vanished. Note also that the amplitude of the frequencies above 125 Hz are as high as those below 50 Hz: They are contributing significantly to the resolving power of this dataset.

## CONCLUSIONS

A high-resolution 3-D program was acquired in eastern Baylor county to image Palo Pinto and deeper targets. A shallow fluvial unit at 1550 ft (470 m) was imaged as a by-product of this effort. The 3-D survey appears to show clearly the exact orientation of this Cisco stream, as evidenced by the agreement with well control. Point bars, mid-channel sand bodies, levees, and overbank deposits are also visible as a result of the unusual bandwidth recorded by this survey.

This study has been compared to work done from subsurface information by the Bureau of Economic Geology in this area. The 3-D data add to this work by providing evidence of a second lowstand episode early during King deposition.

A well was drilled to test one of the point bars identified from the 3-D image. The resulting log, along with the others, validates the seismic data's ability to accurately predict potential reservoir location and orientation in this shallow fluvial unit. This technique has immediate and obvious application to other such units on the Eastern Shelf which are known to be productive.

## REFERENCE CITED

L. F. Brown, Jr., R. F. Solis-Iriarte, and D. A. Johns, 1990, Regional depositional systems tracts, paleography, and sequence stratigraphy, upper Pennsylvanian and lower Permian Strata, North- and West-Central Texas: Bureau of Economic Geology Report of Investigations 197, p. 76–78.

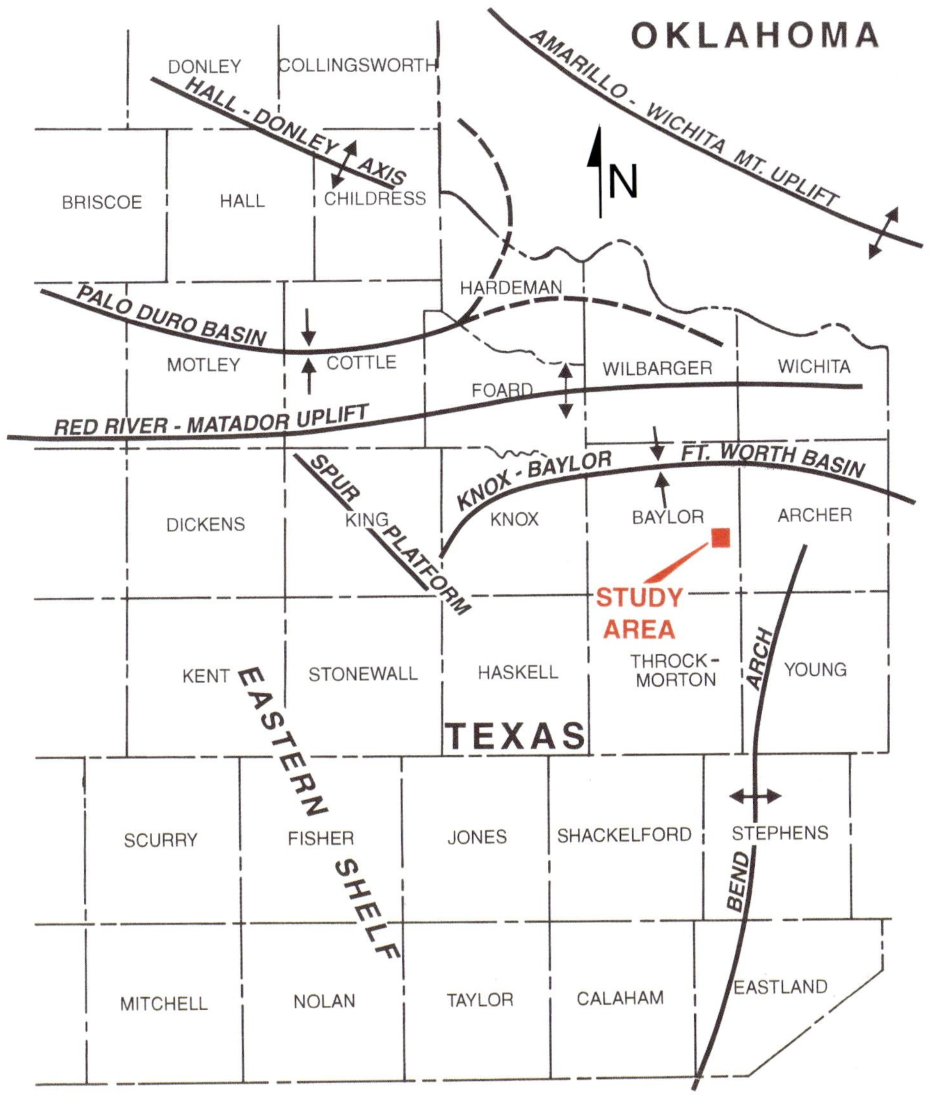


FIGURE 1. Study area location map, showing regional structural features.

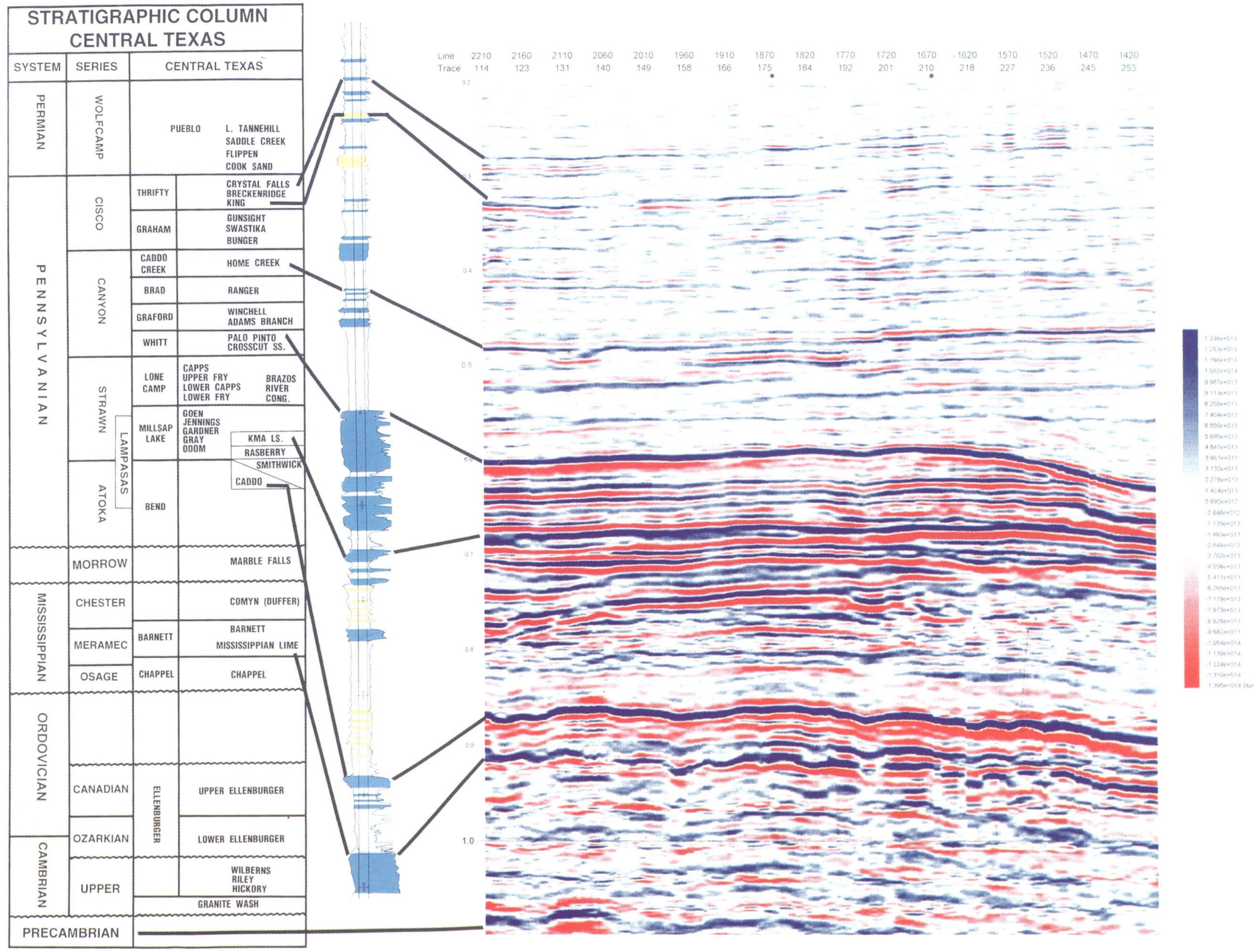


FIGURE 2. Stratigraphic column and type log with representative seismic line. Note the high signal-to-noise character. Also note deep-seated structural features in Caddo–Basement interval. These are the targets of the 3-D survey.

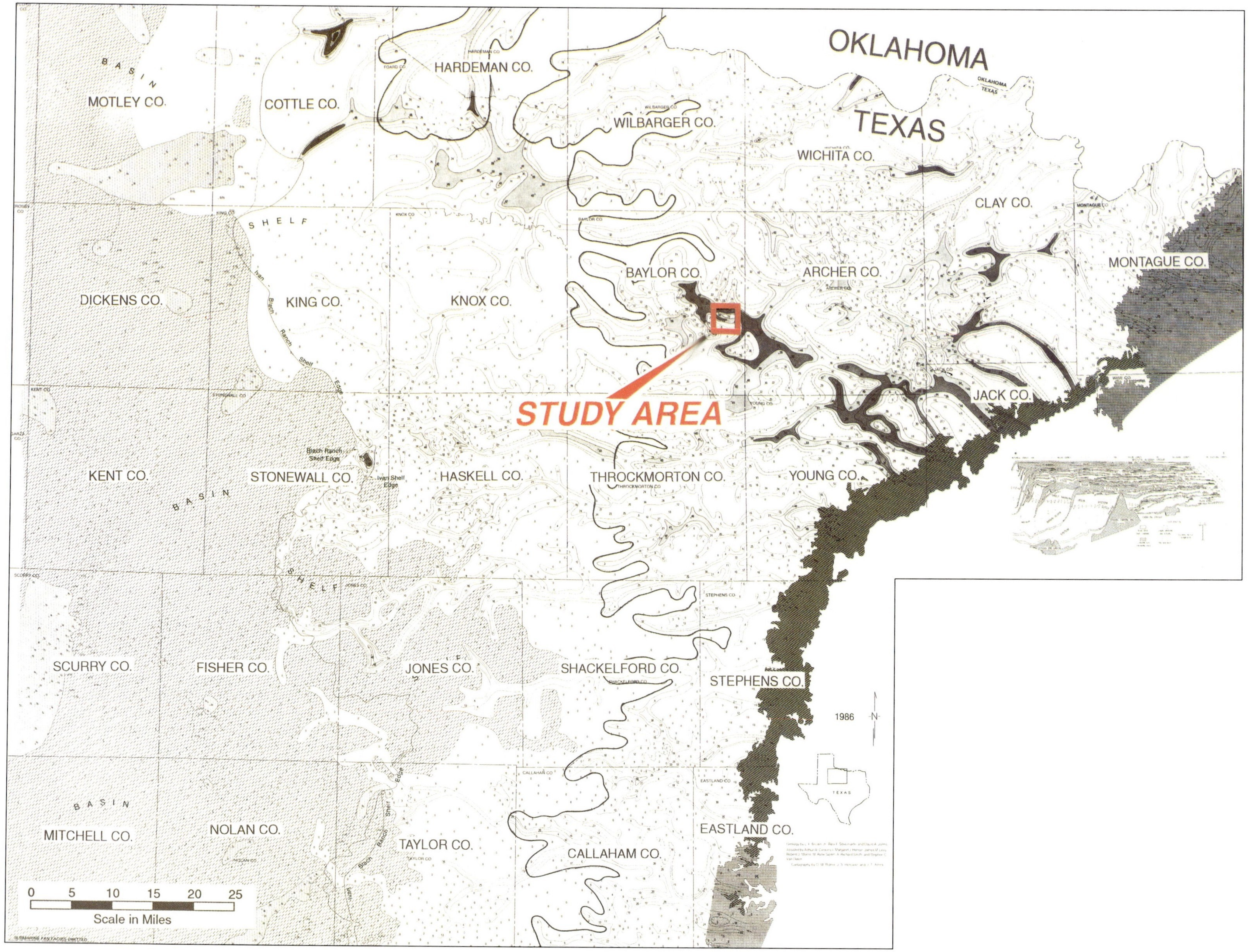


STRATIGRAPHIC POSITION AND PRINCIPAL DEPOSITIONAL SYSTEMS REPRESENTED BY SANDSTONE ISOLITH MAP

| SYSTEM | SERIES | GROUP | FORMATIONS, MEMBERS AND INFORMAL SUBSURFACE SANDSTONE UNITS | MAP INTERVAL: SLOPE-BASIN SYSTEM | MAP INTERVAL: FLUVIAL-DELTAIC SYSTEM |
|---|---|---|---|---|---|
| PERMIAN | WOLFCAMPIAN | CISCO | Coleman Junction Limestone | | |
| | | | Unnamed sandstone | | |
| | | | Sedwick Limestone | | |
| | | | Unnamed sandstone | | |
| | | | Dothan Limestone | | |
| | | | Moutray sandstone | | |
| | | | Camp Colorado Limestone | | |
| | | | Upper Frye sandstone | | |
| | | | Stockwether Limestone | | |
| | | | Lower Frye sandstone | | |
| | | | Lower Stockwether limestone | | |
| | | | Tannehill sandstone | | |
| | | | Saddle Creek Limestone | | |
| | | | Flippen/Bluff Creek sandstones | | |
| ? | ? | | Flippen Limestone | | |
| PENNSYLVANIAN | VIRGILIAN | | Cook sandstone | | |
| | | | Crystal Falls Limestone | | |
| | | | Upper Hope sandstone | | |
| | | | Breckenridge Limestone | | |
| | | | Lower Hope sandstone | | |
| | | | Blach Ranch Limestone | | |
| | | | King sandstone | ▨ | ▨ |
| | | | Ivan Limestone | | |
| | | | Swastika sandstone | | |
| | | | Gunsight Limestone | | |
| | | | McMillan sandstone | | |
| | | | Bunger Limestone | | |
| | | | Unnamed sandstone | | |
| | | | Gonzales Limestone | | |
| | | | Unnamed sandstone | | |
| | | | Salem School Limestone | | |
| | | | Unnamed sandstone | | |
| | MISSOURIAN | CANYON | Home Creek Limestone | | |

EXPLANATION

- Well with net-sandstone value (ft).
- Well with terrigenous shale; sandstone absent.
- Well did not penetrate interval.
- Well with interval obscured by surface casing, or shallow section not logged.
- Well penetrated only part of interval; value is minimum net sandstone (ft).
- Well with shelf limestone, slope-basin limestone, or slope-basin hemipelagic shale.
- Well with limestone buildup.
- Well with limestone buildup buried by terrigenous clastic deposits, either shale (0) or net sandstone in feet (10).
- Well with shelf limestone, slope-basin limestone, or slope-basin hemipelagic shale less than 100 ft above buried limestone buildup.
- Sandstone trend extrapolated; precise location questionable.
- Area of shelf limestone, slope-basin limestone, or slope-basin hemipelagic shale.
- Area of limestone buildup; dashed where buried during deposition of this interval.
- Position of subjacent shelf edge; dashed where approximate.
- Position of overlying shelf edge.
- Inferred fluvial sediment supply.
- Approximate outcrop of Gunsight to Blach Ranch/Breckenridge or equivalent interval; Cretaceous overlap (K).

Net-sandstone contour intervals (ft).

0-40; 40-60; 60-100; 100-140; 140-200; 200-260

Fluvial-deltaic systems: all sandstone thickness values contoured; slope-basin systems: all sandstone values contoured.

FIGURE 3A. Regional sand isolith map for King section, showing depositional environments. Taken from State of Texas Bureau of Economic Geology Report of Investigations No. 197.

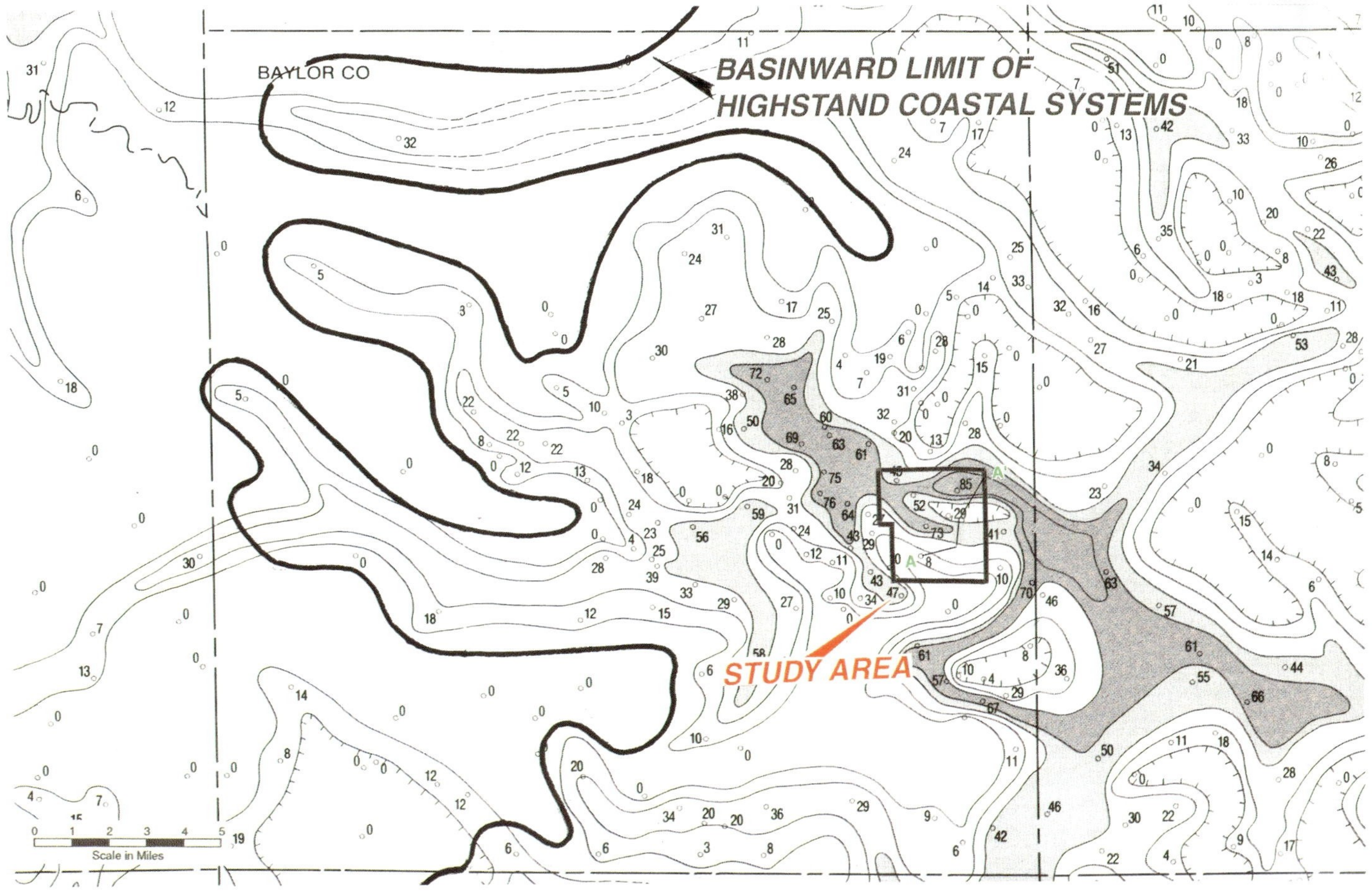


FIGURE 3B. Detail of 3A, showing study and cross section location relative to inferred deltaic system.

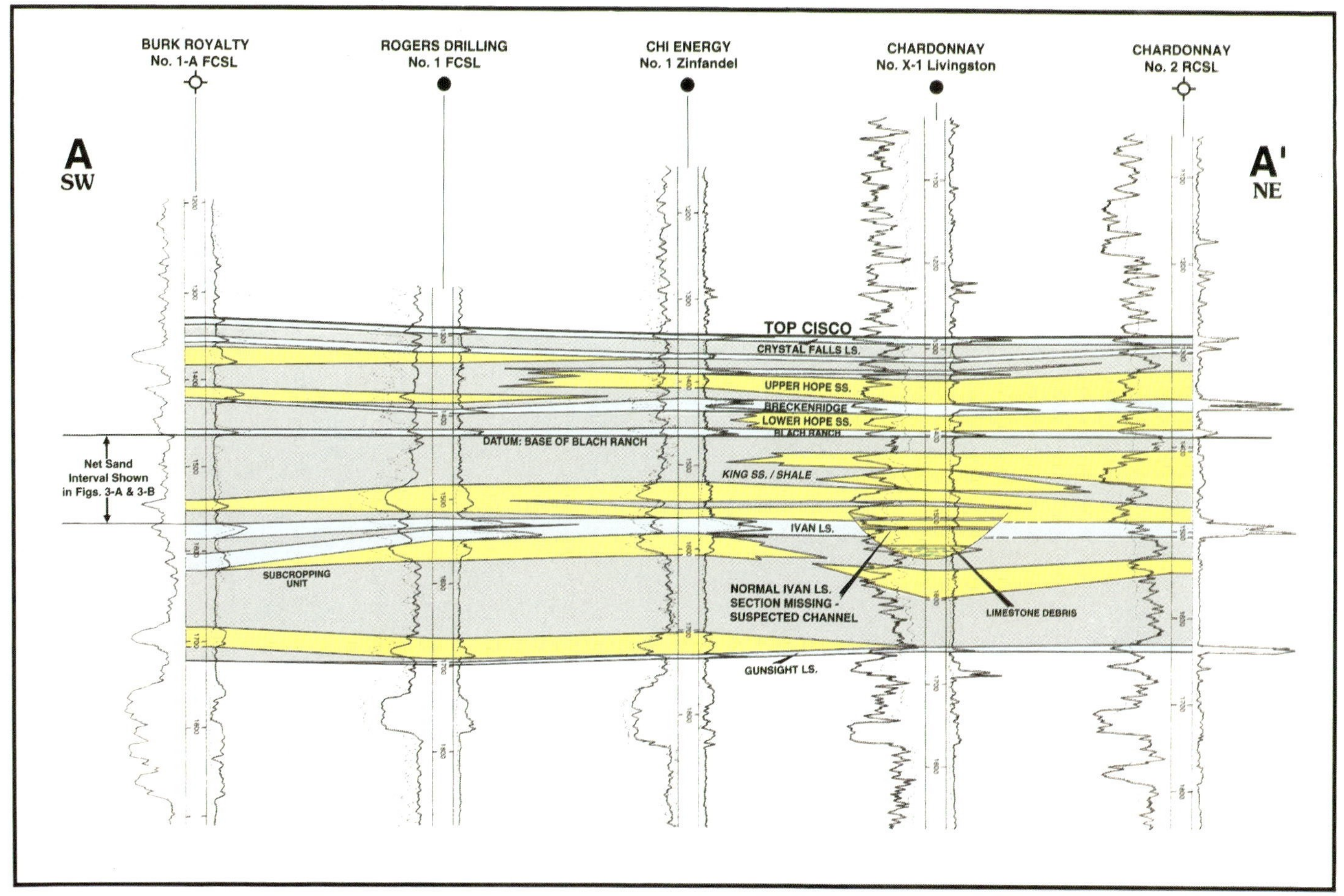


FIGURE 4. Stratigraphic cross section, showing King sand deposition in study area. Note thickening sand towards the northeast.

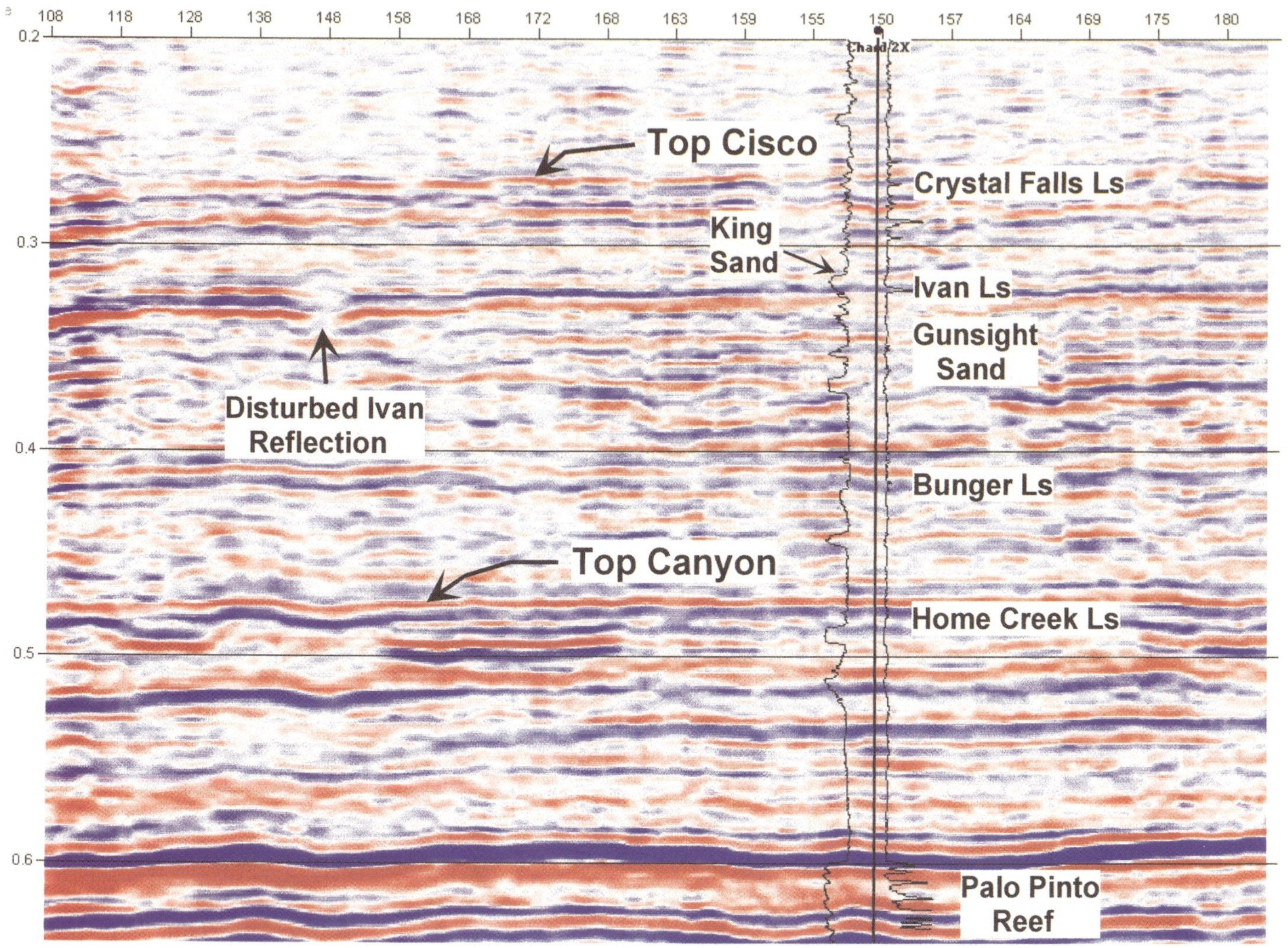


FIGURE 5. Example of disturbed Ivan reflection. Seismic reflectivity has changed due to erosion of the limestone and its replacement by King sandstone. Note the high signal-to-noise ratio and the stratigraphic detail available.

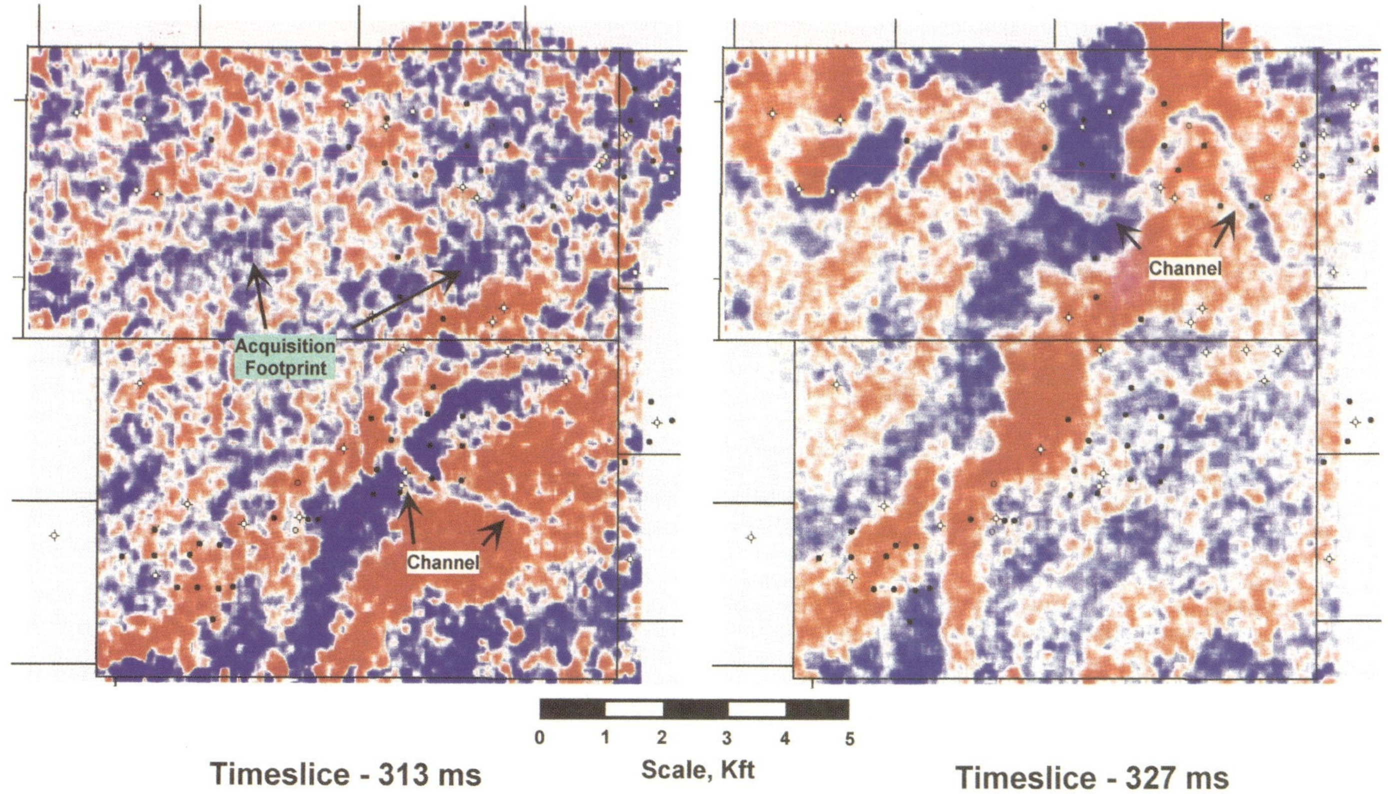


FIGURE 6. Time slices illustrating portions of meandering channel. Note patterns of vertical or horizontal amplitude distortion, probably due to low fold.

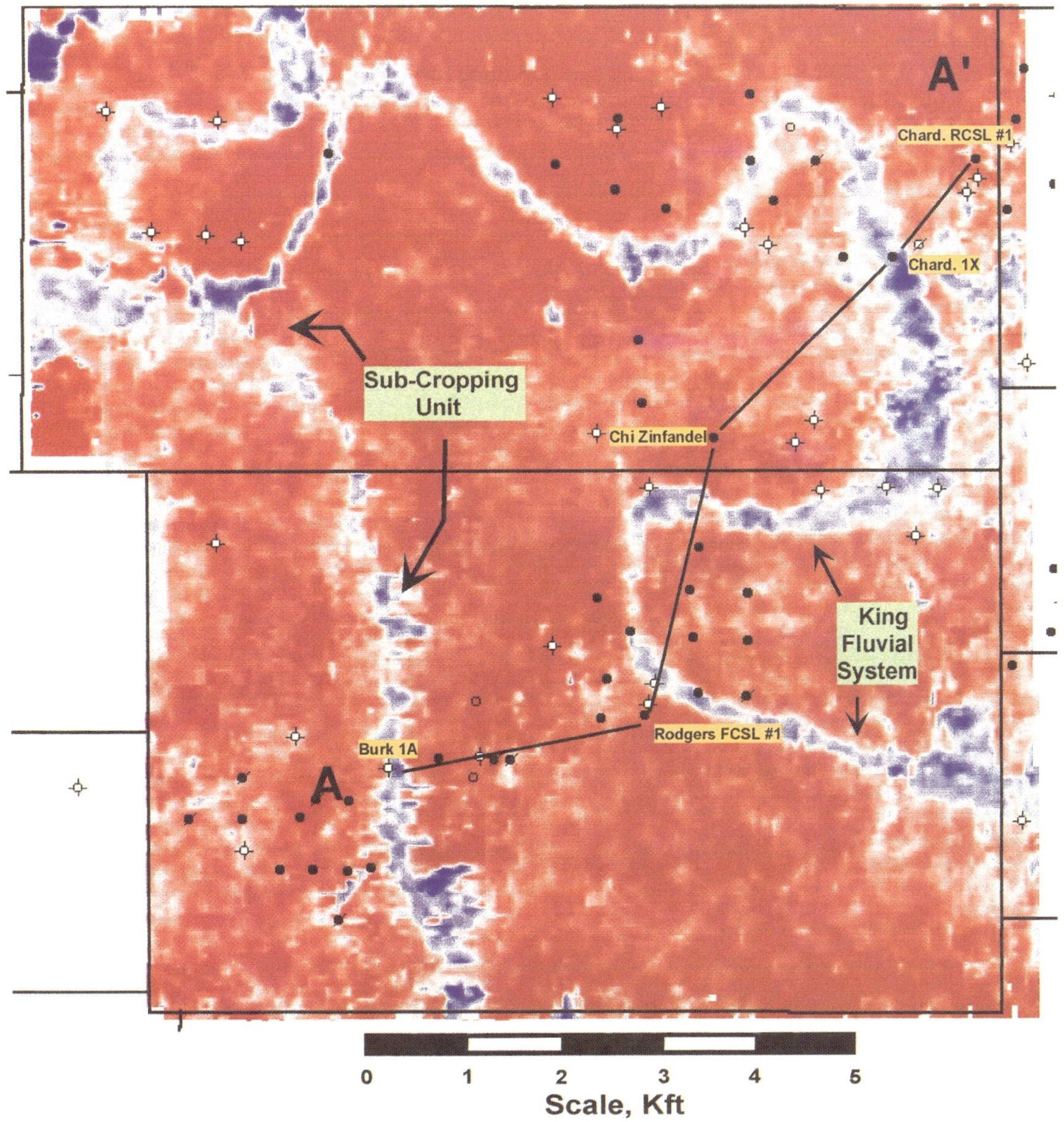


FIGURE 7. Time slice 3 ms under flattened Ivan horizon. Entire channel system is visible. Note well penetrations of channel. No doubt this channel would have been resolved even better with a smaller bin size. Location of cross section of Figure 4 shown. Subcropping unit has been penetrated by Burk Royalty 1A well.

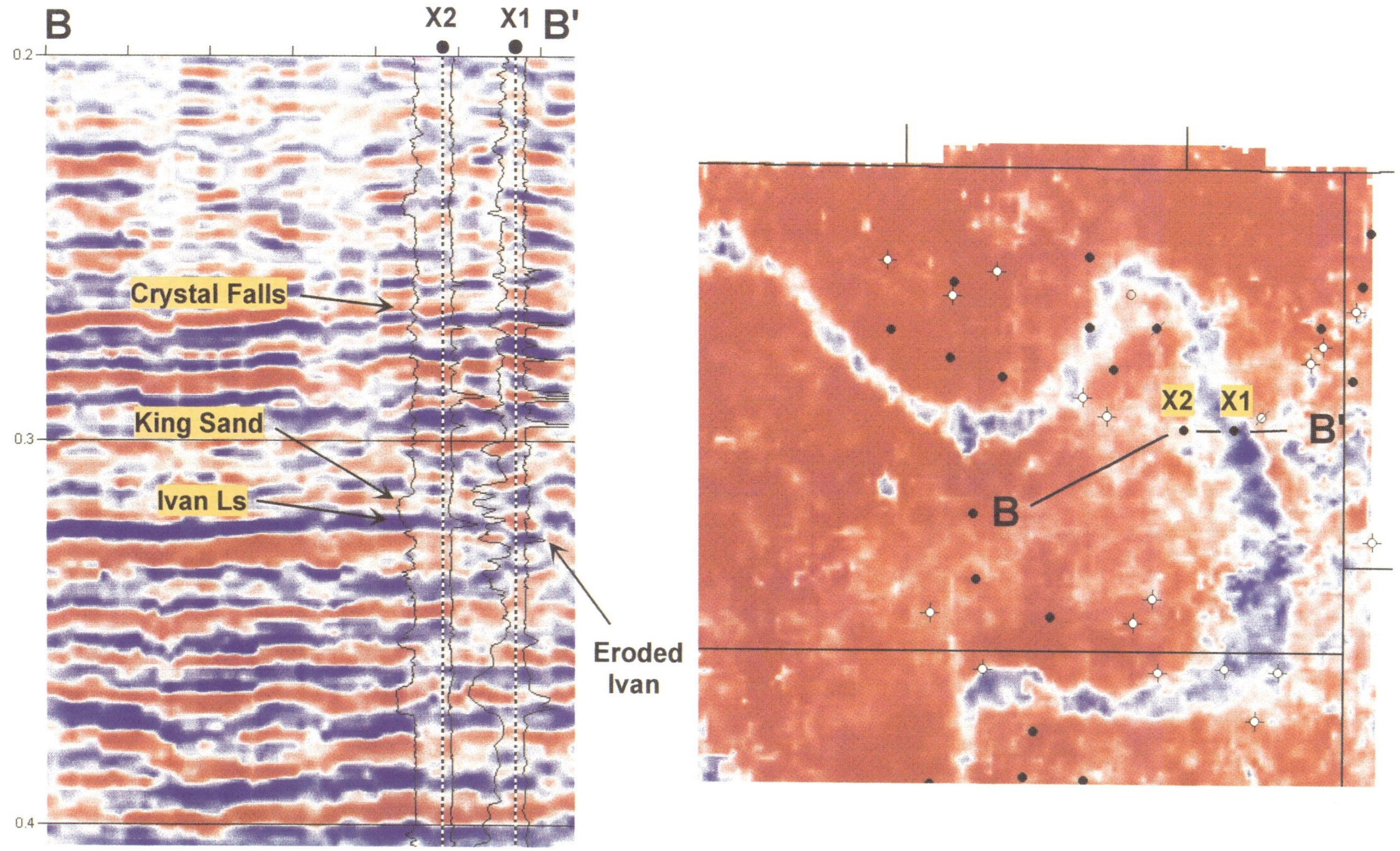


FIGURE 8. Arbitrary line showing a normal Ivan penetration vs. a missing Ivan limestone. The disrupted seismic response of the Ivan is clearly visible.

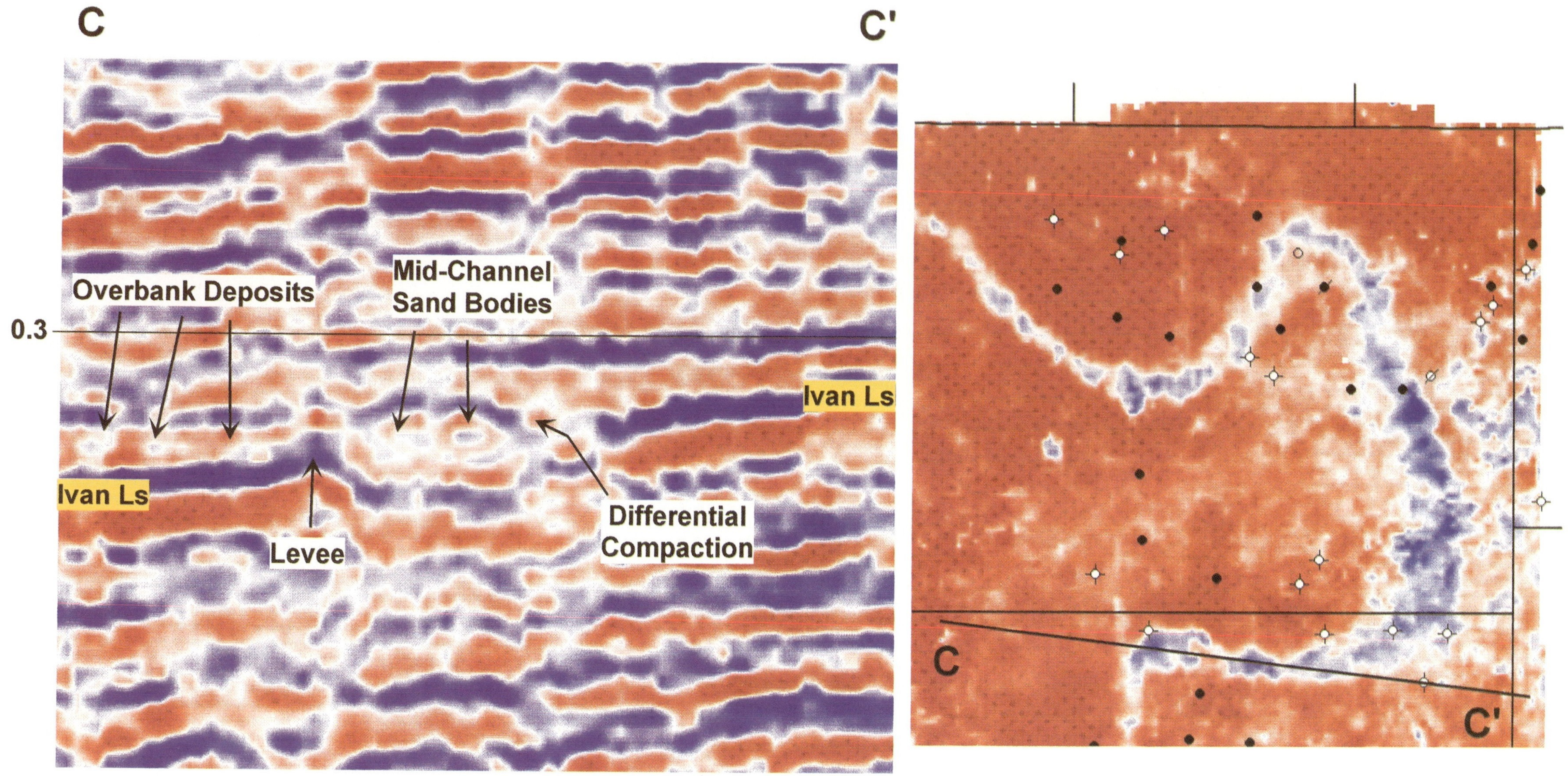


FIGURE 9A. Arbitrary line along channel axis showing mid-channel sand bodies, levee, overbank deposits, and differential compaction.

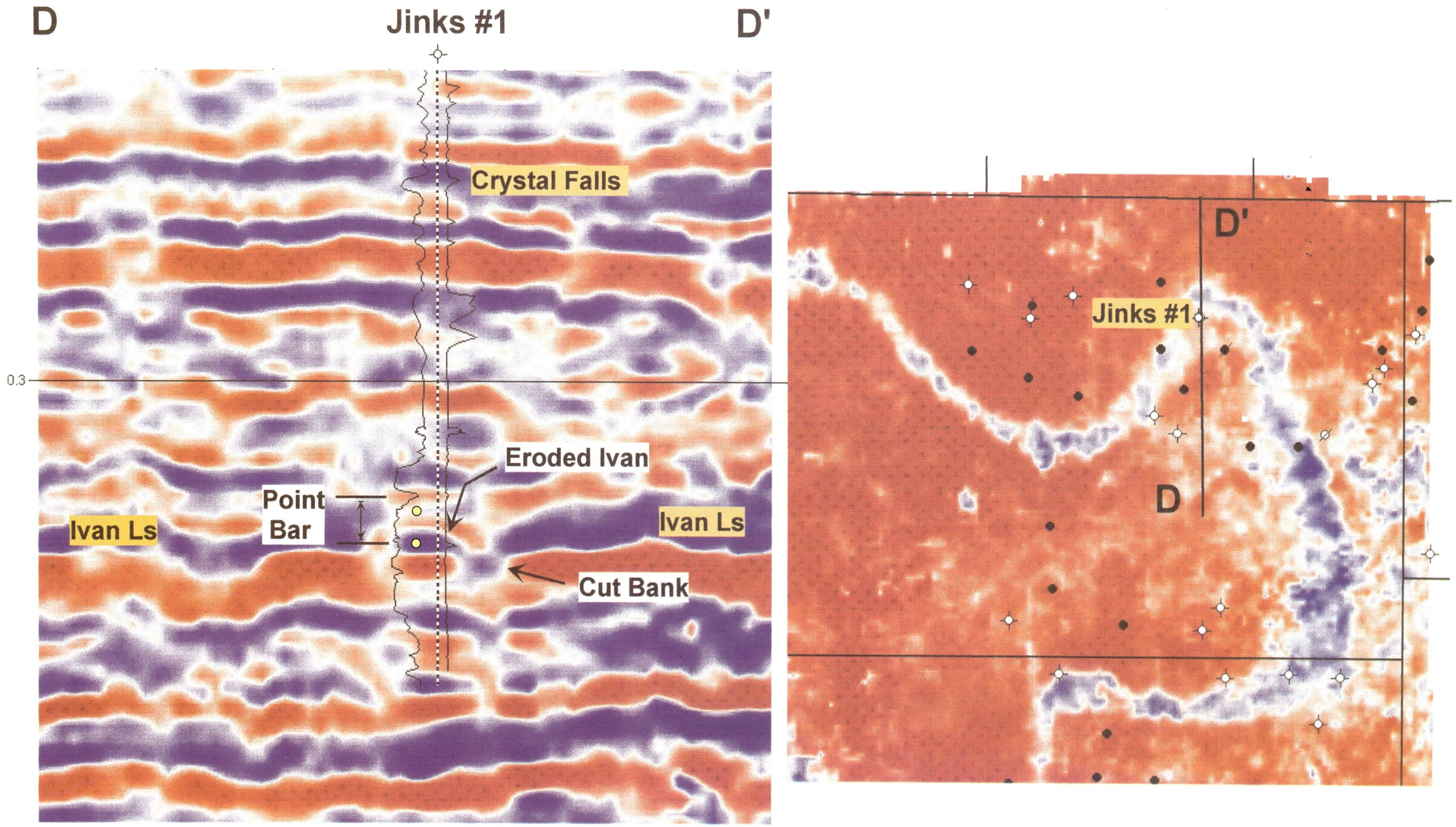


FIGURE 9B. Arbitrary line showing point bar axis and log from test well. Note the close correlation between the log and the seismic response of the sandstone point bar.

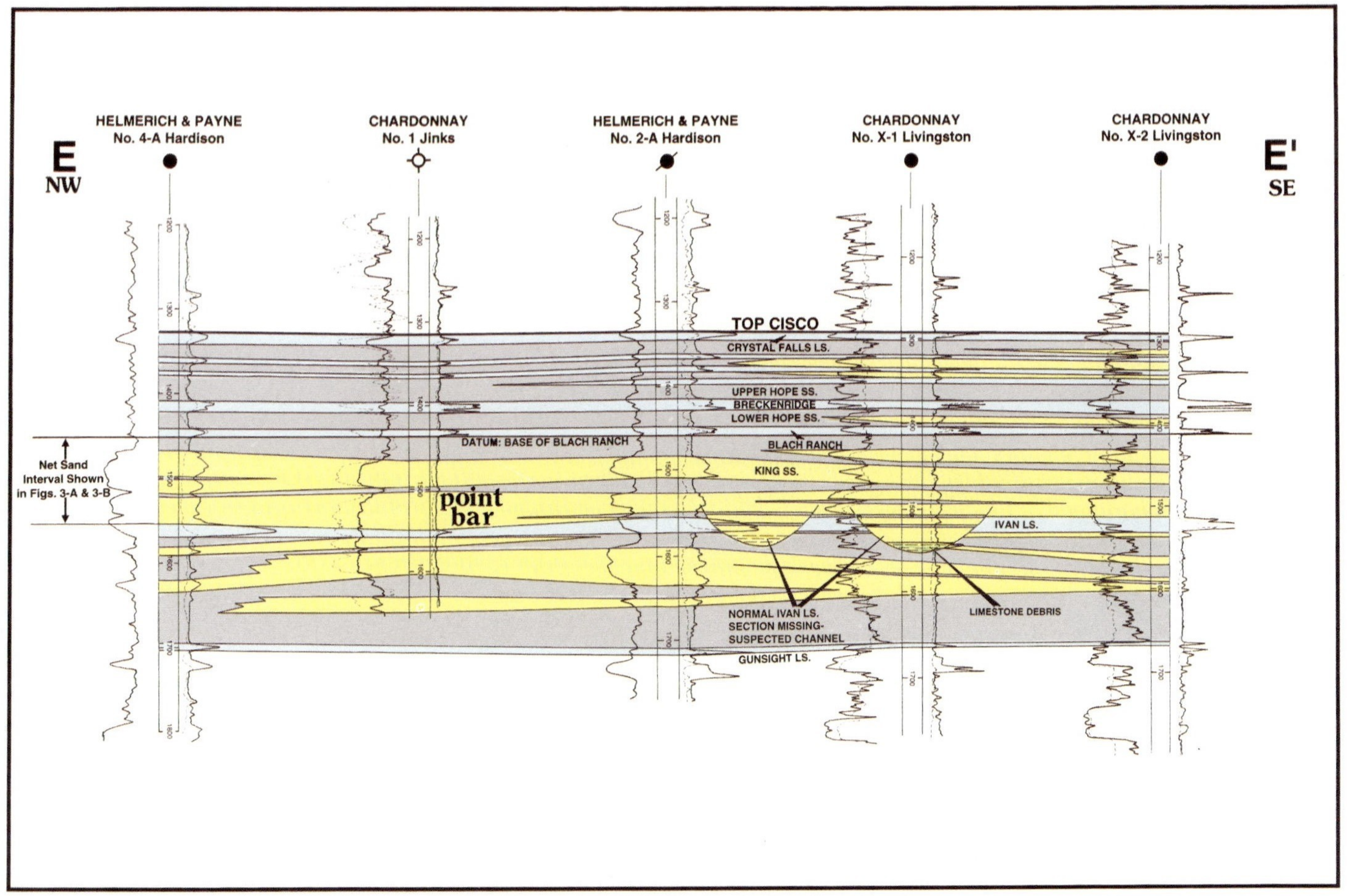


FIGURE 10A. Stratigraphic cross section showing point bar test. The Jinks #1 well shows a massive, well-developed point bar, while the Livingston X1 well has penetrated several stacked sand bodies.

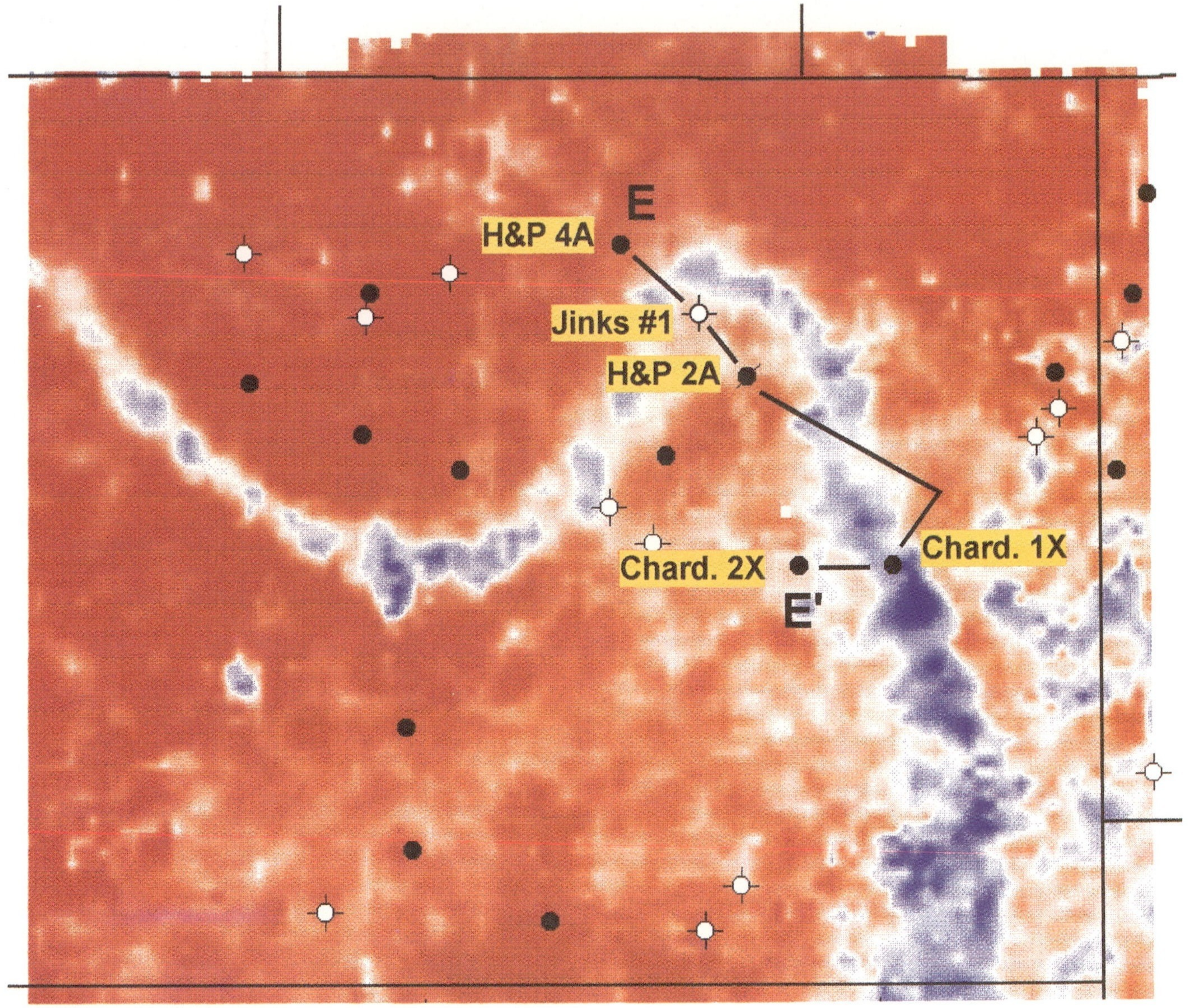


FIGURE 10B. Location of cross section in Figure 10A.

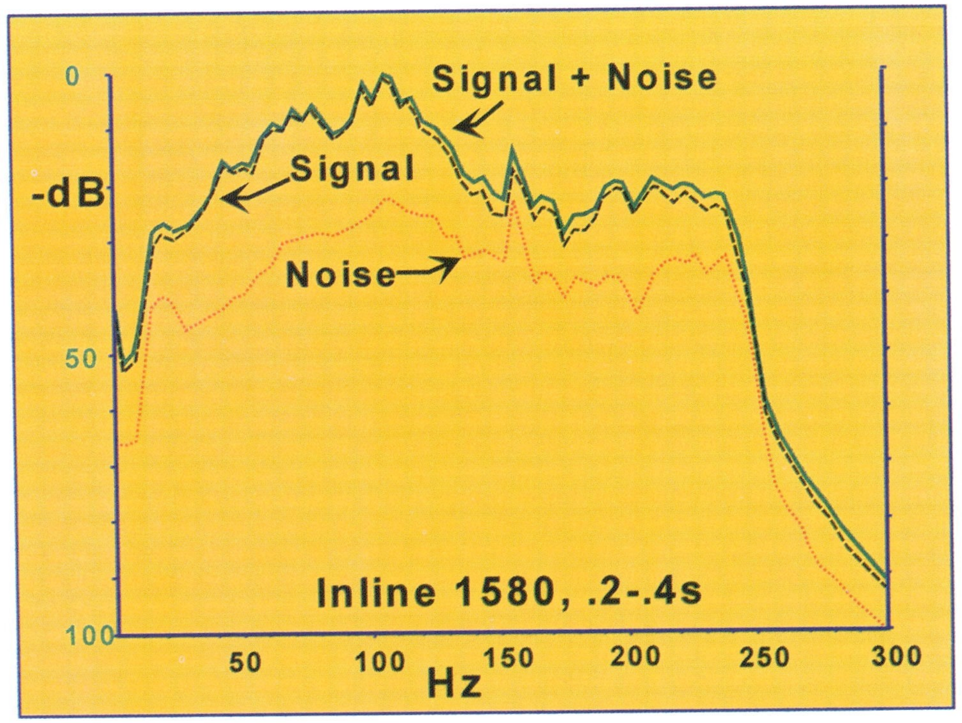


FIGURE 11. Spectral analysis of inline 1580 from 0.2 s–0.4 s, covering the time interval of the fluvial system. Note calculated bandwidth of data is greater than 200 Hz. Dominant bandwidth at this time window of the dataset, estimated by visual inspection of main reflection events, is closer to 110 Hz, which corresponds to the 0-dB position of the spectrum.

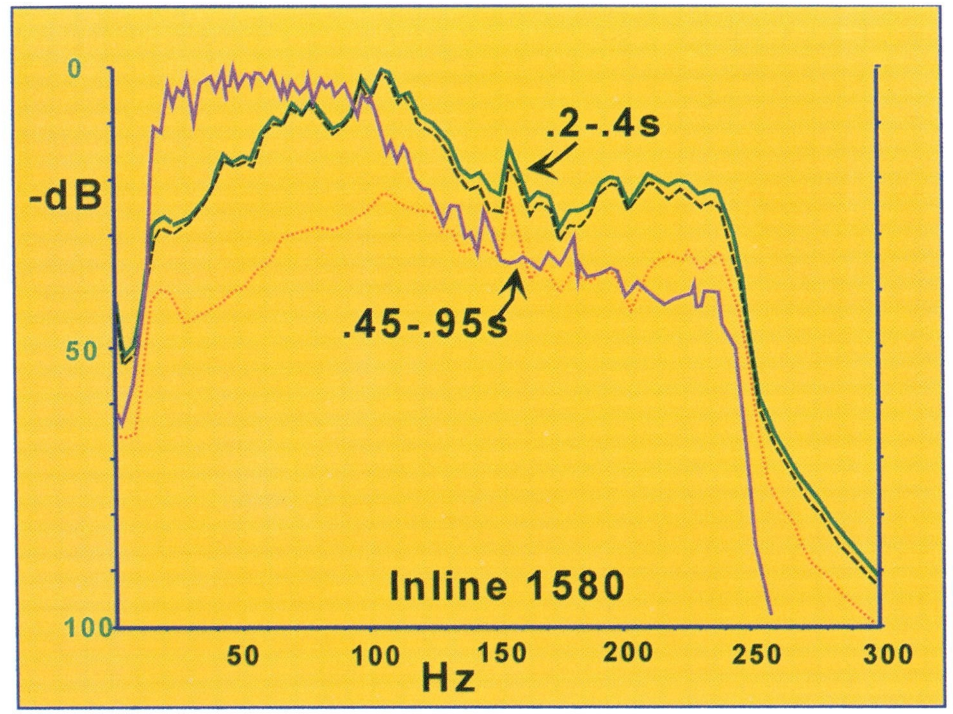


FIGURE 12. Comparison of Figure 11 with deeper data. While the frequencies between 10 and 110 Hz are effectively balanced in amplitude in the deeper data, stratigraphic details can still be seen in the 200 Hz range in this part of the dataset. Signal-to-noise ratio for this survey is excellent, which demonstrates the effective noise rejection results of the data processing.

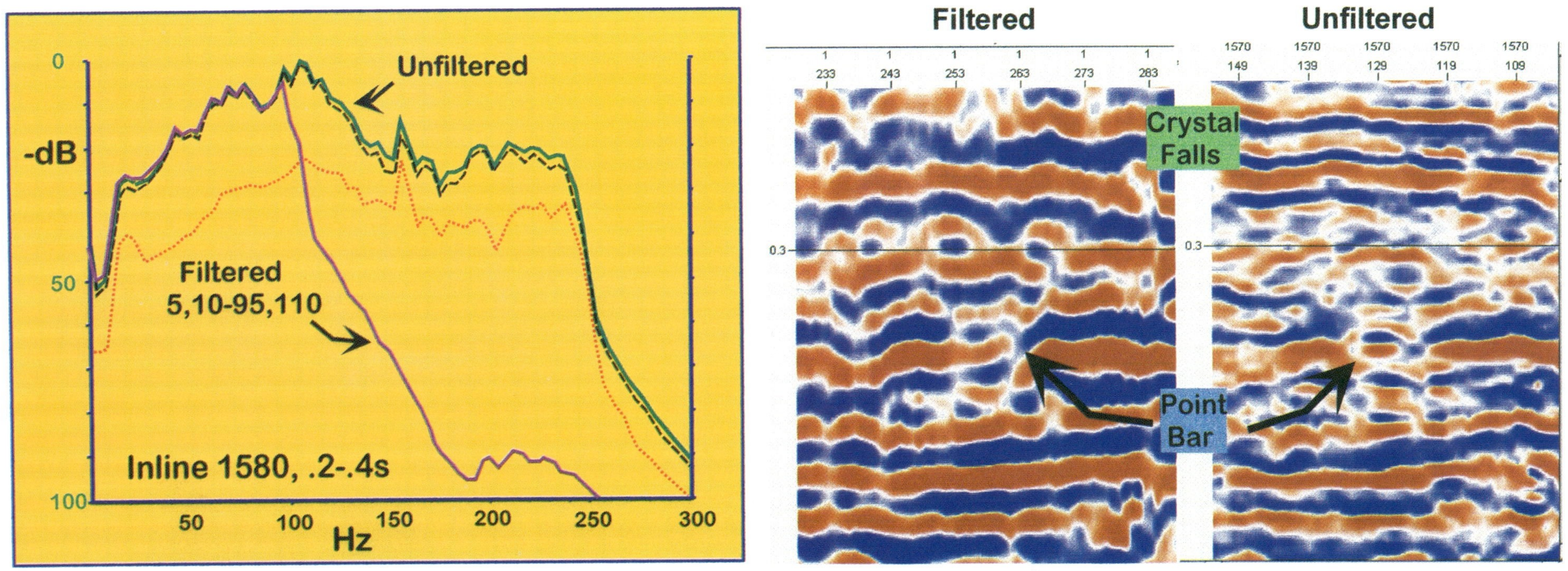


FIGURE 13. Bandpass filter applied to the point bar. This channel might not have been discovered had the seismic source been Vibroseis. These results were possible in large part because of the increased dynamic range of the recording system.

CHAPTER 6

# Impact of 3-D Seismic Interpretation on Reservoir Management in the Apiay–Ariari Oil Fields, Llanos Basin, Colombia

Gutierrez, M., E. Lozano, and M. Sroczynski, Impact of 3-D seismic interpretation on reservoir management in the Apiay-Ariari oil fields, Llanos Basin, Colombia, *in* P. Weimer and T. L. Davis, eds., *AAPG Studies in Geology No. 42* and *SEG Geophysical Developments Series No. 5*, AAPG/SEG, Tulsa, p. 57–68.

Mario Gutierrez*, Eduardo Lozano†, and Michael Sroczynski‡

## ABSTRACT

The Apiay–Ariari oil field of the Llanos Basin, Colombia, is located within a structurally complex area. A 3-D seismic survey was necessary to understand the structural factors controlling the distribution of hydrocarbons within the field. The results of the interpretation were used in conjunction with core, production, and log data to complete an integrated reservoir evaluation. Time structure maps were converted to depth using geostatistical techniques which resulted in a more detailed realization of strike-slip, normal, and reverse faults, consistent with a right lateral shear system. The 3-D interpretation is also much more simplified than the 2-D interpretation. This integrated study realized an increase of field reserves of greater than 60.7 million bbl over those originally calculated. The study also identified several exploration plays, such as that successfully tested by the Gavan-1 well. The ability to fully integrate the new 3-D seismic information with all available data resulted in a more accurate understanding of structural components responsible for the trapping of hydrocarbons within the Apiay–Ariari fields. This knowledge has proven to be essential for proper management of the field development.

## INTRODUCTION

To more closely match production models with actual geologic data, a 3-D seismic survey was conducted over existing oil fields in Colombia. The study area is located in the southwestern portion of the Llanos Basin, west of the Guyana Shield and 100 km (61 mi) south-southeast of Bogotá (Figure 1). The eastern half of the study area consists of the Apiay field proper, while the western half comprises both the Suria and Libertad fields. The entire area contains 48 closely clustered wells. Hydrocarbons in the Apiay field are found in siliclastic upper Cretaceous to Tertiary strata. ECOPETROL discovered the Apiay field in 1981 with the Apiay-1 well, which produced 1673 BOPD, at 25° API, from the Cretaceous K-2 unit. Currently the field produces 30,000 BOPD with a 20% water cut. Cumulative oil production is at 68 million bbl. The OOIP is calculated at 900 million bbl for the Apiay–Ariari area, and the recoverable reserves are 270 million bbl. The Apiay–Ariari area includes the Apiay, Suria, and Reforma–Libertad fields and surrounding areas (Figure 2). The objectives of the study were to (1) obtain a more complete understanding of the regional geologic model, (2) determine a more accurate estimate of the reserves in place, (3) reduce the risk of future development wells, and (4) locate additional potential drilling prospects. To best achieve the objectives, a 3-D seismic survey was acquired and used as part of an integrated reservoir evaluation over the oil fields of the Apiay–Ariari area.

*† ECOPETROL, Reservoir Division, Bogotá, Colombia
‡E & P Services, Western Atlas International, Inc., Houston, TX 77042

## SEISMIC OVERVIEW

ECOPETROL contracted Western Geophysical to acquire and process 125 km² (78 mi²) of 3-D seismic data covering the Apiay–Ariari area. The data were jointly interpreted by an interdisciplinary team of ECOPETROL and Western Geophysical personnel.

The overall topography of this region is gently rolling farmlands. The acquisition of data was slightly hampered by pre-existing production facilities around the Apiay and Suria fields. Table 1 lists a summary of the main parameters used for the seismic acquisition (Figure 3).

Standard 3-D seismic processing was done in Bogotá. In addition, random noise attenuation, instrument phase correction, and time variant spectral whitening were performed in Houston to achieve a dataset suitable for stratigraphic interpretation.

## STRATIGRAPHIC FRAMEWORK

The stratigraphy of the Llanos Basin, Colombia, is typically described with standard lithostratigraphic nomenclature (Reyes, 1992). Nomenclature used in this study is consistent with the operational units of the Llanos Basin as defined by ECOPETROL. These units are bounded by unconformities and/or lithology changes (Figure 4).

**TABLE 1. Summary of 3-D Seismic Survey Geometry**

| | |
|---|---|
| Fold | 20 |
| Lines/swath | 4 |
| Receiver line separation | 500 m (1640 ft) |
| Number of channels | 480 |
| Source line separation | 300 m (984 ft) |
| Receiver interval | 50 m (164 ft) |
| Energy source | Dynamite ( 3 mini holes 3 m [10 ft] deep) |
| Bin size | 25 × 25 m (82 × 82 ft) |

### PALEOZOIC–MESOZOIC STRATA

Paleozoic strata lie unconformably upon Precambrian rocks. The lower units consist of Cambro-Ordovician strata overlying phyllites. The seismic reflection for this boundary is of poor quality. The Mesozoic section in the Apiay area consists of Upper Cretaceous strata, which are Cenomanian–Maastrichtian in age (Castro-Arévalo et al., 1991). ECOPETROL recognizes two operational units in this formation. The older K-2 reservoir is the most economically significant. Core analysis of this unit indicated quartzose, medium- to coarse-grained sandstones, varying locally to conglomeratic. This unit has been interpreted as braided stream channel deposits. Overlying the K-2 is the K-1 unit, which consists of shales with interebedded fine- to medium-grained sandstone intercalations. This unit is interpreted as consisting of a deltaic environment—distributary channels, point bars, and crevasse splay deposits. The top of the K-1 unit is an unconformable surface recognizable on seismic sections. The reservoirs within the K-1 have fair production rates due to heterogeneity associated with low-stratigraphic continuity of the sand bodies and poor petrophysical properties.

### UPPER EOCENE-OLIGOCENE STRATA

The T-1 operational unit is located at the base of the Carbonera Formation and consists of coarse- to fine-grained sandstones, with intercalations of gray-to-green mudstones and siltstones (Figure 4). The T-1 unit has strong facies variations from sand to shaly sand. The T-1 sand produces gas/condensate in the Apiay field, where the reservoir contains sandstones that are extremely porous. The T-1 reflection on the seismic sections varies from very good to fair.

The Carbonera reservoir facies consist of white coarse- to fine-grained sandstones, deposited in a shallow marine environment. The Carbonera sandstone is associated with an excellent seismic reflector.

## PROJECT METHODOLOGY

An integrated approach was implemented, merging geophysics, geology, and reservoir engineering, in order to maximize the information available for interpretation. In an integrated study, the seismic structural map is no longer the final step in the geophysical process. Continuous interaction within a team of geophysicists, geologists, petrophysicists, and reservoir engineers ensures that all results are within each discipline's acceptable parameters. This interactive feedback system ensures that any discrepancies are corrected early in the process, when they are easily changed.

To fulfill the objectives of the study, it was necessary to integrate the interpretation with core, production, and log data. In addition to a structural interpretation, a reservoir characterization, seismic inversion, geostatistical analysis, and evaluation of the spatial variability of the petrophysical heterogeneities were conducted. A discussion of these processes is not included in this paper.

## SEISMIC TO WELL TIES AND DEPTH CONVERSION

The study area contains 48 wells, of which 12 have check-shot velocity surveys. The check-shot surveys were used to calibrate sonic logs. Synthetic seismograms were used to tie the lithologic information at the wells to the seismic reflection data. The gamma ray, density, and sonic logs were first correlated to the synthetic seismograms, and then to the time reflections on the workstation. Optimum correlation was achieved by tying the reverse polarity section of the synthetic to a positive amplitude value (peak) on the seismic data. This process was repeated for each mapped horizon and consistently applied throughout the entire survey (Figures 5 and 6). A summary of the horizons and their seismic responses is listed in Table 2.

Conversion of the time structure maps to depth was accomplished through the use of geostatistical techniques. Kriging estimation techniques differ from traditional estimation techniques in the way in which weighting functions are computed. Kriging methods not only consider the distance between input data, but also the direction between data points. The computation of weights is the result of the minimum mean square error. This method results in higher accuracy than with traditional estimation techniques.

The initial step in the depth conversion is a comparison of subsea depth at each horizon with the corresponding two-way seismic time at each control well (Figures 7–9). The resultant depth map accurately ties each well-control point, and then uses the seismic data to guide the algorithm away from known points. Depth maps generated in this manner tie wells, and honor the dip of the seismic data, without having to generate an intermediate error map.

## STRUCTURAL INTERPRETATION

After review of the inline, crossline, and time slice data, several geologic features were observed. These features were confirmed where possible with well control. Three different types of faults and two types of folds could be identified (Figure 10).

**TABLE 2. Summary of Seismic Horizon Characteristics**

| Horizons | Seismic Response | Time (sec) / Depth (m) (ft) |
|---|---|---|
| Carbonera | Peak | 1.400–1.900 / 3305–4285 (5320–6900) |
| T-1 | Peak | 2.140–2.485 / 5440–6101 (8760–9680) |
| K-1 | Trough | 2.198–2.535 / 5630–6235 (9060–10040) |
| K-2 | Peak | 2.265–2.609 / 5975–6660 (9620–10720) |
| Basement | Peak | 2.300–2.705 / 6035–7290 (9720–11735) |

### FAULTS

Type 1 are high-angle normal faults that cut the entire section from the surface to the basement. These faults strike approximately N60°E and are present over the Suria–Libertad fields and south of the Apiay field. Type 2 are reverse faults that cut from the Tertiary to the basement. These faults strike approximately N30°E and are the boundaries for the Apiay–Suria fields. Type 3 are normal faults that only cut the basement and in some instances extend up to the Carbonera Formation over the Suria field.

### FOLDS

Primary folds are asymmetric, anticlinal folds conforming to the structure of the Apiay–Suria fields. These folds trend approximately N30°E and are bounded to the east by Type 2 faults. Secondary folds are anticlinal and synclinal features that conform to the structures of the Guayuriba area and Libertad–Reforma fields, are generally oriented in an east–west direction, and are associated with all three fault types mentioned previously.

The former 2-D structural model (Figure 11) consists of only contractional fault blocks and basement thrusts (Obando, 1988; Reyes, 1992). The current 3-D structural model is less faulted and is consistent with a simple stress regime (Figure 12). A strain ellipse was used to determine the proper orientation of major structural elements associated with wrench tectonics. The strain ellipse of a right-lateral convergent shear system fits the major structural elements of the Apiay–Ariari area. These elements include a shear zone, normal faults, reverse faults, and folds. This model is also consistent with other regional structural models of the Eastern Cordillera and the Llanos Basin (Covey and Dengo, 1993; Cooper et al., 1995).

## INTERPRETATION RESULTS

The principal improvements of the 3-D interpretation over the 2-D interpretation are as follows: (1) comprehensive regional structural models, (2) external reservoir structural models for each field, (3) better field production boundaries, (4) identification of reservoir sands updip of production, (5) more accurate fault placement, and (6) new locations in untested fault blocks.

A total of five horizons were mapped. Discussion of the above improvements is concentrated on the K-2 level, which is the main pay horizon and accounts for over 80% of the total production to date. The interpretation of the K-2 horizon is described field by field below.

### APIAY FIELD

The interpretation from this study indicates that the Guatiquia and Apiay field areas are a single field, not two separate fields, as they were considered before the 3-D survey. Also, the new interpretation indicates that fault B is at least 457 m (1500 ft) farther south than indicated by the interpretation based on 2-D data (Figure 13). The time and depth maps indicate that the structurally highest point of the field is 0.012 seconds (22 m [72 ft]) higher than at the Apiay-9 well, which is the highest producing point in the Apiay field. Additionally, there was an untested three-way closure upthrown to fault A near the shear zone (Figure 13). This feature added exploration potential to the development aspect of the field study.

**TABLE 3. Comparison of Oil in Place Estimates of the Apiay-Suria Fields**

| Area | Before 3-D (million bbls) | After 3-D (million bbls) | Difference (million bbls) |
|---|---|---|---|
| Apiay–Guatiquia | 149.7 | 298.3 | 148.6 |
| Suria | 236.6 | 277.6 | 41.0 |
| Total | 386.3 | 575.9 | 189.6 |

### SURIA FIELD

The time and depth maps indicate that the structurally highest point of the field is 0.020 seconds (39 m [129 ft]) higher than the Suria-1 well, which is the highest well in the Suria field. This high occurs in a position near a fault that was formerly interpreted as a low area. Also, it may be seen on the new interpretation that the Guayuriba field is joined to the Suria field (Figure 14).

### LIBERTAD–REFORMA FIELDS

The time map indicates that the structural model is more simple than originally interpreted with 2-D data. In addition, there is a new three-way closure west of the Libertad N-2 well (Figure 15).

## RESERVOIR DEVELOPMENT PLAN

The integration of 3-D interpretation results with a numerical simulation for the K-2 reservoir has led to a new development plan for the Apiay and Suria fields. Reevaluation of oil in place before and after the 3-D interpretation is listed in Table 3 (Salamanca et al., 1995).

This new development plan should increase production by 36 million bbl until the year 2013. The recoverable reserves have increased by 60.7 million bbl. Reevaluation of the original reserves before and after the 3-D study is listed in Table 4 (see Figures 16 and 17).

In the future, the K-1 and T-1 reservoirs will be reevaluated using the same procedure.

**TABLE 4. Comparison of Recoverable Reserve Estimates for the Apiay-Suria Fields**

| Area | Before 3-D (million bbls) | After 3-D (million bbls) | Difference (million bbls) |
|---|---|---|---|
| Apiay–Guatiquia | 64.0 | 106.8 | 42.8 |
| Suria | 103.4 | 121.3 | 17.9 |
| Total | 167.4 | 228.1 | 60.7 |

Based on the 3-D study, ECOPETROL drilled one successful well and plans to drill two more wells. The first well was drilled during early 1995 to test a new block located to the east of the Apiay field. The Gavan-1 is currently producing 500 BOPD (Figure 18). The total field reserves of 60.7 million bbl do not include any production from the Gavan-1 well. The remaining two wells (one of which will be horizontal) will be acceleration wells and are scheduled to be drilled at the end of 1995 or in the beginning of 1996.

## SUMMARY

This project was a success because the 3-D seismic volume was truly integrated with all available data by the interdisciplinary team. The ability to describe the structure and faulting with a higher degree of confidence resulted in an increase of recoverable reserves over 60.7 million bbl for the K-2 reservoir (Figures 19 and 20). In addition, a discovery was made in an untested three-way closure based on the new maps. Future development wells will be drilled with reduced risk, resulting in better reservoir management through the minimization of unnecessary capital expenditure.

### *Acknowledgments*

We are grateful for the assistance provided by Trino Salinas and Nestor Meza of ECOPETROL, and Gary Jones, John Sherwood, Roger Murray, Nazih Al-Najjar, Bill Skibbe, Liz Watkins, Linda Noyes, and Janice Sweitzer of Western Atlas.

We thank the management of ECOPETROL, especially Herman Gutierrez P. and Jorge Gualdron S., for supporting this work and granting release of the material for publication.

This manuscript benefited from the review and constructive comments and suggestions of: Cesar Vasquez H. of ECOPETROL and Rod Hotz of Western Atlas.

## REFERENCES CITED

Castro-Arévalo, E., et al., 1991, Determinación de ambientes sedimentarios en la subcuenca Apiay–Ariari, Llanos Orientales, Colombia: Empresa Colombiana de Petroleos, Instituto Colombiano del Petróleo.

Cooper, M. A., F. T. Addison, R. Alvarez, M. Coral, R. H. Graham, A. B. Hayward, S. Howe, J. Martinez, J. Naar, R. Peñas, A. J. Pulham, A. Taborda, 1995, Basin development and tectonic history of the Llanos Basin, Eastern Cordillera, and middle Magdalena Valley, Colombia: AAPG Bulletin, 79, 10, p. 1421–1435.

Covey, C., and C. Dengo, 1993, Structure of the Eastern Cordillera of Colombia: Implication for trap styles and regional tectonics: AAPG Bulletin, 77, 8, p. 1315–1325.

Obando, G., 1988, Contribución a la evolución geologica classificación de cuenca y modelo estructural del área Apiay-Ariari: III Congreso Colombiano del Petróleo, Memorias v. 1, p. 133–139.

Reyes, J. P., 1992, Potential stratigraphic traps in Meta sub-basin Pachaquiaro area, Colombia, South America: Master's thesis, Colorado School of Mines, 110 p.

Salamanca, M. L., et al., 1995, Sísmica 3-D y simulación integradas en administración de yacimientos—un caso práctico: VI Congreso Colombiano del Petróleo.

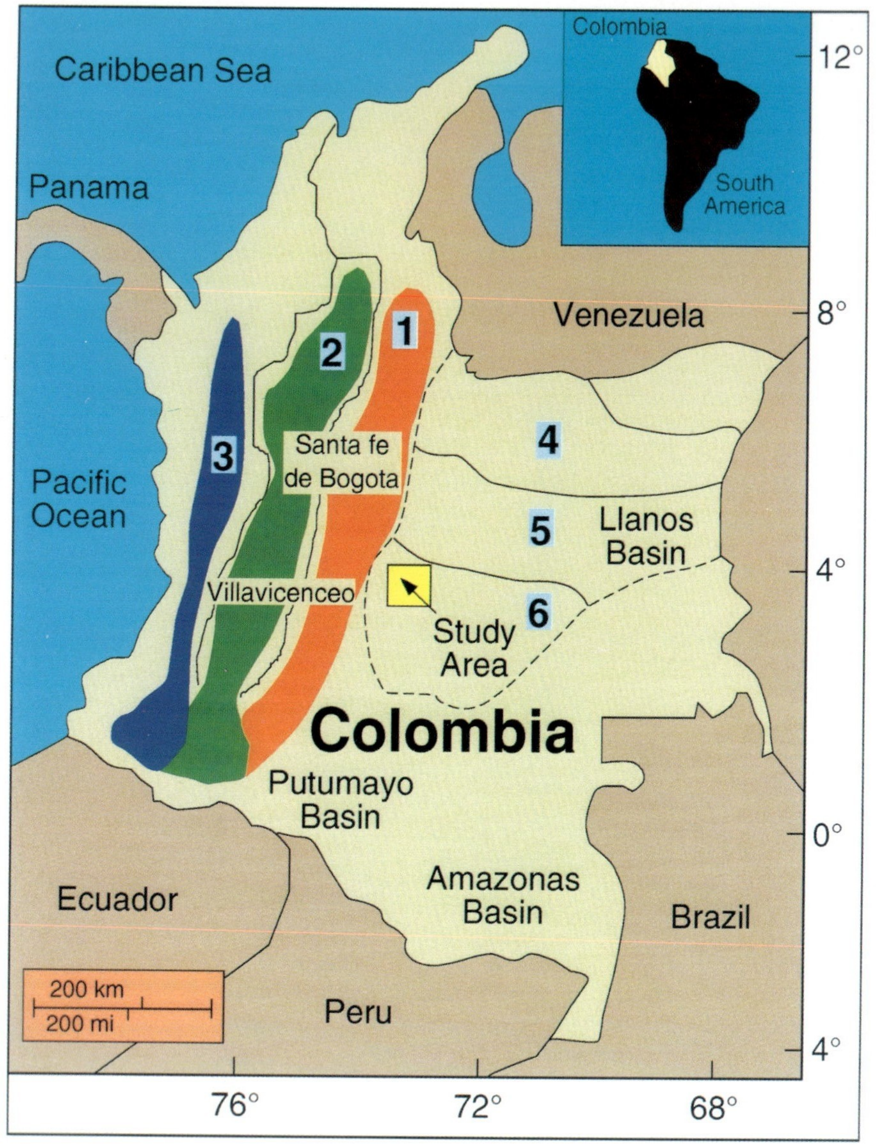


FIGURE 1. Location map of geographic features of Colombia. The Apiay–Pachaquiaro area is in the Meta subbasin of the Llanos Basin. (1) Cordillera Oriental. (2) Cordillera Central. (3) Cordillera Occidental. (4) Arauca subbasin. (5) Casanare subbasin. (6) Meta subbasin.

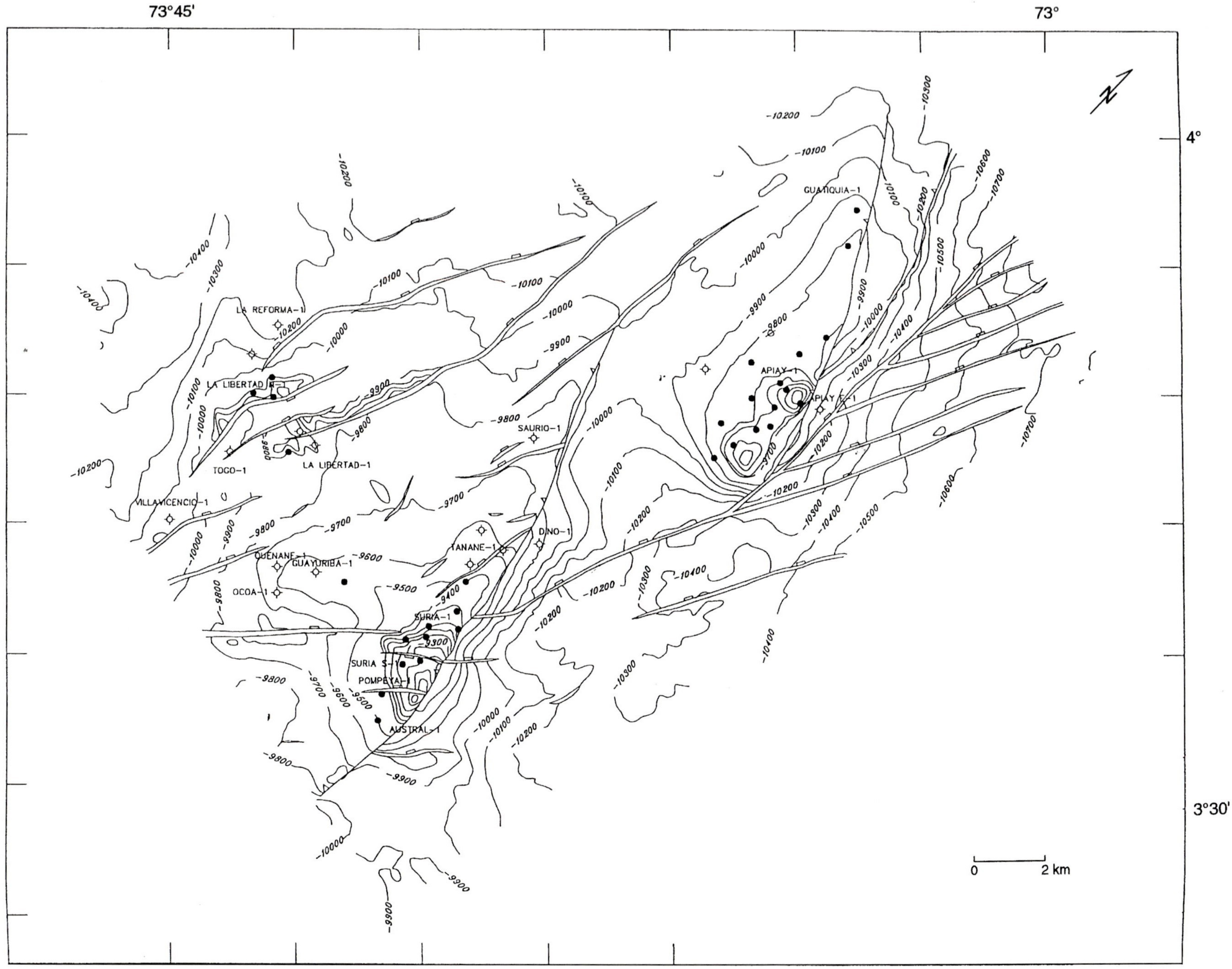


FIGURE 2. K-2 structure map over Apiay–Ariari area. This includes the Apiay, Suria, Reforma, and Libertad fields.

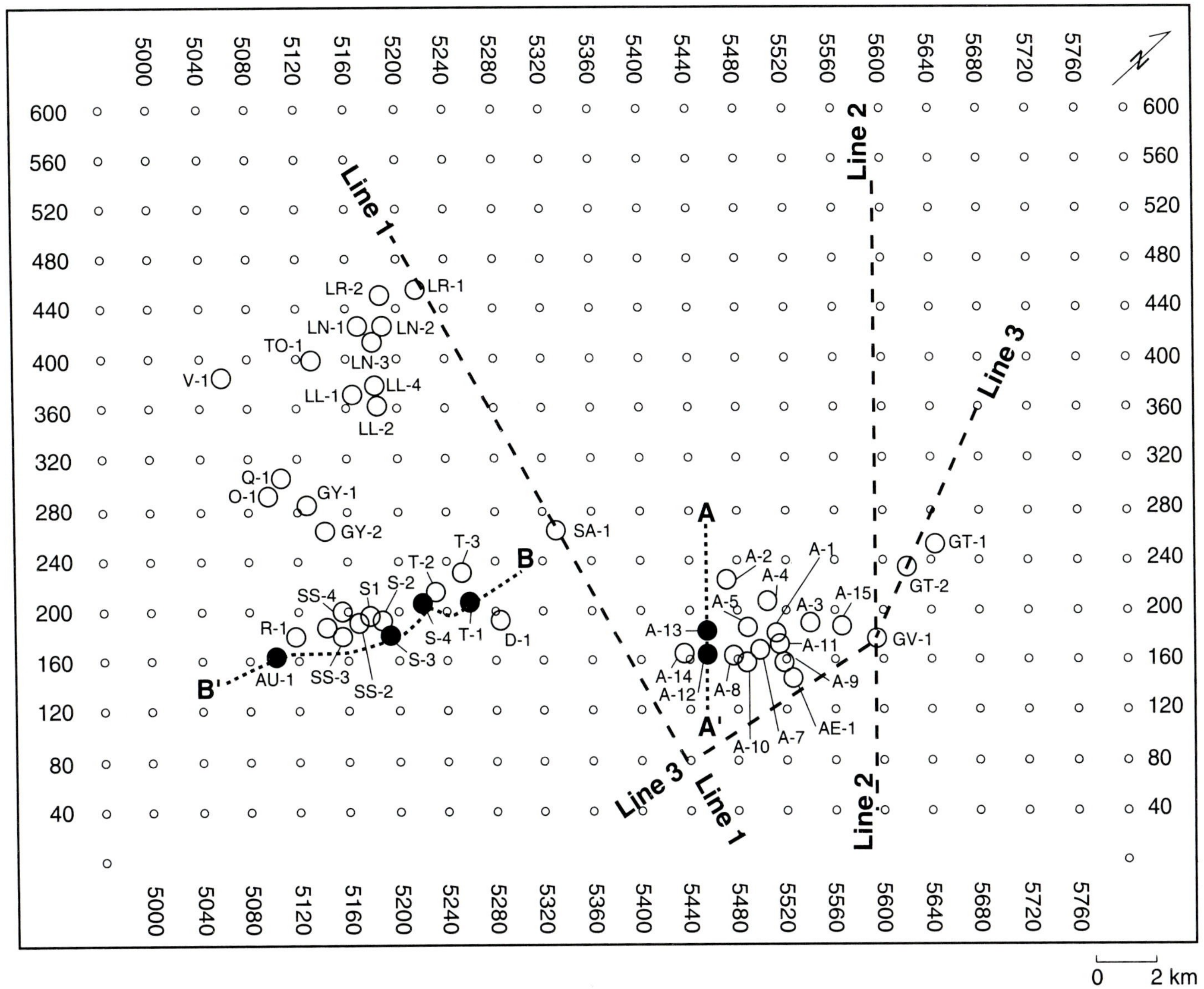


FIGURE 3. Seismic 3-D data, well location, key lines, and cross section map.

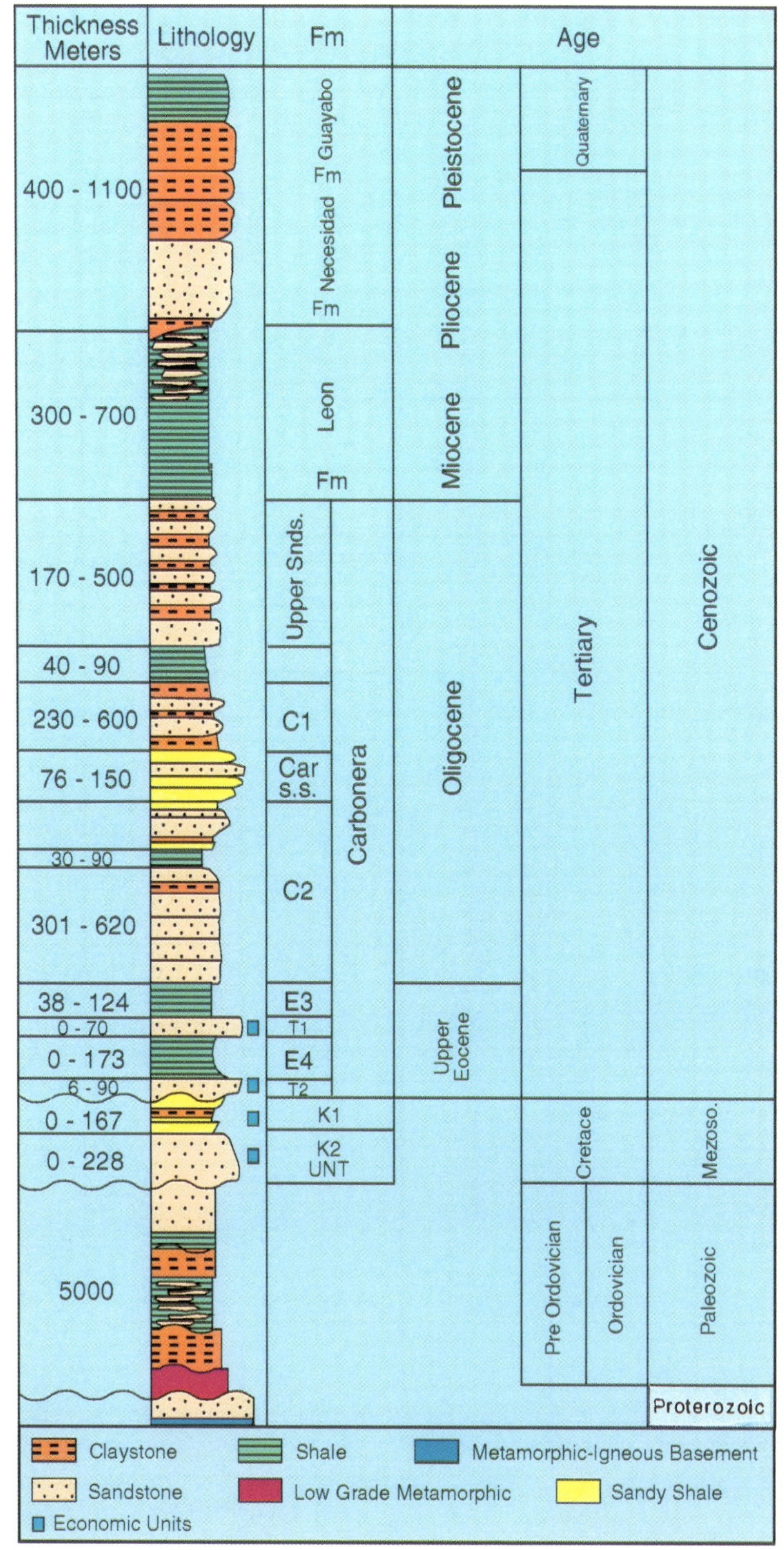


FIGURE 4. Generalized stratigraphic column for the Apiay and Suria fields.

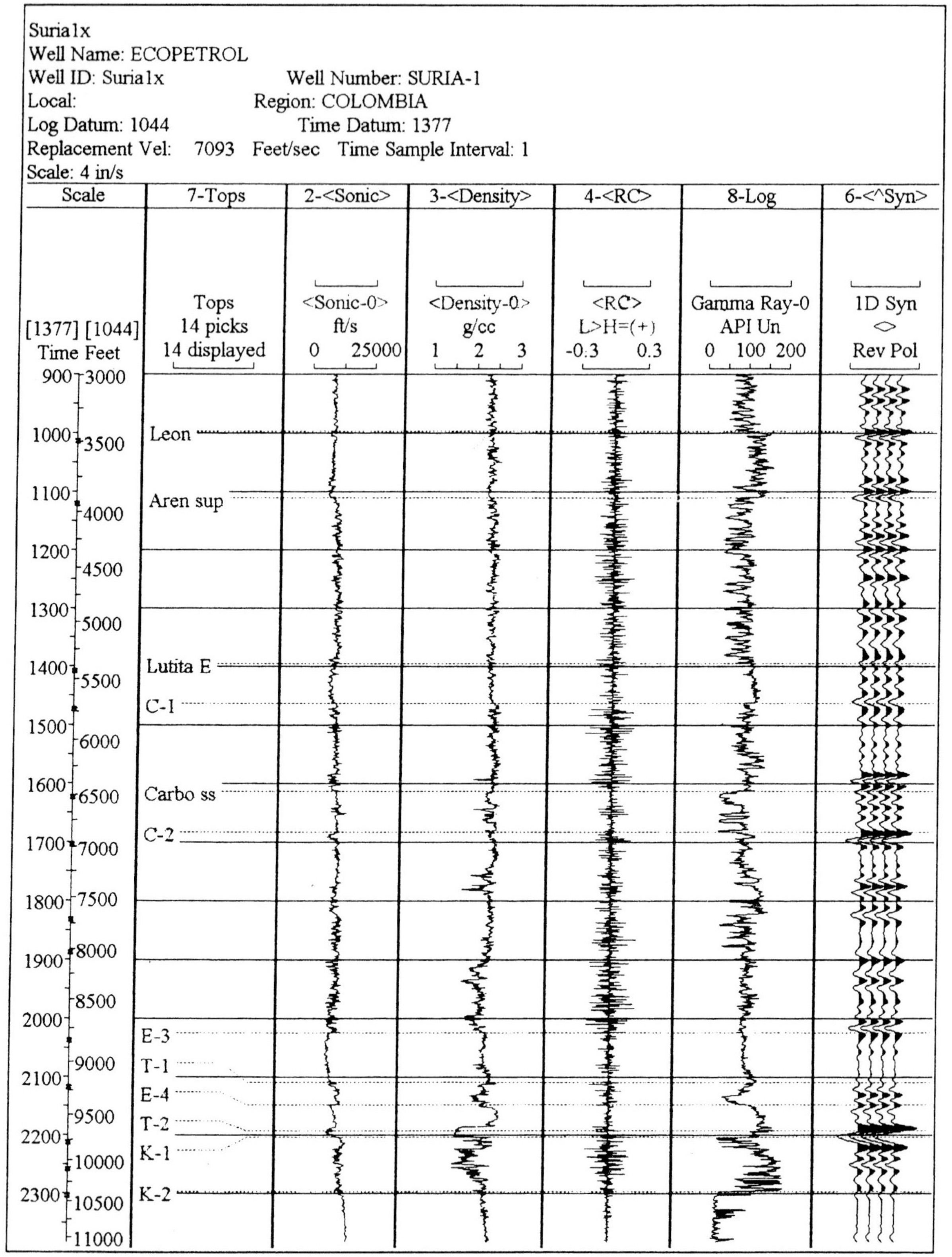


FIGURE 5. Synthetic seismogram of the Suria-1 well.

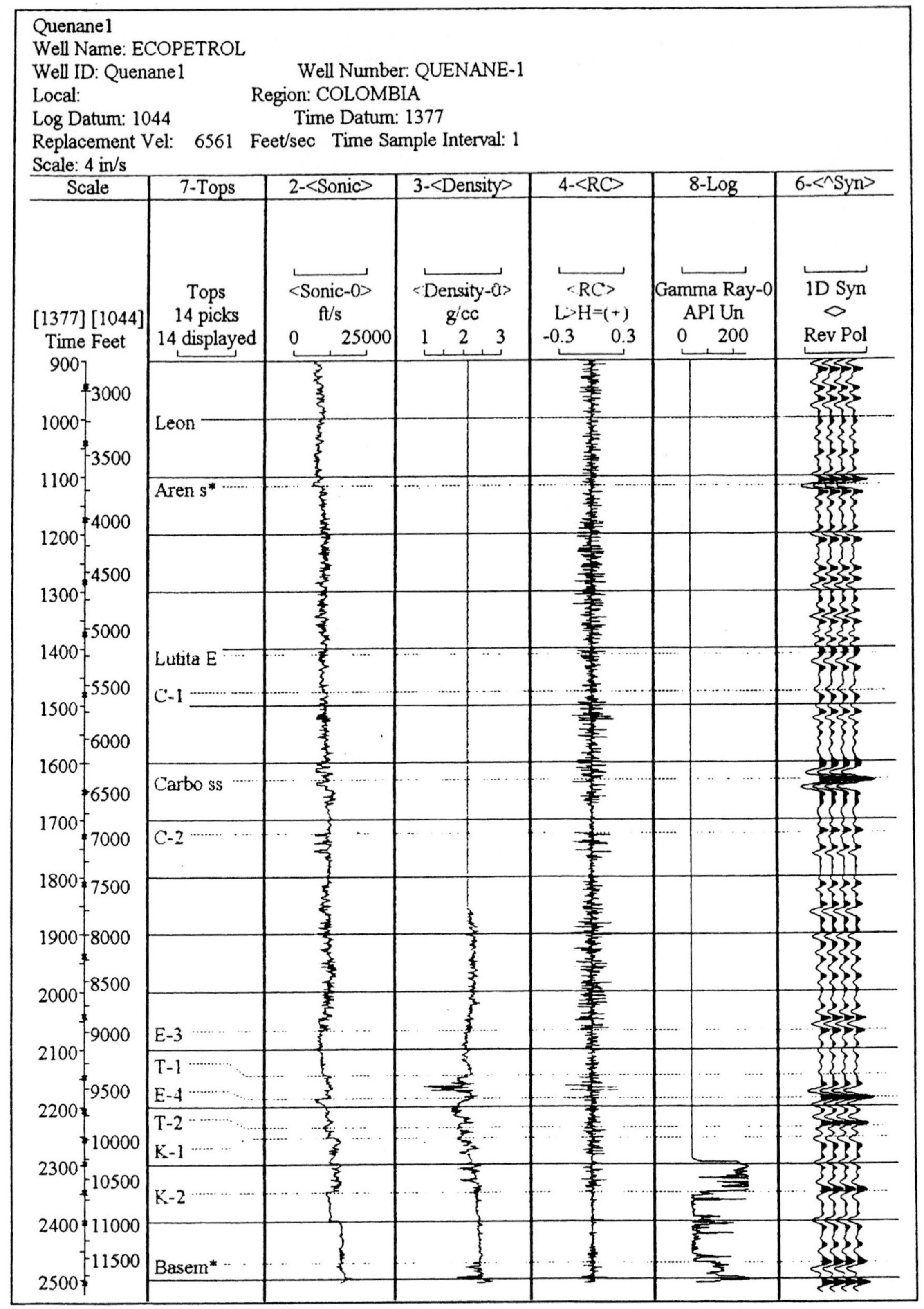


FIGURE 6. Synthetic seismogram of the Quenane-1 well.

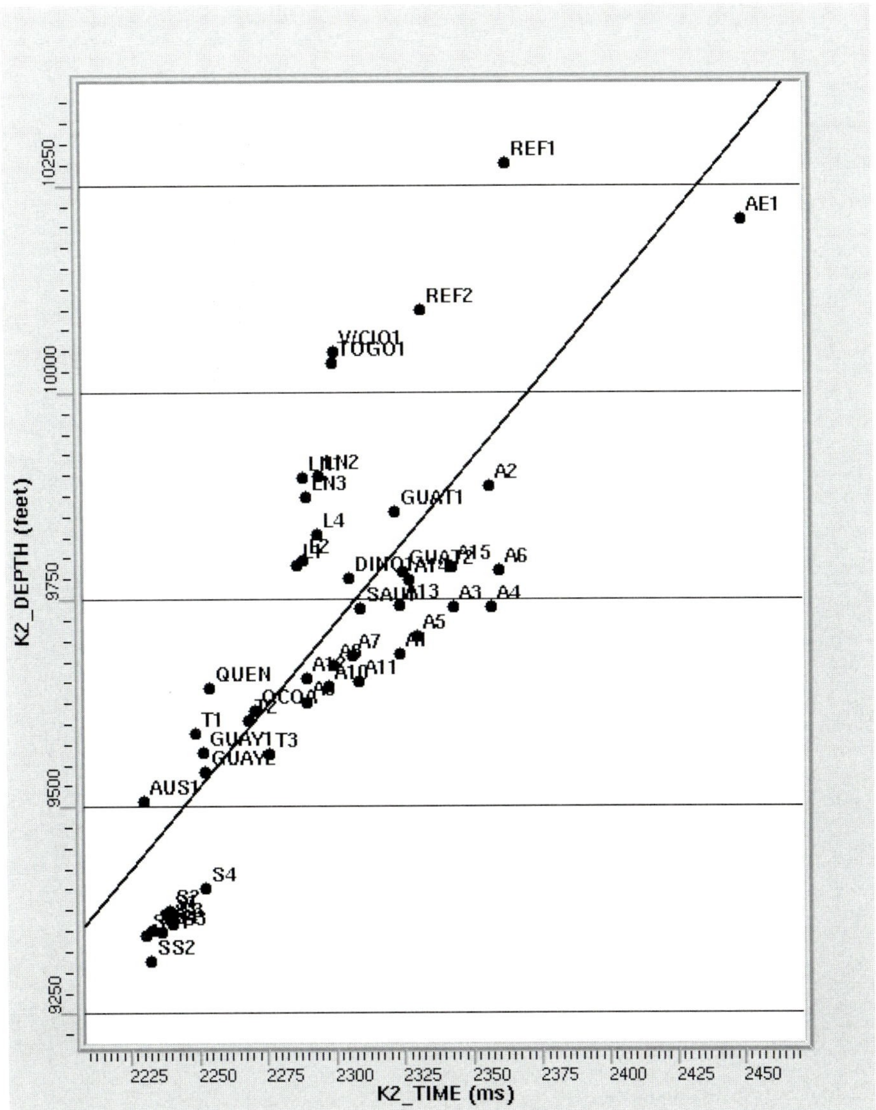

FIGURE 7. Crossplot of 2-way seismic times versus subsea depths for the K-2 level.

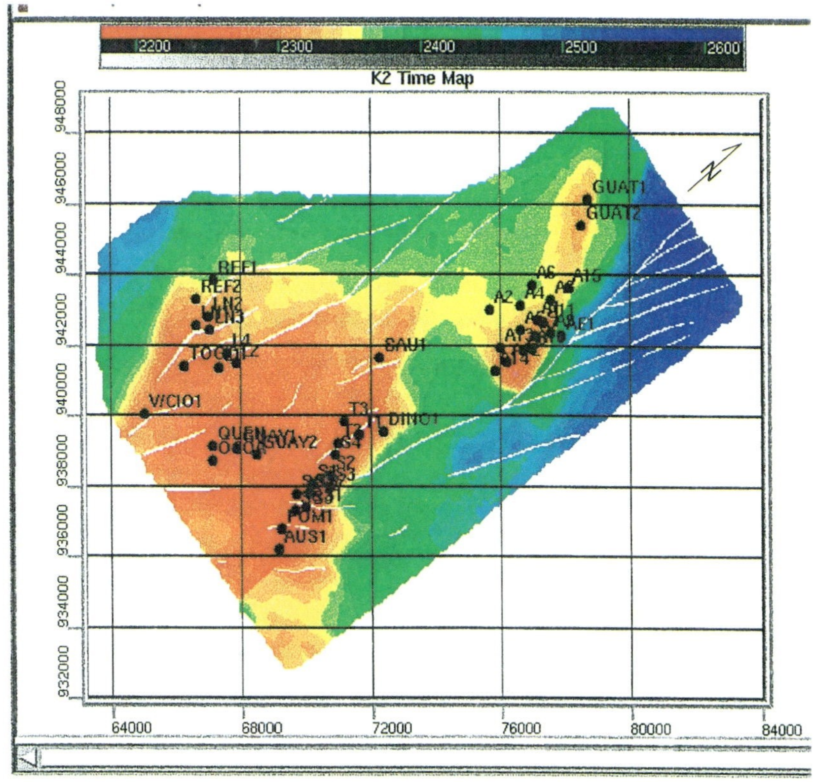


FIGURE 8. K-2 time map.

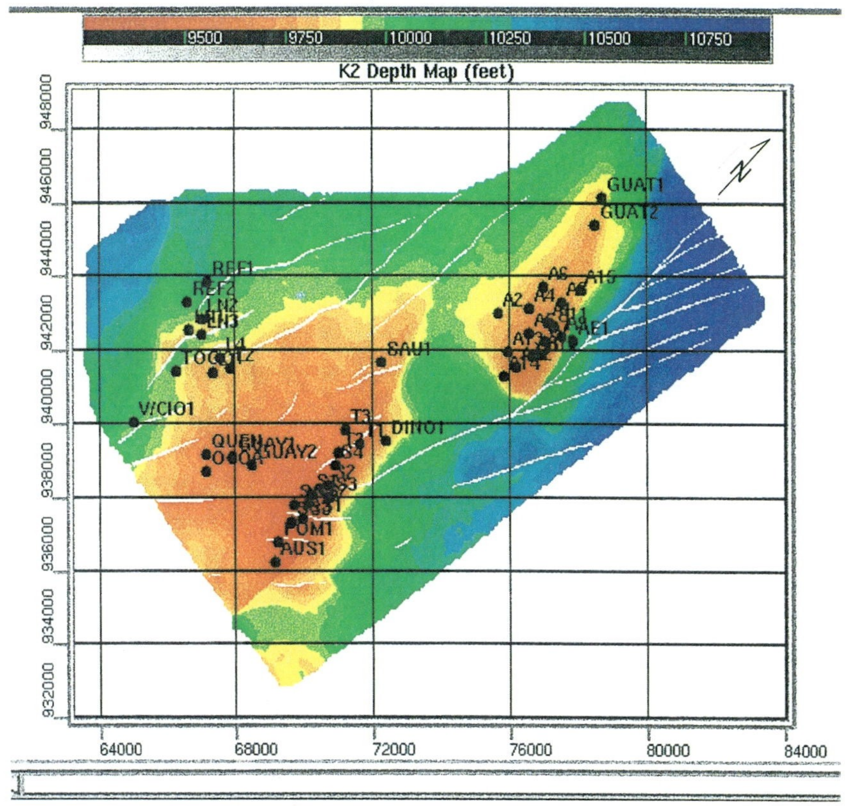


FIGURE 9. K-2 depth map.

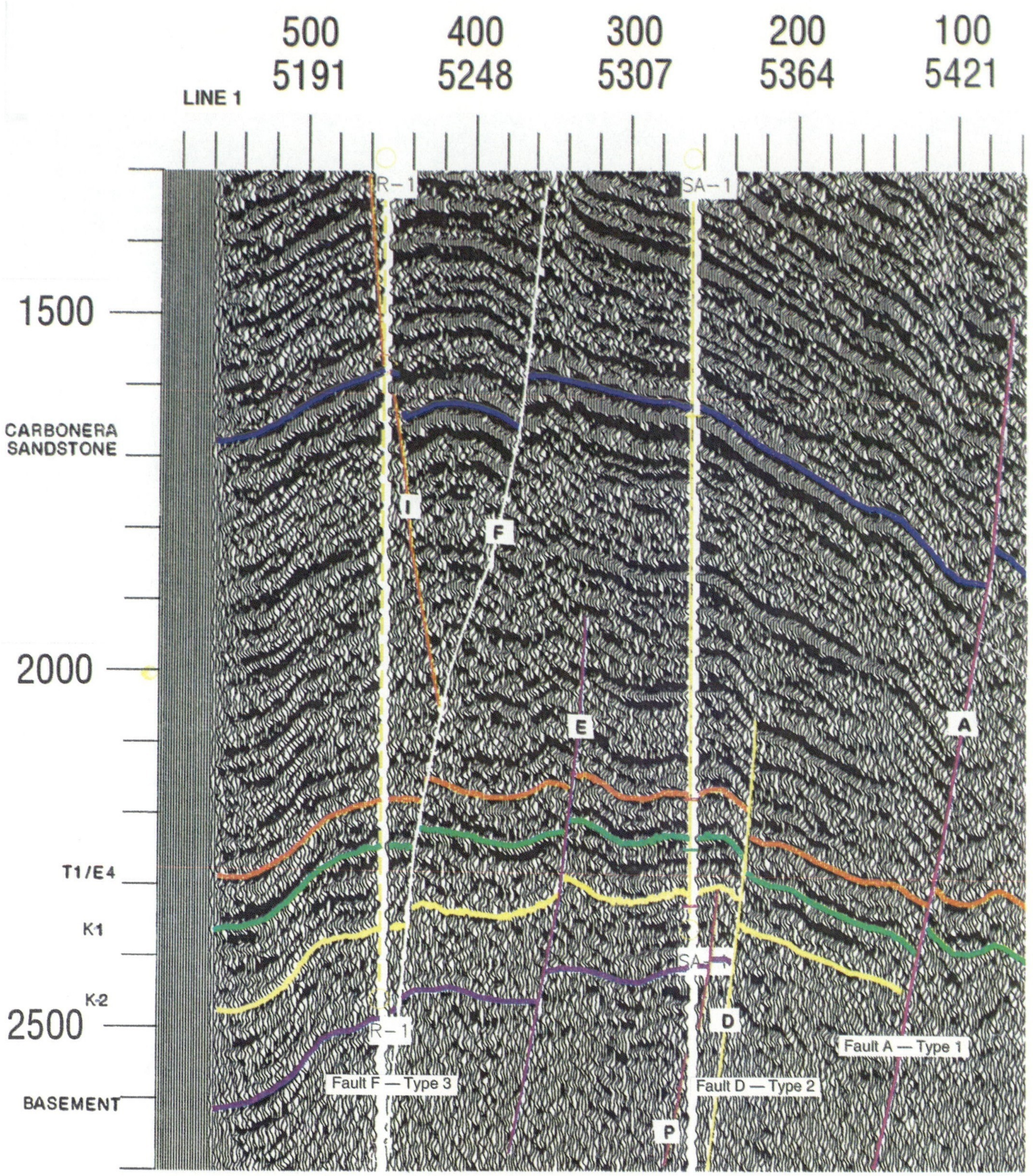


FIGURE 10. Arbitrary Line 1 with examples of three types of faults over the Suria field. (See Figure 3 for location of profile.)

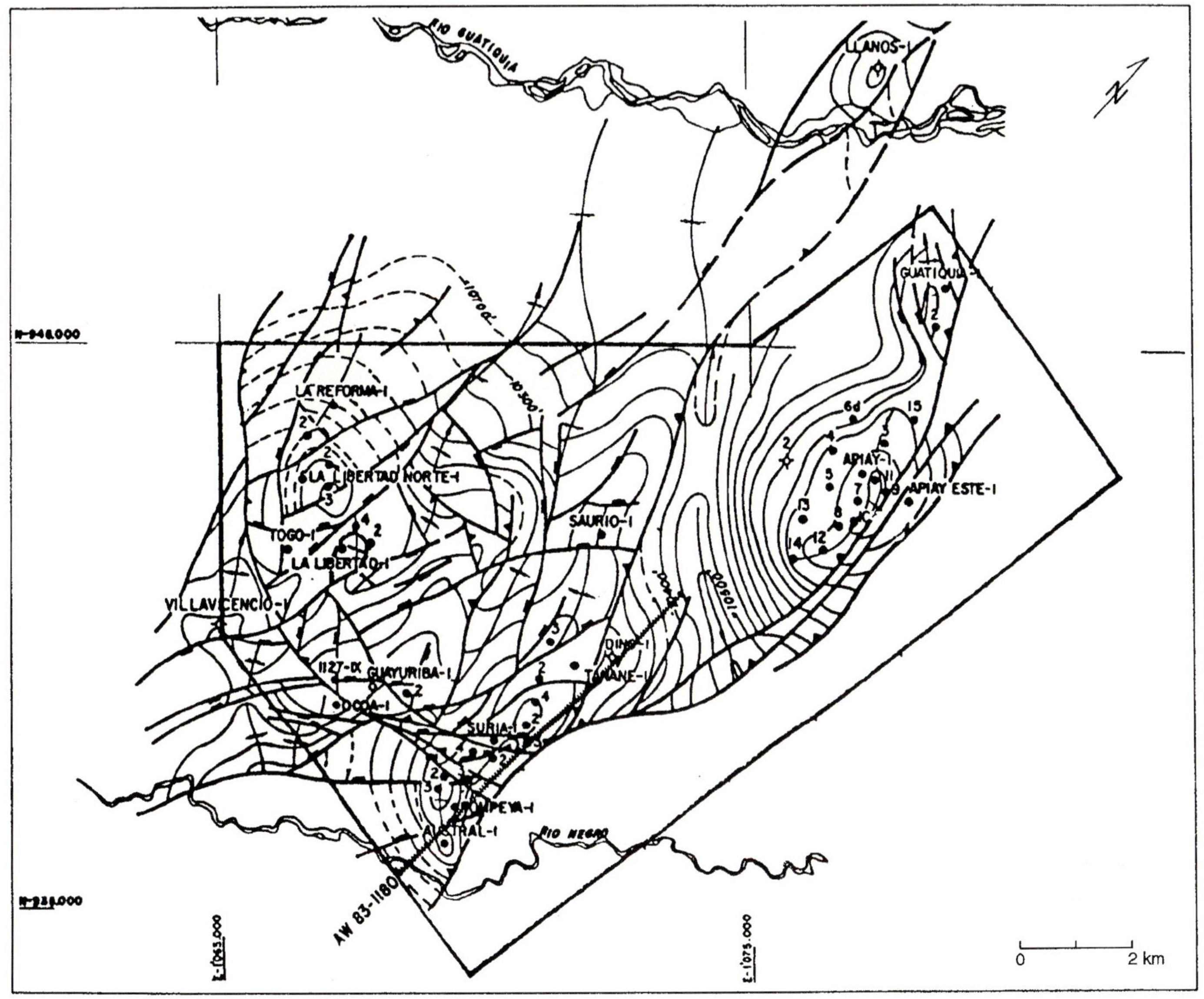


FIGURE 11. A previous interpretation of Apiay–Ariari Area constructed using 2-D data. The outline shows the new 3-D seismic data grid.

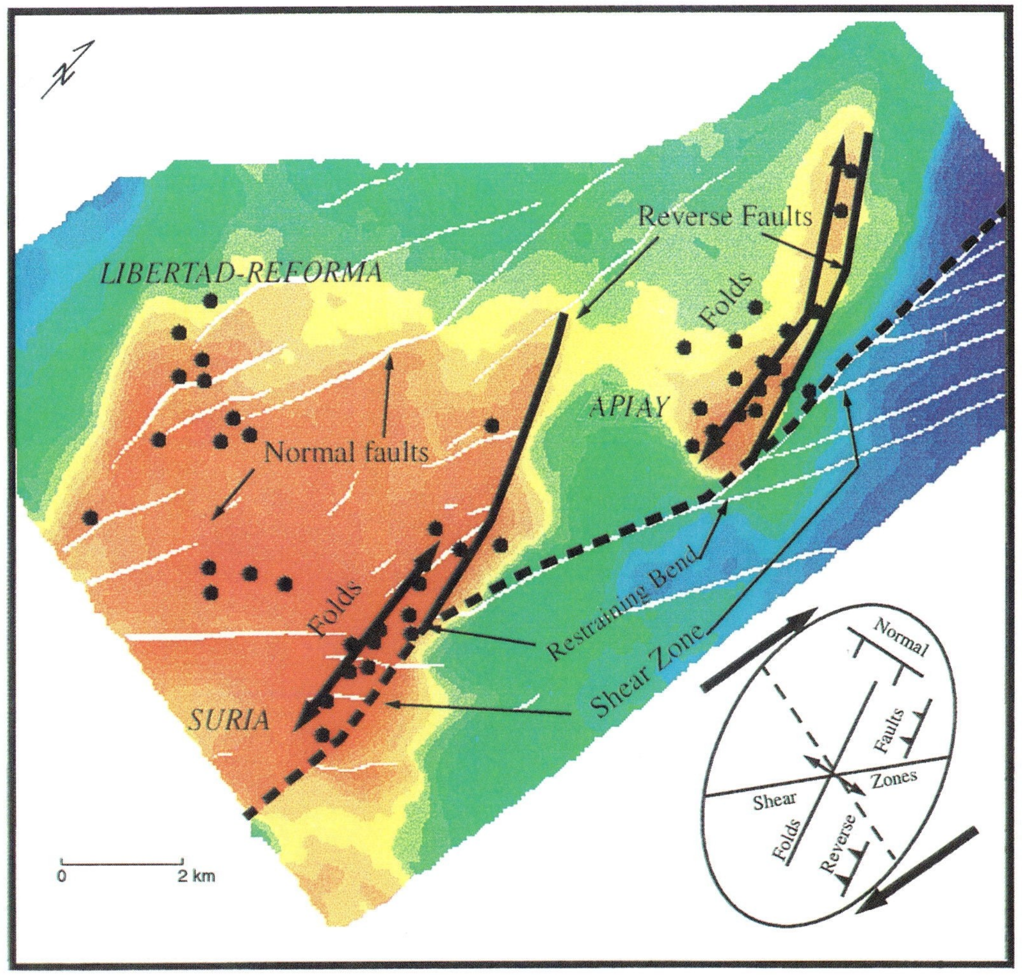


FIGURE 12. K-2 structural map with strain ellipse.

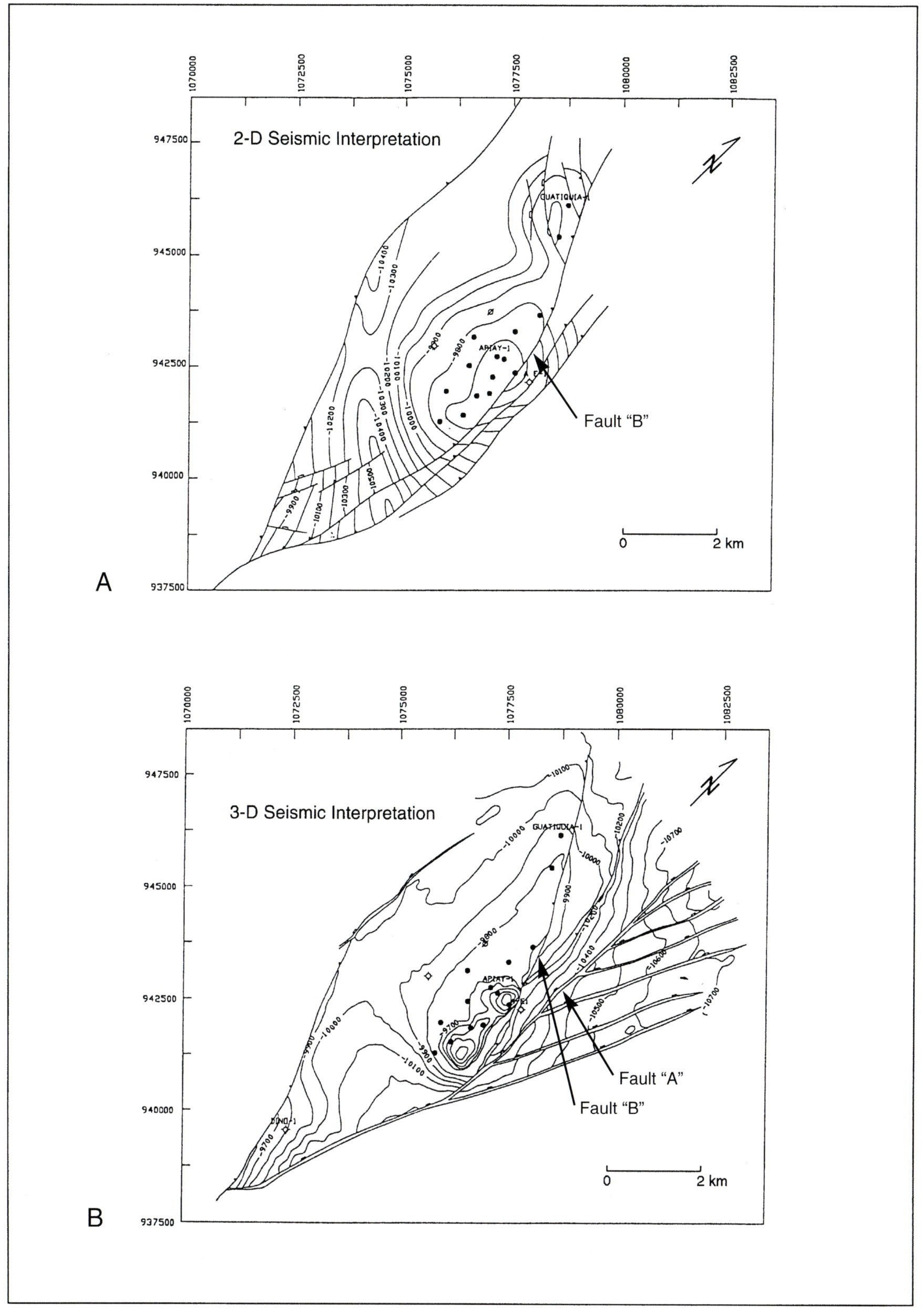


FIGURE 13. Apiay field structure. (A) Based on 2-D seismic interpretation. (B) Interpretation of the 3-D survey.

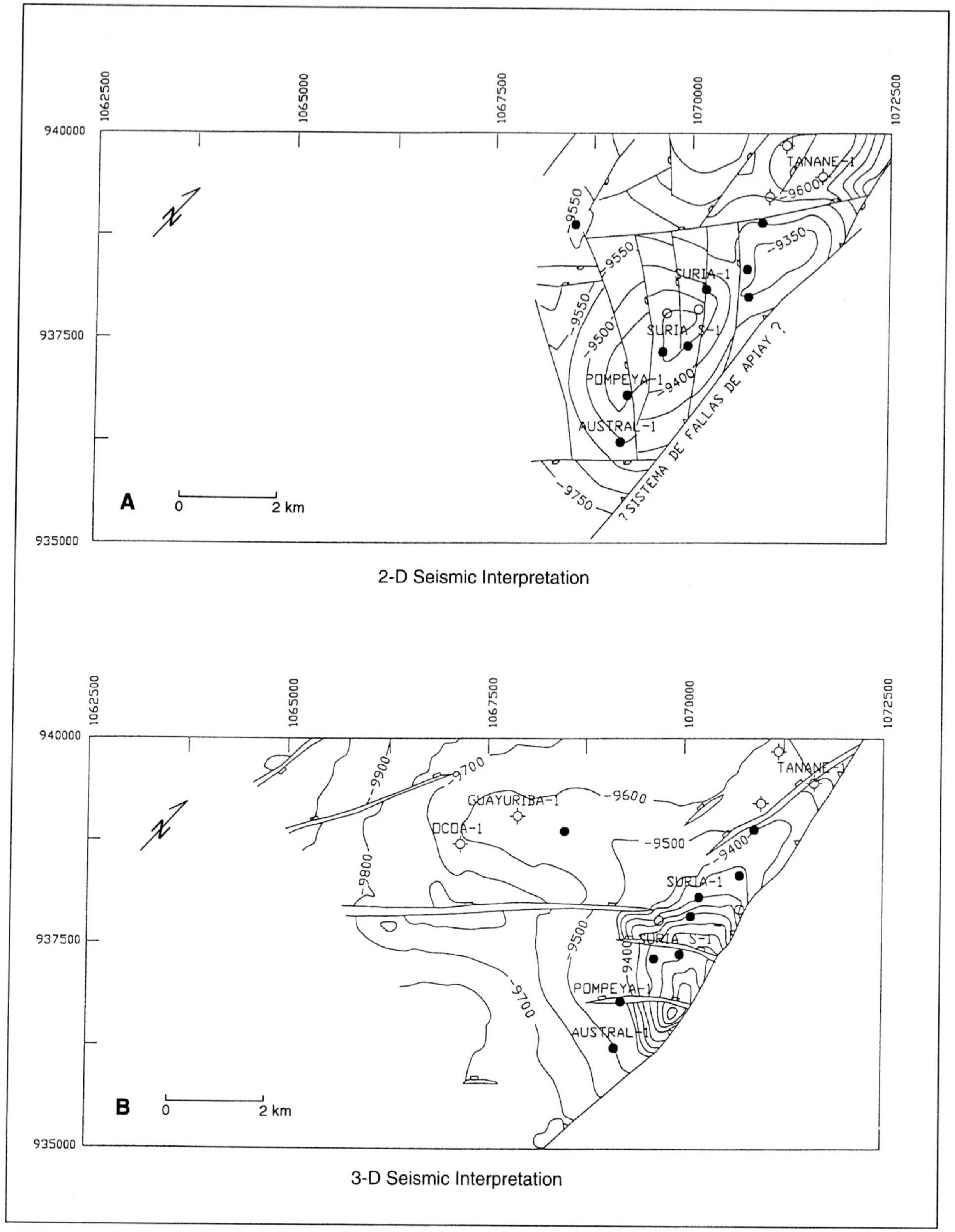


FIGURE 14. Suria field structure. (A) Based on 2-D seismic interpretation. (B) Interpretation of the 3-D survey.

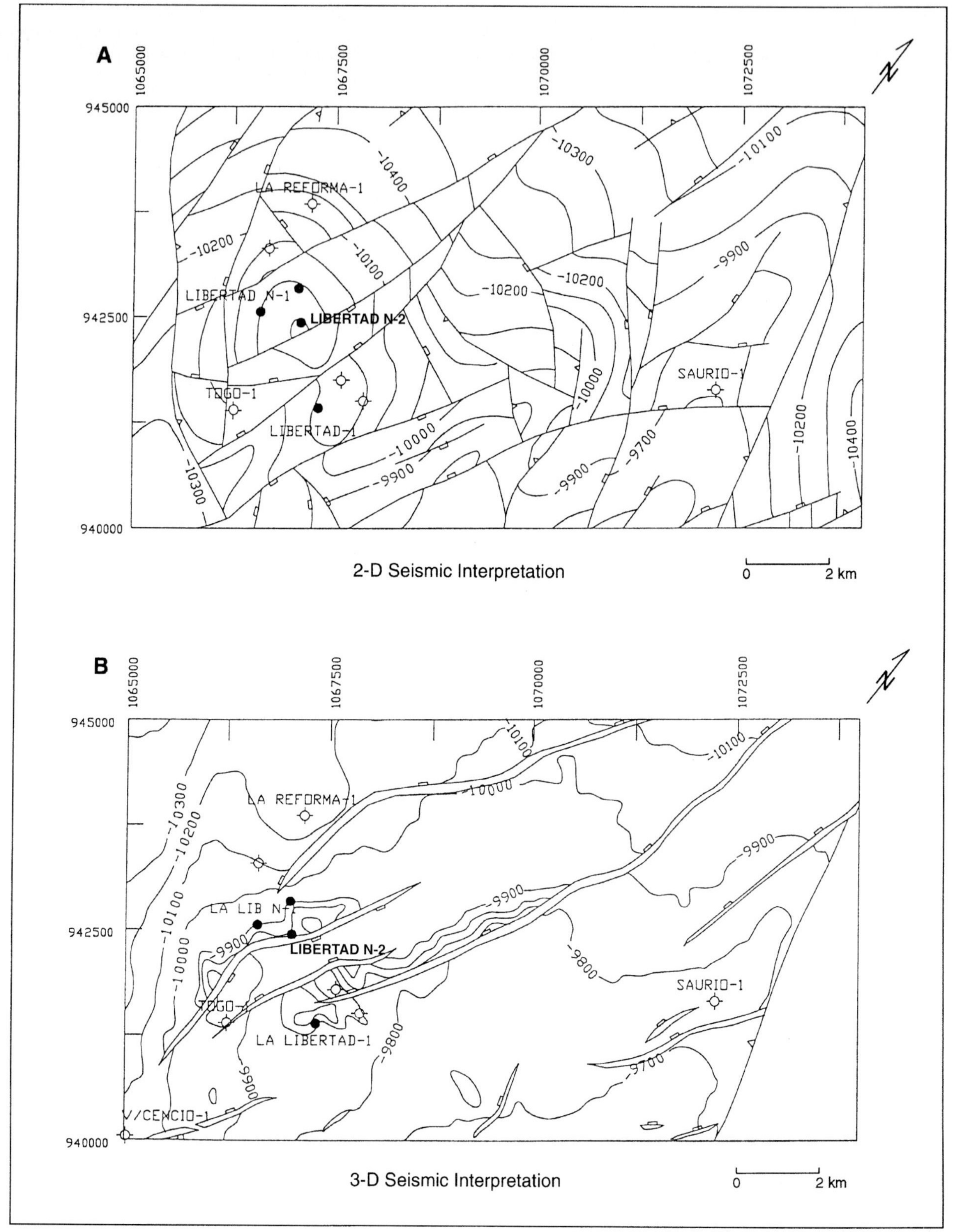


FIGURE 15. Libertad–Reforma fields structure. (A) Based on 2-D seismic interpretation. (B) Interpretation of the 3-D survey.

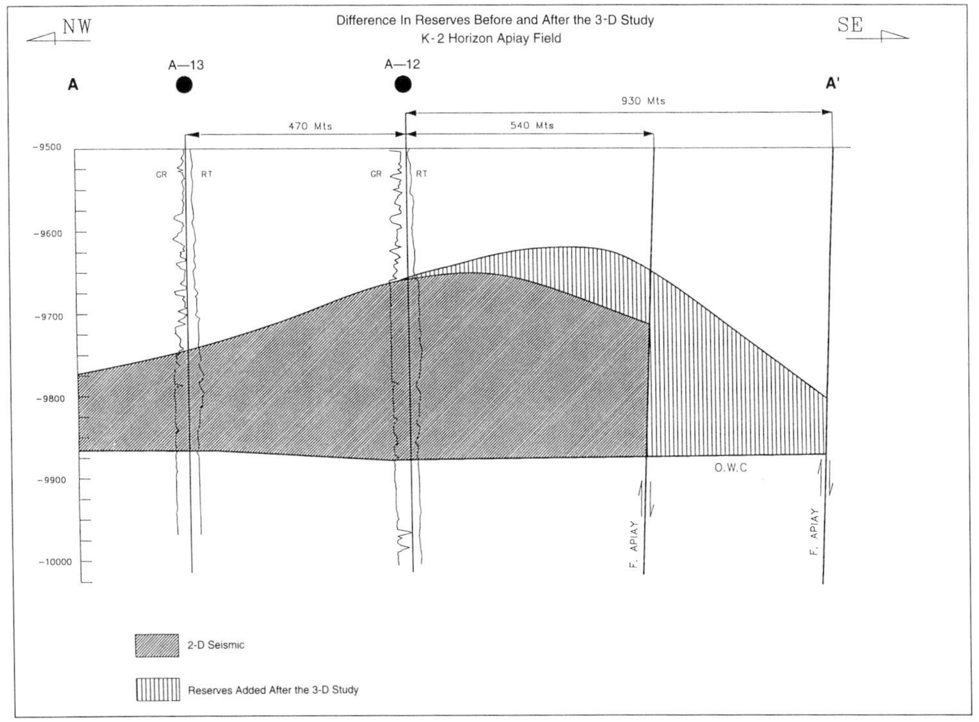


FIGURE 16. Re-evaluation of the original reserves before and after the 3-D study—Apiay area (See Figure 3, cross section A-A'.)

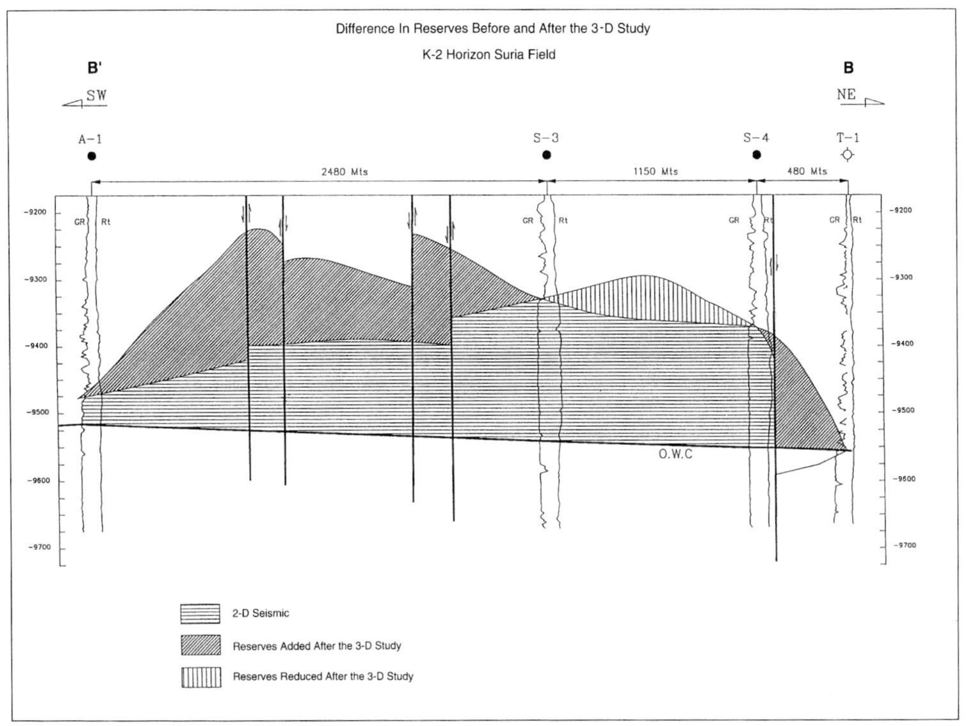


FIGURE 17. Re-evaluation of the original reserves before and after the 3-D study—Suria area (See Figure 3, cross section B-B'.)

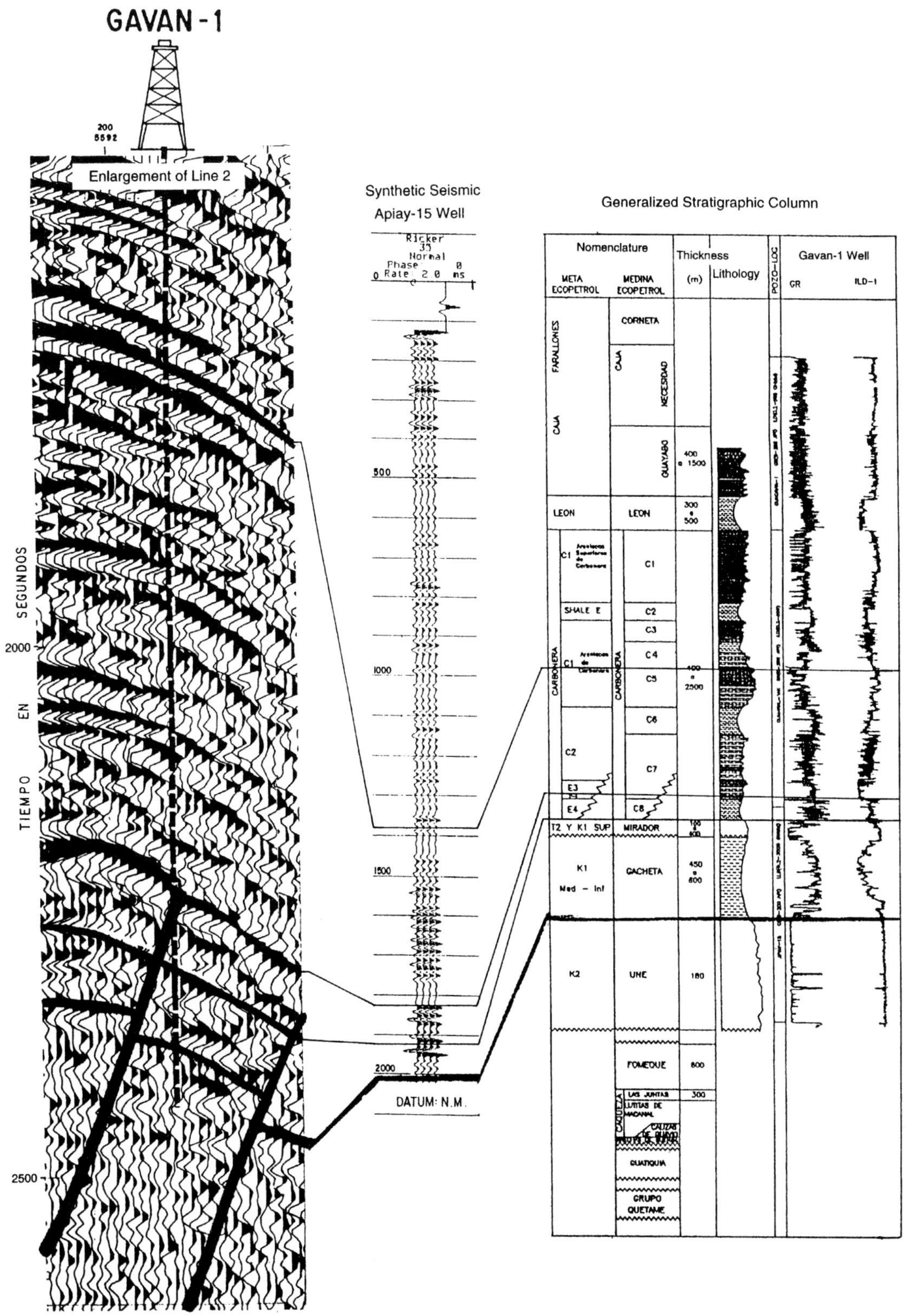


FIGURE 18. Seismic Line 2. Enlarged seismic line showing the Gavan-1 well location, a synthetic seismogram of the Apiay-15 well, and a generalized stratigraphic column. (See Figures 3 and 20.)

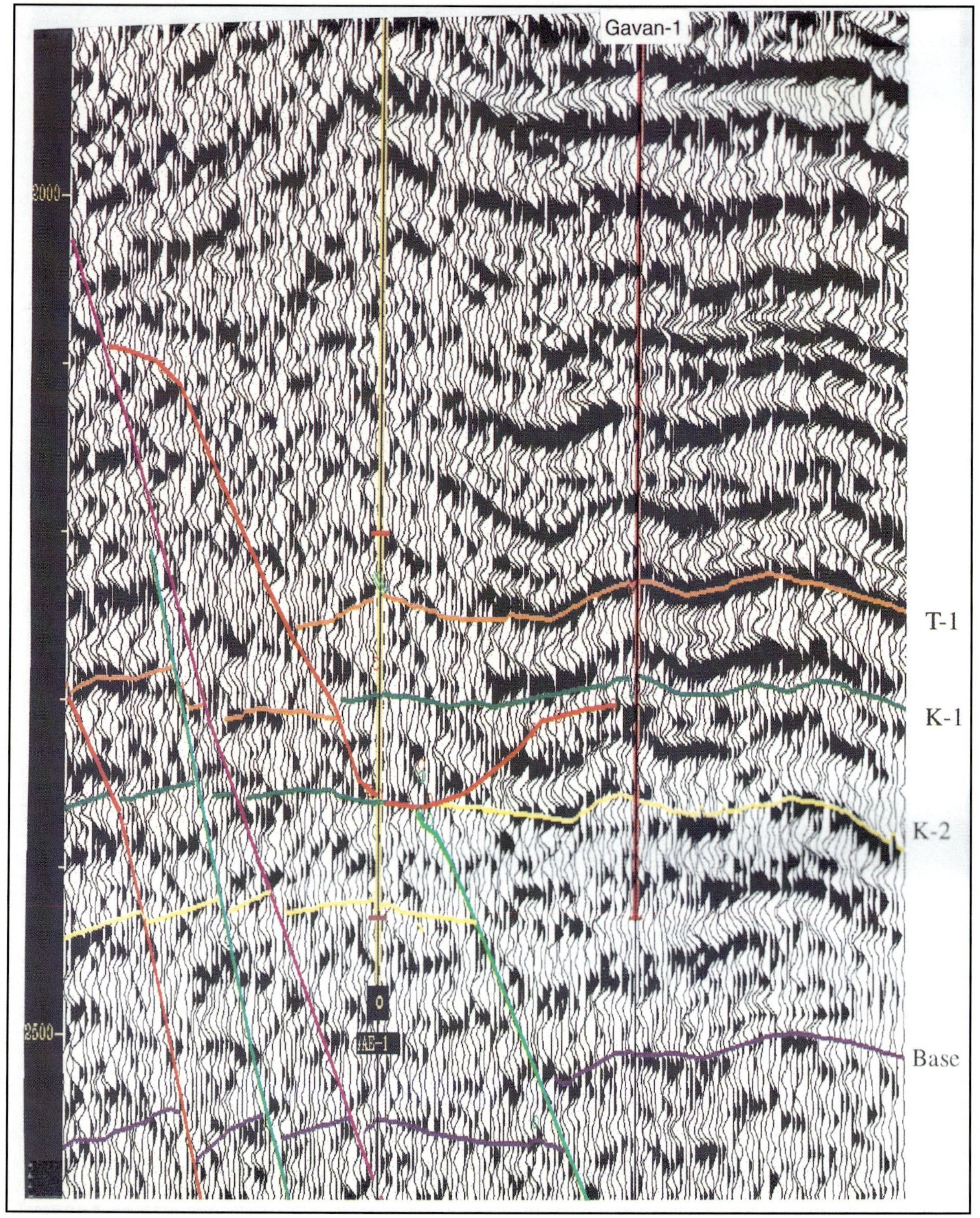


FIGURE 19. Arbitrary Line 3. Seismic section with Gavan-1 location in the strike direction. Note repeated K-1 section in the AE-1 well. (See Figure 3.)

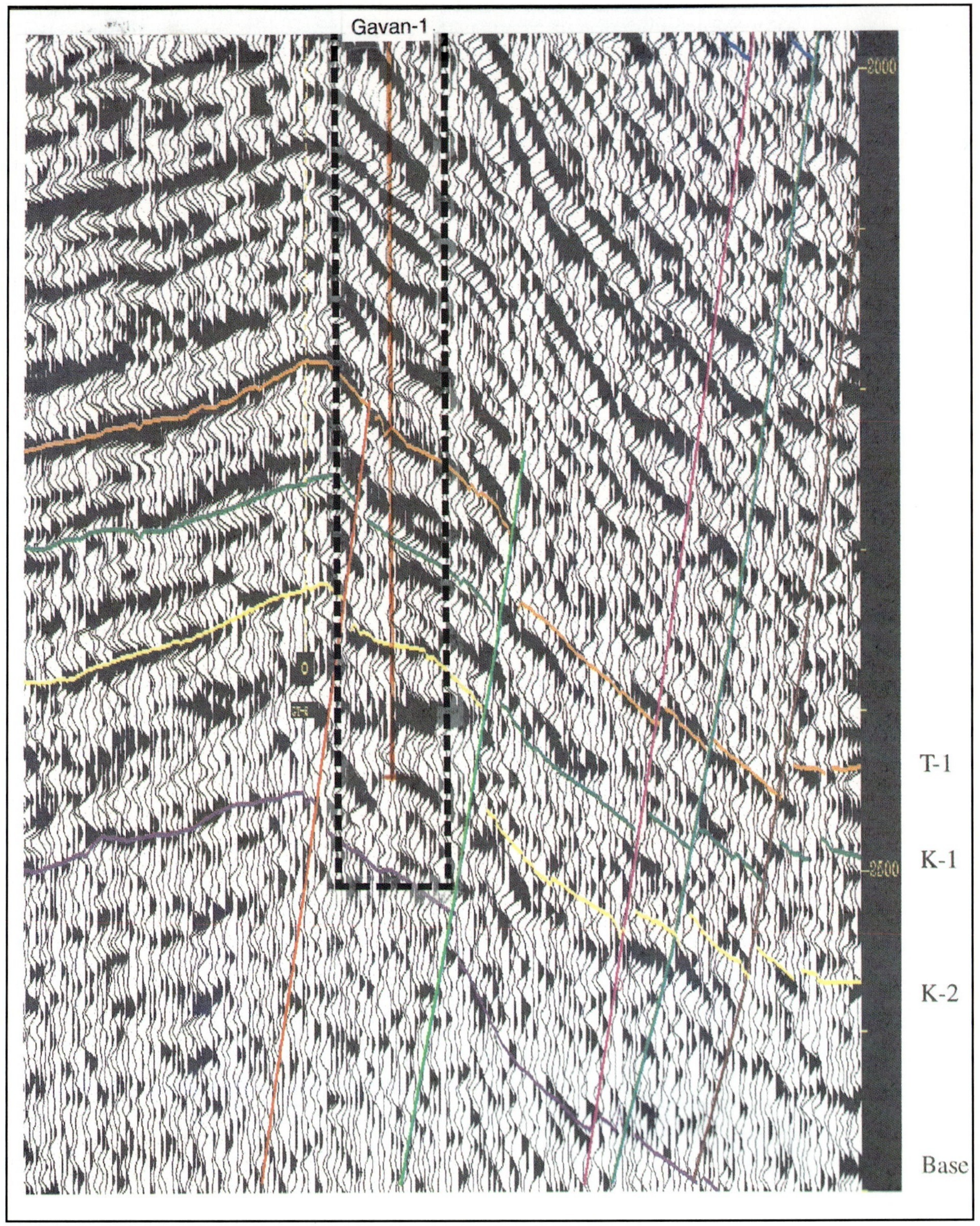


FIGURE 20. Seismic Line 2. (See Figures 3 and 18.)

## CHAPTER 7

# 3-D Seismic Examples from Central Lake Maracaibo, Maraven's Block I Field, Venezuela

Link, M. H., C. K. Taylor, N. G. Muñoz J., E. Bueno, and P. J. Muñoz, 3-D seismic examples from Central Lake Maracaibo, Maraven's Block I Field, Venezuela, *in* P. Weimer and T. L. Davis, eds., *AAPG Studies in Geology No. 42* and *SEG Geophysical Developments Series No. 5*, AAPG/SEG, Tulsa, p. 69–82.

**Martin H. Link[*†], Christopher K. Taylor[*‡], Nicolás G. Muñoz J.[*§], Emilio Bueno[*], and Pedro J. Muñoz[*]**

## ABSTRACT

The structural and stratigraphic framework of Maraven's Block I was re-interpreted using 3-D seismic and existing data as part of an evaluation of the remaining oil potential. More than 1800 MMBO have been produced from Block I in the past 40 years, mainly from structural traps. In order to maintain production levels, it has become increasingly important to define the seismic stratigraphic framework for the area and to accurately locate faults and stratigraphic pinchouts.

The dominant structures are the Icotea fault, its conjugate fault system, and the Eastern Boundary fault. The most prominent fault is the NE-striking Icotea fault, which subdivides the area into two main structural blocks, a graben in the West Flank and a horst in the East Flank. The Icotea fault is a highly complex fault zone with a long history of deformation. It is a nearly continuous fault zone with both vertical and lateral offsets and is locally inverted. Along the eastern flank of the Icotea, prominent reverse-fault bounded upthrown blocks, called the Attic, have developed. Along the western flank, contraction has re-activated listric faults into reverse and thrust faults. Major northwest-striking normal faults delineate a large paleoarch that occurs in the south-center of the East Flank. This phase of faulting produced small horst and graben blocks bounded by normal faults that dip to the northeast and southwest. The Eastern Boundary fault is subparallel to the Icotea fault and is an east-dipping normal fault that has been locally inverted and occurs in a synclinal area of the block. Two play concepts, utilizing (1) horizontal wells in Attic and (2) vertical wells along the Eastern Boundary fault, were successfully tested during this study.

The stratigraphic section includes, from oldest to youngest, pre-Triassic basement rocks; the Jurassic graben-fill Quinta Formation; the Cretaceous Rio Negro, Cogollo Group, La Luna, Colon, and Mito Juan formations; the Paleocene Guasare Formation; the Eocene Misoa Formation; the Miocene La Rosa, Lagunillas, and La Puerta formations; and the Quaternary El Milagro Formation. Only the lower part of the Eocene Misoa Formation (C sands) is preserved in Block I, and most of the Eocene B sands and all of the Pauji were either eroded or not deposited in this area. The main reservoirs occur in the Eocene Misoa Formation and the basal Miocene Santa Barbara member of the Lagunillas Formation. Sedimentation occurred throughout the Eocene and was strongly influenced by tectonism. The Eocene section in the horst block is up to 760 m thick and is bracketed by two major unconformities. The upper angular unconformity places the basal Miocene Santa Barbara member (16–25 Ma) over the Eocene Misoa C sands (45–54 Ma). The lower disconformity (54 Ma) occurs at the top of the Paleocene Guasare Formation. In between, eight seismic sequences occur within the Eocene horst section. The adjacent stratigraphic sections east and west of the horst block are thicker than the East Flank section. The C sands in Block I form a retrogradational clastic sequence deposited as transgressive (70–80%), highstand (10–15%), and lowstand wedge and incised valley fill (10–15%) systems tracts with prominent marine-flooding surfaces separating these systems tracts. The main reservoirs are thick-bedded transgressive sandstone deposits.

[*]Maraven, S. A., Caracas, Venezuela. Current address: P.O. Box 62, Tidewater, OR 97390
[†]Landmark Graphics, Caracas, Venezuela
[‡]Geco Prakla, Caracas, Venezuela
[§]S.A. Consultores CSC, Venezuela

## INTRODUCTION

The use of 3-D seismic data to interpret old producing fields has become very important in field development. When integrated with well, core, and production data, 3-D seismic data provide the best tool to determine the stratigraphic and structural framework for an area. Maraven, S.A., one of Venezuela's three national oil companies, has been using 3-D seismic data to locate new wells by accurately defining faults and stratigraphic terminations/pinchouts and by developing new geologic models to better develop their existing fields. In addition, the mapping of seismic horizons and amplitudes combined with time and horizon slices have been used to determine reservoir trends in fields.

The Maracaibo Basin is an intermontaine basin today, lying between the Perija and Merida Andes in the northwest corner of Venezuela. Maraven's Block I is 10 × 22 km and is in the north-central part of Lake Maracaibo along the Icotea fault (Figure 1). This field has more than 550 wells, which have produced more than 1800 MMBO during the past 40 years, mainly from structural traps. The Block I area is the southern continuation of the Lagunillas field and the northern part of the Lama field (Delgado, 1992). The purposes of these examples are to show the use of 3-D seismic data to (1) define the structure and stratigraphy of this Block by integrating older log, core, and production data, (2) to develop a seismic and sequence stratigraphic model for the Lower Eocene Misoa C sands in the East Flank, and (3) highlight two successful wells that were drilled using 3-D seismic technology. The 3-D seismic data allow, for the first time, the complex fault relationships of the Icotea and Eastern Boundary faults to be resolved and major sequence boundaries and systems tracts to be recognized. The 3-D seismic data were utilized to illustrate some of the various methodologies used to better visualize the reservoir at both a regional and a prospect scale.

## METHODOLOGY

The structure and stratigraphy of Maraven's Block I were re-interpreted using 3-D seismic data integrated with well logs, cores, older 2-D seismic lines, and synthetic seismograms. Both Landmark and Charisma software with Sun workstations were used to construct maps, cross sections, and various 3-D graphic presentations. The Block I 3-D seismic survey was collected and processed during 1990–1991 and covers an area approximately 20.4 km long by 11.6 km wide (Figure 2). The shot lines were recorded perpendicular to the Icotea fault to ensure optimum imaging of the subsurface. Between 250 and 300 time-based logs were combined with the seismic data to correlate horizons. Eight velocity surveys and four synthetic seismograms were used to calibrate and adjust the data. Regional 2-D seismic lines were used to identify the dip and shape of the main fault traces, especially where the quality of the 3-D seismic data was diminished due to the edge effect of the survey. In addition, ten cored wells, containing more than 300 m of core, helped correlate lithofacies to well logs and calibrate different seismic facies.

Time horizons, structures, stratigraphic tops, sequence boundaries, and marine-flooding surfaces were mapped from seismic data in the area. Structures were defined using horizon and dip maps and fault polygons interpreted from seismic lines (Figures 2 and 3). The stratigraphic framework was established by (1) correlating horizons from existing well tops using over 300 wells and by (2) extrapolating tops into areas with little well control. The sequence stratigraphy terminology follows Mitchum (1977), Vail (1987), Posamentier and Vail (1988), and Van Wagoner et al. (1990). Sequences, sequence boundaries, marine-flooding surfaces, systems tracts, and parasequences were interpreted and correlated for the Eocene section in the Block I horst. Seismic and wireline-log cross sections were used together to carefully identify sequence boundaries, marine-flooding surfaces, and faults.

## STRUCTURE

Block I area has a long and complex structural history (Link et al., 1994). The area is on the eastern margin of a Jurassic rift basin that trends north-south (Bartok et al., 1981; Bartok, 1993; Figure 1A). This graben, containing redbeds of Jurassic La Quinta Formation, has been deformed and later eroded (Figures 1B and 4). The dominant structures of Block I are the Icotea fault, its conjugate fault system, and the Eastern Boundary fault (Figures 1–3, 5–9, and 11). These faults follow the older Jurassic basin trends. The most prominent fault is the north-east–striking Icotea fault, which subdivides the area into two main structural blocks, a graben in the West Flank and a horst in the East Flank (Figures 2 and 5). The Icotea fault is a highly complex fault zone with a long history of deformation. Along the eastern flank of the Icotea, prominent reverse-fault bounded blocks, called the Attic, have developed. On the West Flank, contraction has re-activated normal faults into reverse and thrust faults. The Eastern Boundary fault has a trend similar to that of the Icotea and merges with it, farther to the north, outside Block I (Figure 2). This listric fault is downthrown to the east and was later re-activated as a reverse fault (Figures 5 and 6). Northwest-striking normal faults delineate a large paleoarch that occurs in the south-center of the East Flank (Figures 2 and 3). This arch contains smaller horst and graben blocks bounded by normal faults that dip to the northeast and southwest.

The Icotea fault system strikes northeast across the west-central part of Block I (Figure 2). This fault zone contains the main trace of the Icotea and is flanked by several normal and reverse faults that dip to both the east and the west (Figures 2, 3, 5, and 6). The Icotea fault is a relatively linear, continuous narrow zone, about 0.5–1 km wide and more than 100 km long. Along the strike, this narrow fault zone has both anticlinal and synclinal structures (Figures 7, 8, and 11). Vertical displacement along the Icotea fault in the West Flank basin is about 1–2 km, and movement is down to the west (Figures 8 and 11). The amount of transpressional slip on the Icotea fault system could not be determined in Block I. The estimated left slip of 0.8 km is based on offset surface topography and subsurface mapping (Lugo, 1992). Miocene compression created inversion locally along the Icotea and related faults with about 15° of clockwise rotation for the Icotea fault–bounded blocks (Lugo, 1992). The Icotea fault is the boundary between the horst and graben (Figures 2, 3, and 5).

The Icotea fault had been interpreted as a normal fault, strike-slip (Lugo, 1992, among others), thrust fault (Delgado, 1992), and inversion structure (Roberto et al., 1993). We believe that four separate faults occur in close proximity along the Icotea and each can be distinguished on seismic profiles. Each fault system is superimposed upon another: (1) early Eocene high-angle reverse fault(s), followed by (2) early Eocene normal faults, (3) a middle to late Eocene listric fault system with a detachment surface at the Cretaceous Colon level, and (4) late Eocene and early Miocene inversion-related reverse and thrust faults with a strike-slip component (i.e., Attic fault) (see Figures 4, 5, 6, and corresponding structural events [1]–[4]).

We believe the Icotea fault originally formed as an east-dipping reverse fault (Icotea 1) (Figure 5). A west-dipping boundary fault (Icotea 2) developed later over the eastern edge of the Jurassic half-graben, directly overlying and overprinting the earlier reverse or ramp fault. A thick Eocene stratigraphic section was deposited on the downthrown side of the fault, forming the West Flank Basin (Figures 5–8). West-dipping listric normal faults (Icotea 3) formed during and after sedimentation, utilizing the older fault traces. Both antithetic and synthetic faults were associated with the listric faults, which sole out in the underlying Cretaceous Mito Juan/Colon and Paleocene shales (Figures 5, 6, and 8). Later during the early Eocene, transpression was caused by the oblique collision between the Caribbean and the South American plates. The Icotea and its listric fault system were re-activated as reverse and thrust faults (Icotea 4) with some strike-slip component caused by the clockwise rotation of the blocks (Bueno et al., 1993; Roberto et al., 1993). During inversion, west-directed backthrusts occurred during the late Eocene and again in the Miocene. Experimental work and field observation of inversion structures show that thrust and reverse faults commonly form and use existing extensional structures (Williams et al., 1989; McClay and Buchanan, 1991), especially where the original structures had a low-angle dip.

The Attic fault is the eastern most reverse fault of the Icotea system and, where present, always contains the East Flank section (Figures 2–3, 5–9, and 11). It consists of two segments: (1) a northern part, and (2) a southern part, separated by a saddle in the middle, where the Attic fault merges with the main trace of the Icotea fault (Figures 2, 5–7, and 9). The Attic fault verges to the east, and its vertical displacement is 50–100 m. The origin of these fault blocks is probably related to late-stage shortening, where the western corner of the horst block was thrust- or reverse-faulted over itself (Figure 5). The Attic fault may have originated as a west-directed synthetic fault of a listric fault system. During inversion, this fault was re-activated and thrust eastward. Northwest- and northeast-striking cross faults cut the Attic block into a series of smaller fault blocks. These faults are young and cut the Attic and Icotea faults as well as the Eocene unconformity. The use of 3-D seismic data allowed the complexities of the Attic fault blocks to be imaged for the first time.

A large paleoarch occurs in the south-central part of the East Flank of Block I and is characterized by northwest-striking normal faults (Figures 2, 3, 7, 8, 9, and 10). The paleoarch trends northwest and is 6–10 km wide and at least 60 km long, extending into the Lagoven area to the east. This arch may be the northwestern-most continuation of the Merida Arch (Lugo and Mann, 1995). The arch does not appear to extend across, or to have been offset along, the Icotea fault. The younger Icotea and Eastern Boundary fault systems cut the arch (Figure 2). The paleoarch appears to have been a buttress to later fault deformation, as shown by the uplift of the southern and northern Attic fault blocks on the sides of the arch (Figure 9). The northwest-striking faults delineate the arch and are parallel to subparallel to one another and dip to the northeast and southwest (Figures 2, 8, 9, and 10). Individual faults range from 1 to 5 km in length and are curvilinear with some showing oblique slip. Both planar and listric faults occur, and some of these faults have been re-activated as reverse faults during later compression and inversion. Typical secondary normal fault structures such as roll-over and drag folds and antithetic and synthetic faults are associated with these faults (Withjack et al., 1995). Many of these northwest-striking faults are deep-seated and define horsts and grabens on the arch. The Cretaceous to C-5 strata have been folded and faulted uniformly within this arch (Figures 7 and 10). The arch is eroded, with almost all of C-4 and C-5 strata locally removed.

The origin of the paleoarch in Block I is related to the Merida arch and probably represents one of two foreland bulges that developed in front of a series of southwest-advancing thrust nappes related to the collision of the Caribbean and South America plates.

The West Flank or graben area of Block I is very different from the East Flank in both its structure and its stratigraphy. Structurally, the West Flank is downthrown 1–2 kms to the west along the Icotea /Attic fault system (Figures 8, 11). This flank is characterized by three subparallel to parallel prominent northeast-striking fault zones that merge with the Icotea and Attic fault system (Figures 2 and 3). These faults are linear, 12–15 kms long, and mainly have reverse motion with 50–100 m of displacement. At depth these faults merge with the Icotea/Attic fault zones to the east (Figures 5, 6, and 11). These structures are young and cut most of the older northwest-trending structures and the paleoarch, which are poorly imaged in the West Flank. This 3-D seismic interpretation of these faults and their trends is significantly different from the original mapping in the area.

The Eastern Boundary fault is a major north-northeast–striking fault system that occurs on the eastern side of Block I (Figures 1 and 2). The fault zone consists of closely spaced, subparallel normal and reverse faults that bound the east side of the horst block and are subparallel to the Icotea fault (Figures 2, 5, and 6). This fault system merges with the Icotea fault north of Block I, forming a wedge-shaped basin between these two fault systems. This fault system continues to the north and south of Block I (Figure 2); in Block I, it is 16–20 km long with total vertical displacements up to 1 km. The main fault trace is listric and dips to the east-southeast with prominent roll-over fold locally dipping into this fault (Figures 5 and 6). Locally, the fault has been inverted, re-activating the listric fault into a reverse fault, with the Miocene strata being folded and displaced above this fault. The main folds associated with this fault system are to the east of Block I in an adjoining concession. An asymmetric half graben, containing a clastic wedge, occurs in the extreme northeastern part of Block I (Figures 5 and 6) along the Eastern Boundary fault. This clastic wedge thickens in the C-1 to C-4 sands interval to the east-northeast onto the adjacent area. In addition, it onlaps the arch in the south-central part of Block I and is not present in the south-southwest part of the block. The top of the wedge is characterized by an angular unconformity and the base by stratigraphic onlap.

The Tertiary tectonics of Block I are summarized by (1) early Eocene compression, forming reverse faults on both sides of the horst block, followed by uplift and erosion; (2) early Eocene extension and transpression along the side of a Jurassic graben system with C-7 to C-5 sands deposition; (3) compression, inversion, and block rotation in the middle to late Eocene with development of the paleoarch, folding, faulting, uplift, and local erosion followed by (4) stratal onlap and listric faulting over older structures during C-4 to C-1 sands deposition; (5) inversion during the late Eocene, with uplift and erosion; and, (6) following the early Miocene-related Andean orogeny, deposition was followed by inversion with re-activation of some of the older Eocene normal faults into reverse and thrust faults such as occurred along the Attic and Eastern Boundary faults (Figures 5 and 6). Sedimentation occurred throughout the Eocene and was strongly influenced by tectonism, as evidenced by the development of paleoarch and listric faults along with the deposition of clastic wedges and stratal onlap surfaces.

## SEISMIC STRATIGRAPHY AND SEDIMENTOLOGY

The stratigraphic section of the East Flank of Block I includes, from oldest to youngest: pre-Triassic basement rocks; the Jurassic graben La Quinta Formation; the Cretaceous Rio Negro, Cogollo Group, La Luna, Colon, and Mito Juan formations; the Paleocene Guasare Formation; the Eocene Misoa Formation; the Miocene La Rosa, Lagunillas, and La Puerta formations; and the Quaternary El Milagro Formation (Figure 4). Only the lower part of the Eocene Misoa Formation is preserved in Block I; it includes the C-1 to C-7 sands (Figures 5 and 12). Most of the Eocene B sands and all of the Pauji were either eroded or not deposited in this area. The section on the West Flank is thicker than that on the East Flank and, in addition, contains the B-8 and B-9 sands.

The Block I Eocene section is bounded by two major bracketing unconformities, the E1 (25 Ma) and E2 (54 Ma) (Maraven Exploration Regional Study Team, 1995, personal communication). The Eocene stratigraphic section contains two main depositional sequences, the C-1 to C-4 sands and the C-5 to C-7 sands that are separated by an onlap surface at the top of C-5 (Figure 5). At least seven disconformities occur within the Eocene section and allow the section to be subdivided into eight sequences in the horst block (Figures 5 and 12).

The E1 unconformity spans 16–50 Ma and is considered to be 25 Ma (late Oligocene) in this area. It is an angular unconformity and separates the Oligocene to Miocene basal Santa Barbara member (16–25 Ma) from the various Eocene Misoa C sands (45–54 Ma). This unconformity is very distinctive on seismic profiles and truncates all underlying strata (mainly C-3 to C-5 sands). The upper Eocene depositional sequence (C-1 to C-4 sands) is a wedge-shaped section that has stratal onlap onto the older C-5 sand and is shale prone. This sequence thickens to the northeast and is not found in the south-southwestern part of the East Flank. The lower Eocene depositional sequence contains C-5 to C-7 sands, which are subparallel to one another. This section has a relatively uniform thickness and is sand-prone across the East Flank of Block I. The lower E2 unconformity (54 Ma) occurs at the top of the Guasare Formation and is a major disconformity in the area. Incised valleys filled with C-7 sands are locally cut into this surface (Figure 12).

Several attempts at establishing a sequence stratigraphic framework for the Maracaibo Basin (Perdomo and Bot, 1986; Marais-Gilchrist and Higgs, 1993; Pestman et al., 1994; other unpublished studies by BP and TOTAL, among others) have been made. Our study is built on the recognition of major unconformities and seismic markers that have been age-dated by well control and paleontology by Maraven and S. A. Consultores CSC (W. Wornardt, 1995, personal communications) during the past 40 years (Figures 4, 5, and 12). There was no attempt to correlate the unconformities and sequence boundaries with the global sea level curves (Haq et al., 1988) as was done by Marais-Gilchrist and Higgs (1993). Instead, our study tried to integrate the local stratigraphic nomenclature of the Misoa C sand with third- and fourth-order seismic events, namely, the recognition of sequence boundaries, marine-flooding surfaces, and systems tracts at a seismic level and parasequences at a log-correlation level. We have found that several of the major unconformities are time-transgressive in this area and span long periods of time. For example, the E1 unconformity represents a possible 39 Ma erosional marker and/or nondepositional interval (from 16 to 45 Ma). In different parts of the Maracaibo Basin, the age relations of unconformities vary (Pestman et al., 1994), especially where they cut and merge with one another.

The C sands of Block I are up to 760 m thick and subdivided into eight sequences within the horst block (Figures 5 and 12). They consist of transgressive, highstand, and lowstand systems tracts deposited in coastal and shallow-marine environments. Overall, the C sands in Block I form a retrogradational clastic sequence that is sand prone at the base, becoming more shale prone upwards (Figures 12 and 13). This back-stepping pattern is attributed to rapid subsidence and net southwestward basin transgression during deposition of the C sands. Afterwards, the B sands were deposited as a progradational outbuilding sequence and are locally preserved in the West Flank. Facies and formation names change dramatically from southwest to northeast. The Misoa Formation varies in facies characteristics in both space and time. The paralic and shallow-marine Misoa Formation is transitional into the fluvial Mirador in the southwest, and may be laterally equivalent with the deep-marine late Paleocene to early Eocene Trujillo Formation and middle(?) to late Eocene Pauji to the northeast (Figure 13). Block I is located in the Misoa part of these facies associations (Figures 13 and 14). Regionally, structure controlled sedimentation in the Maracaibo Basin. The ancestral Orinoco River was forced to cross several paleoarches that formed as foreland bulges subparallel to series of southwest-directed thrust nappes.

The Eocene section on the East Flank horst contains nine sequence boundaries (SB) and eight marine-flooding surfaces (MFS) that define eight sequences (Figures 5 and 12). Sequence boundaries occur at the top of Eocene unconformity E1 (SB1), within lower C-2 sand (SB2), middle C-3 sand (SB3), within the C-4 sand (SB4 and SB5), middle C-5 sand (SB6), middle C-6 sand (SB7), lower C-6 sand (SB8), and at the top of Guasare unconformity E2 (SB9). Each of these disconformities or unconformities has a corresponding maximum marine-flooding surface. The Eocene sections east of the Eastern Boundary fault and west of the Icotea fault system have not been correlated in this study and are apparently different.

The transgressive systems tracts make up about 70–80% of the C sands and are sandstone prone at the base (C-7 and C-6 sands), becoming more shale prone in the upper C sands (C-2 and C-3 sands). C-4 and C-5 sands contain nearly equal amounts of shale and sandstone (Figure 12). The transgressive systems tracts consist of stacked marine channels and bars arranged into thinning- and fining-upward or blocky intervals (Figure 12). There are two types: (1) thin basal transgressive sheet sands that are 10–30 m thick, and (2) thicker transgressive retrogradational intervals 100–200 m thick (Figures 12, 14, and 15). Both types of transgressive systems tracts are commonly overlain by thinner highstand systems tracts. The base of each transgressive systems tract is characterized by a sequence boundary and its top by a marine-flooding surface. Each systems tract consists of an erosional channel or channels characterized by bell-shaped log patterns that become more shaly upward. The seismic facies of the transgressive systems tract consists of discontinuous reflections of variable amplitude and are best viewed in horizon slices. The impedance contrast between the relatively thick sandstone intervals and the underlying and overlying shales usually produces several reflections. Downlapping seismic events occur locally at the lower sequence boundary, with parallel to subparallel reflections occurring at the upper marine-flooding surface. Horizon- and flattened time-slice maps show north-trending linear amplitude events interpreted to be channels and bar complexes (Figure 15). Net sand and oil accumulation maps mirror these amplitude trends. The transgressive systems tracts are the best reservoirs in the field area, with porosities in the 18–25% range and permeabilities up to several hundred md (Muñoz et al., 1994).

Highstand systems tracts comprise about 10–15 percent of the C sands and are mainly shale prone (Figure 12). Individual highstand cycles are 10–30 m thick and characterized at the base by a marine-flooding surface and at the top by a sequence boundary. The base of this systems tract is predominantly shale and is transitional upward into units with alternating sand and shale. On logs, this system tract is characterized by a shale, which forms seals over the transgressive systems tracts in these retrograding systems. At least seven highstand systems tracts are recognized in the horst block of Block I. Seismically, the highstand systems tracts occur as both continuous and discontinuous, subparallel reflections of variable amplitude. No distinctive toplap or baselap patterns were visible on seismic studies in this facies, probably due to the factor that the seismic resolution was below minimum thickness values of this facies for these depths. This is considered a nonreservoir facies in this area.

Lowstand systems tracts make up 10–15% of the Misoa Formation in the block. Both incised valley fills and lowstand wedges are recognized (Figure 12). The incised valley fills are found at the top of the Guasare, where the lower C-7 sands infill erosional channels. The incised valleys are less than 100 m thick, trend north-northeast, contain the coarsest sands in the system, and generally contain water in the field area. They have a blocky log signature, and core studies show tidal bundles characteristic of subtidal channels (well VLA-186). They are overlain by a transgressive systems tract in this area. At least two, and possibly three, lowstand wedges (above SB4 and SB6) occur in Block I and are best developed in the northern part of the block (Figure 12). They are 30–60 m thick and are characterized by thickening- and coarsening-upward log patterns of mixed sandstone and shale. The upper lowstand wedge (above SB4) onlaps and pinches out against the paleoarch. On seismic profile, the incised valleys can be locally seen in cross section, where they are enhanced by structure. Both the lowstand wedges and the incised valleys downlap sequence boundaries. Both facies consist of discontinuous reflections of variable amplitude. These two subfacies are locally productive in Block I, with porosities in the 12–20% range and less than 100 md.

In Block I, the main reservoirs are transgressive deposits of the Misoa C sands and the basal Santa Barbara member (Figures 4 and 12). Reservoir trends and facies associations were determined using logs, cores, and amplitude and flattened horizon time-slice maps (Figures 14 and 15). The composite geologic and depositional model (Figure 14) for the Eocene Misoa C sands recognizes north-trending marine channels and bar systems deposited mainly subtidally in front of a back stepping tidally influenced delta system (Figures 14 and 15). Water injection, production, petrophysical, and geochemical data shows similar north–south trends (Muñoz et al., 1995) that match the seismic amplitude patterns in this area. Recently found microfossils in marine-flooding surfaces in the lower C-4 sands suggest inner neritic depths for Block I (W. Wornardt, 1995, personal communications). Modern examples of tidally-influenced transgressive deltas that may be similar to the Misoa C sand-rich system include the subtidal part of the Colorado River delta in Mexico, the Mahakam delta in Indonesia, and the Fly delta in Papua New Guinea (Meckel, 1975; Galloway and Hobday, 1983). In the Eocene B sands in the onshore Lagunillas field, Maguregui and Tyler (1991) also documented a tidal deltaic paleoenvironment setting (Figure 1).

## EASTERN BOUNDARY FAULT PROSPECT

Using 3-D seismic data, the Eastern Boundary fault was remapped and a new structural model was developed. Older 2-D data coverage was inadequate in this area, and no wells had been drilled in this area of Maraven's acreage during the past 36 years. The use of 3-D seismic data allowed this complex structure to be imaged for the first time and the locations of previous wells to be better evaluated.

### NEW CONCEPT FROM 3-D SEISMIC DATA

The Eastern Boundary fault is interpreted to be a narrow, complex, north-northeast striking fault zone within a syncline. It is more than

10 km long and 0.5–1 km wide (Figures 16 and 17). Originally, the fault formed as an east-dipping listric fault; it was later inverted, forming a small and distinctive anticline cut by one or more northeast-striking antithetic faults that dip to the west and by several northwest-northeast–striking normal and reverse cross faults. The main structure can be subdivided in most places into three distinctive parts: (1) roll-over fold, (2) upthrown antithetic, tilted-fault block, and (3) downthrown, antithetic fault block (Figures 17 and 18). The inversion structure clearly formed post-Eocene, as seen seismically by the folding and uplift of the Eocene unconformity. Much of the fault zone is outlined by distinctive higher-amplitude seismic events, which allowed this trend to be accurately mapped and potential reservoirs to be calibrated (Figures 19 and 20). All the previous wells in this area were drilled into the upthrown and/or the downthrown parts of the main antithetic fault(s) that dipped into the Eastern Boundary fault (Figure 18). These wells had mixed results based on the close proximity of reservoirs to sealing faults. Based on the new mapping, a roll-over fold prospect with four-way closure was identified in a fault block, approximately 0.8 km long by 0.5 km wide, that was bounded by faults on all sides (Figures 16–20).

### RESULTS

Three wells were drilled during late 1994 and early 1995 (VLA1131; its twin, VLA1147; and VLA1159) to test the prospect. A 1000-m-thick hydrocarbon-bearing zone, including the Miocene Lagunillas Formation, the Santa Barbara member of the La Rosa Formation, and the Eocene Misoa C-1 to C-6 sands (Figure 4), was encountered with original pressures in the fault block. Gas and condensate were tested in the C-1 to C-5 levels and are believed to be responsible for the higher-amplitude seismic events within this fault zone. Oil was found in the Miocene strata and the Misoa C-5 and C-6 sands. Initial flow rates were 1400 BOPD from the C-6 sand, 200–300 BOPD from the C-5 sand, and 8 MCFGD from the C-3 sand. A checkshot survey and synthetic seismogram were used to calibrate the 3-D seismic data, and better identify the reservoirs and the amplitude events (Figure 20). A major sequence boundary or angular discordance recognized on the 3-D seismic data at the top of the C-5 sand (Figure 19) may allow the Eocene reservoirs here to be developed as two separate reservoir units (C-6/C-5 oil-bearing unit and C-1 to C-4 gas-bearing units).

Based on the success of these wells in this synclinal area, several new well localities and prospects have been identified. In addition, further testing of the existing reservoirs and mapping of the multiple reservoir intervals of this area are planned to optimize field development.

## VLA1035 HORIZONTAL WELL IN THE ATTIC

This development well was drilled in 1992 and required a horizontal section to drain the reservoirs from an Attic fault block, which have been producing oil for more than 40 years (Figures 21 and 22). The percentage of water in the produced oil zone in this block had increased from 20% in 1960 to over 85% in 1991. Water-coning problems in the field had rendered any additional vertical wells uneconomical because of the steep dip of the fault block and the strong water drive. The remedy was to drill a horizontal well in the upper zone of the producing structure, targeting a thick continuous sand at the top of the oil column. The 3-D seismic data coverage played an important role in selecting the well location by documenting existing faults and their positions, stratigraphic and amplitude trends, and terminations and pinchouts of strata. In addition, the use of closely spaced (40 m) inlines and random well-tie lines allowed accurate mapping of the dip and strike of the C-7 reservoir, which was critical for locating the drainhole within the productive interval. The strike of the reservoir horizon was parallel to the sealing Icotea fault and dipping steeply to the east. The structure consisted of an upthrown reverse fault block, or "pop-up" structure, bounded on the west by the Icotea fault and to the east by the Attic reverse fault (Figure 21). The Attic fault block here is only 0.6 km wide and less than one km long. The horizontal well was drilled parallel to the strike of the reservoir in an updip position above the existing oil-water contact. The well trace followed a high-amplitude reflection calibrated to be part of the upper C-7 reservoir target that was interpreted to be a channel, based on sedimentologic studies (Figure 23).

### RESULTS

The VLA1035 well was Maraven's first horizontal well in Lake Maracaibo and successfully utilized a new drilling technique combined with 3-D seismic data to better develop existing field areas (Belloso et al., 1994). More than 300 m of horizontal pay was drilled in the well in continuous channel sand complex estimated to have a real vertical thickness of only 12 m. The well originally tested more than 2400 BOPD, has produced more than 800,000 BO between 1992 to 1995, and is still producing more than 700 BOPD.

Based on the success of this well, four more horizontal wells were drilled in the northern Attic region of Block I, mainly as re-entry localities (Figure 22). In addition, several new horizontal and highly deviated wells are planned for areas that require maximum penetration of highly dipping strata in the Attic fault zone in the near future.

## CONCLUSIONS

(1) Further development of older producing fields, such as Maraven's Block I, can be greatly improved by using 3-D seismic data combined with all other available data. Remapping of stratigraphic tops, horizons, and faults, combined with production data, allows the evaluation of earlier wells and the accurate location of any future wells and/or injectors. In particular, the precise location of fault and stratigraphic terminations can be accurately determined and mapped.

(2) A new, more complete structural model has reduced the necessity for using over half of the original faults recognized in the field. Complex faults, such as the Attic, Icotea, and Eastern Boundary faults were found to consist of several fault traces and to be locally compartmentalized. Stratigraphic tops and horizons were combined with sequence boundaries, marine-flooding surfaces, and depositional sequences to better characterize reservoir trends and facies in the field. Predictive depositional models were made utilizing all available data. The interpretation of seismic horizons, various derivations, and fault polygon maps have led to the complete revision of the structural and stratigraphic models for Block I.

(3) Using 3-D seismic and log displays in the form of cross sections, fence diagrams, seismic cube and chair displays, isometric projections, and time and horizon slices greatly improved confidence in drilling any prospect or infill locality. The combined graphic representation of seismic, log, and map data allows greater visualization of prospects and concepts to be presented to management.

(4) 3-D seismic data were used in Block I to better locate wells and to develop new exploration and production concepts as shown in the two successful wells drilled: the VLA1131 discovery along the Eastern Boundary fault and the VLA1035 horizontal well in the Attic fault zone. Both wells tested concepts for developing new areas in an old producing field and led to drilling of other wells using these new concepts.

## ADDENDUM

In the year following the writing of this chapter, Maraven tested the three wells described in the VLA1131 Discovery section. The wells' (VLA1131, VLA1147, and VLA1159) initial production totalled 1392 BOPD; VLA1131 tested 264 BOPD, VLA1147 produced 825 BOPD, and VLA1159 tested 305 BOPD from single horizons. Each of these wells will develop other horizons through their life to better exploit and develop the thick potential stratigraphic section present in each well.


### *Acknowledgments*

We thank the many people who have contributed to this project, particularly the Maraven Production of Segregación Lagomar Group; Maraven Exploration Regional Study Team; Maraven/Amoco Cretaceous project, S.A. Consultores CSC (Muñoz et al, 1994); Maraven Sedimentology (E. Sampson and F. Chacartegui) and Geochemistry groups (L. Mompart); log correlation and petrophysical mapping by S.A. Consultores CSC (P. Jam, J. Delgado, and A. J. Guerrero); initial sequence stratigraphy interpretations by R. M. Mitchum and paleontologic studies by W. W. Wornardt. P. Bartok, S. K. Ghos, P. J. Pestman, and P. Weimer reviewed various drafts of this manuscript and earlier abstracts. Maraven Exploration Regional Study Team kindly provided the age and stratigraphic relationships in the Maracaibo Basin and discussed their original work with us. Some of the figures used in this paper were drafted by C. A. M. Edo and S.A. Consultores CSC. Maraven S. A. supported this project and kindly gave permission that it be published.


## REFERENCES CITED


Bartok, P., T. J. A. Reijers, and I. Juhasz, 1981, Lower Cretaceous Cogollo Group, Maracaibo Basin, Venezuela: Sedimentology, diagenesis, and petrophysics: AAPG Bulletin, v. 65, p. 1110–1134.

Bartok, P., 1993, Prebreakup geology of the Gulf of Mexico-Caribbean: Its relation to Triassic and Jurassic rift systems of the region: Tectonics, v. 12, p. 441–459.

Belloso, L., F. Chacartegui, B. Cortiula, F. Escorcia, T. Mata, E. Sampson, R. Casco, J. Husband, G. Monsegui, C. Taylor, B. Lesso, and T. Suárez, 1994, Teamwork renews an old field with a horizontal well: Schlumberger Oilfield Review, v. 6, p. 59–69.

Bueno, E., J. Zubizarreta, J. Pinto, C. Taylor, and E. Prieto, 1993, Interpretación estructural de la Falla Icotea, en el Lago de Maracaibo, Venezuela (Abs.): AAPG Bulletin, v. 77, p. 309.

Delgado, I., 1992, Lama Field–Venezuela, Maracaibo Basin, Zulia State, *in* N. H. Foster and E. Q. Beaumont, eds., Structural Traps III, Treatise of Petroleum Geology, Atlas of Oil & Gas Fields: AAPG, p. 271–294.

Galloway, W. E., and D. K. Hobday, 1983, Terrigenous clastic depositional systems: Springer-Verlag, New York, 423 p.

Haq, B. U., J. Hardenbol, and P. R. Vail, 1988, Mesozoic and Cenozoic chronostratigraphy and cycles of sea-level change, *in* C. K. Wilgus, H. W. Posamentier, C. A. Ross, C. G. St. C. Kendall, and B. S. Hastings, eds., Sea-level changes: an integrated approach: SEPM Special Publication 42, p. 71–108.

Link, M. H., C. K. Taylor, E. Bueno, and R. M. Mitchum, 1994, Structure, stratigraphy, and sequence stratigraphy, Maraven's Block I: A seismic perspective, Maracaibo Basin, Venezuela, *in* VII Congreso Venezolano de Geofísica, Memorias 1994, p. 401–408.

Lugo, J., 1992, Historia tectónica a lo largo de los sistemas de fallas de Icotea y Pueblo Viejo, Cuenca de Maracaibo: VI Congreso Venezolano de Geofísica, p. 118–124.

Lugo, J., and P. Mann, 1995, Jurassic–Eocene tectonic evolution of Maracaibo Basin, Venezuela, *in* A. J. Tankard, R. Suárez Soruco, and H. J. Welsink, eds., Petroleum Basins of South America, AAPG Memoir 62, p. 699–725.

Maguregui, J., and N. Tyler, 1991, Evolution of middle Eocene tidal-dominated sandstones, Lagunillas field, Maracaibo Basin, Western Venezuela, *in* A. D. Miall and N. Tyler, eds., The three-dimensional facies architecture of terrigenous clastic sediments and its implications for hydrocarbon discovery and recovery: SEPM Concepts in Sedimentology and Paleontology, v. 3, p. 233–244.

Marais-Gilchrist, G. and R. Higgs, 1993, Secuencia estratigráfica de la Formación Misoa (Eoceno), Lago de Maracaibo, Venezuela (Abs.): AAPG Bulletin, v. 77, p. 332.

McClay, K. R., and P. G. Buchanan, 1991, Thrust faults in inverted extensional basins, *in* K. R. McClay, ed., Thrust tectonics: Chapman & Hall, p. 93–104.

Meckel, L. D., 1975, Holocene sand bodies in the Colorado Delta area, northern Gulf of California, *in* M. L. Broussard, ed., Deltas models for exploration, Houston Geological Society, p. 239–265.

Mitchum, R. M., 1977, Seismic stratigraphy and global changes of sea level. Part I: Glossary of terms used in seismic stratigraphy, *in* C. E. Payton, ed., Seismic Stratigraphy—Applications to Hydrocarbon Exploration: AAPG Memoir 26, p. 205–212.

Muñoz, N. G., L. Mompart, and S. C. Talukdar, 1995, Application of oil gas–chromatography in reservoir compartmentalization in a mature Venezuelan oil field (Abs.): AAPG Annual meeting volume, Houston, TX.

Muñoz, N. G., M. H. Link, J. Delgado, C. K. Taylor, R. M. Mitchum, R. M. and P. J. Muñoz, 1994, Integrated sedimentological and seismic studies, Eocene Misoa C Sands, Maraven's Block I, Maracaibo Basin, Venezuela, *in* VII Congreso Venezolano de Geofísica, Memorias 1994, Caracas, Venezuela, p. 259–266.

Perdomo, J. L., and P. Bot, 1986, Eocene seismic stratigraphy in the Maracaibo Basin, *in* III Congreso Venezolano de Geofísica, Caracas, p. 474–483.

Pestman, P. J., P. Bartok, S. K. Ghosh, and L. Melendez, 1994, Influencia tectónica y de fluctuaciones del mar en sistemas fluviales, con ejemplos del Eoceno en la cuenca de Maracaibo, Venezuela, *in* Simposia Bolivariano, Exploración Petrolera en las Cuencas Subandinas: Memoria, Puerto La Cruz, p. 65–84.

Posamentier, H. W., and P. R. Vail, 1988, Eustatic controls on clastic deposition II—sequence and system tract models, *in* C. K. Wilgus, H. W. Posamentier, C. A. Ross, C. G. St. C. Kendall, and B. S. Hastings, eds., Sea-Level Changes: An Integrated Approach: SEPM Special Publications 42, p. 125–154.

Roberto, M., B. Duval, and C. Cramez, 1993, Inversión tectónica en el área de Maracaibo, Venezuela: AAPG Bulletin, v. 77, p. 343.

Vail, P. R., 1987, Seismic stratigraphy interpretation using sequence stratigraphy, Part I: Seismic stratigraphy interpretation procedure, *in* A. W. Bally, ed., Atlas of Seismic Stratigraphy: AAPG Studies in Geology 27, v. 1, p. 1–10.

Van Wagoner, J. C., R. M. Mitchum, Jr., K. M. Campion, and V. D. Rahmanian, 1990, Siliciclastic sequence stratigraphy in well logs, cores, and outcrops: concepts for high-resolution correlation of time and facies: AAPG Methods in Exploration series 7, 55 p.

Williams, G. D., C. M. Powell, and M. A. Cooper, 1989, Geometry and kinematics of inversion tectonics, *in* M. A. Cooper, and G. D. Williams, eds., Inversion Tectonics: Geological Society London Special Publication 44, p. 3–15.

Withjack, M. O., Q. T. Islam, and P. R. La Pointe, 1995, Normal faults and their hanging wall deformation: an experiment study: AAPG Bulletin, v. 79, p. 1–18.

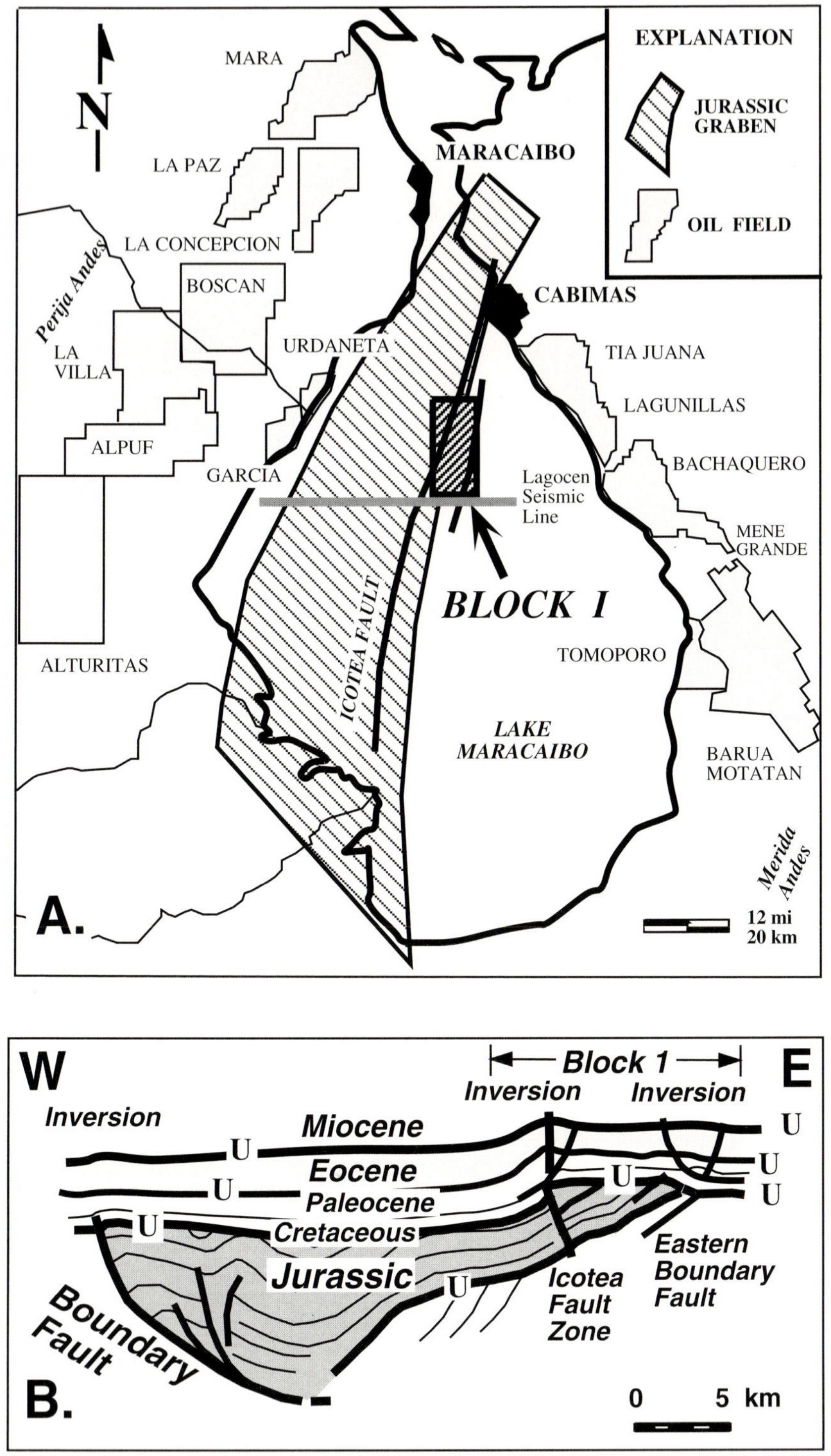


FIGURE 1. Index map showing Lake Maracaibo. (A) Map of major onshore fields, trace of the Icotea and Eastern Boundary faults, location of Lagocen regional seismic line 15, Jurassic graben outline (Bartok et al., 1981), and position of Block I. (B) Sketch of west–east Lagocen 15 seismic line, highlighting the Jurassic basin and Icotea and Eastern Boundary fault-bounded horst of Block I, and major unconformities (U).

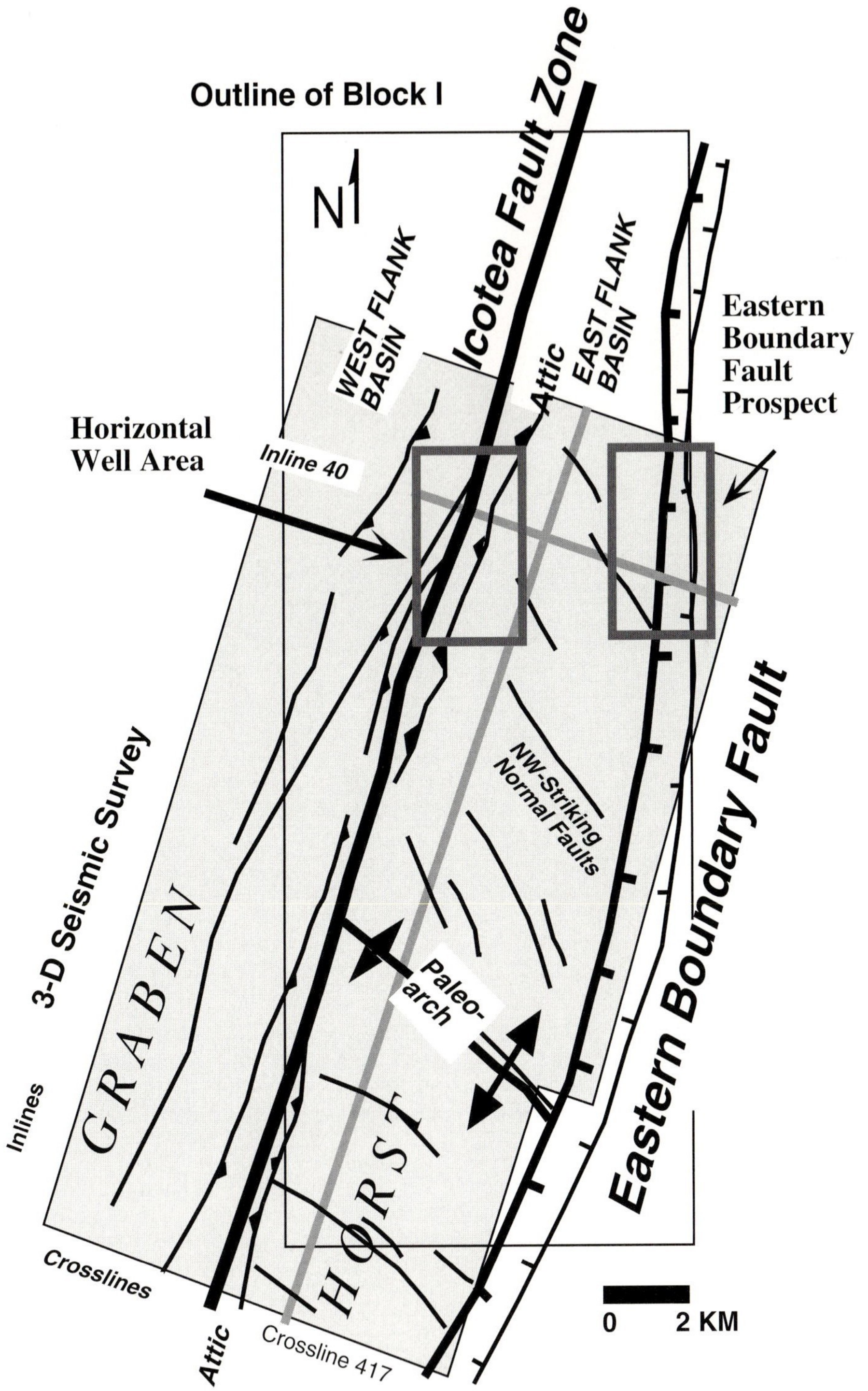


FIGURE 2. Simplified structure map at the top of the Guasare Formation for Block I, modified from the Guasare first derivative or dip map shown in Figure 3. Major faults; the location of the paleoarch; Block I and seismic survey outlines; the horst and graben; and West and East Flank basins are shown. The position of inline 40 and crossline 417 are indicated. The location of Horizontal Well and the Eastern Boundary Fault Prospect areas are highlighted with a box.

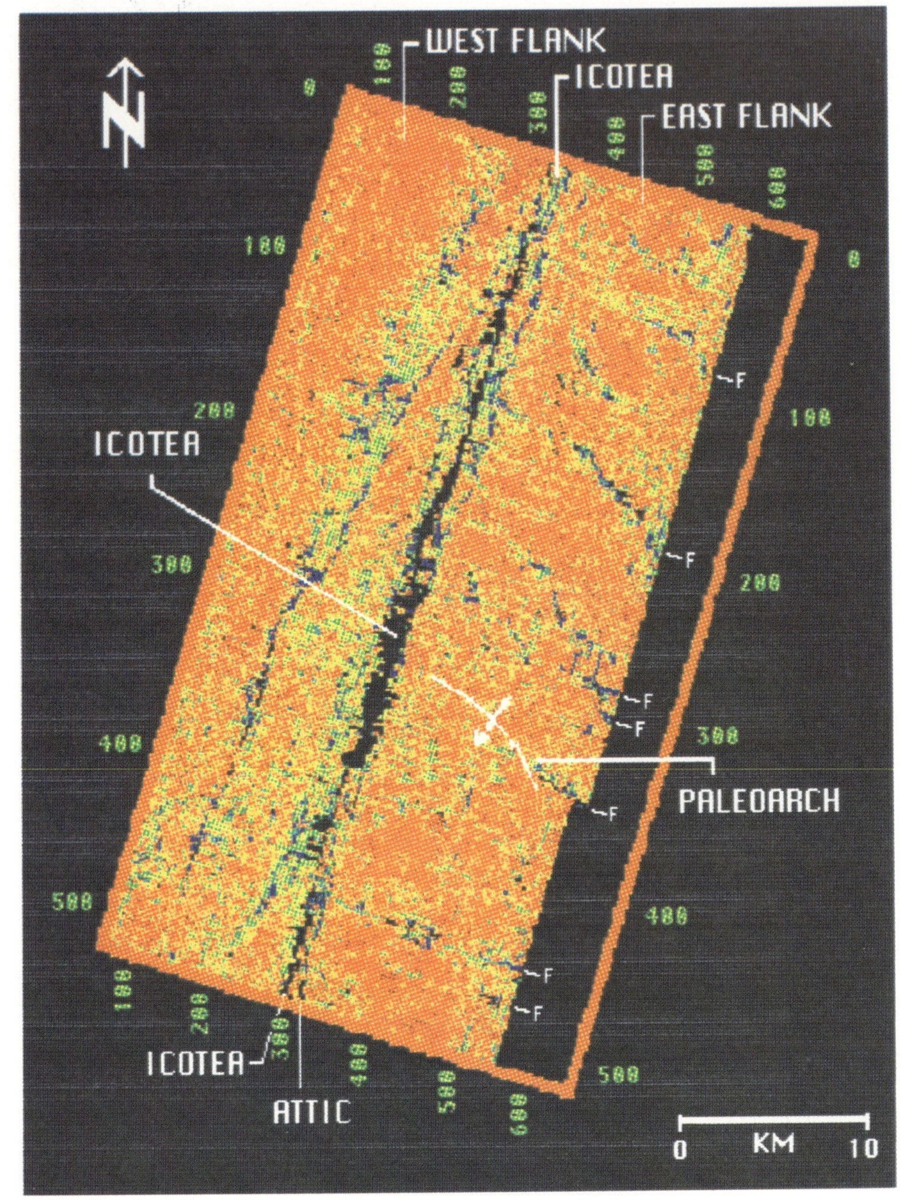


FIGURE 3. Dip or first derivative map on the top of the Guasare Formation for the entire block, showing major fault (F) patterns, position of the paleoarch, West and East flanks, and inline and crossline numbers.

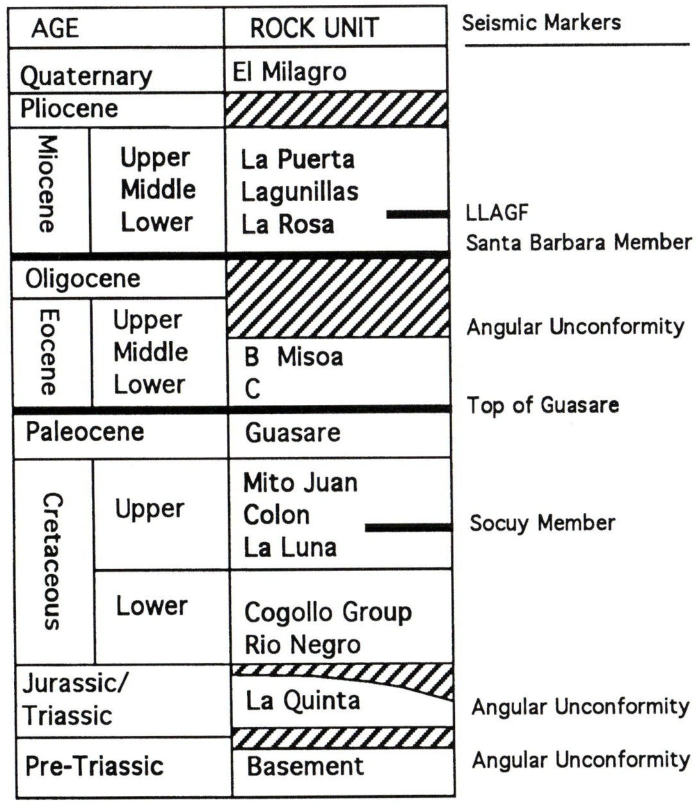


FIGURE 4. Block I stratigraphic section, highlighting ages, rock units (formations), unconformities, and major seismic markers.

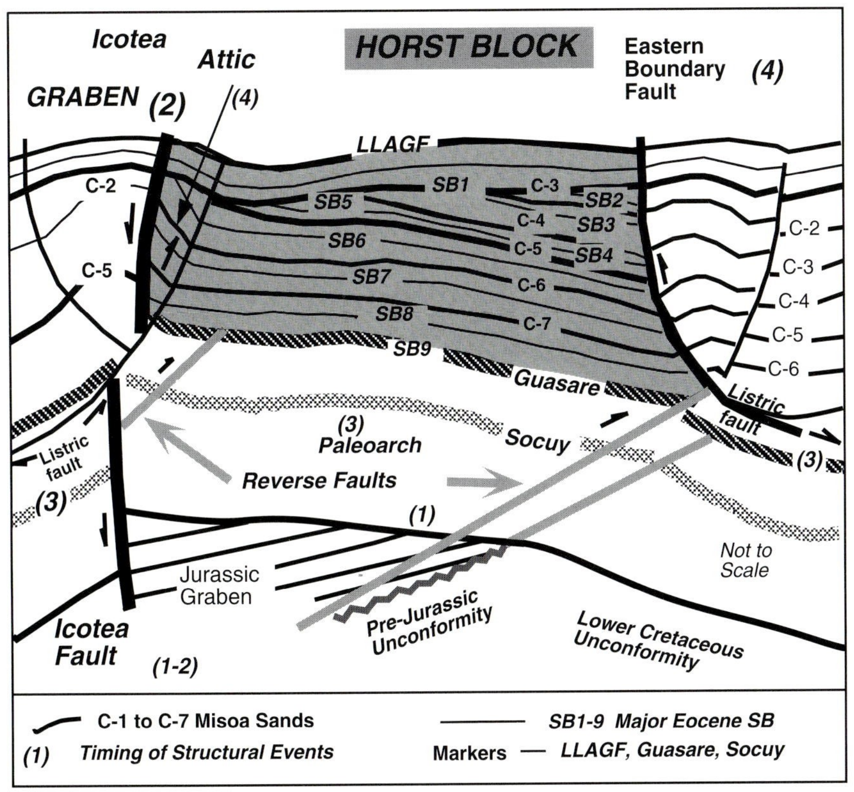


FIGURE 5. Geologic sketch of the horst area of Block I, showing the unconformities, major structures, sequence boundaries, various C sand members of the Misoa Formation, and the major stratigraphic relations for the East Flank of Block I. Timing of structural features from oldest to youngest is indicated with (1)–(4). See Figure 6 for seismic line interpretation.

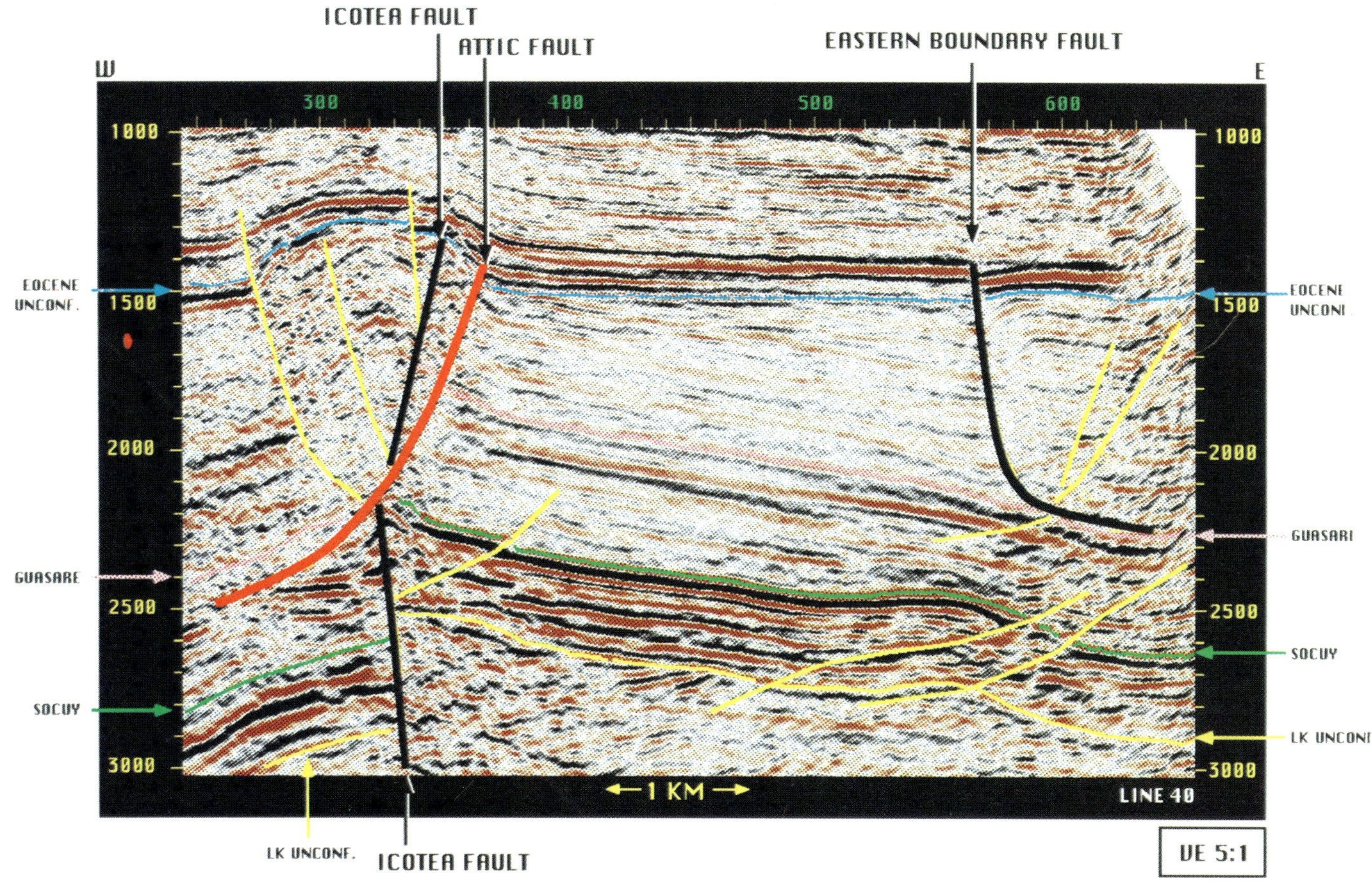


FIGURE 6. West–east inline 40 across the East Flank of Block I, highlighting the major structures, unconformities, and stratigraphic relationships. See Figure 2 for location of inline; Figure 4 for seismic markers; and Figure 5 for schematic interpretation of faults, the various C sands, and sequence boundary localities. Seismic markers include the Lower Cretaceous (LK) unconformity, Upper Cretaceous Socuy member, top of the Paleocene Guasare Formation, and top of Eocene Unconformity (see Figure 4).

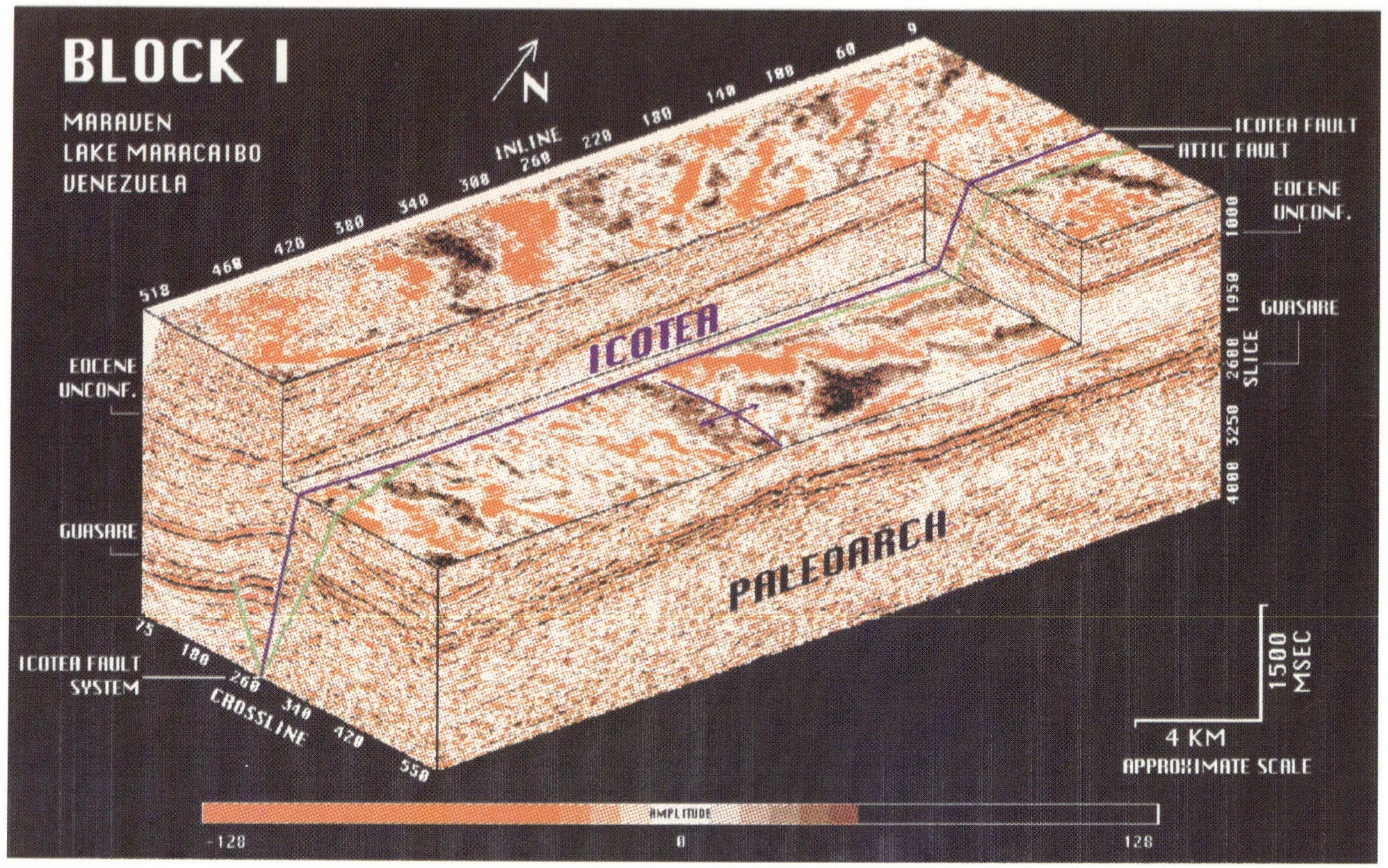


FIGURE 7. Chair display of the complete 3-D seismic survey, looking northwest at the East Flank of Block I. The Icotea fault, the southern and northern Attic blocks, northwest-striking faults that delineate the paleoarch, inline and crossline localities, top of Guasare and Eocene unconformity seismic markers, and depth in time (ms) (see slice) are indicated.

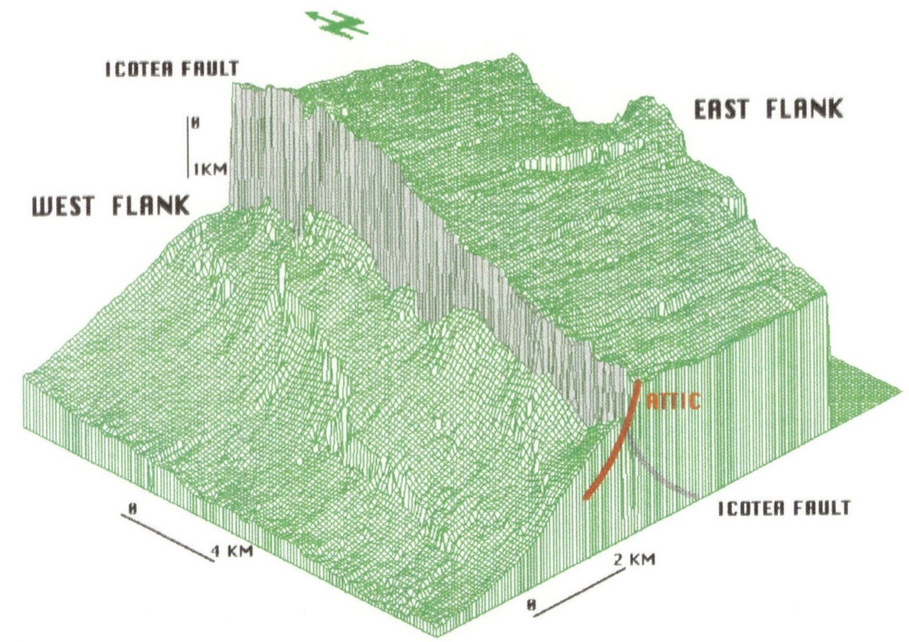


FIGURE 8. A 3-D isometric projection on the top of the Guasare Formation in the West Flank, looking northeast and highlighting the Icotea and the subparallel reverse faults of the West Flank that merge with the Icotea fault forming triangle- to wedge-shaped fault blocks. The southeast corner of the West Flank adjacent to the Icotea fault is down-dropped about 1/2–1 km to the west, whereas the northeast corner adjacent to the Icotea fault has a displacement of almost 2 km.

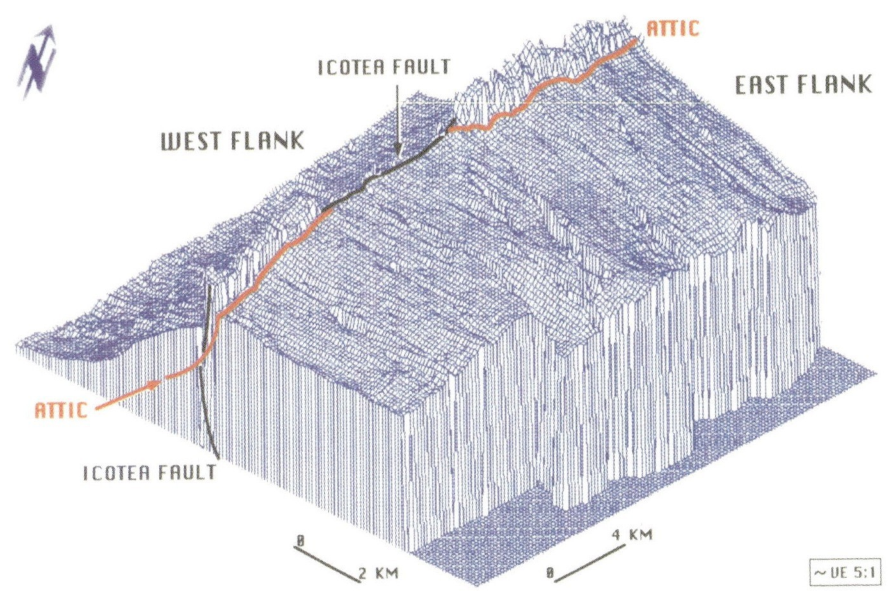


FIGURE 9. A 3-D isometric projection on the top of the Guasare Formation in the East Flank, looking northwest to highlight the Icotea and Attic faults. The northern and southern upthrown Attic blocks are bounded by reverse faults that dip to the west and are separated by a saddle in the middle. The saddle coincides with the paleoarch, which occurs near the south-center of the East Flank and is delineated by northwest-striking normal faults that dip both to the northeast and southwest. The two Attic fault blocks may have formed due to shortening of thicker section on sides of the paleoarch during later fault deformation.

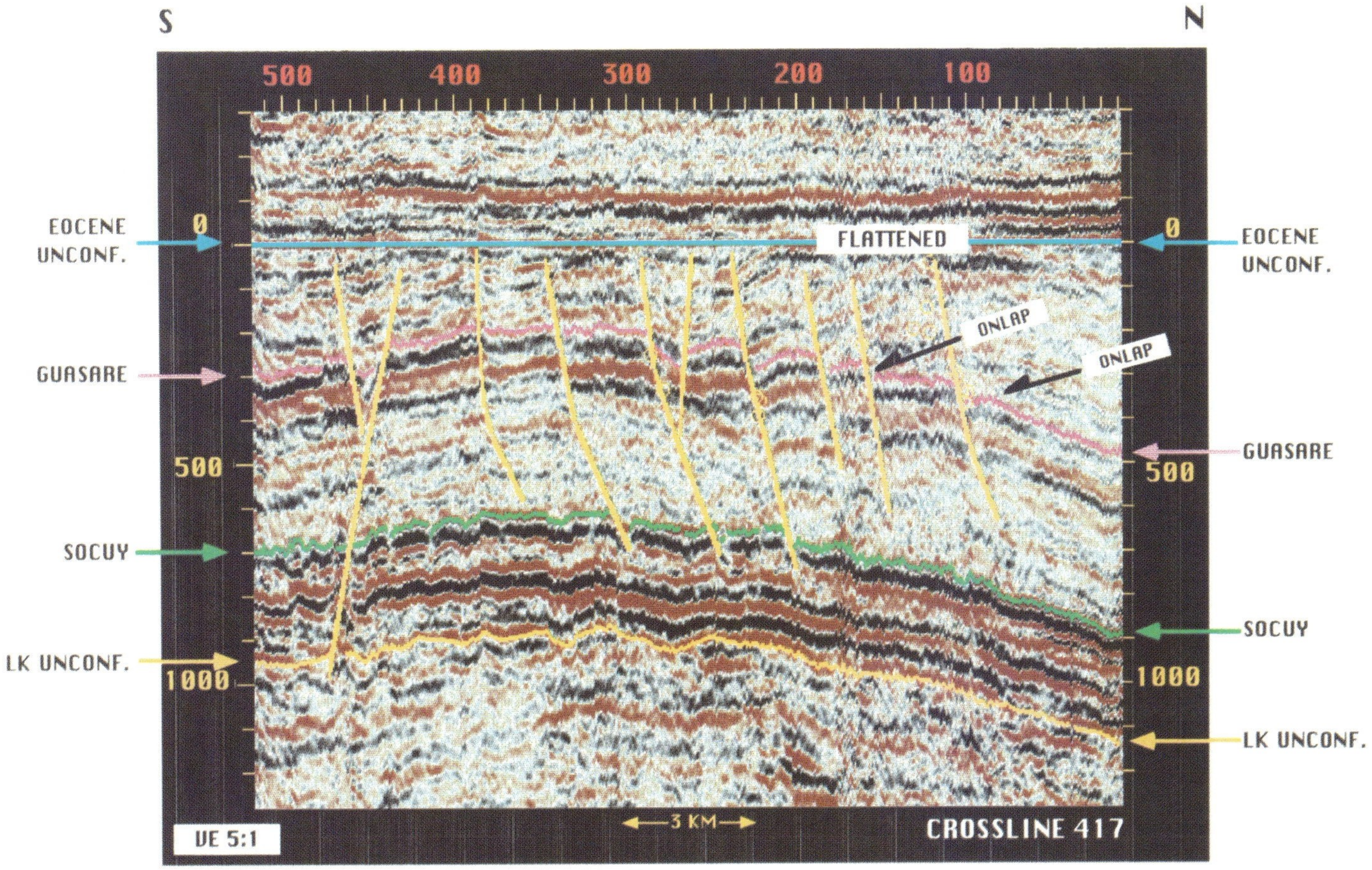


FIGURE 10. North–south crossline 417 across the East Flank, flattened on the top of Eocene Unconformity, and highlighting the paleoarch and its northwest-striking faults (F). See Figure 2 for location of crossline. Seismic markers include the Lower Cretaceous (LK) unconformity, Upper Cretaceous Socuy member, top of the Paleocene Guasare Formation, and top of the Eocene Unconformity (see Figure 4). North-to-south onlap (arrows) can be seen above the Guasare marker, starting just above the 500 ms depth that is flattened on the Eocene unconformity.

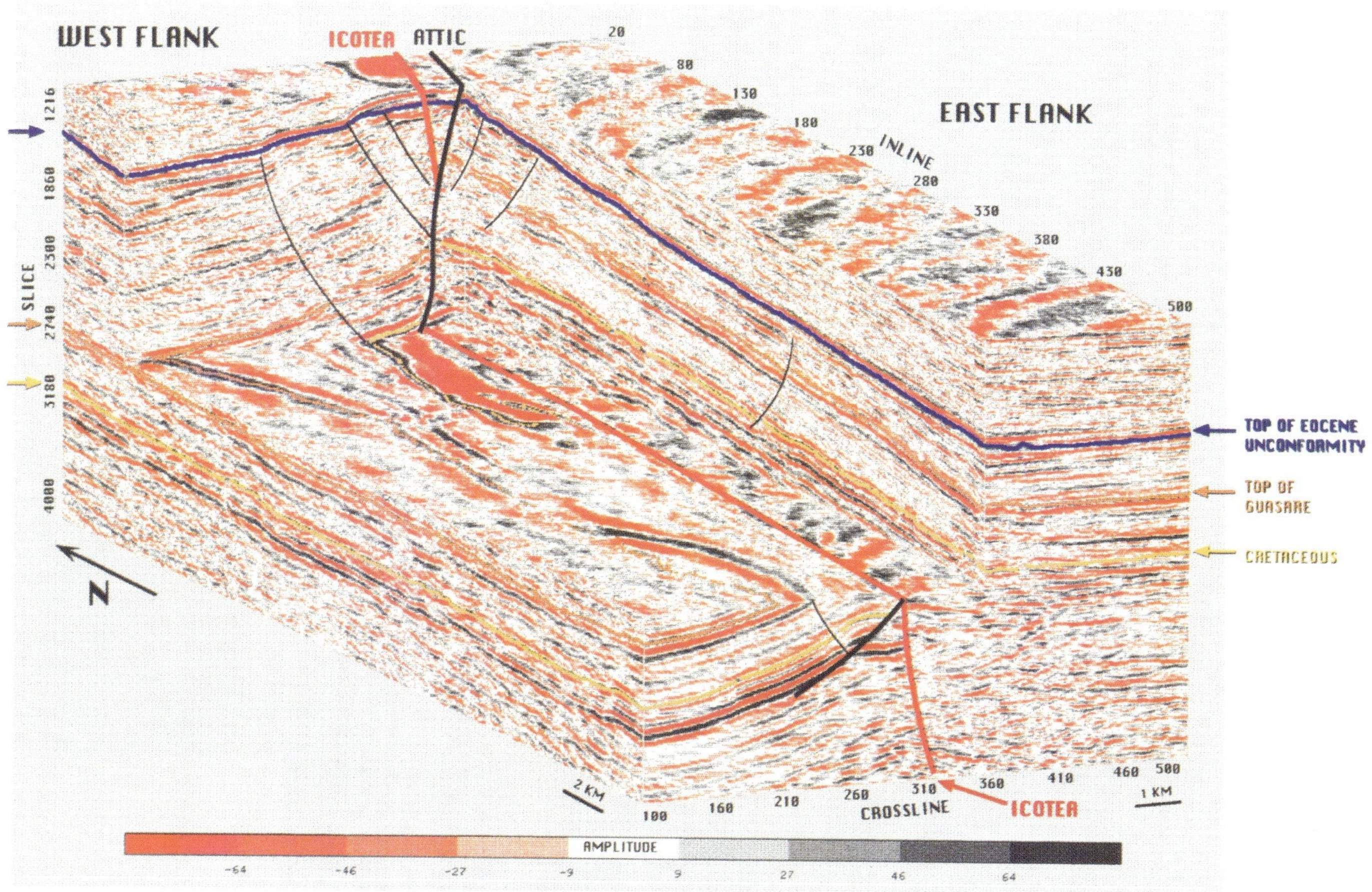


FIGURE 11. Chair display of the 3-D seismic survey, looking northeast at the West Flank of Block I. The Icotea fault strikes northeast across the chair display and is characterized by a nearly vertical trace that is offset and displaced by the Attic fault (best seen in the upper central part of the chair display). Northeast-striking antithetic reverse faults are subparallel to and dip into the Attic/Icotea faults. The Eocene section is 1–1/2 times thicker on the West Flank than on the East Flank. Attic blocks subparallel the Icotea fault to the east. Inline and crossline localities; top of Cretaceous, Guasare, and Eocene unconformity seismic markers; vertical scale in time (ms) (see slice 4000, 3188, 2748) are indicated.

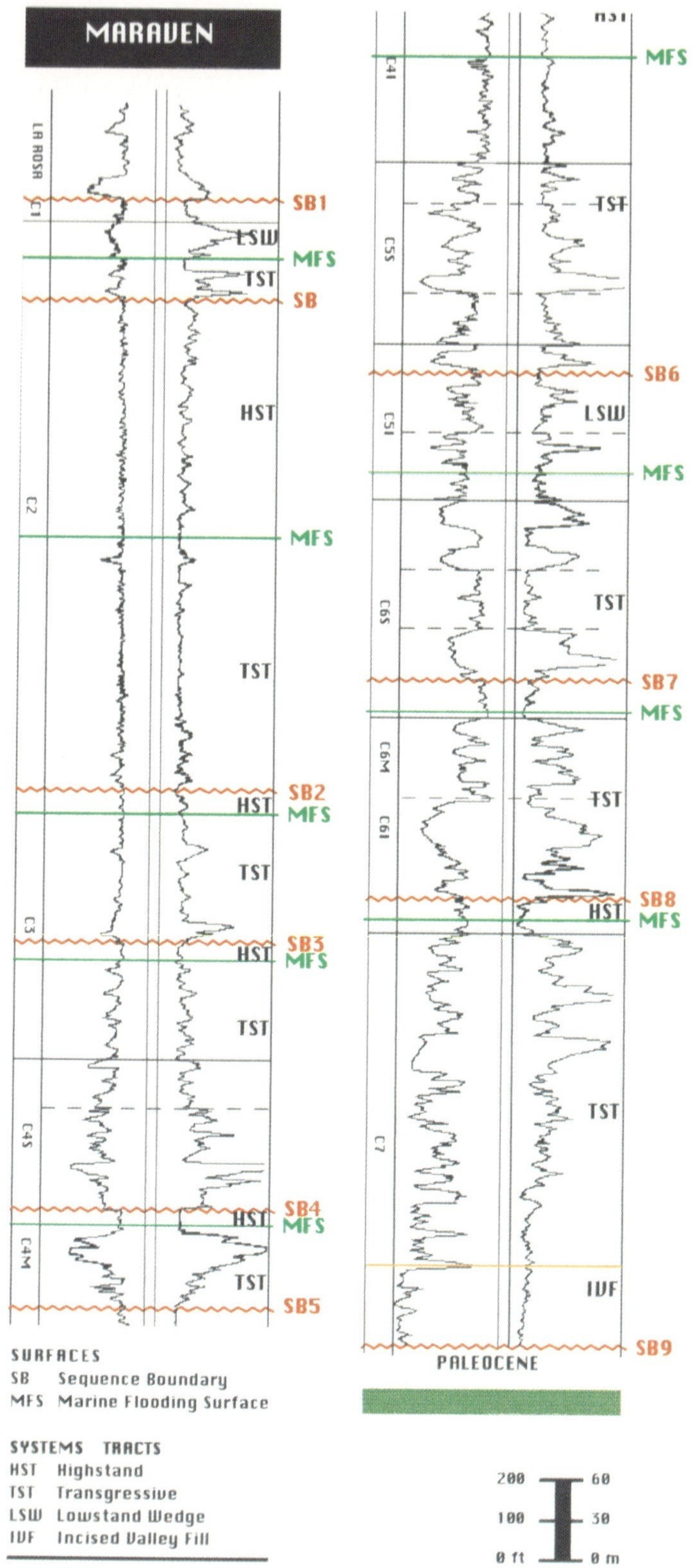


FIGURE 12. Type log and sequence stratigraphic diagram for the Eocene Misoa C sands in the East Flank of Block I, showing tops, sequence boundaries, marine-flooding surfaces, and systems tracts. This stratigraphic log is a composite log for the area east of the Eastern Boundary fault.

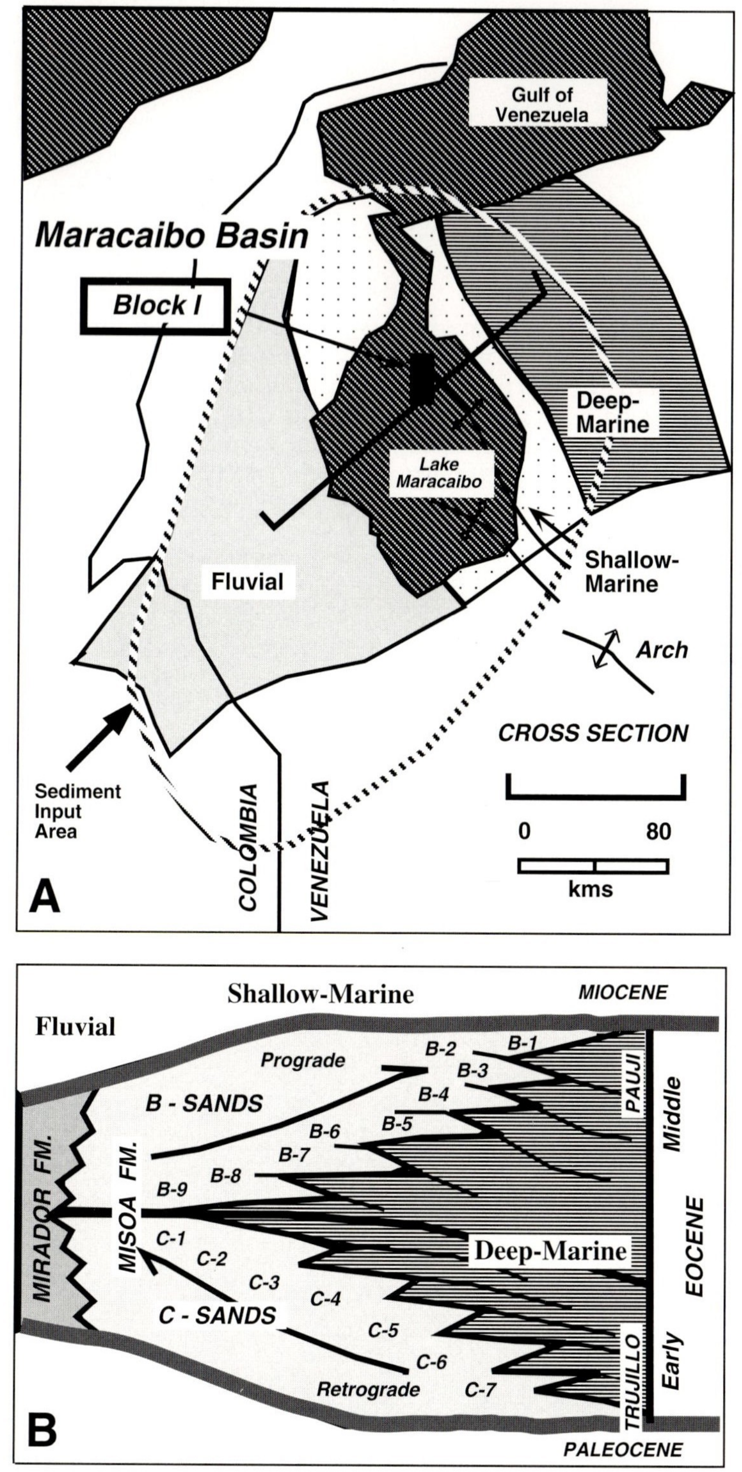


FIGURE 13. Stratigraphic model for the Eocene strata of the Maracaibo Basin. (A) Simplified early to middle Eocene paleogeographic map, showing southwest to northeast transition from fluvial to shallow-marine to deep-marine facies relations; and (B) southwest to northeast stratigraphic cross section, highlighting the transition from the Mirador to the Misoa to the Trujillo formations and from the nonmarine to the deep-marine facies relations, the various B and C sands of the Misoa Formation, and the retrograde (back-stepping) of the C sands vs. progradation (out-building) stacking patterns of the B sands.

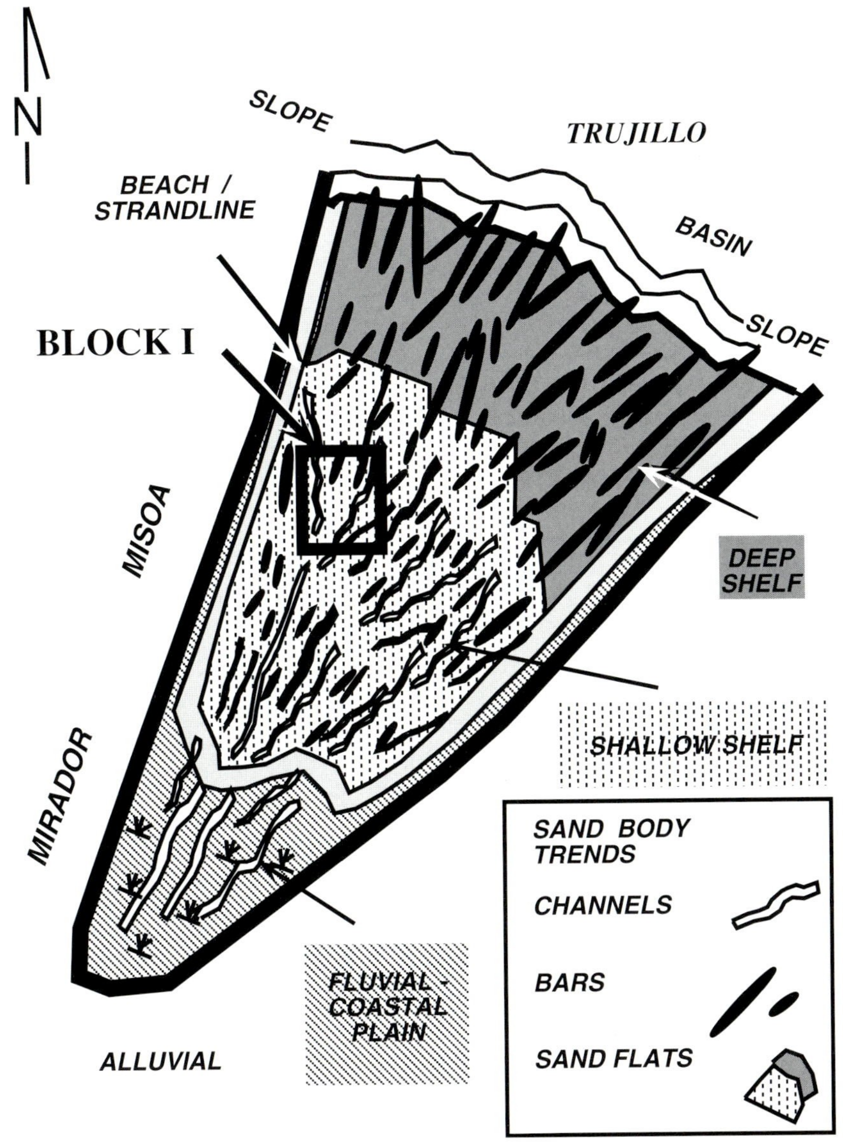


FIGURE 14. Depositional model for the Misoa and Mirador formations at C-5 to C-7 time. A retrograde back-stepping pattern of sand-prone sedimentation occurred during lower Misoa deposition. Shallow-marine facies (Misoa) were deposited and are transitional from nonmarine units (Mirador) to the southwest to deep-marine facies (Trujillo) to the northeast. Depositional trends are aligned northwest, north, and northeast and follow the original confining shape of the northeast-trending basin.

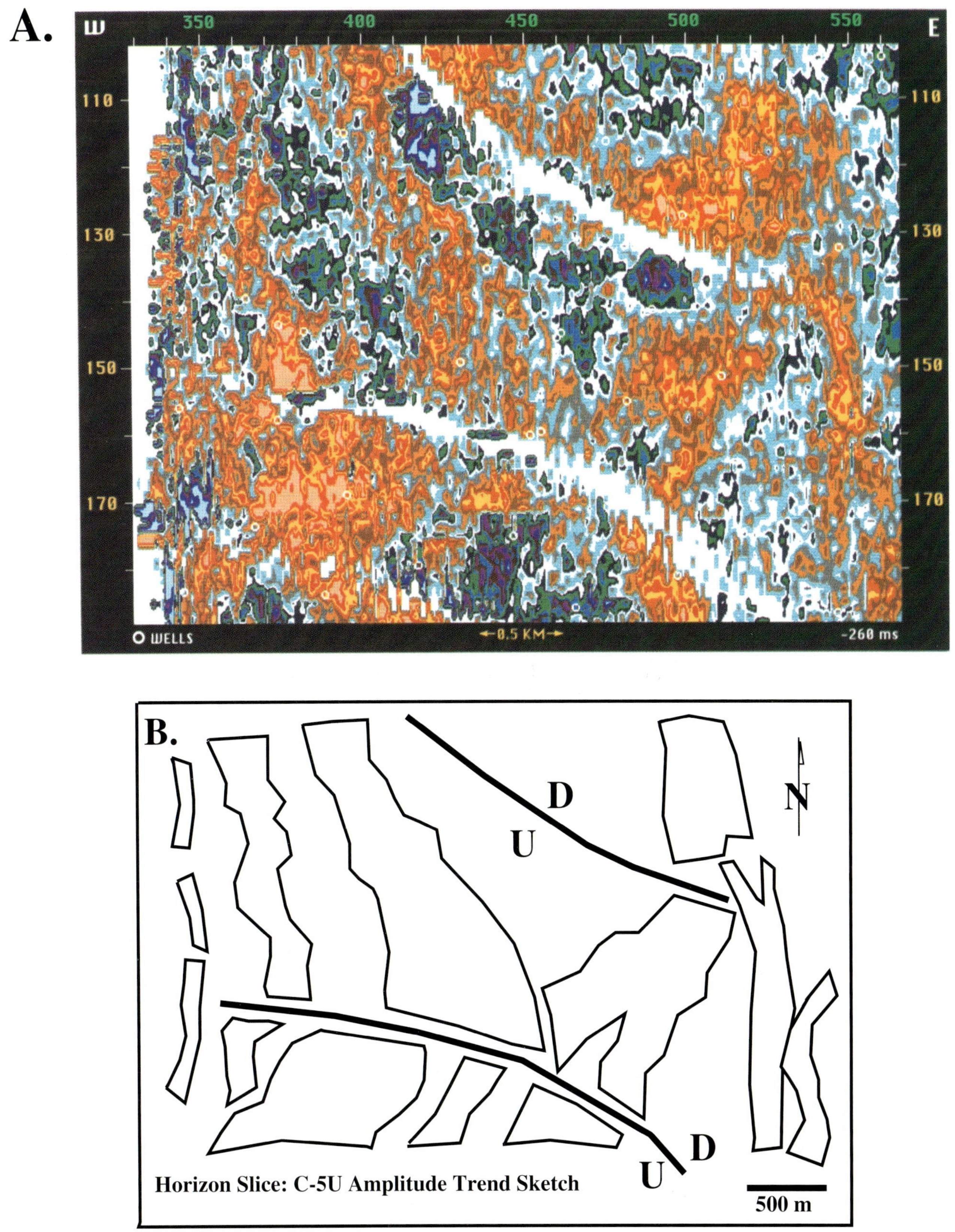


FIGURE 15. Flattened horizon-time slice and sketch at the C-5U horizon level. (A) Horizon-time slice taken at 260 ms above the top of Guasare. The north-trending linear amplitudes are cut by major northwest-striking faults (white fault traces). (B) Sketch of the north-trending amplitude patterns interpreted from these amplitudes and other core and log data to be subtidal marine channels. Fault abbreviations: D = down; U = up.

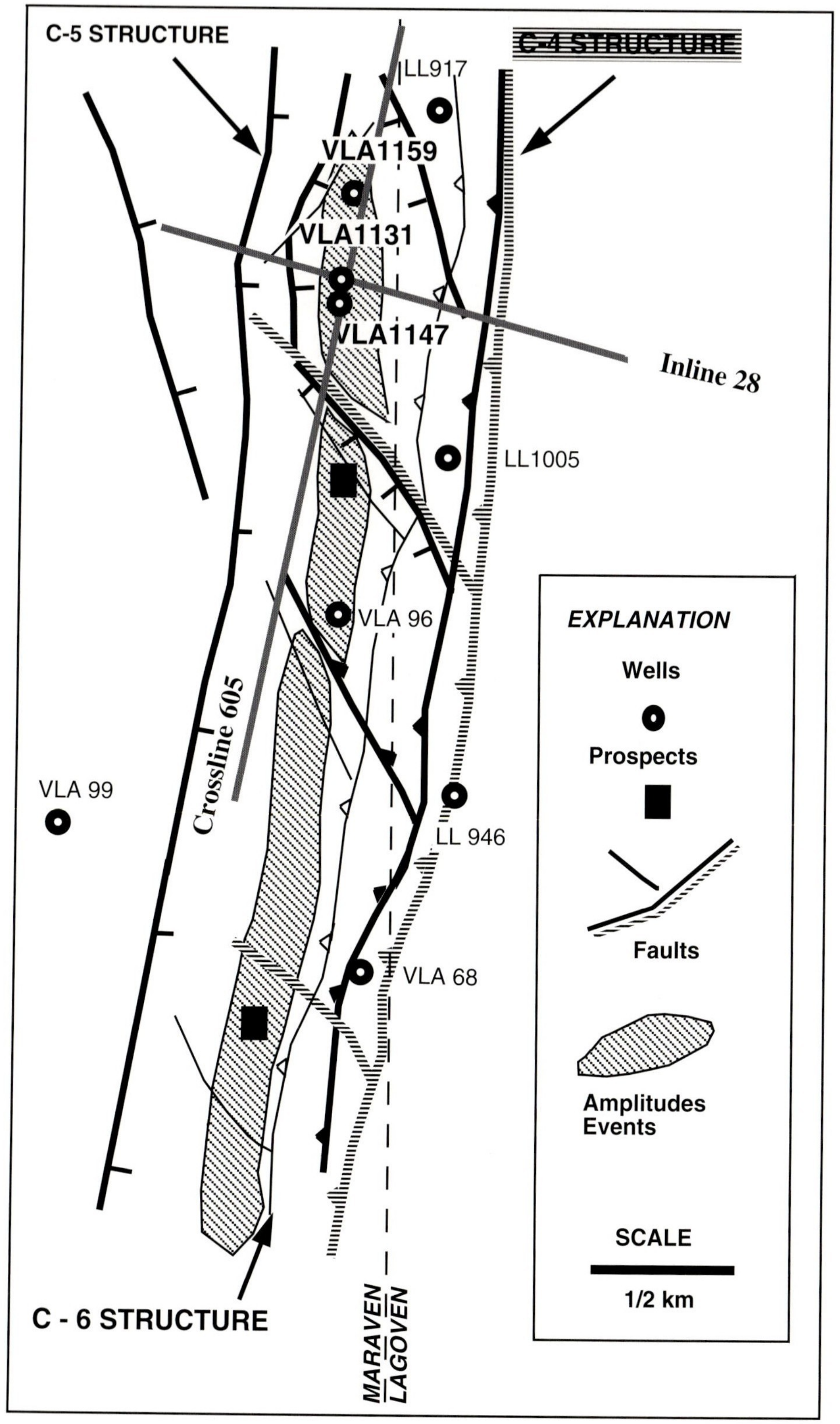


FIGURE 16. Simplified structure map of the Eastern Boundary fault, showing wells, prospects, trends of seismic amplitude events, and major fault patterns at the C-6, C-5, and C-4 levels. The positions of inline 28 and crossline 605 are shown in Figures 18, 19, and 20.

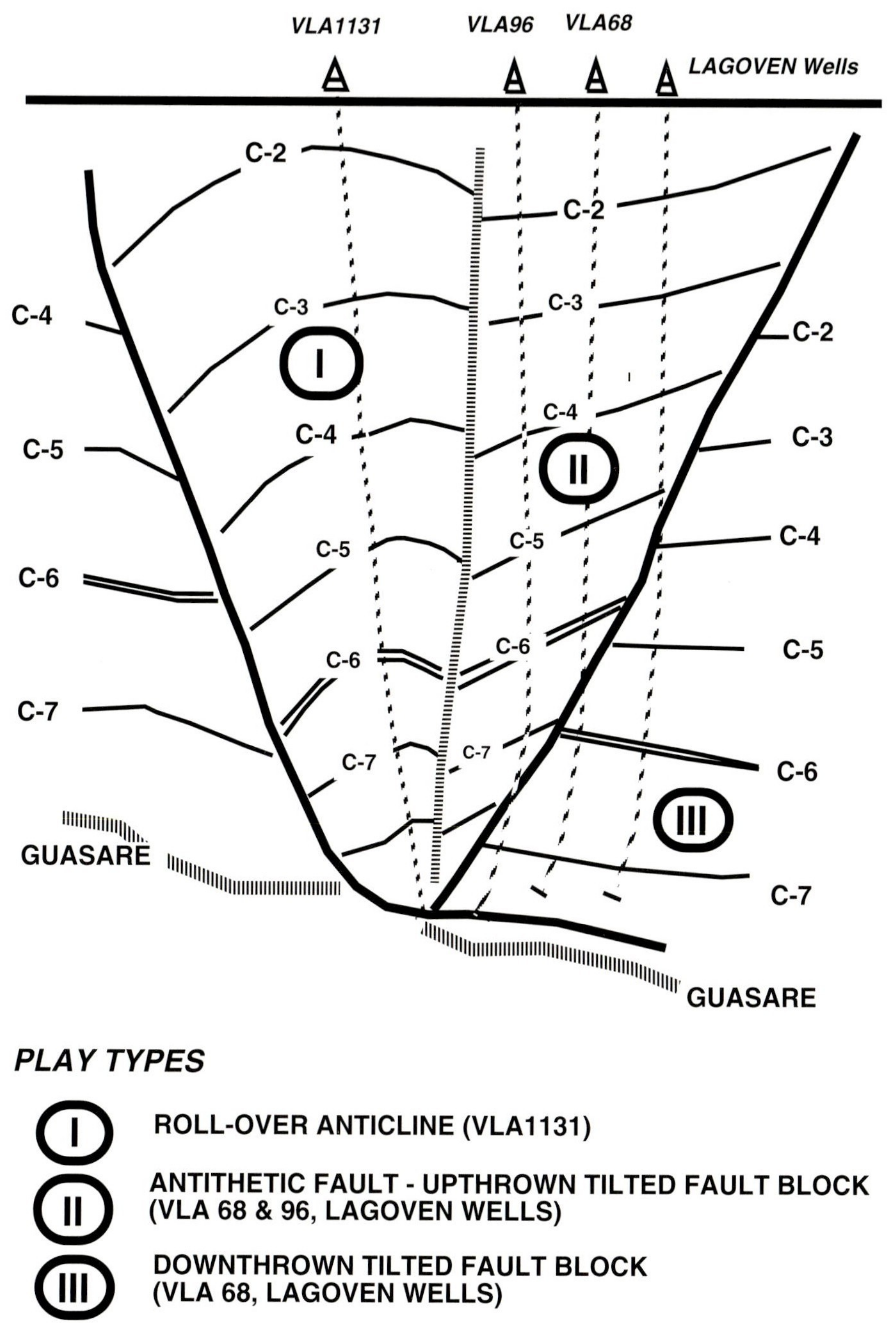


FIGURE 17. Composite structural model for the Eastern Boundary fault based on sketches made from west-east seismic lines, highlighting the compartmentalization of the narrow structure into (1) a roll-over fold, (2) upper antithetic, tilted-fault block, and (3) a lower antithetic, tilted-fault block. Horizons and the relative positions of key wells are indicated (see Figure 16).

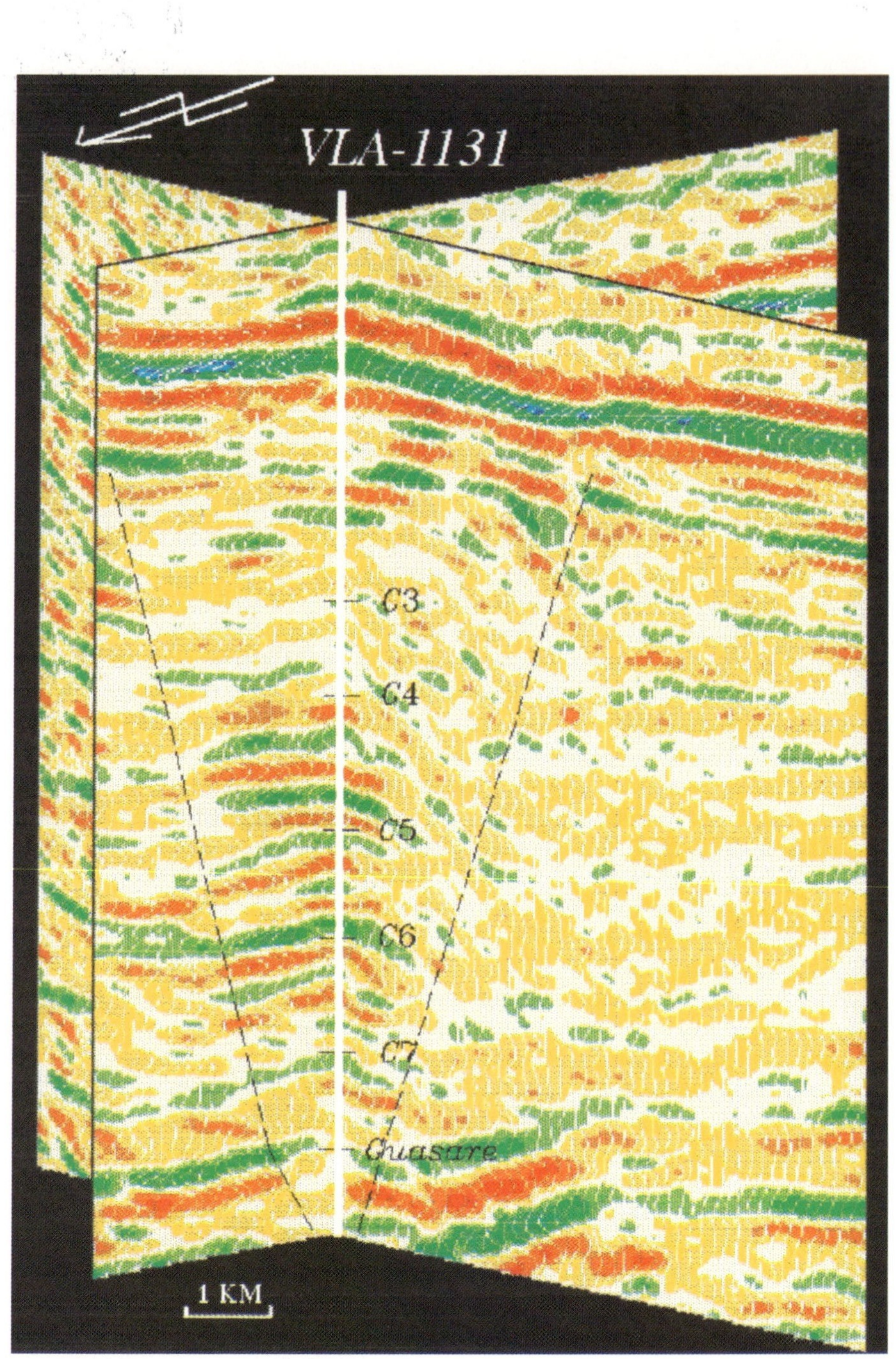


FIGURE 18. Fence diagram of the VLA1131 well, showing the structural closure, faults, and unconformity at the top of the C-5 horizon. West-east inline 28 and north-south crossline 605 make up the two sides of the fence diagram. See Figure 16 for location of seismic lines. The high-amplitude seismic events outline the C-4 to C-6 reservoirs.

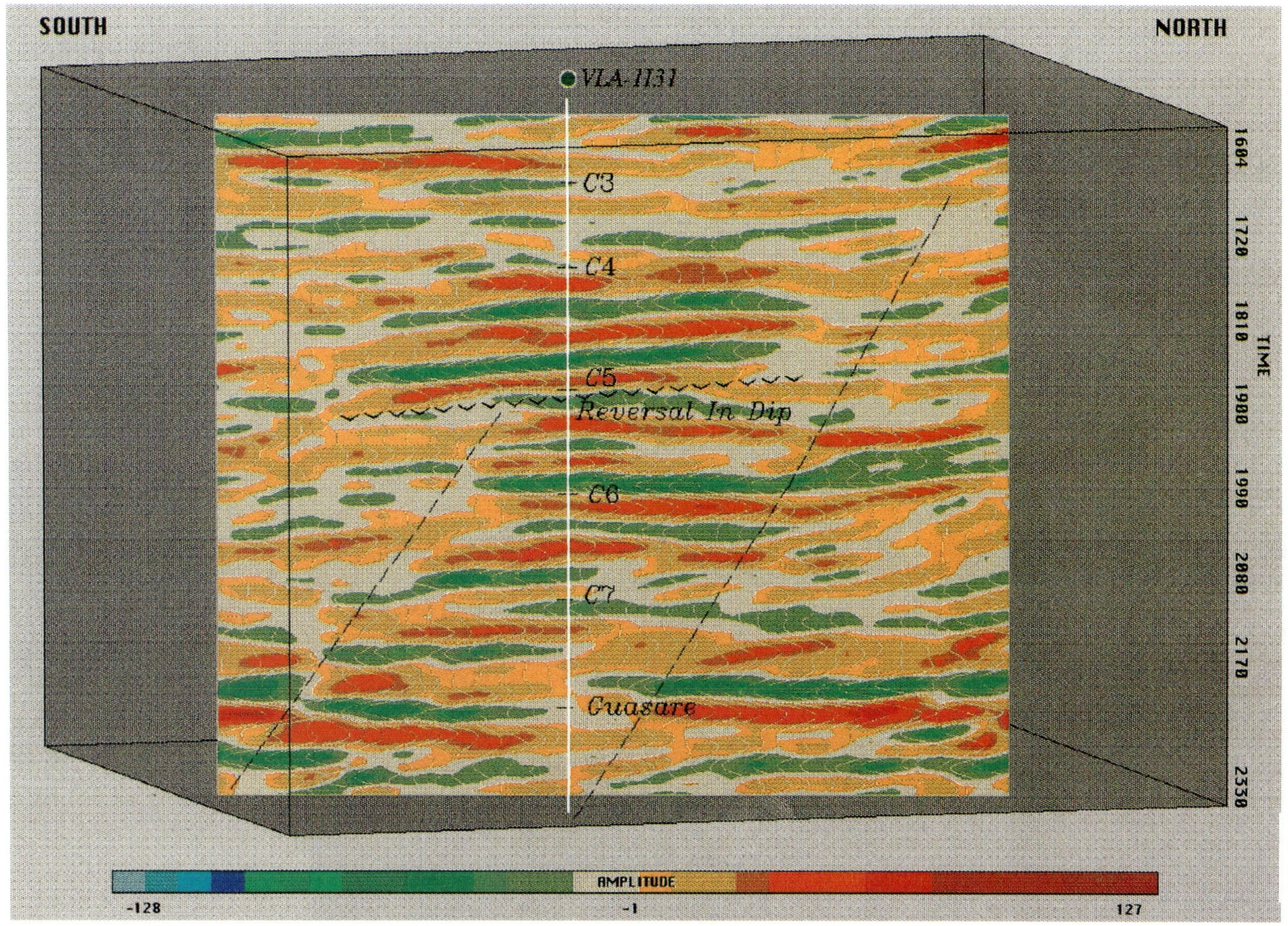


FIGURE 19. North-south seismic crossline 605, showing the structure and amplitude patterns for part of the Eastern Boundary fault. Top of C-5 unconformity, major faults, and stratigraphic tops are indicated. Note the reversal of dips below the unconformity.

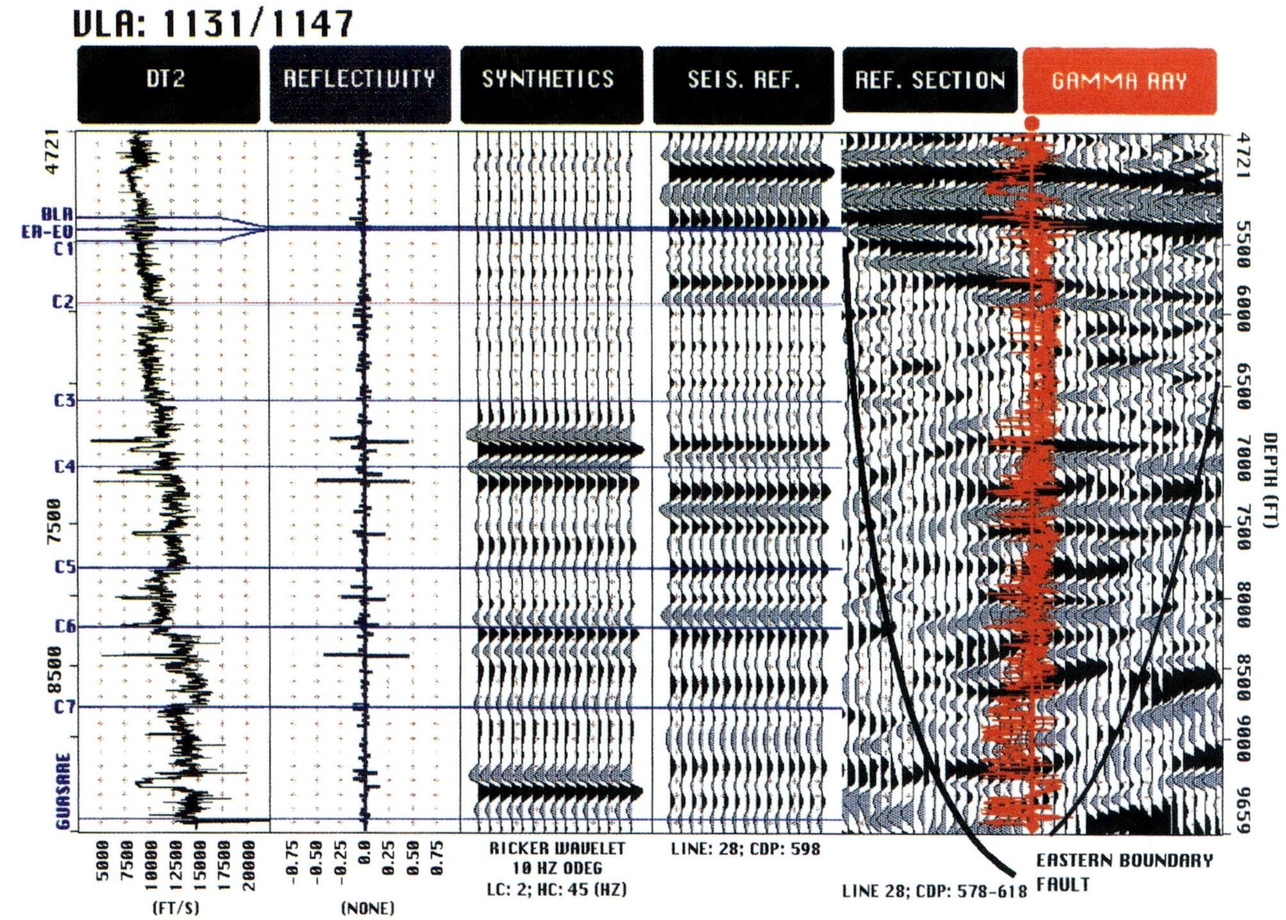


FIGURE 20. Synthetic seismogram, gamma ray log, and west-east inline 28, showing the structure and log character for the VLA1131 discovery well and the VLA1147 twin well. This display shows, from left to right: sonic log, reflectivity, synthetic seismogram, seismic reference trace (trace at well location), and an interpreted section of inline 28 from the 3-D seismic survey with gamma ray log overlain at the well location. Other abbreviations: BLR = basal La Rosa Formation, ER-EO = Eocene Unconformity, and CDP = common depth point.

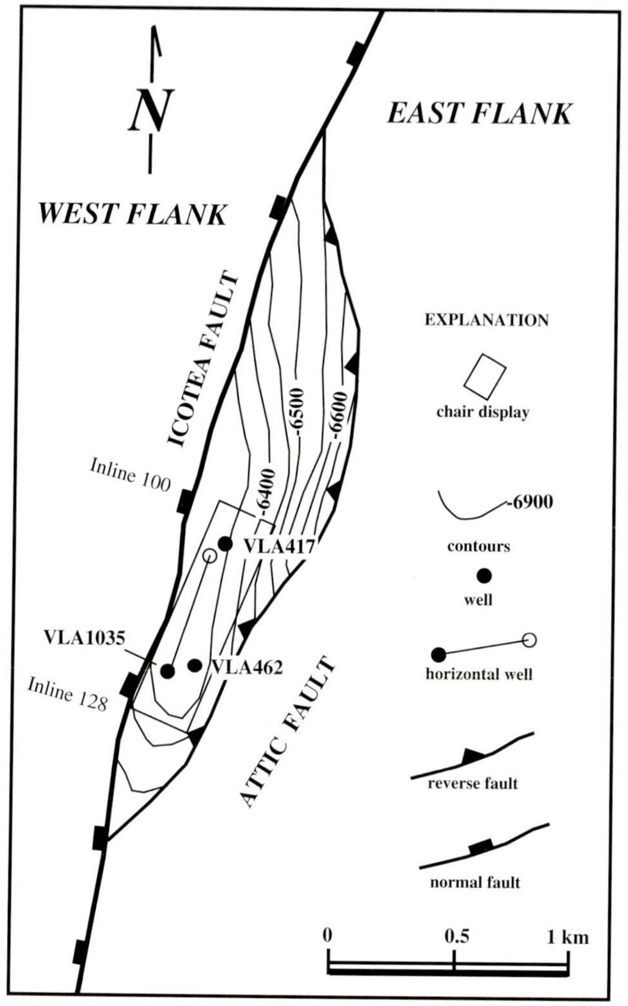


FIGURE 21. Index map for the VLA1035 well in the northern Attic fault block. Contours and structures are on the top of the C-7 sand. Key wells and seismic lines, position of the chair display, and isometric projection area for this part of the northern Attic fault zone are indicated.

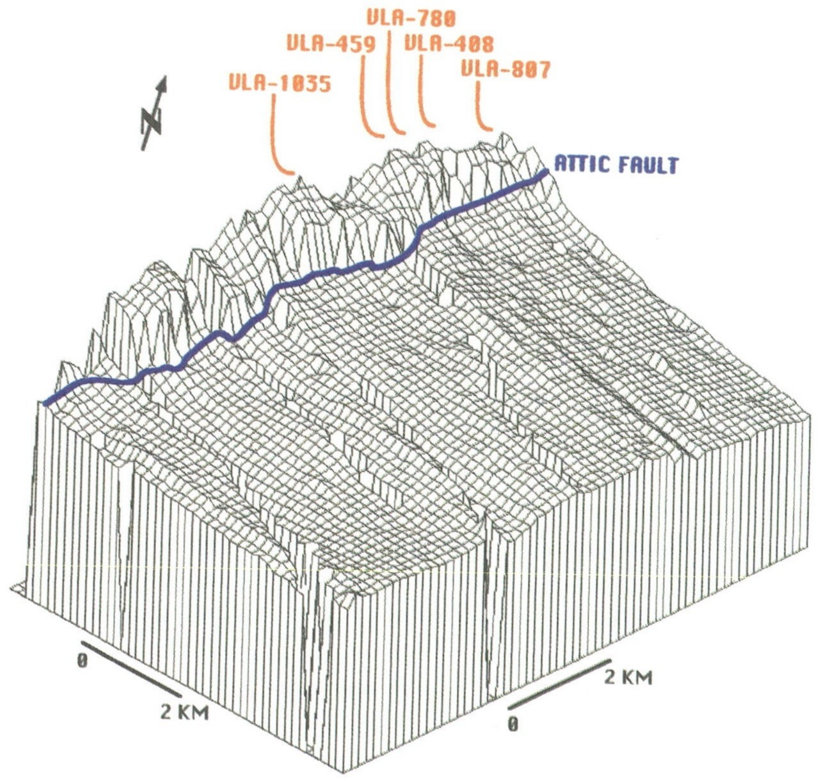


FIGURE 22. Isometric projection of the northern Attic fault block at the top of the Guasare Formation, showing the location of the VLA1035 well and four other horizontal wells drilled between 1992 and 1995. The success of the VLA1035 well led to the drilling of the VLA408, 459, 780, and 807 re-entries as horizontal or highly deviated wells in this fault block.

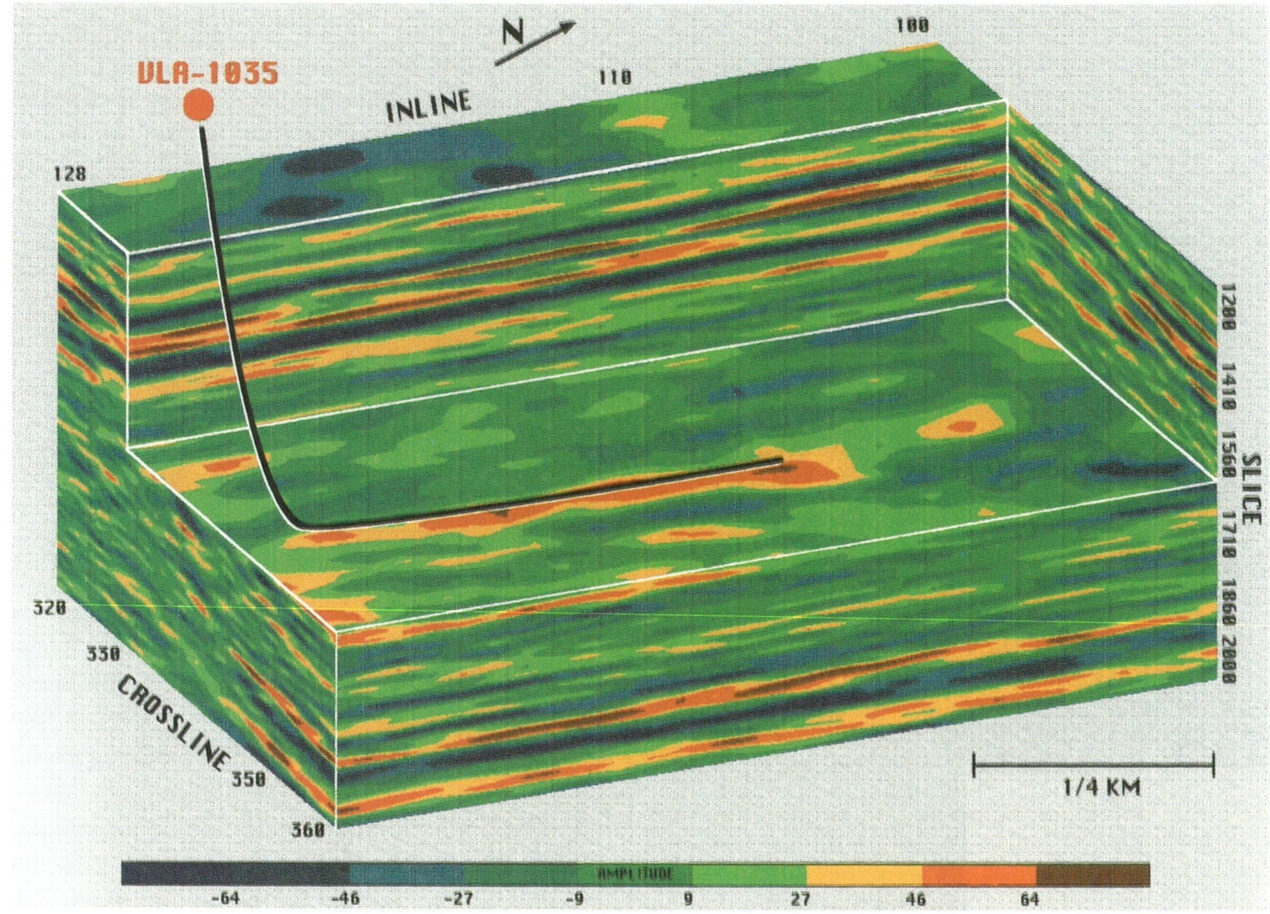


FIGURE 23. Chair display of part of the northern Attic block and the VLA1035 well, highlighting the amplitude trends at the top of the C-7/lower C-6 horizon. These amplitude patterns are interpreted to be a map view of north-trending channels as viewed as a chair–time slice and their discontinuous lenticular channel cross sections in an west-east inline direction. See Figure 21 for location of chair display.

# CHAPTER 8

Mackie, S. I. and C. M. Gumley, The Dirkala South Oil Discovery: Focusing on Cost-Efficient 3-D Seismic Reservoir Delineation, Cooper/Eromanga Basin, Central Australia, *in* P. Weimer and T. L. Davis, eds., *AAPG Studies in Geology 42* and *SEG Geophysical Developments Series No. 5*, AAPG/SEG, Tulsa, p. 83–90.

# The Dirkala South Oil Discovery: Focusing on Cost-Efficient 3-D Seismic Reservoir Delineation, Cooper/Eromanga Basin, Central Australia

S. I. Mackie* and C. M. Gumley**

## ABSTRACT

The Dirkala Field is located in the southern Murta block of Petroleum Exploration Licenses (PELs) 5 and 6 in the southern Cooper and Eromanga Basins of central Australia. Excellent oil production from a single reservoir sandstone in the Jurassic Birkhead Formation in Dirkala 1 had indicated a potentially larger resource than could be mapped volumetrically. The hypothesis that the resource was stratigraphically trapped led to the need to define the fluvial sand reservoir seismically and thereby prepare for future development.

A small (16 km$^2$) 3-D seismic survey was acquired over the area in December 1992. The project was designed not only to evaluate the limits of the Birkhead sand but also to evaluate the cost efficiency of recording such small 3-D surveys in the basin.

Interpretation of the dataset integrated with seismic modeling and seismic attribute analysis delineated a thin Birkhead fluvial channel sand reservoir. Geological pay mapping matched volumetric estimates from production performance data. Structural mapping showed that Dirkala 1 was optimally placed and that no further development drilling was justifiable.

Seismic characteristics comparable with those of the Dirkala 1 Birkhead reservoir were noted in another area of the survey, beyond field limits. This led to the proposal to drill an exploration well, Dirkala South 1, which discovered a new oil pool in the Birkhead Formation. A post-well audit of the pre-drill modeling confirmed that the seismic response could be used to determine the presence of the Birkhead channel sand reservoir.

The acquisition of the Dirkala 3-D seismic survey demonstrated the feasibility of conducting small 3-D seismic surveys to identify subtle stratigraphically trapped Eromanga Basin reservoirs at lower cost and risk than appraisal/development drilling based on 2-D seismic data.

## BACKGROUND

The Dirkala Field is located within the southern Murta block of PELs 5 and 6 in the southern Cooper and Eromanga Basins (Armstrong and Barr, 1982; Thornton, 1979) of central Australia (Figure 1) in an area whose surface geomorphology comprises sand dune fields and interdune clay pans. The field produces oil from the Jurassic Birkhead Formation and gas from the Permian Epsilon Formation (Figure 2). In 1992 the remaining oil reserves in the Birkhead reservoir were estimated at 290 million stock tank barrels (MSTB) (55 mL) and Dirkala 1, the crestal and only oil well, was producing at a rate of 100 BOPD (19 kLOPD). Production data indicated a potentially larger resource than mapped, which required an additional development well for efficient drainage. Geological analysis suggested that the oil was stratigraphically trapped, highlighting the need to seismically define the reservoir.

Available 2-D seismic data were reviewed. Small amplitude anomalies detected near Dirkala 1 in the Birkhead Formation (Birkhead) and underlying Hutton Sandstone (Hutton) interval were believed to be indicative of sand development, but the available data were too sparse (1.0 × 0.5 km grid) to adequately resolve the possible reservoir distribution. A 16-km$^2$ 3-D seismic survey was proposed to assist in determining the geometry of the reservoir and, if possible, in characterizing reservoir properties (Figure 3).

The survey was also proposed to evaluate the strategic and economic implications of recording small, low-cost (<AUS$200,000) 3-D surveys in the Eromanga Basin. There are several small (one- or two-well) Jurassic oil fields in the basin. This survey was a test case to determine whether small surveys can delineate producing reservoirs cost-efficiently. Prior to Dirkala, only much larger and more expensive 3-D surveys had been acquired.

Acquisition was undertaken over three days in December 1992, using four swaths of six lines, each with 96 channels per line. There was a three-line overlap, so the data were recorded at a nominal 24 fold with no duplicate raypaths. Source and receiver lines were perpendicular to each other, with the source lines designed to be acquired in the interdune zones, where feasible, for environmental reasons. The data were processed using a standard 2-D processing stream, with the addition of 3-D shot-receiver azimuth dip moveout (DMO) and a single-pass 3-D migration. The project was completed in April 1993. The data volume was then loaded onto a seismic interpretation workstation and interpreted over the following year.

## SEISMIC INTERPRETATION AND MODELING

### STRUCTURAL INTERPRETATION

Although not the primary focus, a review of previous structural interpretations was conducted. Previous interpretations indicated a lunate structure (Figure 3) with basement fault control. The time interpretation was straightforward, as key reflectors could be readily identified and tied to the stratigraphy. A comparison with the previous 2-D interpretation, however, showed shifts in time highs at both the Jurassic and Permian levels (Figure 4), demonstrating the better resolution of 3-D DMO and migration on the low-relief unfaulted structure.

For the low-relief Dirkala structure, velocity analysis was critical to the structural interpretation. Velocity models were produced both from average and interval well velocities and from average and interval stacking and migration velocities. These models were compared in an effort to understand the sensitivity of depth to velocity and hence to reduce the risk involved in the structural analysis.

### STRATIGRAPHIC INTERPRETATION

Dirkala 1 was the only well in the area that had encountered Jurassic pay. Consequently, the stratigraphic interpretation centered on understanding the reservoir seismic response and comparing it with that of other wells. To understand the seismic wavelet correctly at and around the reservoir level, various bandpass filters were convolved with the reflection coefficients from the Birkhead/Hutton interface. As the bandpass was widened (increasing the higher-frequency content), the peak within the Birkhead split into two peaks (Figure 5). Closer investigation indicated that the first peak appeared to be related to the velocity–density contrast at the shale–sand interface at the top of the Birkhead sand. The second peak was generated by the velocity–density contrast at the shale–sand interface at the top of the Hutton. The single peak produced by convolving the reflection coefficient series with a 10- to 55-Hz bandpass wavelet (which was the frequency content of the old 2-D data) failed to resolve the Birkhead sand thickness. The single peak was generated by the addition of the peak generated from the top of the Birkhead sand interface, together with constructive interference from the peak generated at the top of Hutton interface and destructive interference from the trough produced by both interfaces. The actual peak in the 3-D seismic data, at the Dirkala 1 location, is strong and front-loaded with a longer, less defined, trailing peak. From this analysis it was concluded that the signal within the 3-D seismic data (at this level) was very close to zero phase and could be represented by a higher frequency bandpass wavelet.

Investigation of the 3-D data showed areas where similar splits in the seismic peak occurred. Closer investigation using local well

*Santos Ltd., 60 Edward Street, Brisbane, Queensland, 4000, Australia
**Santos Ltd., 101 Grenfell Street, Adelaide, South Australia, 5000, Australia

control, however, indicated that the split was related to a shale section within the Hutton, i.e., the first peak relates to the top Hutton interface and the second peak to a shale–sand interface within the Hutton. This peak splitting was separated, in an areal sense, from that around Dirkala 1. The split in the Birkhead peak relating to the presence of Birkhead sand was mapped around Dirkala 1. This map shows an elongate feature striking northwest–southeast, with minor cross trends at 90° (Figure 6). Dipmeter data over the Birkhead section in Dirkala 1 (Scotford, 1986) are interpreted to demonstrate that the pay sand's dominant depositional direction is from southeast to northwest. This trend is also evident from other dipmeter interpretations in the region. This depositional direction corresponded well with the observed seismic data and enhanced confidence in seismically interpreting the Birkhead sand distribution from the 3-D data.

## SEISMIC MODELING

A series of seismic models was constructed using the higher-frequency bandpass wavelet. These models were based on a modification of the Birkhead aimed at explaining the apparent split in the peak reflector in the Birkhead at the higher frequency. The thickness of the Birkhead section in Dirkala 3, where no sand is present, was comparable with that in Dirkala 1. Dirkala 3 was used as the control well for a "no sand" case.

Using these various 1-D models, a synthetic seismic section was constructed using a 2-D geological model with varying sand thickness. Figure 7 shows the input 2-D model and the interpolated synthetic section compared with the real seismic data. It was concluded from this simple modeling exercise that the split in the Birkhead peak indicated a section of Birkhead sand at least 10 m thick (in order to be resolved seismically; see Sheriff and Geldart, 1982). Dirkala 1 has a Birkhead sequence of 22.5 m of shale, 6 m of sand, and 6 m of shale. Where the lower peak (Hutton) weakens and the Birkhead peak strengthens, it can be interpreted from the 2-D model that the Birkhead stratigraphic section may be 8 m of shale, 15 m of sand, and 1.5 m of shale. A stratigraphic sequence of 18 m of shale and 16.5 m of sand causes the lower peak (Hutton) to disappear altogether. This represents the case where the Birkhead sand is in direct contact with the Hutton sand.

## GEOLOGICAL MODEL AND PAY MAPPING

The regional model for the deposition of Birkhead sands consists of a series of high-sinuosity channels transporting sediment down a low-gradient paleoslope (Marcus, 1983). The sediment was sourced from the area of the Dunoon Horst and was transported towards a shallow lake in the Pando and Wooloo troughs (Figure 1). The channel and overbank sediments were deposited on a very sandy paleoslope (Hutton) with lacustrine influence (Birkhead) and are interpreted to have been controlled by climate. During stages of no flow, the sands are interpreted to have been reworked by wind, forming small aeolian mounds. Lateral and down-dip seals were created on the edges on the channel belt by lacustrine deposits. Over time, the area was inundated by the shallow lake, depositing a thick shale sequence that produced the top seal of the Birkhead shale.

Using this depositional model and the results of the seismic modeling, it was possible to infer the distribution and thickness of the Birkhead pay sand being accessed by Dirkala 1. The areal distribution was defined by the presence of the split in the Birkhead peak reflector, and the reservoir thickness was related to the time thickness of the subsequent interpeak trough. Volumetric mapping indicated a total original oil in place (OOIP) comparable with the estimated OOIP based on production data, confirming that the modeling was plausible. Dirkala 1 has been drilled at the structural crest of a confined stratigraphic pool, and production information indicated that this single well should be able to produce the total accumulation. The 3-D data and interpretation had saved the cost of drilling an unnecessary development well.

Interpretation also showed that the Birkhead reservoir sand was present farther south of Dirkala 1. Approximately 1.5 km south of Dirkala 1, a small untested structural closure that had the interpreted thickest accumulation of the Birkhead pay sand at its crest was identified. The Dirkala South prospect was interpreted to be a stratigraphic trap similar to Dirkala 1 and to have a maximum potential to contain approximately 4 million bbl (764 ML) OOIP. Trap configuration, reservoir presence, and characteristics, along with the lateral and vertical seals, were interpreted from the modeling and seismic attributes. Analysis of the migration pathway for the oil indicated Permian sourcing south of the prospect (Heath et al., 1989). The hydrocarbons had migrated up faults into the Hutton and, as there was no regional seal, continued into the Birkhead sand. Seismic modeling demonstrated that these migration windows could be identified by the lack of the top Hutton seismic reflector. Several such areas were outlined on the 3-D seismic in and around the prospect.

## DISCOVERY

The risks involved in trapping hydrocarbons (i.e., closure, reservoir, seal, and hydrocarbon charge) were minimized enough to allow reward to outweigh risk, and hence to warrant the drilling of this exploration target.

Dirkala South 1 was drilled in April 1994 and reached a total depth of 2149 m KB in pre-Permian sediments. The primary objective of the well, the Birkhead sand, was present and was intersected 1.5 m lower than prognosed. An 18.3-m core was cut over the Birkhead/Hutton sand interval. Drill stem test (DST) 1 over the interval 1626 m–1639 m KB flowed 49.4° API oil at 1080 BOPD (206 kLOPD) and recovered 60 bbl (11.5 kL) of oil and 15.5 bbl (3.0 kL) of water. The well also discovered another new oil pool in the Murta Member of the Mooga Formation when DST 2, over the interval 1376 m–1402 m KB, recovered 17.8 bbl (3.4 kL) of 46.9° API oil. A new gas pool discovery was also made when DST 3B, over the upper Patchawarra Formation 65-4 sand (1964 m–1979 m KB), flowed gas at 1.48 MMCFGD (42 $km^3$GD) with 73 BWPD (13.9 kLWPD). A Repeat Formation Tester (RFT) survey of the Epsilon Formation 62-2 sand showed it to be gas-saturated but depleted via production from the main Dirkala Field.

Final log and core analysis showed oil over 5 m in the Birkhead Formation. An oil–water contact (OWC) was present in the Birkhead Formation at –1608 m SS, 9 m lower than the lowest known oil (LKO) in Dirkala 1 (Figure 8). The well was cased and suspended for completion as a Birkhead Formation producer and as a future Murta Member oil well, with the possibility of conversion to Patchawarra Formation gas production at a later date.

## POST WELL AUDIT

The preferred model at the Dirkala South 1 location was that of the Birkhead channel sand with a gradational top and a shale below, which gave rise to a weak upper peak and a strong lower peak. The model forecast a total Birkhead thickness of 30 m, consisting of a 5-m shale, a gradational top to the reservoir sand with 12 m of silty sand above 8 m of good-porosity sand, and a 5-m basal shale. This could be expressed in the form of a ratio, viz., 5:12:8:5. The well actually encountered a 24.5 m Birkhead section, with a breakdown of 6.5:9: 5:4. This result approximates the prognosed ratio. The aim of the audit was to determine whether the thickness of the good porosity Birkhead sand could be estimated from the seismic data. Several key geophysical picks, based on the impedance logs and correlatable from well to well, were used for the geophysical analysis (Table 1).

The upper peak within the Birkhead package is interpreted to be generated by the top of the Birkhead sand. This is demonstrated by the Dirkala 1 synthetic (Figure 9). Underlying the Birkhead sand is a basal Birkhead shale, which is resolvable only as a minor trough (due to its thickness). The trailing peak is generated by the contrast at the top of the massive Hutton sand, about 8 m below the true stratigraphic top Hutton. The response of the Dirkala South 1 synthetic (Figure 10) is similar to that of Dirkala 1. The peak-to-peak separation, however, is greater than that generated solely from the thinning of the overlying Birkhead shale.

The subtle differences between the two wells are shown in Figure 11 and are detailed below. The upper Birkhead shale is much thicker in Dirkala 1 and has a noticeable low-impedance package above (Figures 9 and 11). This separates the top Birkhead trough response, with a minor peak above it, from the sharp upper boundary of the low-impedance package. Dirkala South 1 has a much thinner

**TABLE 1. Legend for Post-Well Audit Seismic Modeling***

| Abbreviation | Description |
|---|---|
| base Dn | A change in impedance (from low to high) at the base of the Namur Sandstone |
| Birkhead | The top of the Birkhead Formation |
| ts | The top of the Birkhead Formation sand package |
| tgs | The top of the good sand development at the base of the Birkhead Formation sand package |
| bgs | The base of the good sand development at the base of the Birkhead Formation sand package |
| Hutton | The geologic pick at the top of the Hutton Sandstone |
| f2s | The onset of the massive clean Hutton sand, marked by a decrease in acoustic impedance |
| s2f | A change in impedance (from low to high) within the Hutton Sandstone |
| f2f | A change in impedance (from low to high) above top Toolachee Formation |

*See Figures 9–15.

Birkhead shale, and the low-impedance package above the shale has a gradational upper boundary. Therefore, the top Birkhead is not a trough proper but a minor trough within a peak doublet (Figures 10 and 11). In both wells the top of the Birkhead sand is represented by a characteristic peak, the difference being that the top of the sand is more gradational and thicker in Dirkala South 1 (Figure 11).

The Birkhead sand–basal Birkhead shale interface imaged in Dirkala 1 generates a minor trough, superimposed on the peak doublet. This response is generated, as the sand–shale layers are thin and destructively interfere. The lower peak is the response from the massive sand about 8 m below the true top of the Hutton (Figures 9 and 11). It is possible in Dirkala 1 to infer the thickness of the Birkhead sand, as the top of the sand generates a peak and the base is constrained by the trough from the basal shale and the onset of the trailing peak from the massive Hutton sand. At Dirkala South 1, the basal Birkhead shale is slightly thicker and, together with the upper Hutton, generates a true seismic trough. In both Dirkala 1 and Dirkala South 1, the trough is seismically shifted upwards in relation to the impedance logs due to interference. At Dirkala 1, the basal shale has a gradational base, and hence a trailing peak is not generated and does not destructively interfere. At Dirkala South 1, there is an interval of about 18 m from the true top of the Hutton to the massive Hutton, which generates the second peak (Figure 11). The thin, very clean sand (low impedance) immediately below the basal Birkhead shale in Dirkala South 1 (Figure 10) generates a peak–trough response that interferes and does not contribute. The thickness of the alternating sand–shale sequence in the top 18 m of the Hutton does not generate any recognizable seismic response.

The response at Dirkala South 1 is therefore generated by a combination of a thinner top Birkhead shale plus a thickening of the lower-quality top Hutton section. The top of the thicker basal Birkhead shale is seismically resolvable as a trough, and the extra 18 m of sand–shale at the top of the Hutton, above the massive, clean sandstone, contributes to the trough–peak separation below the Birkhead. These observations were tested by splicing the Dirkala South 1 Birkhead logs (top Birkhead to top Hutton) into the Dirkala 1 logs (Figure 12). Here the response was as previously modeled, i.e., without the complication of the varying top Hutton. The separation in peaks is due to the thinner top Birkhead shale and the better development of the Birkhead sand. The intervening trough is generated from the basal Birkhead shale.

The observation that the Birkhead sand could be identified from the peak/trough relationship was tested using interpolation modeling. The results from the modeling are displayed in Figure 13, which has been flattened on the top Birkhead pick. The top of the Birkhead sand unit is recognizable as a peak, followed by a trough representing the basal Birkhead shale. The Birkhead sand package thins towards Dirkala 3, as does the top Hutton section, between the true top and the top of the massive sand, until the Dirkala 3 response is achieved (Figure 14). The Dirkala 3 response is clearly different, with no true top peak generated for the top of the Birkhead sand, as the sand is of poor quality and virtually absent. The main peak is generated from the top of the massive Hutton sand, which is about 10.5 m below the top of the true Hutton. The intervening trough, seen in both Dirkala 1 and Dirkala South 1, is not developed. This response is significantly different from the wells containing the Birkhead sands. The results of the modeling can be related to the actual seismic line tying the wells flattened on the Birkhead pick (Figure 15).

The audit shows that the presence of a split in the Birkhead peak response indicates the presence of sand (if the seismic response is tied to well control). It is also possible to estimate the thickness of the section between the base of the top of the Birkhead shale and the base of the basal Birkhead shale from the seismic response.

## ECONOMICS

The original acquisition of the 3-D seismic dataset had two primary objectives. The first was to delineate and characterize the Birkhead reservoir sand. The success of that objective was demonstrated by the discovery of oil in Dirkala South 1 in a reservoir section similar to that prognosed. The second objective was to determine the cost efficiency of small 3-D surveys in the Eromanga Basin.

The cost of the acquisition, processing, and interpretation of the Dirkala 3-D dataset was AUS$226k. The major results of the survey to date have been the prevention of drilling an unnecessary development well and the new field discovery at Dirkala South. As the discovery was made in an area of the basin where production infrastructure is present, production began almost immediately and at very little incremental cost. Using simple discounted cash flow methods, the Dirkala 3-D seismic survey shows a positive return on investment based solely on the incremental value of the Birkhead oil discovery, without taking into account any contribution from the Permian gas discovery or from the extension of the Dirkala 1 Birkhead oil pool or the new Murta oil pool. This demonstrates that the second objective of the survey was satisfied.

## CONCLUSIONS

This review demonstrates the positive use of 3-D seismic data in the Eromanga Basin—especially that such a small amount of low-cost 3-D seismic data over Jurassic oil pools in the basin can be a very cost efficient appraisal and development tool. The most likely use of such 3-D data is during a field's appraisal stage, because without seismically calibrated well control, lateral prediction of stratigraphy is likely to be complex.

The interpretation of the Dirkala 3-D dataset has led to several positive outcomes. The interpretation of the survey delineated the productive Birkhead reservoir around Dirkala 1, which resolved the questions of ultimate recoverable reserves and further development drilling. It also successfully delineated another subtle, stratigraphically trapped hydrocarbon pool. Combined with seismic modeling, the 3-D data were used to achieve the characterization of a relatively thin stratigraphic reservoir sand development in a cost-efficient way.


### *Acknowledgments*

The authors would like to thank the Joint Venture parties in the Murta Block of PELs 5 and 6 (Santos Ltd., Delhi Petroleum Pty. Ltd., Boral Energy Resources Ltd., and Crusader Resources NL) for permission to publish this paper.


## REFERENCES CITED


Armstrong, J. D., and T. M. Barr, 1982, The Eromanga Basin, *in* P. S. Moore, and T. J. Mount, compilers, Eromanga Basin Symposium, Summary Papers: Geological Society of Australia and Petroleum Exploration Society of Australia, p. 20–42

Heath, R. S., S. McIntyre, and N. Gibbins, 1989, A Permian origin for Jurassic reservoired oil in the Eromanga Basin, *in* B. J. O'Neil, ed., The Cooper and Eromanga Basins, Australia: Proceedings of Petroleum Exploration Society of Australia, Society of Petroleum Engineers, Australian Society of Exploration Geophysicists (SA Branches), Adelaide, p. 405–416.

Marcus, G., 1983, The stratigraphy, paleogeography and hydrocarbon potential of the Birkhead Formation in South Australia: Flinders University, Adelaide, South Australia, B.Sc. Thesis.

Scotford, G. L., 1986, Dirkala 1 well completion report: Delhi Petroleum (unpublished). Available from Department of Mines and Energy, Adelaide, South Australia.

Sheriff, R. E. , and L. P. Geldart, 1982, Exploration Seismology, v. 1, Cambridge, Mass., Cambridge University Press, p. 117–122.

Thornton, R. C. N., 1979, Regional stratigraphic analysis of the Gidgealpa Group, Southern Cooper Basin, Australia: Department of Mines and Energy, Geological Survey of South Australia, Bulletin 49.

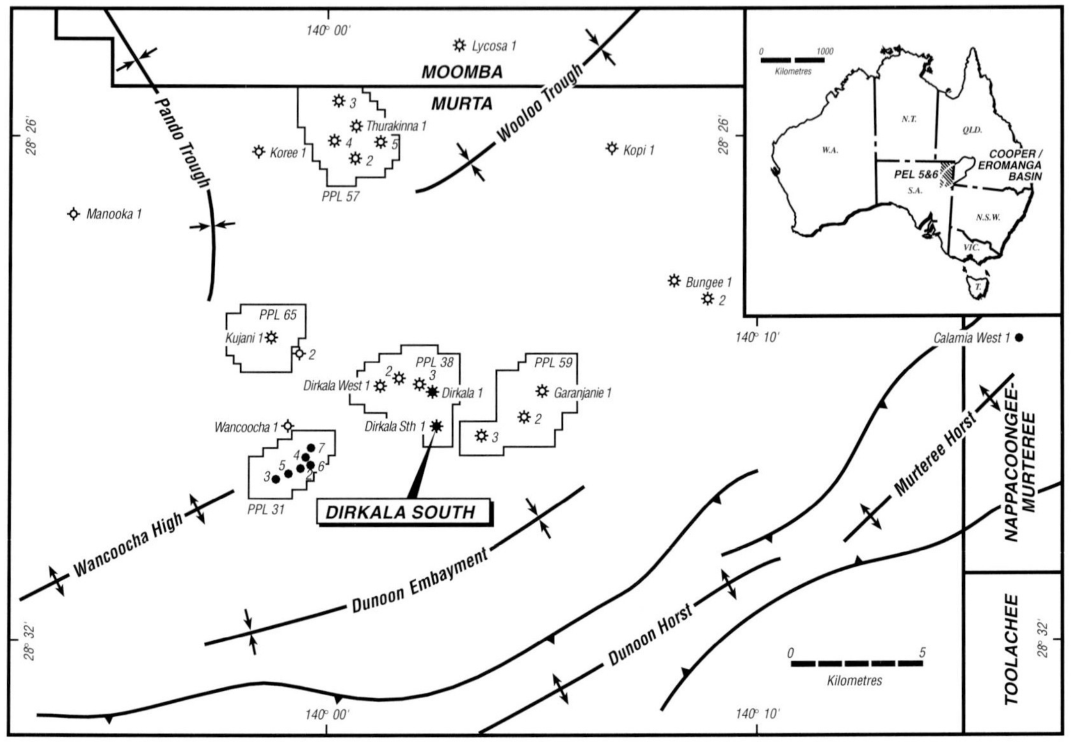


FIGURE 1. Location map of Dirkala South 1 and southern Murta Block, Cooper/Eromanga Basin, Central Australia.

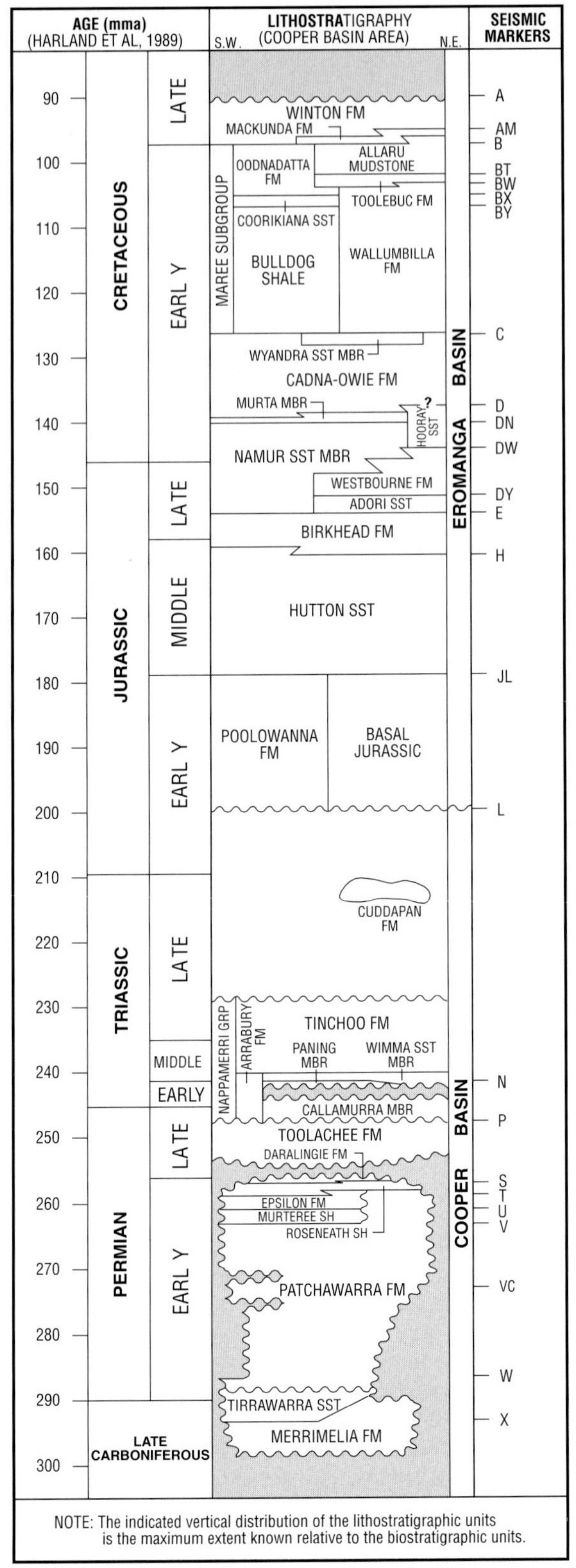


FIGURE 2. Cooper/Eromanga Basin lithostratigraphic table.

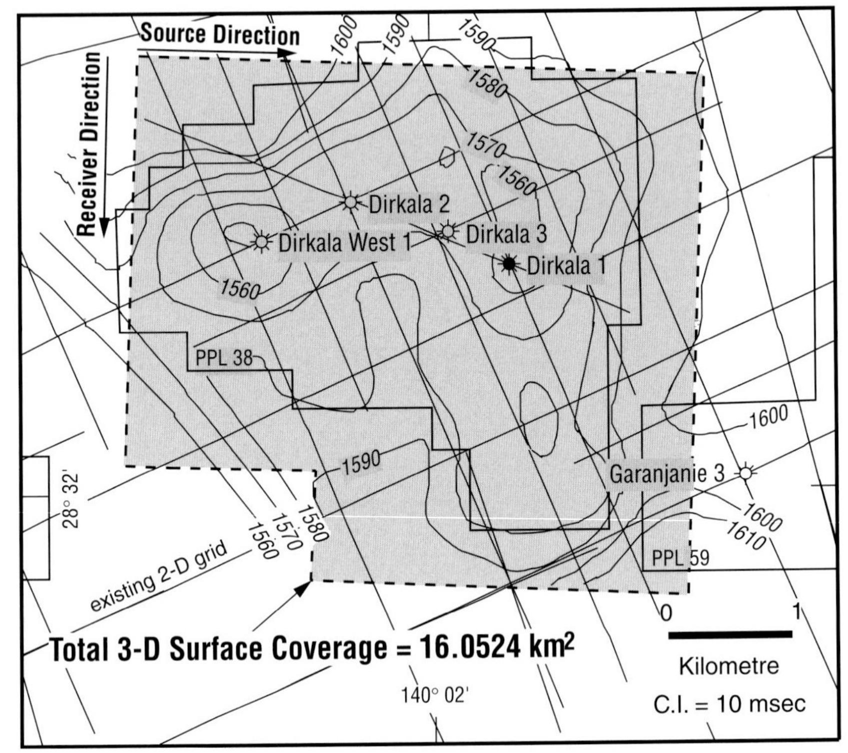


FIGURE 3. Time structure map of Patchawarra Formation (2-D interpretation), showing the 3-D area outline and existing 2-D grid.

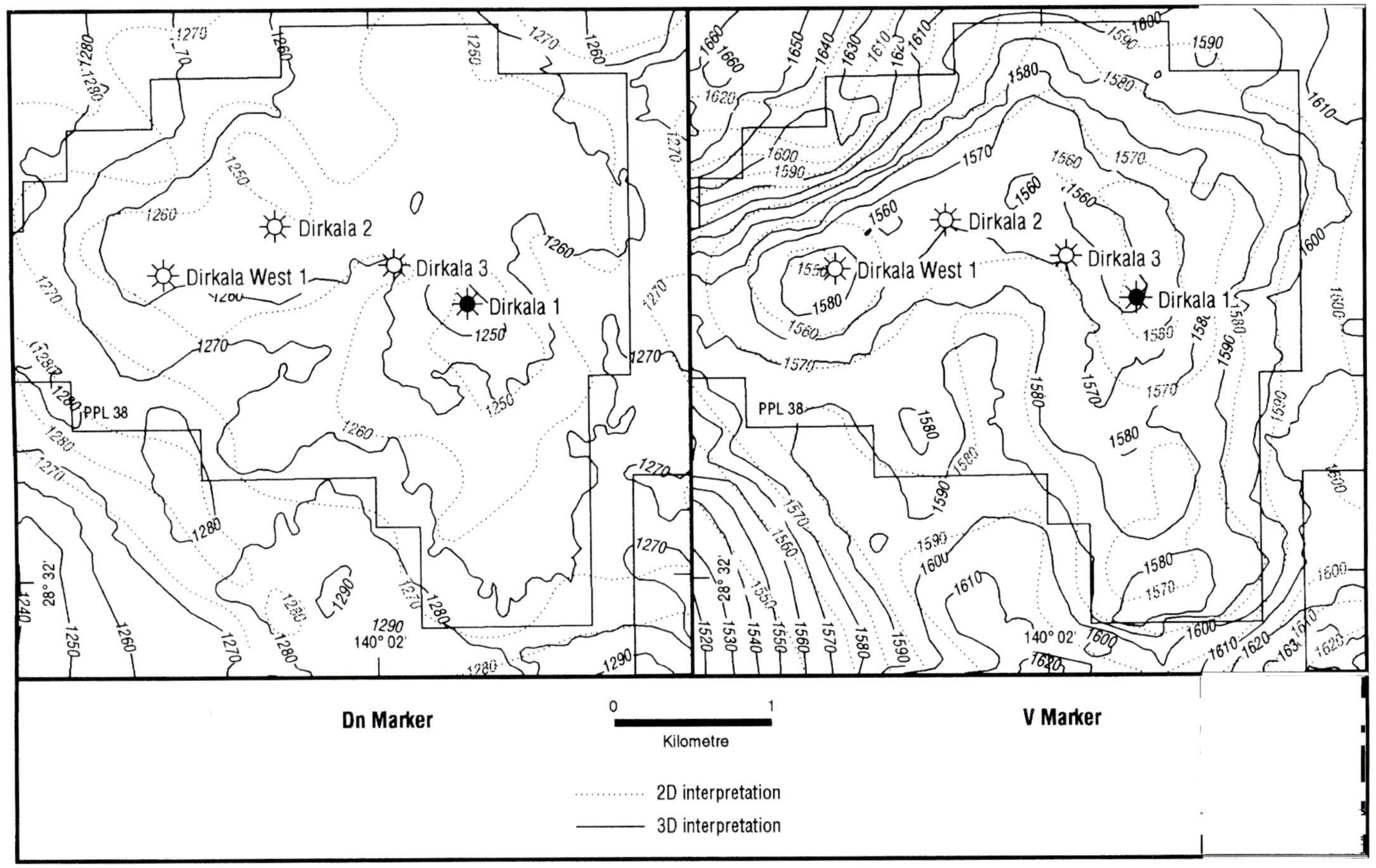


FIGURE 4. Comparison of 2-D and 3-D time interpretation.

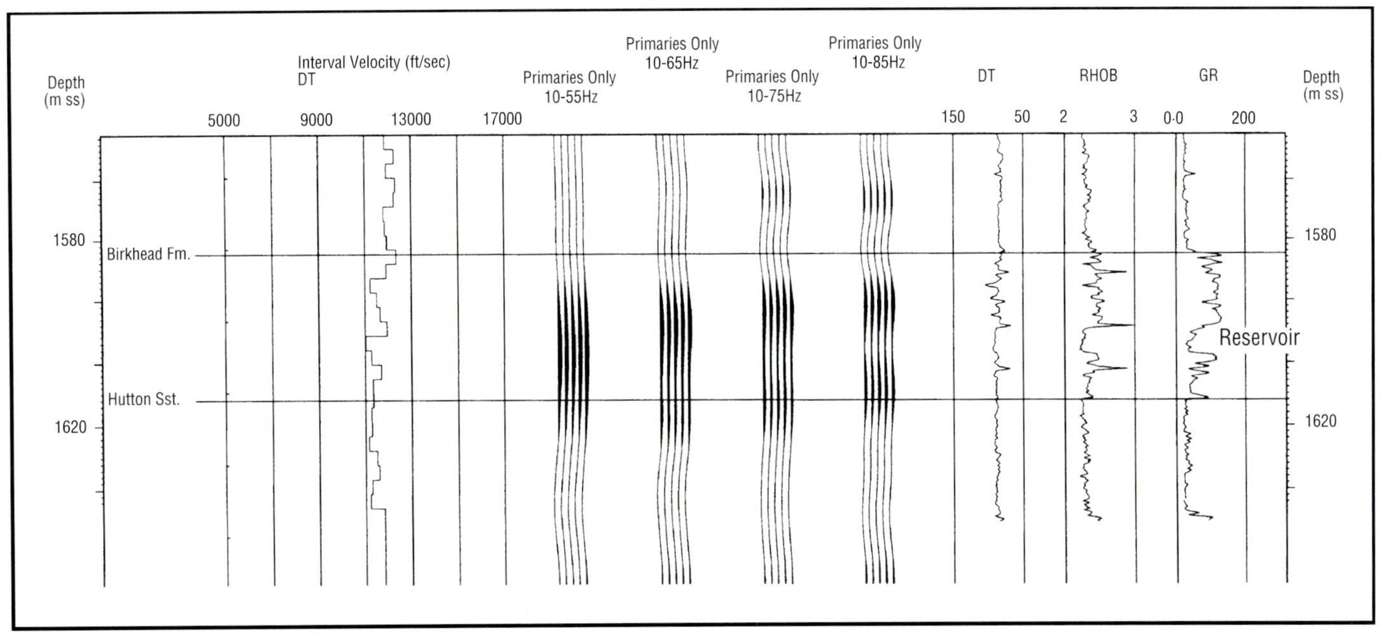


FIGURE 5. Synthetic seismogram for Dirkala 1.

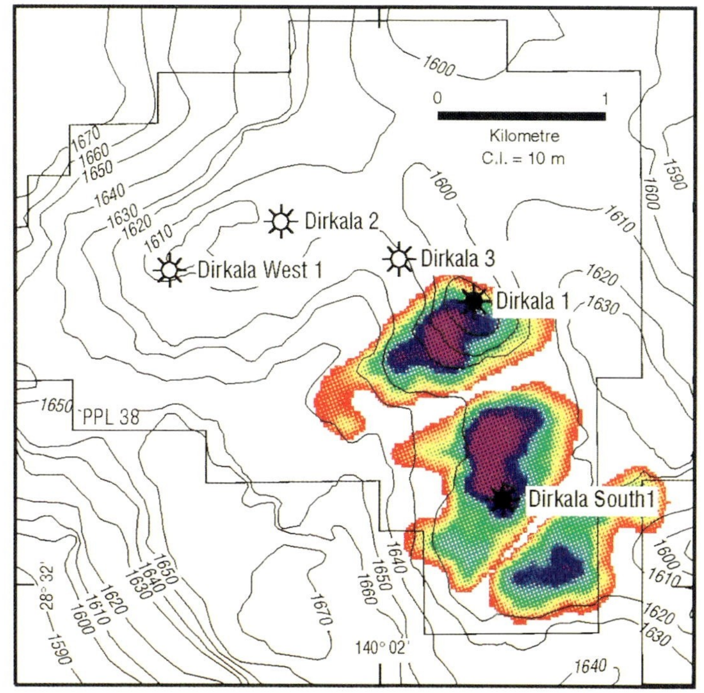


FIGURE 6. Depth structure map of Birkhead Formation with Birkhead sand thickness underlay derived from the time thickness of the split in the seismic peak.

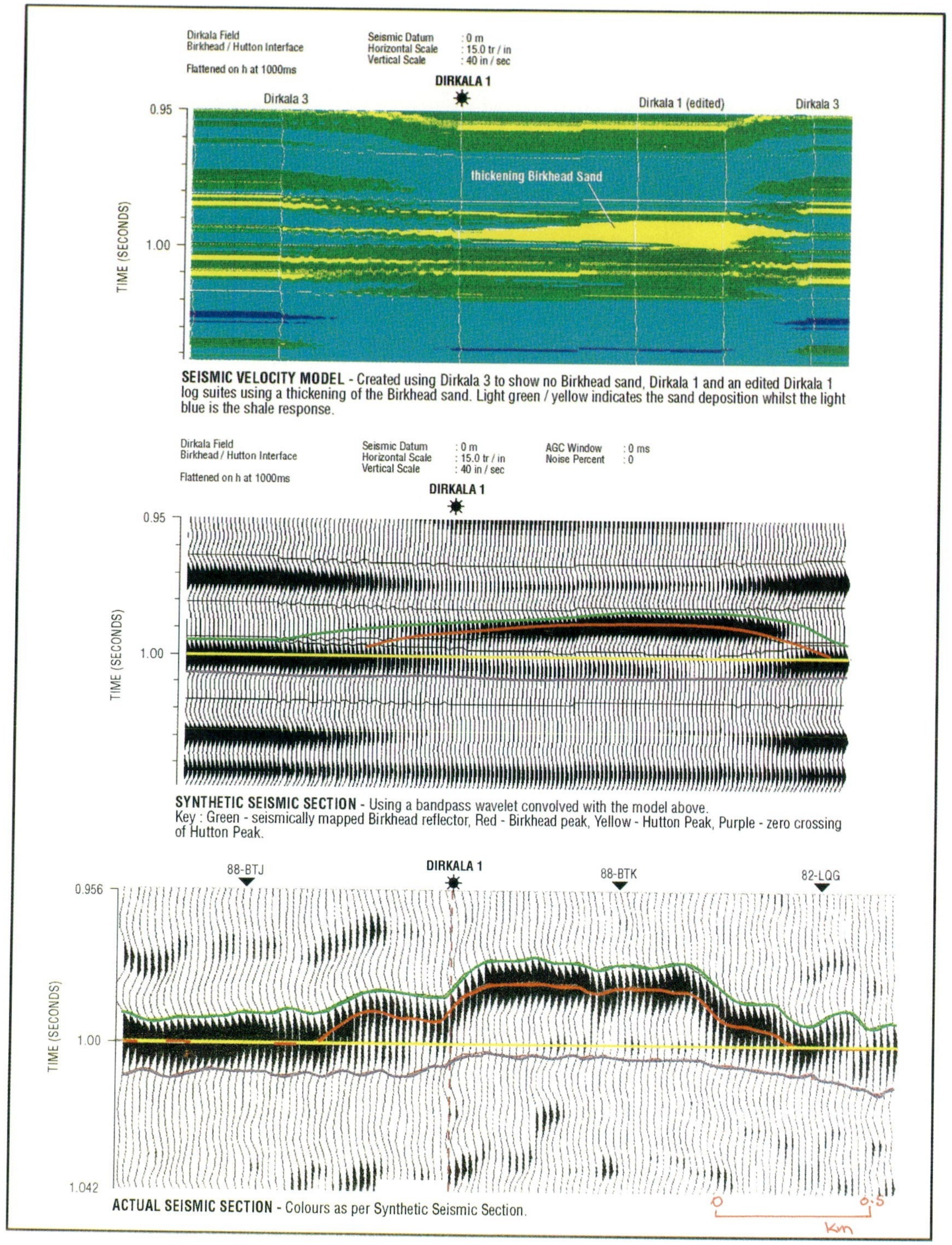


FIGURE 7. Seismic modeling of the Birkhead Formation, showing model, synthetic seismic section, and real seismic section.

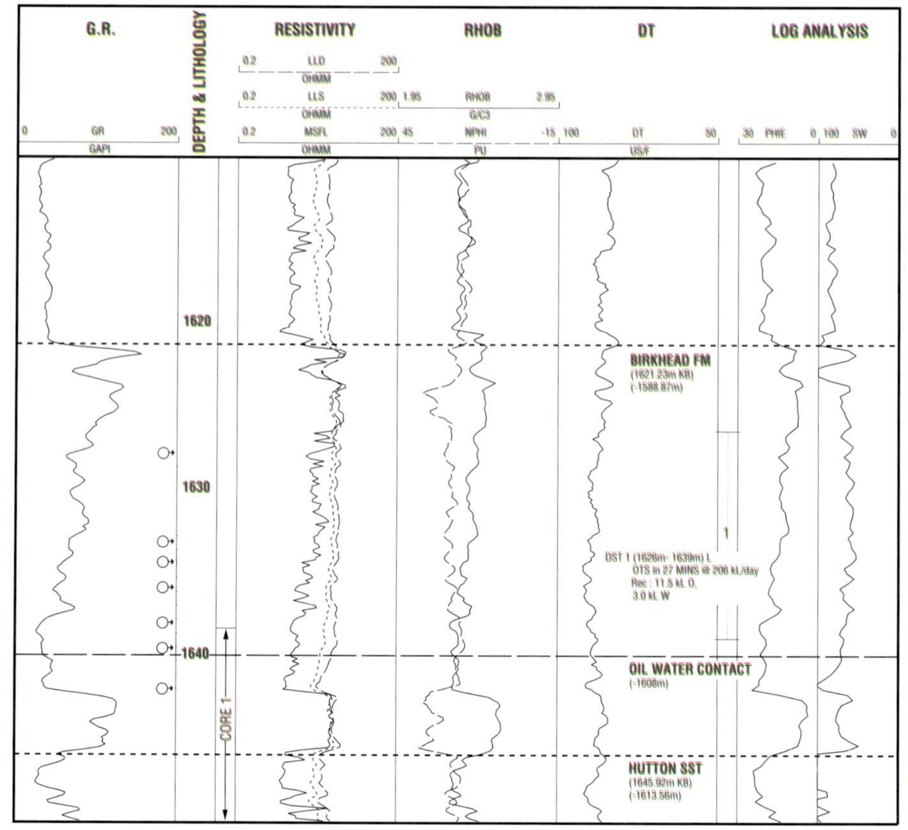


FIGURE 8. Dirkala South 1 petrophysical logs.

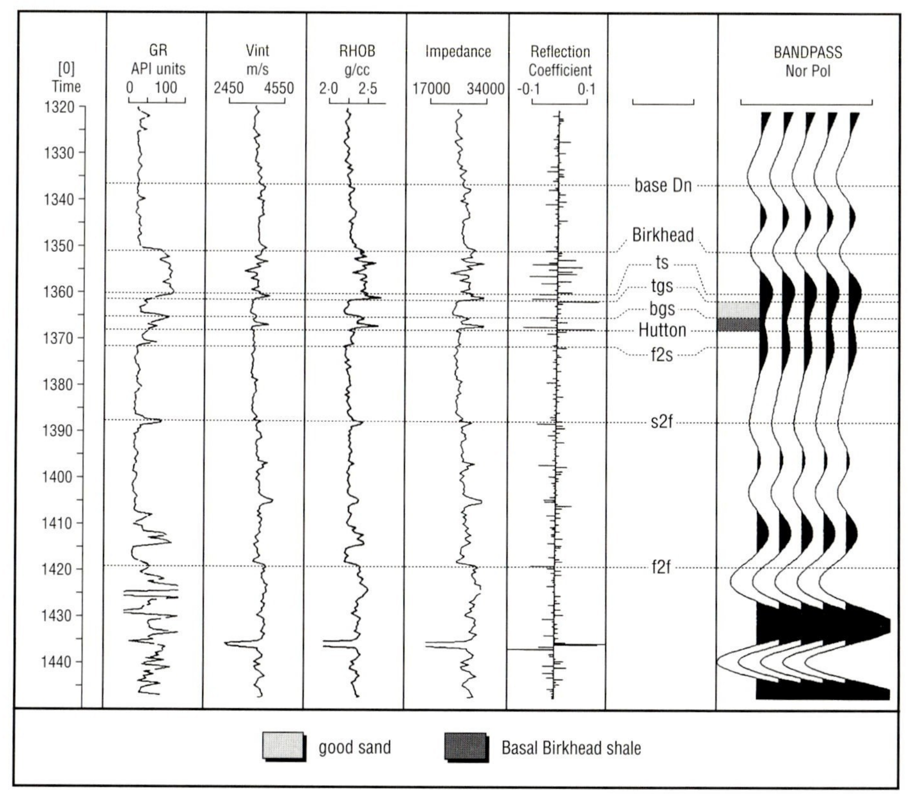


FIGURE 9. Dirkala 1 synthetic seismogram.

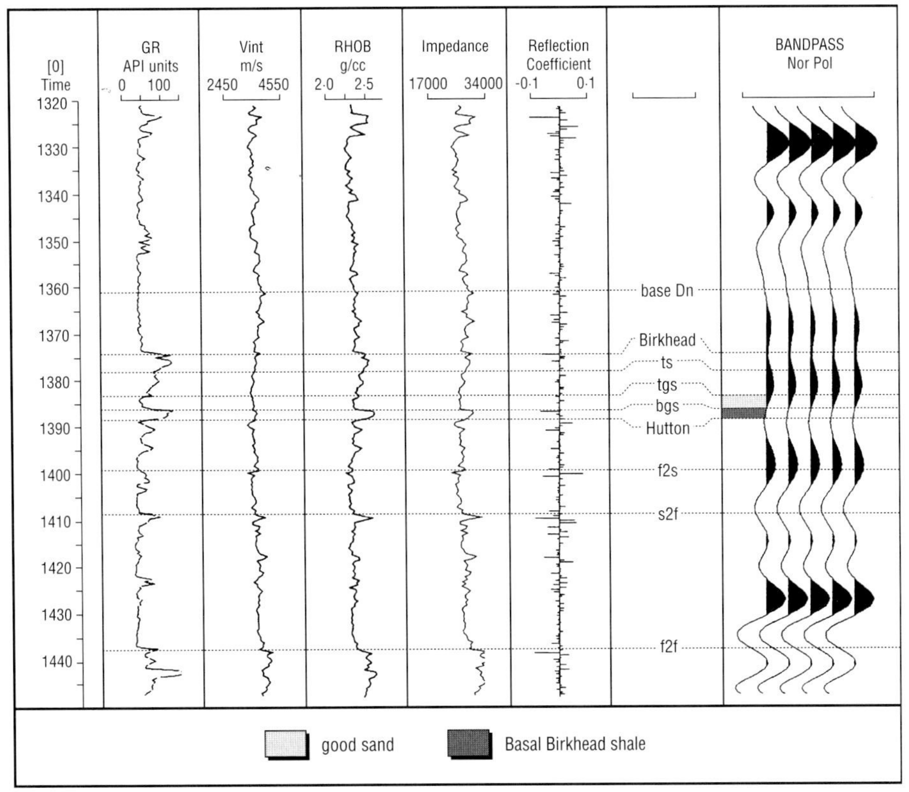


FIGURE 10. Dirkala South 1 synthetic seismogram.

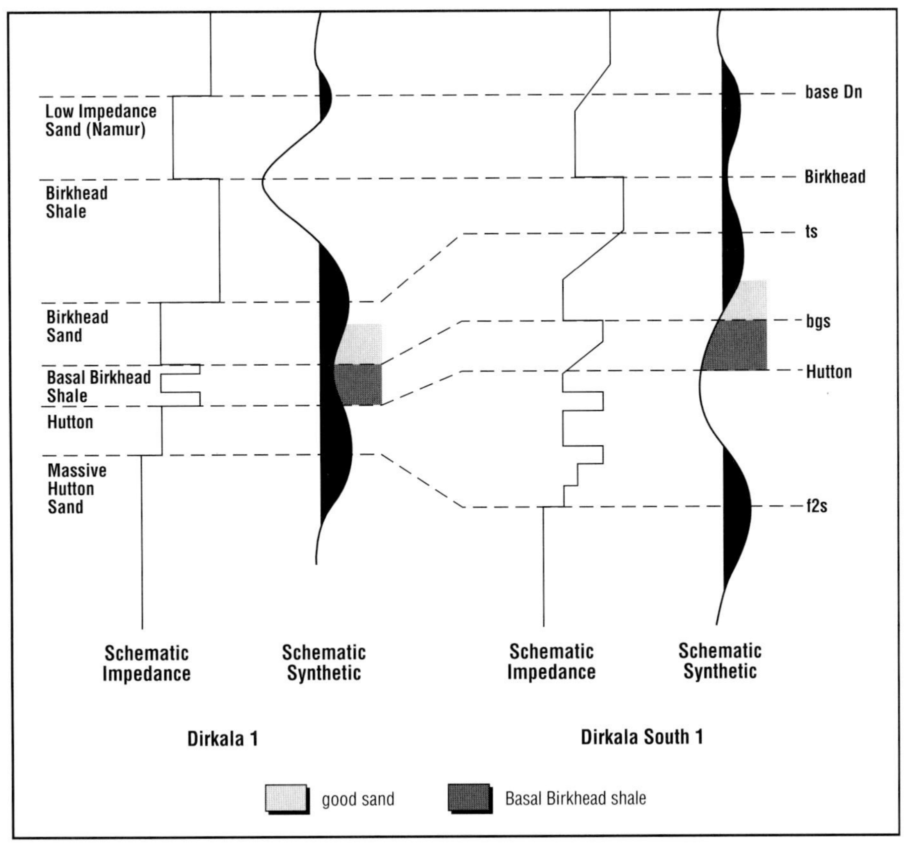


FIGURE 11. Schematic comparison of Dirkala 1 and Dirkala South 1 seismic response.

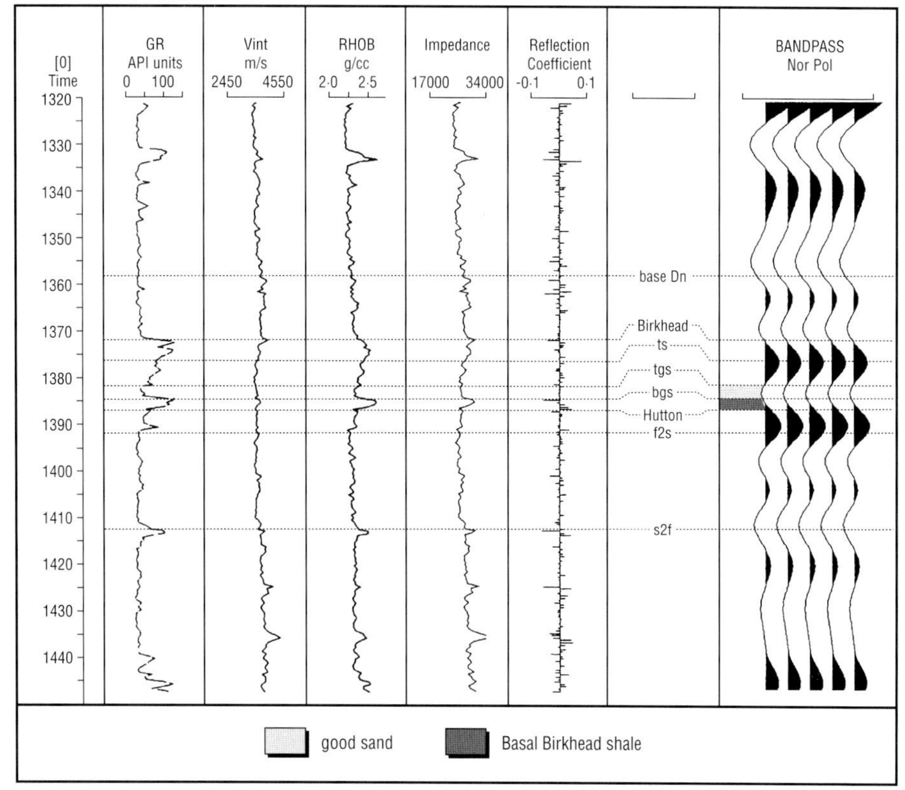


FIGURE 12. Modified Dirkala 1 synthetic seismogram. Dirkala 1 logs above the top Birkhead and below the top Hutton. Dirkala South 1 logs between the top Birkhead and the top Hutton.

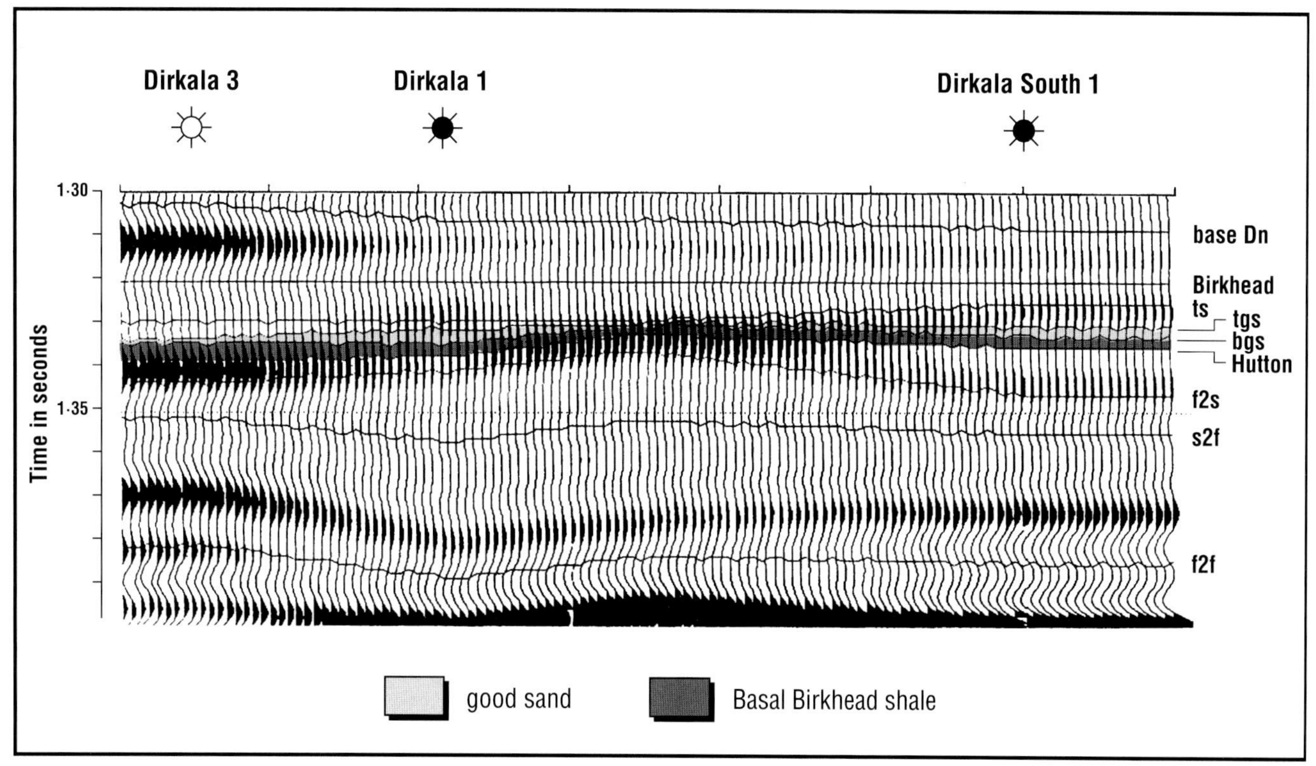


FIGURE 13. Interpolation model of Birkhead sand.

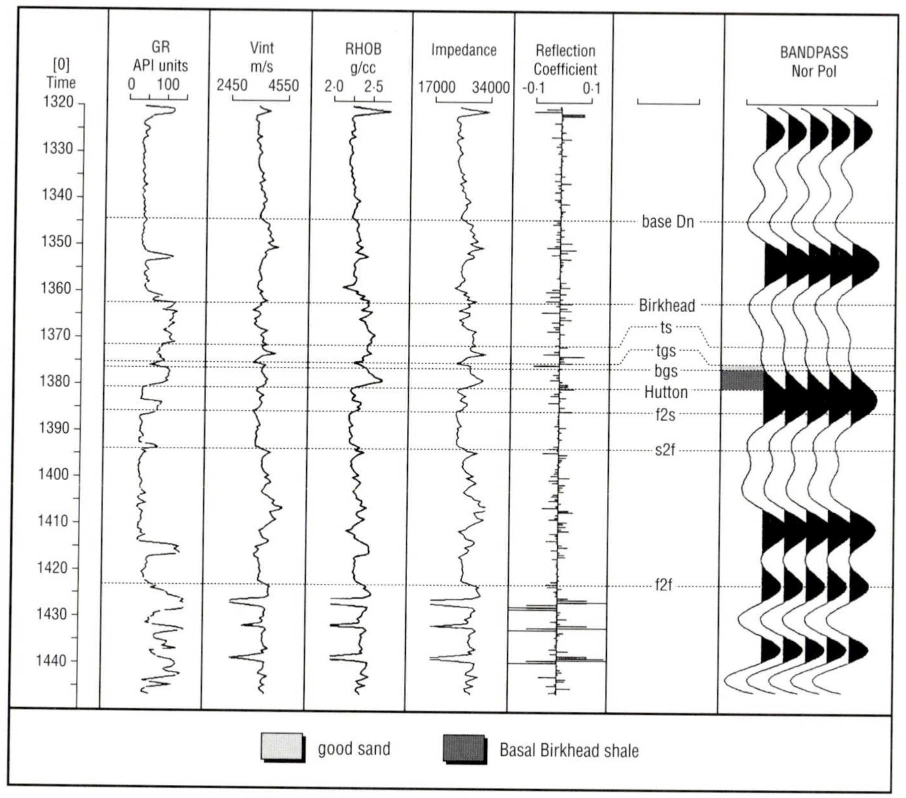


FIGURE 14. Dirkala 3 synthetic seismogram.

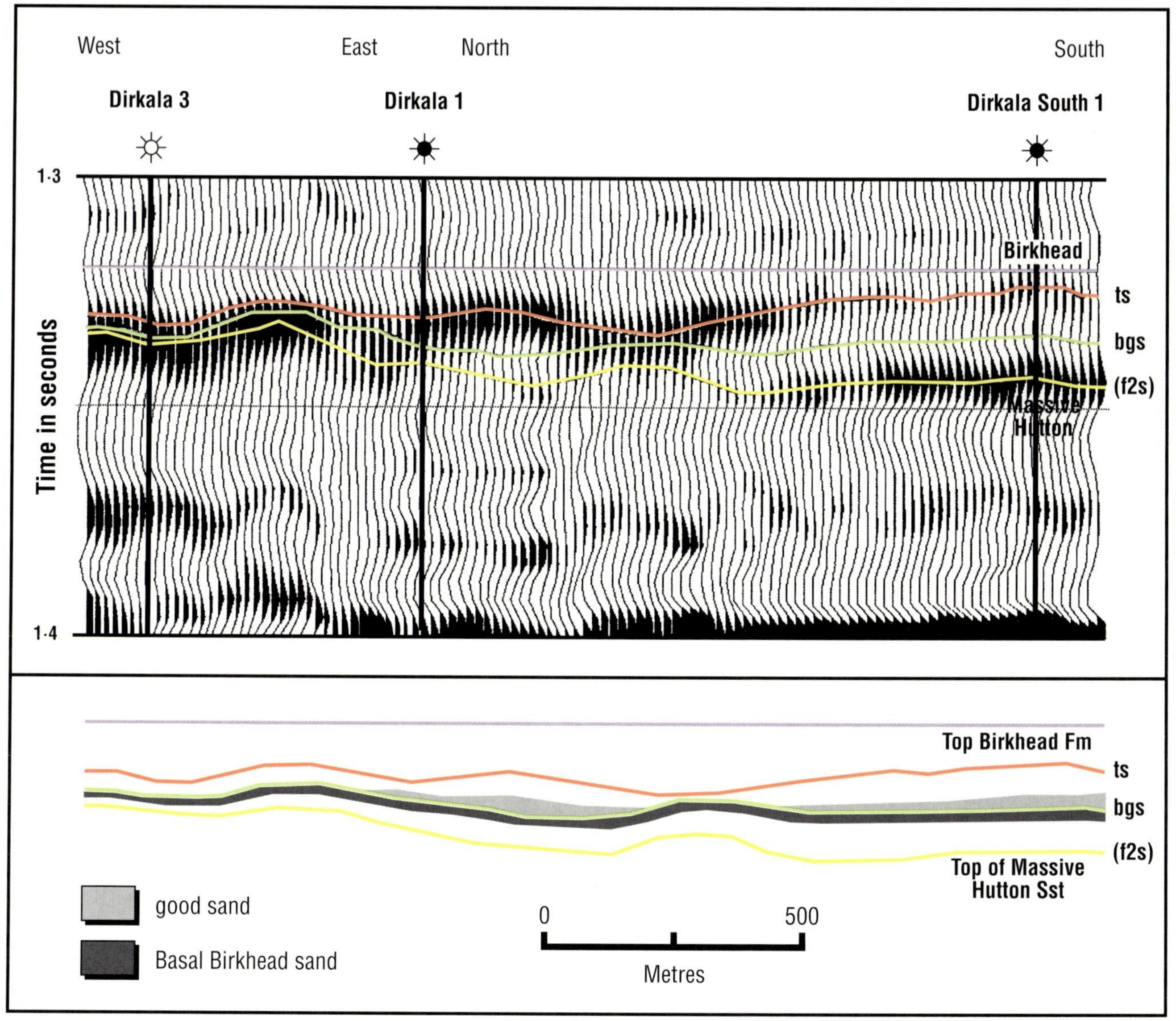


FIGURE 15. Arbitrary seismic traverse between Dirkala 3, Dirkala 1, and Dirkala South 1, flattened on the Birkhead, showing current interpretation.

# CHAPTER 9

# 3-D Evaluation of the Ping Hu Field, East China Sea

Silverman, M. R., Qin Wang, B. L. Byrd, Zhao Jiong Kun, Zhang Hai Ying, Chen Ming, and P. Weimer, 3-D Evaluation of the Ping Hu Field, East China Sea, *in* P. Weimer and T. L. Davis, eds., *AAPG Studies in Geology No. 42* and *SEG Geophysical Developments Series No. 5*, AAPG/SEG, Tulsa, p. 91–102.

**Matthew R. Silverman,* Qin Wang,* Bruce L. Byrd,* Zhao Jiong Kun†, Zhang Hai Ying,† Chen Ming†, and Paul Weimer‡**

## ABSTRACT

Ping Hu Field, discovered in 1982 in the East China Sea, 365 km offshore China, consists of two structural closures on a complex, faulted anticline. Chinese authorities have planned for the field to provide natural gas to Shanghai in the 21st century. Five wells have been drilled, and more than 2400 km of 2-D and 118 km² of 3-D seismic data have been interpreted. Both structures were identified by 2-D data and drilled prior to acquisition of the 3-D seismic data.

Interpretation included mapping key reservoirs and fault analysis. Reservoir composition, distribution, and lateral variations were described. These variations were determined by studying geophysical attributes of the data, which included instantaneous amplitude, frequency, phase, and acoustic impedance. Seismic data were inverted to acoustic impedance (density and velocity) and transformed into reservoir parameters. Crossplot regression analyses of petrophysical parameters were performed, and relationships between pairs of values of porosity, acoustic impedance, permeability, and water saturation were derived.

An integrated geological–geophysical–reservoir engineering evaluation was required for accurate hydrocarbons-in-place calculations. The interpretation was utilized for field development planning, but implementation has been delayed due to distance from shore relative to field size.

## INTRODUCTION

Ping Hu Field (Figure 1) is located in the Xihu Depression of the East China Sea and is bounded on the west by the Haijiao Uplift. The field is approximately 440 km southeast of Shanghai. Discovered in 1982, the field is expected to provide gas to the rapidly growing Shanghai region. Five wells, designated Ping Hu 1 through 5 (PH-1, PH-2, etc.) have been drilled to a maximum depth of over 4600 m. All had high flow rates of oil, gas, and condensate.

More than 2400 line km of 2-D seismic data and approximately 118 km² of 3-D seismic data have been acquired, processed, and interpreted (Figure 2). Seismic data were acquired in several stages, beginning with a 394-km, 2 × 2-km survey in 1982. PH-1, the discovery well, was drilled on the larger Fangheting (F) structure following this survey. In 1985, the Bajiaoting (B) structure was discovered by drilling PH-2. A 1917-km, 1 × 1-km survey was completed from 1985 to 1987, and PH-3 was drilled to confirm the Fangheting structure. PH-4 was drilled on the crest of the Fangheting structure in 1988 following a 132-km, 0.5 × 0.5-km detailed 2-D seismic survey. In 1987, a 3-D survey of 55 km² was acquired on the Bajiaoting structure. Finally, in 1991, a 63-km² 3-D survey was acquired for the Fangheting structure. PH-5 was drilled as a delineation well, based on the 3-D data, in 1992.

*Gustavson Associates, Inc., 5757 Central Ave., Suite D, Boulder, CO 80301
†Shanghai Petroleum Corporation, 1985 Hongqiao Road, Shanghai 200335 China
‡University of Colorado, Department of Geological Sciences, University of Colorado, Boulder, Co 80309

## GEOLOGY

Ping Hu is located on a faulted, complex anticline (Figure 2), immediately east of the Ping Hu main (normal) fault. The field consists of two principal structures: the larger Fangheting structure is 7 km south of the smaller Bajiaoting structure. Closure in the Fangheting structure is about 26 km² in the Hua Gang Formation (oil-bearing reservoirs) and about 22 km² in the Ping Hu Formation (gas-bearing reservoirs). At Bajiaoting, closure is about 4 km² in the Hua Gang and about 7 km² in the Ping Hu. Both structures are complicated by additional minor normal faults and cross faults that may control the later extent of individual reservoirs.

The dominant trapping mechanism for the field is structural and may include several subtle fault-related traps that either restrict the reservoir or control the oil/water contact with their displacement. Combination structural–stratigraphic traps in individual reservoirs and traps in isolated, lenticular sandstones are likely to occur through the field area.

Strata penetrated in the field range in age from Eocene to Quaternary; basement is Cretaceous andesite (Figure 3). Seismic data indicate that the total thickness of the sedimentary section is about 12,000 m. All the hydrocarbons are contained in the Oligocene Hua Gang Formation and the Eocene Ping Hu Formation. The two formations are separated by an unconformity. Marine shales provide viable regional seals, whereas fluvial–deltaic and littoral sandstones are the primary reservoir facies (Figure 4).

The Eocene Ping Hu Formation includes gas and condensate reservoir targets. The unit is characterized by basal volcaniclastics and sandstones overlain by argillaceous sandstones with interbedded dark gray mudstone and minor coals. The succession grades upward into marine facies. The top of the unit is an unconformity (sequence boundary) that has been mapped regionally. Strata were deposited as fluvial, paludal, and littoral sediments, including tidal channels, sandy-to-mixed tidal flats, and littoral sandstones. Cores indicate that common sedimentary characteristics in the Ping Hu Formation are well-developed sandstone facies with fining-upward sequences and cross-bedding. Some sandstones contain bidirectional cross-bedding indicative of tidal channels. The formation ranges in total thickness from about 750 to 1700 m and includes 12 sandstone reservoirs (P1–P12). The lateral continuity in P1, P3, P8, and P11 is good; in all the other reservoirs it is poor to fair. Based on pressure data, log characteristics, and other factors, the Ping Hu has been divided into upper and lower members; the dividing line is the base of the P8 reservoir. The Upper Ping Hu sandstones are fine- to medium-grained. Thin sections from cores indicate that secondary porosity is dominant and usually 17 to 19%. Average permeabilities are about 50 md. Lithological characteristics of the lower Ping Hu are similar, but porosity and permeability average only about 12 to 14% and 10 md, respectively. Conflicts between core and log data render acceptance of these values tentative. The pressure gradient is normal in the upper Ping Hu but about 1.3 times higher than normal in the lower pays.

Condensate concentration decreases with depth, and no significant $H_2S$ is present. In summary, most of these reservoirs (P12 has much lower permeability) should exhibit fair to good production characteristics, but poor lateral continuity presents correlation and production problems in certain zones.

The Oligocene Hua Gang Formation is a clastic shallow lacustrine and fluvial facies, typical of upper delta plain sedimentation. The lithology includes interbedded sandstone, siltstone, and mudstone, with occasional intercalated carbonates and coal seams. An erosional unconformity occurs at the top of the unit. The sandstones were deposited in a complex of lacustrine environments, including river channels and shallow-lake fan deltas. The Hua Gang ranges in total thickness from about 550 to 800 m and includes eight sandstone reservoirs (H1–H8). These sandstones are medium- to fine-grained and quartose with good lateral continuity. Intergranular porosity is often 20 to 25%, and permeabilities in excess of 200 md are common. Average log porosity and permeability are 20% and 341 md, respectively. Lateral continuity is good for most reservoirs, but the H6 (a very important reservoir) may not be continuous throughout the structural closure. The pressure gradient is normal, and edge-water and bottom-water will both be encountered. No significant gas cap is present, and viscosity, density, pour point, and sulphur content of the crude oil are all low. In summary, these reservoirs should exhibit good to excellent production characteristics.

Regional studies indicate that the Paleocene fluvio-lacustrine shales deposited in the deepest trough of the Zhe Dong Depression are responsible for the hydrocarbons generated in the area. Apparently, this section has yet to be penetrated by any well. Hydrocarbon source rocks for both the Hua Gang and Ping Hu formations are thought to be interbedded organic-rich, thermally mature shales. Some source rock contribution from thin coals in both units is also anticipated, but is unquantified. Vertical distribution of hydrocarbons (light oil in the Hua Gang, gas and condensate in the Ping Hu below) reflects a normal geochemical/geothermal history. Variations in the source rock characteristics (kerogen type, total organic carbon, and geothermal maturity, for example) of the two units and the timing of expulsion and migration are not known.

Hydrocarbon seals in both the Hua Gang and the Ping Hu sandstone reservoirs are interbedded low-permeability shales. Many of the reservoirs are lenticular, and the intraformational shales serve both as vertical seals and as elements in combination structural–stratigraphic traps. (Reserves in the more continuous sandstones may be trapped purely by structure.) Marine shales provide viable regional seals. Deltaic shales of the Hua Gang and Ping Hu formations are also acceptable seals. The Ping Hu formation is dominated by shales, with sandstone reservoir composing a minority fraction of the overall stratigraphic thickness. On the other hand, the Hua Gang contains several massive sandstones, and vertical seals in this coarser-grained, dominantly sandstone unit may not be complete. This is especially problematic in the Hua Gang, where faults may not be sealing if permeable sandstones are located on opposite sides of the fault. There is no reason to suspect that this provides cause for major concern, but more evaluation is required, especially on the eastern side of the Fangheting closure.

Relatively full suites of logs were run on all wells. Well PH-5 was logged by different engineers and equipment from these who logged the earlier four wells. There is some discrepancy in the logs related to the use of different types of equipment. Extensive formation evaluation from logs and cores suggests two possible problem areas. First, the high water saturations (Sw) indicate a possible problem with log calculations. If the high Sw interpretation is correct, related production problems of these reservoirs may be anticipated. If the interpretation is incorrect, confidence in the reserve estimates and other reservoir parameters is reduced. Second, there are apparent inconsistencies in correlations in both the Hua Gang and Ping Hu formations. These discrepancies occur between well correlations, as indicated by seismic ties between the wells. However, questions about reservoir continuity will have a serious impact on reserve estimation and projected productivity.

## SEISMIC INTERPRETATION

The integrated 3-D interpretation of the Ping Hu Field included structural mapping, detailed fault interpretation, and analysis of seismic amplitudes from the 55-km$^2$ 3-D survey of the B Structure (acquired and processed in 1987–1988) and the 63-km$^2$ survey over the F Structure (acquired and processed in 1991). From a structural standpoint, both surveys tied quite well, with fault trends and seismic reflections tying across the boundary of these two surveys (Figure 5). However, due to seismic processing differences in deconvolution, filtering, and gain recovery, the seismic amplitudes for a particular reflection varied considerably from one survey to the other. Because of this situation, it was not possible to construct an extracted amplitude map from a particular reservoir horizon over the combined B and F structures. Separate amplitude maps from these individual surveys were constructed to depict the extent of seismic amplitude anomalies.

The 3-D structural interpretation and seismic reservoir characterization of the Ping Hu Field was accomplished. Seismic horizons were calibrated to the existing well control with synthetic seismograms, and time-depth relationships were derived from VSP surveys on the PH-5 well. Seismic picks were made on actual reservoir zones where possible.

The Hua Gang formation is predominantly oil bearing, and generally exhibits simple anticlinal closure with little faulting. A map on the H-2 horizon (Figure 6) illustrates the closure which is present across both the F and B structures. Seismic amplitudes from the H-2 reservoir zone (Figure 7) in the F Structure were converted to impedance utilizing seismic inversion methodologies. Seismically derived impedance values for the H-2 reservoir zone (Figure 8) were calibrated to the values derived from wells drilled in the Ping Hu Field. Utilizing this calibration, and the relationship between impedance and porosity derived from well data, a relationship between porosity and impedance was derived (Figure 9). Using this relationship, and integrating the seismic values for impedance, a seismically derived porosity map was created on the H-2 horizon (Figure 10).

This same procedure was used to calculate seismically derived porosity for the gas-prone P-8 horizon. Figure 11 depicts the structure configuration of the P-8 reservoir horizon within the Ping Hu formation. The seismic impedance map created from seismic inversion for the P-8 on the F Structure is illustrated in Figure 12, and the seismically derived porosity map for this reservoir is shown in Figure 13.

Utilizing the methodology described above, it was possible to determine structural configuration of important reservoirs in the Ping Hu Field, as well as variations in porosity of these reservoirs. Using this seismically derived data and integrating borehole-derived engineering data, a more accurate determination of in-place reserves could be made for the H-2 and P-8 reservoirs.

Additionally, using Society of Petroleum Evaluation Engineers (SPEE) criteria in conjunction with the 3-D seismic data, a more optimistic assessment of hydrocarbon reserves was made. This is particularly important, because this field is marginally economic based on its distance from shore, the presence of gas which would require a long subsea gas pipeline for full exploitation, and the relatively small initial reserve estimation.

## RESULTS

Detailed mapping revealed the Ping Hu Field to be a structurally complex, wrench-faulted anticline with considerable variations in reservoir quality across the structure. A 3-D visualization of the Ping Hu Field F Structure is illustrated in Figure 14. Analysis of amplitude changes in reservoir zones in the Hua Gang and Ping Hu formations suggest lithological changes might be responsible for some of these variations.

For example, the high-potential H-6 reservoir zone seen in the PH-5 well is interpreted to be a delta front sandstone body, which is laterally limited to the east and west. This reservoir is draped over the north end of the Ping Hu F Structure, which creates a reversal and the hydrocarbon trap. This reservoir is not observed in the PH-2 well to the north on the B Structure or to the south of the PH-4 well. The seismic event correlative with the H-6 reservoir appears to correspond to this interpretation (Figure 15).

A number of high amplitude anomalies within the upper Ping Hu formation appear to be present to the east of, and terminated updip by, the Fahn fault. These zones are interpreted to be possible pay zones within the Ping Hu Formations which have not been penetrated by any of the five wells currently drilled in the Ping Hu Field. Additional reserves can be assigned to these zones under the criteria established by SPEE.

Alternatively, Wang (1995) has suggested that amplitude anomalies in the Ping Hu data are related to thin (0.5 to 5 meters) coal beds and/or data acquisition and processing problems. These anomalies appear on both the structural crest and in low positions, throughout the Ping Hu Formation. There is little agreement among Chinese researchers and foreign oil company geophysicists about the cause of this distribution of anomalies. Wang's reservoir calibration, especially to the PH-5 well, and 2-D seismic modeling suggest coals are the primary reason. In this well, high-amplitude anomalies correspond with intervals in which coal seams are present. On the other hand, the abnormally narrow vertical appearance of high amplitudes, or *striping*, can be created during stacking and may be caused by navigation errors. Ultimately, the correct interpretation of these anomalies is dependent on reprocessing or acquisition of a better dataset.

## SUMMARY

In summary, the Ping Hu Field has been mapped in detail utilizing a combination of two 3-D seismic surveys acquired over the study area. The integration of this quality 3-D seismic base with geological and engineering data has provided a means for an accurate determination of hydrocarbon reserves in-place. Additional potential reserves were identified by utilizing this integrated approach.


### *Acknowledgments*

The authors gratefully acknowledge the cooperation of Shanghai Petroleum, Gustavson Associates, the Asian Development Bank, and the Department of Geological Sciences at the University of Colorado, Boulder.


## REFERENCE CITED


Wang Qin, 1995, Reservoir Delineation Using 3-D Seismic Data of the Ping Hu Field, East China Sea, unpublished M.S. thesis, University of Colorado, Boulder.

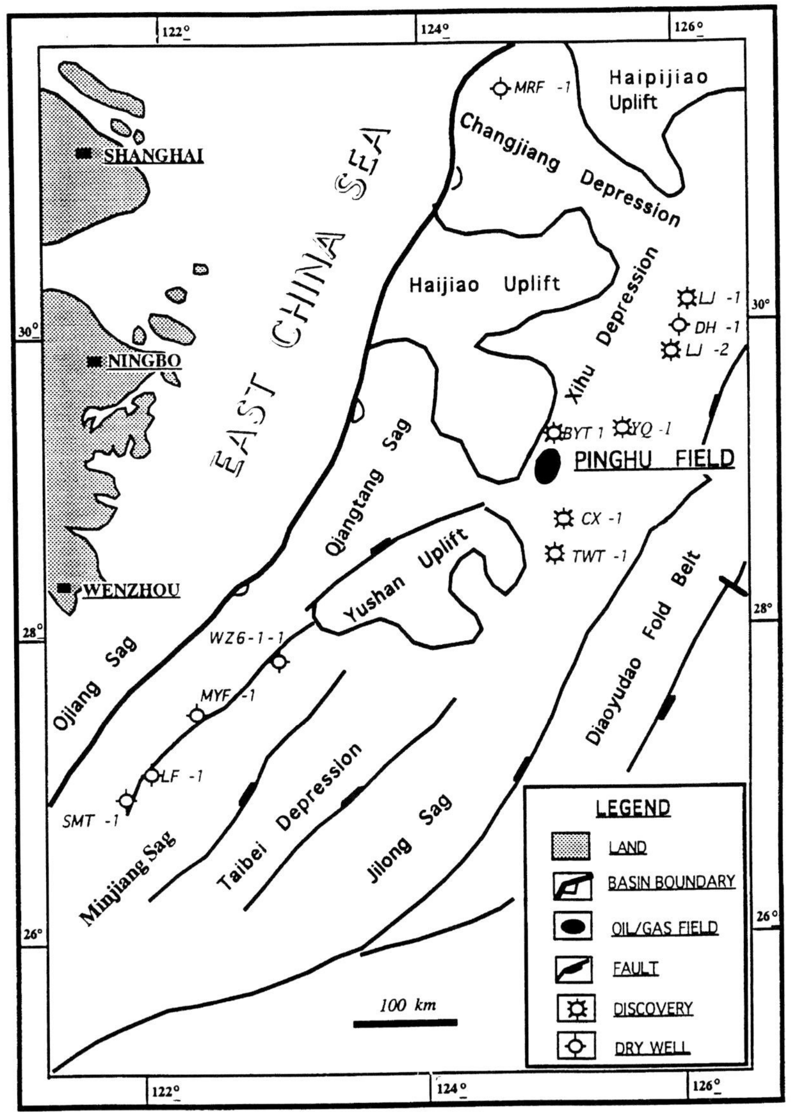


FIGURE 1. Map showing regional tectonic features of the East China Sea and the location of the Ping Hu Field area, offshore the People's Republic of China. This gas, condensate, and oil field lies in 90 m of water. Other petroleum discoveries are shown.

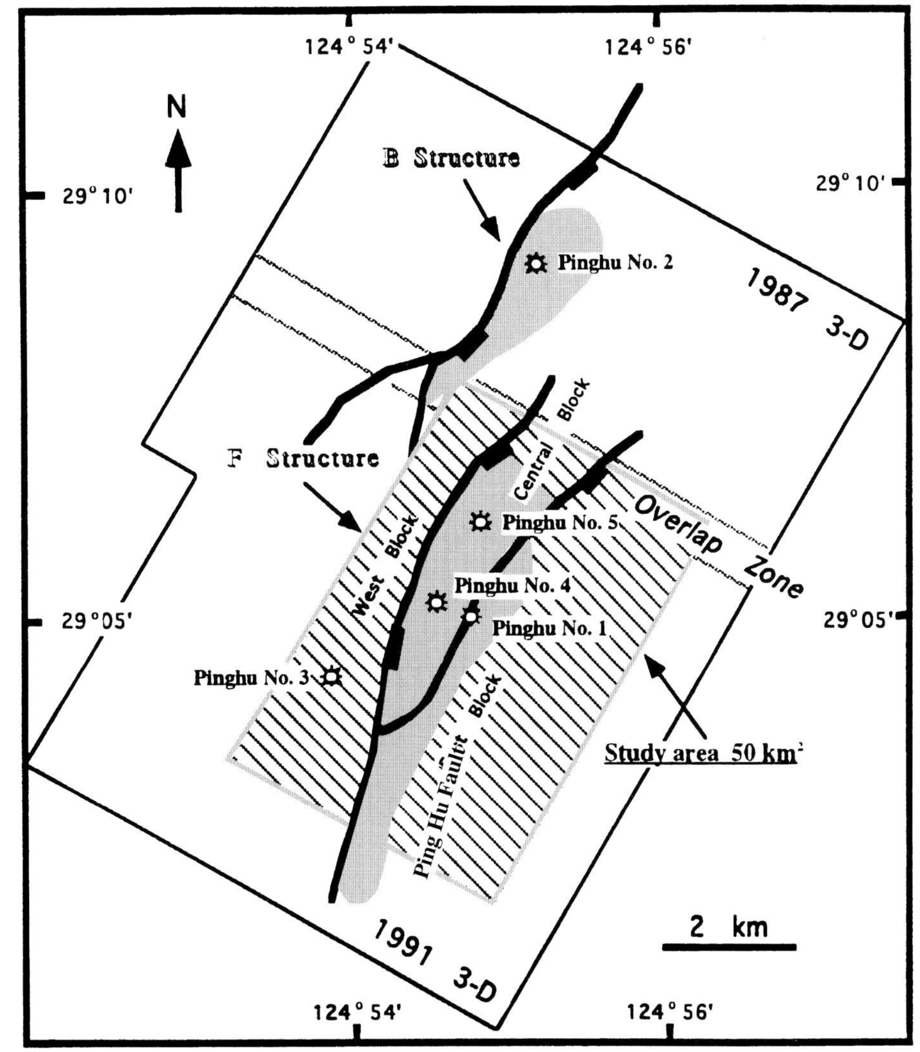


FIGURE 2. General structural configuration of the Ping Hu Field; Fangheting (F), and Bajiaoting (B) structures. The 3-D dataset was acquired in 1987 (B Structure) and 1991 (F Structure) and covers approximately 118 km².

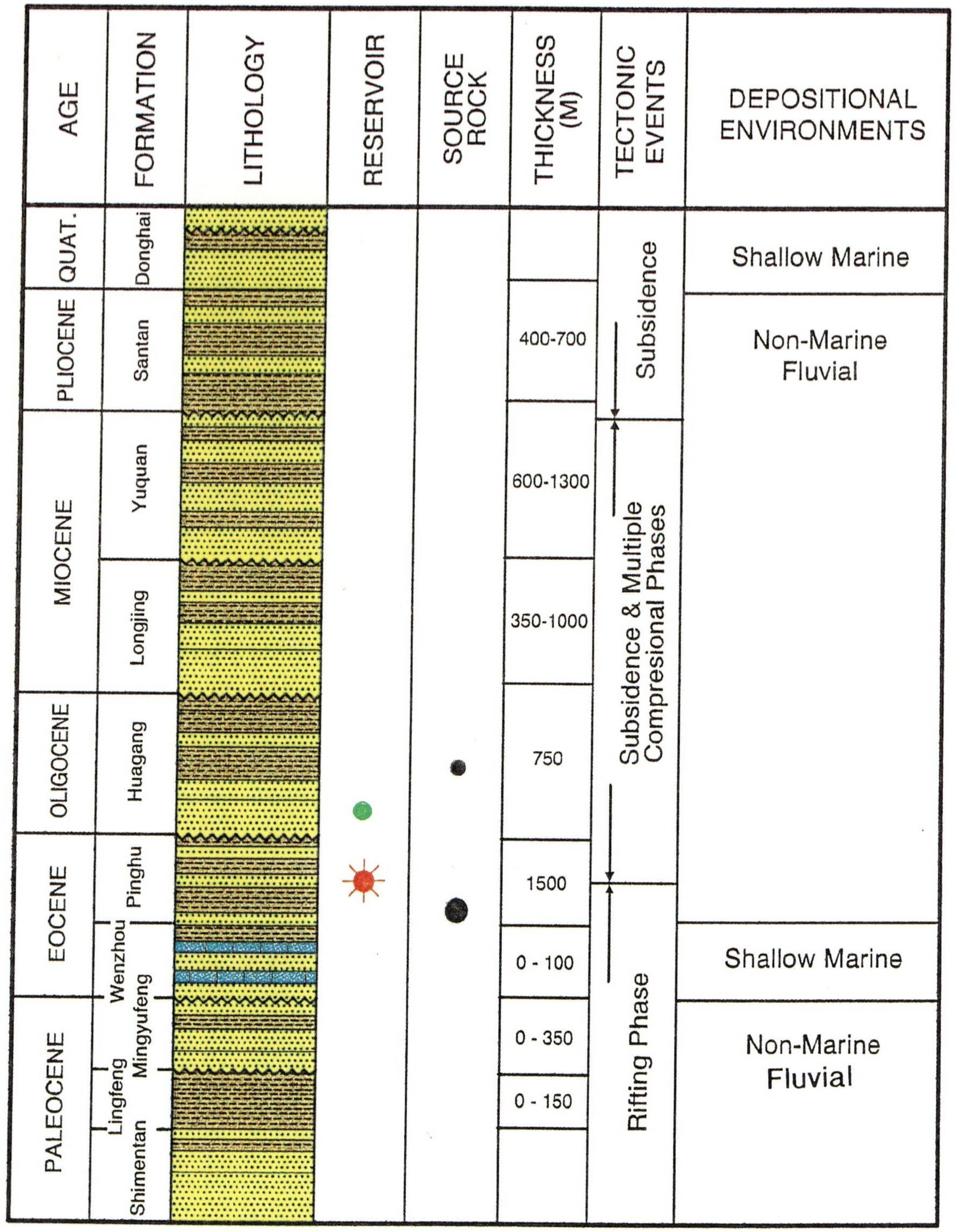


FIGURE 3. Generalized lithostratigraphy of the Ping Hu Field area, East China Sea Basin. Shallow- and non-marine strata dominate the Xihu depression of this Cenozoic basin. Principal reservoir rocks are Lower Tertiary fluvial and deltaic sandstones. Oligocene coals and Eocene shales are potential source rocks. Note that strata older than the Ping Hu are not present in the field itself. Wells PH-2 and PH-3 were drilled to Cretaceous basement.

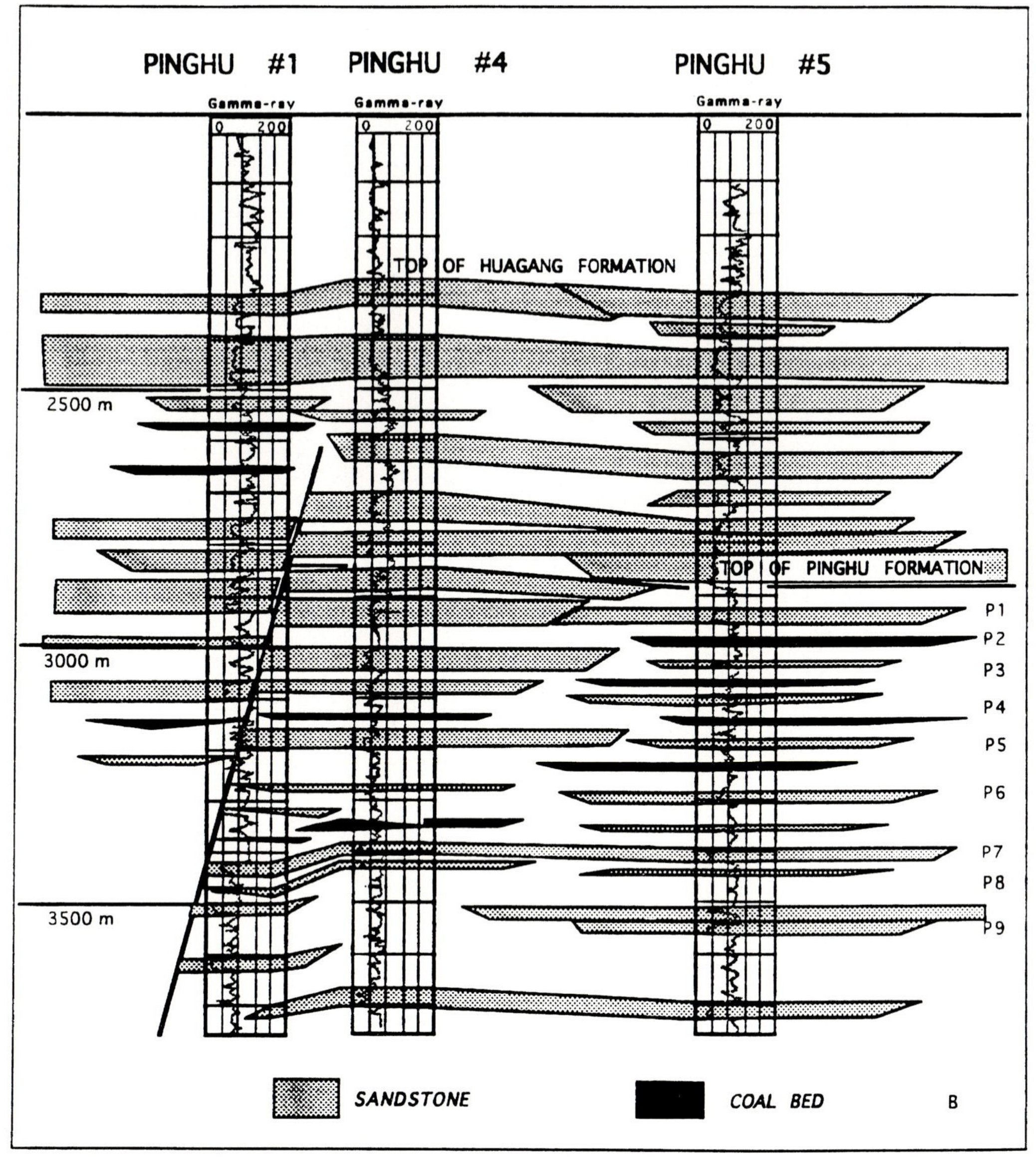


FIGURE 4. Correlation of reservoir sandstones in PH-1, PH-4, and PH-5. See Figure 2 for well locations. The Hua Gang reservoirs include thick, coarse-grained sandstones with thin shale interbeds. The Ping Hu Formation reservoirs are interbedded medium- to fine-grained sandstones, shales, and coals. Sandstone and coal are highlighted. The rest is shale.

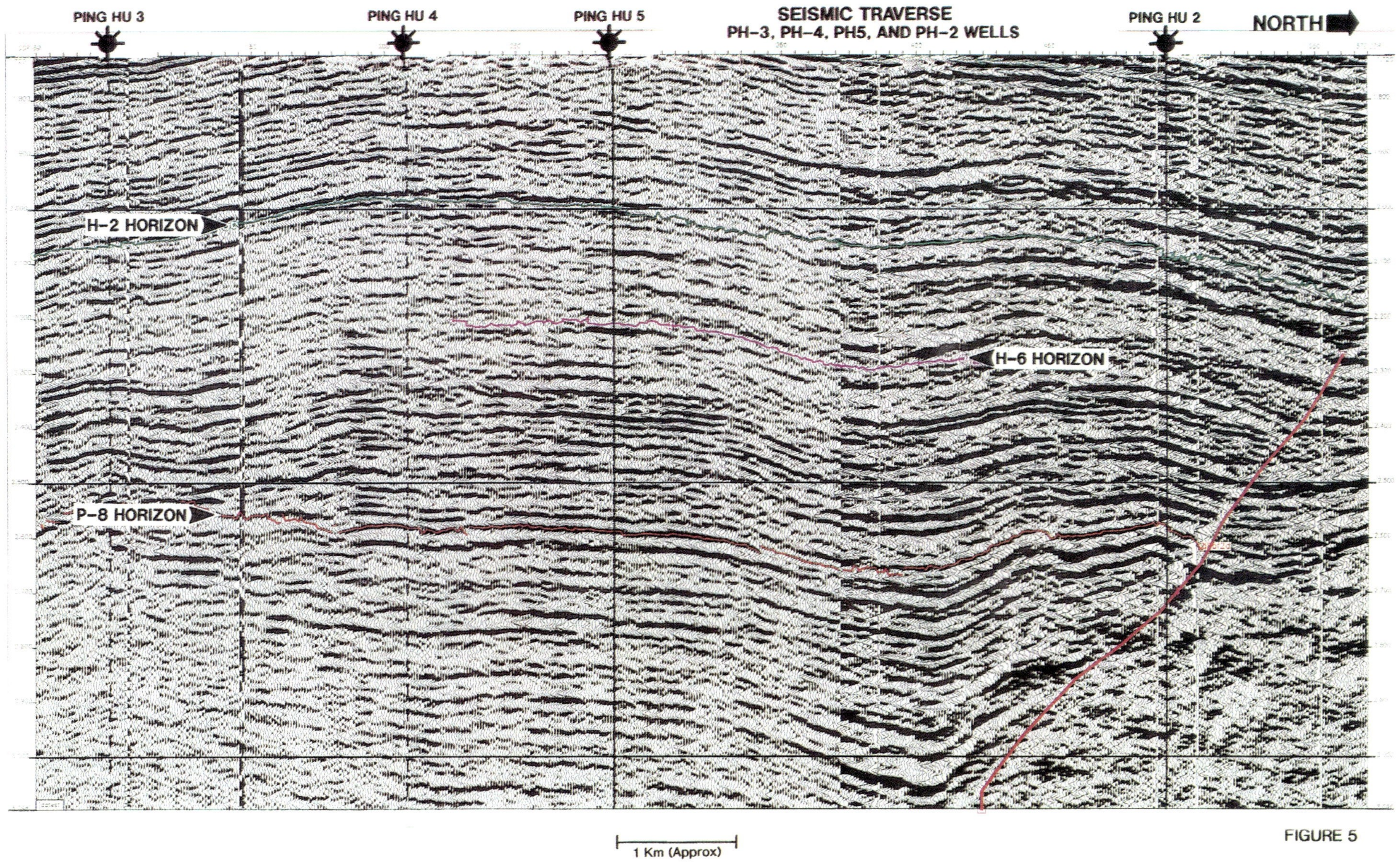


FIGURE 5. Seismic traverse from Ping Hu 3-D seismic survey through PH-3, PH-4, PH-5, and PH-2 wells. This north-south extracted seismic traverse illustrates fair to good continuity of events across the juncture of the F Structure and B Structure 3-D surveys. See Figure 2 for location of profile.

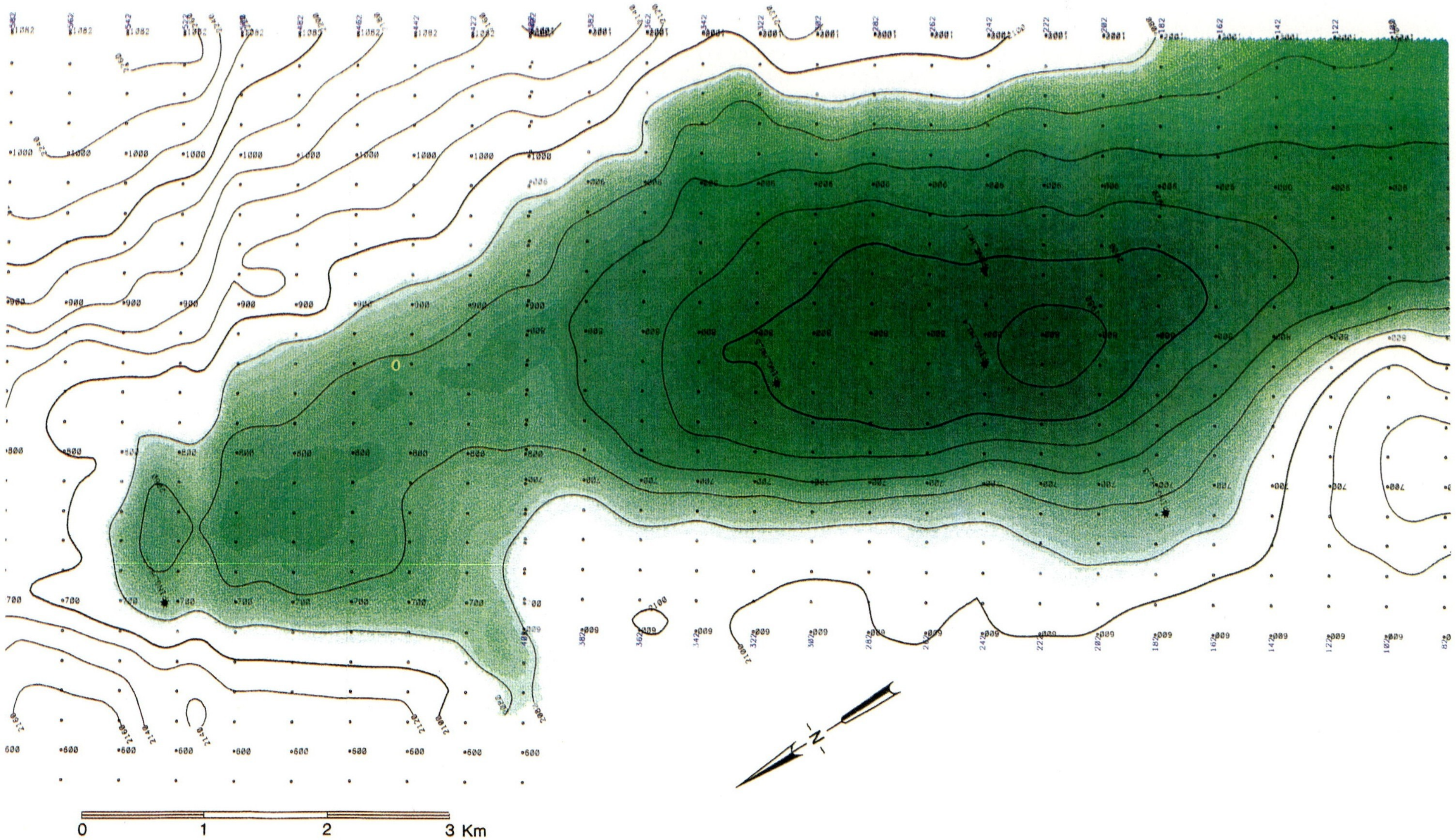


FIGURE 6. Time structure map of the H-2 reservoir zone—Hua Gang Formation. The H-2 is a predominantly oil-bearing zone at a depth of 2250–2400 m. Structure exhibits anticlinal closure with little evidence of faulting.

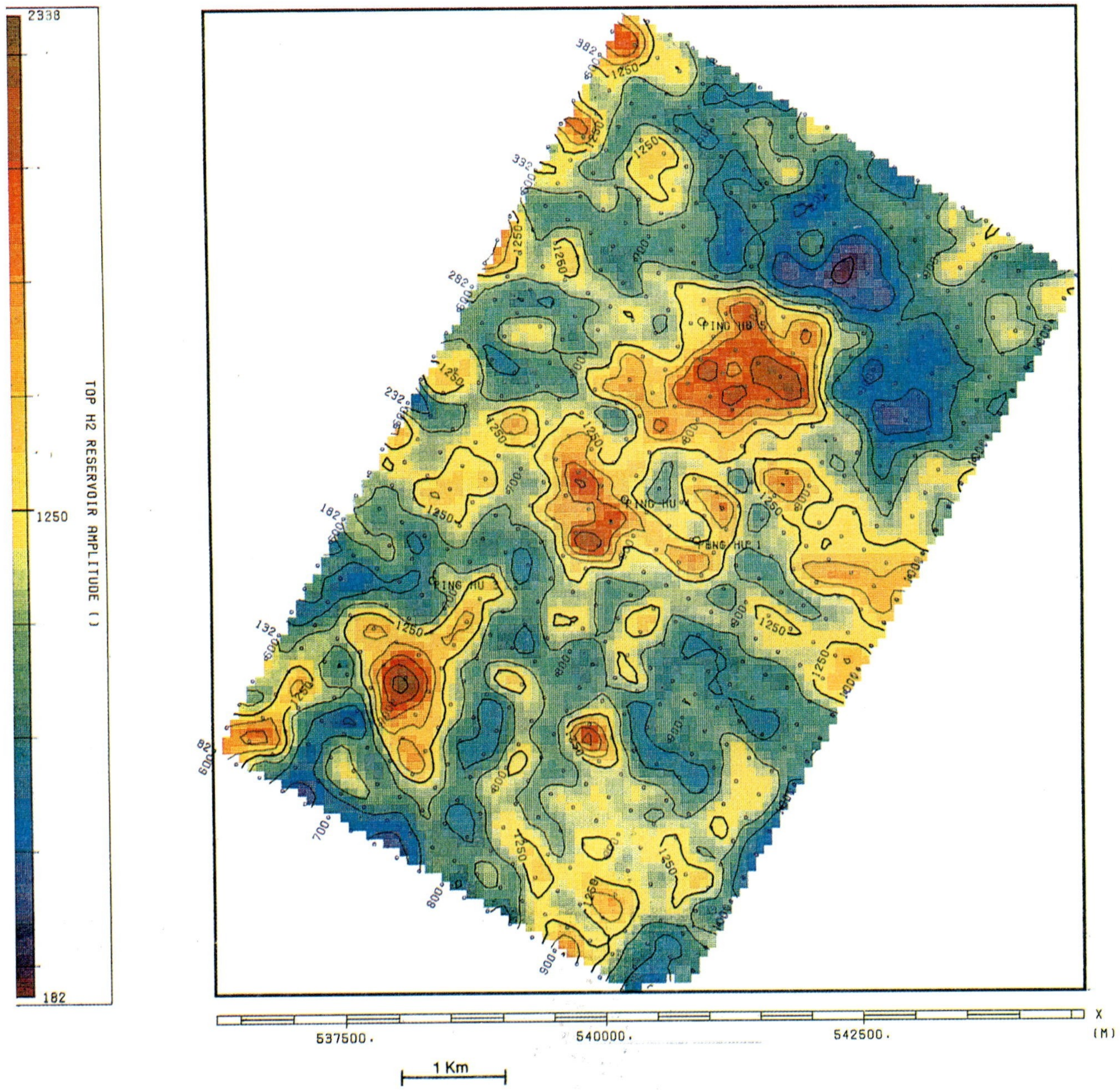


FIGURE 7. Extracted amplitude map from seismic picks on the H-2 reservoir zone, Ping Hu F Structure. The relatively continuous nature of the H-2 seismic event over the F Structure provides for a good calibration to reservoir properties.

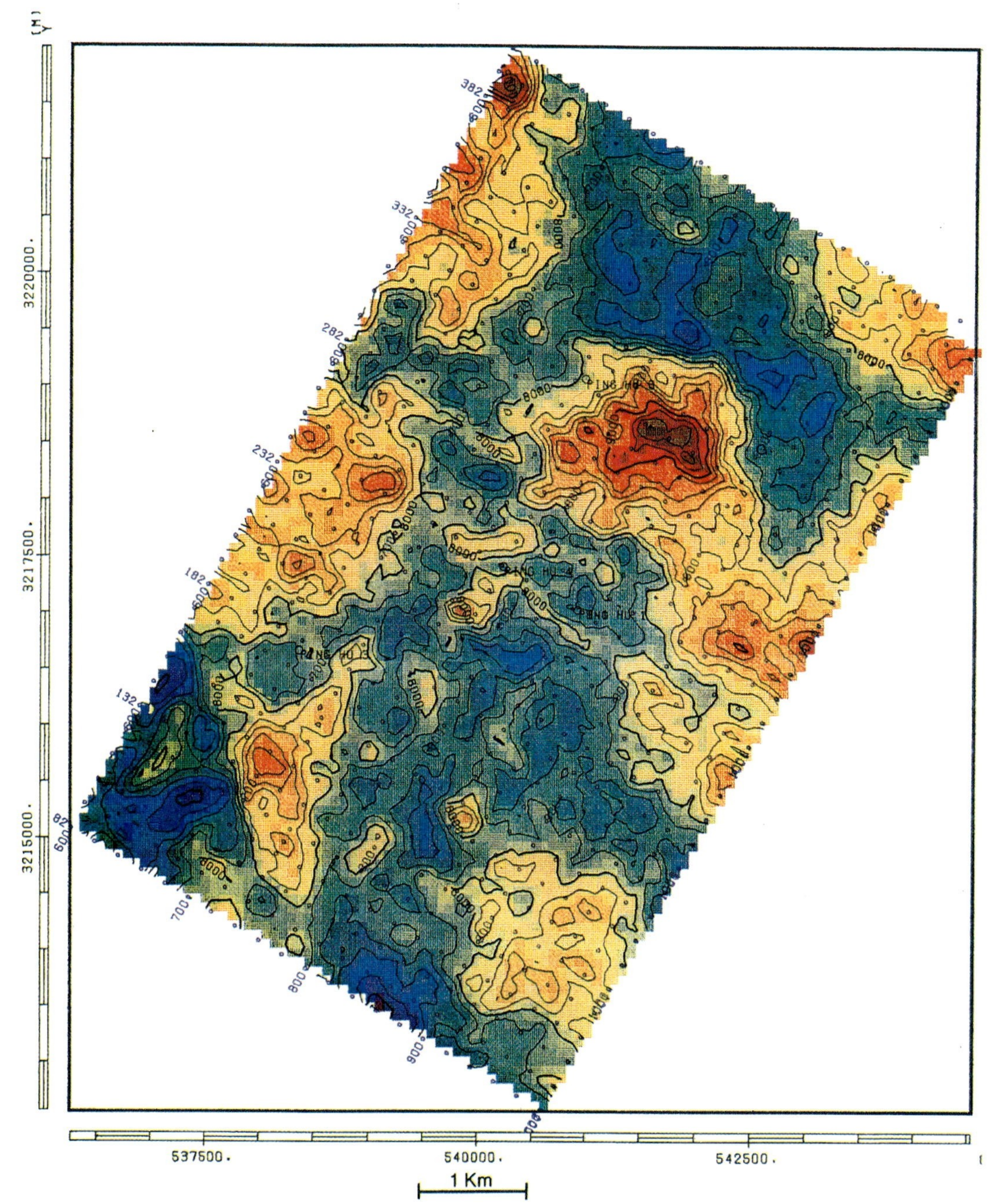


FIGURE 8. Seismically derived impedance for H-2 reservoir zone, Ping Hu F Structure. Impedance is defined as the product of seismic velocity and bulk density.

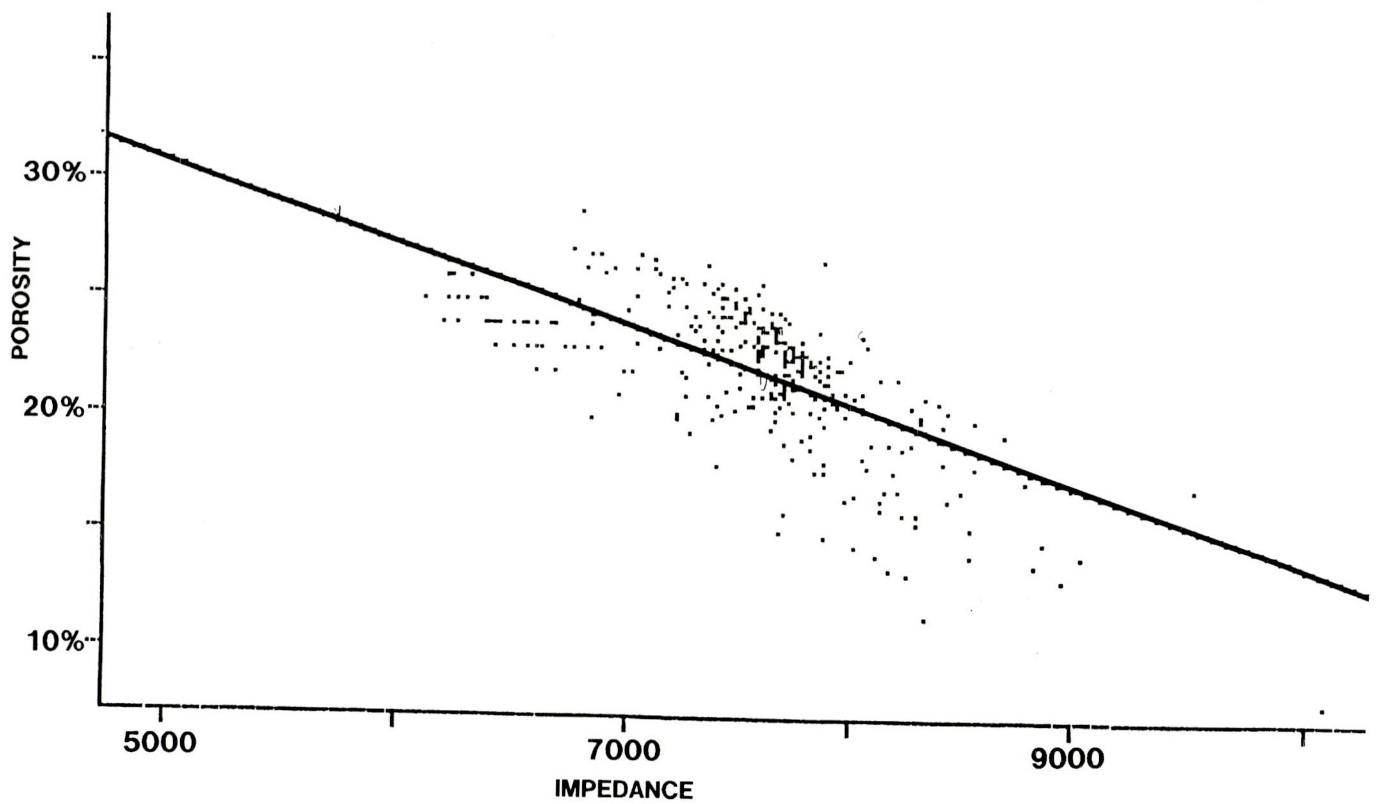


FIGURE 9. Seismically derived relationship between porosity and impedance, H-2 reservoir zone. Values were obtained by first calibrating measured seismic impedance with reservoir porosity at the well bore. Additional porosity values were then calculated away from the well bore using values of seismic impedance.

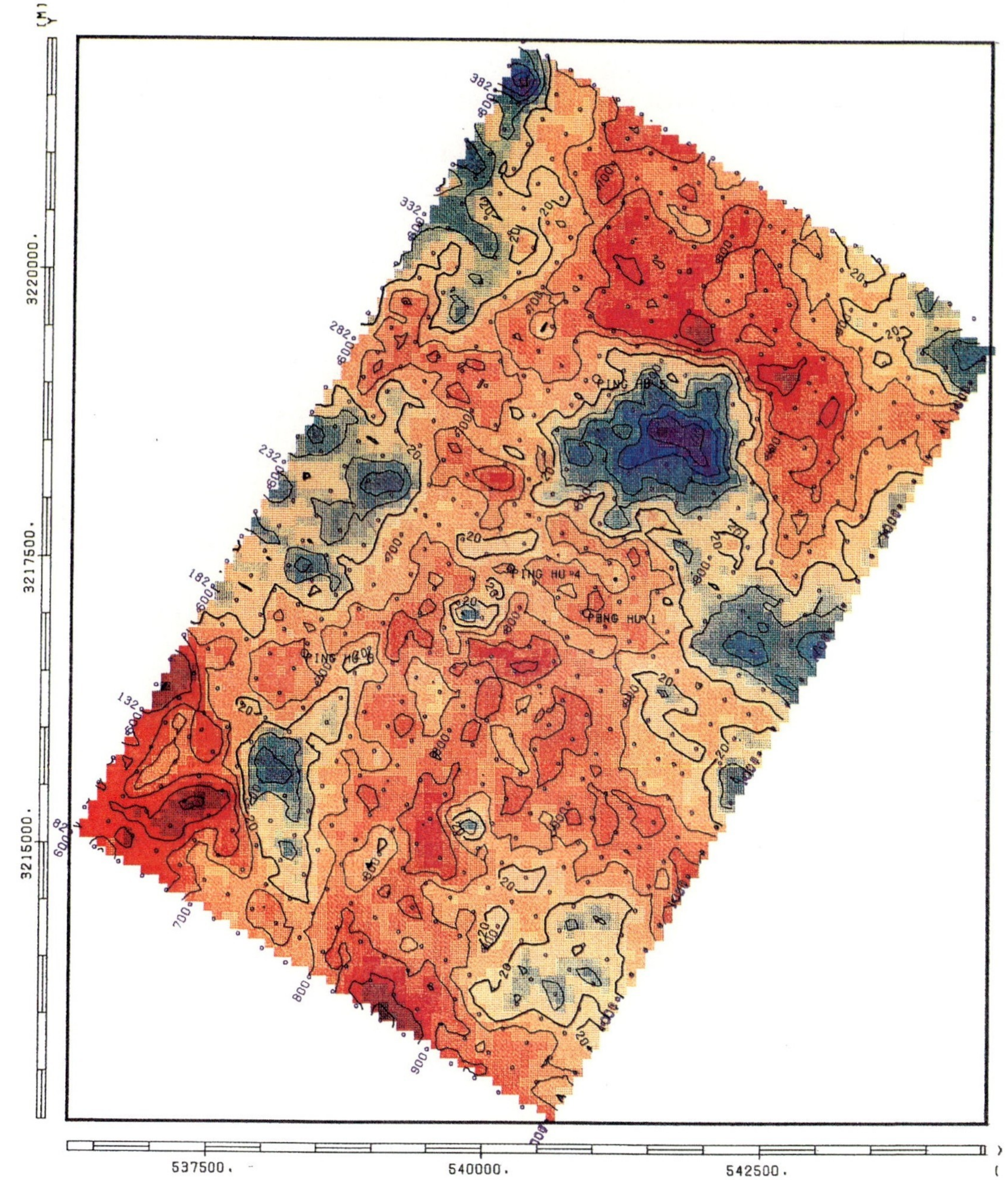


FIGURE 10

FIGURE 10. Seismically derived porosity map for H-2 reservoir zone—Ping Hu F Structure. Using the relationship derived in Figure 9, porosities were calculated from the 3-D dataset away from the borehole across faults and downdip.

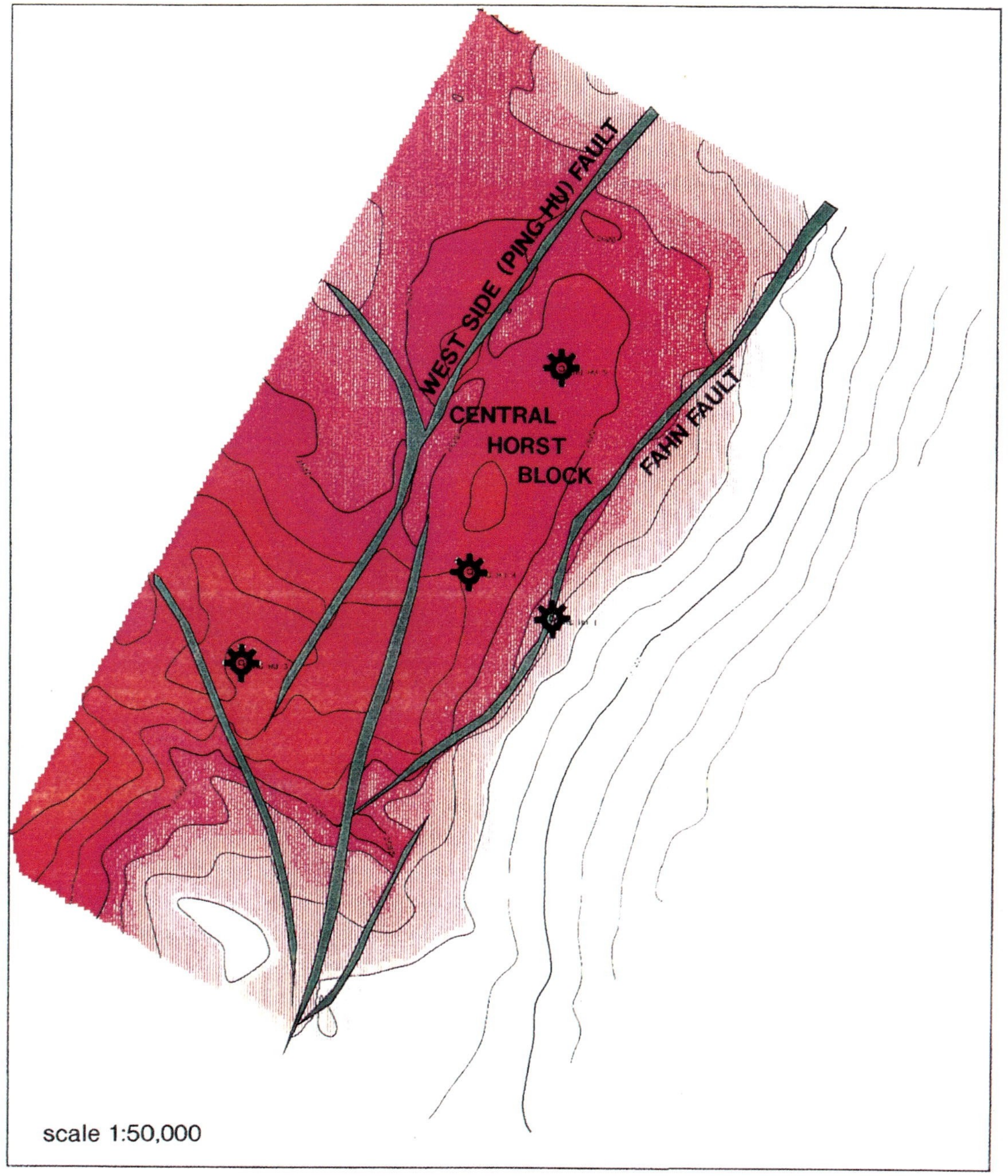


FIGURE 11. Time-structure map on P-8 reservoir zone, Ping Hu F Structure. The P-8 zone is predominantly gas bearing. Significant faulting cuts through the P-8 reservoir and may be responsible for some of the reservoir compartmentalization throughout the Ping Hu Formation.

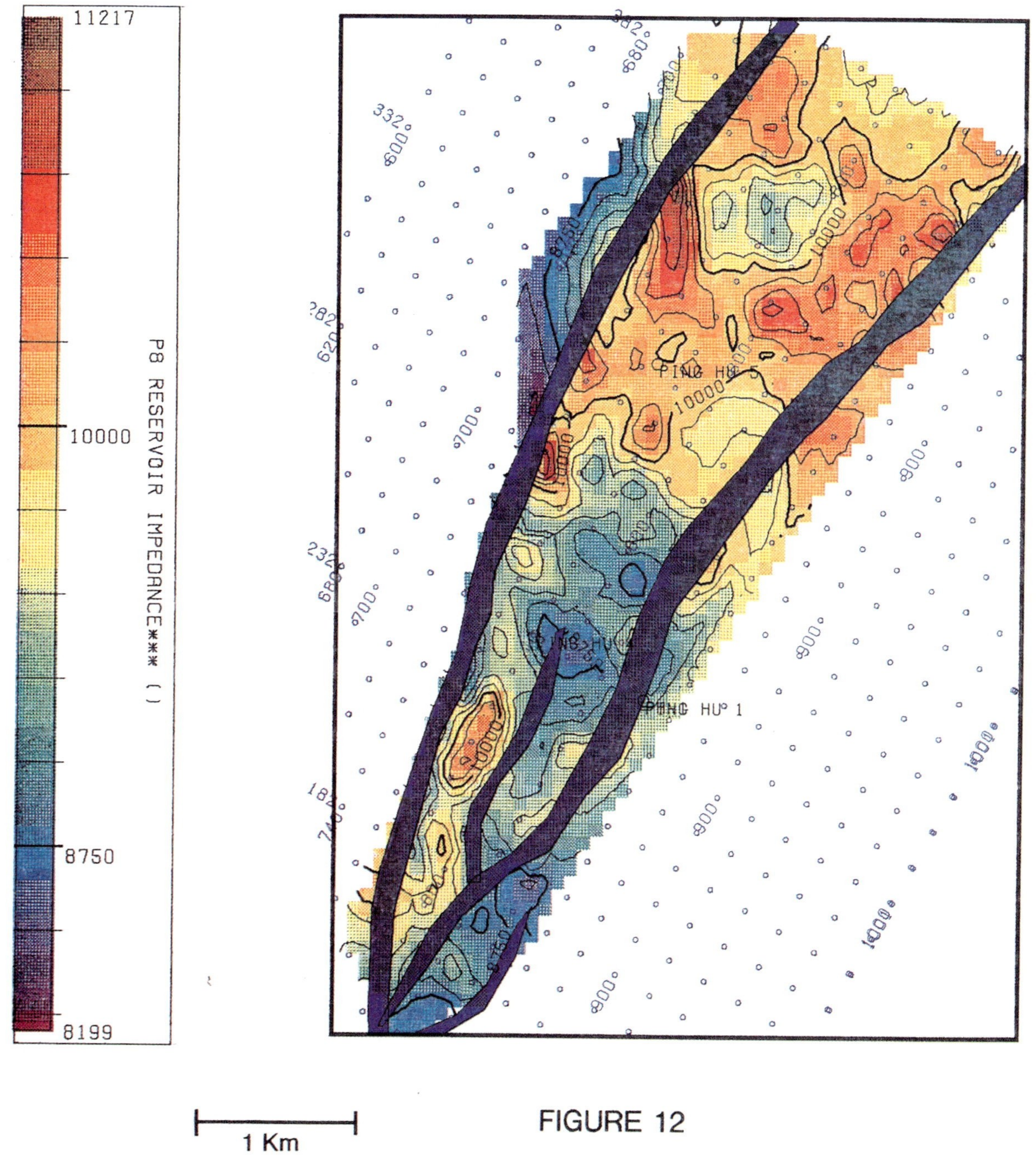


FIGURE 12. Seismically derived impedance for P-8 reservoir zone, Ping Hu F Structure. The P-8 zone is predominantly gas bearing and provides a good seismic pick across most of the Ping Hu structure.

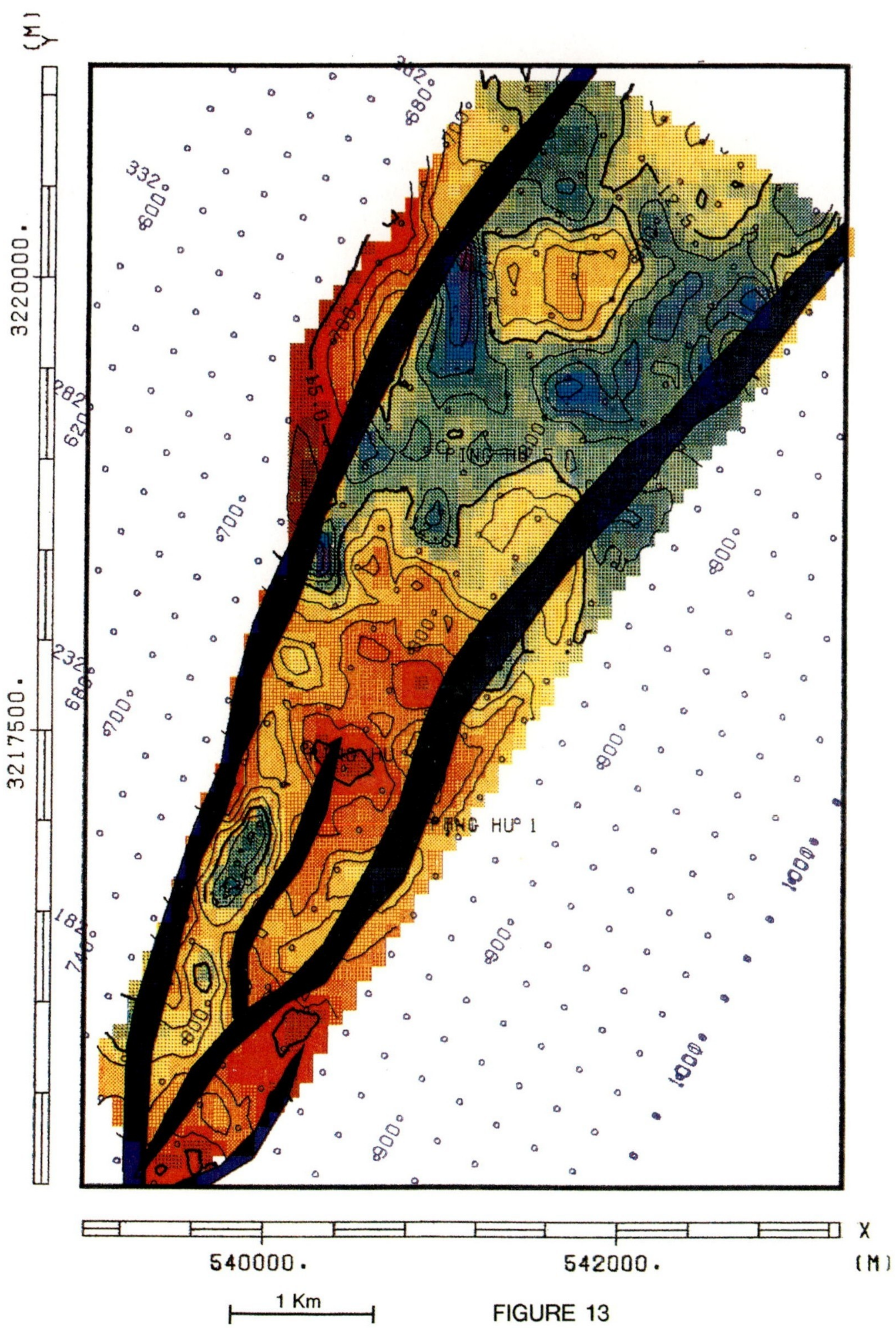


FIGURE 13. Seismically derived porosity for P-8 reservoir zone, Ping Hu F Structure. Calculated porosities for this horizon range from 10 to 16%.

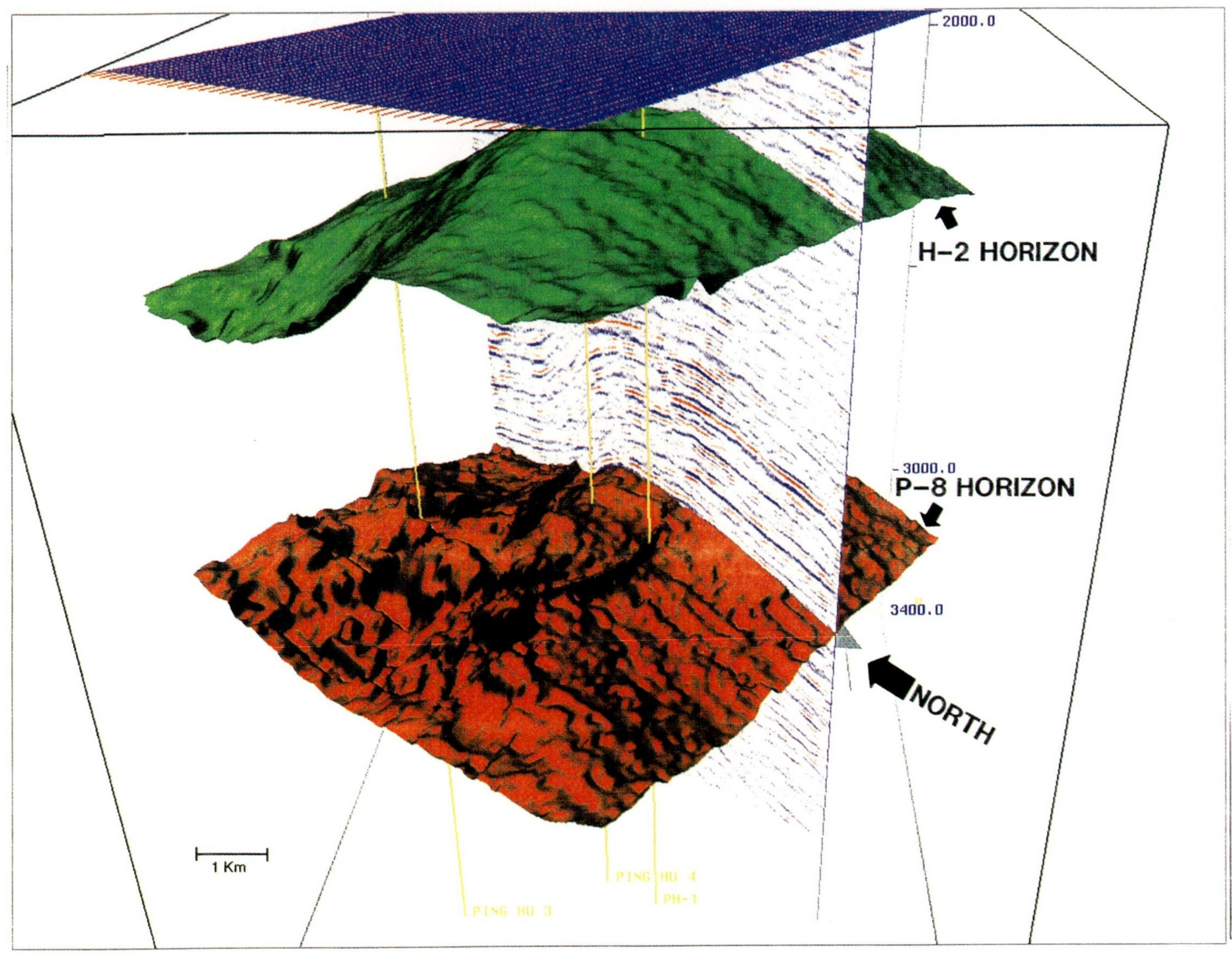


FIGURE 14. 3-D visualization of the H-2 (green) and the P-8 (red) reservoir horizons in the "F" Structure of the Ping Hu Field. Faulting becomes more intense and complex in the deeper Ping Hu reservoir horizons (zones P-1 through P-12). The upper Hua Gang reservoirs exhibit simple anticlinal structure with four-way closure.

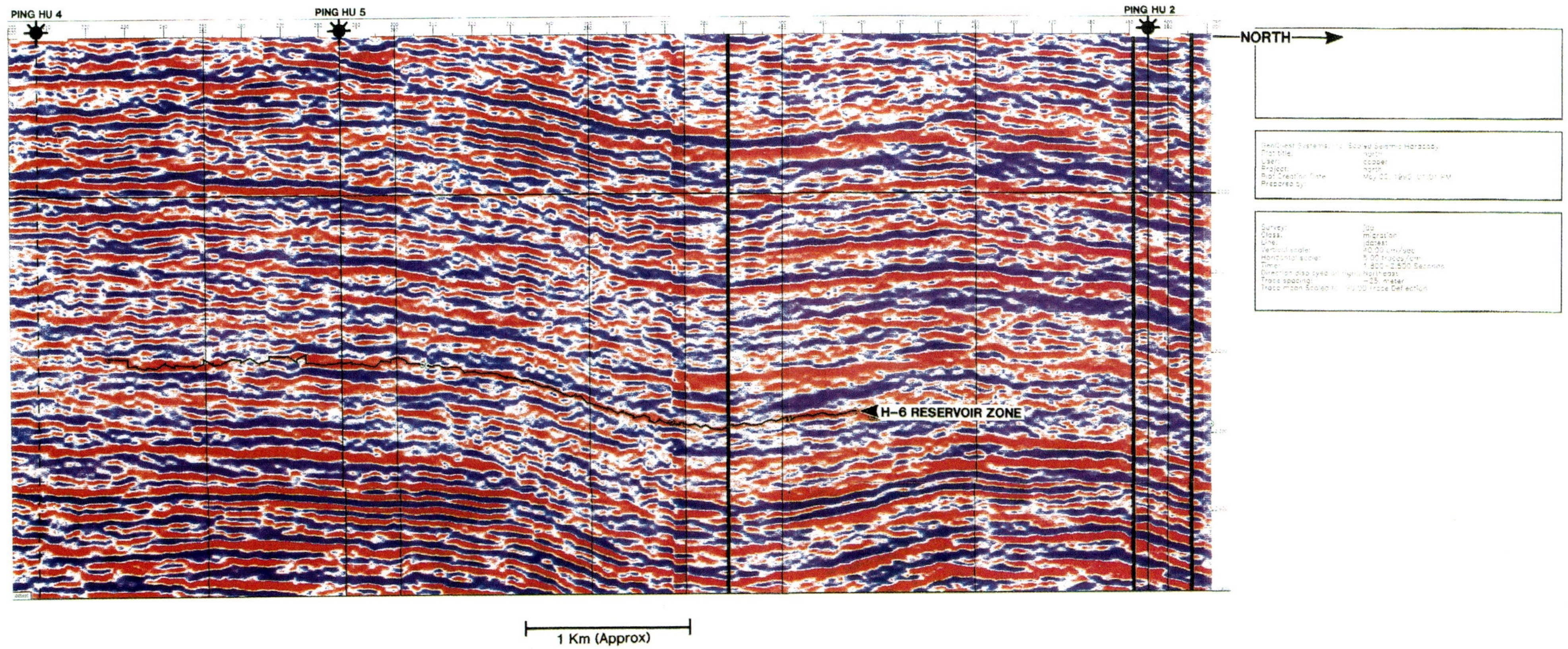


FIGURE 15. Seismic traverse from PH-5 to PH-2 wells, illustrating the distribution of the H-6 reservoir zone. This important oil reservoir is present in the PH-5 well and is interpreted to be a delta front sandstone body. Note pinchout and termination of this zone to the north—this reservoir is not present in the PH-2 well. See Figure 2 for location of profile.

# CHAPTER 10

Saudi Arabian Oil Company, Impact of 3-D seismic data on development drilling, Ghinah and Umm Jurf Fields, Central Saudi Arabia, *in* P. Weimer and T. L. Davis, eds., *AAPG Studies in Geology No. 42* and *SEG Geophysical Developments Series No. 5*, AAPG/SEG, Tulsa, p. 103–106.

# Impact of 3-D Seismic Data on Development Drilling, Ghinah and Umm Jurf Fields, Central Saudi Arabia

**Saudi Arabian Oil Company, compiled and edited by S. C. Simms***

## ABSTRACT

The Saudi Arabian Oil Company has reduced development drilling costs and increased oil reserves by selecting well locations using 3-D seismic data in the Ghinah and Umm Jurf Fields, Central Saudi Arabia. 3-D seismic data provide more accurate structural interpretation of strike-slip structures, and more important, provide stratigraphic predictability within a heterogeneous braided stream reservoir. The success rate for wells with high flow rates (>2000 BOPD) has increased from 43% to over 90% since 3-D seismic data have been used to locate development wells. In addition, the cumulative reserves of the two fields have increased 400%.

## INTRODUCTION

The Saudi Arabian Oil Company has discovered 13 oil and gas fields in central Saudi Arabia since 1989, including the Ghinah and Umm Jurf fields. These fields are located 100 to 200 km south of Riyadh and about 100 km east of the Arabian Shield (Figure 1). All the fields produce from the Permian-Carboniferous Unayzah formation, a complexly interbedded, multilateral, multistory sequence of braided stream sandstones and flood-plain siltstones. While the sandstone facies can be highly productive, the siltstone facies is nonproductive. Early in the development of several of the fields it was recognized that the sandstone and siltstone could be differentiated using 2-D seismic data (Simms, 1993; Al-Hauwaj et al., 1993; Al-Husseini and Chimblo, 1994). However, the spatial distribution of the sands and silts were too complex to predict using 2-D seismic data, and numerous dry holes and poor producers were drilled.

A 3-D seismic program was initiated in 1992 in an effort to reduce drilling costs and optimize well locations. 3-D seismic data was acquired for approximately 900 $km^2$ including the 300 $km^2$ survey discussed in this paper.

## STRUCTURE AND STRATIGRAPHIC SECTION

All of the fields discovered to date were drilled as structural prospects which are draped over Paleozoic fault blocks. The Ghinah field was drilled on what was thought to be a doubly plunging anticline trending about 30° northwest (Figure 2). This en echelon structure lies west of a north-trending, right lateral strike-slip fault (Simms, 1994). The strike-slip fault and en echelon fault to the west are shown in Figure 3, an east-to-west 3-D seismic section through the field.

Prior to acquisition of the 3-D seismic survey, the two southernmost wells were wells A and D. Fourteen wells had been drilled to the north of well D, and less than half of these were considered to be good producers. Well D, on the southern nose of the Ghinah field, encountered a predominantly siltstone section with several thin sandstones and was considered a poor producer. Well A, drilled on what was thought to be a separate structure to the south, also encountered siltstone above the oil/water contact, and although oil shows were present in a thin sandstone, it was classified a dry hole. A stratigraphic section through these wells (Figure 4) indicates a rather poor prognosis for additional drilling in the vicinity. In fact, no further drilling was anticipated until the 3-D survey was acquired.

## RESULTS OF THE 3-D SEISMIC SURVEY

The 3-D Unayzah structure map (Figure 5) is significantly different from the 2-D structure map (Figure 2). The east flank of the field dips very steeply over the north-trending strike-slip fault, and the oil/water contact (not shown) is up to three kilometers east of the contact as mapped using 2-D seismic. In addition, the 3-D structure map indicates that only one structure is present and that the area between A and D is above the oil/water contact.

A 3-D seismic section between wells A and D (Figure 4) reveals dramatic changes in seismic character, which were interpreted to be braided stream sands. A horizon slice through the reservoir (Figure 6) shows a southeast-trending braided stream system with the characteristic delta-shaped alluvial plain. Wells B and C were drilled shortly after acquisition of the 3-D survey and, as anticipated, encountered thick, stacked channel sands (Figure 7). Well B was designated the discovery well of the Umm Jurf field. More than 90% of the wells drilled in this area are considered good producers (Al-Hajhog et al., 1994).

Although the success rate for drilling wells with high flow rates more than doubled after acquisition of the 3-D survey, much of this success can be attributed to simply drilling in a more sand-prone area—in this case, on the flanks of the en echelon structure rather than on the crest. As commonly occurs, a topographic high was present over the structural high during the deposition of the Unayzah, and the braided streams flowed to the southeast around the high. A modern analog of the horizon slice (Figure 8) is seen on a Landsat image of the field (Figure 9). A topographic high is present over the structural high, and the major drainage system flows to the southeast in nearly the same location as the Unayzah channel system. A 3-D view of the basement structure (Figure 10) shows the en echelon fault and structural high dying to the south. Superimposed on this image are probable paths of braided streams deposited on a topographic surface controlled by basement structure.

## DISCUSSION

Drilling is still in progress within the Ghinah and Umm Jurf fields. All proposed locations are evaluated for reservoir quality using 3-D seismic data prior to drilling. Drilling subsequent to acquiring the 3-D survey has resulted in a fourfold increase in the cumulative reserves of the Ghinah and Umm Jurf fields. The success rate for drilling wells with high flow rates (>2000 BOPD) has increased from 43% to over 90%. As drilling has progressed, it has become evident that the Umm Jurf field is a stratigraphic trap rather than a structural trap, with braided stream sands absent up-dip to the west. The productive limits of the field have been extended south of the original 3-D survey, and another 3-D seismic survey has recently been acquired.

## REFERENCES CITED

Al-Hajhog, J., S. C. Simms, and C. Heine, 1994, Staying ahead of the drill bit: 3-D seismic guides the development drilling at the Ghinah and Umm Jurf Fields, central Saudi Arabia: Presented at the Middle East Geosciences Conference, GEO '94.

Al-Hauwaj, A. Y., S. C. Simms, J. Al-Hajhog, 1993, Impact of 3-D seismic on facies analysis of clastic reservoirs, Ghinah Field, central Saudi Arabia: Organization of Arab Exporting Countries—Institut Fransais du Petrole Joint Workshop, Paper 2.

Al-Husseini, M. I., and R. D. Chimblo, 1994, 3-D seismology in the Arabian Gulf region: Manama, Bahrein, Middle East Petroleum Geosciences, GEO '94, p. 93–102.

Simms, S. C., 1993, Integration of 3-D seismic data with reservoir modeling of a stratigraphically complex reservoir, central Saudi Arabia: Presented at the American Association of Petroleum Geologists International Conference, The Hague.

Simms, S. C., 1994, Structural style of recently discovered oil fields, central Saudi Arabia: Middle East Petroleum Geosciences, GEO '94, p. 861–866.

*Saudi Arabian Oil Company, Dhahran, Saudi Arabia

FIGURE 1. Regional map of the Ghinah and Umm Jurf fields, central Saudi Arabia.

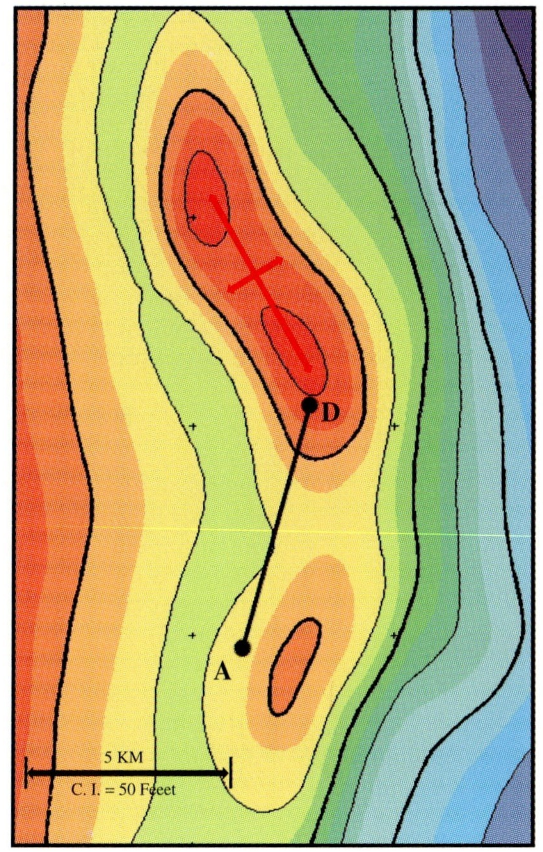


FIGURE 2. Structure map of the Ghinah field based on 2-D seismic data.

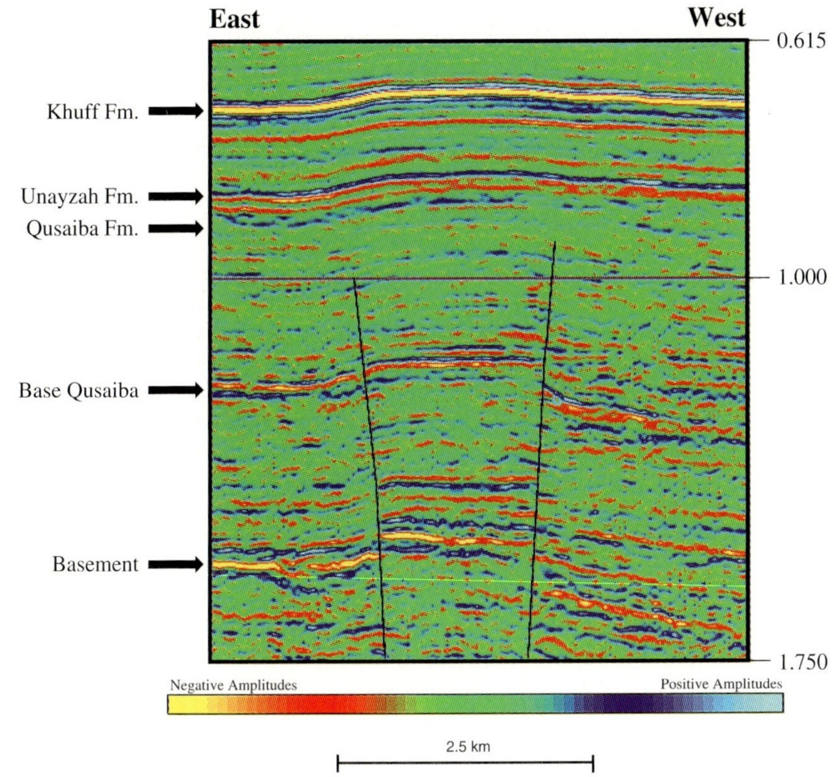


FIGURE 3. 3-D seismic section across the Ghinah field. Note the drape of the Unayzah reservoir over the basement faults.

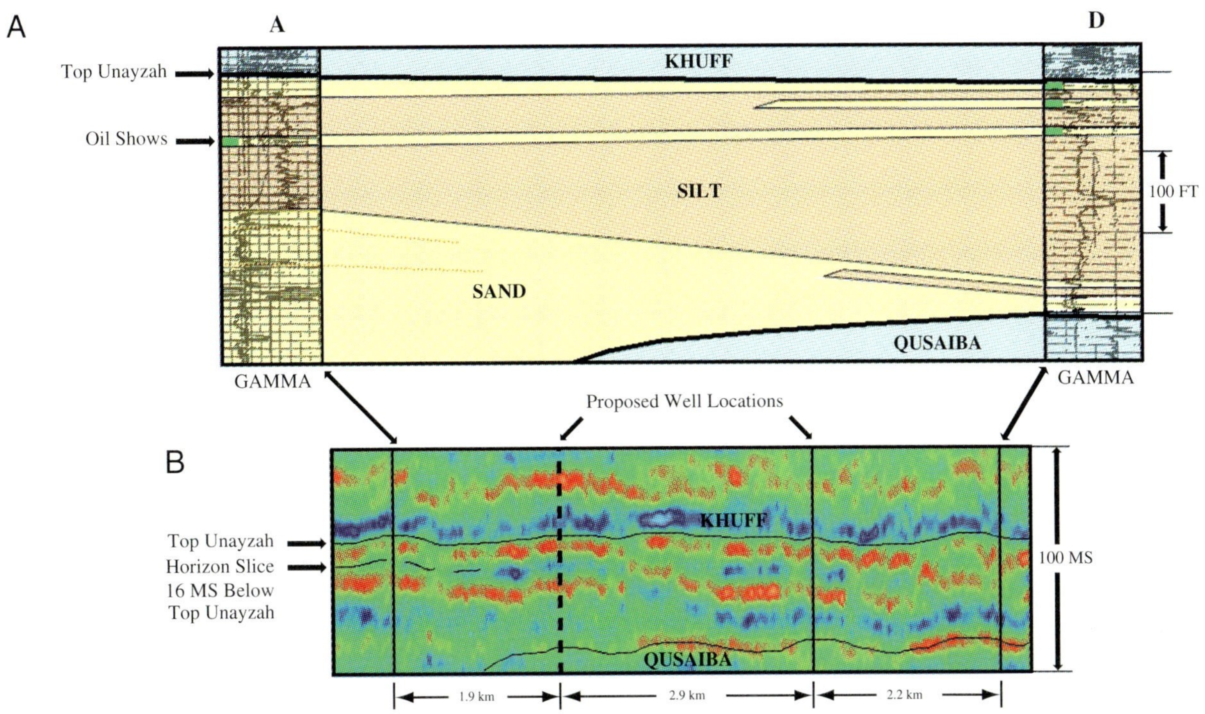


FIGURE 4. (A) Well log cross section across the Unayzah reservoir. Location of section shown in Figure 2. (B) 3-D seismic section across well log section acquired before the drilling of infill wells.

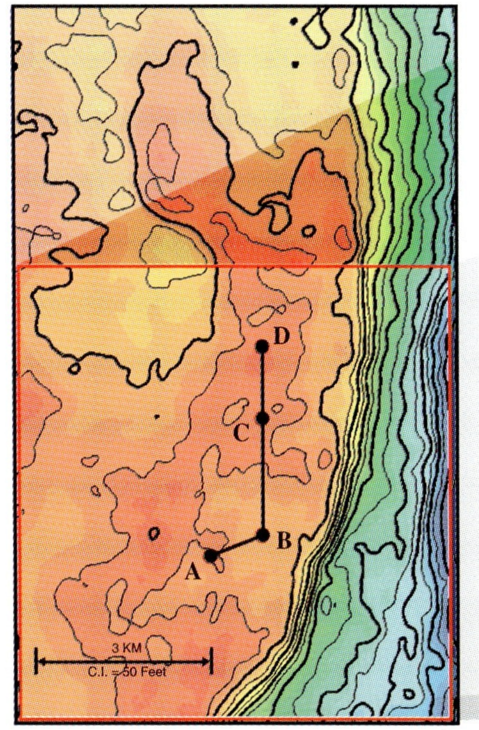


FIGURE 5. Structure map of the Ghinah field based on 3-D seismic data.

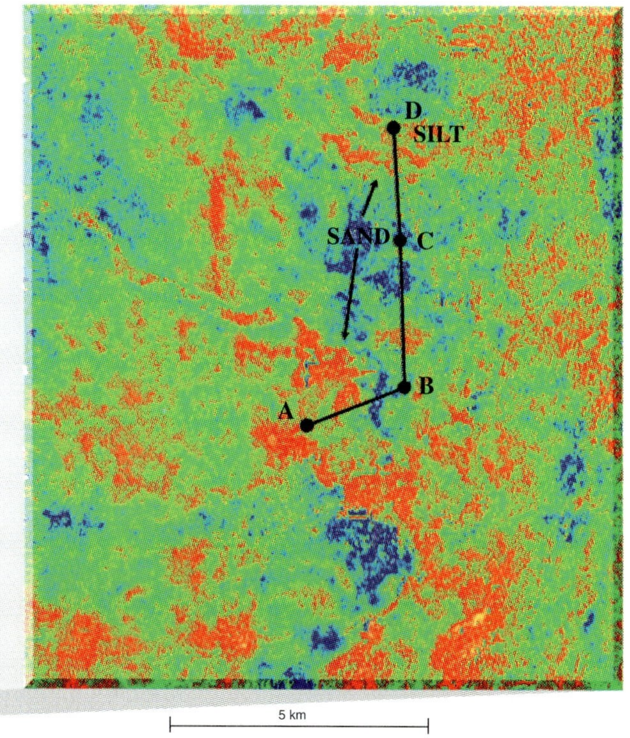


FIGURE 6. Horizontal slice 16 ms below the top Unayzah formation, showing interpreted southeast-trending braided stream system.

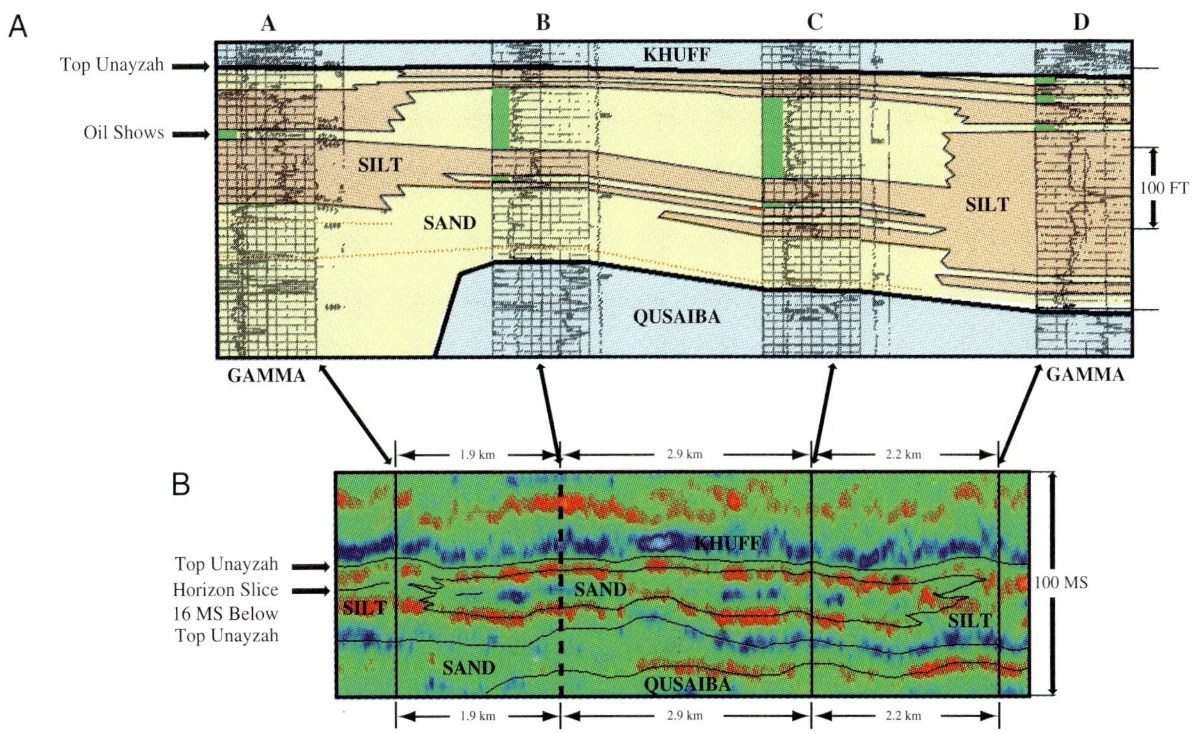


FIGURE 7. (A) Well log cross section across the Unayzah reservoir with infill wells B and C. Location of section shown in Figure 2. (B) Interpreted 3-D seismic section after drilling infill wells B and C.

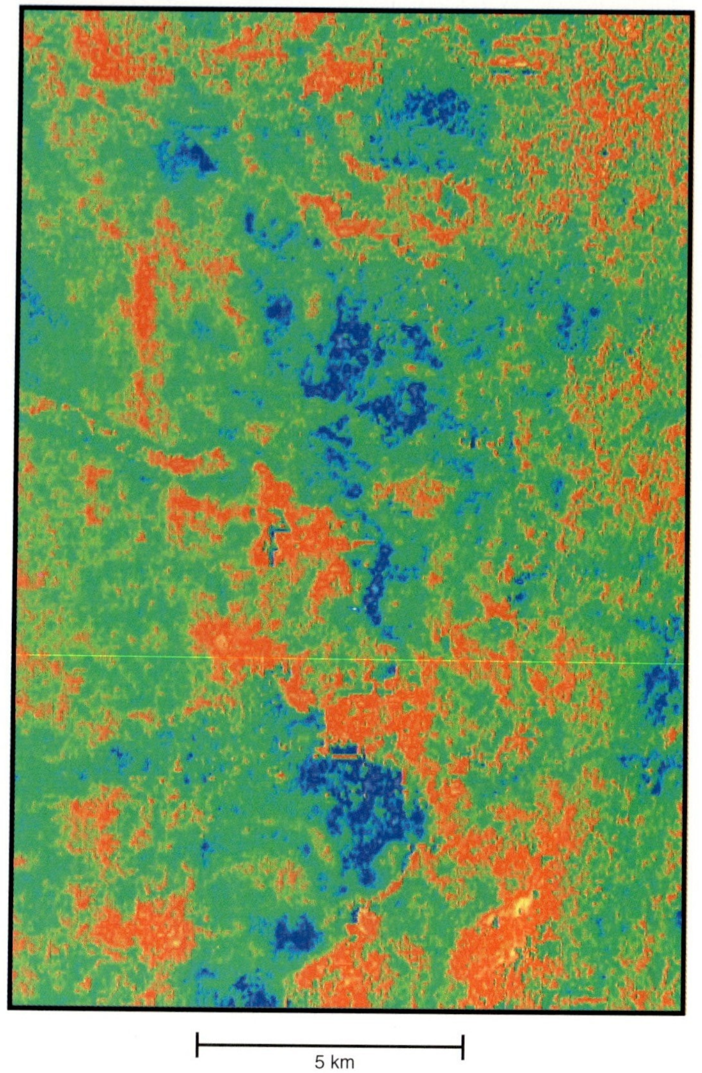


FIGURE 8. Horizon slice 16 ms below top Unayzah Formation.

FIGURE 9. Landsat image of the Ghinah oil field, central Saudi Arabia.

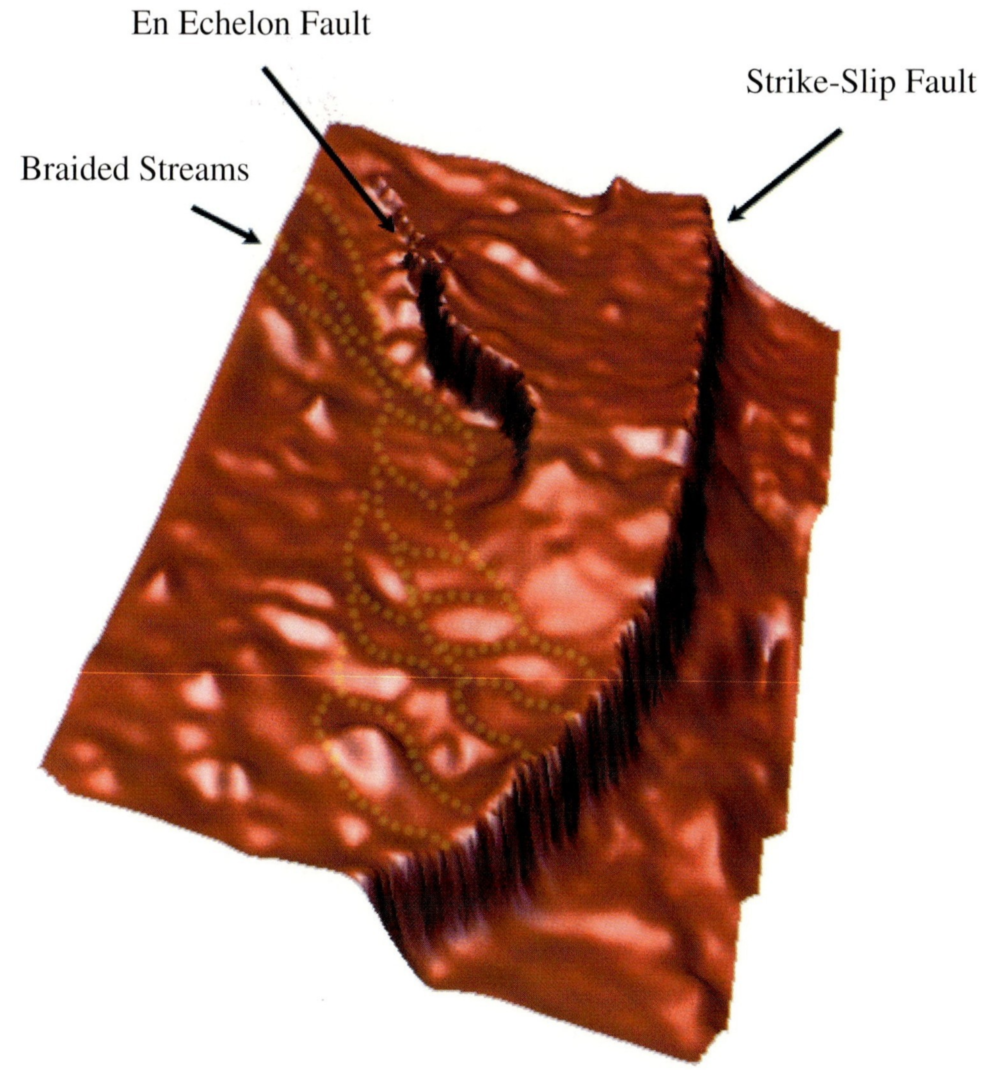


FIGURE 10. 3-D perspective of the basement structure of the Ghinah field, with braided streams developed on flanks of the en echelon structure.

# CHAPTER 11

Dorn, G., K. Tubman, D. Cooke, and R. O'Connor, Geophysical Reservoir Characterization of Pickerill Field, North Sea, Using 3-D Seismic and Well Data, *in* P. Weimer and T. L. Davis, eds., *AAPG Studies in Geology No. 42* and *SEG Geophysical Development Series No. 5* AAPG/SEG, Tulsa, p. 107–122.

# Geophysical Reservoir Characterization of Pickerill Field, North Sea, Using 3-D Seismic and Well Data

Geoffrey A. Dorn*, Kenneth M. Tubman*, Dennis Cooke†, and Rob O'Connor§

## ABSTRACT

Pickerill Field is a relatively thin, highly faulted gas reservoir in the southern gas basin of the North Sea. The 100- to 250-ft (30-to 76-m) thick, Permian Rotliegend reservoir consists primarily of thin dune and interdune deposits overlying generally poorer-quality fluvial sands. There is a rapid lateral variation in reservoir quality due to facies changes and compartmentalization due to diagenesis associated with faults.

A combination of petrophysics and geophysics was used to develop seismic criteria that could be used to optimize the location of development wells. An analysis of horizon attributes from the 3-D seismic survey produced a detailed reservoir fault map. Analysis of log data, seismic modeling, and horizon attributes produced an estimated reservoir porosity map. These have been used to help optimize the position of development wells in the field.

A detailed interpretation was made of several horizons in the 3-D seismic survey, including the Top Rotliegend (top reservoir) reflection. A set of seismic horizon attributes, based on horizon structure and reflection amplitude, were generated at the Top Rotliegend. Reservoir faults with throws as small as 15 ft (5 m) were interpreted and mapped. Since many faults were sealed by diagenesis subsequent to faulting, these results have helped identify potential compartmentalization and have allowed development wells to be positioned away from these potential barriers to flow.

Synthetic seismic modeling using log data from exploration wells indicated a linear relationship between reservoir reflection amplitude and average reservoir porosity. Log-based wavelet extraction was used to correct phase errors in the seismic data. Phase-corrected reflection amplitude from the Top Rotliegend reflection was correlated with porosity at exploration wells. The resulting empirical amplitude-porosity relationship has been used to successfully predict gross reservoir porosity in several wells drilled since the work was concluded.

## INTRODUCTION

Pickerill Field is located in Blocks 48/11a, 48/11b, 48/12b, and 48/17b in the UK portion of the southern gas basin of the North Sea (Figure 1).

Figure 2 is a schematic lithologic cross section from northwest to southeast through the field. The target is a relatively thin (~125 ft, or 38 m) Rotliegend section consisting of higher-quality aeolian sands overlying poorer-quality fluvial sands. A thin cemented to partially cemented Weissliegendes overlies an aeolian dune complex, which provides the highest-quality reservoir facies. A fluvial sandstone sequence of variable thickness underlies the aeolian dunes and provides lower-quality reservoir. The gross reservoir interval thickens from northwest to southeast, with most of the thickening occurring in the lower-quality fluvial sands.

Figure 3 is a generalized stratigraphic column, showing a fairly typical stratigraphic succession for this portion of the North Sea. The early Permian-age Rotliegend reservoir is overlain by Zechstein evaporites of varying thickness. Immediately above the reservoir sands is a basal Zechstein anhydrite which ranges in thickness from 175 to 250 ft (53 to 69 m), as determined from well data. Within the Zechstein sequence, the Plattendolomit, a major reflection event on the seismic data, is faulted and rafted so portions of the reservoir are overlain by Plattendolomit while other portions are not.

The reservoir itself is highly faulted, with two northwest-to-southeast en-echelon fault zones connected by a fault transfer zone consisting of discontinuous and variably oriented faults. Many of the faults have been reactivated several times during different periods of structural deformation. A particular concern in Pickerill Field is the presence and extent of structurally controlled cementation. Compressional shear fractures in the West Sole Field, to the north of Pickerill (Figure 1), are filled with anhydrite and quartz, and extensional fractures are filled with anhydrite (Glennie et al., 1978; Winter and King, 1991). Anhydrite-filled fractures are also present in the Amethyst Field to the northwest of Pickerill.

Drilling of the first development well in Pickerill (48/11b-A1Z) demonstrated that at least some of the faults in Pickerill Field are diagenetically altered. These altered zones may form barriers to flow within the reservoir, even when throw along the faults is smaller than the pay thickness.

Previous studies in the neighboring Amethyst Field (Garland, 1991) suggested that an empirical relationship might exist between the seismic reflection strength at the Top Rotliegend and the reservoir quality. Exploration drilling results in Pickerill Field indicated a rapid lateral variation in reservoir porosity.

The objective of this study was to develop criteria that could be used to optimize the location of development wells planned for the field. To accomplish this, the study was designed to provide a detailed fault map in the reservoir interval, an estimate of the gross reservoir thickness, and an indicator of reservoir quality. The fault map would be used to avoid flow barriers presented by diagenetically altered faults and fracture zones. Detailed estimates of porosity and thickness not only would help volumetric determinations but also would provide a means of optimizing well locations for field development.

Numerous technical problems had to be overcome to achieve these objectives. There are poor data zones in the 3-D seismic survey, particularly in areas where there is significant thickening of the overlying Zechstein sequence. In these poor data zones there is often no coherent reflection from the Top Rotliegend (top reservoir). The Top Rotliegend reflection changes from a single peak to a doublet in some areas, causing problems with horizon autotracking. Edges of the doublet zone could cause anomalies on the Top Rotliegend which might be misinterpreted as small faults. The change from a single event to a doublet could also complicate any attempt to use the amplitudes at the Top Rotliegend as a porosity predictor.

The 3-D seismic volume was not processed to zero phase, so a phase correction had to be determined and applied to the data. There was concern that the overlying Plattendolomit, which was present over portions of the reservoir, would cause amplitude and frequency anomalies on the Top Rotliegend reflection, complicating the fault interpretation, thickness determination, and reservoir quality estimation.

The northwest portion of the reservoir was known from drilling to be sufficiently thin so that reflections from the top and base of the reservoir would interfere with each other. Reflections from an anhydrite immediately above the reservoir sand and the Top Rotliegend would also interfere with each other. These interference effects could influence both the thickness determination and the porosity estimation. Finally, well control from which to develop any relationship between reservoir properties and seismic amplitudes was limited.

## HORIZON INTERPRETATION

"Seed" picks were made on a grid of inlines and crosslines. Figure 4 shows the seed picks on the Top Rotliegend in the northwest half of the survey area. The density of the picks varies somewhat with the structural complexity. A combination of volume autotracking and interpolation were used to infill the interpretation. Each horizon was then enhanced or "snapped" to the exact feature being picked. This resulted in the preliminary Top Rotliegend horizon interpretation shown in Figure 5.

The preliminary horizons were adequate for structural interpretation and time-to-depth conversion. However, cycle skipping occurred in some areas due to a low signal-to-noise ratio or due to the Top Rotliegend peak's splitting into a doublet. Figure 6 shows an example

*ARCO Exploration and Production Technology, 2300 W. Plano Pkwy, Plano, Texas, 75075
†ARCO Indonesia Inc., Jakarta, Indonesia 12910
§ARCO British Ltd., Guildford, Surrey, GU1-1UE, United Kingdom

of this on inline 500. These cycle skips may cause anomalies in the horizon attributes which would affect the fault trace interpretation and the porosity estimation. The enhanced horizon picks were edited and re-enhanced to eliminate the majority of these cycle skips. The results of applying this process to inline 500 are shown in Figure 7.

The final time structure map for the Top Rotliegend is shown in Figure 8. Comparing this final map to the preliminary time structure map (Figure 5) reveals a number of areas where some of the detail in the structural interpretation has changed due to the cleanup. However, the gross structural picture remains unchanged.

Even with the cleaned-up picks, the boundary of the doublet zone could cause time and amplitude anomalies on attribute maps. The Plattendolomit above the Top Rotliegend (Figure 7) is "rafted" into blocks that could cause amplitude anomalies at the Top Rotliegend. There are also areas where there is no coherent reflection from the Top Rotliegend (poor data zones). The boundaries of these regions (doublet zone, the Plattendolomit, and poor data zones) were picked and mapped during the cleanup process. The doublet boundary is shown in Figure 9; the boundary of the poor data zone is shown in Figure 10; and the Plattendolomit boundary is shown in Figure 11.

## FAULT TRACE INTERPRETATION

The interpretation of detailed fault trace patterns from horizon attribute maps began in the mid-1980s and is well documented in the literature. Fault traces appear as lineaments on a variety of horizon attributes, including dip magnitude and azimuth (Dalley et al., 1989), residual structure (Denham and Nelson, 1986), and reflection amplitude maps. The general techniques used here have been successfully applied over the last ten years in a variety of fields (e.g., see Rijks and Jauffred, 1991; Hoetz and Watters, 1992; and Bouvier et al., 1989).

A suite of four horizon attributes were used in the fault trace interpretation process: dip magnitude, dip azimuth, residual structure, and reflection amplitude. These attributes, calculated at the Top Rotliegend horizon, are shown in Figures 12 through 15. Several attributes were used because a fault is not necessarily visible on any one attribute but may appear on others. The horizon was interpreted across all intersecting fault surfaces so the appearance of all faults would be consistent on any given attribute, regardless of the amount of throw.

Dip magnitude (Figure 12) is the steepness of dip calculated at each data point on the interpreted horizon. Faults appear on this attribute as lineaments of anomalous dip magnitude. These lineaments, along the Top Rotliegend horizon, are typically steep dip (red) lineaments. Dip azimuth (Figure 13) is the direction of the dip at each point on the horizon. Faults appear as narrow lineaments of anomalous dip azimuth. Due to the relative geometry of the horizon and the fault surface, a fault may appear on either dip magnitude or dip azimuth, or both.

The residual horizon structure (Figure 14) is the difference between the interpreted horizon structure and a "regional" or smoothed version of the horizon. The smoothed version of the horizon is a low-pass spatial filtered version of the original structure. A fault will appear on the residual horizon structure map as a narrow lineament of positive residual adjacent to negative residual.

The final attribute used is reflection amplitude (Figure 15). Faults frequently appear as low amplitude anomalies (blue lineaments) on reflection amplitude maps, since there is little coherent energy reflected to the surface from the fault.

These attributes were calculated and displayed interactively on the workstation. The fault interpretation was performed interactively, adjusting the color mapping of each attribute as smaller faults were interpreted.

Three-dimensional displays were also used to understand the structure and to aid the fault trace interpretation. Figure 16 shows a large portion of the northern half of the reservoir, with several well locations. The structure of the surface is given by the structure of the interpreted horizon. The color on the surface is dip magnitude. The color scale used is the same as that shown for dip magnitude in Figure 12. The display also uses a light source to cast highlights and shadows on the surface. The interpreter interactively adjusts the position and orientation of the light source and the interpreter's viewpoint. These 3-D displays were a valuable aid in highlighting very small throw faults in structurally complex areas.

Preliminary fault trace interpretations were refined by extracting traverse sections from the seismic volume in directions normal to the strike of the fault traces. These are the vertical sections that will best image the fault. The traverses used in the northern half of the survey are shown in red in Figure 17; fault traces are shown in black. The traverse section along the traverse path highlighted in blue on Figure 17, is shown in Figure 18.The traverse is displayed with markers where it intersects each preliminary fault trace.

Although such traverse sections are a useful tool for performing quality control on the fault trace interpretation, they must be used with some judgment. The horizon attribute maps will image faults with throws that are less than one sample in the seismic data. As a result, some of the fault traces that are interpreted from the attribute maps may not be visible even on a traverse that is perpendicular to the fault trace. In such cases, the interpreter must make a judgment based on what is imaged in the attributes and on a knowledge of the geology.

The preliminary fault traces were also compared with the doublet boundary (Figure 9) and the Plattendolomit boundary (Figure 11) to eliminate any false traces that were associated with anomalies caused by either of these two boundaries. Confidence levels were assigned to the interpreted fault traces based on the interpreter's assessment of the quality control traverses and the poor data zones mapped in Figure 10. Throw values were obtained at several points along each interpreted fault.

The final fault map is shown in Figure 19. Estimated throw values are posted in feet, and dashed fault traces indicate a lower level of confidence, while solid fault traces indicate higher confidence.

## RESERVOIR CHARACTERIZATION

The analysis of the seismic data for a relationship with reservoir quality involved several steps. First, seismic modeling was needed to ascertain whether a relationship would exist in an ideal case for the range of physical properties (thickness, porosity, etc.) expected for this reservoir. Second, extraction of the seismic wavelet was required to assess the phase of the seismic data. The use of zero-phase seismic data optimizes the possibility of detecting a seismic amplitude–reservoir porosity relationship. This is particularly important for reflections from a seismically thin reservoir, where seismic tuning effects could be anticipated (i.e., reservoir thickness is less than 1/4 wavelength in the seismic data).

If the wavelet extraction demonstrated that the seismic data were not zero phase, a phase correction would have to be applied to the data. Finally, since the seismic data were time migrated, and the well data were in the depth domain, a final map migration correction (Hubral correction) was required to obtain the corrected locations of the wells with respect to the interpreted Top Reservoir reflection.

Each of these steps will be discussed in the following sections.

## MODELING

The generation of synthetic seismograms and modeling was an essential part of the analysis leading to an estimation of the porosity. Modeling results suggested that an amplitude-versus-porosity relationship should be present in the data. Modeling also helped develop an understanding of the relative importance of the parameters that could affect the amplitude at the Top Rotliegend reflection and helped explain changes in the character of the Top Rotliegend reflection.

Figure 20 is a plot of a typical suite of logs obtained in one of the exploration wells in the Pickerill Field. The logs (gamma ray, density, velocity, impedance, and neutron porosity) cover the interval from within the Zechstein down through the Basal Anhydrite, the Rotliegend reservoir, and into the underlying Carboniferous section. Features of particular note, relating to the seismic response, are the sharp impedance contrasts at the Zechstein/Basal Anhydrite boundary and at the Basal Anhydrite/Rotliegend boundary and the gradational impedance boundary at the base of the reservoir. Modeling based on these data suggests that significant reflections will occur at the Zechstein/Basal Anhydrite and the Basal Anhydrite/Rotliegend boundaries. The gradational boundary at the base of the Rotliegend will give rise to only a weak reflection.

The possible amplitude effects due to tuning were of particular concern. The Anhydrite is sufficiently thin (Figure 20) that reflections from the Top Anhydrite and the Top Rotliegend will be in tuning (will interfere) at the dominant frequency in the seismic data. Figure 21 is a crossplot of reflection amplitude at Top Rotliegend versus the isochron between the Top Rotliegend and the Top Anhydrite from a large portion of the 3-D survey, including a portion of the reservoir. On this crossplot, each dot represents a data point from a single trace. The envelope of the scatter plot clearly describes a statistical tuning curve (Brown, 1988). Fortunately, data from exploration wells suggest that both the thickness and the impedance of the Anhydrite over the reservoir itself are relatively constant. Although there is clearly a tuning effect present, it is a second-order effect on the amplitude variation over the reservoir proper.

The Rotliegend reservoir is also sufficiently thin that reflections from its top and base would also potentially produce a tuning effect. Since the impedance contrast at the base of the Rotliegend is gradational, tuning effects due to changing reservoir thickness would have a small effect on the amplitudes measured at the Top Rotliegend.

Figure 22 is an example crossplot of acoustic impedance versus log-derived porosity in the Rotliegend interval from one of the wells in the Pickerill field. The linear relationship shown is typical for wells in the reservoir. The reflection amplitude at the Top Rotliegend is a function of the acoustic impedance contrast between the Anhydrite and the Rotliegend. Since the thickness and acoustic impedance of the Anhydrite are relatively constant over the field, the dominant effect on the reflection strength is due to the variation in acoustic impedance in the Rotliegend. Finally, the linear relationship between porosity and acoustic impedance in the Rotliegend shown in Figure 22 suggests that there should be a linear relationship between reflection amplitude and reservoir porosity.

Synthetic seismograms were created at ten wells in the field area using a zero-phase wavelet and impedance logs from the wells. Figure 23 is a plot of the average reservoir porosity (log-derived porosity averaged over the thickness of the reservoir in each well) versus the synthetic reflection amplitude at the Top Rotliegend. The analysis shows a good linear relationship between amplitude and porosity, consistent with the other modeling results.

## WAVELET EXTRACTION

In the empirical search for a relationship between horizon attributes and reservoir properties, crossplots were made between a variety of seismic attributes and petrophysical properties obtained from well data. An initial crossplot of amplitude versus porosity is shown in Figure 24. The data are a poor fit to a linear trend and the scatter is relatively large. Since the relationship predicted by the modeling study (Figure 23) is much tighter than that obtained in Figure 24, an effort was made to further reduce the scatter.

One step in this effort was the extraction of wavelets from the seismic volume at a number of exploration well locations. Wavelet extraction allows the interpreter to determine the phase of the seismic data and the spatial variability of the phase. There was concern that the data might not be zero phase and that the phase might not be constant over the 3-D survey area. With a non-zero phase, the reflection character would be more complex, with larger wavelet side lobes creating more complicated interference effects. If the phase were variable laterally in the 3-D survey, the desired log-seismic ties would be difficult to establish. By correcting the phase, if necessary, to achieve a zero-phase volume, the vertical and horizontal resolution of the data would be optimized, and the chances of finding an amplitude-porosity relationship would be improved (e.g., Poggiagliolmi and Allred, 1994).

Wavelets were extracted at a number of wells along a traverse through the center of the field. The traverse was chosen to avoid as many potential complications (i.e., poor data, Top Rotliegend doublet, Plattendolomit) as possible. The phase could be corrected to zero provided it remained roughly constant. The wells in which wavelet analysis was done were: 48/11b-4, 48/11b-6, 48/11a-7, 48/11a-9, 48/11b-A1Z, 48/12b-4, and 48/17a-8. Wavelet extraction was also attempted in 48/11b-10 to confirm the phase far to the northwest; however, the poor data quality in that region precluded reliable analysis.

The wavelet extraction process uses impedance logs and a section of seismic data around each well location to determine amplitude and phase as functions of frequency (Nyman et al., 1987). Amplitude and phase spectra are determined separately, with the ability to smooth and hand-edit each at a number of stages of the process.

The amplitude spectrum is determined by calculating the average spectrum from traces near the well location. A compensated spectrum is then calculated by dividing the amplitude by the spectrum of the reflectivity sequence defined by the well log data. Both the average amplitude spectrum and the spectrum of the reflectivity can be smoothed before the spectral division. Figure 25 shows an example of the average, reflectivity, and compensated spectra.

The reflectivity series is filtered to have a spectral bandwidth similar to that of the seismic data. A correlation is performed between the resulting series and each seismic trace within the spatial window surrounding the well location. The phase spectrum of the wavelet is determined from an average of the correlograms. Smoothing and editing is performed on both the correlograms themselves and the resulting phase spectrum. The smoothed phase spectrum (Figure 25D) is combined with the compensated amplitude spectrum to yield the extracted wavelet. An example is shown in Figure 26.

In this study, the spatial window was typically about 15 traces. The exact number of traces used at each well depended on the nature of the data in the vicinity of that well. Spatial windows encompassing faults decrease the correlation of the seismic traces with the synthetic traces. Smaller windows would be used in this case to eliminate the fault region from the analysis. The window was similar for both the amplitude and the phase determination. The temporal windows for the correlations were chosen to be as long as possible while ensuring that events on the synthetic traces matched events observed in the seismic data. Consistency and quality of the wavelet extractions were checked by varying both the spatial and the temporal windows.

An average wavelet was needed to determine an operator for correcting the phase of the seismic volume. Individual phase spectra from different wells were calculated separately and then averaged. This approach allowed examination of not just the average phase spectrum but also individual deviations from it. The resulting average amplitude and phase spectra are shown in Figure 27. The appearance of these spectra is as anticipated. The amplitude spectrum retains the notch in the middle frequencies, and the phase spectrum shows some variation, with the phase approximately –100°.

The average wavelet derived from these spectra is shown in Figure 28. The energy is shifted towards the back of the signal. The trailing peak exhibits the "notch" observed in most of the wavelets from the individual wells.

## PHASE CORRECTION

A dephasing operator was calculated from the average wavelet and applied to the data volume. The operator was designed to shift the average phase spectrum (Figure 27) to flat zero phase. To characterize the effectiveness of this operator across the field, phase and wavelet analyses were performed again at each of the wells, this time using the phase-corrected seismic data. Both variable and constant phase calculations were used. The variable phase analysis was the same as that discussed above. The constant phase analysis consists of correlating the reflectivity series and the seismic data for a variety of phase shift values. The shifts that yield the highest correlation are taken as the best fit (see Figure 29).

Figure 30 shows the residual phase for each of the wells. The values are generally less than 10°. The most notable exception is 48/11b-A1Z. This well was very close to a number of faults. It was not possible to get enough seismic traces for a reliable wavelet calculation without crossing a fault. This made correlation between the well reflectivity and the seismic traces very poor. This well was not included in the calculations of the average wavelet.

The dephasing operator had some interesting effects on the Top Rotliegend reflection amplitude. Figures 31A and 31B show the reflection amplitude before and after phase correction. The gross character of the map has not changed, but some of the subtle features have.

Another effect of the dephasing operation was a change in the doublet Top Rotliegend reflection. Figures 32A and 32B show inline 1150 before and after dephasing. The doublet nature of the Top Rotliegend is evident in the original data. The same area exhibits a single, strong reflection in the post–phase correction data. Subsequent modeling demonstrated that synthetic seismograms exhibit the doublet if they are generated using the extracted –100° wavelet and the anhydrite section above the Rotliegend exceeds a certain thickness. The doublet was due to interference between reflections of the non-zero-phase wavelet from the Top Anhydrite and the Top Rotliegend. When the data are modeled using a zero-phase wavelet, the doublet is eliminated.

## POROSITY-AMPLITUDE RELATIONSHIP

A variety of techniques have been used in the literature to relate reservoir porosity or the product of porosity and thickness to seismic attributes. The particular approach that should be used depends on the nature of the relationship between the various reservoir parameters that affect the seismic data. For a thin reservoir, such as occurs at Pickerill Field, the approach chosen was to search for a simple direct relationship among reflection amplitude and porosity. Modeling showed that such a relationship should exist, and log data suggested that the tuning effects caused by the overlying anhydrite layer would be minimal because of the relatively constant thickness of the anhydrite over the reservoir.

Once the phase of the seismic data was corrected, an effort was made to find a linear relationship between the reflection amplitude at the Top Rotliegend reflector and average log-derived porosity in the reservoir using data from the exploration wells. The approach used was similar to that described by Stanulonis and Tran (1992).

A porosity-amplitude crossplot was made using the phase-corrected seismic data. Since the seismic data were not depth migrated, a Hubral correction was applied to each of the well locations prior to wavelet and amplitude analysis. This adjustment to the well locations was required to correctly position the wells with respect to the time-migrated seismic data.

Figure 33 shows the resulting porosity-amplitude relationship. All available wells are included on this plot except four: 48/11-1, 48/11b-10, 48/11a-11, and 48/17a-8. Wells 48/11-1, 48/11b-10, and 48/11a-11 were not included because they are within the poor data zones, making the amplitudes unreliable. Well 48/17a-8 is off the field to the southeast.

The elimination of single points from the dataset has little effect on the fit or the value of the slope. The absolute prediction error is less well constrained. The prediction error with 95% confidence is 3.8% porosity. The prediction error with 80% confidence is 2.2% porosity.

The statistics of the porosity-amplitude trend provide insight as to how best to use the information. Predictions of absolute porosity on a global scale are difficult because of the large error bars associated with absolute prediction. The slope is well defined, so the relationship is useful for determining the best porosity in a local area. It is useful to know the slope, not just that there is some trend, because it provides an indication of the tradeoff between location, from a reservoir drainage standpoint, and porosity, and it can indicate how much porosity can be gained by moving a certain distance from the best drainage location.

The porosity-amplitude trend can be combined with the reflection amplitude map to produce an estimated porosity map of the entire field. This map is shown in Figure 34. The estimated porosity can then be combined with interpreted Top Rotliegend structure and rendered in 3-D displays with existing or proposed well locations (Figures 35A and 35B). This visually integrates the structural and porosity data in one display.

Figure 36 is a plot of the gross reservoir porosity versus the reflection amplitude from the phase-corrected volume (similar to Figure 33), including four new data points from wells drilled after this study was concluded. In all cases, the porosity obtained in the new wells fits with the established amplitude-porosity trend.

## SUMMARY AND CONCLUSIONS

The primary goal of this study was to provide guidance for the placement of development wells to be drilled in Pickerill Field. Previous drilling had shown that some of the faults could act as barriers to flow, even with throws that are significantly less than the thickness of the reservoir. Exploration drilling had also shown that porosity varied rapidly laterally within the reservoir. Finally, a more detailed estimate of the thickness of the reservoir than could be obtained from the wells alone would also help development planning.

The detailed fault interpretation from horizon attribute surfaces resulted in a map of faults at the top of the reservoir with throws greater than about 15 ft (5 m). A porosity–reflection amplitude relationship was established over the reservoir, providing porosity estimates for local area optimization of well locations. It was not possible to obtain a seismically derived estimate of the variation in thickness of the reservoir because of the gradational character of the impedance change between the Rotliegend and the Carboniferous.

The recommended use of these results in well positioning was to determine well locations based on:

- Reservoir optimization requirements
- Structure
  - faults
  - elevation
- Local area porosity estimates
- Drilling considerations

This procedure has been used to help position several wells which have been drilled in the field since the study was concluded. The gross reservoir porosities obtained in the new wells conform well with the established amplitude-porosity trend.

## REFERENCES CITED


Bouvier, J. D., C. H. Kaars-Sijpesteijn, D. F. Kluesner, C. C. Onyejekwe, and R. C. Van Der Pal, 1989, Three-dimensional seismic interpretation and fault sealing investigations, Nun River Field, Nigeria: AAPG Bulletin, v. 73, n. 11, p. 1397–1414.

Brown, A., 1988, Interpretation of three-dimensional seismic data: AAPG Memoir 42, 2nd Edition, 253 p.

Dalley, R. M., E. C. A. Gevers, G. M. Stampfli, D. J. Davies, C. N. Gastaldi, P. A. Ruijtenberg, and G. J. O. Vermeer, 1989, Dip and azimuth displays for 3D seismic interpretation: First Break, v. 7, n. 3, p. 86–95.

Denham, J. I., and H. R. Nelson, Jr., 1986, Map displays from an interactive interpretation: Geophysics, v. 51, n. 10, p. 1999–2006.

Garland, C. R., 1991, The Amethyst Field, Blocks 47/8a, 47/9a, 47/13a, 47/14a, 47/15a, U.K. North Sea, *in* I. L. Abbotts United Kingdom Oil and Gas Fields: 25 Years: Commemorative Volume, Geologic Society Memoir 14, p. 387–393.

Glennie, K. W., G. C. Mudd, and P. J. C. Nategaal, 1978, Depositional environment and diagenesis of Permian Rotliegendes sandstones in Leman Bank and Sole Pit areas of the U.K. southern North Sea: Journal of the Geologic Society of London, v. 135, p. 25–34.

Hoetz, H. L. J. G., and D. G. Watters, 1992, Seismic horizon amplitude mapping for the Annerveen Gasfield, The Netherlands: First Break, v. 10, n. 2, p. 41–45.

Nyman, D.C., M. J. Parry, and R. D. Knight, 1987, Seismic wavelet estimation using well control: 57th Annual Meeting of the Society of Exploration Geophysicists, October 11–15, New Orleans, LA.

Poggiagliolmi, E., and R. D. Allred, 1994, Detailed reservoir definition by integration of well and 3-D seismic data using space adaptive wavelet processing: The Leading Edge, July 1994, p. 749–754.

Rijks, E. J. H., and J. C. E. M. Jauffred, 1991, Attribute extraction: An important application in any detailed 3-D interpretation study: The Leading Edge, Sept. 1991, p. 11–19.

Stanulonis, S. F., and H. V. Tran, 1992, Method to determine porosity-thickness directly from 3-D seismic amplitude within the Lisburne carbonate pool, Prudhoe Bay: The Leading Edge, January 1992, pp. 14–20.

Winter, D. A., and B. King, 1991, The West Sole Field, Block 48/6, UK North Sea, *in* I. L. Abbotts, ed., United Kingdom Oil and Gas Fields: 25 Years Commemorative Volume, Geologic Society Memoir 14, p. 517–523.

FIGURE 1. Location of Pickerill Field in the southern North Sea.

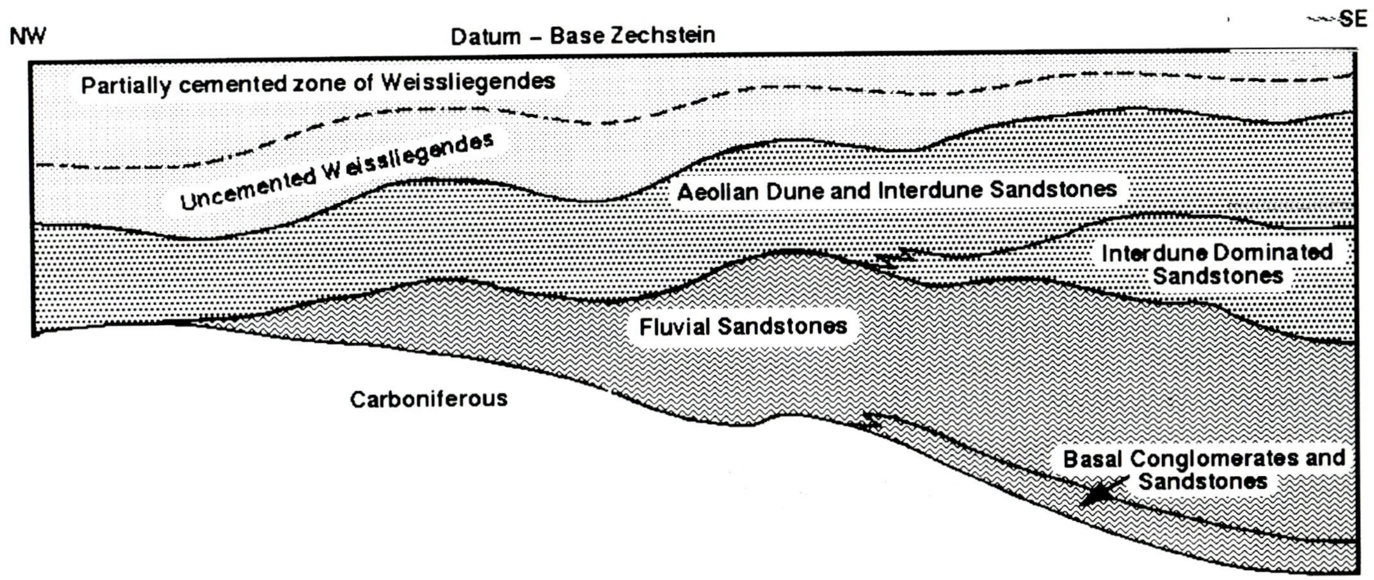


FIGURE 2. Schematic lithologic cross section through Pickerill Field datumed on the Base Zechstein or Top Rotliegend.

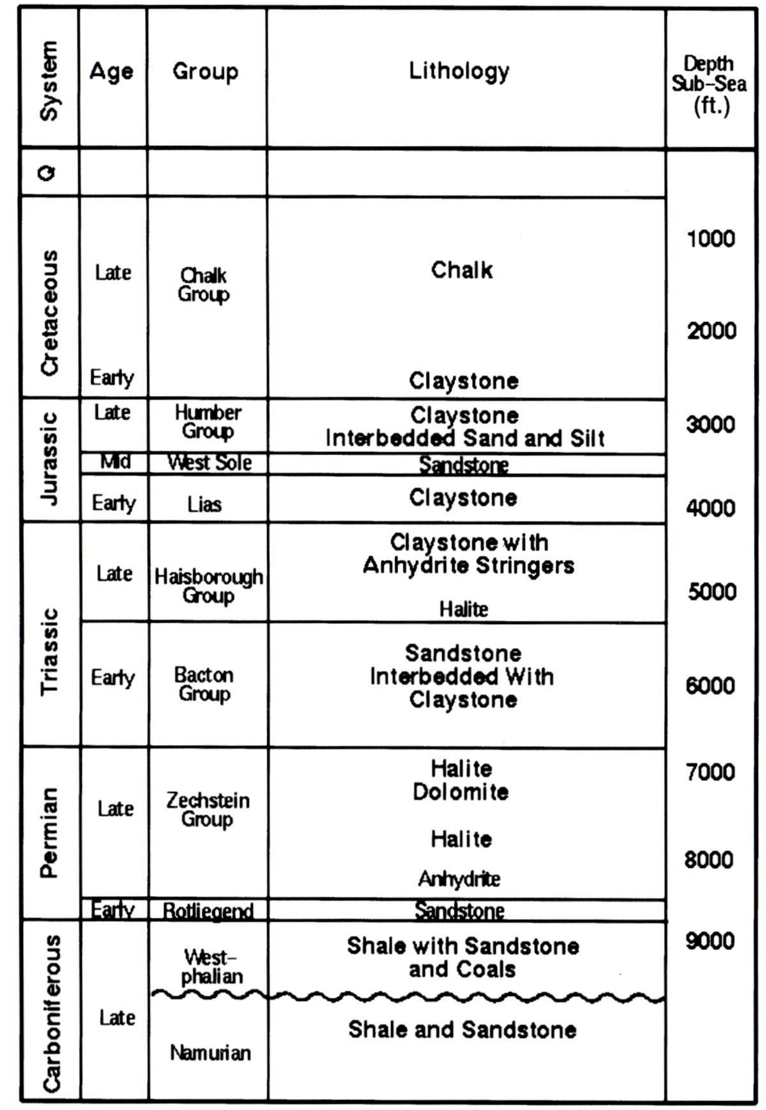


FIGURE 3. Generalized stratigraphic column for the Pickerill Field area, southern North Sea gas basin.

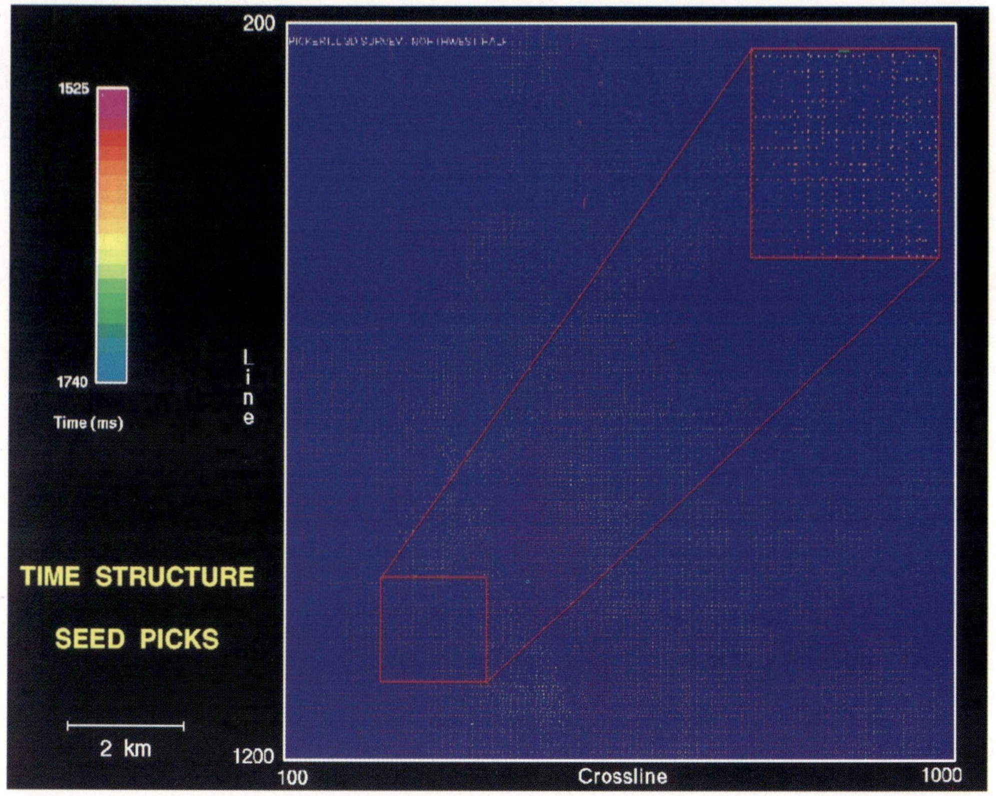


FIGURE 4. Seed picks on the Top Rotliegend in the northwest half of the 3-D survey area. Two-way time to the picks is shown in color.

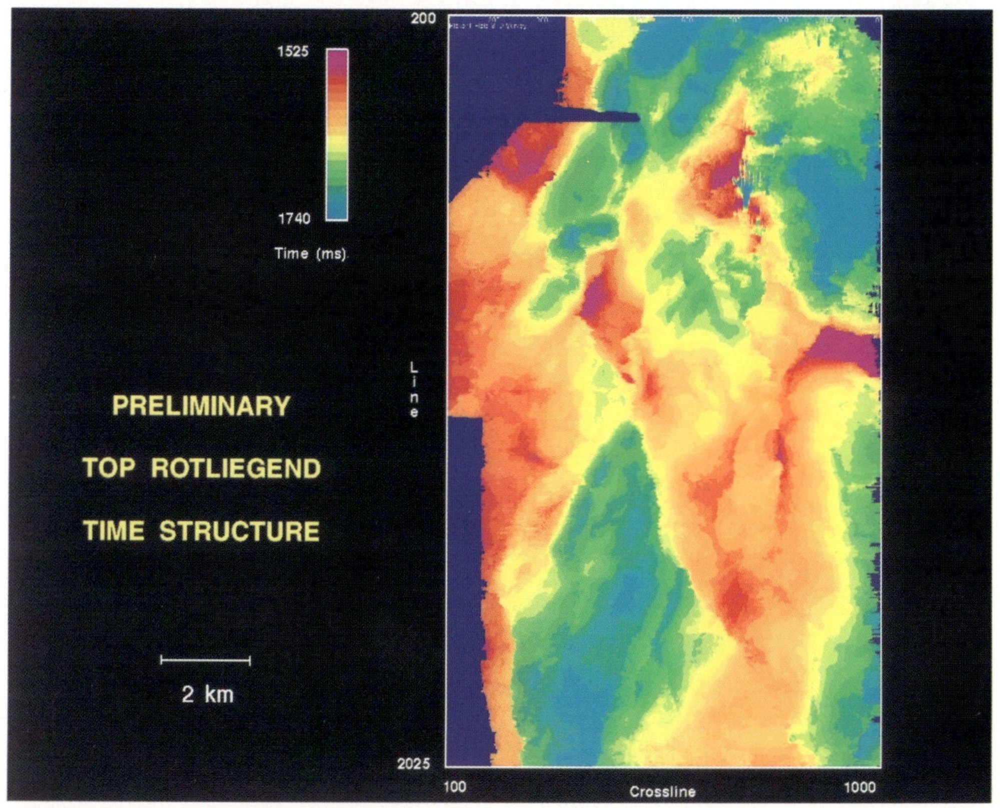


FIGURE 5. Preliminary time structure map of the Top Rotliegend. This version of the horizon was obtained from the seed picks using a combination of autotracking and interpolation.

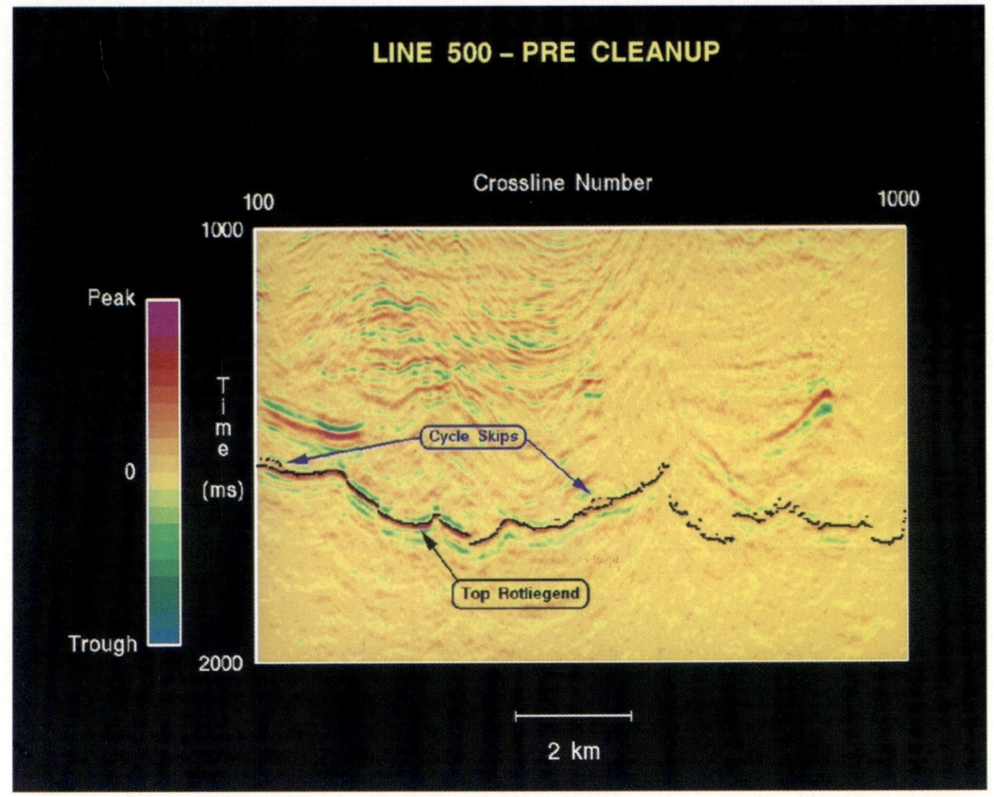


FIGURE 6. Inline 500 from the 3-D seismic survey, showing the preliminary picks for the Top Rotliegend horizon. Cycle skips have occurred where the Top Rotliegend reflection splits into a doublet and where the signal-to-noise ratio is low.

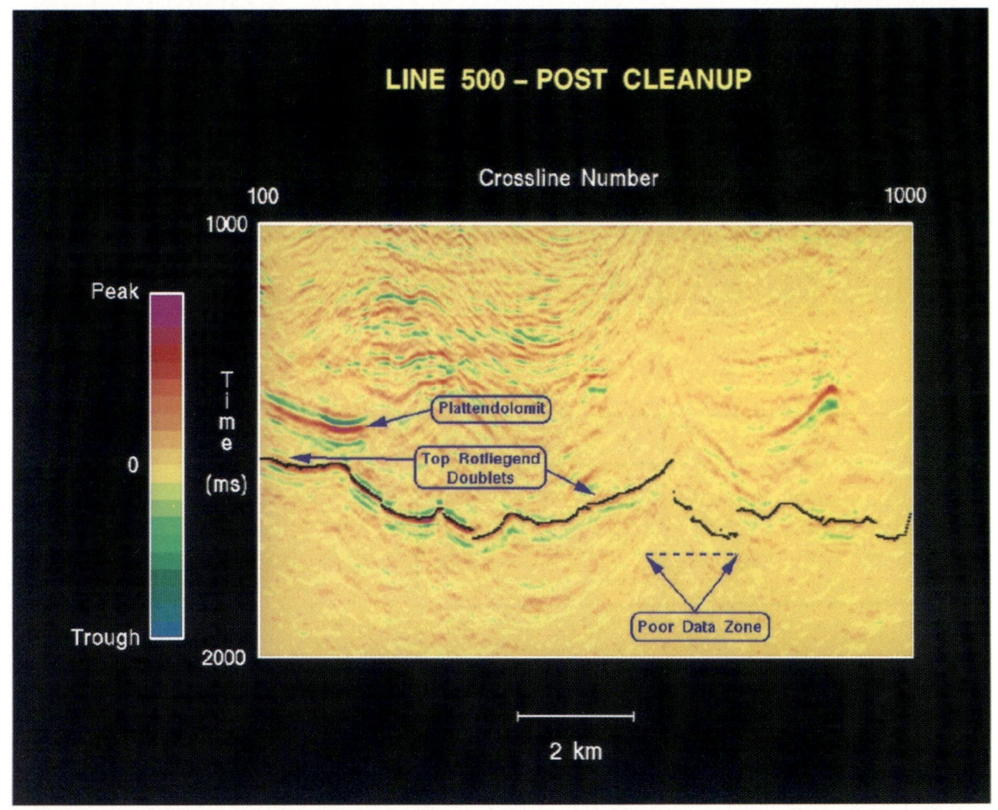


FIGURE 7. Inline 500 from the 3-D seismic survey, showing the final picks for the Top Rotliegend horizon. The cycle skips have been eliminated by editing and re-enhancing the horizon.

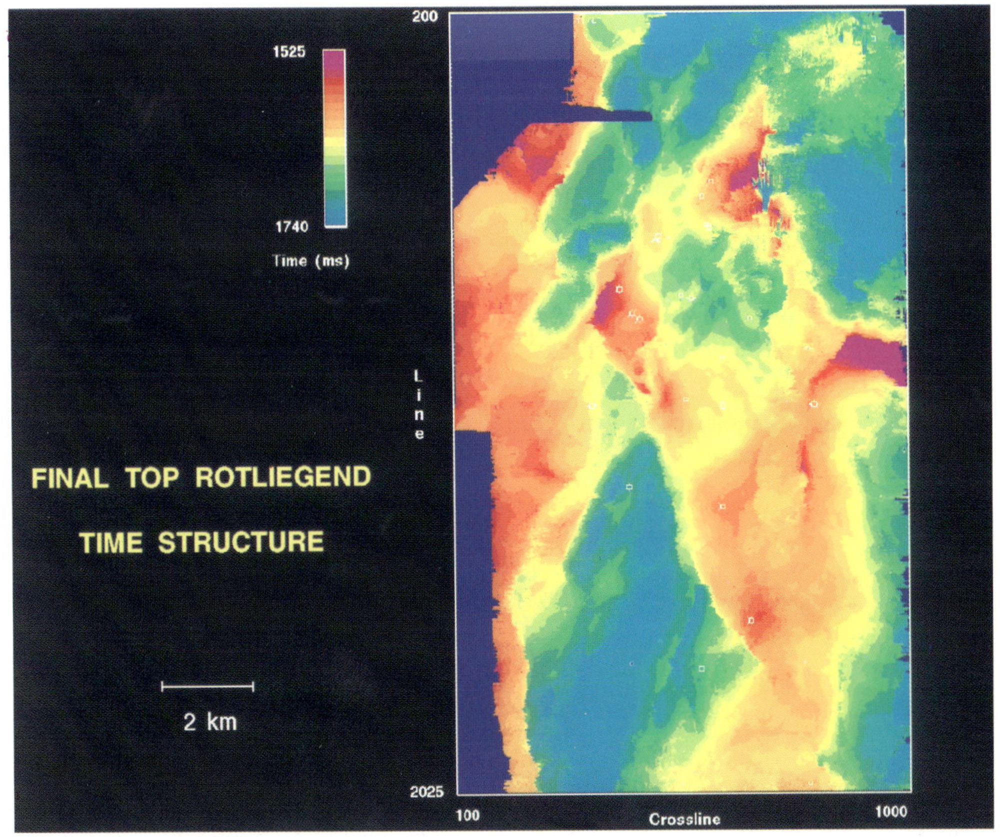


FIGURE 8. Final time structure map of the Top Rotliegend.

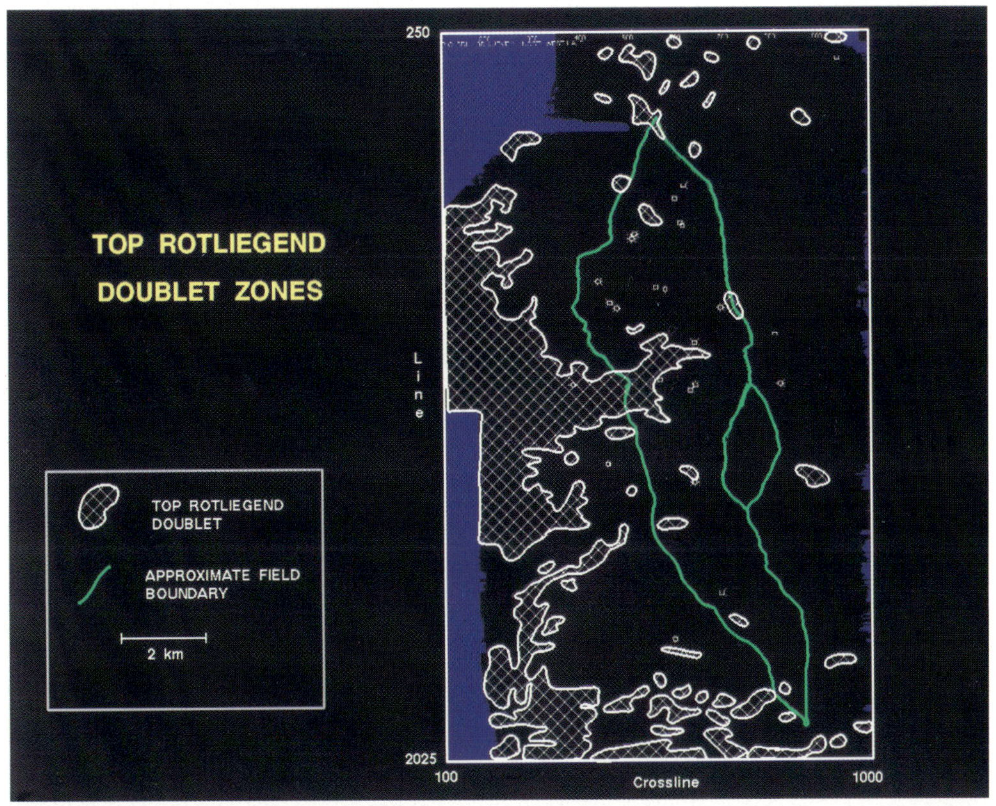


FIGURE 9. Boundary of the areas in which the Top Rotliegend event is a doublet.

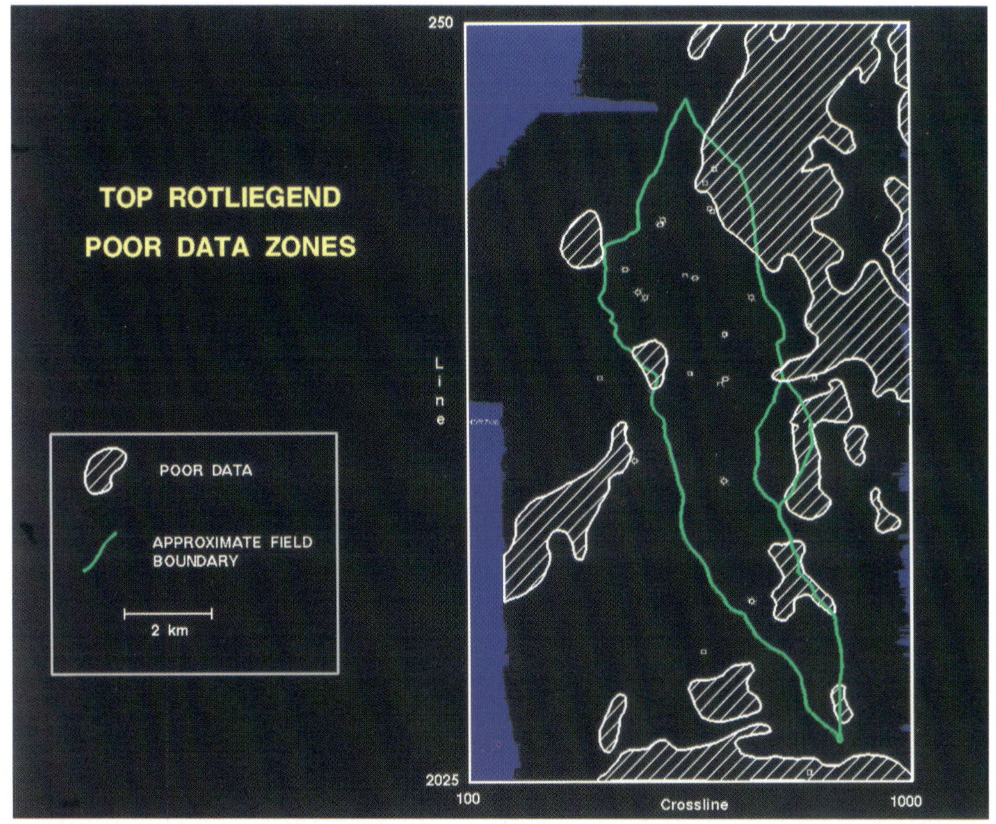


FIGURE 10. Boundary of the areas in which there is very poor quality data at the Top Rotliegend.

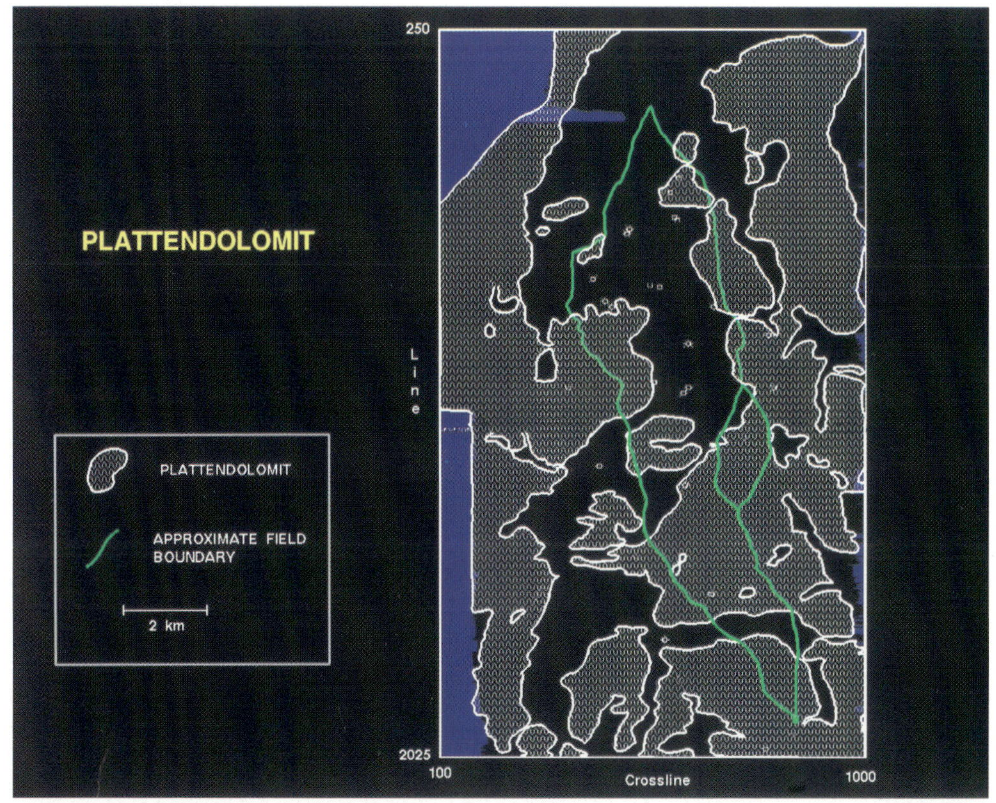


FIGURE 11. Boundary of the Plattendolomit rafts.

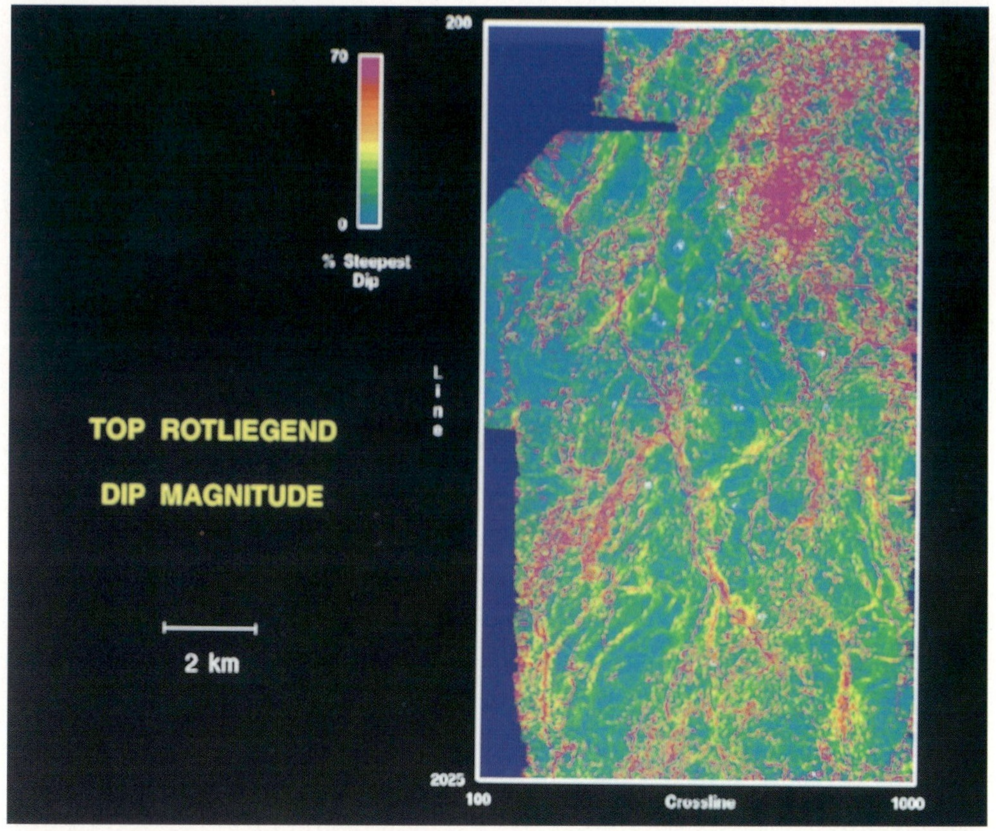


FIGURE 12. Dip magnitude at the Top Rotliegend.

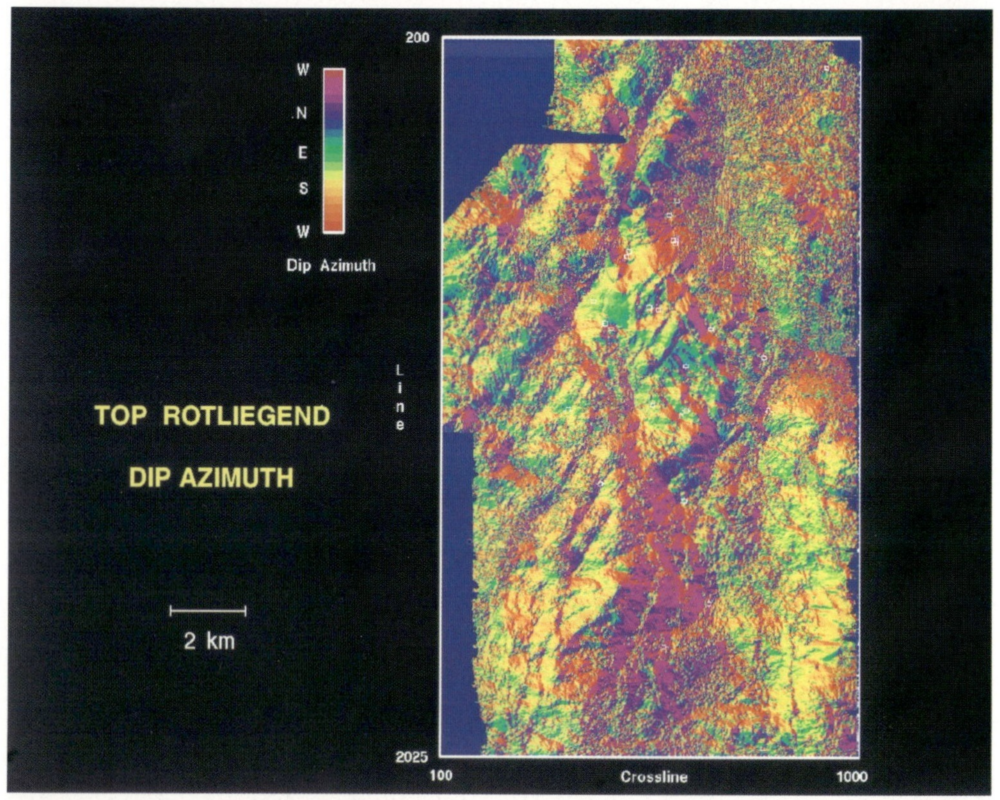


FIGURE 13. Dip azimuth at the Top Rotliegend.

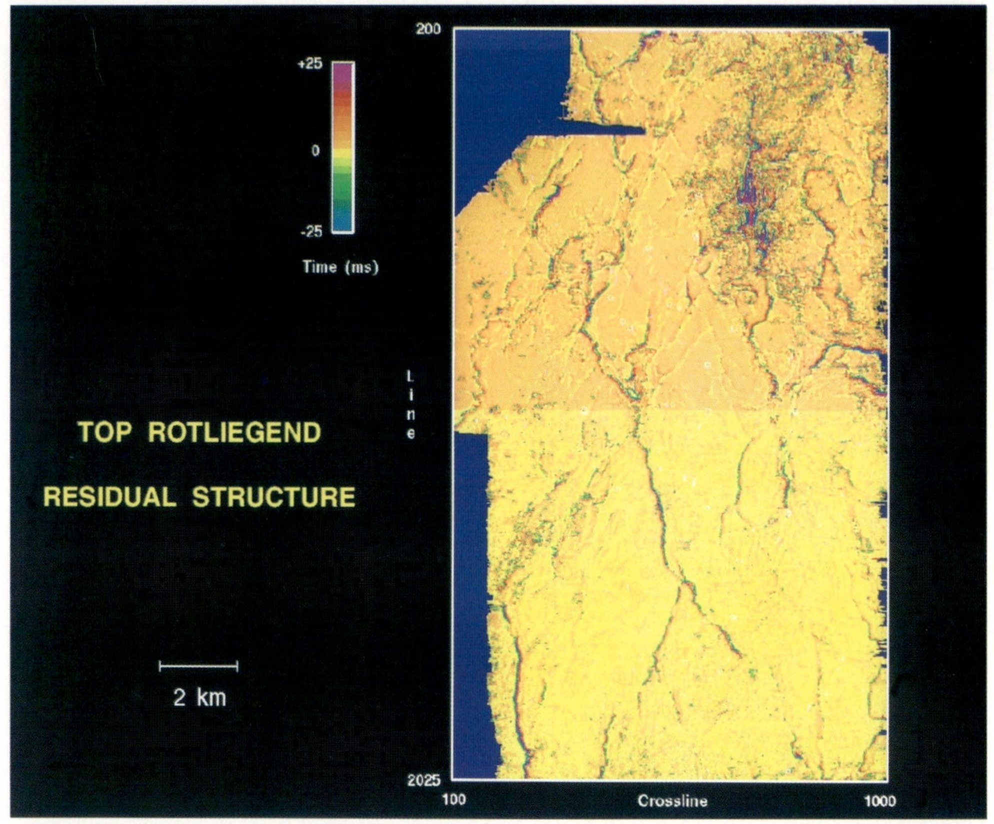


FIGURE 14. Residual horizon structure at the Top Rotliegend.

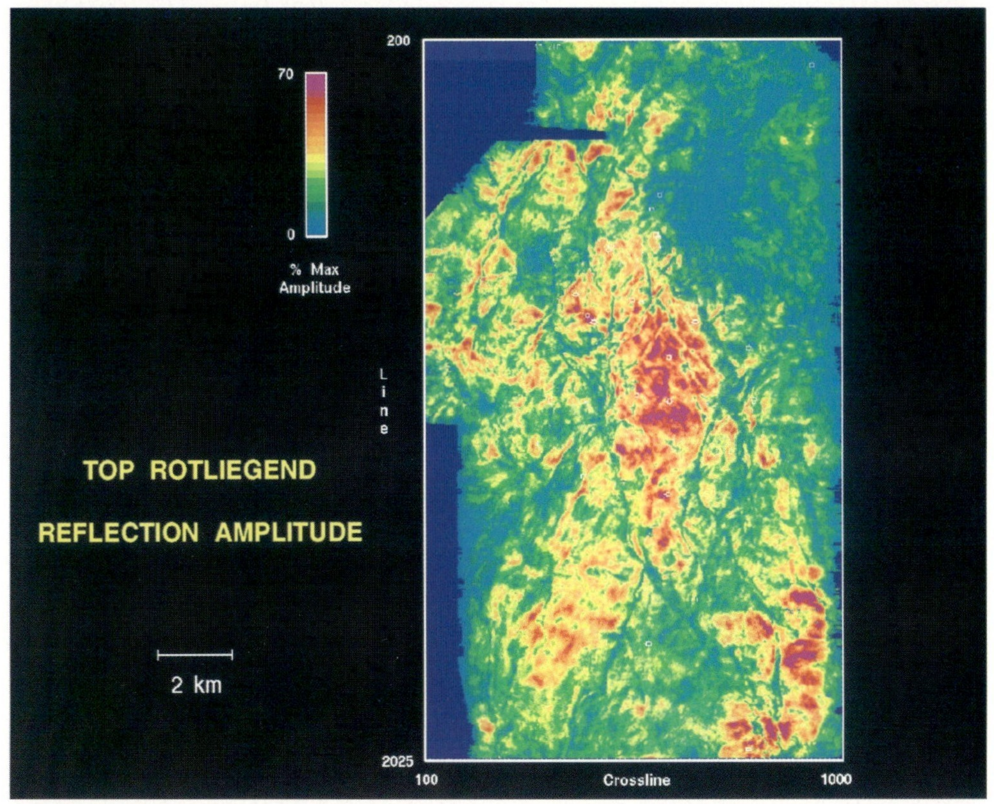


FIGURE 15. Reflection amplitude (of the peak) at the Top Rotliegend.

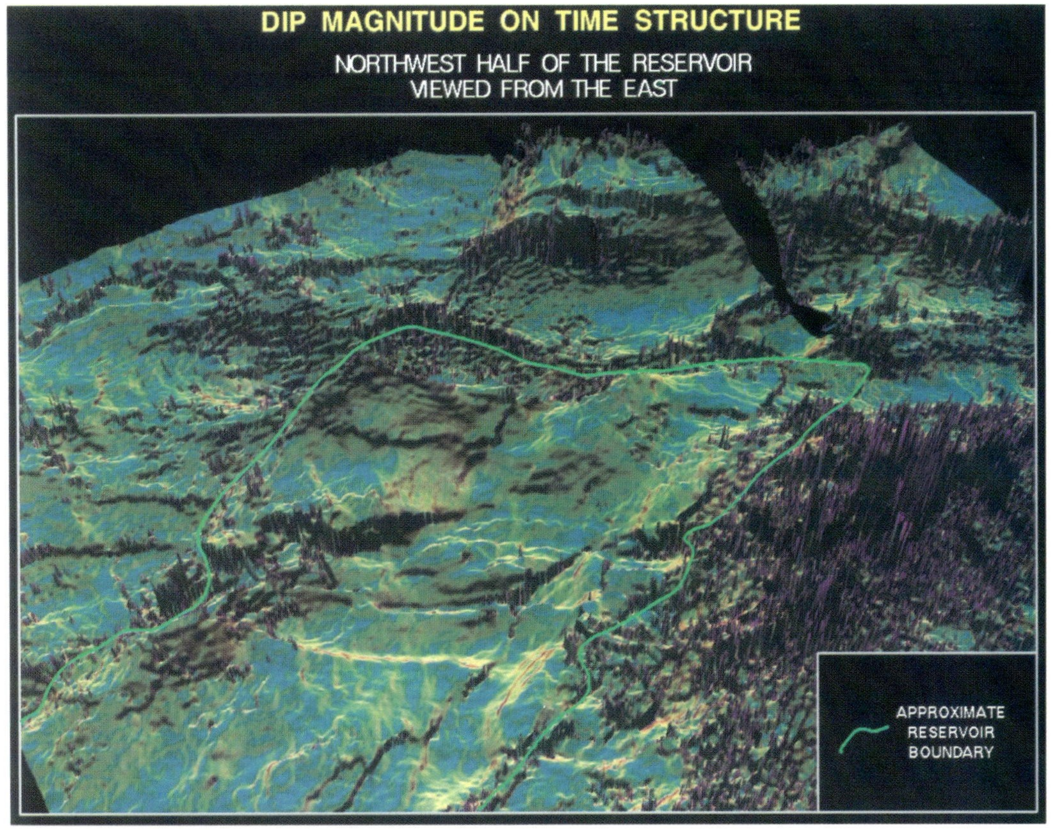


FIGURE 16. Three-dimensional display of the interpreted Top Rotliegend horizon in the northwest portion of the survey, with dip magnitude shown in color. View is from the east.

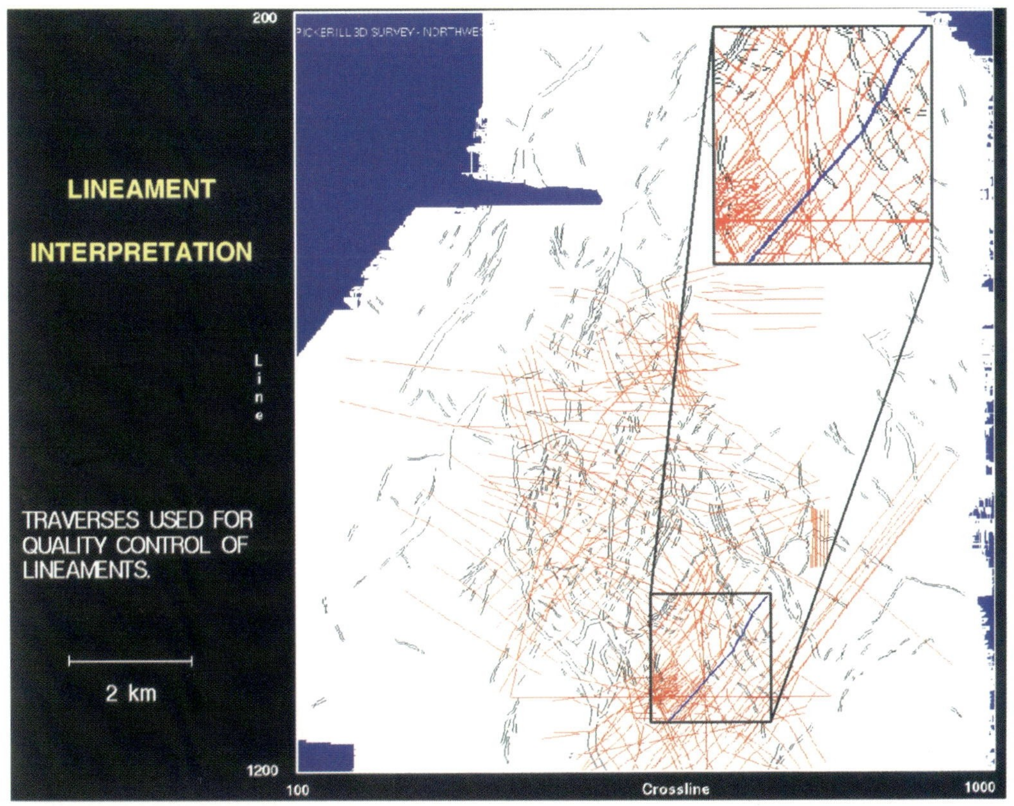


FIGURE 17. Preliminary fault trace interpretation (black) at the Top Rotliegend. The traverses used to refine the fault interpretation are shown in red (and blue).

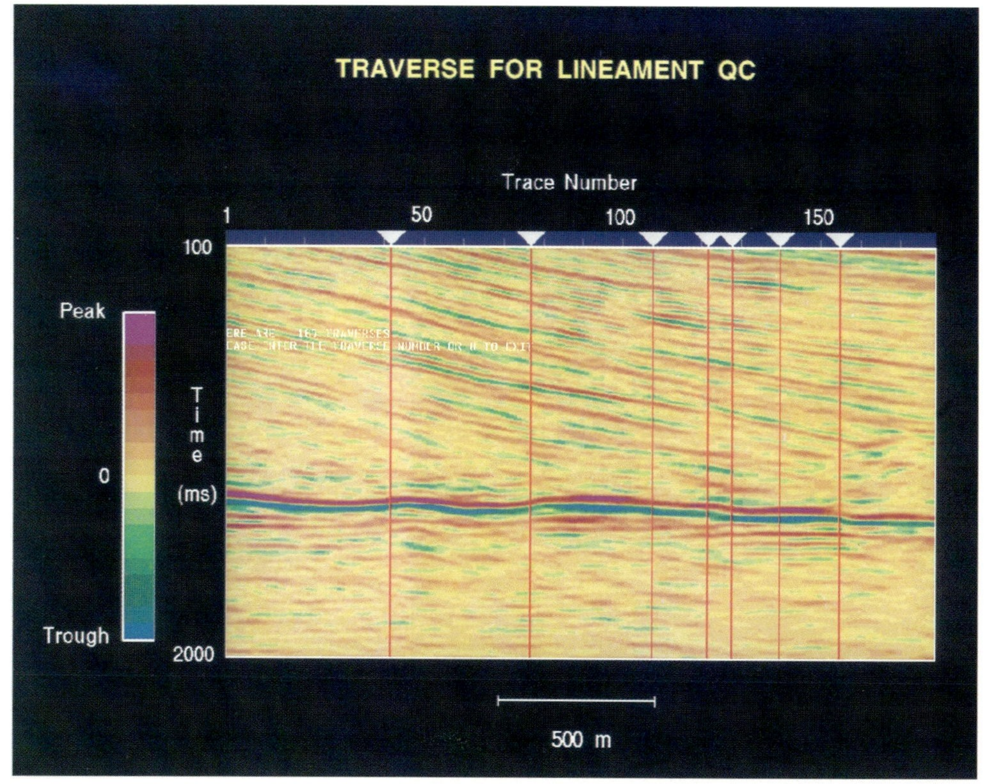


FIGURE 18. Example of a traverse selected to cut several fault traces in a direction normal to the strike of the faults. This is the traverse highlighted in blue on Figure 17. Markers indicate where the traverse crosses interpreted fault traces.

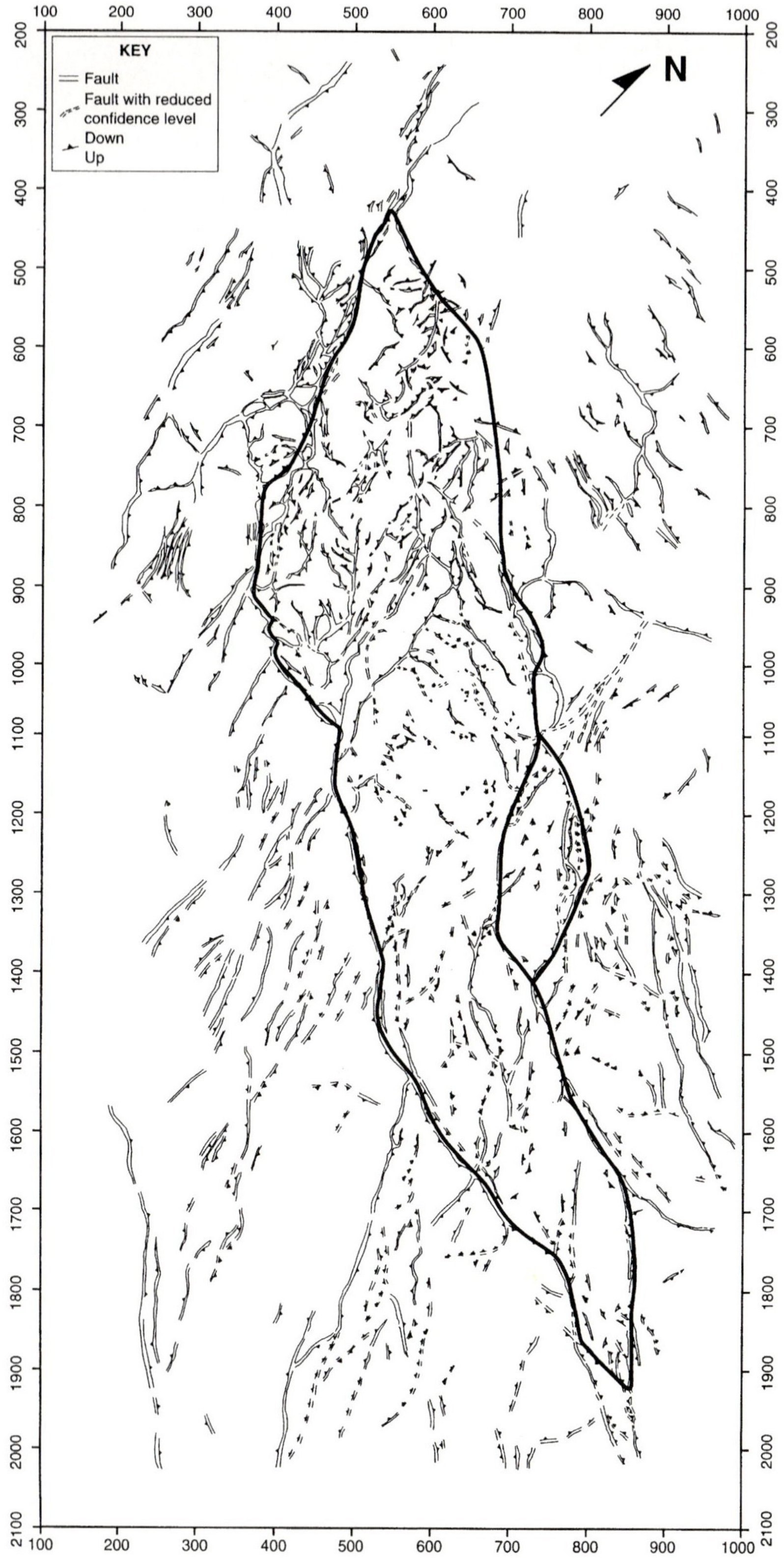


FIGURE 19. Final fault trace map at the Top Rotliegend. The estimated throws are posted in feet.

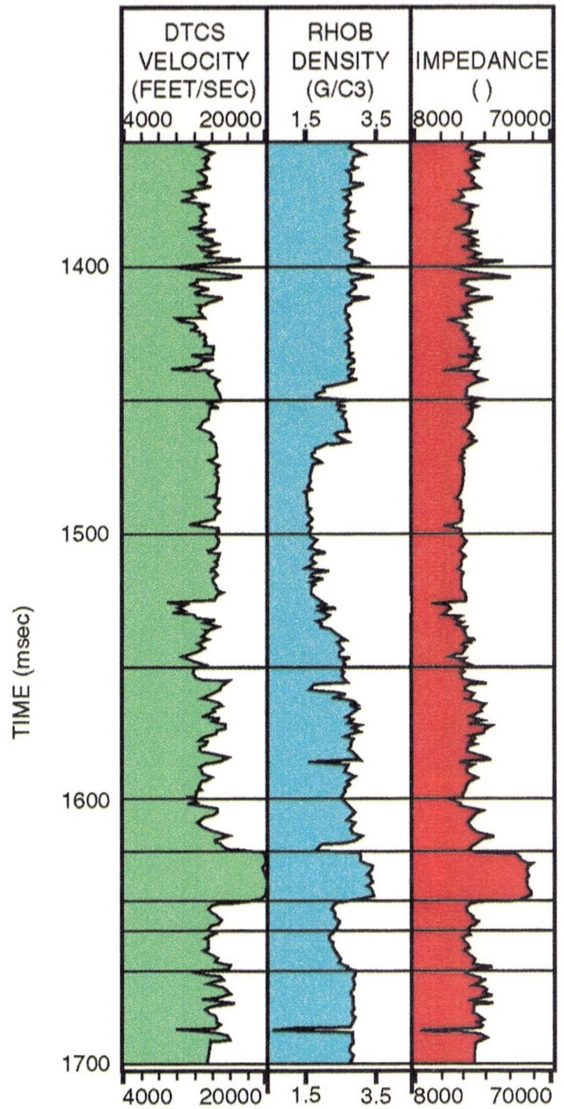


FIGURE 20. A typical suite of logs from one of the exploration wells in the Pickerill field.

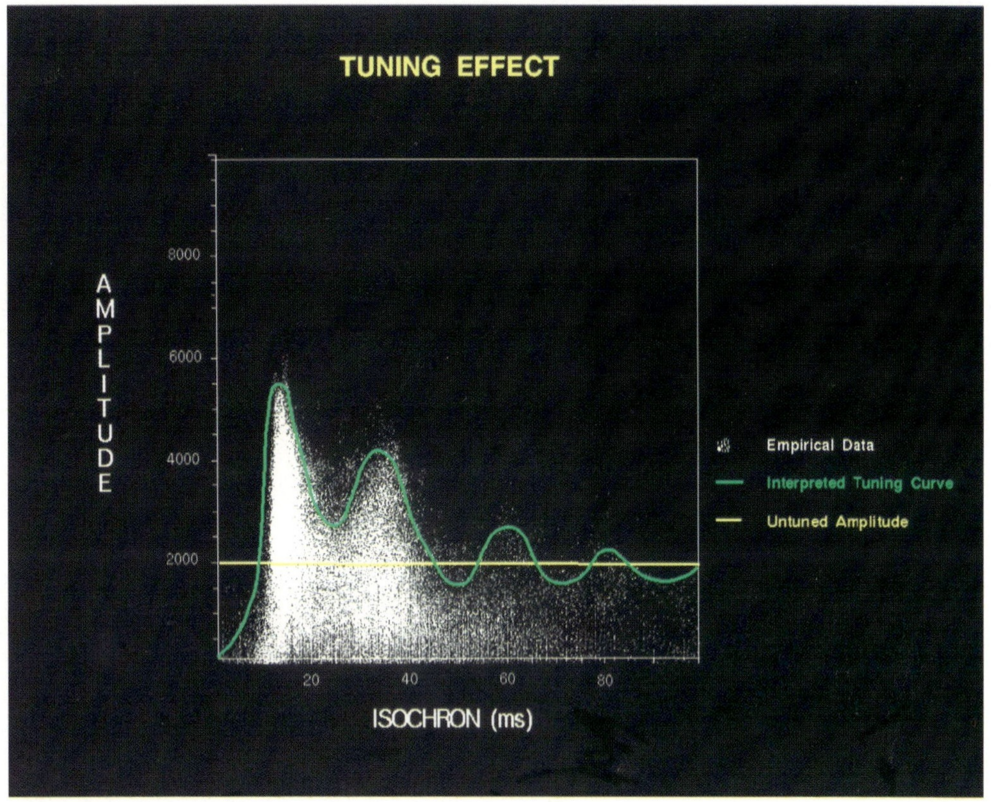


FIGURE 21. Top Rotliegend reflection amplitude vs. the isochron between the Top Rotliegend and the Top Anhydrite.

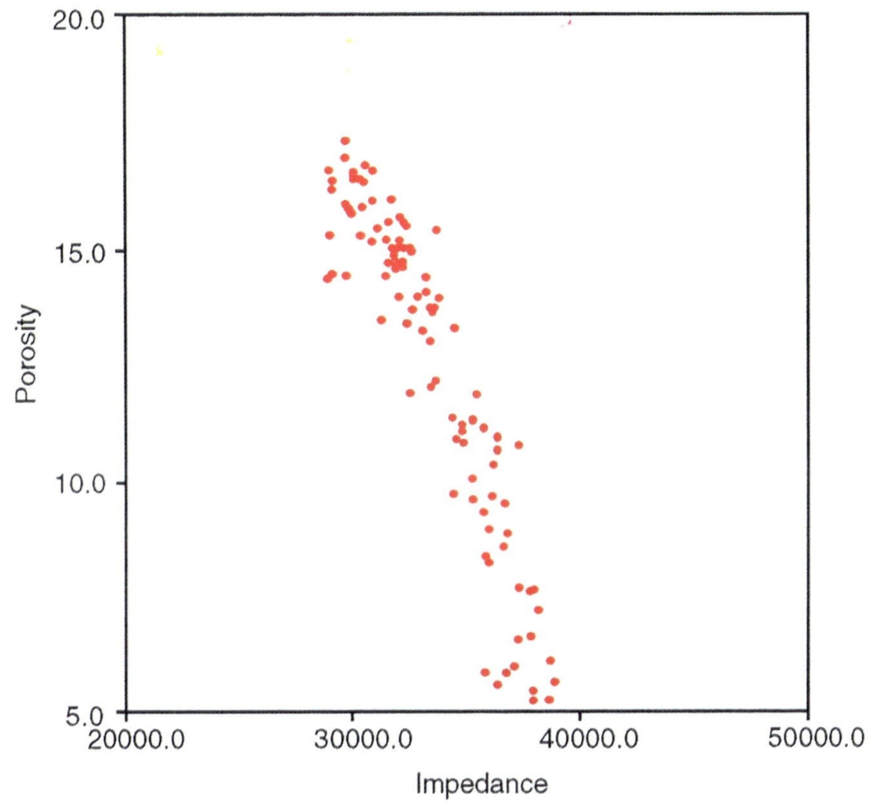


FIGURE 22. Acoustic impedance vs. log-derived porosity in the Rotliegend interval from an exploration well in the Pickerill Field.

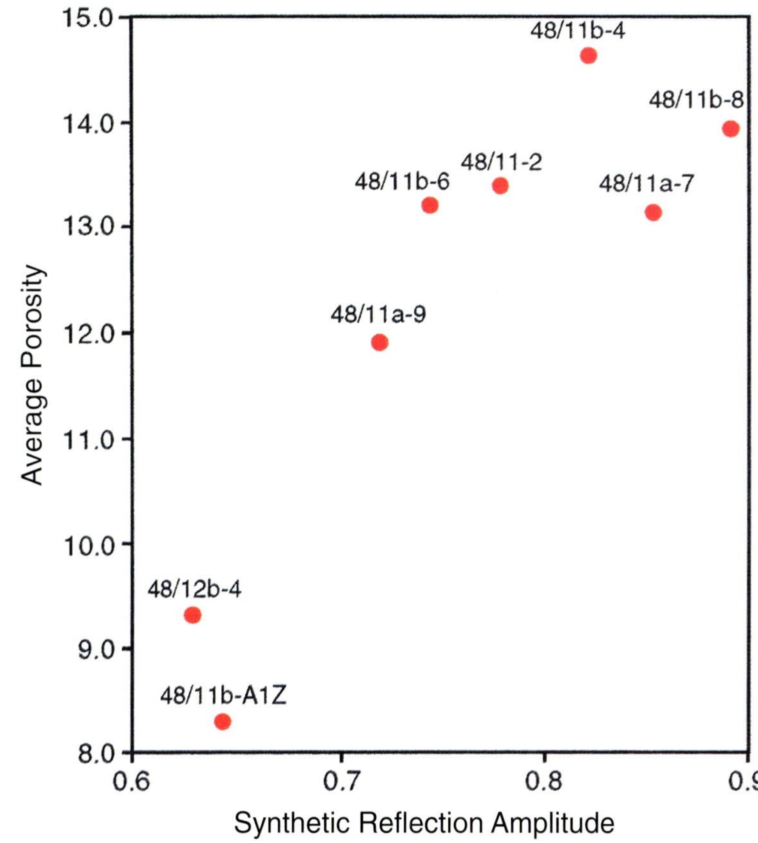


FIGURE 23. Average reservoir porosity vs. synthetic reflection amplitude at the Top Rotliegend.

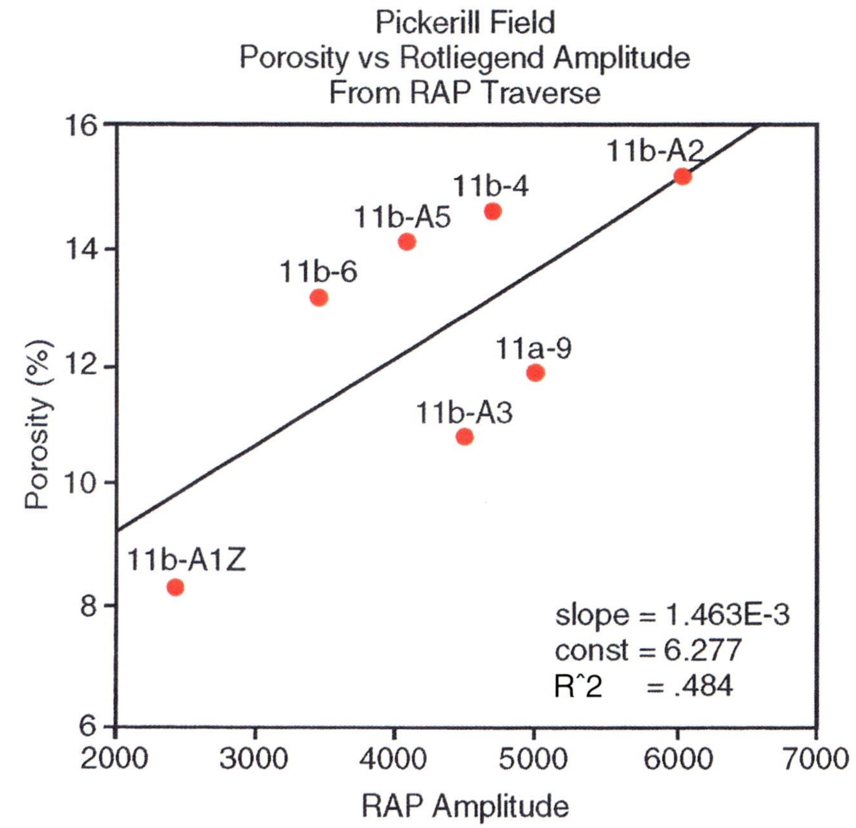


FIGURE 24. Porosity vs. Rotliegend amplitude from the Relative Amplitude Processed (RAP) seismic volume.

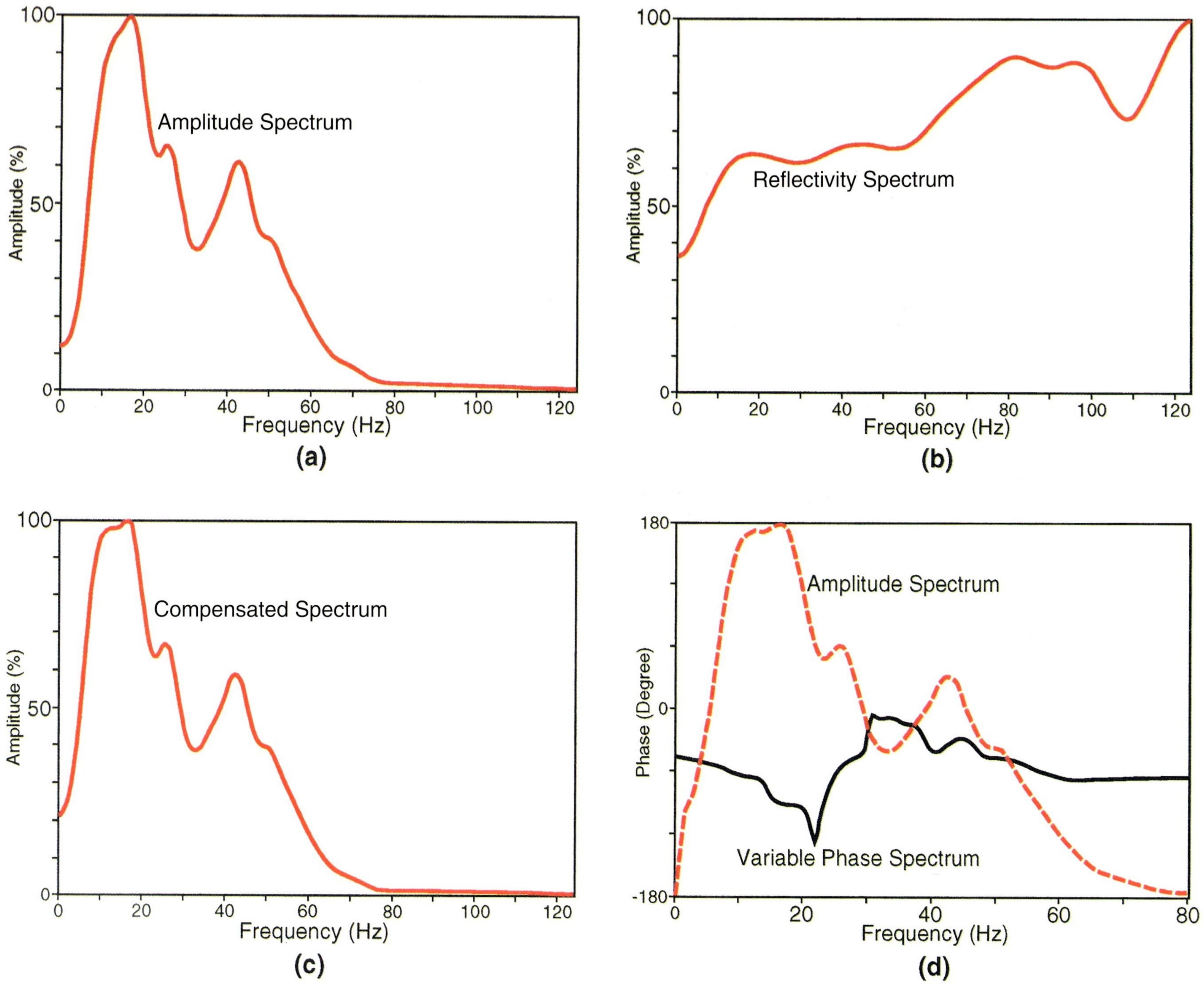


FIGURE 25. Plots used in the extraction of a wavelet at well 48/12b-4. (a) Amplitude spectrum near the well. (b) Reflectivity spectrum from the well. (c) Compensated spectrum from the well. (d) Amplitude and phase spectra for the extracted wavelet.

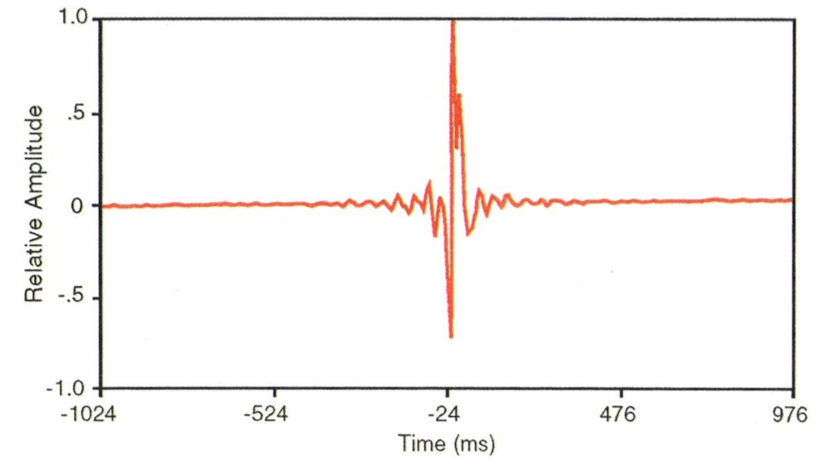


FIGURE 26. Wavelet extracted at well 48/12b-4.

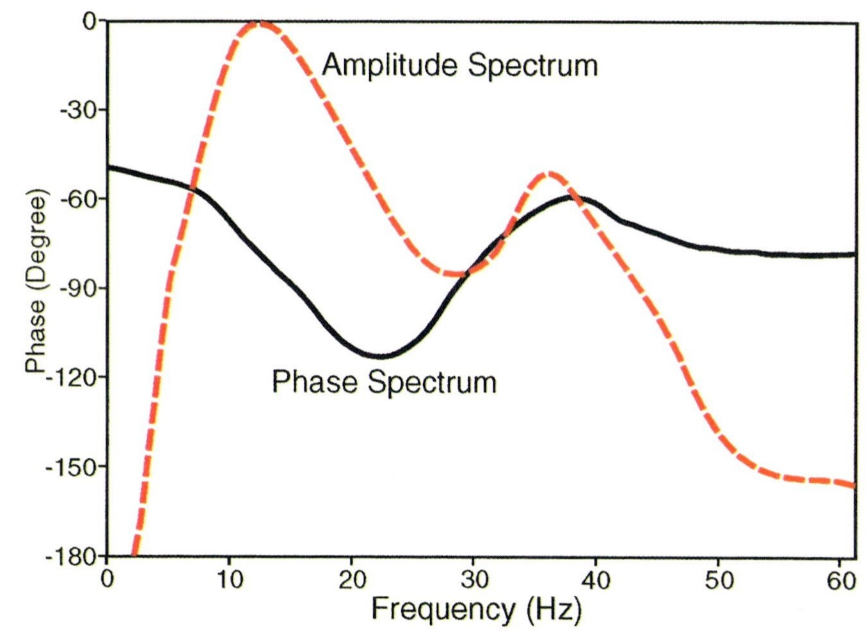


FIGURE 27. Average compensated amplitude and phase spectra.

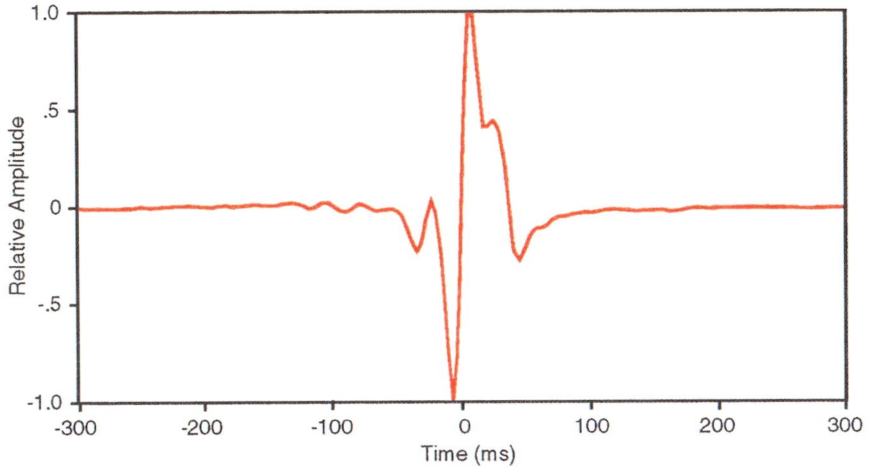


FIGURE 28. Average wavelet.

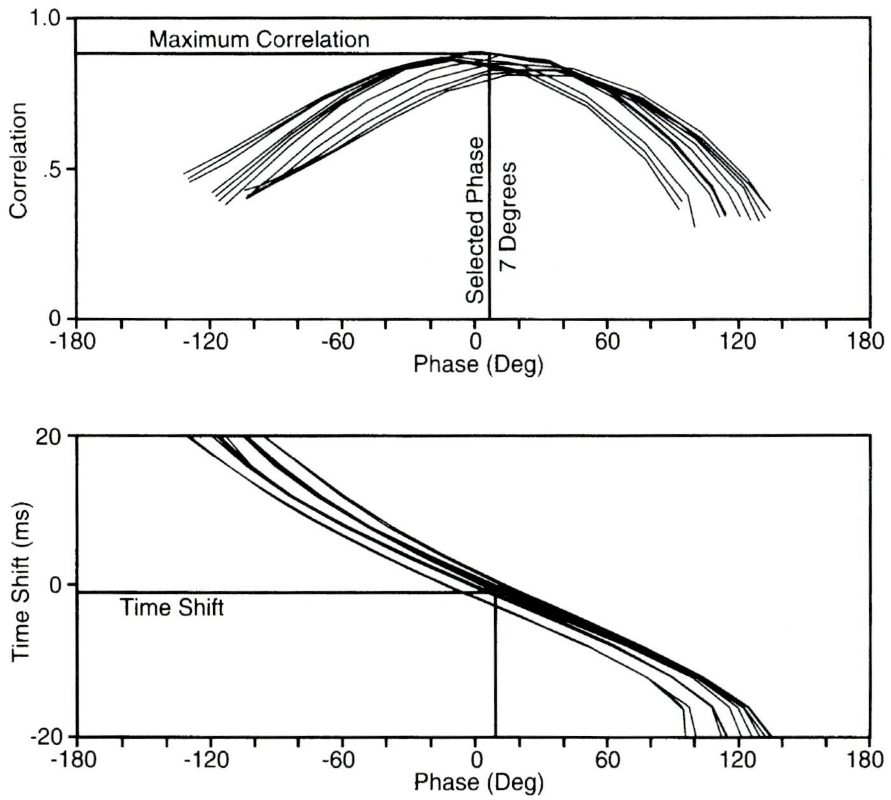


FIGURE 29. Constant phase analysis from well 48/12b-4 to check residual phase, after phase correction operator has been applied.

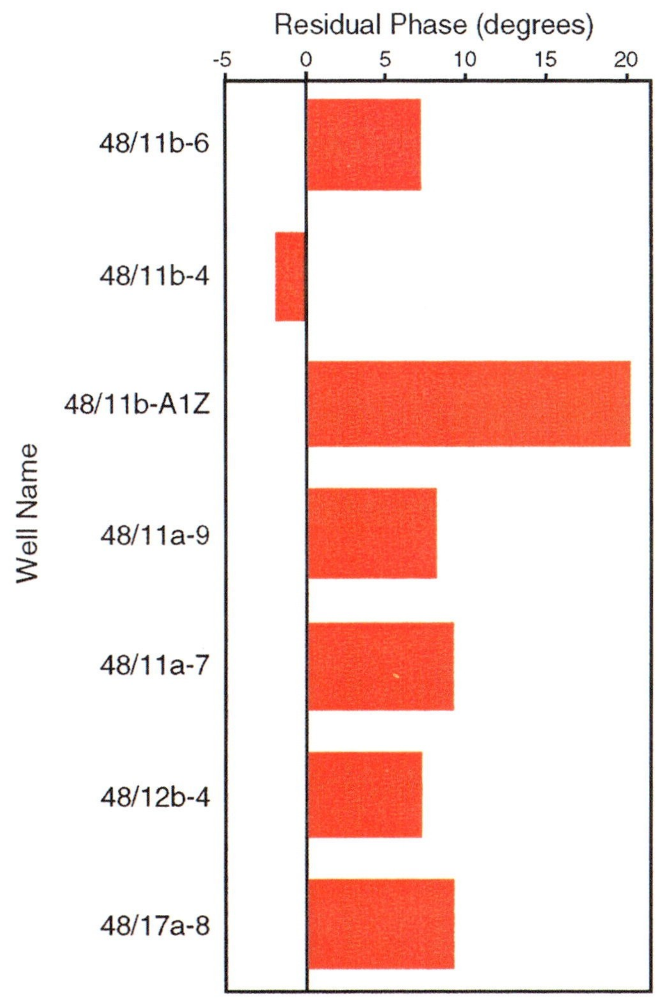


FIGURE 30. Residual phase in each well after phase correction.

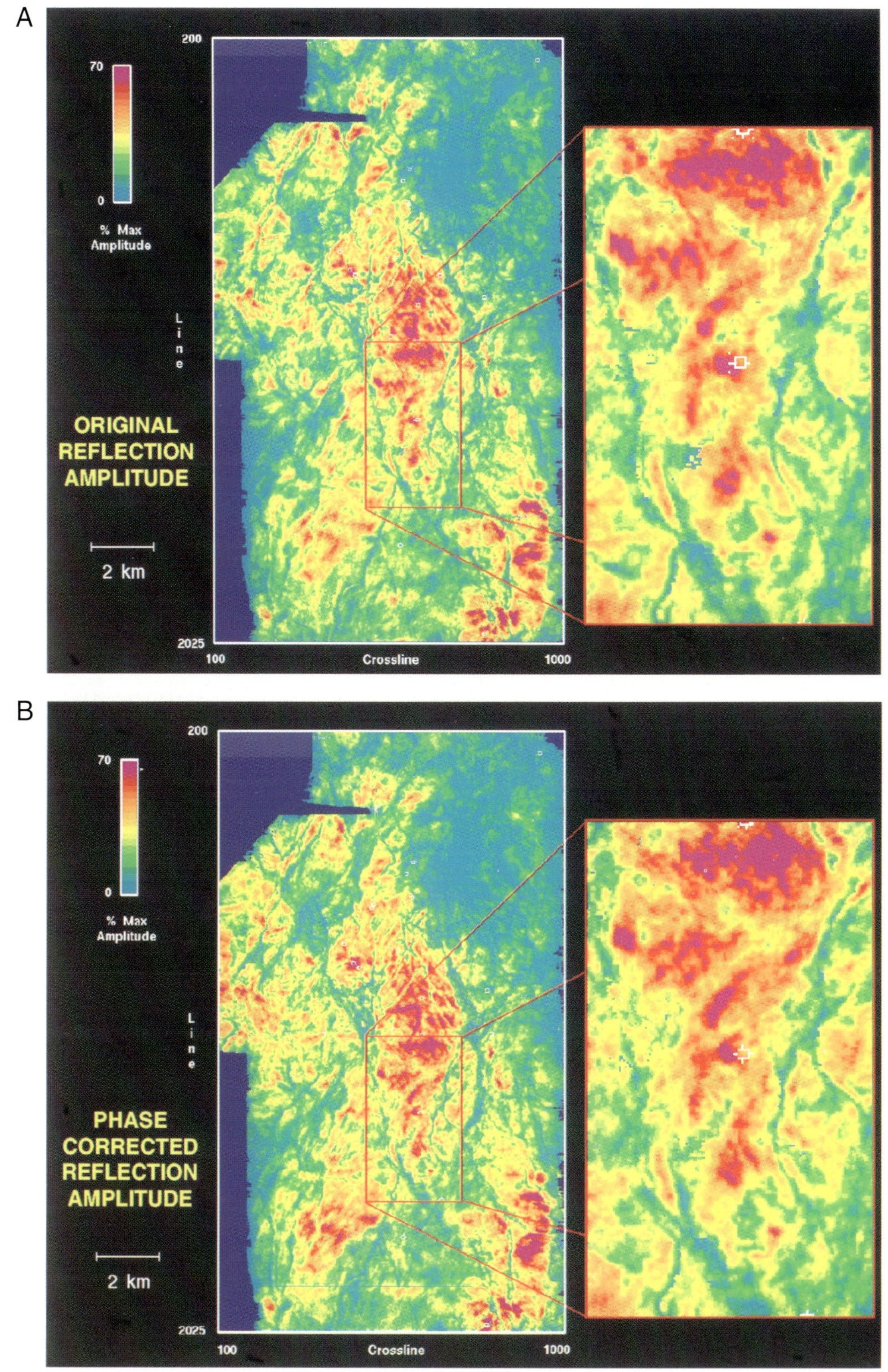


FIGURE 31. Top Rotliegend reflection amplitude. (A) Before phase correction. (B) After phase correction.

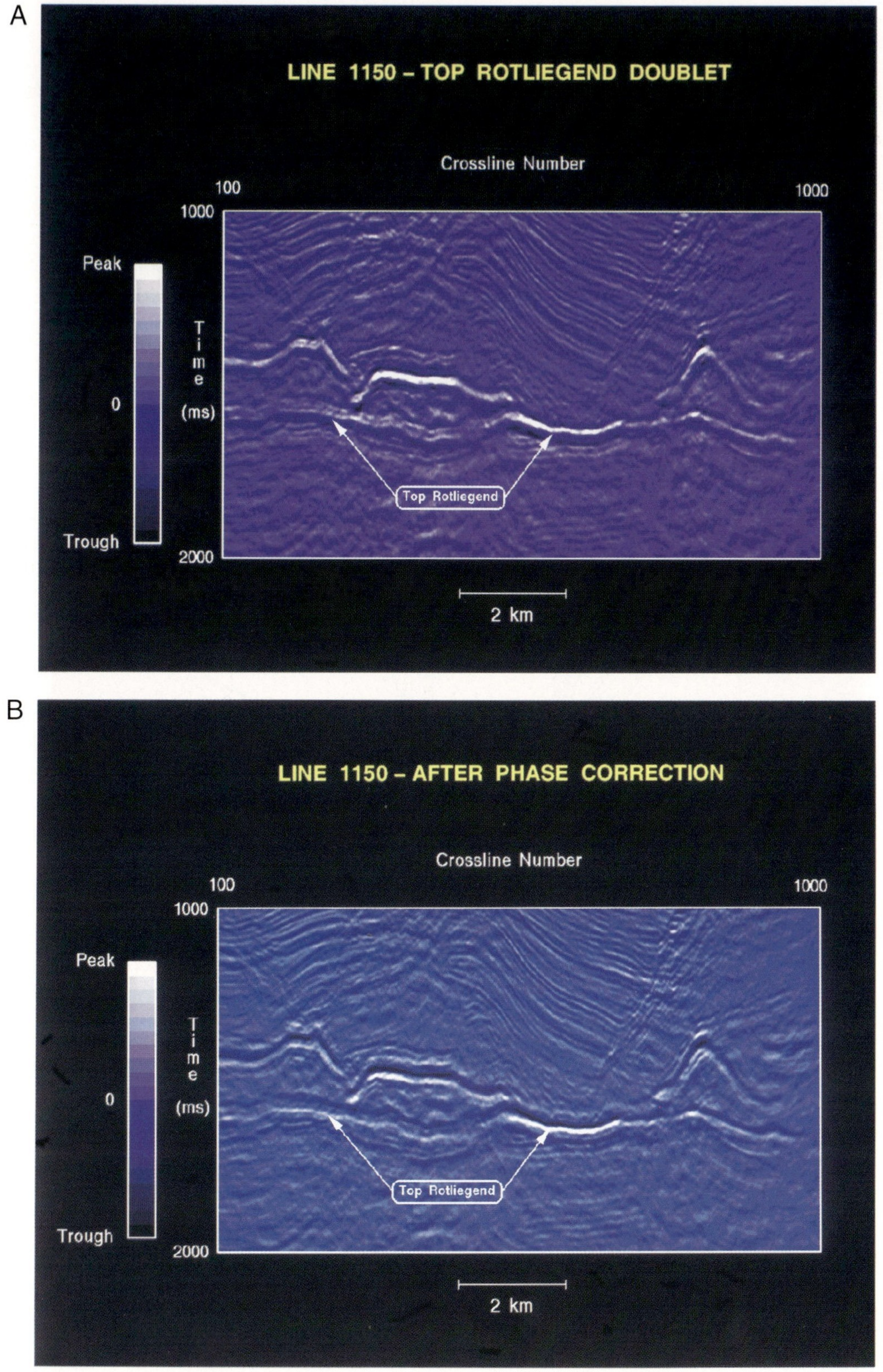


FIGURE 32. Inline 1150. (A) Before phase correction (doublet Top Rotliegend arrival). (B) After phase correction (doublet has been eliminated).

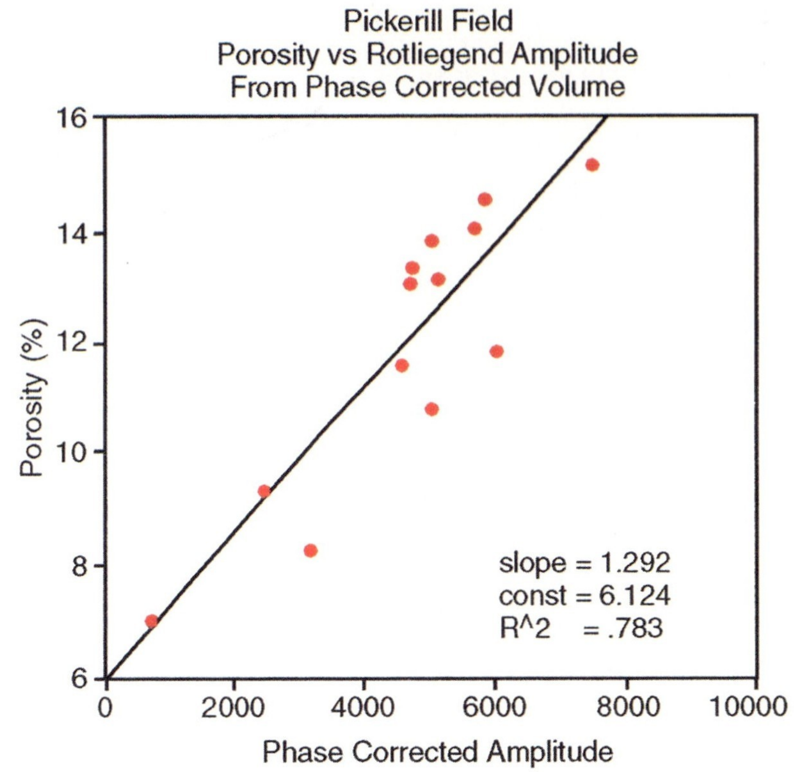


FIGURE 33. Porosity-amplitude relationship. The red points indicate wells used in the analysis. Shows Pickerill Field: Porosity vs. Rotliegend amplitude from phase-corrected volume.

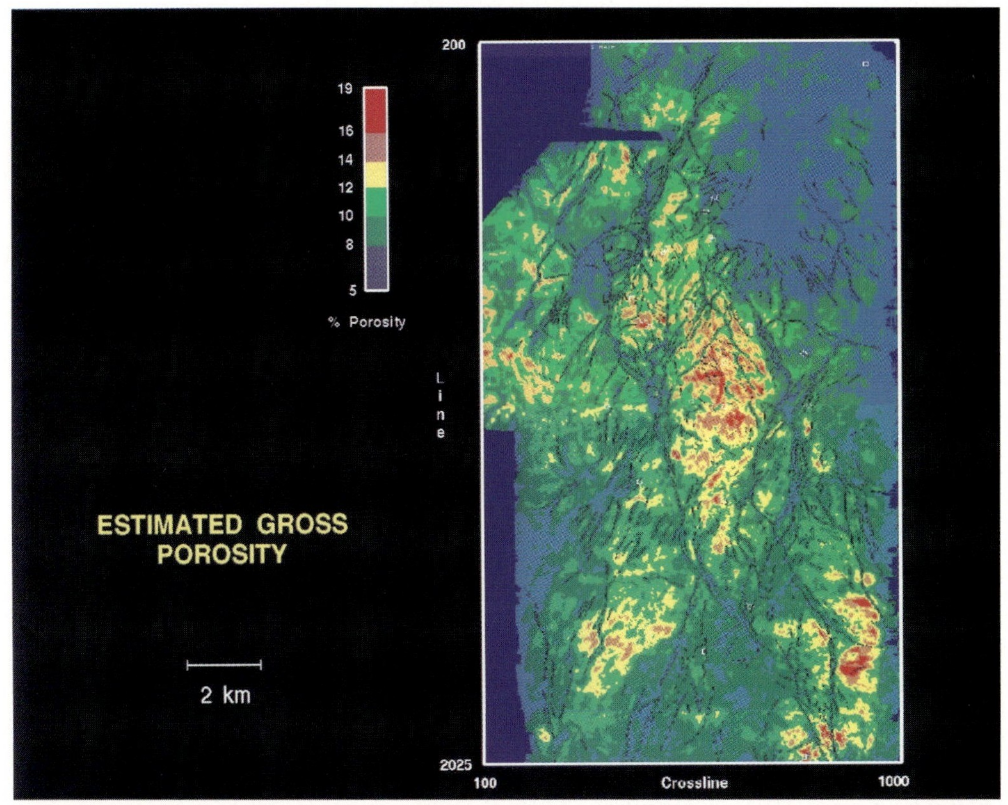


FIGURE 34. Porosity map derived from the Top Rotliegend reflection amplitude and the porosity-amplitude relationship.

A

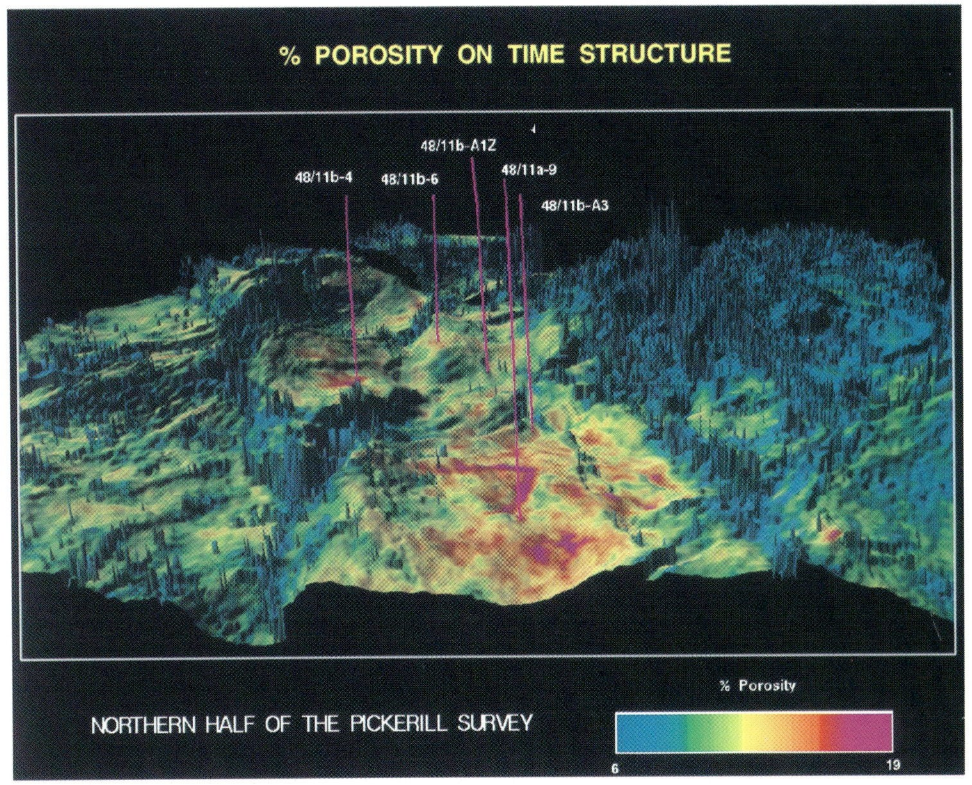


B

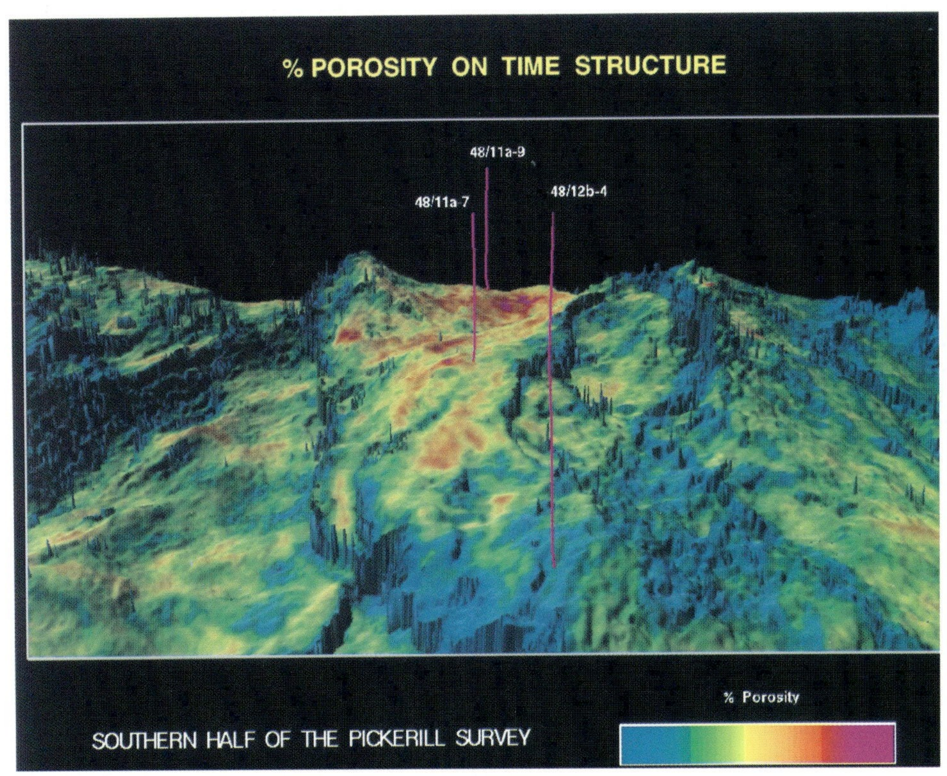


FIGURE 35. 3-D rendering of porosity (color) mapped on time structure. (A) Northern half of the field. (B) Southern half of the field.

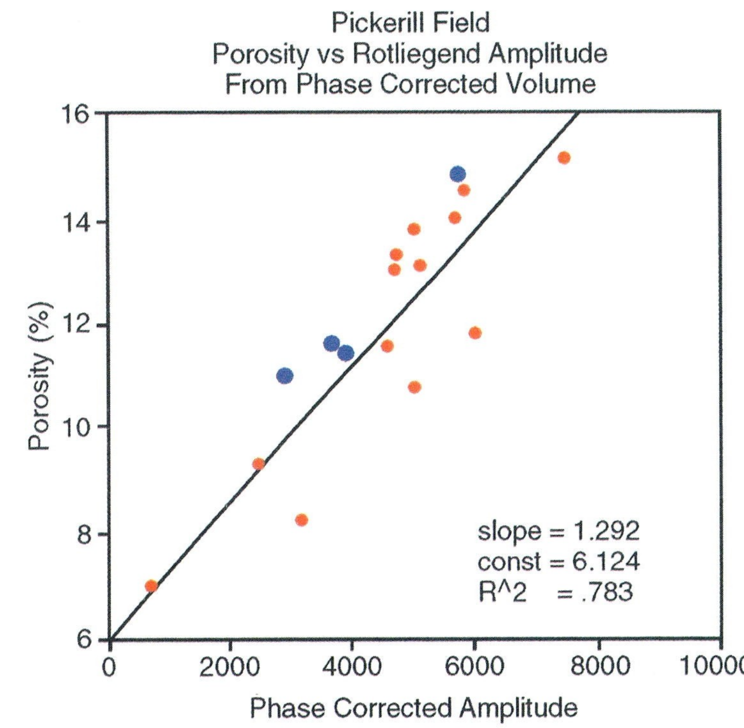


FIGURE 36. Porosity-amplitude crossplot. The red points indicate wells used in the analysis; the blue points are from wells drilled after the analysis. Shows Pickerill Field: Porosity vs. Rotliegend amplitude from phase-corrected volume.

# CHAPTER 12

Chapin, M. A., G. M. Tiller, and M. J. Mahaffie, 3-D architecture modeling using high-resolution seismic data and sparse well control: example from the Mars "Pink" Reservoir, Mississippi Canyon Area, Gulf of Mexico, *in* P. Weimer and T. L. Davis, eds., *AAPG Studies in Geology No. 42* and *SEG Geophysical Developments Series No. 5*, AAPG/SEG, Tulsa, p. 123–132.

# 3-D Architecture Modeling Using High-Resolution Seismic Data and Sparse Well Control: Example from the Mars "Pink" Reservoir, Mississippi Canyon Area, Gulf of Mexico

M. A. Chapin, G. M. Tiller, and M. J. Mahaffie*


## ABSTRACT

Most deep-water development projects are planned using high-quality 3-D seismic data and sparse well control. Economic considerations require large reservoir volumes to be drained with relatively few wells. We have used 3-D seismic data to constrain large-scale, deterministic reservoir bodies in a 3-D architecture model of Pliocene turbidite sands of the "E," or "Pink," reservoir, Prospect Mars, Mississippi Canyon Areas 763 and 807, Gulf of Mexico. A geological interpretation derived from 3-D seismic data and three wells was linked to 3-D architecture models through seismic inversion, resulting in a reservoir rock property distribution incorporating all available data. High-resolution reprocessing of a high-quality marine seismic dataset resulted in the ability to deterministically map sedimentary reservoir bodies. Distinguishing subtle stratigraphic shingles from faults was accomplished by detailed, loop-level mapping and was important to characterize the different types of reservoir compartments. Seismic inversion was used to detune the seismic amplitude, adjust the sand-body thickness, and update the rock properties. This modeling project illustrates how high-quality seismic data and architecture models can be combined in a pre-development phase of a prospect, in order to optimize well placement.


## INTRODUCTION

Economic considerations of the deep-water turbidite play—in the Gulf of Mexico and elsewhere—require large reservoir volumes to be drained by relatively few, very expensive wells. Deep-water development projects to date have been planned on the basis of high-quality 3-D seismic data and sparse well control. An in-house software package allows integration among seismic interpretation, well log information, geophysical modeling, 3-D geologic architecture modeling, and reservoir simulation. The link between 3-D seismic data, well control, and the 3-D architecture model are demonstrated here for Pliocene turbidite sands of the "E," or "Pink," reservoir, Prospect Mars, Mississippi Canyon Areas 763 and 807, Gulf of Mexico. This information was used to better understand reservoir compartmentalization for development well planning.

The Mars discovery lies 210 km (130 mi) south-southeast of New Orleans, in 950 m (3100 ft) water depth (Figure 1). Potentially productive intervals at Mars include Miocene and Pliocene turbidite sands deposited in a narrow salt withdrawal mini-basin (Figure 2; see also Mahaffie, 1994). The "Pink" reservoir, arbitrarily named for the seismic horizon mapping color, comprises unconsolidated Pliocene turbidite sands, covers approximately 12 km$^2$ (3000 acres) and is at a depth of approximately 3500 m (11,500 ft) (Figure 3). A seismic line across the structural dip shows the Pink sands in the upper portion of nested, onlapping sequences (Figure 4). The Pink sand onlaps and pinches out to the north and west and is truncated by a major fault overlying deep-rooted salt to the south (Figure 4).

Three well penetrations were used for the bulk of this study (Figure 5). Preliminary seismic mapping of gross sand-body geometries showed four primary stratigraphic elements, termed "shingles" and interpreted as turbidite sand lobes (Figures 5 and 6). However, more complex reservoir compartmentalization was suggested by early interpretations of repeat formation tester (RFT) pressure measurements and geochemical analysis of RFT oil samples (Figure 7). Because of these indications of compartmentalization, more detailed mapping and characterization were initiated.

## HIGH-RESOLUTION 3-D SEISMIC DATA AND MAPPING

The seismic data were first reprocessed to boost the maximum frequency content from 40 to 60 Hz (Figure 8). The higher-frequency data showed clear separation of the "West Shingle" from the "Main Pink," whereas previously the West Shingle was resolved only as a subtle shift in loop geometry. Interpretation of the enhanced-resolution dataset allowed subtle stratigraphic and structural features to be mapped. In addition to the primary stratigraphic elements identified earlier, the higher-frequency data indicated other stratigraphic shingles within the Main Pink level (Figure 9). Subtle, synsedimentary faults were mapped in the south, and related to subsidence on the large southern boundary fault (Figure 10). Distinguishing between these small faults and the stratigraphic shingles within the Main Pink required careful mapping of each stratigraphic element. To confirm fault interpretations, horizons under the Pink level were also mapped to check for consistent fault offsets. Seismic attribute maps show the locations of some of the major structural and stratigraphic elements (Figures 11 and 12). Seismic time slice and flattened horizon slices were also useful in confirming and checking consistency of many of the interpretations. Most of the actual picking and mapping, however, was done in profile.

## 3-D SEISMIC-CONSTRAINED ARCHITECTURE MODELING

3-D architecture models were built on the basis of the seismic mapping. Architecture modeling has gained some popularity, but its application to date has been mainly in generating stochastic realizations of rock properties or facies in the presence of significant well control, and with very little in input of deterministic, meso-scale sedimentary bodies, as we have mapped from high-resolution seismic data (e.g., Budding et al., 1992; Massonat et al., 1992). Two architecture models were built (Figure 13). One, the "Simple" model, contained the four stratigraphic elements originally observed and generally agreed upon (Figure 14). The second, or "Complex," model incorporated more interpretive, fine-scale stratigraphic shingles within the Main Pink (Figure 15).

The procedure for constructing the architecture model was as follows: An initial model was built using top and base structure maps and an interval thickness map derived from seismic data. This step requires that velocity analysis and time-depth conversions have been completed. Polygon outlines of zero-edges of stratigraphic shingles were picked from the seismic data volume, with estimates of their approximate maximum thickness and their time-stratigraphic stacking sequence. Available well control was incorporated, with several deterministic markers picked, and rock property logs were blocked according to those markers.

Several benefits of the 3-D architecture modeling are immediate: (1) assurance of consistency of maps among themselves and with the seismic data, (2) consistency of the rock property model with the geologic interpretation of stratigraphy, and (3) the ability to quickly

*Shell Exploration and Production Technology Company, Bellaire Technology Center, P.O. Box 481, Houston, TX 77001

update and modify the model and the maps as additional well control is obtained. Ultimately, the model is used for reservoir development planning and flow simulation.

## SEISMIC MODELING AND ROCK PROPERTY ASSIGNMENT

Key rock properties for seismic modeling are velocity and density (Figure 16). Seismic inversion was used to give a detailed blocked model of sand-body thickness and acoustic impedance which, loop for loop, best matched the seismic traces. The a priori rock properties and thicknesses from the well control were used to constrain the inversion, which otherwise would be ambiguous. Figure 17 shows the initial velocity model. Figure 18 shows a seismic inversion acoustic impedance model, with stratigraphic shingles, and the resulting correspondence between the actual seismic data and a synthetic trace over the architecture model. Seismic inversion, therefore, acts to detune the seismic amplitude, adjust layer thickness, and update acoustic rock properties within the model. Because inversion solutions are nonunique, the adjustments are limited and are guided by user-specified constraints.

In these unconsolidated turbidite sands, reservoir rock properties such as sand fraction, porosity, and permeability are correlated to acoustic properties velocity and density (Figure 19). These relationships were used to distribute reservoir properties within the architecture model using a spatial cross-correlation similar to kriging with external drift (Deutsch and Journel, 1992). Postmodeling interval velocity maps and derived sand fraction maps and permeability maps are shown in Figure 20.

## DISCUSSION

The combination of rock-property and architecture information in a 3-D model allows creation of various diagnostic maps that are more informative than are amplitude or net sand maps (Figure 21). Measures such as permeability feet and connected-pore volume are more important for optimizing placement of individual, high-performance wells.

As stated earlier, these stratigraphic shingles are interpreted as laterally offsetting turbidite lobes, similar to large-scale compensation cycles. The dimensions of these lobes range from 3000 to 6000 m (10,000 to 20,000 ft) in length and 1500 to 3000 m (5000 to 10,000 ft) in width. Their occurrence is consistent with the ponded nature of the seismic sequence, as described for other deep-water Gulf of Mexico reservoirs by Mahaffie (1994), McGee et al. (1994), and Holman and Robertson (1994).

The sand bodies shown here are relatively large scale. Reservoir rock properties vary on a much finer scale. Fine-scale models have been constructed for detailed rock-property modeling and reservoir simulation. However, the fine-scale models are not discussed in detail here, because our purpose is to emphasize the link among seismic interpretation, quantitative rock property information contained in the seismic data, and reservoir architecture and properties. It is important that the rock-property character derived from the seismic models be passed on consistently to the fine-scale realizations.

## CONCLUDING REMARKS

A geological interpretation derived from 3-D seismic mapping and a few wells was linked to 3-D architecture models through seismic inversion, resulting in a 3-D rock-property distribution of a reservoir incorporating all available data. Important details from this study include the following:

(1) High resolution reprocessing of a high-quality marine seismic dataset resulted in the ability to deterministically map sedimentary reservoir bodies.

(2) Distinguishing subtle stratigraphic shingles from faults was accomplished by detailed, loop-level mapping and was important to characterize the different types of reservoir compartments.

(3) Seismic inversion was used to detune the seismic amplitude, adjust sand-body thickness, and update the rock properties.

(4) This modeling project illustrates how high-quality seismic data and architecture models can be combined in a predevelopment phase of a prospect to optimize well placement.

New well information is now being obtained by Mars platform pre-drills penetrating the Pink horizon en route to deeper pay horizons. Conclusions derived from the RFT pressure data were originally used to justify detailed analysis but are now suspected to be inconclusive and perhaps reside within the error bars of the tool measurements. However, sand distribution from the new well penetrations and detailed geochemical analysis from oil samples show compartmentalization similar to that mapped on the basis of the high-resolution seismic data. Well planning is in progress for the Pink level and will initially target the Main Pink reservoir, where the largest connected volumes occur. Well paths have not been finalized at the time of this submission.

### *Acknowledgments*

The seismic interpretation system (3DI), inversion software (FLINT), and 3-D geological modeling package (DepSim) were developed by Shell E&P Technology Company at Bellaire Technology Center, Houston. Many people at Shell contributed to the ideas presented here and to the interpretation of the Mars prospect. Hua Zhu and Howard Chan worked with Gottfried Tiller on the architecture modeling program. Johan Neiuwoudt, Eldon Sims, and Alan Cohen were instrumental in developing and applying the seismic inversion procedures. Justin Freeman provided petrophysical analysis of the log data and established procedures for correlating acoustic and reservoir properties. Roger Shew helped tie core data to the reservoir model. Members of the Mars development team providing input to this study include Bob Barnard, Larry Franceware, Chris Lerch, Bill Henry, and John Kendrick. Paul Weimer and Bob Hardage provided helpful and timely reviews of the manuscript. We thank Shell E&P Technology Company, Shell Offshore, Inc., and British Petroleum, Inc., for permission to publish this paper.

## REFERENCES CITED

Budding, M. C., A. H. M. Paardekam, and S. J. van Rossem, 1992, 3-D connectivity and architecture in sandstone reservoirs: SPE paper 22342, presented at the SPE International Meeting on Petroleum Engineering, held in Bejing, China, March 24–27.

Deutsch, C. V., and Journel, A. G., 1992, GSLIB: Geostatistical Software Library and User's Guide: Oxford University Press, New York.

Holman, W. E., and S. S. Robertson, 1994, Field Development, Depositional Model, and Production Performance of the Turbiditic "J" Sands at Prospect Bullwinkle, Green Canyon 65 Field, Outer-Shelf Gulf of Mexico, *in* P. Weimer, A. H. Bouma, and B. F. Perkins, eds., Submarine Fans and Turbidite Systems: Sequence Stratigraphy, Reservoir Architecture, and Production Characteristics—Gulf of Mexico and International: 15th Annual Gulf Coast SEPM Research Conference, Houston, p. 139–150.

Mahaffie, M. J., 1994, Reservoir classification for turbidite intervals at the Mars discovery, Mississippi Canyon 807, Gulf of Mexico, *in* P. Weimer, A. H. Bouma, and B. F. Perkins, eds., Submarine Fans and Turbidite Systems: Sequence Stratigraphy, Reservoir Architecture, and Production Characteristics—Gulf of Mexico and International: 15th Annual Gulf Coast SEPM Research Conference, Houston, p. 233–244.

Massonat, G. J., F. G. Alabert, and C. B. Giudicelli, 1992, Anguille Marine, a deepsea-fan reservoir offshore Gabon: from geology to stochastic modelling: SPE paper 24709, presented at the 67th Annual Technical Conference and Exhibition of the Society of Petroleum Engineers, Washington, D.C., October 4–7.

McGee, D. T., P. W. Bilinski, P. S. Gary, D. S. Pfeiffer, and J. L. Sheiman, 1994, Geologic models and reservoir geometries of Auger Field, Deepwater Gulf of Mexico, *in* P. Weimer, A. H. Bouma, and B. F. Perkins, eds., Submarine Fans and Turbidite Systems: Sequence Stratigraphy, Reservoir Architecture, and Production Characteristics—Gulf of Mexico and International: 15th Annual Gulf Coast SEPM Research Conference, Houston, p. 245–256.

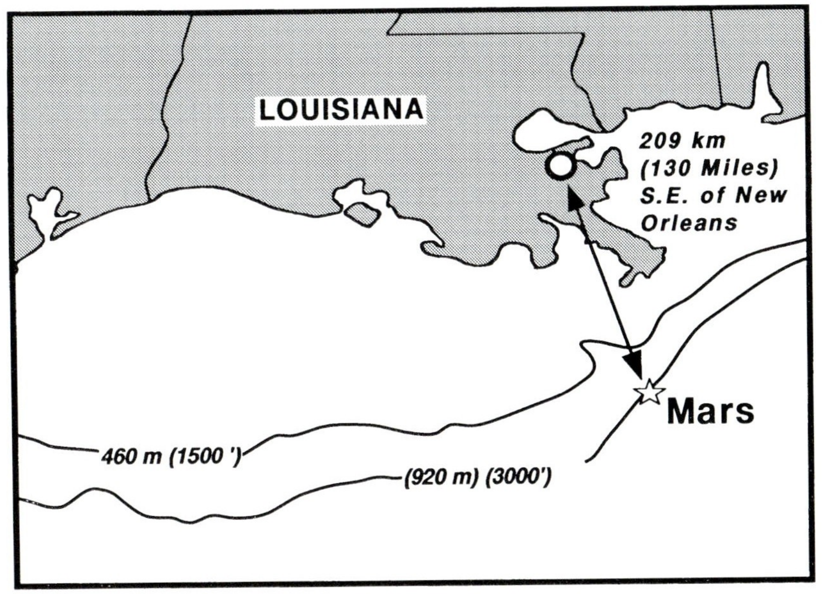


FIGURE 1. Location map, Mars Development Project, Gulf of Mexico.

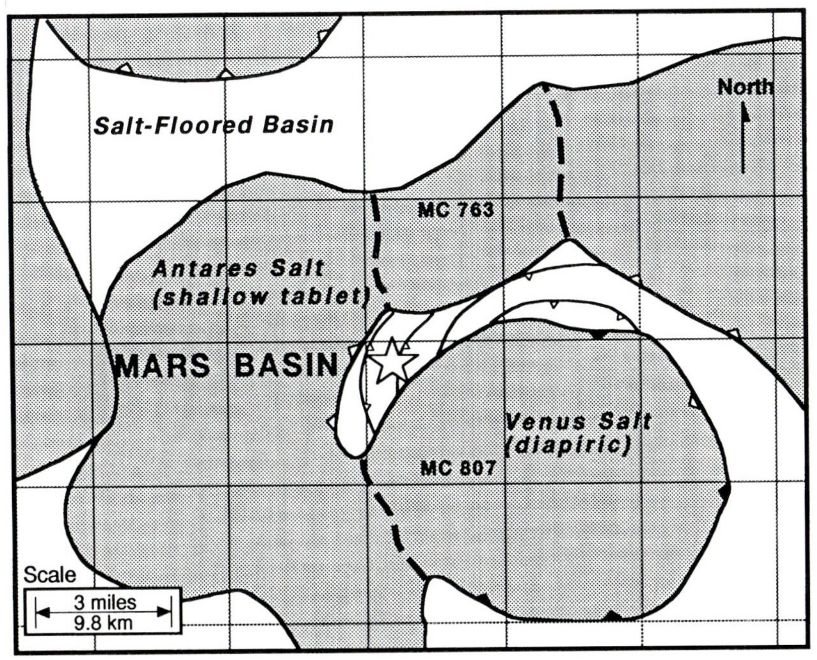


FIGURE 2. Mars basin setting. (Star indicates Mars development.) Mars basin fill includes Miocene through Recent deep-water sediments deposited in a salt-withdrawal minibasin between the Antares salt , a shallow tablet overhanging to the south, and the Venus salt, a deeper, mushroom-shaped diapir.

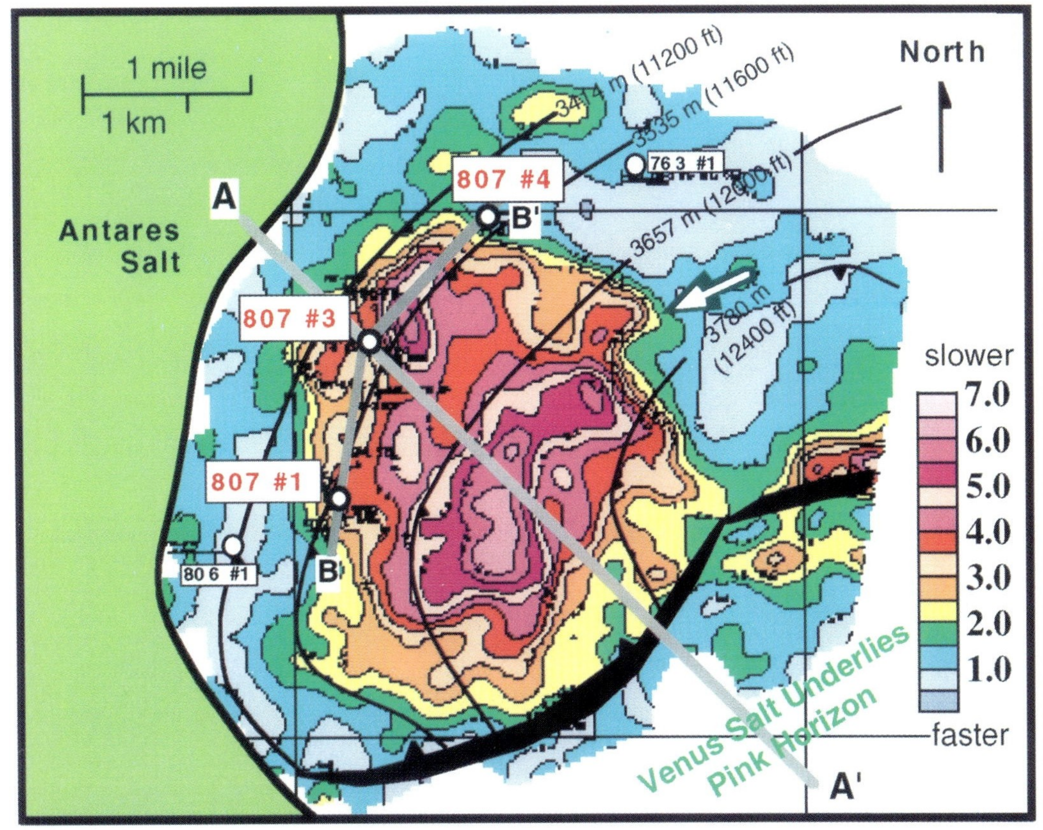


FIGURE 3. Amplitude and structure maps of the Mars "Pink" horizon, with key wells and seismic traverses.

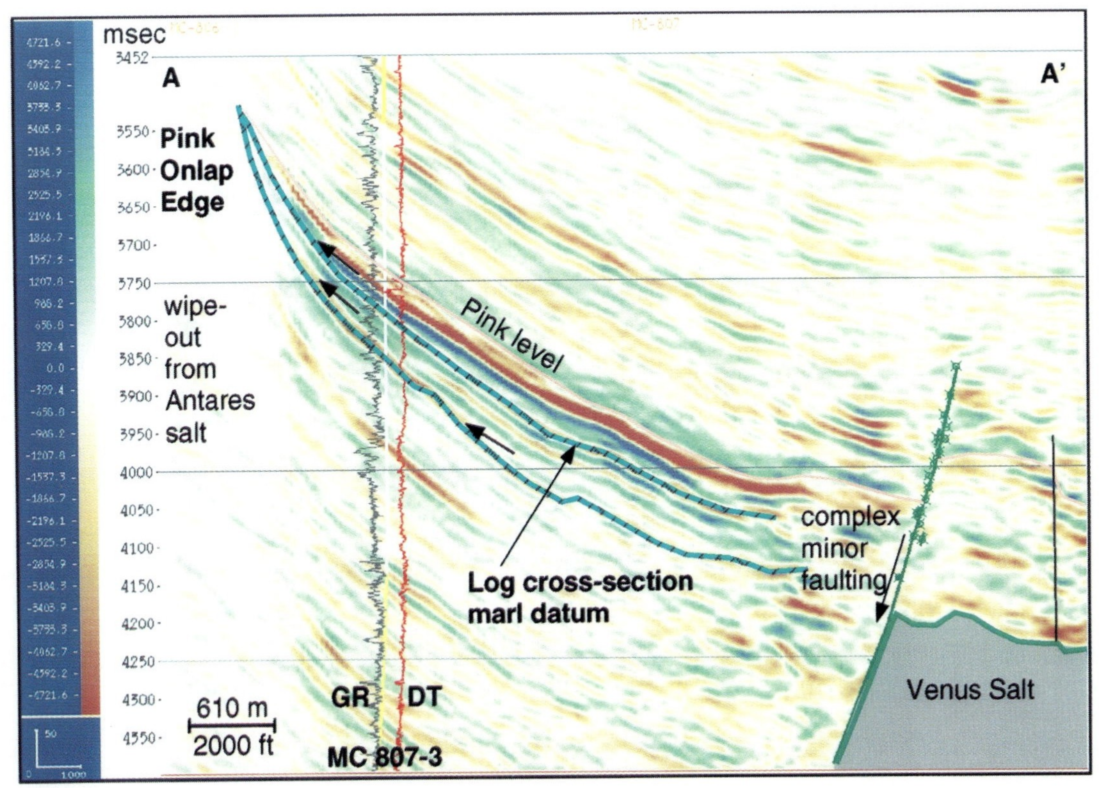


FIGURE 4. Key seismic traverse A-A´ across structural dip, showing setting of the Pink horizon. Red = negative amplitude, blue = positive amplitude. Note the expansion of the Pink sequence into the north-dipping growth fault associated with the Venus salt. Traverse location shown on Figure 3.

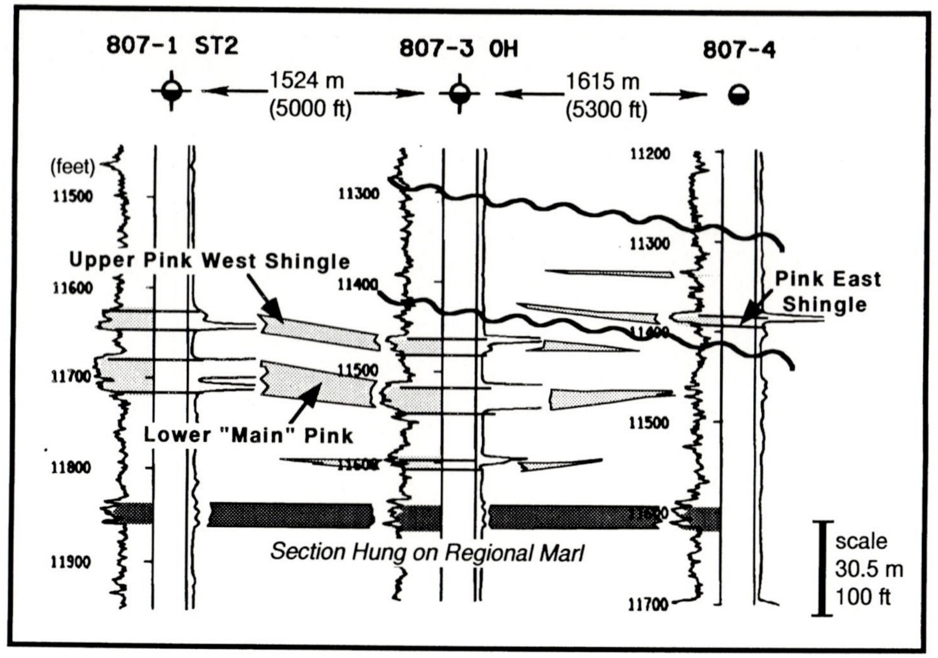


FIGURE 5. Cross section of three key wells used for this study, showing major stratigraphic elements and correlations (locations given in Figure 3). Three principal stratigraphic shingles are shown by the well control. An additional stratigraphic shingle is present to the southwest but has not been penetrated by wells.

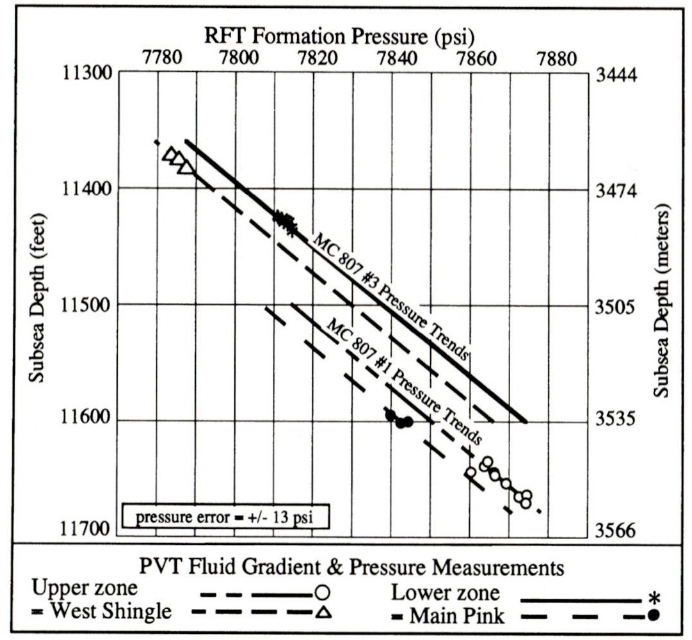


FIGURE 7. Repeat formation tester (RFT) pressure vs. depth plots for two wells. The points represent pressure measurements, and the lines represent extrapolated fluid pressure gradients derived from analysis of RFT oil samples. Although the lower and upper zones appear to correlate stratigraphically from seismic and well logs, none of the zones appeared to be in pressure continuity from these preliminary RFT data.

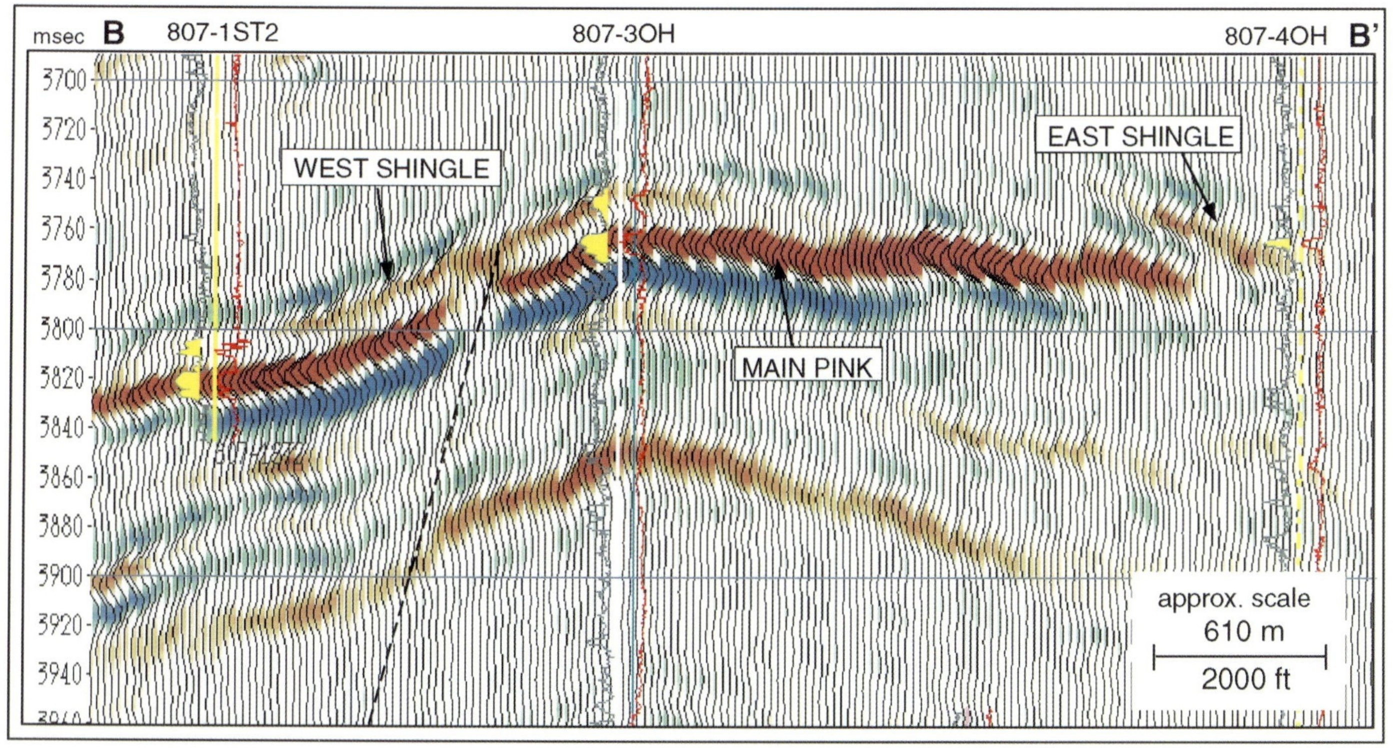


FIGURE 6. Seismic traverse B-B´ tying well control and showing the three principal stratigraphic shingles. The left log curve is gamma ray and the right curve is sonic velocity. All seismic data shown is integrated, that is, the top and base of acoustically slow sands ideally correspond to seismic zero crossings.

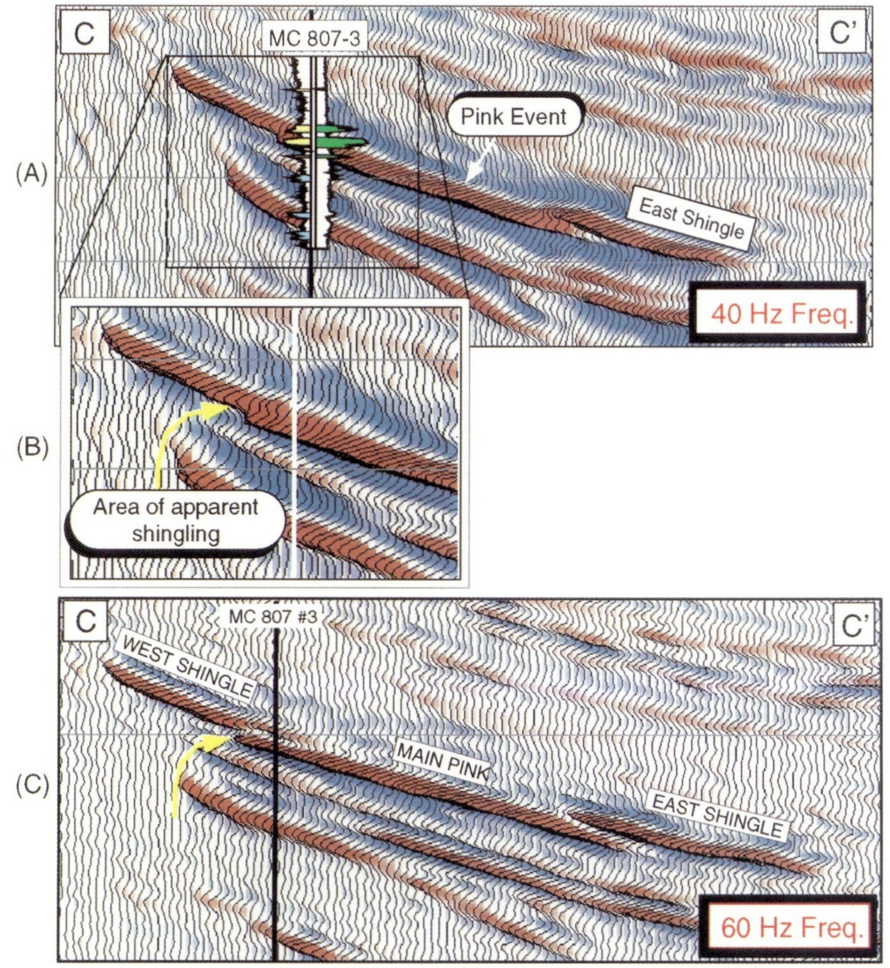


FIGURE 8. Seismic traverse C-C´ illustrating the effects of boosting the high-frequency content of the data on resolving stratigraphic complexity (location on Figure 11). (A) 40-Hz dataset with well tie (well colors: yellow = sand, green = oil pay, blue = marl). (B) Detail shows shift in seismic loop shape corresponding to West Shingle on the 40-Hz dataset. (C) Reprocessed, 60-Hz dataset, showing clear separation of the West Shingle. All other seismic data shown in this paper are from the 60-Hz dataset. The 40-Hz comparison is important because it shows the influence of stratigraphic features slightly below seismic resolution on loop shape.

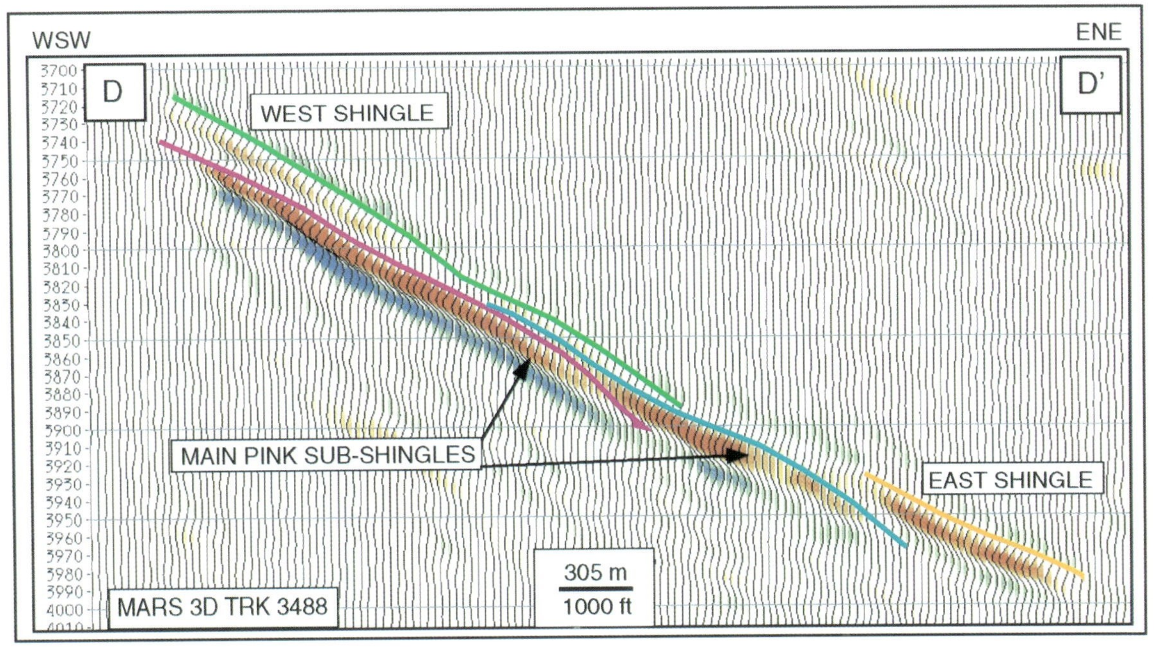


FIGURE 9. High-resolution reprocessing showed additional stratigraphic complexity in the Main Pink level, previously thought to be a single sand body. Traverse location shown on Figure 11.

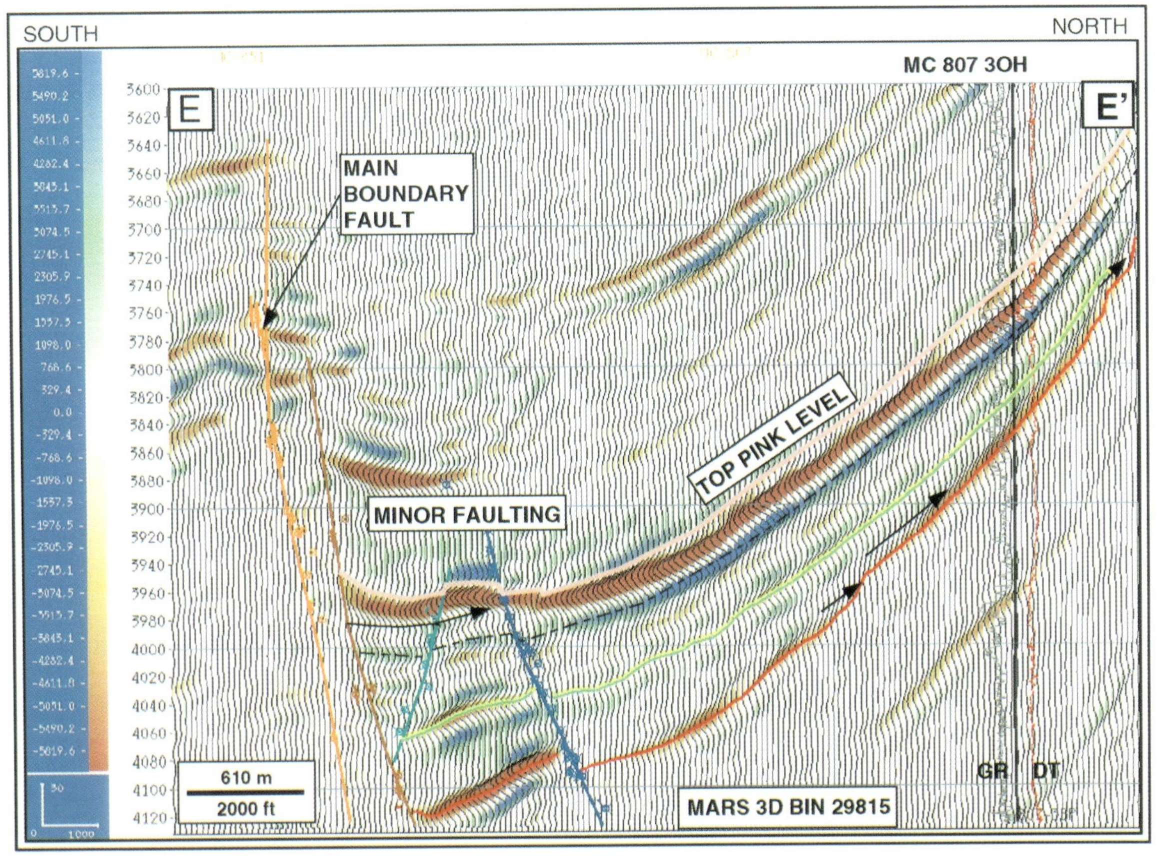


FIGURE 10. Subtle, syndepositional faults were also enhanced on the high-resolution dataset. This faulting was abundant in the south, close to the principal growth fault. Traverse location shown on Figure 11.

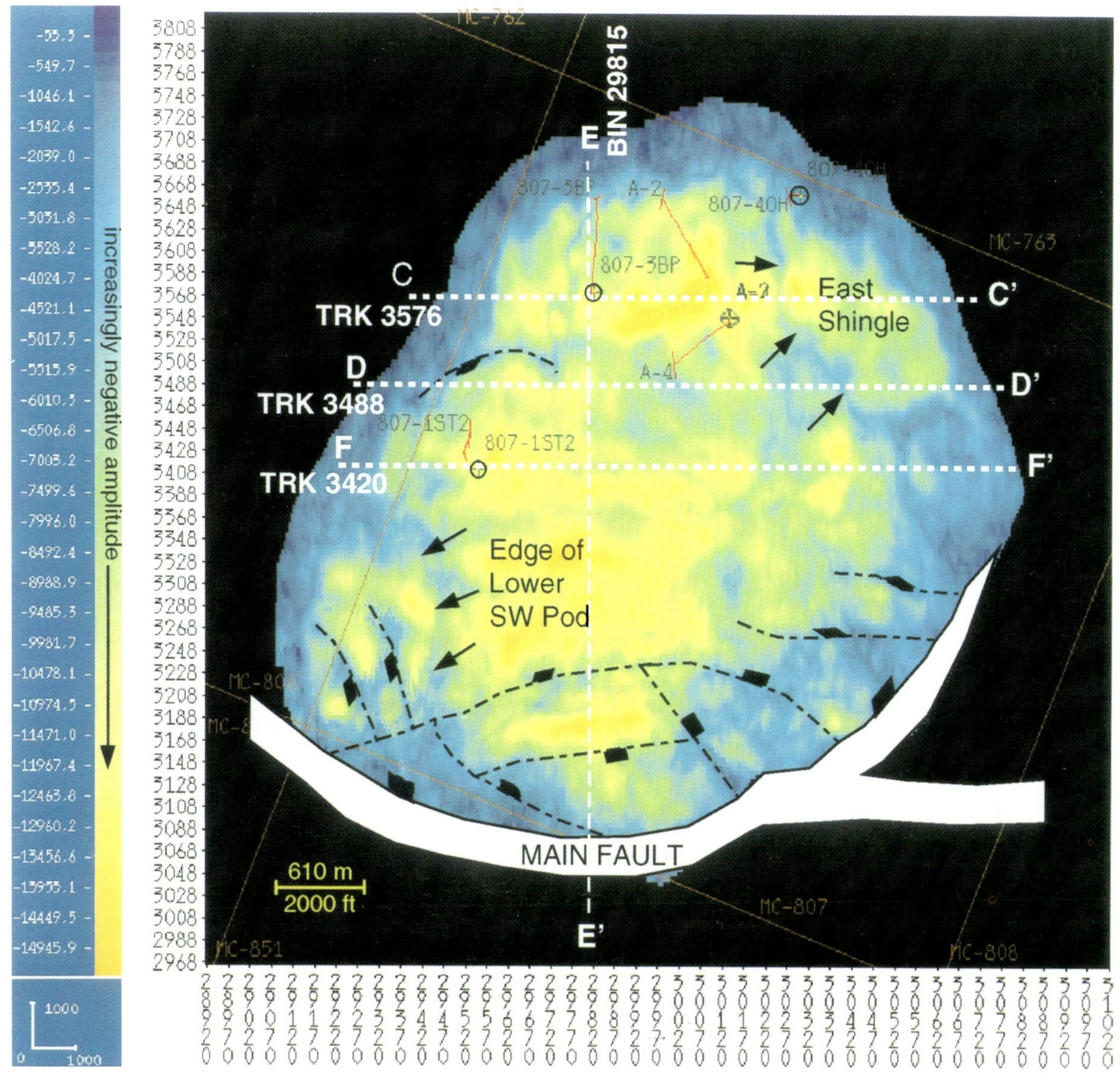


FIGURE 11. Seismic amplitude map of the Pink reservoir horizon. Dim zones demarcate some structural and stratigraphic features. A possible water level is indicated by the downdip conformance of high amplitude area to structural contours in the East Shingle (compare with Figure 3). The Main Pink reservoir does not display structural conformance of amplitude overall because the sand is pinching out to the east and because of complex faulting downdip to the southeast.

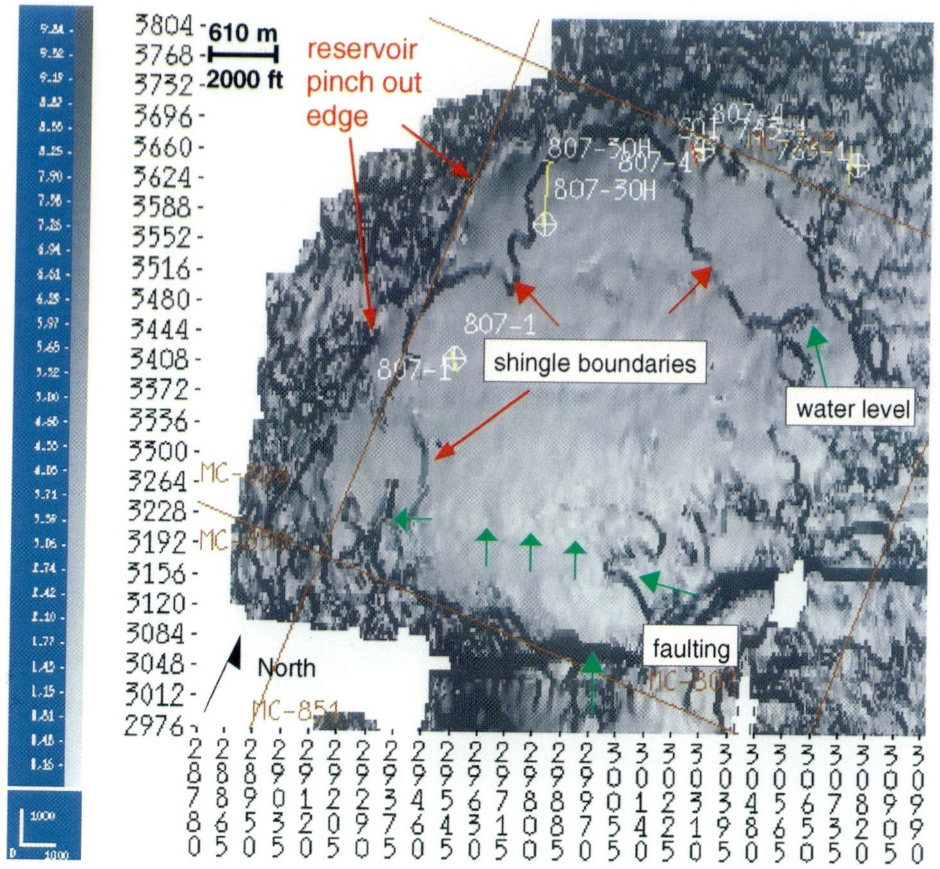


FIGURE 12. Seismic dip map, showing some stratigraphic and structural features. Dark gray = high local gradient, light gray = low local gradient.

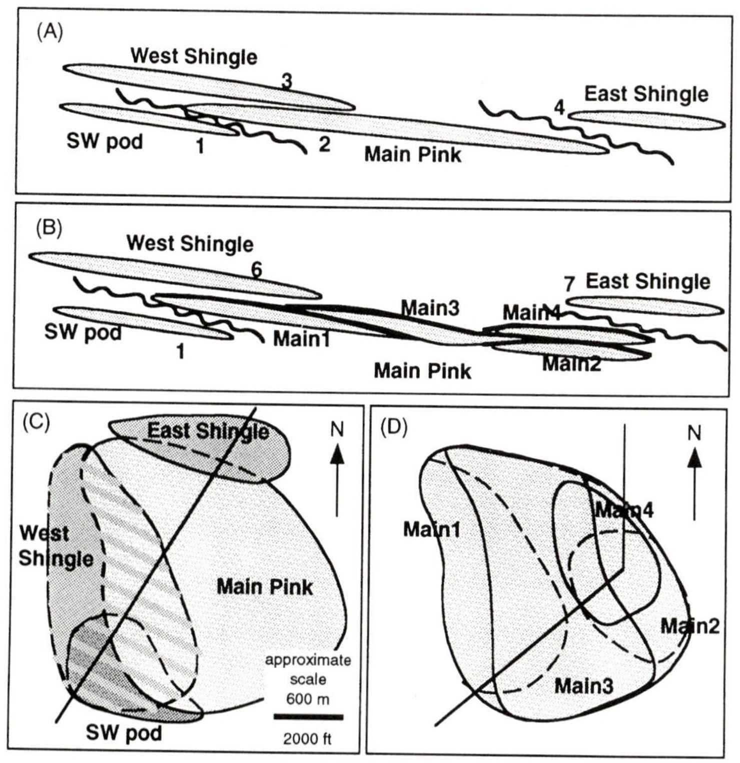


FIGURE 13. Schematic drawings of two architecture models. (A) Simple model cross section. (B) Complex model cross section. (C) Plan view of four principal sand bodies of the Simple model. (D) Smaller-scale shingles within the Main Pink in the Complex model.

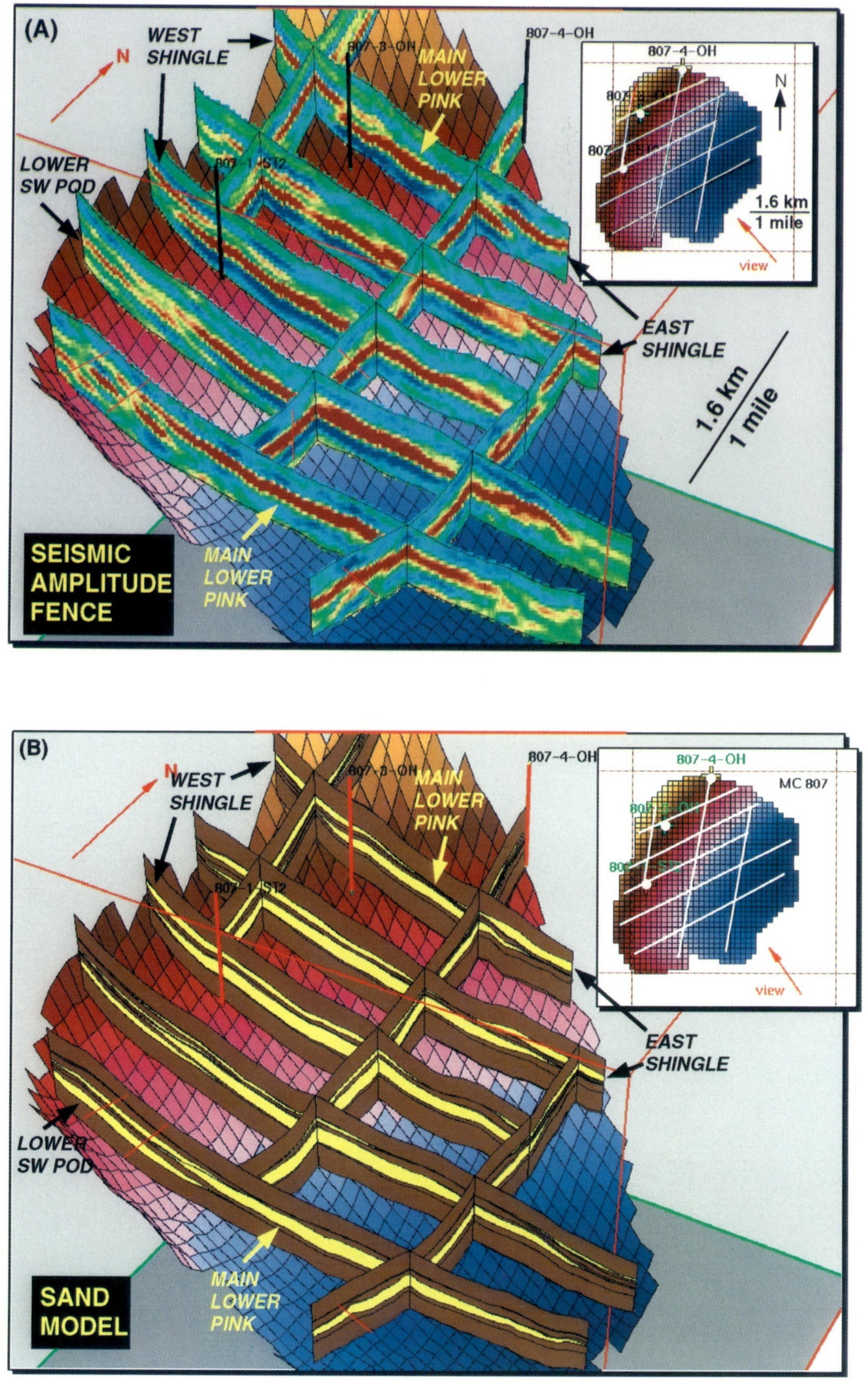


FIGURE 14. (A) A seismic amplitude fence diagram is displayed in a 3-D architecture model and clearly illustrates the principal sand bodies. (B) 3-D sand-body architecture of the Simple model.

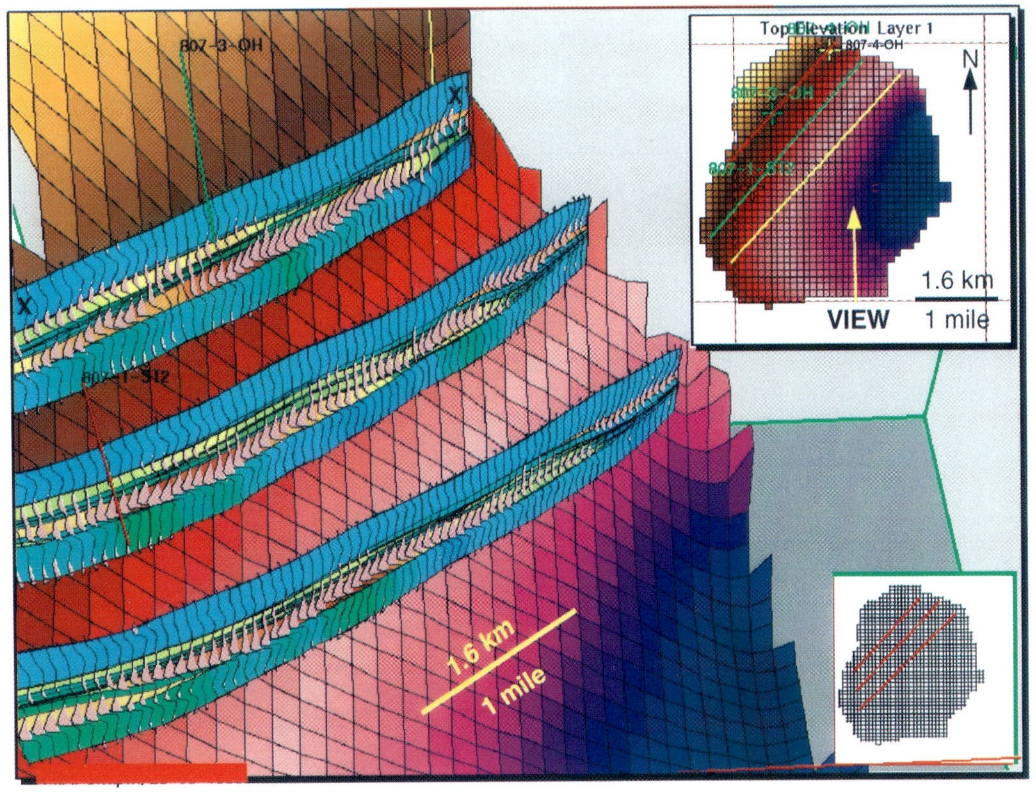


FIGURE 15. Seismic data and architecture of some traverses through the Complex model. The seismic traces shown are the actual seismic data. The underlying architecture model is color coded by interval velocity according to the scale shown in Figure 16. Traverse X-X′ is shown in more detail in Figure 16.

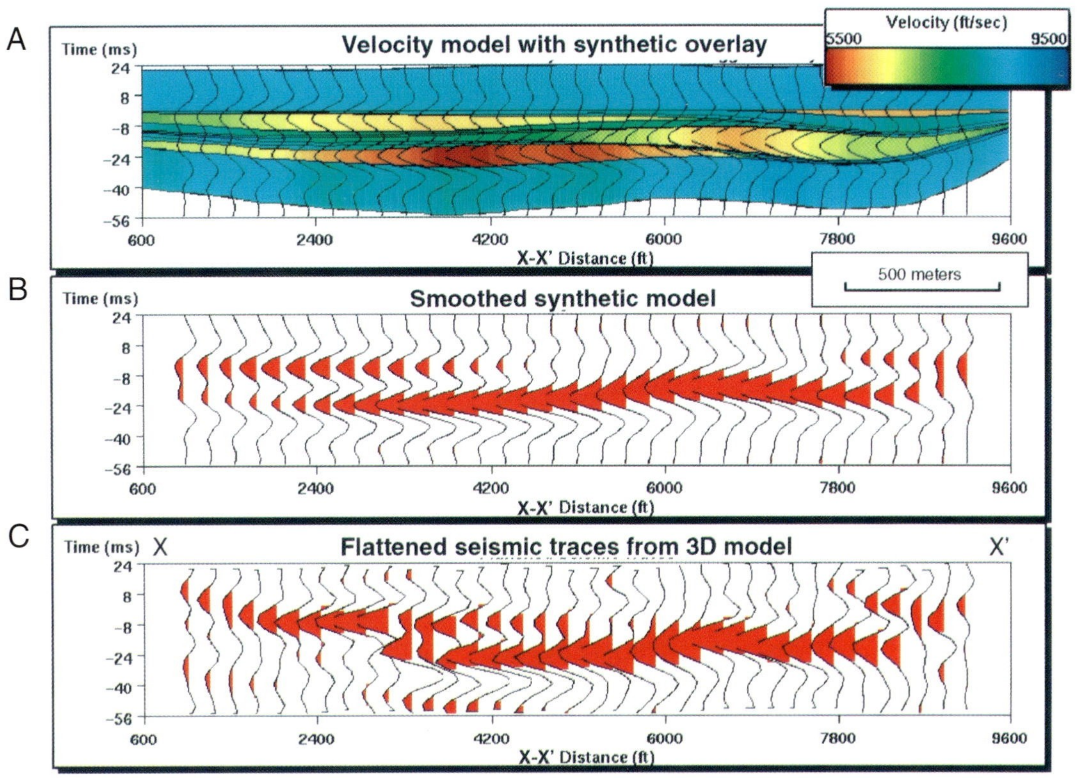


FIGURE 16. (A) Architecture model coded by interval velocity.
(B) Synthetic seismic traces generated by that model. (C) Actual seismic traces of cross section X-X′ (from Figure 15). Acoustic models utilized both velocity and density, but velocity is displayed here, because velocity variations are more dramatic.

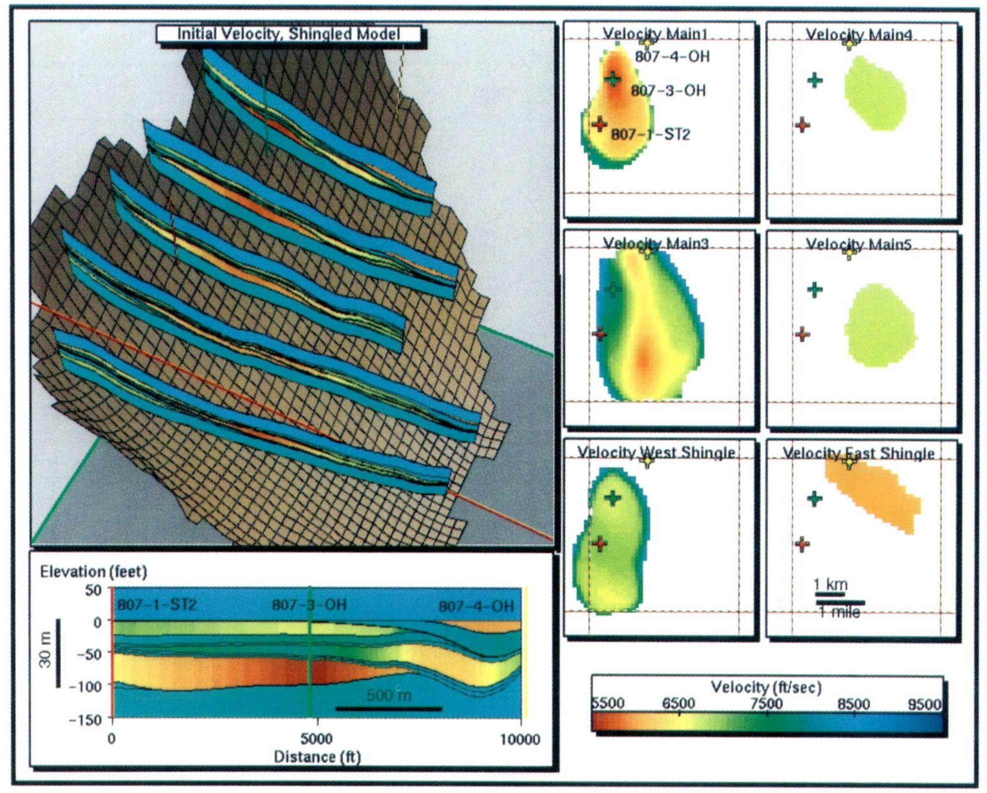


FIGURE 17. Initial velocity model showing stratigraphic shingles of the Complex model. The cross section ties the wells. Blocked velocities from well logs were interpolated by kriging in those shingles with a well penetration. Average Pink sand velocity was used for those shingles with no well penetrations.

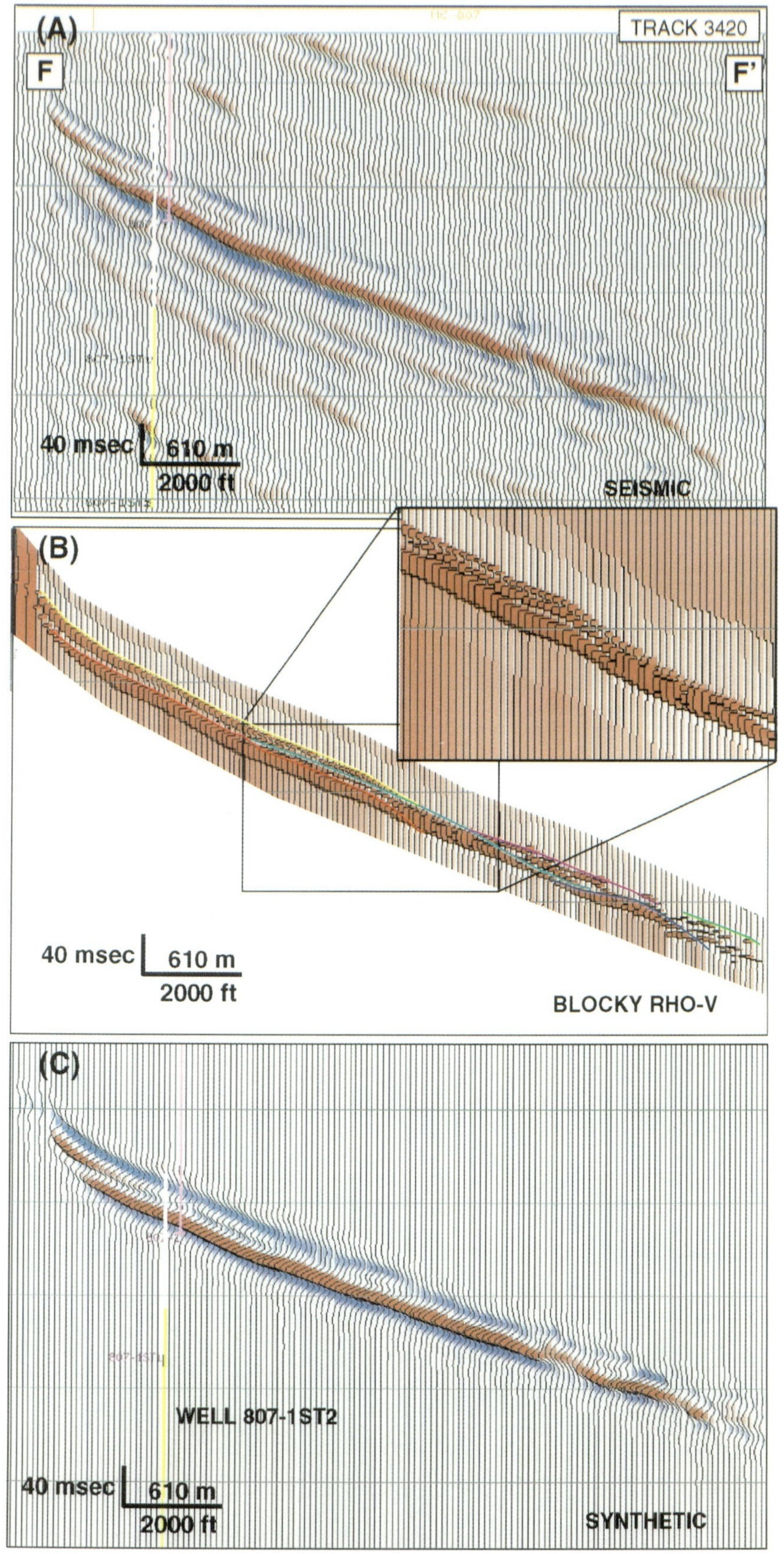


FIGURE 18. Using seismic inversion to aid architecture and property modeling. (A) Seismic data (traverse shown on Figure 11). (B) Blocked, acoustic impedance model of inverted seismic data showing stratigraphic shingles. (C) Synthetic seismogram generated from blocked impedance model, using a wavelet generated from the seismic data at points of well control.

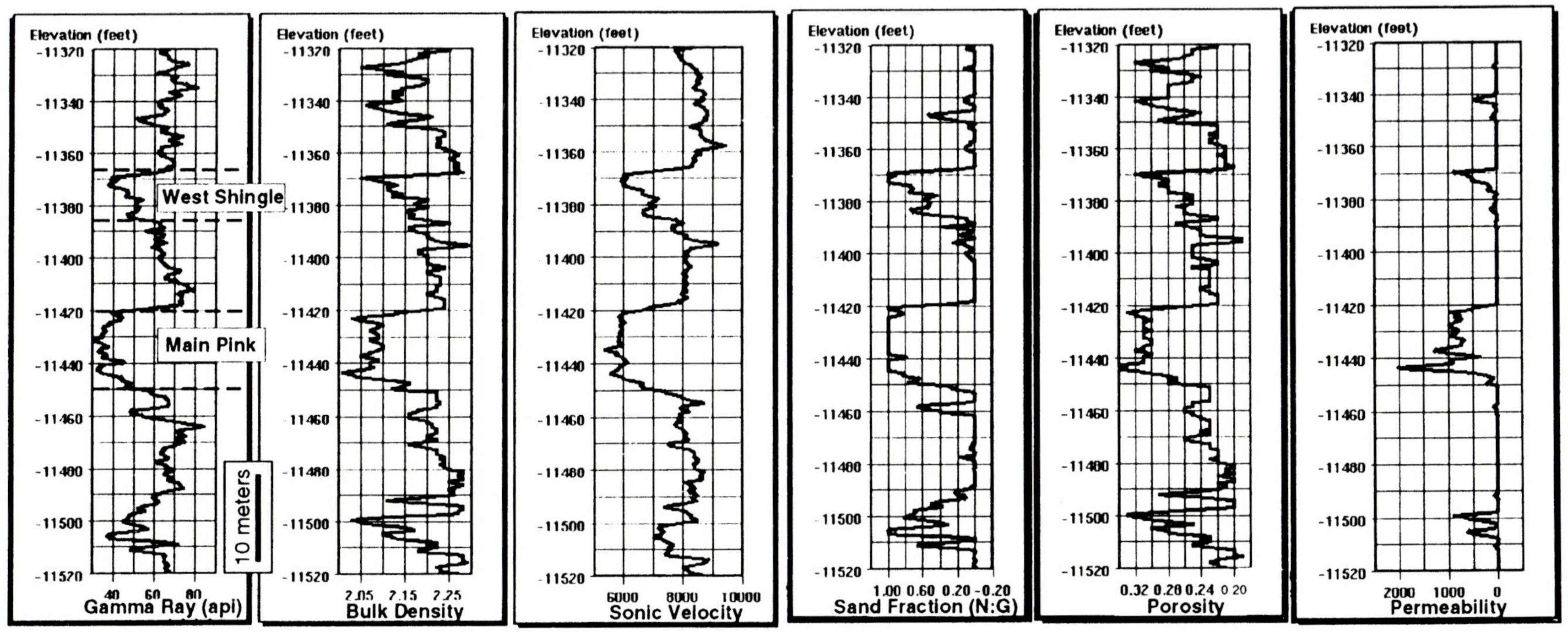


FIGURE 19. Acoustic and reservoir properties for the MC 807-3 well. Sand fraction and permeability are derived from other log measurements and locally calibrated to core.

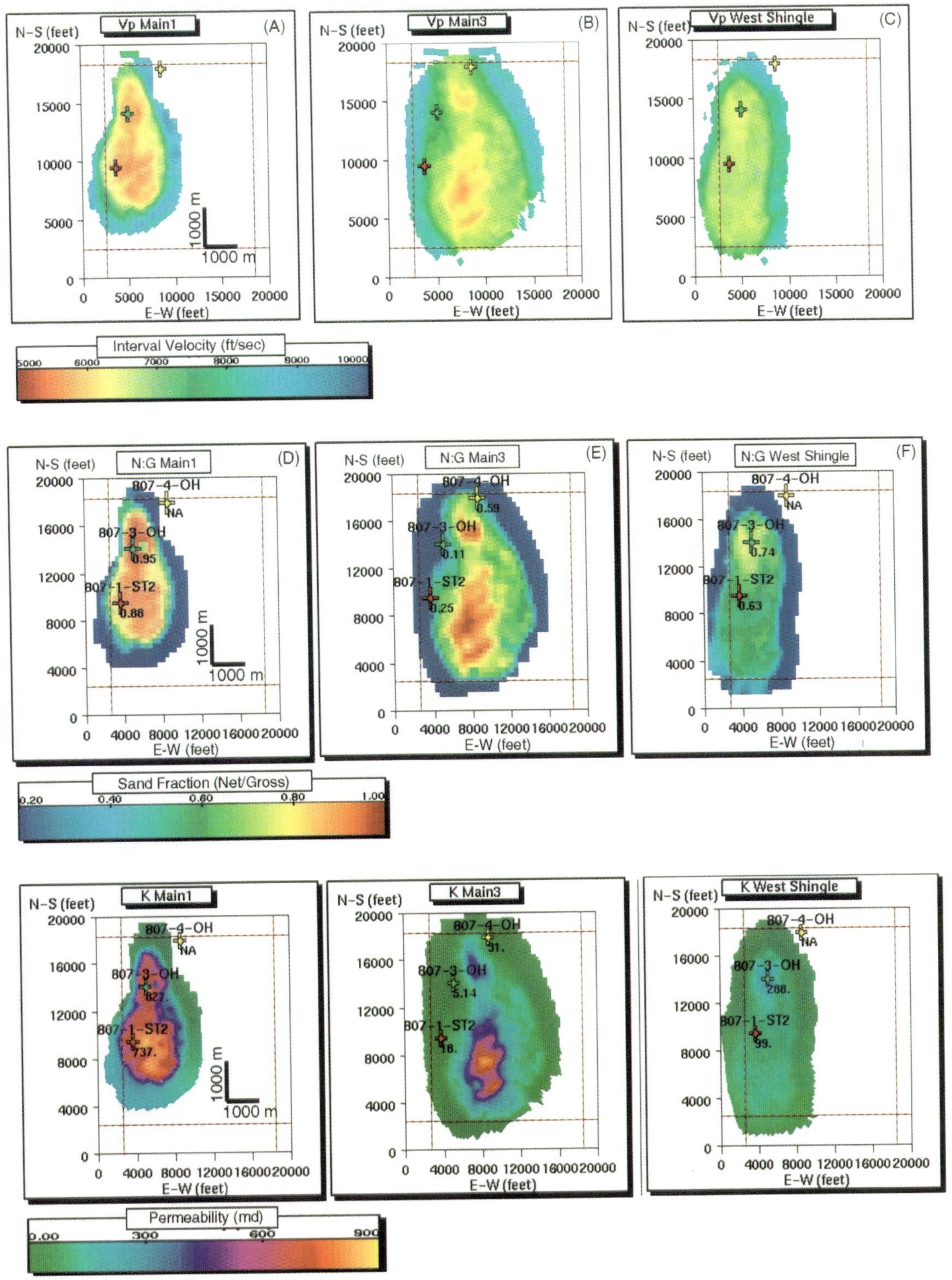

Vp Main1
(A)
Vp Main3
(B)
Vp West Shingle
(C)
N-S (feet)
E-W (feet)
1000 m
Interval Velocity (ft/sec)
N:G Main1
(D)
N:G Main3
(E)
N:G West Shingle
(F)
807-4-OH
807-3-OH
807-1-ST2
Sand Fraction (Net/Gross)
K Main1
K Main3
K West Shingle
Permeability (md)


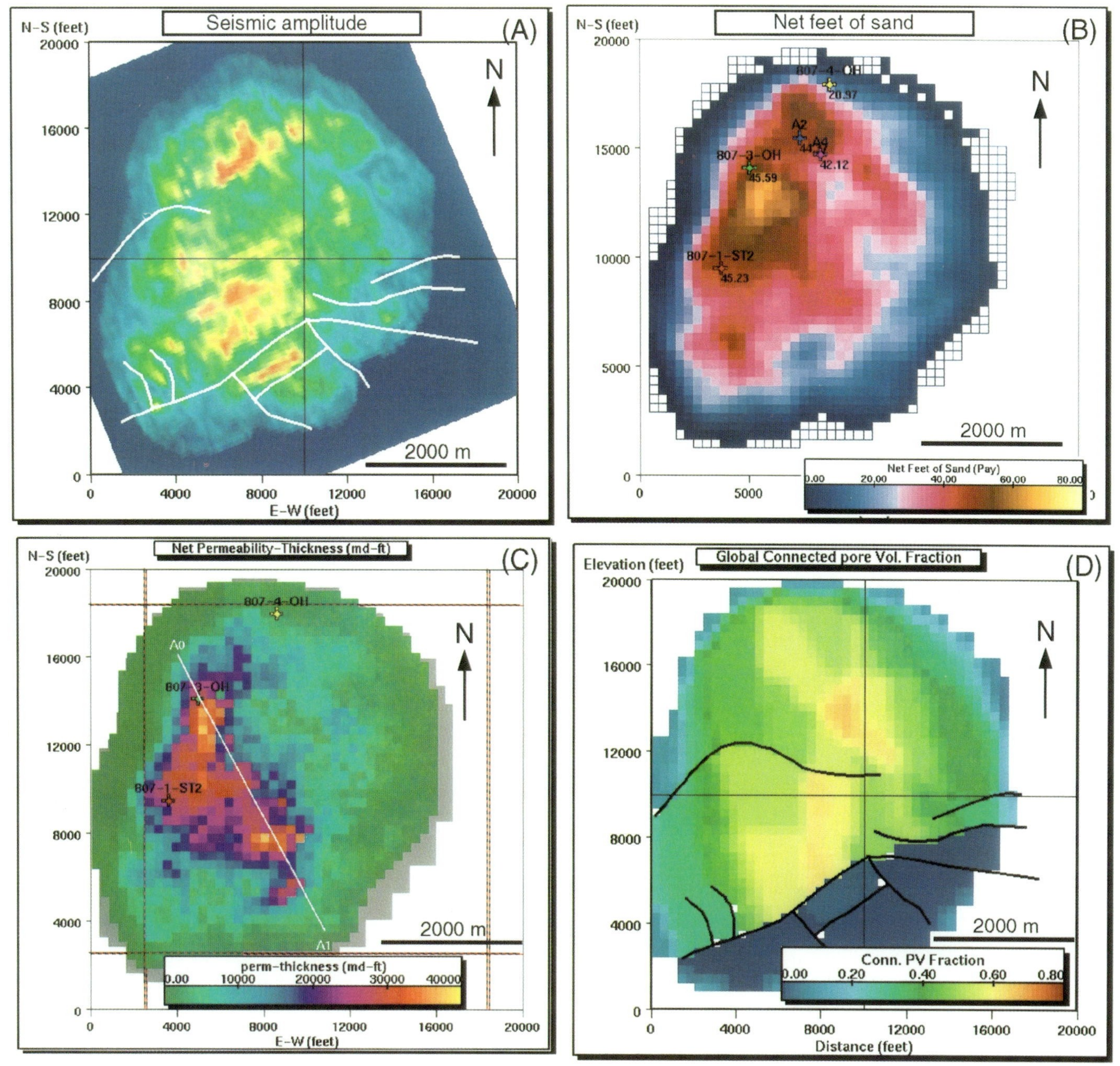

Seismic amplitude
(A)
N
2000 m
N-S (feet)
E-W (feet)
Net feet of sand
(B)
807-4-OH
807-3-OH
807-1-ST2
Net Feet of Sand (Pay)
Net Permeability-Thickness (md-ft)
(C)
perm-thickness (md-ft)
Global Connected pore Vol. Fraction
(D)
Elevation (feet)
Distance (feet)
Conn. PV Fraction

## CHAPTER 13

Rafalowski, J. W., B. W. Regel, D. L. Jordan, and D. O. Lucidi, Green Canyon Block 205 Lithofacies, seismic facies, and reservoir architecture, *in* P. Weimer and T. L. Davis, eds., *AAPG Studies in Geology 42* and *SEG Geophysical Developments Series No. 5*, AAPG/SEG, Tulsa, p. 133–142.

# Green Canyon Block 205 Lithofacies, Seismic Facies, and Reservoir Architecture*

John W. Rafalowski, Bernard W. Regel, Douglas L. Jordan, and Douglas O. Lucidi†

## ABSTRACT

The Green Canyon Block 205 prospect has five primary reservoirs, which range in age from late Pliocene to early Pleistocene. The N1 and N3 sands, which contain 60% of the resource in the prospect, were deposited primarily as turbidites in a middle- to lower-slope environment within bathyal water depths. Key uncertainties which affect fluid displacement paths and displacement efficiencies were defined during a detailed reservoir characterization. They include (1) facies type and distribution, (2) reservoir architecture, and (3) reservoir continuity. A probabilistic approach allowed the manipulation of a facies based geologic model to quantify the range of these and other reservoir uncertainties.

Geologic and geophysical data indicate that the N1 and N3 sand lithofacies and their associated subenvironments are distinctly different. The N1 sand was deposited as a sand-rich, early lowstand fan by a series of low-density turbidity flows. Facies range from massive sand to thin-bedded turbidites, and subenvironments include channel, channel margin, levee, interchannel/overbank, and possibly fan fringe. The N3 sand was deposited as a very sand-rich, middle-to-late lowstand fan by a series of high-density turbidity flows. It is composed of a massive sand facies within an amalgamated channel complex.

A lateral offset stacking pattern, which became progressively younger to the west, is the dominant architectural element for both the N1 and N3 reservoirs. Relative to the N1 sand, the N3 sand exhibits a greater degree of overlap between successive depositional units and therefore has a higher probability of amalgamation.

Drill stem test (DST)-type curve analyses indicate the presence of transmissibility restrictions in both sands, which correspond to discontinuities seen in the seismic data. In the N1 sand, the analyses indicate that fluid flow may not be inhibited between channel and thin-bedded facies but may be restricted between laterally offset depositional units. Seismically defined amalgamation surfaces in the N3 sand do not impede fluid flow within the radius of investigation of the DST.

## INTRODUCTION

The Green Canyon Block 205 (GC 205) prospect is located 241 km (150 mi) southwest of New Orleans, Louisiana, in water depths ranging from 762 to 855 m (2500 to 2800 ft) (Figure 1). The prospect is situated on Green Canyon Blocks 160, 161, and 205 and has an estimated 350–500 MMBO in place from five late Pliocene to early Pleistocene reservoirs. Because of the number and range of project uncertainties, a probabilistic analysis was performed during 1993 and early 1994 to support decision-making in the development scenario screening phase. A comprehensive reservoir characterization was one of the primary processes utilized to assess the economic viability of developing the prospect. A critical aspect of the characterization was the determination of the key uncertainties within the reservoirs. Uncertainties which affect fluid displacement paths and displacement efficiencies are: (1) facies type and distribution, (2) reservoir architecture, and (3) reservoir continuity. A facies-based geologic model was built to allow the manipulation of these and other reservoir uncertainties, and each facies defined was assigned reservoir property values (porosity, permeability, etc.). Thickness, aquifer support ratios, and continuity were varied in the geologic model to create cases representing the 10, 50, and 90% probabilities of occurrence. Additional heterogeneity was introduced geostatistically, and a reservoir simulator was utilized to determine the low, medium, and high case depletion schedules for each development option being considered. The ultimate goal was to create depletion schedules for a wide variety of geologic conditions to yield cumulative probability curves from which the expected value reserves could be derived. Ongoing geotechnical studies continue to enhance the geologic model, including the definition of the range of alternate realizations. This paper focuses on the source of the key uncertainties for the early Pleistocene N1 and N3 sands, because they contain approximately 60% of the resource in the field.

The N1 and N3 sands (N sands) have been penetrated by eight wells (Figures 2 and 3) at depths ranging from –3509 to –4184 m (–11,513 to –13,728 ft) subsea. Whole core was obtained from two wells (Exxon GC 205 #1 ST1 and Chevron GC 161 #1 ST3) in the N1 and N3 sands, and DSTs were performed over portions of the cored intervals. Standard log suites were obtained for all eight penetrations and were supplemented by Formation MicroScanner (FMS) images for five of the wells. Lithofacies and associated subenvironments were identified using data derived from these sources. A proprietary 3-D seismic survey was utilized to extrapolate those facies to the unpenetrated portions of the reservoirs. In general, amplitude values reflect not only reservoir fluid content but also sand quality, and the detuned composite amplitude extractions were utilized to determine reservoir morphology, facies distribution, and architecture. These data were integrated with the core and test data to relate flow behavior to reservoir facies and to predict flow restrictions.

## DEPOSITIONAL MODEL AND GEOLOGIC SETTING

The N sands were deposited primarily as turbidites in bathyal water depths within an intraslope basin. Sediment was derived from the shelf to the north during early Pleistocene sea level lowstands. Salt-cored paleobathymetric highs focused sediment from the shelf to the slope, and in the field area those highs caused localized dip reversals or gradient reductions. This in turn caused sediment transport velocity to decrease and release the coarser-grained components from turbulence. The N1 sand was deposited by low-density turbidity flows as a relatively sand-rich (>70% sand), early lowstand fan. The N3 sand was deposited as a middle-to-late lowstand fan by a series of high-density turbidity flows in the form of a very sand-rich (>0.85 net-to-gross) amalgamated channel sand system.

The N sands occur on the east flank of a north-south–trending salt-cored ridge. Hydrocarbons are trapped by a combination of structure and stratigraphic onlap, with oil column heights approaching 366 m (1200 ft) (Figure 2). The structural trapping element is a northeast-dipping, northwest-trending normal fault system. Seismic data indicate that deposition was partially influenced by this fault, as the thickest and sandiest sediment is found on the downthrown side. However, the dominant process which affected reservoir architecture involved successive deposition of active younger units on the flanks of previously active units. Turbidites were transported from the north onto a gently sloping basin covered with hemipelagic clays, and sediment volumes were only sufficient to partially cover the basin. The positive feature created by the initial sedimentation caused the synclinal axis to shift westward, and subsequent deposition occurred on the flanks of that initial accumulation. This process occurred repeatedly, resulting in the youngest deposition's taking place at the western edge of Blocks 161 and 205. Figure 4 shows diagrammatically the interpreted lateral offset stacking patterns which resulted from these processes. Aquifer support and continuity for this reservoir morphology, therefore, depend upon the amount of juxtaposition of progressively younger depositional units against the preceding units and the degree of amalgamation of individual sand beds between those units. Relative to the N1 sand, the N3 sand exhibits a greater degree of overlap between successive offset depositional units.

## RESERVOIR LITHOFACIES

Geologic and geophysical data indicate that the N1 and N3 sand lithofacies and associated subenvironments are distinctly different. These were defined using whole core, FMS images, and various open-hole logs. Bedding structures and faunal evidence present in

*This article was originally published in *Submarine Fans and Turbidite Systems: Sequence Stratigraphy, Reservoir Architecture, and Production Characteristics*, P. Weimer, A. H. Bouma, and B. L. Perkins, eds., 1994, 15th GCS-SEPM Annual Research Conference. The article is reprinted here by permission of the Gulf Coast Section Society of Economic Paleontologists and Mineralogists Foundation.

†Chevron U.S.A. Production Company, 935 Gravier Street, New Orleans, LA 70112

whole core were used to narrow the range of interpreted subenvironments in which the facies were deposited.

### N1 SAND

Figure 3 is a north-northwest–south-southwest true stratigraphic thickness (TST) cross section, which shows the N sand distribution across the three-block study area. The N1 sand, which was deposited on the 1.86-million-year sequence boundary, is easily correlated along an 8-km (5-mi) traverse. The TST ranges from 15 to 42 m (48 to 137 ft), and facies types range from massive channel sands (Chevron GC 205 #2 and #2 ST1 wells) to various combinations of interbedded sands and shales and thin-bedded turbidites (Exxon GC 205 #1 and ST1, all Chevron GC 161 #1 wells, and BP GC 160 #1). Bouma (1992) indicates that thin-bedded turbidites can be found in various subenvironments, including the top of a channel fill, channel margin, levee, interchannel/overbank, and possibly fan fringe. The facies present in the N1 sand are interpreted to have been deposited in most of these subenvironments. Interpretations based on core and log information are further substantiated by 3-D seismic data and are presented in the subsequent portions of this paper.

Penetrations of the N1 sand exhibit an overall log character that fines upward (with the exception of the Chevron GC 205 #2 ST1 well) and has a sharp base. 15 m (48 ft) TST of unconsolidated, interbedded, and laminated sand and shale were obtained from the N1 sand whole core in the Chevron GC 161 #1 ST3 well (Figure 5A). There are 275 individual sand beds present, ranging from less than 1.5 to 76.0 cm (0.6 to 30 in.) thick, with over 75% between 1.5 and 6.0 cm (0.6 and 2.4 in.) thick. The individual sand units are subarkosic, moderately to well sorted, and very fine grained to fine grained. The cored interval has a 0.70 net-sand/gross-interval (net/gross) value. Porosities for the individual sand units range between 28 and 35% and permeabilities from 200 md to 1.7 d. Sandstone interbeds show ripple cross-laminae and planar laminae, and there is a substantial amount of bioturbation within the thin shale units as well as within some of the sand beds. Sandstone beds more than 8.0 cm (>2.0 in.) thick were probably deposited by low-density turbidity flows and represent incomplete Bouma-sequence turbidites. The thinner (<2.0 in., or <8.0 cm) ripple-laminated sands and shales may represent deposition either as the distal portions of low-density turbidity currents or by reworking of sands by vigorous bottom currents (Schweller et al., 1994). The bioturbation indicates that deposition was slow enough to allow benthic fauna to rework the sediment. The degree of bioturbation most likely enhances the vertical connectivity between individual sand beds but decreases the permeability in the horizontal direction due to lithology mixing. High net/gross values were possibly created from the low-density turbidity flows through a flow-stripping process. Slight seafloor irregularities may have caused the sandier components to be released from turbulence and deposited while the clay and silt components were transported further downslope (Schweller et al., 1994). This process may also account in part for the high permeability values measured in the core (J. D. Schuppers, 1994, personal communications).

### N3 SAND

The N3 sand is correlative, but not entirely consistent in thickness, along the same 5-mi (8 km) traverse (Figure 3). The N3 sand TST ranges from 2 to 34 m (7 to 112 ft) with the thinnest units occurring where it onlaps the salt-cored structural highs to the northwest (British Petroleum GC 160 #1) and south (Chevron GC 205 #2 wells). The N3 log character shows multiple blocky and massive sand facies separated by few shales. It exhibits a sharp base, and the uppermost beds contain thin shales interbedded with thicker sands. Figure 5B illustrates that whole core from the Chevron GC 161 #1 ST3 well comprises 30 m (99 ft) (TST) of unconsolidated massive subarkosic sand interrupted by few persistent shale beds. The sand is predominantly fine grained but ranges from very fine to medium, with a 0.95 net/gross value. Porosities vary between 27 and 33% and permeabilities from 1.0 to 4.0 d. Based on whole core and FMS images, the N3 can be divided into two parts: a very massive to parallel laminated lower section (lower two thirds) with minor amounts of shale rip-up clasts, and an upper section with massive to parallel laminated sands, ripple cross-bedding, and bioturbation within thin shale layers. The character of the massive lower section and the presence of the shale clasts indicates rapid deposition by suspension fallout from a series of high-density turbidity flows. Parallel laminations are a result of shearing between grains during traction-carpet deposition. Ripple cross-bedding and minor bioturbation within the thin shales of the upper portion indicate periods of lower depositional energy followed the high-energy gravity flows that deposited the more massive sands (Schweller et al., 1994).

## SEISMIC FACIES DISTRIBUTION, RESERVOIR ARCHITECTURE, AND FLOW BEHAVIOR

A proprietary 3-D seismic survey was acquired to determine the seismic facies distribution and architecture of the N1 and N3 reservoirs. These data were integrated with the core and DST analyses to relate flow behavior to the reservoir facies and architectural elements of each sand and to predict flow restrictions. Stacking patterns, the presence (or lack) of levees, and the degree of amalgamation between sands are only a few of the architectural elements which affect the connectivity of a reservoir. Seismic character and attributes were utilized to determine these elements and to extrapolate lithofacies from well control to the unpenetrated portions of the N1 and N3 reservoirs.

### N1 SAND

The N1 sand detuned composite amplitude extraction (Figure 6A) exhibits high-amplitude, slightly sinuous, and elongate geometric features which trend northwest to southeast. Generally, the downdip limits of the amplitude anomaly conform with the probable oil-water contact constrained by the GC 161 #1 ST1 and ST3 wells. The possible exception occurs at the British Petroleum #1 well in Block 160, which has approximately 5 ft (1.5 m) of apparent water sand at the base. This well, which penetrated the eastern edge of the high-amplitude feature, found better-developed and thicker sand interbeds than the Chevron GC 161 #1, ST1, ST2, and ST3 and Exxon 205 #1 wells did (Figure 3). Log analysis indicates an overall net/gross value of 0.85. Because of the thicker interbedded facies, combined with the proximity to the sinuous geometry of the amplitude extraction, the N1 sand is interpreted to have been deposited in a channel margin subenvironment at this location.

The Chevron GC 205 #2 and #2 ST1 wells penetrated near the axial portion of a similar linear feature 8 km (5 mi) to the south. The N1 sand at those locations consists of massive, thick accumulations of clean sand with net/gross values ranging from 0.90 to 0.98. The #2 ST1 well contains two coarsening-upward sand units separated by a thin shale unit. The amplitude geometry and massive sand facies developed at these locations are interpreted as a channel subenvironment.

The Exxon GC 205 #1, ST1, and all Chevron GC 161 #1 wells penetrated various amounts of thin-bedded turbidites in the N1 sand with net/gross values of 0.70 or greater. Seismic line #1 is an east-west profile through the Exxon GC 205 #1 discovery well (Figure 7A). The N1 sand at this penetration is composed of 26 m (84 ft) (TST) of interbedded and laminated sand and shale, with a net/gross value of 0.75. Updip of this well, the N1 sand is represented by a higher-amplitude, slightly mounded reflection which corresponds to the channel features seen on the extraction. Levees are not obvious (or not preserved), and basal scour is minimal. Interruptions or discontinuities in the seismic event appear to be limited to thinning between the higher-amplitude mounds. These thinner and lower-amplitude segments are interpreted to be interbedded to thin-bedded sands and shale facies deposited in interchannel/overbank, levee, or channel margin subenvironments as part of a sand-rich, low-relief channel-levee system. The high net sand content of the system may account for the lack of levees. Reading and Richards (1994) indicate that higher sand content leads to a decrease in levee stability and increased channel switching.

The lack of obvious seismically defined levees could lead to the alternate interpretation that the thin-bedded turbidites are part of a fan fringe subenvironment. In this scenario, Blocks 161 and 205 are inferred to be located at the distal end of a channel-levee complex where sands flowed from the channel mouth to form a proximal lobe dominated by areally restricted channelized turbidite facies. The thin-bedded penetrations represent the edge of one (or more) of these fan lobes.

Line #2 is a seismic profile oriented along the Chevron GC 161 #1 ST3 well path (Figure 7B). The N1 sand at this location consists of 15 m (48 ft) (TST) of fining-upward, interbedded, and laminated sand, silts, and shales. With a 0.70 net/gross value, this is the poorest development of N1 sand observed to date. A small area of decreasing amplitude and minor offset occurs slightly updip of the penetration point. This amplitude decrease with vertical offset may represent the position at which a younger, laterally offset depositional unit onlaps the previously active unit. Updip of this position and to the northwest, amplitude values increase and erosional downcutting is not obvious. This high-amplitude area is interpreted to contain thicker, cleaner, more massive sand and is possibly indicative of unconfined channel development. The absence of well-developed levees leads to the formation of a low-sinuosity–to-braided system of unconfined channels that switch constantly across the fan surface (Reading and Richards, 1994). The N1 sand detuned composite amplitude map (Figure 6A) exhibits patterns which are suggestive of a low-sinuosity system.

A map of the areal facies distribution for the N1 sand is presented in Figure 8A. This interpretation was made by integrating the lithofacies from the wells; the thicknesses were derived from the seismic and well data and the geometries from the detuned composite amplitude maps. Whole-core data and well logs narrowed the range of depositional subenvironments to those most suitable for use as a depositional model guideline. Inspection of numerous seismic profiles was necessary to "fine-tune" the morphologic interpretation. These maps show five facies, "A" through "E," with "A" being massive sand and "E" being 100% shale. Facies "B" through "D" were defined as various combinations of interbedded sands and shales and thin-bedded turbidites. A geologic model was built using these maps, and reservoir properties were assigned to each facies. In effect, the most favorable reservoir properties correspond to the "A" facies (highest net/gross values, permeabilities, etc.) and the least favorable to "E." The facies maps presented here represent the dominant facies for each sand, additional heterogeneity was introduced into the model geostatistically. Three different N1 facies geostatistical realizations were vertically stacked together to compensate for the probability of having more than one facies present at any single location.

Type-curve interpretations of DST data taken from two wells in the N1 sand (Figures 9A, B, and Table 1) were utilized to relate flow behavior to facies type and reservoir architecture. These data indicate transmissibility restrictions in the vicinity of the well bores. For the GC 161 #1 ST3 test, a strong transmissibility restriction is seen within 73 m (240 ft) of the penetration point. This restriction corresponds to the area of lower amplitude with minor vertical offset previously discussed and shown on seismic line 2 (Figure 7B). The most likely reason for the discontinuity is that the overlap between the laterally offset depositional units is minimal in the immediate vicinity of the well bore, thus affording very little amalgamation and connectivity. A less likely possibility is that the amplitude decrease is representative of a locally sand-poor levee.

The discontinuity can also be seen as a lineation on an edge detection map for the N1 sand (Figure 10A). An edge detection map is a Landmark product which displays an extracted attribute—in this case, the response to differences in dip across a horizon (Landmark Graphics Corporation, 1994). The resultant map shows lineations that demarcate faults and more subtle depositional features, such as the edges of lobes or channels. The lineation attributable to the discontinuity trends northwest-southeast and also extends southward to the vicinity of the GC 205 #1 well. For the GC 205 #1 test, a moderate transmissibility restriction is seen 175 m (575 ft) from the penetration point. Seismic line #1 (Figure 7A) shows a minor dip change updip of the GC 205 #1 well within this approximate distance, and this corresponds to the lineation seen on the edge detection map west of the well. This lineation is considered to have the same origin as that for the 161 #1 ST3 well. In this case, the two successive depositional units probably have a higher degree of amalgamation. The surface between these units, therefore, locally acts as a fluid baffle.

The N1 edge detection map indicates that additional lineations exist within the 423-m (1390-ft) radius of investigation of the GC 205 #1 well bore. These may correspond to dip changes caused by thinner interchannel/overbank, channel margin, or leveed areas which lie between mounded channel features. Because no other flow restrictions were interpreted to occur within that radius of investigation, it is probable that these lineations did not impede flow between mounds. As the N1 sand is interpreted to have been deposited in a relatively unconfined manner in a sand-rich depositional system, areas which contain interchannel/overbank or levee subenvironments are not necessarily discontinuities to fluid flow. The massive sands deposited by high-density turbidites—in this case the mounded, unconfined channels—may grade laterally into the thin-bedded facies with minimal flow restrictions.

## N3 SAND

An amplitude extraction of the N3 sand (Figure 6B) shows a linear and braided pattern of northwest-southeast–trending anomalous amplitudes. These features originate in Block 117, decrease in strength in Block 160, and resume a high-amplitude character in west central Block 161. That character continues through most of the west-central portion of Block 205. The downdip limits of the amplitude anomaly conform with the oil-water contact established at –4053 m (–13,298 ft subsea) in the GC 161 #1 ST3 well.

The N3 sand at the GC 205 #1 location comprises of 23 m (77 ft) TST of massive sand deposited in a shingled stacking pattern, as shown on seismic line #1 in Figure 7A. As with the N1 sand, deposition began eastward of the penetration point and progressed from east to west as progressively younger units were deposited on the flanks of the previous unit. When westward lateral accommodation space was depleted (possibly due to a slight escarpment caused by the syndepositional fault), vertical stacking became the predominant depositional pattern (Figure 4). The N3 sand exhibits much greater overlap between successive depositional units than the N1 sand, and this affords a greater probability of amalgamation and reservoir communication.

Whole-core and open-hole logs from the GC 161 #1 ST3 well N3 sand display the best sand development penetrated to date in the prospect. The 99 ft (30 m) TST of blocky and massive sand is represented seismically on line #2 by reflections which are mounded and partially amalgamated (Figure 7B). This seismic profile, as well as others, indicates that the N3 sand has a higher degree of basal scour than the N1 sand. The areal amplitude distribution, the stacking patterns, and the massive sand lithofacies indicate that the N3 sand is a package of amalgamated channel sands deposited by high-density turbidity flows. The areal facies interpretation for the N3 sand is shown in Figure 8B.

Interpretation of a DST (Figure 9C) conducted in the N3 sand at the GC 161 #1 ST3 location indicates that a transmissibility restriction was encountered at 335 m (1100 ft). This corresponds with faulting updip and to the west as seen on seismic line #2 (Figure 7B). The N3 edge detection map (Figure 10B) indicates that lineations near the penetration were not restrictions to flow. These lineations correspond to dip changes between successive sand units which appear to be amalgamated. This implies that within the radius of investigation of the DST, the N3 massive channel sands have a high degree of connectivity across seismically defined amalgamation surfaces. A high degree of connectivity was experienced within similar amalgamated channel sands in the K-16 reservoirs at South Timbalier Block 295. This was evidenced by stable pressures during the production period (Mason, 1992).

**TABLE 1. DST Summary Showing Flow Restrictions, Radii of Investigation, and General Facies**

| Well | Sand | Facies | $K_0$ (md) | $r_{inv}$ (ft) | L (ft) |
|---|---|---|---|---|---|
| GC 205 1 | N1 | Laminated | 565 | 1390 | 575 |
| GC 161 1 ST3 | N1 | Laminated | 235 | 954 | 250 and 460 |
| GC 161 1 ST3 | N3 | Massive | 3000 | 2050 | 1100 |

$K_0$ = effective oil permeability
$r_{inv}$ = radius of investigation
L = distance to flow restriction

## SUMMARY

Quantifying key uncertainties within turbidite reservoirs is an essential part of risk mitigation and proper economic evaluation for high-cost deep-water projects. Key sources of uncertainty addressed for the GC 205 prospect include lithofacies, seismic facies, reservoir architecture, and reservoir continuity. These were defined by integrating whole-core, well log, 3-D seismic, and DST data for the N1 and N3 sands. Facies distribution maps were created for each sand and used as initial input to build a facies-based geologic model for a probabilistic analysis of the reservoirs. This methodology allowed the manipulation of various uncertainties within the geologic model to cover the range of possibilities for each development option considered.

The N1 sand is interpreted to be an early lowstand fan deposited by low-density turbidity flows in a relatively unconfined manner onto a gently sloping, smooth basin floor with minimal basal scour. Well control indicates the N1 unit to have a high net sand content (>70% sand), with facies types which range from massive to laminated. Subenvironments include channel, channel margin, levee, interchannel/overbank, and possibly fan fringe. Deposition progressed from east to west as successively younger units were emplaced on the flanks of the preceding units, resulting in lateral offset stacking patterns. Type-curve matching of DST data indicates that flow restrictions exist updip from existing penetrations, corresponding directly to discontinuities seen on seismic profiles and edge detection maps. The most likely reason for the restrictions is either the nonamalgamated overlap between two successive depositional units or a locally sand-poor levee. The N3 sand is interpreted to be a middle-to-late lowstand fan deposited by a series of high-density turbidity flows. It is part of a very sand-rich (>85% sand) complex of massive channel sands which are manifested by shingled, amalgamated, and mounded architecture. DST analysis indicates that these architectural elements do not behave as flow restrictions within the radius of investigation of the well bore.

*Acknowledgments*

We thank the many colleagues who contributed to this paper, particularly M. E. Levy, J. P. Logan, D. S. McCormick, W. J. Schweller, and C. E. Stelting. Additional gratitude is extended to G. F. Foster, A. E. Young, and C. A. Pinion for graphics preparation. We wish to express our appreciation to Chevron U.S.A. and Exxon U.S.A. for permission to publish this material.

## REFERENCES CITED

Bouma, A. H., 1992, Locations and characteristics of thin-bedded turbidites in passive margin setting submarine fans: Gulf Coast Association of Geologic Societies, v. XLII, p. 425–432.

Landmark Graphics Corporation, 1994, Seisworks/3-D Reference, p. 128.

Mason, E. P., 1992, Reservoir geology and production performance of turbidite sands at South Timbalier 295 Field, offshore Louisiana: Gulf Coast Association of Geologic Societies, v. 42, p. 267–278.

Reading, H. G., and M. Richards, 1994, Turbidite systems in deep-water basin margins classified by grain size and feeder systems: American Association of Petroleum Geologists Bulletin, v. 78, n. 5, p. 792–822.

Schweller, W. J., M. E. Levy, D. S. McCormick, C. E. Stelting, J. W. Rafalowski, B. W. Regel, and J. P. Logan, 1994, Analysis of Green Canyon 205 reservoir stratigraphy of N1, 2, and 3 intervals from cores, FMS images and wireline logs: Technical Memorandum, Chevron U.S.A., 14 p.

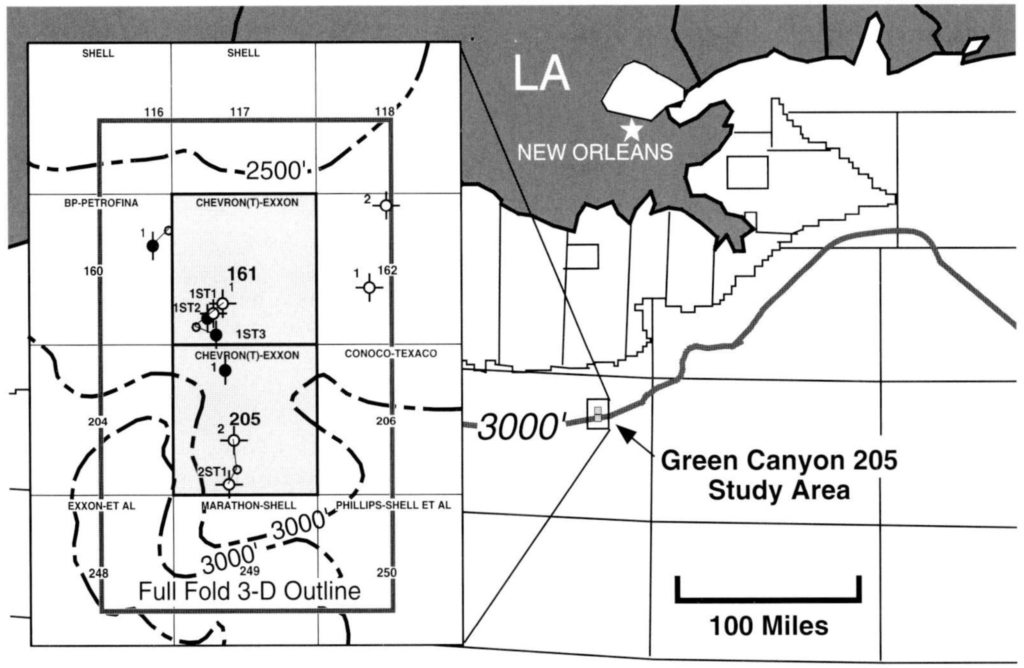


FIGURE 1. Green Canyon Block 205 location map.

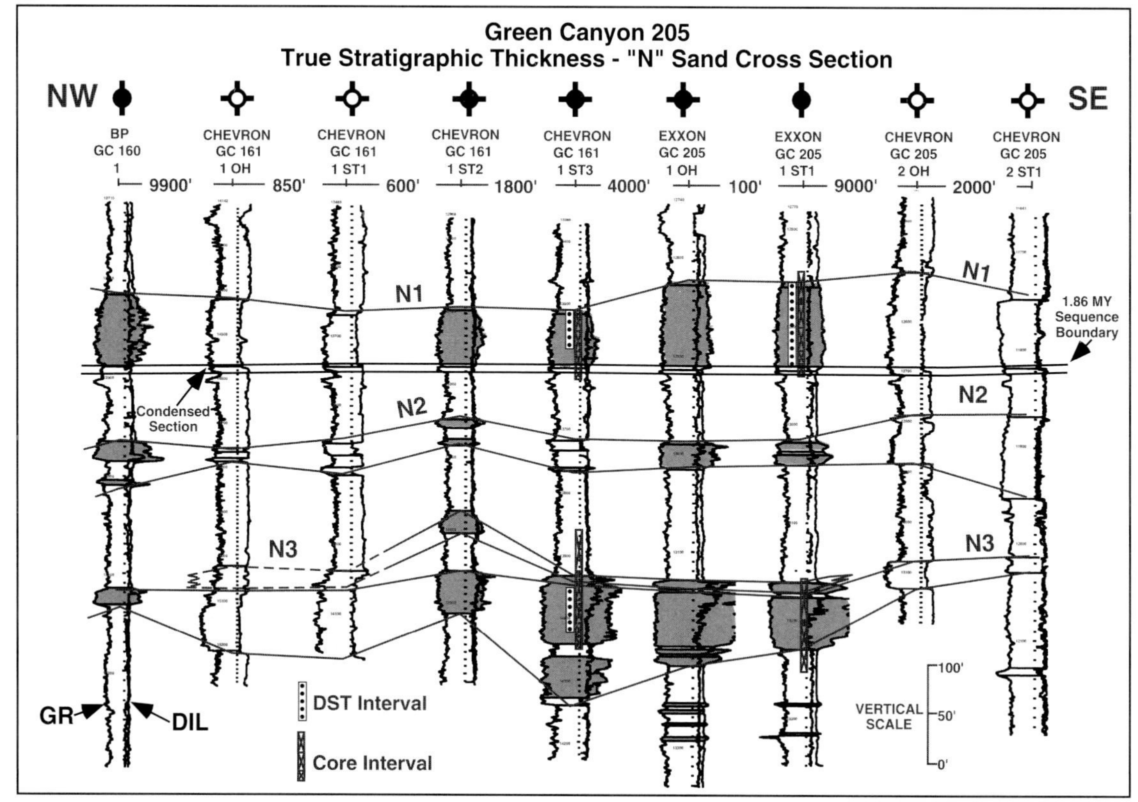


FIGURE 3. True stratigraphic thickness "N" sand cross section, showing "N" sand distribution and sand development.

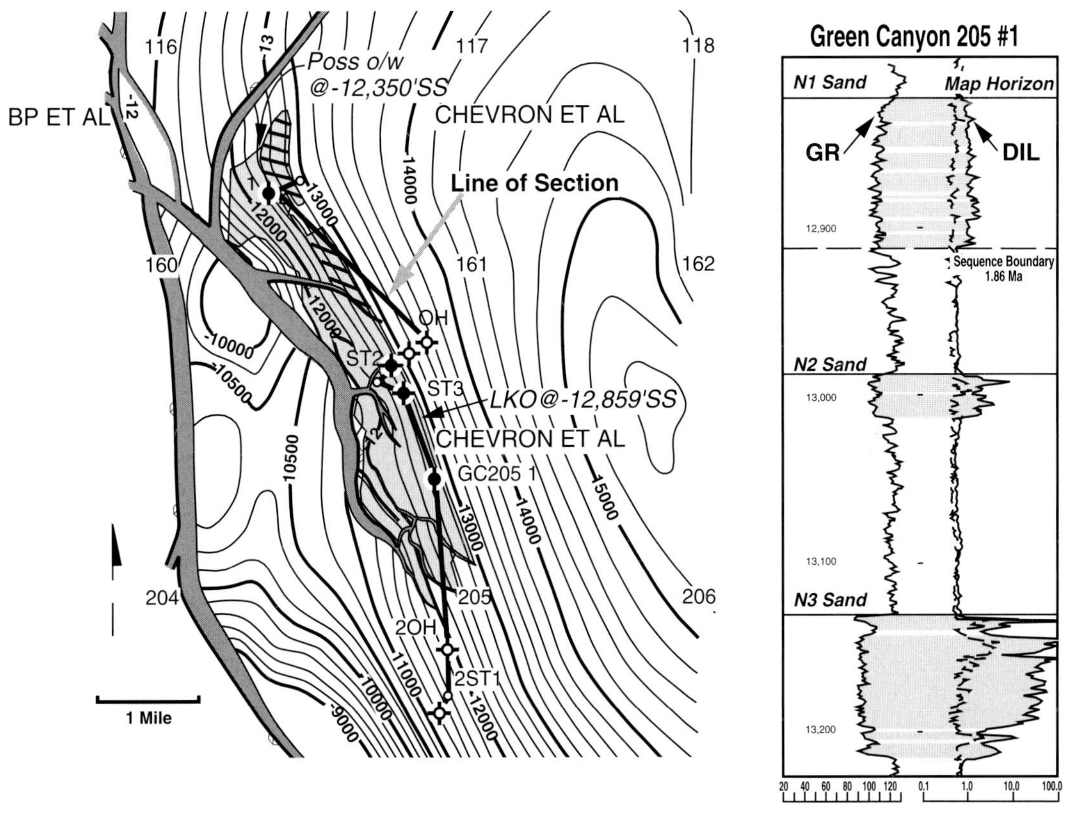


FIGURE 2. N1 structure, showing "N" sand true stratigraphic thickness line of section and general outline of the N1 sand reservoir.

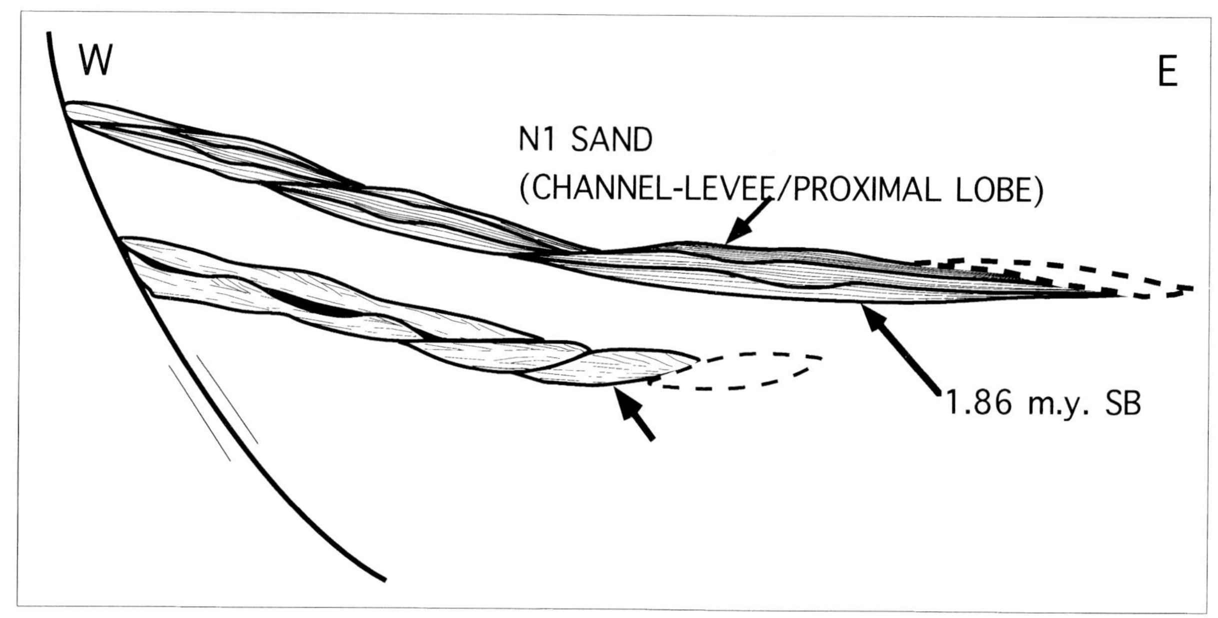


FIGURE 4. Diagram showing interpreted lateral offset stacking patterns of the N sands. The N3 sand exhibits a greater degree of overlap between depositional units and has a higher probability of amalgamation than the N1 sand.

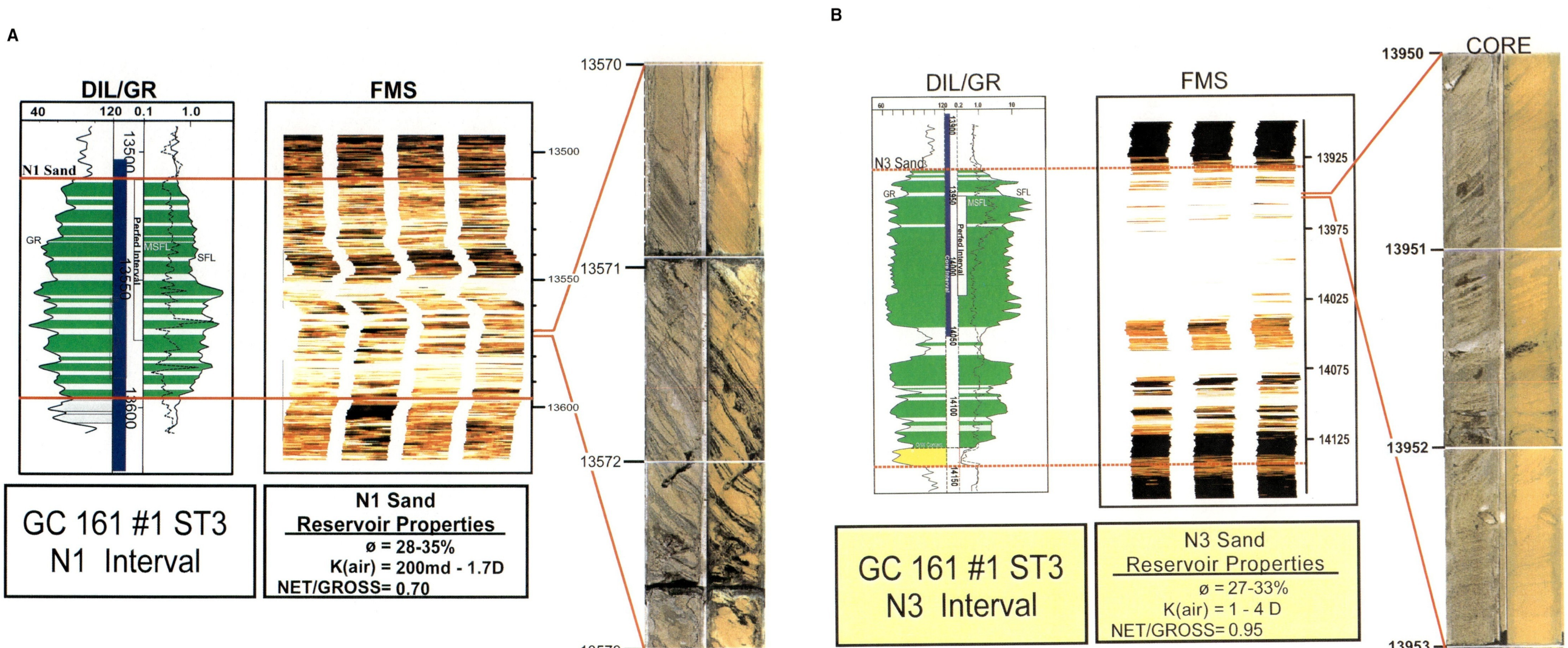


FIGURE 5. (A) N1 sand DIL/GR, FMS, whole core, and reservoir properties. FMS shows thin-bedded nature of sand, and core shows bioturbation and planar and ripple laminations. (B) N3 sand DIL/GR, FMS, whole core, and reservoir properties. Core shows massive sand facies with rip-up clast overlain by planar lamination and ripple cross-bedding.

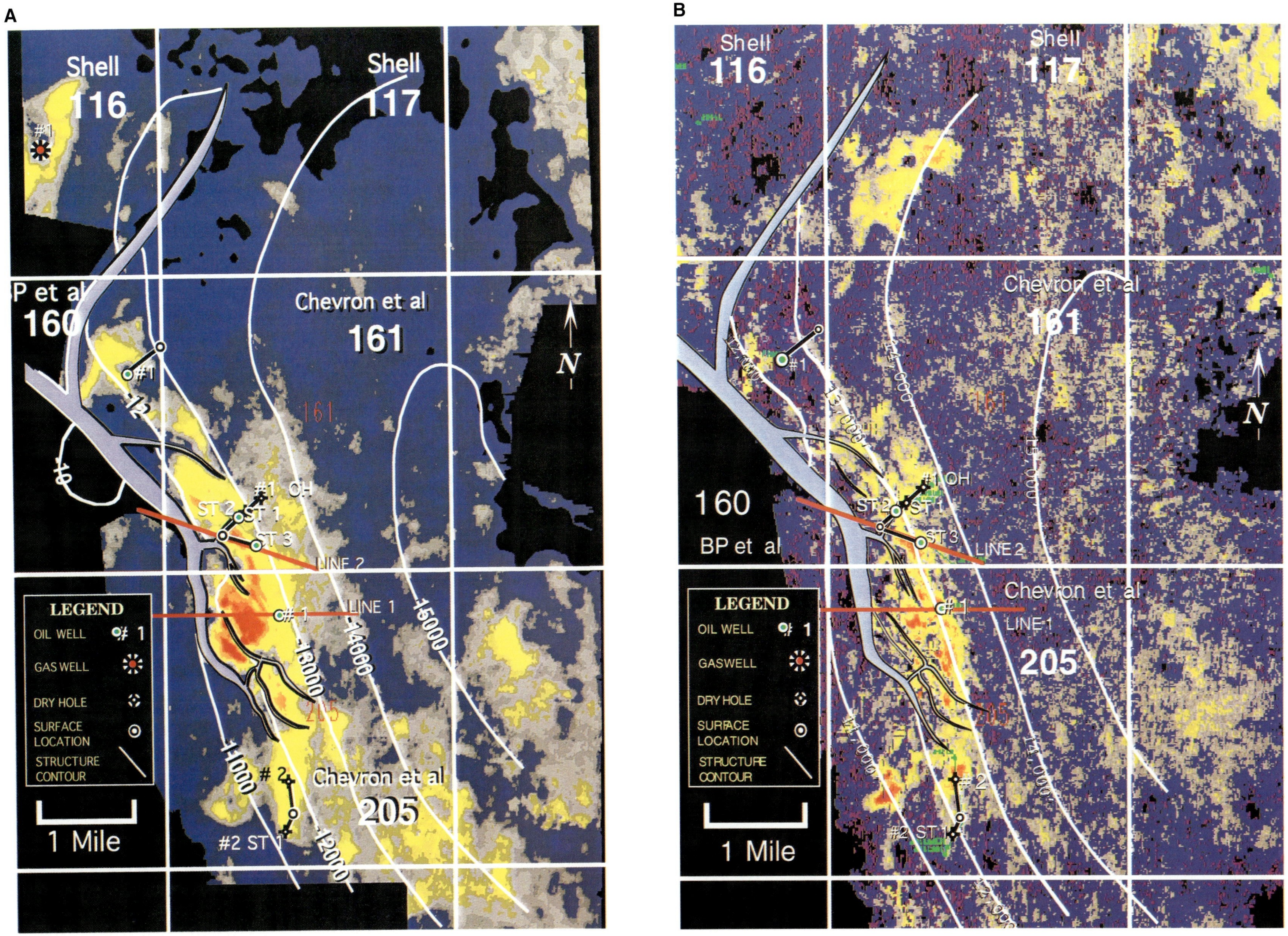


FIGURE 6. (A) N1 sand structure and detuned composite amplitude extraction show the downdip limit of the amplitude conforms to structure. (B) N3 sand structure and detuned composite amplitude extraction show the downdip limit of the amplitude conforms to structure and the oil/water contact seen in the GC 161 #1 ST3 well.

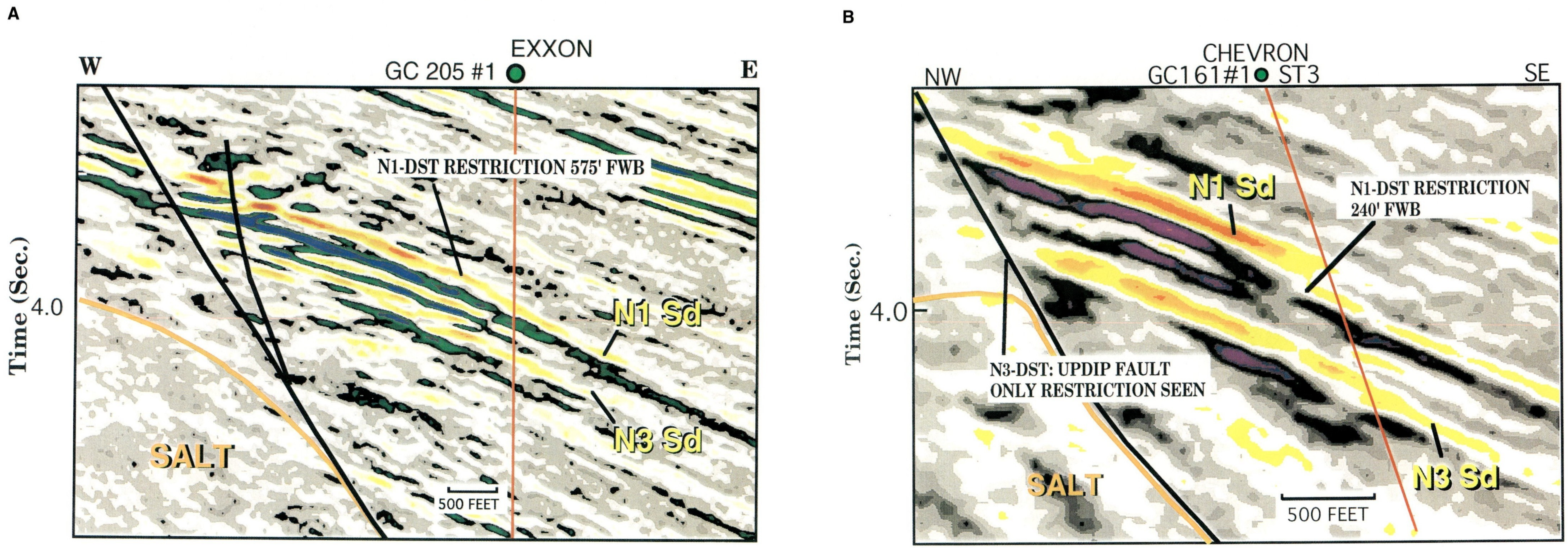


FIGURE 7. (A) Line 1 is an east-west seismic profile through the Exxon GC 205 #1 well, showing N1 and N3 sands seismic expression and DST flow restrictions. (B) Line 2 is a northwest-southeast seismic profile through the Chevron GC 161 #1 ST3 well, showing N1 and N3 sands seismic expression and DST flow restrictions.

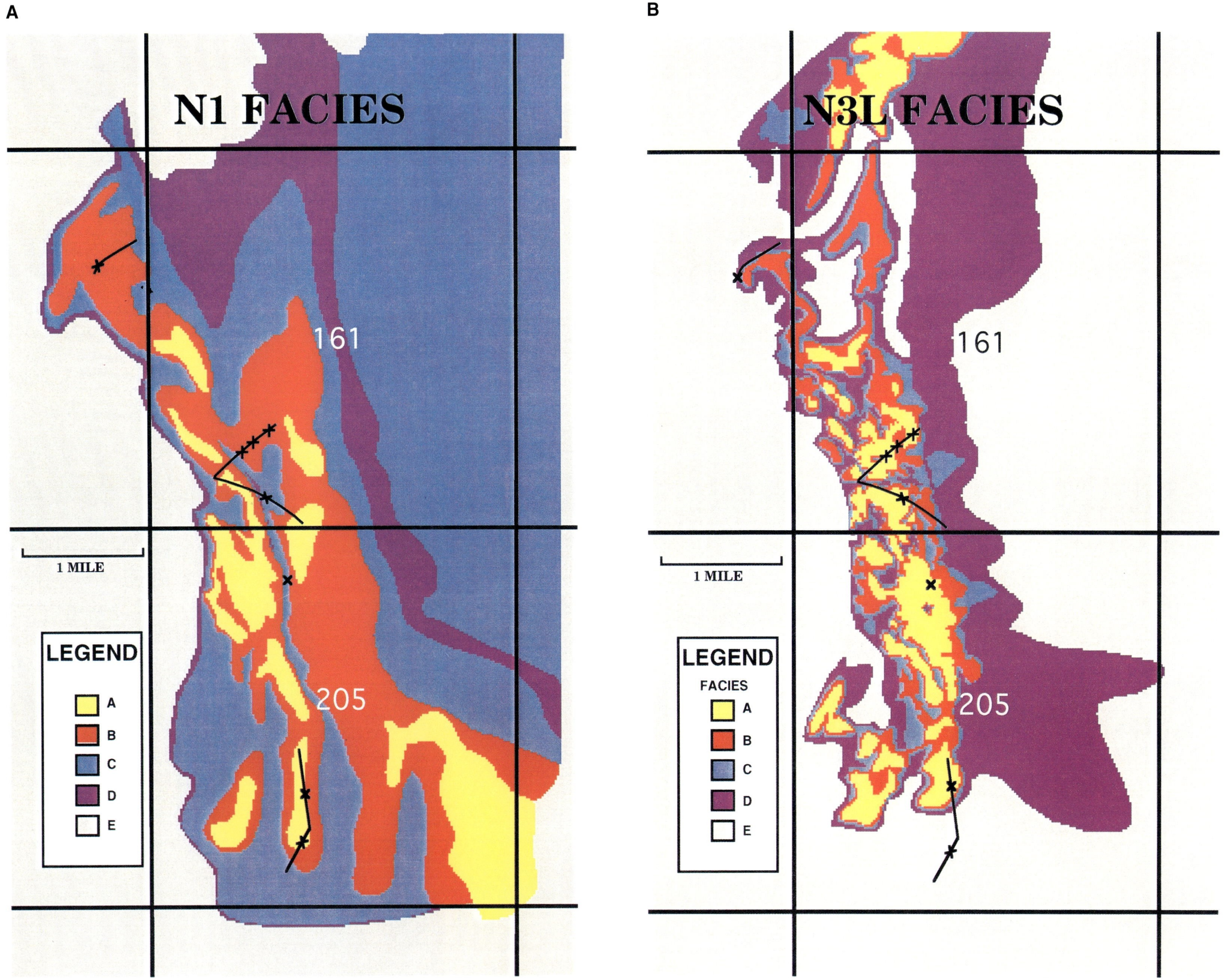


FIGURE 8. (A) N1 sand facies distribution map. (B) N3 sand facies distribution map.

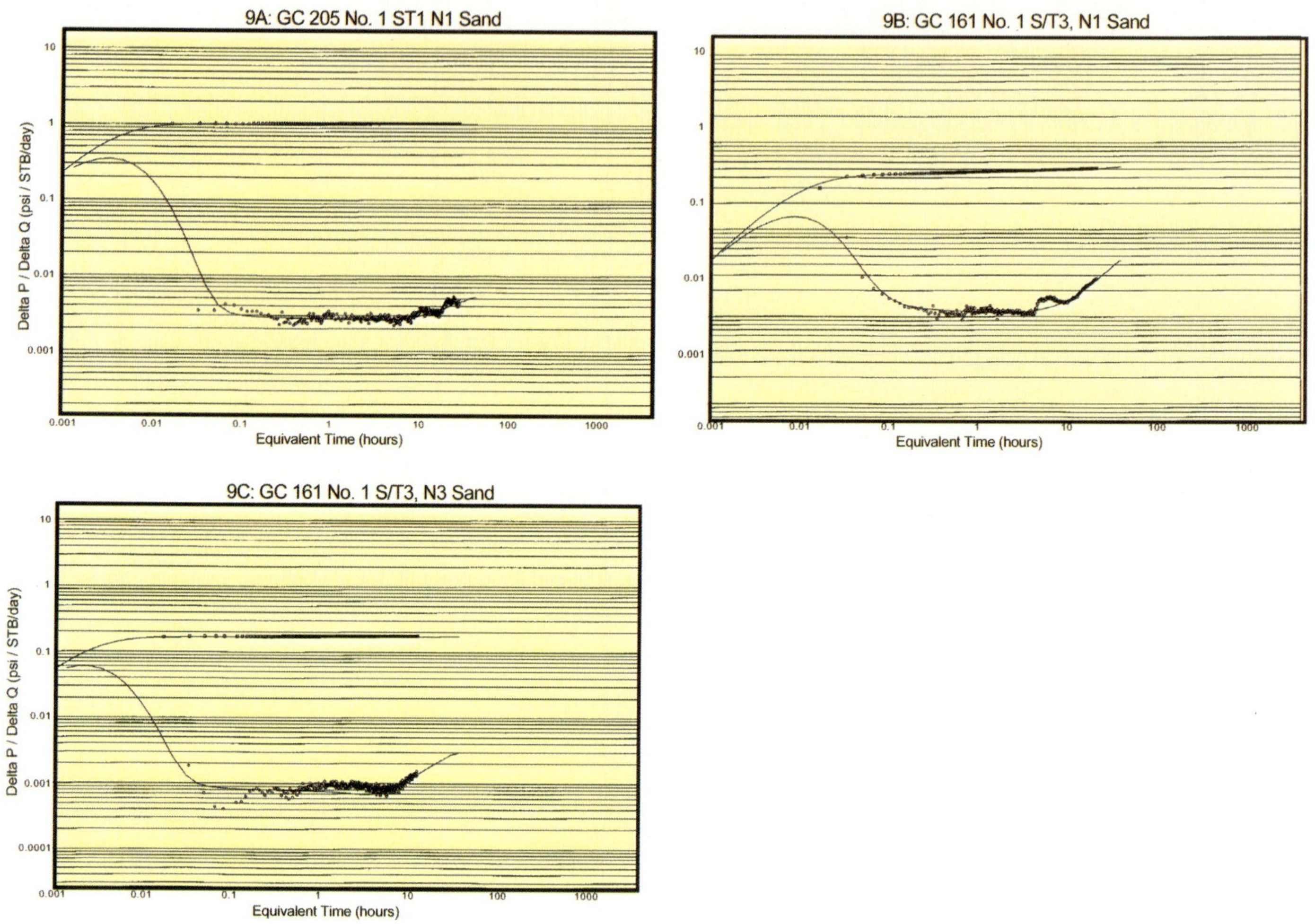


FIGURE 9. DST type curves from Exxon GC 205 #1 ST1 and Chevron GC 161 #1 ST3 wells. The modeled response shows transmissibility restrictions exist within the radii of the well bores.

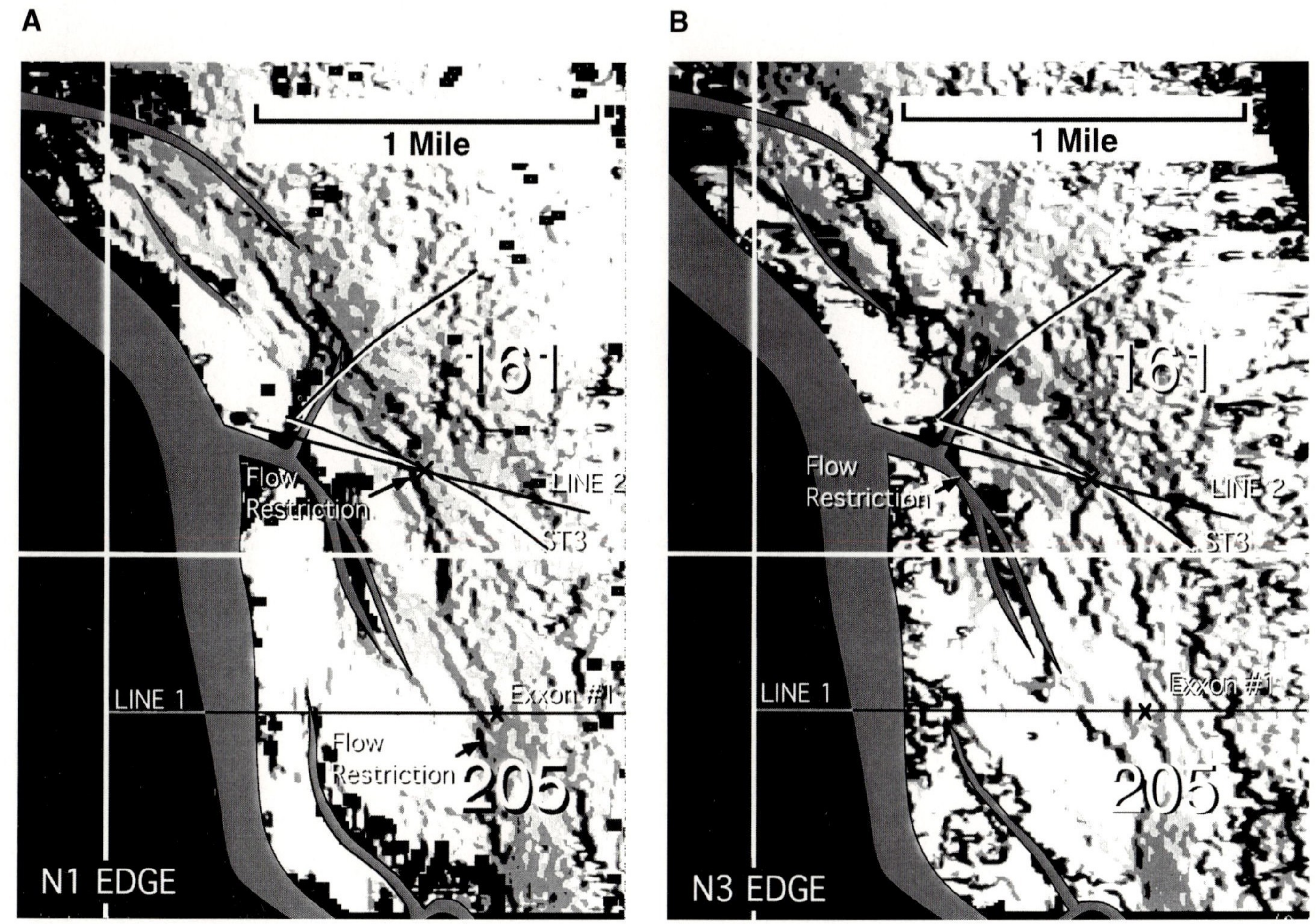


FIGURE 10. (A) N1 sand edge detection map shows lineations which correspond to discontinuities caused by dip differences between laterally offset depositional units. (B) N3 edge detection map shows differences in dip across the horizon which correspond to dip changes between amalgamated channel sands.

Risch, D. L., B. E. Donaldson, and C. K. Taylor, Deep-water facies analysis using 3-D seismic sequence stratigraphy and workstation techniques: an example from Plio-Pleistocene strata, Northern Gulf of Mexico, *in* P. Weimer and T. L. Davis, eds., *AAPG Studies in Geology No. 42* and *SEG Geophysical Developments Series No. 5*, AAPG/SEG, Tulsa p. 143-148.

CHAPTER 14

# Deep-Water Facies Analysis Using 3-D Seismic Sequence Stratigraphy and Workstation Techniques: An Example from Plio-Pleistocene Strata, Northern Gulf of Mexico

David L. Risch*, Barry E. Donaldson†, and Chris K. Taylor‡

## ABSTRACT

Deep-water lithofacies predictions were made in the Green Canyon area of offshore Louisiana using seismic facies analysis of 3-D data within a sequence stratigraphic framework. The versatility of the computer workstation and the high lateral resolution of the 3-D seismic data allowed us to identify stratigraphic boundaries and distinguish internal characteristics. Multiple flattened horizon slices were excellent for viewing the detailed geomorphic shapes and positions of depositional systems. Lithofacies maps of stratigraphic intervals were made quickly from seismic facies maps using volume attribute analysis and calibrated with well logs and selected seismic profiles.

## INTRODUCTION

Twelve federal lease blocks (120 mi$^2$, [307 km$^2$]) were selected in the Green Canyon Outer Continental Shelf (OCS) lease area of offshore Louisiana to interpret the deep-water stratigraphy from a 3-D seismic survey (Figure 1). Interpretation of 2-D seismic data provided the chronostratigraphic framework of the sequences when integrated with the biostratigraphic data in 13 nearby wells. Present and paleo bathymetry are greater than 2000 ft (610 m) in this region of the continental slope.

Sequence stratigraphy is a method of describing sediments deposited in a variety of environments during one cycle of relative sea level rise and fall (Van Wagoner et al., 1987). The cycle is divided into three chronostratigraphic units that relate to different positions of sea level during each cycle; these being the lowstand, transgressive, and highstand systems tracts. In deep-water slope and basin environments, the transgressive and highstand systems tracts are very thin and are commonly below conventional seismic resolution (less than 30 ms). The bulk of the sediments comprise the lowstand systems tract.

The lowstand systems tract is further subdivided. A basal unit, or basin-floor fan (BFF) was deposited during a rapid fall in sea level. The slope fan (SF) complex follows as sea level falls to a minimum and sedimentation occurs primarily on the slope. The prograding wedge (PW) complex is typified by delta progradation on the shelf edge during the initial slow sea level rise.

The deposits comprise channel/levee complexes, debris flows, slumps, and sheet sands. The deep-water sands are generally well sorted and can have porosities as high as 28–35% and permeabilities of 200–4000 md (Sangree et al., 1988; White et al., 1992; Rafalowski et al., 1994).

These lowstand depositional complexes have predictable lithofacies for each basin environment and diagnostic well log and seismic signatures. Biostratigraphic data were used to age-date the sequences, interpret paleo-environments, and identify relative marine deepening events within sequences that are associated with fossil abundance curve peaks. Our lithofacies interpretations were based on integrating well logs, seismic, and biostratigraphic data in the context of sequence stratigraphy.

## GEOLOGICAL SETTING

The structure and stratigraphy in the Green Canyon area are intimately tied to the copious amount of sediments supplied by the mud-rich Mississippi River and salt movements (Pulham, 1993; Weimer et al., 1994). The thick Pleistocene sediments caused salt to remobilize into tongues, sheets, massifs, and diapirs. Bathymetric lows developed where salt evacuated and acted as depocenters for additional lowstand sediments. Hence, a cycle of sediment loading and salt withdrawal created numerous isolated minibasins with complex geological histories. The interpretation of this complex pattern of sedimentation and salt movements necessitated an integrated sequence stratigraphic analysis.

The study area consists of a north-south trough flanked by shallow salt on the east and buried salt on the west (Figure 2). Salt withdrawal, minibasin tilting, and autocyclic switching of sedimentation have produced stacked and laterally overlapping depositional systems. A 10,000-ft (3050-m) thick accumulation of lowstand SF and PW sediments was deposited by successive turbidite flows and slumps around the salt during the Pleistocene period.

## INTERPRETATION

Stratigraphic boundaries and internal depositional units were identified based on their characteristic seismic and well log patterns. We relied on recognizing components of a sequence from various seismic facies, because progradational patterns are absent on the regionally gentle dipping shelf and slope. In addition, local structural events in individual minibasins result in stratal patterns that are independent of the regional trends and cannot be used to identify sequence boundaries.

A deep-water sequence is predominantly composed of the lower SF and, to a lesser amount, upper PW complexes. Figure 3 shows an example of the 3-D seismic character of stratigraphic boundaries and internal stratal geometries of the SF and PW. The SF complex is characterized by chaotic/hummocky seismic reflections indicative of slumping, debris flows, or small, seismically unresolved, stacked channel/levee systems. The overlying PW is typified by parallel, continuous to semicontinuous seismic reflections. A less common basal unit, the BFF, is usually less than 100 ms thick and has limited areal extent and bidirectional downlaps; the top has a high-moderate amplitude reflection. The sequence boundary is marked by the base of the SF chaotic seismic character and/or top of the PW parallel reflections.

The stacked channel/levee system in the middle of Figure 3 was interpreted using three methods, described from the least to the most sophisticated. One method was to construct isochron maps of the SF unit, recognizing that the thickest part is associated with the channels (Figure 4). This map yielded a crude estimate of the location of the channels centered along the axis of the north-south structural trough. Another procedure was to view a series of time slices starting near the top of the SF at its structurally highest point in the north. This method rendered good views of the channel but only in small segments from north to south, because of the southerly structural dip (Figures 5A and 5B). The superior technique was to areally flatten the top of the SF and create parallel horizon slices at 20-ms intervals (Figure 6). The complete detailed shape of the channel/levee system was revealed on each horizon slice (amplitude map). Figure 7 is an amplitude map sliced at 80 ms below the flattened top of the SF. The high-amplitude (black) levees outline the lower-amplitude (gray) channel fill. The flanking overbank deposits range from low- to

*Schlumberger/Geco-Prakla, Houston, TX 77077. Now with BHP Petroleum (Americas) 1360 Post Oak Blvd., Suite 500, Houston, TX 77056.
†Schlumberger/Geco-Prakla, Houston, TX 77077. Now with Jason Geosystems, Houston, TX 77077.
‡Schlumberger/Geco-Prakla, Houston, TX 77077

high-amplitude reflections, probably indicating the degree of sand-proneness. Weimer (1990) interpreted the higher-amplitude seismic reflections of the levee facies in the Mississippi Fan to generally correspond to coarser-grained sediments. Hackbarth and Shew (1994) showed a similar relationship with cores and shallow-penetration 3-D seismic data in a late Pleistocene channel/levee system in the Gulf of Mexico.

The trend of deposition was not always north to south, as shown by two older channels seen in Figure 8. These channels are from a horizon slice below the 1.4-Ma sequence boundary (red horizon). The change in channel orientation could have been in response to slight gradient changes that occurred as nearby salt adjusted for sediment loading or to the topography created by the deposition of underlying sediments.

## FACIES MAPS

Generating seismic facies maps manually is a time- and labor-intensive procedure; however, we rather quickly produced an alternative seismic facies map based on volume attribute analysis of the SF, using a unique function that calculates trace attributes within an interpreted interval. The reflection heterogeneity function calculates the length of each trace within an interval over a specified map area to measure the lateral changes in reflection patterns (i.e., chaotic, continuous). The function is dependent upon on a trace's frequency and amplitude, and each calculation was normalized to account for variations in interval thicknesses. This functionality can be used to locate stratigraphic discontinuities such as slumps, faults, fluids in the sediment, and lithofacies changes.

The SF lithofacies interpretation in Figure 9 is based on interpreting selected seismic profiles and well logs through distinct zones on the reflection heterogeneity map. The general shape of the channel/levee system is evident as a red band running from north to south and is associated with the chaotic (hummocky) reflections seen in cross section on Figure 10. Generally, hemipelagic and overbank silts and muds are interpreted in the low-reflection heterogeneity (green) map areas and on the seismic lines as low- to moderate-amplitude, parallel to subparallel reflections (Figure 10). The log character for the well in Figure 10 also indicated a mud-rich section distal to the channel system. The channel system becomes diffuse in the vicinity of the eastern flanking salt (Figure 11). We interpret this behavior to have been caused by slumping of the channel. A sandier overbank deposit associated with high heterogeneity (inline 724, crossline 1000) was interpreted from the seismic profile (Figure 12) and gamma ray log (Figure 13).

## SUMMARY

This study provided us with procedures and techniques for identifying, interpreting, and mapping turbidite-related sediments with 3-D seismic data on an interactive workstation. The high spatial resolution of the 3-D seismic data and the ability to volume-flatten and horizon-slice through the SF intervals allowed us to clearly image the exact shapes and positions of stacked channel/levee systems. This technique is of great benefit for reservoir continuity studies as well as exploration. In addition, horizon flattening can be used to identify misinterpreted horizons in areas where depositional trends abruptly disappear. Our lithofacies mapping was simplified by using the unique volume functionality of reflection heterogeneity to measure seismic reflection pattern variations. We related seismic facies to lithofacies by examining selected seismic profiles and well logs and applying the depositional model for deep-water systems.

### *Acknowledgments*

The authors wish to thank Geco-Prakla for permission to publish this paper. Special thanks to Karen El-Tawil for her support and to Jerry Watson for his review of this study. We also want to thank Doug McDonnell for technical support with the workstation.

## REFERENCES CITED

Hackbarth, C. J., and R. D. Shew, 1994, Morphology and stratigraphy of a mid-Pleistocene turbidite leveed channel from seismic, core and log data, Northeastern Gulf of Mexico, *in* P. Weimer, A. H. Bouma, and B. F. Perkins, eds., Submarine Fans and Turbidite Systems, GCSEPM 15th Annual Research Conference, proceedings published by Earth Enterprises, Austin, TX, p. 127–133.

Pulham, A. J., 1993, Variations in slope deposition, Plio-Pleistocene, offshore Louisiana, northeast Gulf of Mexico, *in* P. Weimer and H. Posamentier, eds., Siliciclastic sequence stratigraphy; recent developments and applications: American Association of Petroleum Geologists, Memoir 58, p. 199–233.

Rafalowski, J. W., B. W. Regel, D. L. Jordan, D. O. Lucidi, 1994, Green Canyon Block 205 lithofacies, seismic facies, and reservoir architecture, *in* P. Weimer, A. H. Bouma, and B. F. Perkins, eds., Submarine Fans and Turbidite Systems: GCSEPM 15th Annual Research Conference, proceedings published by Earth Enterprises, Austin, TX, p. 293–306.

Sangree, J. B., P. R. Vail, and R. M. Sneider, 1988, Evolution of facies interpretation of the shelf-slope: application of the new eustatic framework to the Gulf of Mexico: Offshore Technology Conference paper 5695, p. 133–144.

Van Wagoner, J., R. M. Mitchum, Jr., H. W. Posamentier, and P. R. Vail, 1987, Seismic stratigraphy interpretation using sequence stratigraphy. Part 2: Key definitions of sequence stratigraphy, *in* A. W. Balley, ed., Atlas of Seismic Stratigraphy: AAPG Studies in Geology 27, v. 1, p. 11–14.

Weimer, P., 1990, Sequence stratigraphy, facies geometries, and depositional history of the Mississippi Fan, Gulf of Mexico: American AAPG Bulletin, v. 74, p. 425–453.

Weimer, P., P. Varnai, Z. Acosta, F. M. Budhijanto, R. E. Martinez, A. F. Navarro, M. G. Rowan, B. McBride, and T. Villamil, 1994, Sequence stratigraphy of neogene turbidite systems, northern Green Canyon and Ewing Bank, northern Gulf of Mexico *in* P. Weimer, A. H. Bouma, and B. F. Perkins, eds., Submarine Fans and Turbidite Systems: GCSEPM 15th Annual Research Conference, proceedings published by Earth Enterprises, Austin, TX, p. 383–399.

White, C. D., Bradburn, F. R., Brown, R. L., and Thieme, M. A., 1992, Reservoir potential of thin-bedded turbidites: Prospect Tahoe: Society of Petroleum Engineers Annual Conference paper 24875, p. 105–117.

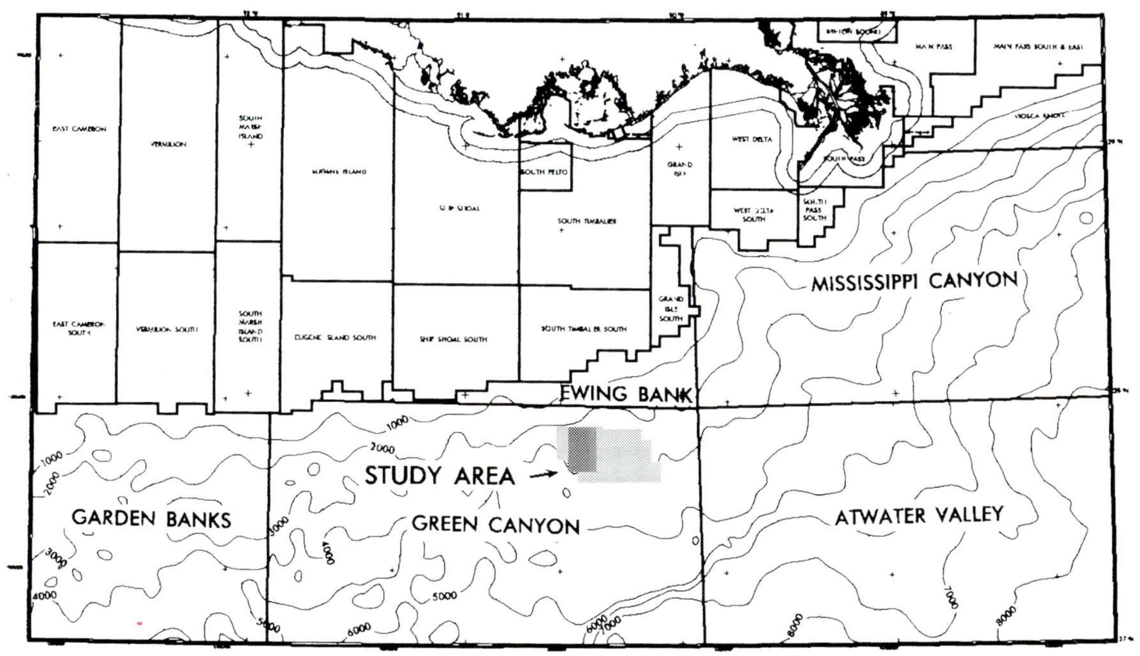


FIGURE 1. The study area includes 12 federal lease blocks extracted from a 43-block 3-D seismic survey on the slope, Green Canyon, northern Gulf of Mexico.

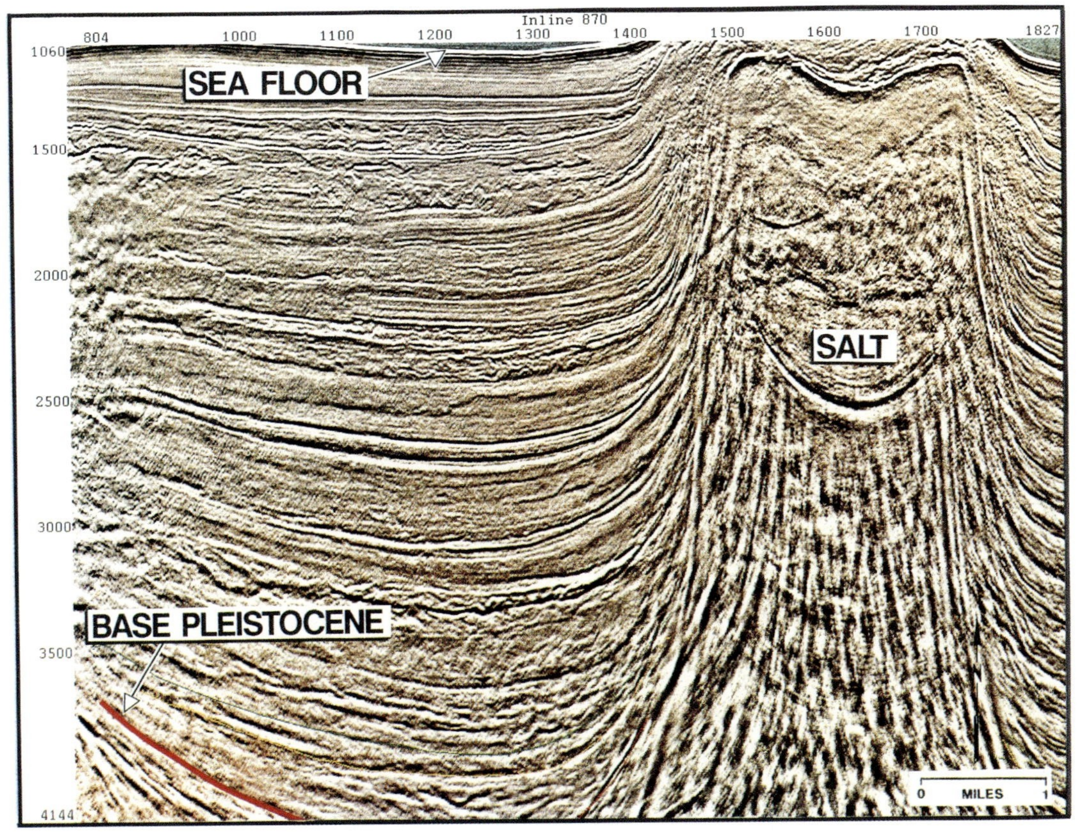


FIGURE 2. East-west 3-D line showing the structure and cyclic stratigraphy in the study area. The Pleistocene sediments are 10,000 ft (3050 m) thick. Location of profile shown in Figure 5B.

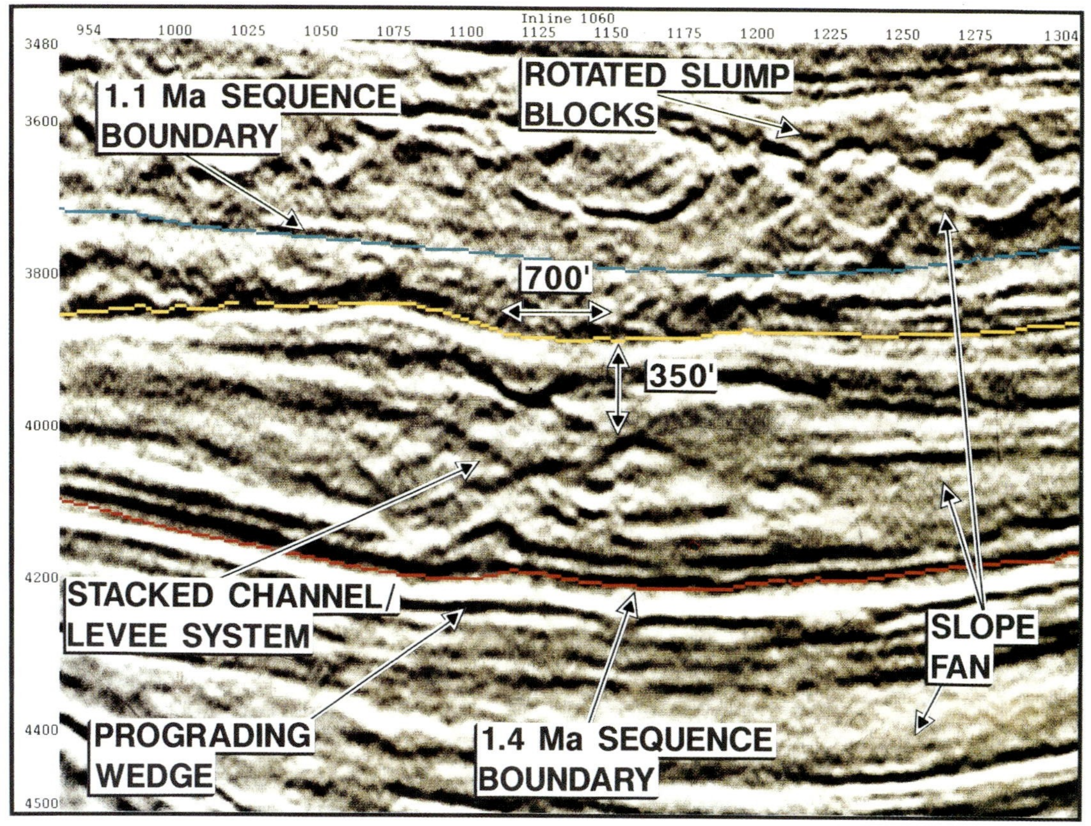


FIGURE 3. Example of 3-D seismic sequence stratigraphic boundaries and depositional units. Note the resolution of the channel at 3900 ms. Location of profile shown in Figure 5A.

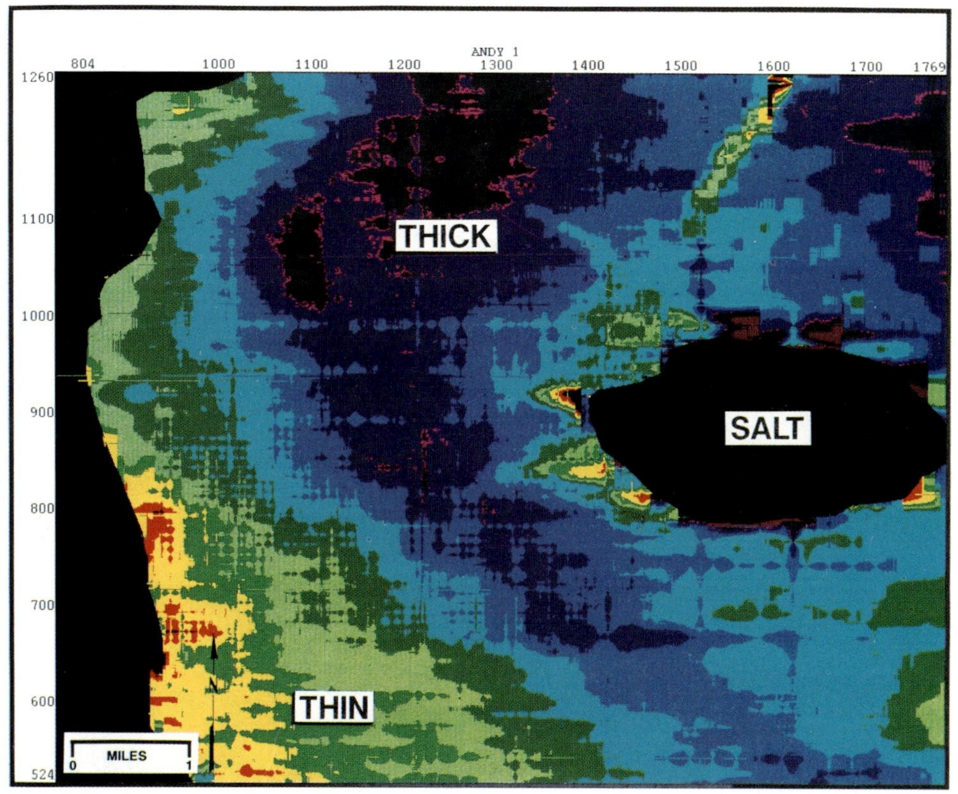


FIGURE 4. Isochron thickness of slope fan interval shown in Figure 3. The top is shown in yellow, the base in red.

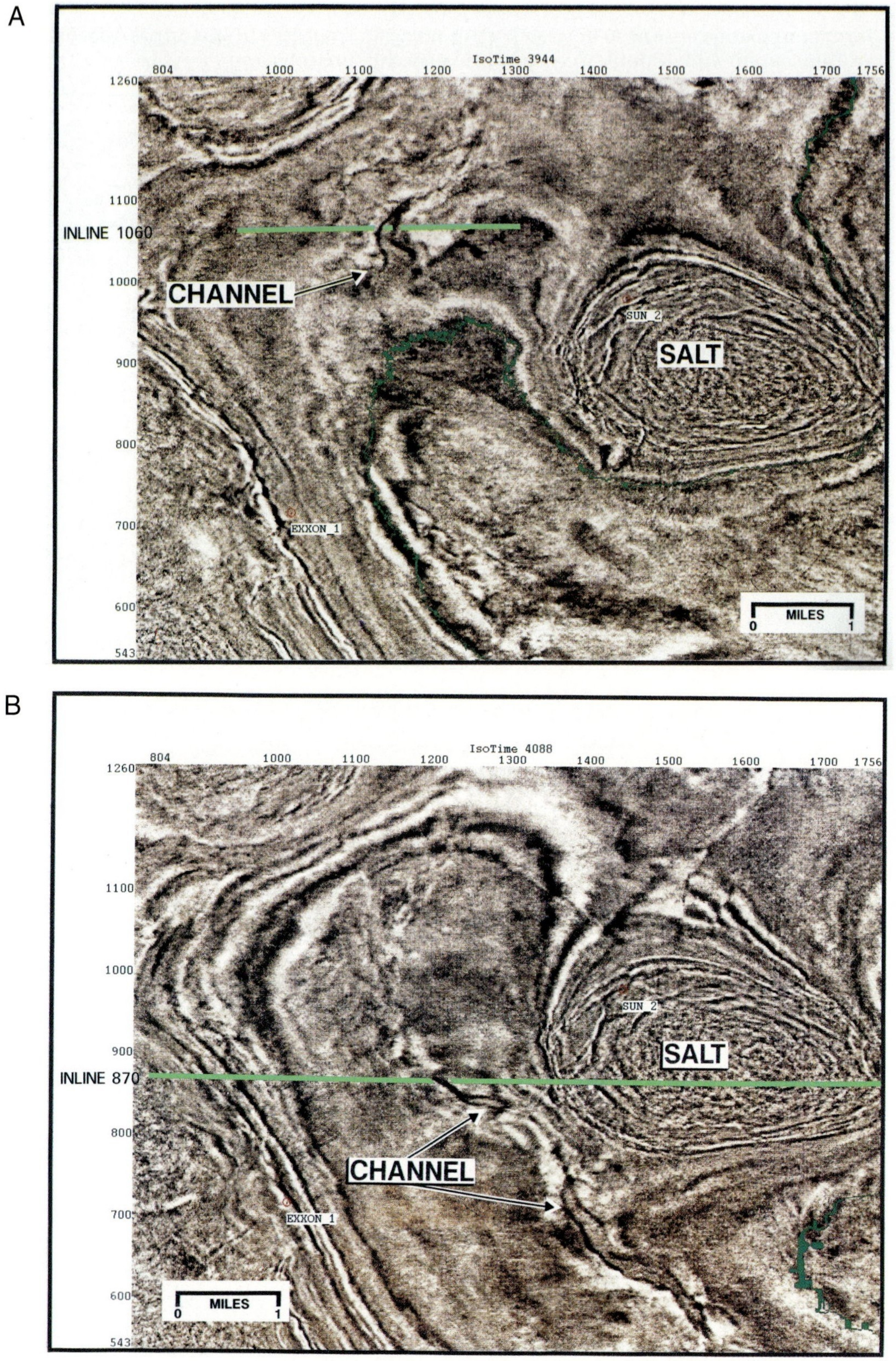


FIGURE 5. (A) 3-D time slice at 3944 ms, showing a northern portion of the channel. (B) 3-D time slice at 4088 ms, showing a southern portion of the channel.

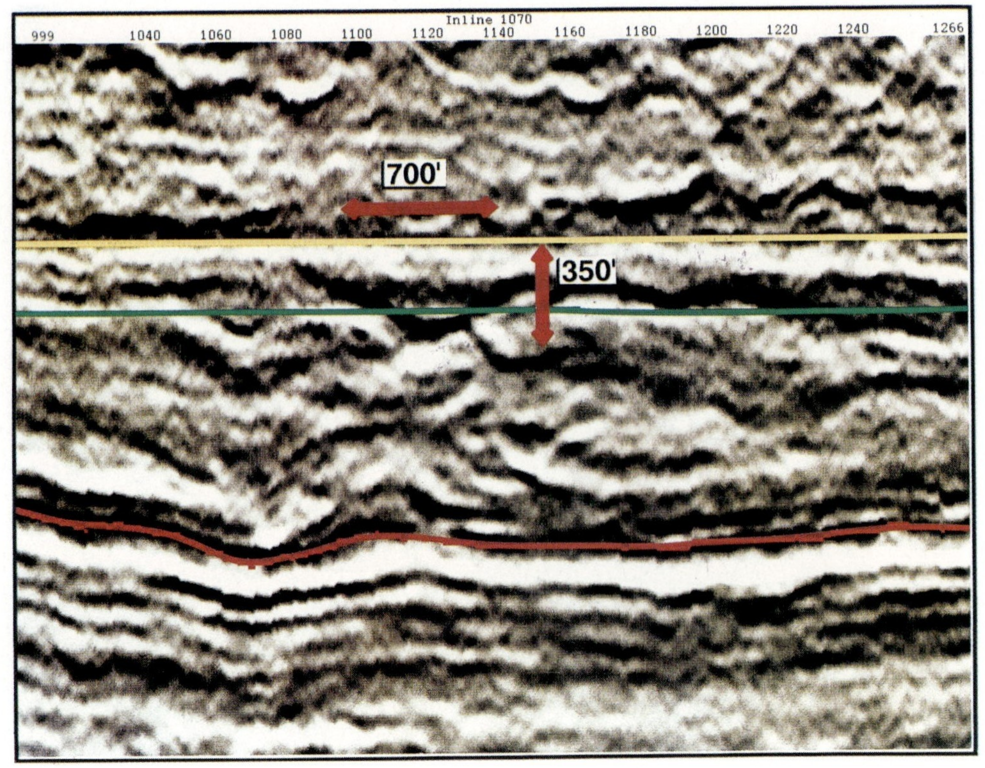


FIGURE 6. East-west line illustrating volume flattening on the SF top (yellow) to remove structural dip, then parallel slicing through the channel (green). The resulting horizon slice is shown in Figure 7 with the location of the profile.

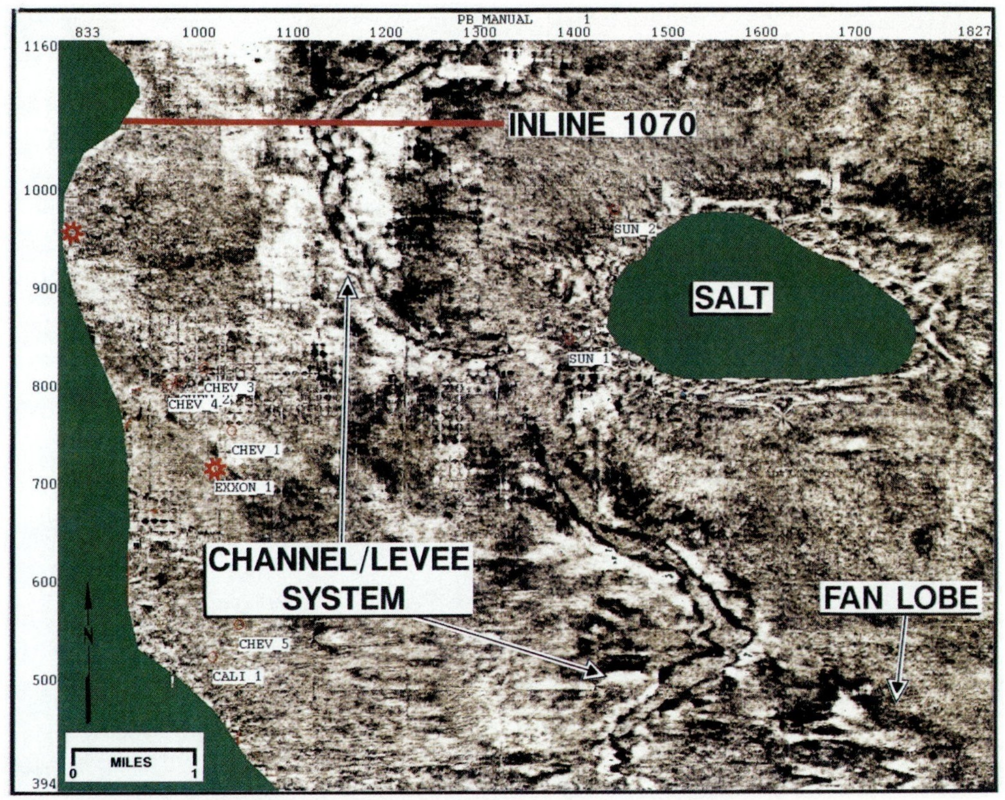


FIGURE 7. Volume-flattened horizon slice through the channel shown in Figures 3 and 6.

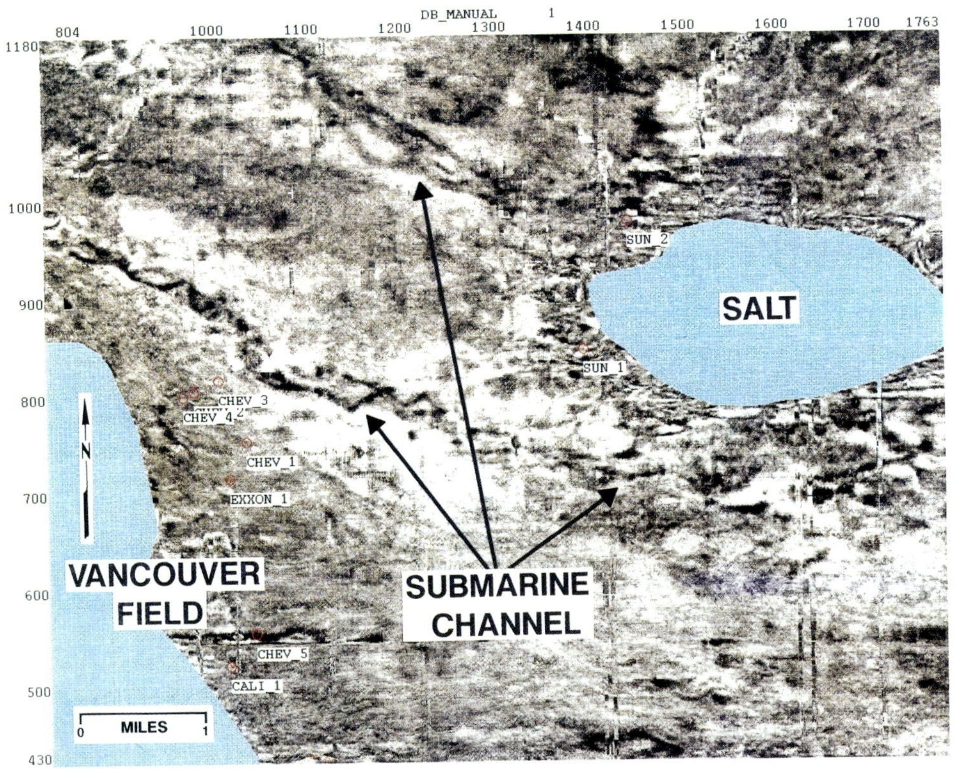


FIGURE 8. Northwest-southwest channel systems from a sequence below the sequence illustrated in Figure 7 (below red horizon).

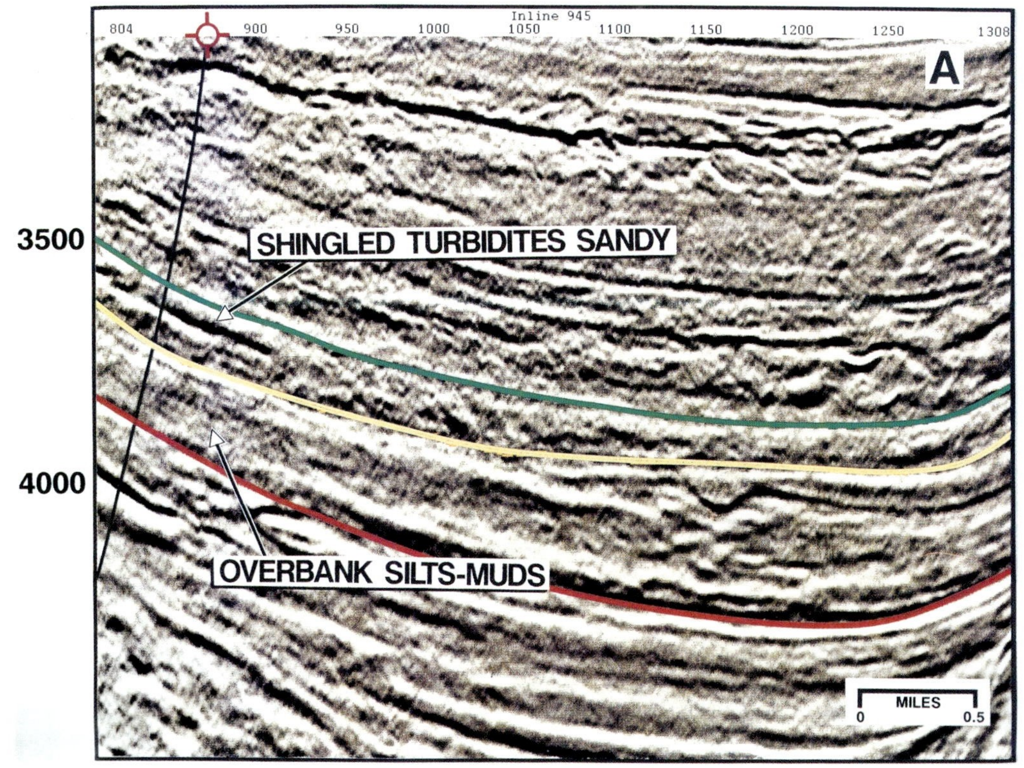


FIGURE 10. Profile A, showing the high-heterogeneity channel system (right) and low-heterogeneity overbank silts and muds (left). See Figure 9 for location of profile. Location of well shown in Figures 7 and 9.

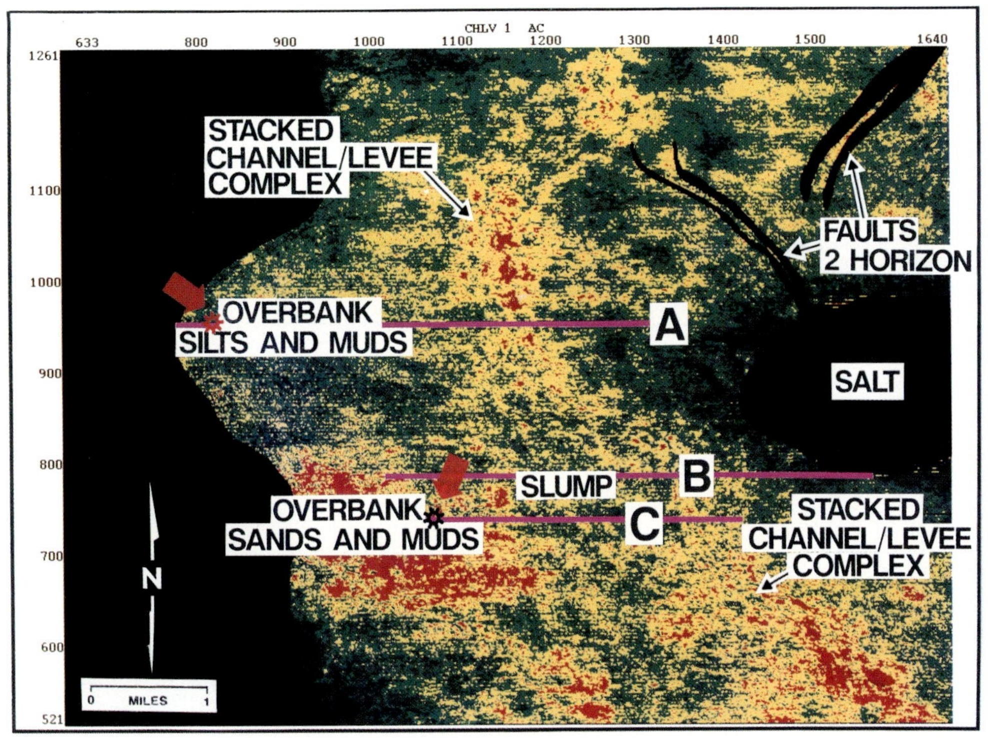


FIGURE 9. Reflection heterogeneity map with lithofacies interpretation. Red and yellow indicate high degree of dissimilarity among nearby traces; green indicates similarity. The channel system in Figure 7 is revealed as red and yellow due to its chaotic/hummocky seismic character. Seismic profiles A, B, and C, used for lithofacies interpretation, are shown in Figures 10–12.

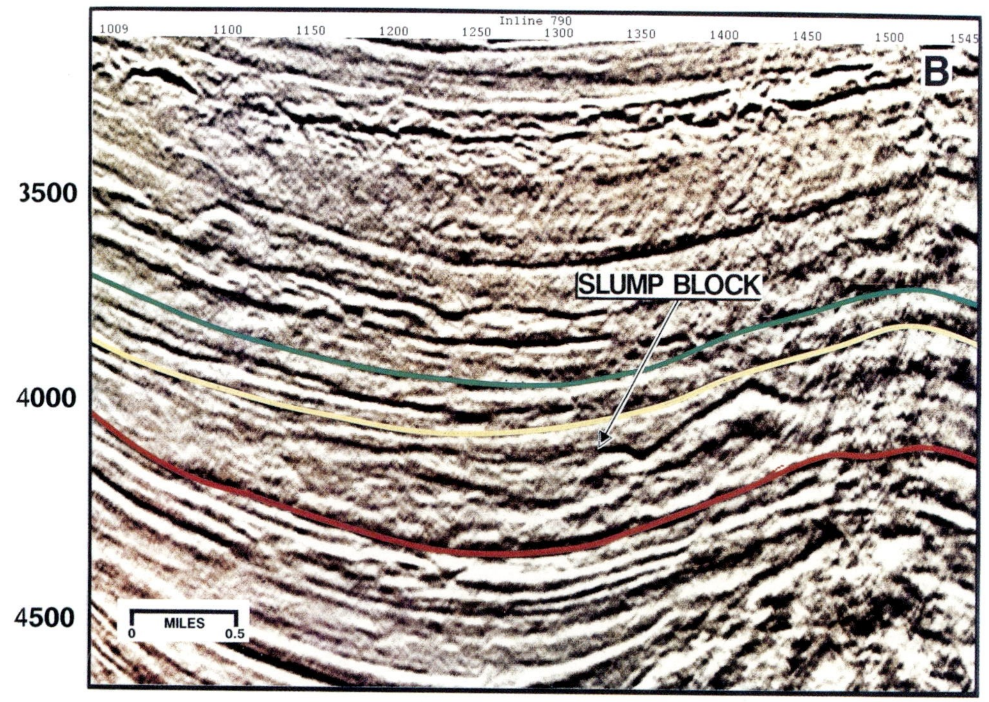


FIGURE 11. Profile B, showing the slumped channel system related to a mixed heterogeneity response. See Figure 9 for location.

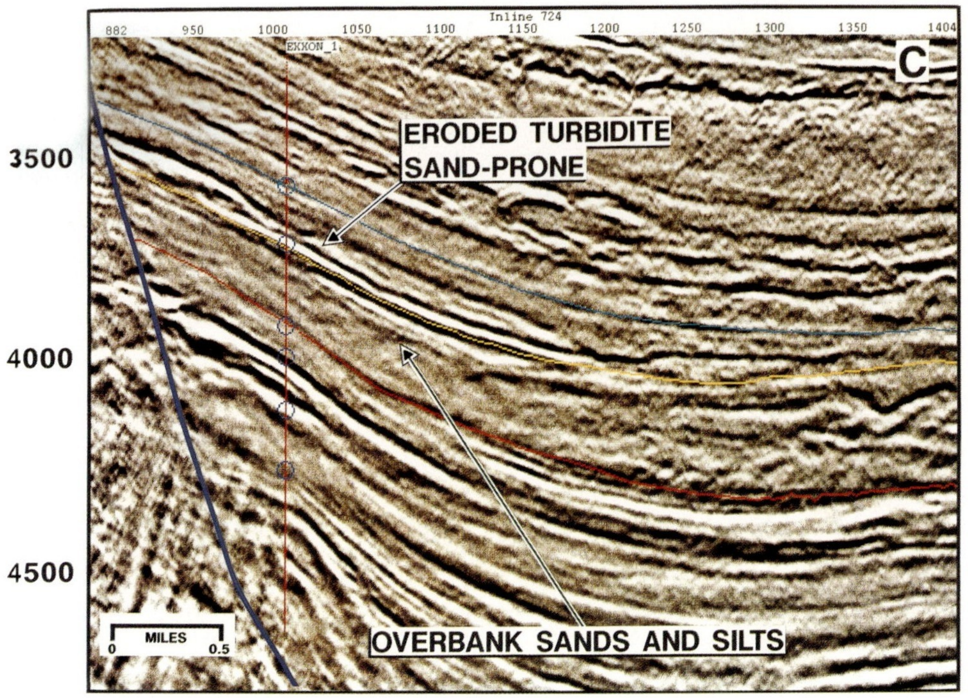


FIGURE 12. Profile C, showing stacked channels (right) and seismic response of overbank silts and interbedded sands (left). Well log shown in Figure 13, location in Figures 7 and 9.

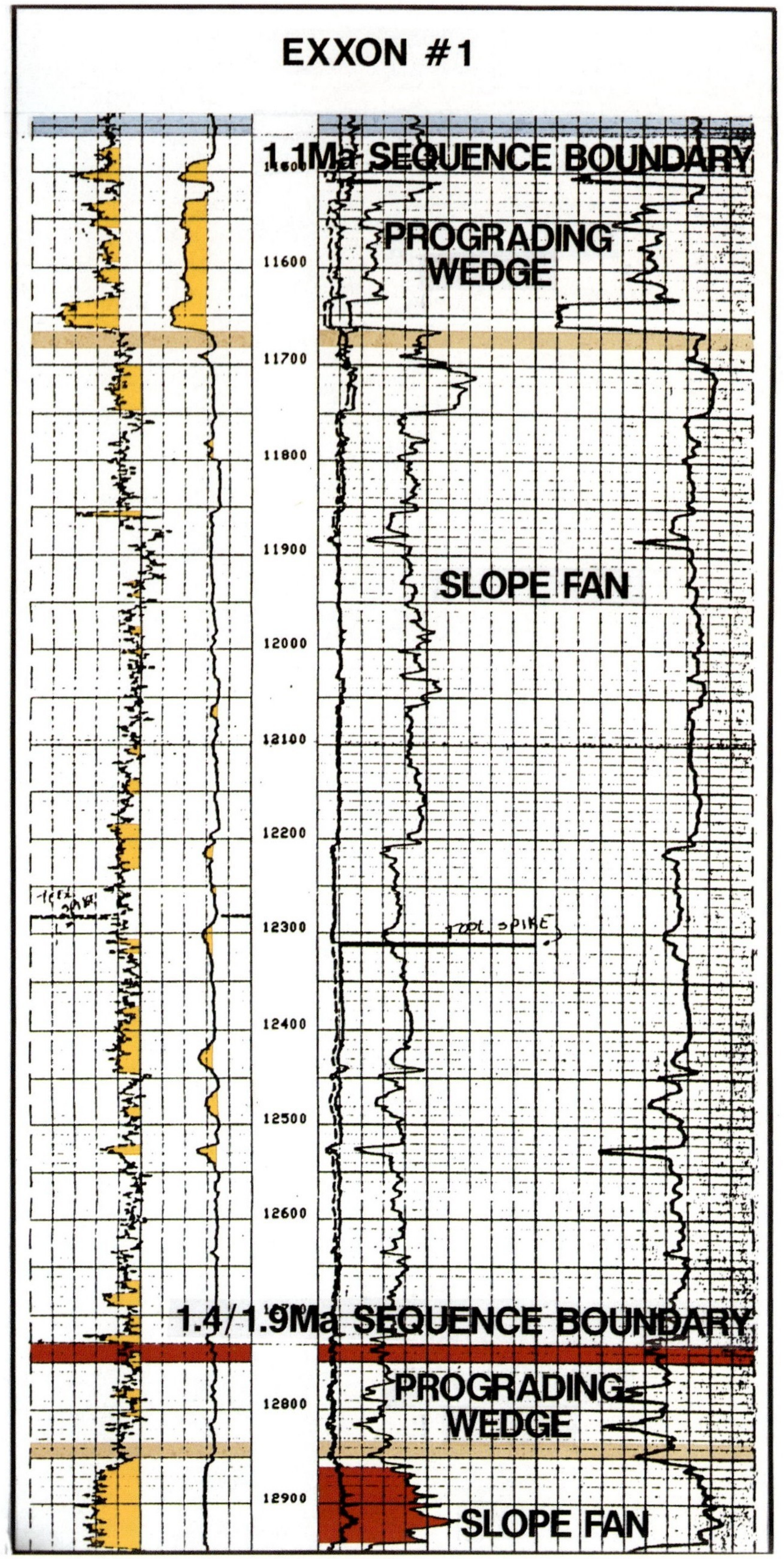


FIGURE 13. Well log shown on profile C (Figure 12). The 1.1-Ma SF is dominated by silts with numerous thin-bedded overbank sands. The petroleum-charged sands below the 1.4-Ma sequence boundary belong to the southern northwest-southwest channel system in Figure 8. See Figures 7 and 9 for location of well.

Lewis, C., A. P. Laferriere, and J. R. Schwalbach, 3-D interpretation using seismic attributes and forward modeling: case study of a deep-water channel system in the North Sea, *in* P. Weimer and T. L. Davis, eds., *AAPG Studies in Geology No. 42* and *SEG Geophysical Developments Series No. 5*, AAPG/SEG, Tulsa, p. 149–154.

CHAPTER 15

# 3-D Interpretation Using Seismic Attributes and Forward Modeling: Case Study of a Deep-Water Channel System in the North Sea

Catherine Lewis*, Alan P. Laferriere†, and Jon R. Schwalbach‡

## ABSTRACT

For a stratigraphic sequence in the Tertiary section of the North Sea Central Graben, seismic attributes have enhanced the 3-D seismic interpretation by imaging a deep-water channel system. Although sands in the channels are seismically invisible, chalk debris flows within the channels appear as high-amplitude reflections, or "bright spots" in the attributes, that mark the axes of the system. Our interpretation of the depositional units initially suggested that the patchy but slightly dendritic pattern of bright spots was part of a distributary channel system. Forward modeling of the seismic attributes indicated that the bright spots do not indicate the presence of hydrocarbons and that individual oil sands cannot be located with the seismic data. Instead, the channels that are prone to sand are delineated by the chalk debris flows that are visible on the seismic attribute maps.

## INTRODUCTION

Strategies for reservoir description are changing as the result of large volumes of 3-D seismic data's becoming readily available to geoscientists. Stratigraphy between well locations that was once only a geological extrapolation is now being imaged by 3-D seismic data. These large 3-D datasets contain enormous potential for better reservoir description, but the sheer volume can be overwhelming to the seismic interpreter. One technique for reducing the complexity of 3-D visualization to the more familiar 2-D map without losing valuable detail is the use of seismic attributes. Attributes extracted within specific intervals can focus the interpreter's eye on the stratigraphy of a specific sequence and help identify subtle patterns that might otherwise go unnoticed.

### PREVIOUS USES OF SEISMIC ATTRIBUTES

Seismic attributes have been used to enhance resolution of subtle structural features such as the intricate distribution of en-echelon faults (Rijks and Jauffred, 1991). Where reservoir facies have an impedance contrast with the surrounding strata, seismic amplitudes can be used to map the distribution and quality of the reservoir itself (Stanulonis and Tran, 1992; Neff, 1992). Amplitudes from horizon slices have highlighted bifurcating and meandering channels (Brown, 1986). A seismic attribute called detuned composite amplitude, extracted from a single-cycle high-amplitude event, has been used to map net pay thickness of gas zones (Brown, 1992). However, many reservoirs do not have bright amplitudes as direct hydrocarbon indicators (DHIs), and many hydrocarbon-bearing facies are more than a single seismic cycle thick. In this paper we present an example of using additional seismic attributes that do not depend on a DHI to identify a sandy channel facies in a seismic interval that is several cycles in thickness.

*Exxon Production Research Company, P.O. Box 2189, Houston, TX 77252
†Exxon Exploration Company, P.O. Box 4778, Houston, TX 77210
‡Exxon Company, U.S.A., P.O. Box 5025, Thousand Oaks, CA 91359

## GEOLOGIC OBJECTIVE AND SEISMIC INTERPRETATION

The geologic setting is a recently discovered oil field in the North Sea Central Graben. A generalized stratigraphic chart for the area is shown in Figure 1. A variety of deep-water fan and channel lithologies make up the Paleocene interval. The Andrew Member of the Lista Formation consists of thickly bedded debris flows and amalgamated deep-water channels 15 to 20 ft (4.5 to 6 m) thick. These were interpreted as being part of a deep-water channel complex, deposited in a proximal to mid-fan setting. The Andrew sands are overlain by basinal shales of the Lista Formation.

The structure is a domal uplift that affects the Cretaceous and lower Tertiary section (Figures 2 and 3). Uplift probably resulted from a combination of Zechstein salt diapirism and structural inversion of underlying extensional faults. Since the structure is simple, with little internal faulting, this is an ideal area for applying seismic attributes from a 3-D survey to facies description.

In the evaluation phase of this field, four wells were drilled on the crest of the structure. Although the wells were drilled to other primary pay zones, a potential pay zone was discovered in a sequence contained within the upper Lista Formation. In the lower channel facies of this sequence, Andrew sands containing oil are interbedded with shales and chalk (Figure 4). These chalk beds are interpreted as debris flows from core. The channel facies is overlain by an upper facies of basinal shale within the same sequence.

A 3-D seismic survey over the new field was acquired in 1988, prior to reservoir development. Figure 3 shows a portion of the seismic interpretation. The interval of interest is the sequence that lies between the Top Andrew Member and the Base Forties unconformity. The Base Forties unconformity is an erosional surface that can be clearly mapped across the survey area as a strong trough with stratigraphic terminations above and below the unconformity. The lower sequence boundary was initially identified on the well logs. Although it was difficult to follow as a distinct event on the seismic data, it was interpreted across the structure by picking stratigraphic terminations.

The objective of this investigation is to locate the oil sands in this particular sequence within the structural closure. The oil sands do not have an impedance contrast that would produce high-amplitude reflections, and the facies boundary between the upper shale facies and the lower channel facies cannot be followed on the seismic data. The geologic problem is complicated because the distribution and stacking patterns of the chalk debris flows, sands, and shales are poorly known, with only four well penetrations. Thickness changes of the sequence, as well as internal thickness changes of the channel facies within the sequence, add to this complexity. Our primary tool for addressing the problem is seismic attribute analysis of the stratigraphic interval, interpreted within a sequence framework. We use the well logs and core to guide the interpretation of the seismic attributes by making a seismic wedge model of the geology that is specifically designed to calibrate seismic attributes.

## PROCEDURES FOR SEISMIC ATTRIBUTE INTERPRETATION

The strategy for locating the oil sands using seismic attributes involves two procedures. The first procedure is a visual analysis of the seismic attributes, making maps of the attributes and looking for patterns that conform to our understanding of the depositional stratigraphy. A qualitative interpretation of anomalies in the seismic attributes can be made at this point. The second procedure involves forward modeling of the seismic attributes, using the well log information, and calibration of the attributes to the geologic thicknesses in the model. This procedure provides a quantitative constraint on the meaning of the attribute anomalies.

### MAPS OF SEISMIC ATTRIBUTES

Seismic attributes must be interpreted in the context of the sequence stratigraphic framework. Attributes can be extracted for a sequence that is interpreted on the seismic data and maps of the attributes interpreted visually for patterns that appear to be depositional geometries. The seismic attributes were extracted from the seismic interval between the Base Forties unconformity and the Top Andrew Member. The three interval attributes are:

(1) Integrated Absolute Amplitude (IAA). This is the sum of the absolute values of all the amplitude samples in the interval (Sonneland et al., 1989).
(2) Integrated Negative Amplitude (IAN). This is the absolute value of the sum of all the negative amplitude samples in the interval. For the polarity of our data display, these are the peaks.
(3) Integrated Positive Amplitude (IAP). This is the sum of all the positive amplitude samples in the interval. For the polarity of our data display, these are the troughs.

The maps of the seismic attributes show dendritic patterns, but the areas of high amplitude captured by the attributes (red and yellow on the maps) have a spotty distribution that is somewhat difficult to connect laterally (Figures 5–7). We have interpreted these as a deep-water distributary channel system. We have outlined what we believe are the axes of sand channels on the maps, although the spotty nature of the pattern makes distinct channels difficult to trace. However, a channel system, such as the one we have interpreted on the attribute maps, is consistent with the sequence stratigraphic interpretation indicating a lowstand fan. The seismic attributes have, in fact, enhanced our ability to delineate the channel system, because on individual seismic lines from the 3-D survey, we could not either recognize the channel geometries or correlate the amplitude anomalies (Figure 3).

## FORWARD MODELING OF SEISMIC ATTRIBUTES

Forward modeling of the seismic attributes was used to enhance our understanding of the spotty nature of the dendritic patterns in the attribute anomalies that we see in map view. The forward model was based on the lithologic information obtained from well logs and cores in the four wells. The model extended the existing well control by creating a series of wedges in which the thicknesses of chalks and oil sands were varied (Figure 8). Because the chalks and sands are gravity flow deposits, their thicknesses and stacking patterns may be highly variable. Erosion from the unconformity at the top of the sequence created the wedge geometry of the model, mimicking the isopach changes that were apparent on the real seismic data.

The first four of the seismic wedges are shown in Figure 8, with the facies boundaries indicated on the synthetic traces. The seismic character of the interval varies with the changing facies character, and the lower sequence boundary even changes polarity. In the full model, the isochron ranges from zero up to 70 ms, with up to seven peaks and troughs and many doublets in the seismic character. This is the type and range of variation in seismic character that is seen in the real seismic data across the field.

From the interpreted sequence interval on the model, the seismic attributes IAA, IAN, and IAP were extracted. These are the same attributes that were extracted from the real 3-D seismic survey. They were calibrated by regressing the cumulative thicknesses of the oil sands and chalks in the model against the attributes (Figure 9). The squared correlation coefficient $r^2$ represents the amount of the variation in one variable, such as IAA, that is explained by the regression on the other variable, such as chalk thickness. The regressions for total chalk thickness versus IAA, IAN, and IAP have $r^2$ values of 54%; in other words, 54% of the variations in IAA, IAN, and IAP is produced by changes in chalk thickness for the interval. On the other hand, the regressions for oil sand thickness versus IAA, IAN, and IAP have $r^2$ values of only 5%; in other words, only 5% of the variation in these attributes is produced by changes in oil sand thickness within the sequence. It is clear that the high amplitudes in the attributes IAA, IAN, and IAP are coming from the chalk debris flows, not from the oil sands.

## DISCUSSION OF INTERPRETATION

In the context of our sequence stratigraphic model and the recognition from cores that the chalks are debris flows, the spotty nature of the dendritic pattern in the seismic attributes can be interpreted as local debris flows containing chalk confined to deep-water channels. The chalk debris flows light up the channels in the seismic amplitudes, which are captured by the seismic attributes. Because of the low correlation between the oil sands and the seismic attributes, we cannot locate individual oil sands within the lower channel facies, but we can map the general channel configuration and high-grade areas that are likely to contain sand, and thus prospective for finding oil.

If explorationists were limited to sequence stratigraphy as a tool to guide the interpretation of the seismic attributes, they might be tempted to interpret the high-amplitude areas as high-porosity oil sands within the channel system. The quantitative analysis of the attributes using forward modeling, which honors the well log and core data, steers the interpretation away from this conclusion. On the other hand, without the sequence stratigraphic framework, the interpreter would see only random debris flows with no connection to channels. Both tools must be used to draw a useful conclusion from the seismic-attribute image of the sequence.

## CONCLUSIONS

Seismic attributes extracted on a sequence interval can be a valuable tool for enhancing subtle stratigraphic features. 3-D seismic data make it possible to look at seismic attributes in map view. What was impossible to see on individual lines of the 3-D survey was captured analytically by the attributes. The interpretation was guided by a sequence stratigraphic framework and depositional facies interpretation for the area. Forward modeling of the seismic attributes was vital to avoiding a misinterpretation of high-amplitude areas as oil DHIs in an area where, instead, chalk debris flows generate high-amplitude events.


### *Acknowledgments*

The authors would like to thank Esso Exploration and Production (UK) Ltd. and Shell UK Exploration and Production for their contributions and for permission to publish this paper.


## REFERENCES CITED


Brown, A. R., 1986, Interpretation of three dimensional seismic data: AAPG Memoir 42, p. 72–83, 97.

Brown, A. R., 1992, Net producible gas sand mapping by interactive analysis of seismic time and amplitude, *in* R. E. Sheriff, ed., Reservoir: Geophysics: Society of Exploration Geophysicists, p. 189–206.

Neff, D. B., 1992, Estimated pay mapping using three dimensional seismic data and incremental pay thickness modeling, *in* R. E. Sheriff, ed,: Reservoir Geophysics: Society of Exploration Geophysicists, p. 207–214.

Rijks, E. H. J., and J. C. E. M. Jauffred, 1991, Attribute extraction: an important application in any detailed 3-D interpretation study: The Leading Edge of Exploration, v. 10, n. 9, p. 1119.

Sonneland, L., O. Brakved, M. Olsen, and G. Snyder, 1989, Application of seismic wavefield attributes in reservoir characterization: SEG Expanded Abstracts, 59th Annual International SEG Meeting, p. 813–817.

Stanulonis, S. F., and H. V. Tran, 1992, Method to determine porosity-thickness directly from 3-D seismic amplitude within the Lisburne carbonate pool, Prudhoe Bay: The Leading Edge of Exploration, v. 11, n. 1, p. 14–20.

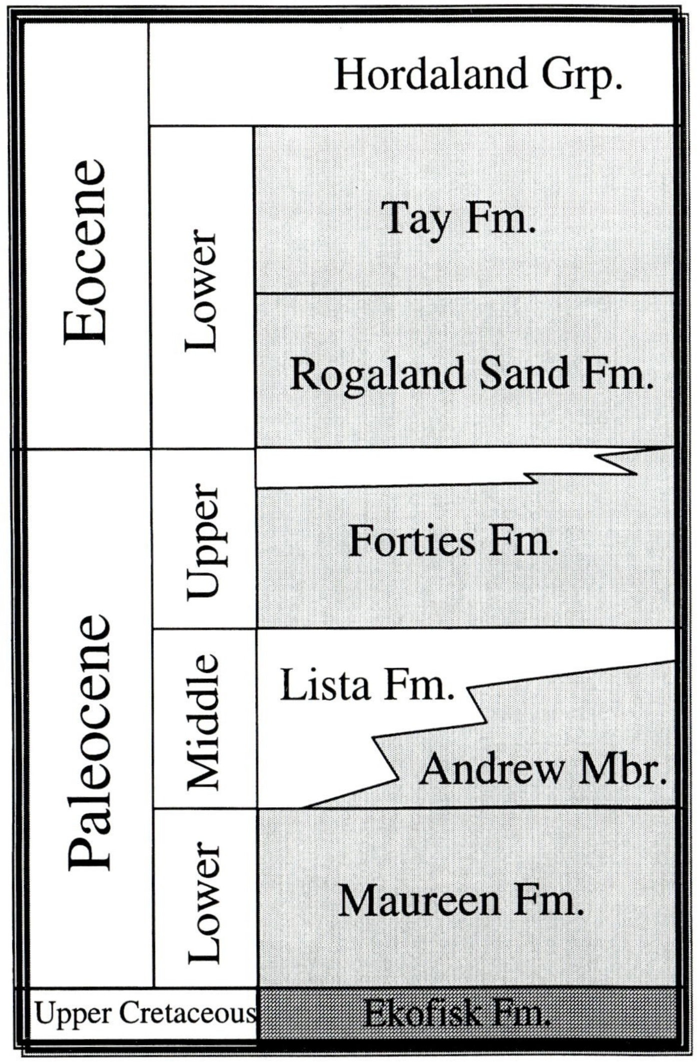


FIGURE 1. Stratigraphic chart of the North Sea Central Graben. The Lista Formation contains the sequence that was modeled and interpreted using seismic attributes.

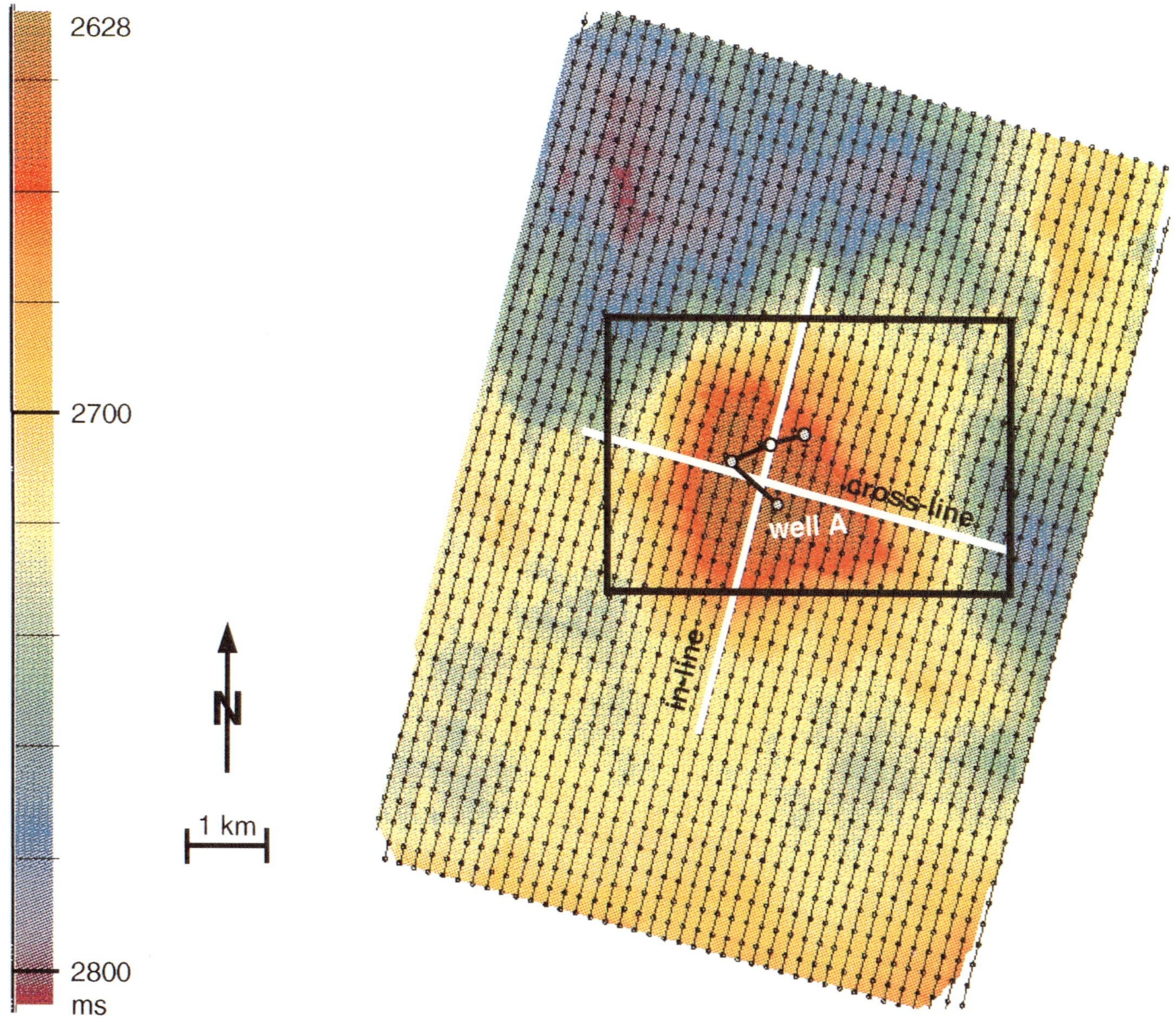


FIGURE 2. Time structure map of the domal closure, with well and line locations. Box encloses area where attributes were extracted.

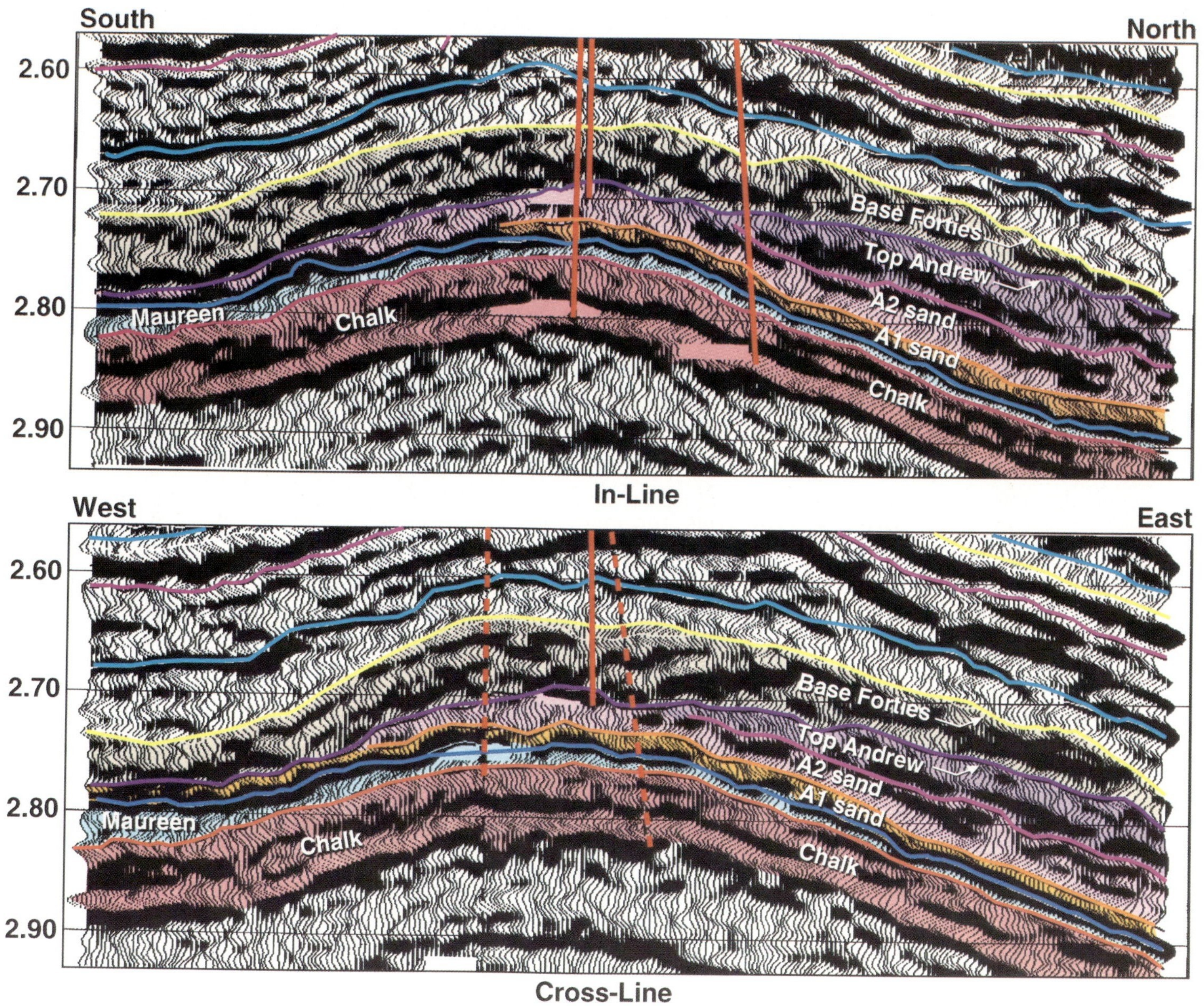


FIGURE 3. Inline and crossline across domal closure. The sequence containing the channel system identified by seismic attributes lies below the Base Forties (yellow) and above the Top Andrew (blue). Well logs for the sequence from well A are shown in Figure 4.

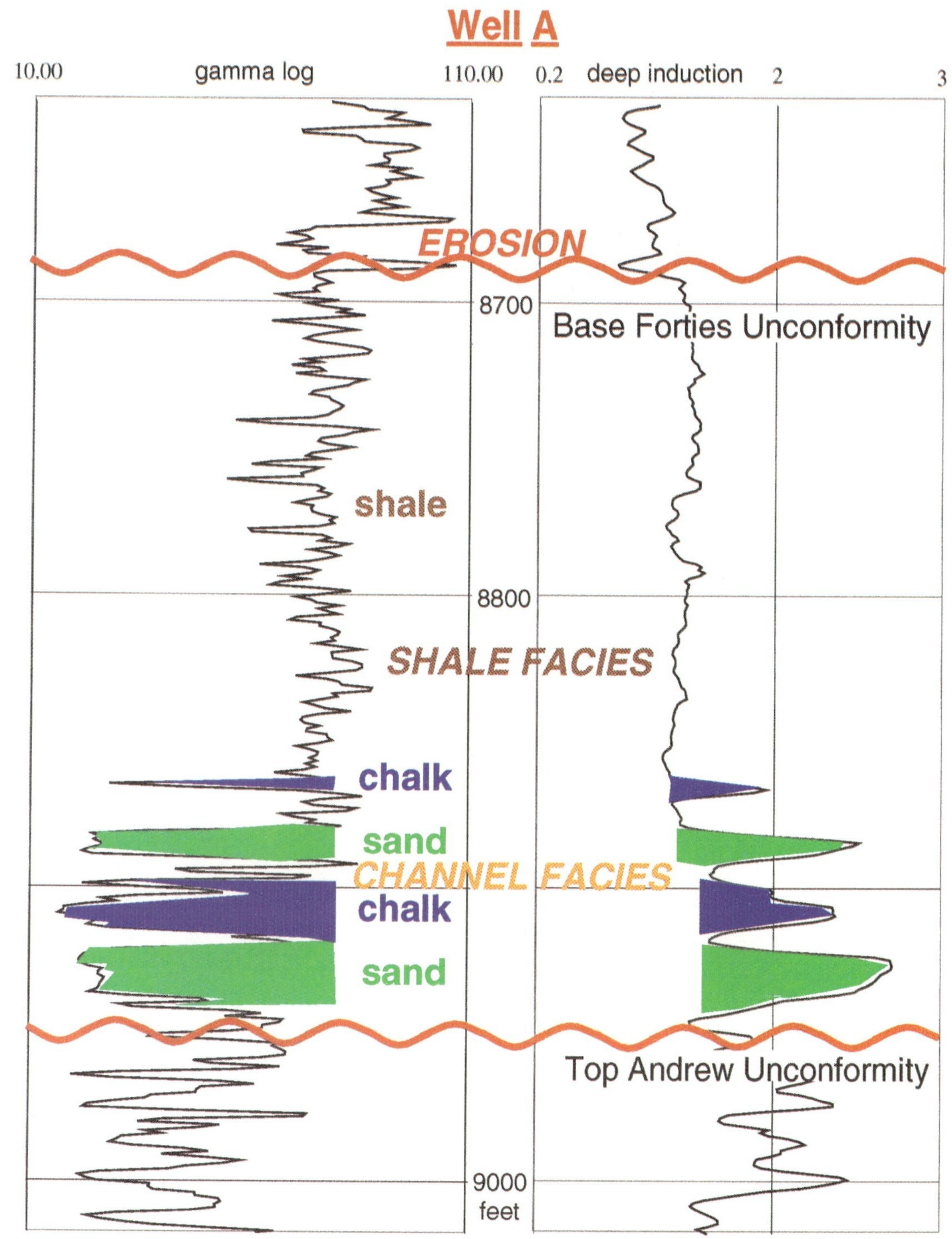


FIGURE 4. Well logs from well A, showing the sequence that was modeled and analyzed with seismic attributes. Chalk beds are interbedded with sands and shales in the lower channel facies.

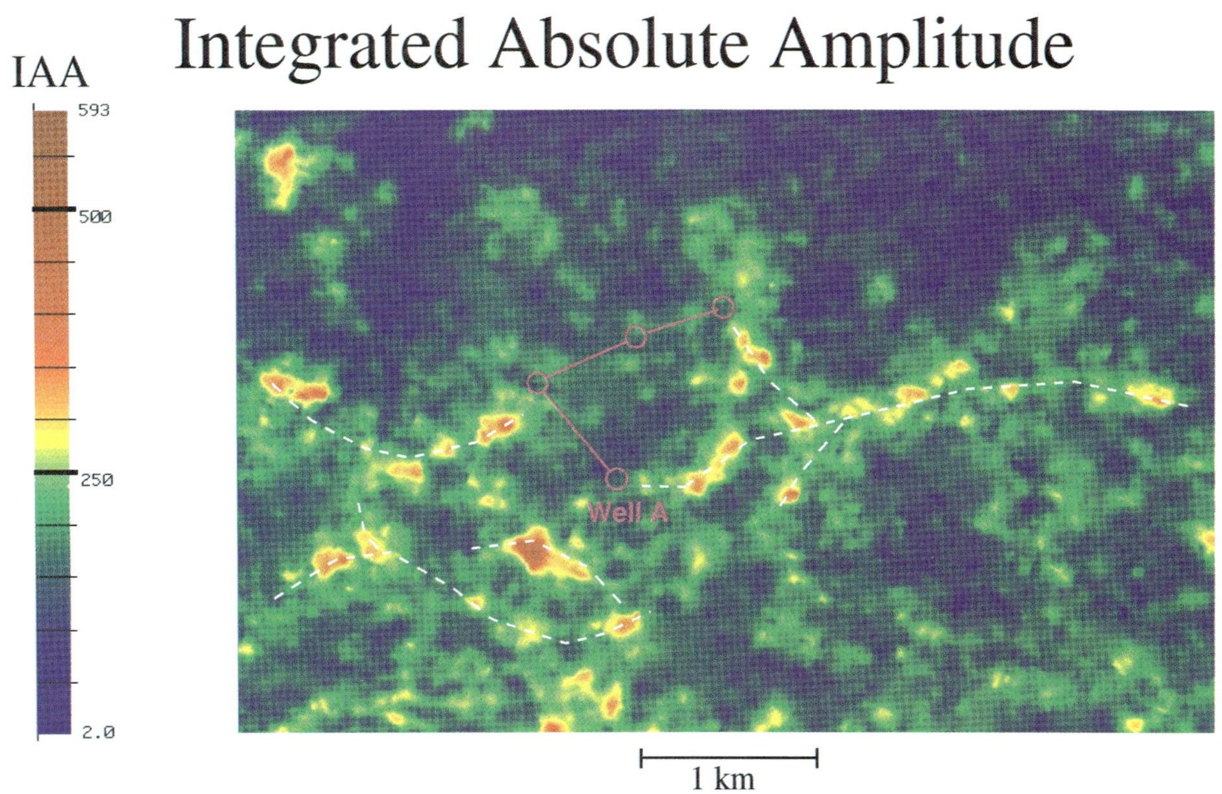


FIGURE 5. Attribute map showing the spotty, dendritic pattern that is interpreted to be a deep-water channel system. The chalk debris flows are confined to the sand channels and create high-amplitude anomalies along the channel axis. This is the IAA, which is the sum of the absolute values of all the amplitude samples in the interval.

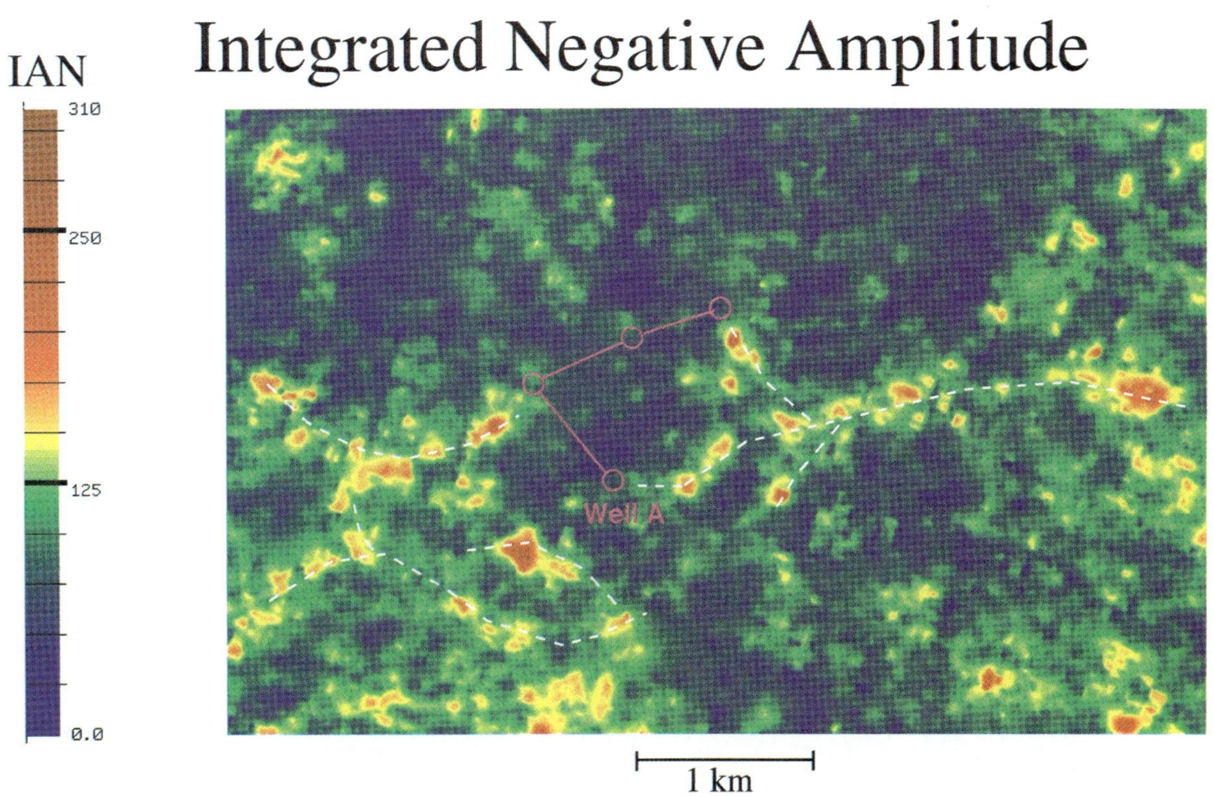


FIGURE 6. Attribute map showing the spotty, dendritic pattern that is interpreted to be a deep-water channel system. This is the IAN, which is the absolute value of the sum of all the negative amplitude samples in the interval.

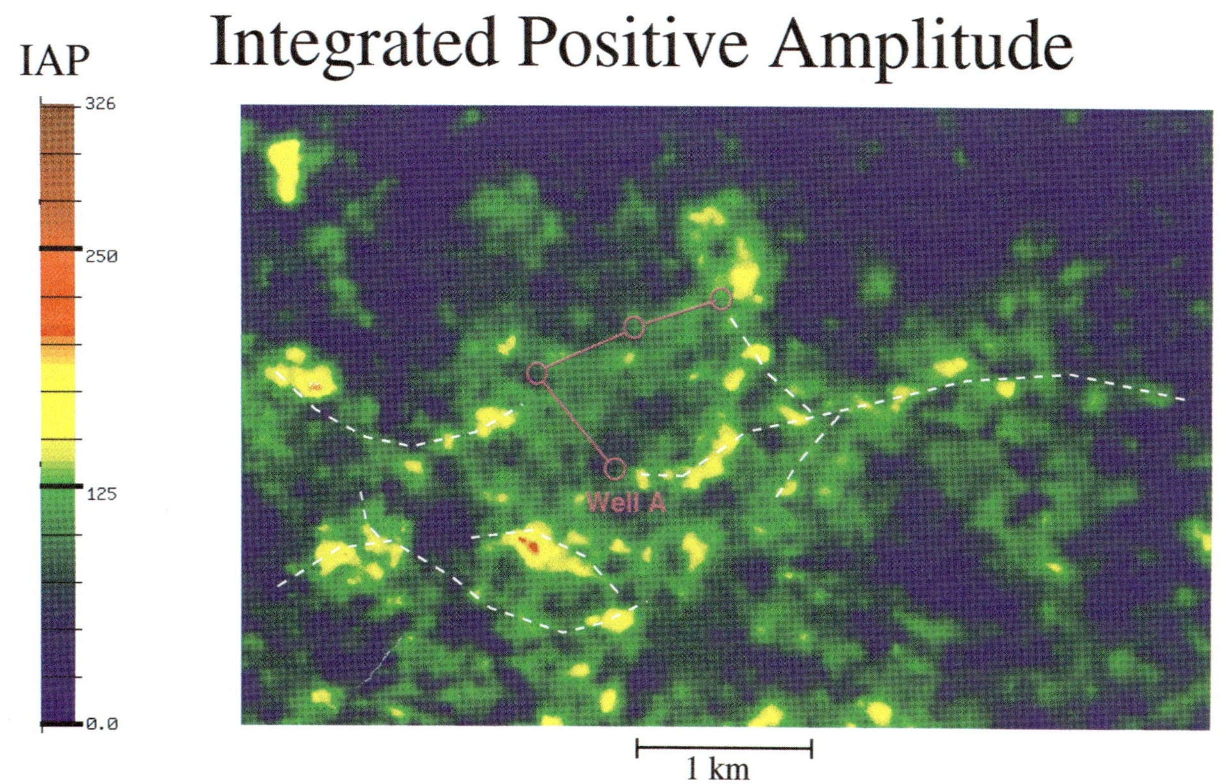


FIGURE 7. Attribute map showing the spotty, dendritic pattern that is interpreted to be a deep-water channel system. This is the IAP, which is the sum of all the positive amplitude samples in the interval.

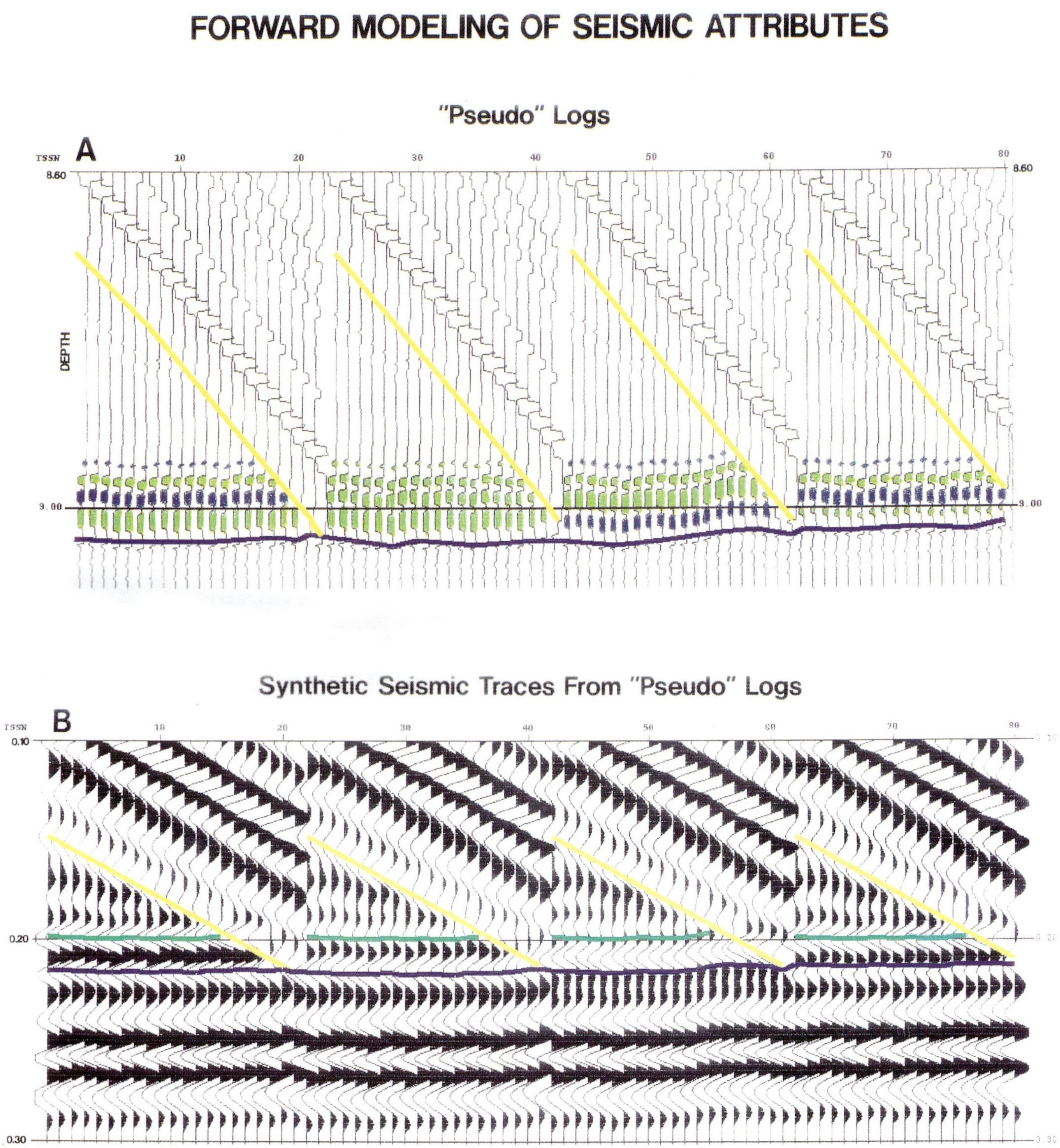


FIGURE 8. Wedge-type models in which the thicknesses of oil sands and chalks are varied, between the Base Forties (yellow) and the Top Andrew (blue). The seismic character of the sequence is highly variable.

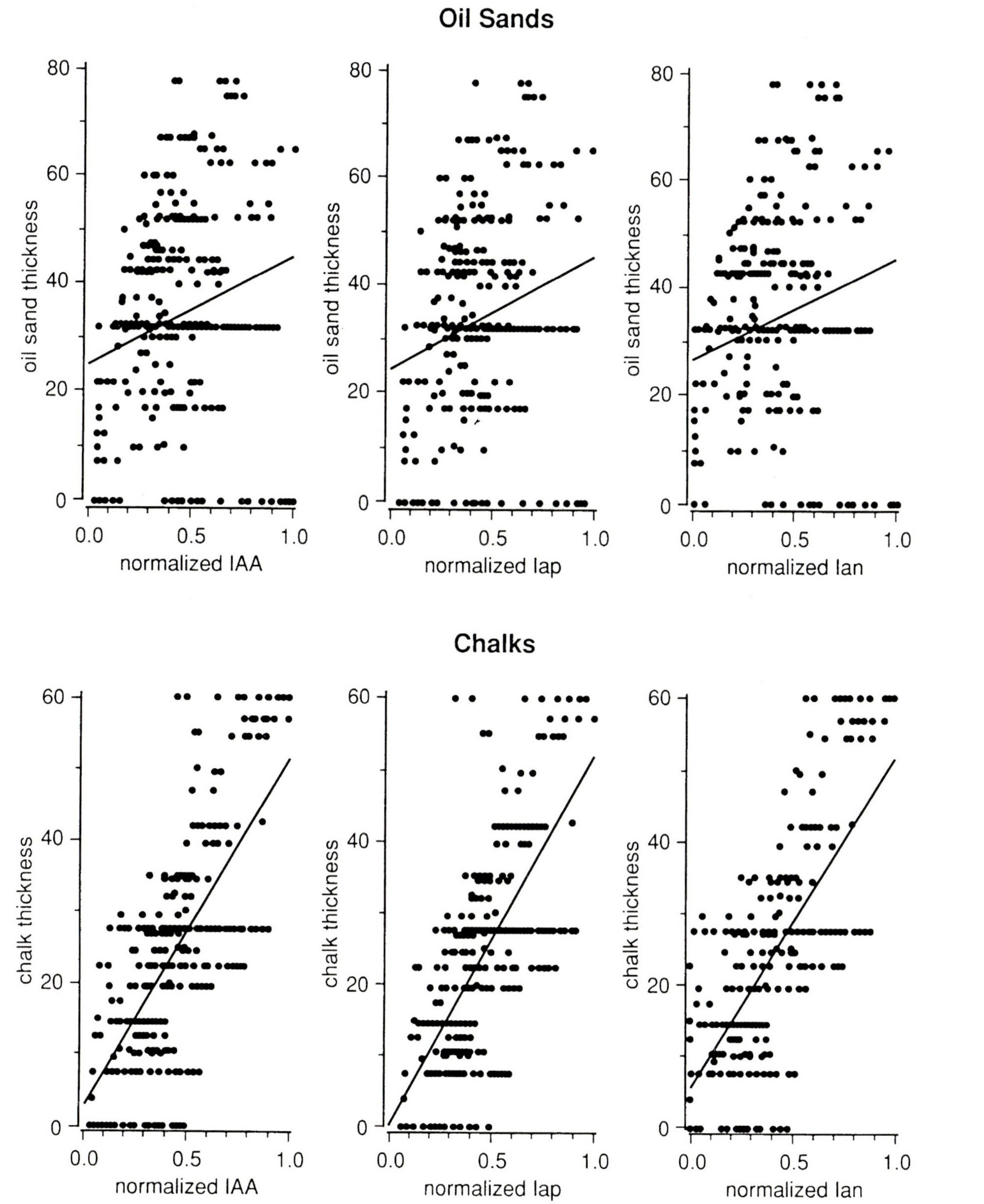


FIGURE 9. Calibration of seismic attributes IAA, IAN, and IAP versus total chalk thickness and total oil sand thickness. The attributes are uncorrelated with oil sand ($r^2 = 5\%$) but are roughly correlated with chalk thickness ($r^2 = 54\%$). The lines are least-squares regression lines.

Jumper, S. C., and H. W. Pardue, A 3-D case history in the Horseshoe Atoll area of Scurry County, Texas, *in* P. Weimer and T. L. Davis, eds., *AAPG Studies in Geology No. 42 and SEG Geophysical Developments Series No. 5*, AAPG/SEG, Tulsa, p. 155–160.

## CHAPTER 16

# A 3-D Case History in the Horseshoe Atoll Area of Scurry County, Texas

**Stephen C. Jumper and Howell W. Pardue***

## ABSTRACT

This case history presents design considerations, processing techniques, and drilling results from a 3-D seismic survey acquired in the pinnacle reef trend of Scurry County, Texas. Pennsylvanian age reefs in this area have long been lucrative but elusive exploration objectives. Their small areal extent and random location pattern have often frustrated explorationists interpreting conventional 2-D seismic data. Through the example shown here, it becomes apparent that pinnacle reef objectives can be located with much-improved confidence using the 3-D seismic method.

The survey has approximately 9 mi$^2$ (23 km$^2$) of surface coverage, more than adequate for a proper migration aperture at the objective horizon. The reefs in question are reasonably identified with a 3-D grid consisting of 110 × 110-ft (33 × 33-m) bin spacing. Economic considerations and accessibility constraints greatly affect the design of the survey. Processing techniques include 3-D surface-consistent statics, post-stack FX (spacial deconvolution) filtering to reduce random noise, and 3-D migration.

Interpretation of the migrated 3-D volume identified several possible drilling locations. Time slice presentations at the pinnacle reef level clearly identified three subsequent producing locations and helped reconcile two previous nonproductive wells. Geophysical and economic success in this survey show that the 3-D seismic technique can solve several of the problems previously encountered using conventional 2-D seismic data.

## INTRODUCTION

The objective of this cost-effective 3-D survey was to locate Pennsylvanian age pinnacle reefs in Scurry County, Texas. These reefs are seaward, or southeast, of the Horseshoe Atoll, a major northeast-to-southwest–trending reef body underlying Scurry County. Discovery of the atoll in the early 1950s led to the development of the Kelly-Snyder oil field. Producing wells drilled on 40-acre spacing delineate the extent of the feature.

Since the development of the Kelly-Snyder oil field, exploration has moved southeast of the atoll in search of small reef bodies. Although the pinnacle reefs have an overall northeast-to-southwest trend, their size and random pattern makes locating the reefs extremely difficult. Exploration methods based on pure geologic data are virtually useless due to the random nature of the pinnacles. The 2-D seismic method has a slightly better success rate but is difficult to apply because of the small areal extent and the isolated occurrence of the reefs. The reefs are located with improved accuracy using the 3-D seismic technique.

*Dawson Geophysical Company, 208 S. Marienfeld, Midland, TX 79701

## HISTORY

The Pinnacle Reef Trend of Scurry County, Texas, was the focal point of Permian Basin 3-D seismic exploration in 1989 and 1990. The geologic setting and exploration objective are ideal for the use of small, inexpensive 3-D seismic surveys. The zone of interest is isolated to a single horizon consisting of small, isolated features set in a relatively simple geologic framework. The issue is not to resolve a complex image but rather to locate the tops of these features. This is efficiently accomplished with the acquisition of a low-fold survey with common midpoint (CMP) bin spacings of 110 × 110 ft (33 × 33 m) and processed through 3-D migration to properly position the reflected energy. The signal-to-noise ratio of this area is good and the processing problems relatively simple.

Prior to the use of the 3-D method, the primary exploration tool was subsurface geology coupled with 2-D seismic data. If and when a reef was detected, multiple 2-D lines were acquired in a wagon wheel fashion over the reef image. This prospect evaluation technique is difficult, with no assurances that the top of the feature is properly located. Several miles of 2-D seismic lines must be acquired in a small area, which makes the process more expensive than it might seem on first appearance. The small 3-D survey is more efficient and comparable economically.

The use of 3-D seismic acquisition in the Permian Basin dates back to the early 1970s. These early surveys were expensive and limited in configuration by the number of recording channels available at that time. Limited recording channels and processing capabilities constrained the recording geometries to small, parallel swath configurations. Since these early Scurry County surveys, there has been a vast expansion in the number of recording channels available, as well as increased processing capabilities. The use of 3-D surveys has expanded throughout the Permian Basin and into more complex geologic settings.

Typical surveys in 1995 are larger, utilize 4 to 5 times the number of recording channels, and are acquired on a more dense surface grid than they were in 1990. As seismic crews become larger and more efficient, the cost of acquisition has begun to decline. Subsurface coverage continues to intensify, and data quality becomes better as the use of the method expands, causing complex as well as subtle features to be more easily resolved.

## DESIGN CONSIDERATIONS

The intent of the project was to obtain nominal 12-fold subsurface coverage at a depth of approximately 8000 ft (2500 m). Surface coverage is 9 mi$^2$ (23 km$^2$). The survey design employed a crossline method, in which source and geophone lines lie perpendicular to one another. The source and geophone lines are spaced every half-mile to take advantage of existing roads and trails which lie along section and half-section lines (Figure 1). The half-mile line spacing helped to reduce permit cost and to increase accessibility to as many surface locations as possible.

The data were recorded with a 240-channel Texas Instruments DFS-V, instantaneous floating point, master/slave recording system. Geophone and source intervals are 220 ft (67 m), resulting in actual CMP bin sizes of 110 × 110 ft (33 × 33 m). The design called for five active geophone lines of 48 channels per line over a 2 × 2-mile (3.2 × 3.2-km) patch. The method of recording produced adequate coverage at the pinnacle reef horizon.

Existing surface problems forced changes in the actual implementation of the project design because the northeast corner of the survey lies within the city limits of Snyder, Texas. Source and geophone lines had to be repositioned to accommodate access constraints.

## DATA PROCESSING

Basic processing concepts were employed throughout the processing sequence. Early processing procedures included 3-D geometry definition, signal processing, and gain recovery. The data were then gathered into their respective CMP locations.

The sparsely acquired survey presented several potential problems pertaining to normal moveout correction. The design of the survey resulted in low-fold CMPs that are rich in source-receiver azimuth distribution but poor in offset distribution. Several CMPs had to be summed on common offsets to properly estimate stacking velocities. The low-fold survey restricts the use of 3-D dip moveout (DMO) techniques and multi-azimuth velocity analysis.

A 3-D surface-consistent statics routine successfully solved the residual statics, and the CMPs were stacked. Post-stack processing included 2-D FX deconvolution to reduce random noise that resulted from the low-fold nature of the survey. The data were then migrated with a two-pass, 2-D migration technique.

## DATA INTERPRETATION

The interpretation of the dataset was performed by a third party utilizing a 3-D interpretive workstation. Vertical displays of the data as well as time slices were incorporated into the interpretation. A selected time slice near the top of the objective horizon is shown in Figure 2. The reef bodies of interest are shaded in red. Three producing wells drilled after the completion of the survey are indicated by yellow dots. The red feature in the northwest corner is the edge of the Horseshoe Atoll. Local dip is to the southeast.

Figure 3 shows an east-west–oriented vertical line passing through a reef body. This reef body is the southernmost producing well shown in the time slice. The red line indicates the interface between the Pennsylvanian age carbonate and the overlying shale section. The reef body is seen in the center of the line.

A northwest-to-southeast extracted line through the same reef body is shown in Figure 4. This line is along the local dip axis. Notice the sharp fore reef flank as well as the edge of the atoll on the left side.

Figure 5 is a computer-generated time contour map at the top of the Pennsylvanian carbonate. It is similar in nature to the time slice showing the isolated reef bodies southeast of the main reef. A three-dimensional projection of the contour map, shown in Figure 6, indicates the relief of each feature.

## CONCLUSION

The data were interpreted on a 3-D workstation and with paper displays. Time slice displays indicated the presence of several possible pinnacle reef locations. Existing well control in the survey area confirmed the validity of the dataset. Subsequent to the interpretation, three new producing wells have been drilled.

Drilling results show that the low-cost, low-fold survey was adequate to identify pinnacle reefs which had been elusive when other exploration techniques were used. This type of survey required caution during processing to properly stack the data. The technique can be improved by using closer line spacing and a larger number of recording channels. This would be an alternative when higher fold or better shallow horizon coverage is required.


### *Acknowledgements*

Dawson Geophysical Company wishes to express its gratitude to HAT Oil and Gas and Paladin Petroleum for releasing this dataset for publication. The data were acquired and processed by Dawson Geophysical Company for HAT and Paladin; the interpretation was performed by a third party.

FIGURE 1. Surface location map showing source and receiver lines. The city of Snyder, Texas lies in the northeast corner.

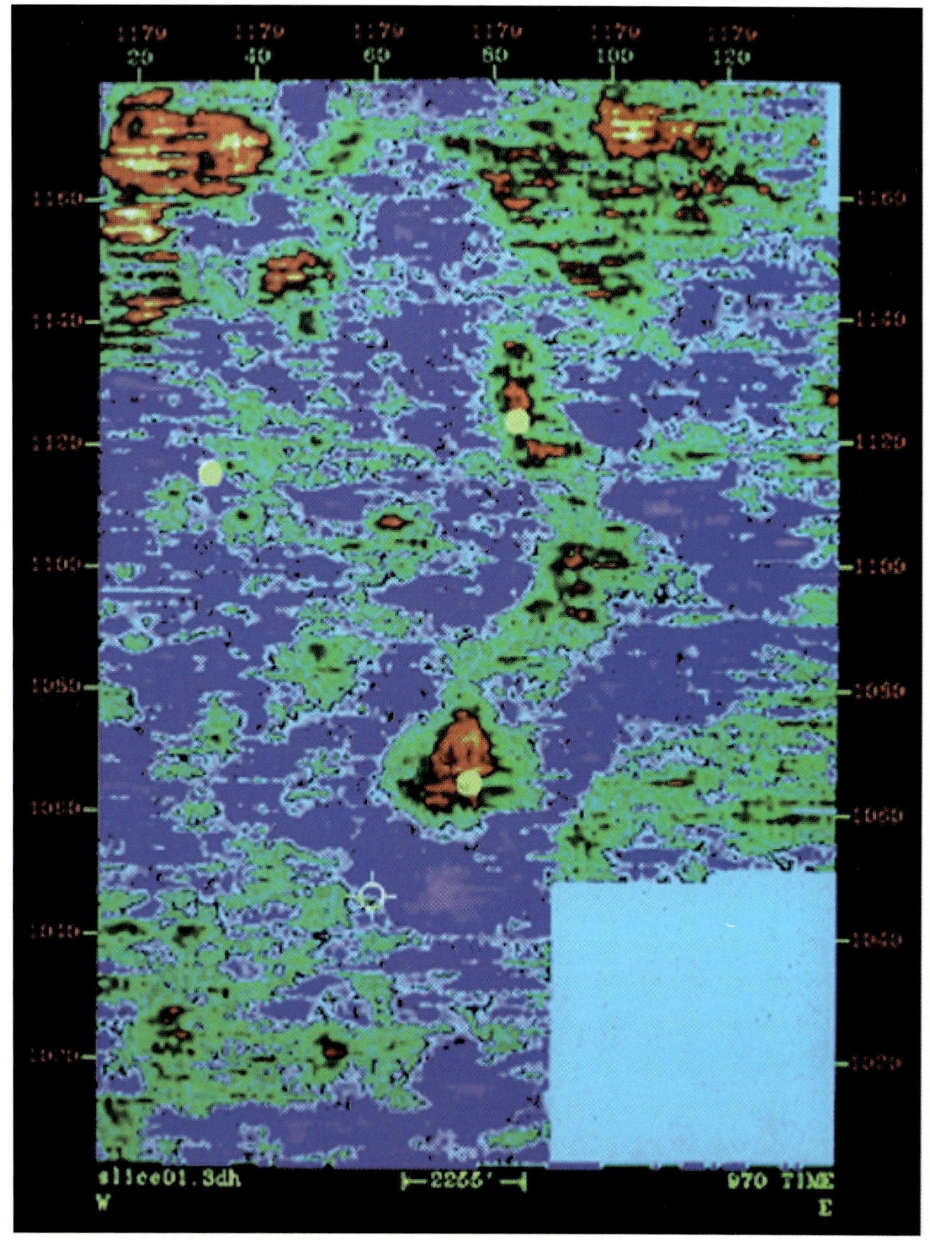


FIGURE 2. Time slice at 970 ms, approximately the top of the Pennsylvanian carbonate. The reef bodies are shaded red; producing wells are indicated by yellow dots.

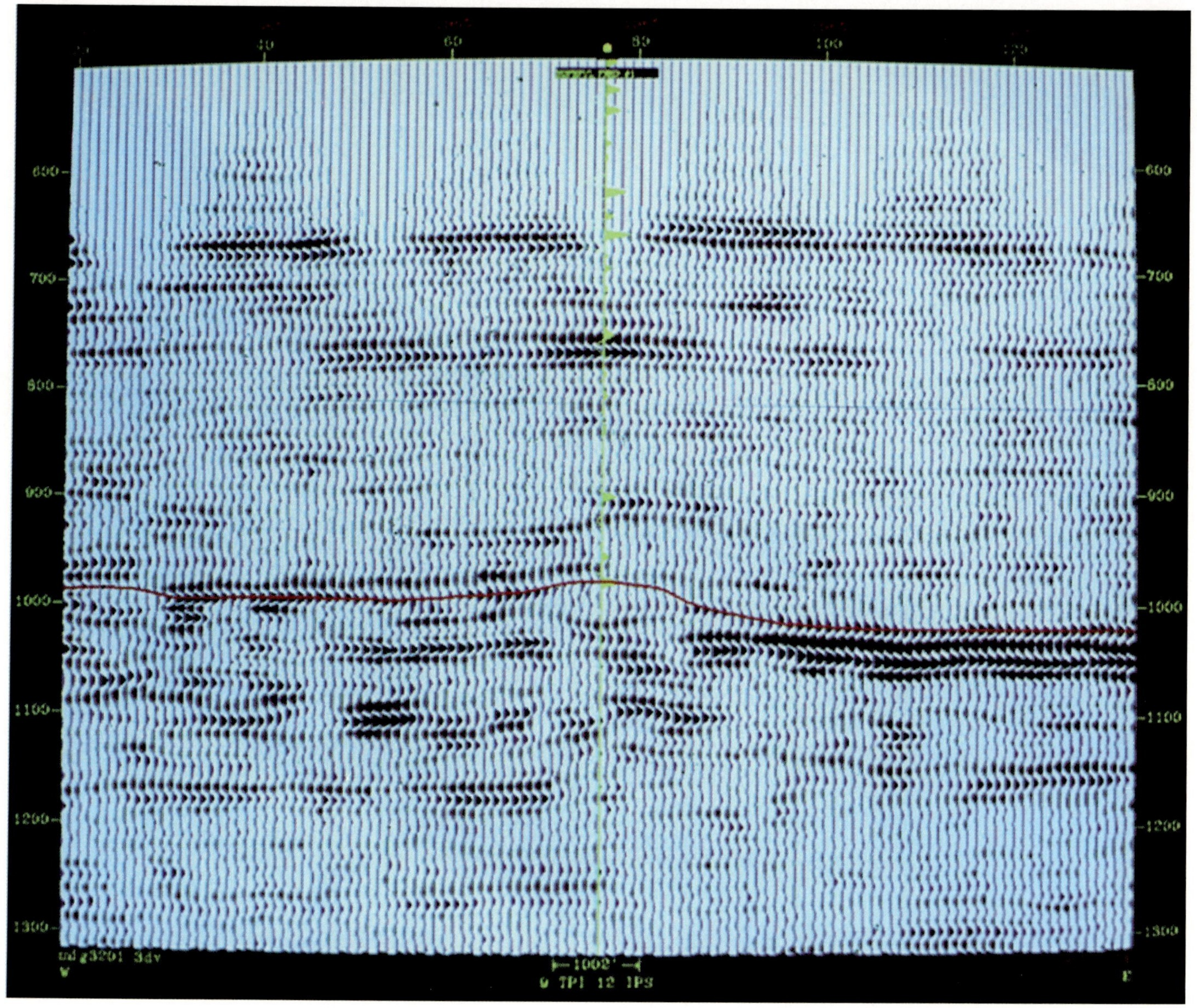


FIGURE 3. East-west extracted profile. The red line indicates the top of the Pennsylvanian carbonate. Note the isolated reef body in the center of the line.

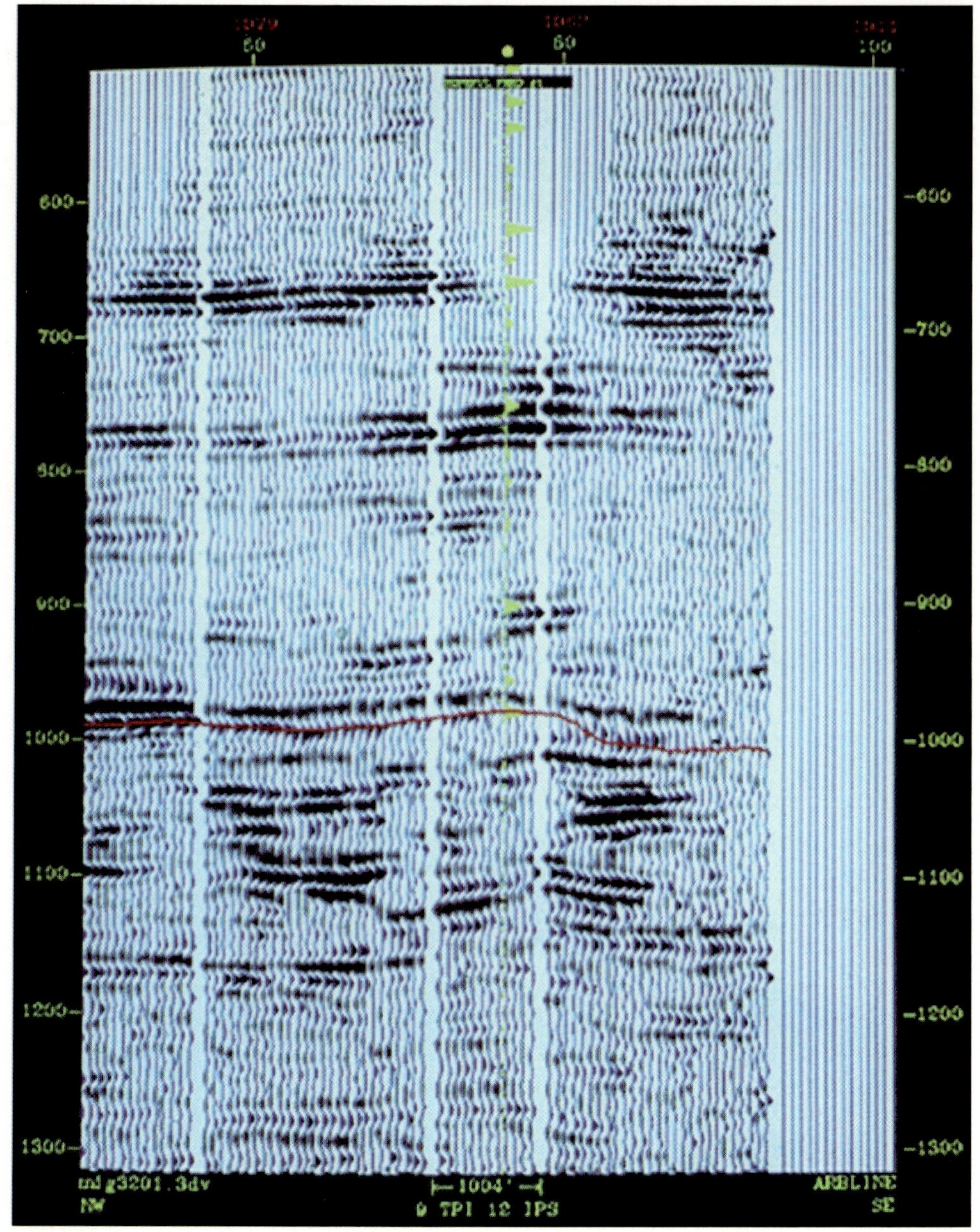


FIGURE 4. Northwest-to-southeast extracted line passing through the same reef body as in Figure 2. This line is along the dip axis away from the Atoll on the left side of the line.

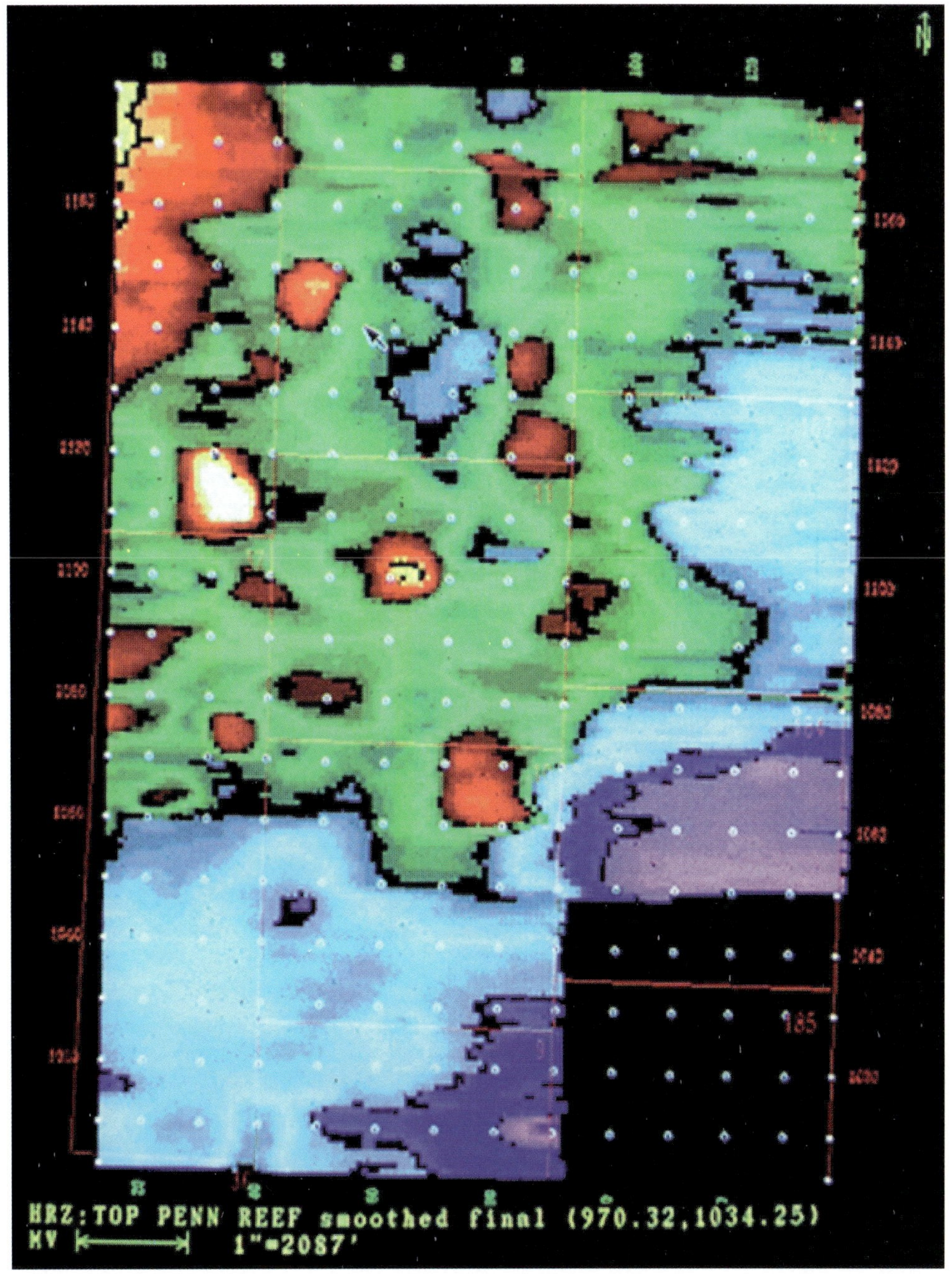


FIGURE 5. Computer-generated time contour map at the top of the Pennsylvanian carbonate. Note the similarity to the time slice in Figure 2.

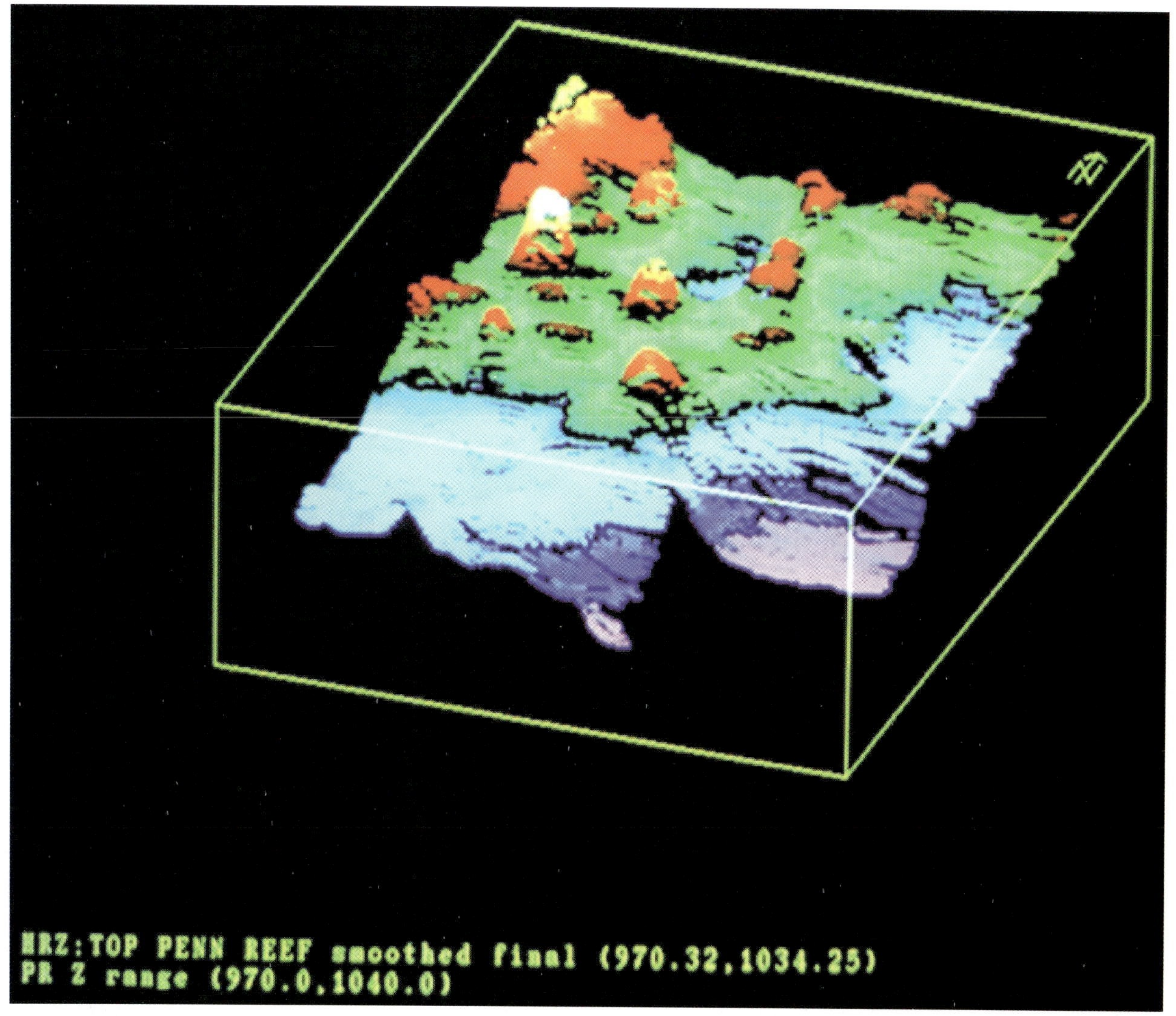


FIGURE 6. Three-dimensional projection of time contour map. The small isolated reef bodies are easily identified.

CHAPTER 17

# Stratigraphic Analysis of 3-D and 2-D Seismic Data to Delineate Porous Carbonate Debris Flows in Permian Strata Along the Northwestern Margin of the Midland Basin, West Texas, U.S.A.

Pacht, J. A., L. Brooks, and F. Messa, Stratigraphic analysis of 3-D and 2-D seismic data to delineate porous carbonate debris flows in Permian strata along the northwestern margin of the Midland Basin, West Texas, U.S.A., *in* P. Weimer and T. L. Davis, eds., *AAPG Studies in Geology No. 42* and *SEG Geophysical Developments Series No. 5*, AAPG/SEG, Tulsa, p. 161–170.

Jory A. Pacht*, Larry Brooks†, and Frank Messa†

## ABSTRACT

Porous carbonate debris flows are important plays in Leonardian (Permian) strata along the northwestern margin of the Midland Basin. Delineation of these strata, however, is difficult, and detailed stratigraphic analysis of both 2-D and 3-D seismic data is important in reducing risk. Debris flows are common in all carbonate slope and base-of-slope environments. However, many flows exhibit a muddy matrix and low permeability. Porous debris flows are best developed during lowstand time. When sea level fell to a point at or below the shelf margin, sand to boulder-sized clasts created by reef-front erosion were funneled through slope gullies onto the base of the slope.

Large debris flows exhibit well-defined mounds that downlap onto the sequence boundary. Many of these flows, however, are too thin to exhibit discrete reflections. In the study area, porous debris flows were distinguished by areas of lower amplitude and lower average weighted frequency than surrounding strata. The porous debris flows occurred downdip of major slump scars. Once sequence boundaries and, where present, mounded reflections that downlap onto them were delineated on a subregional basis using 2-D data, they were interpreted throughout a smaller 3-D dataset. Horizon amplitude and average weighted frequency maps were constructed and used to delineate amplitude and frequency variations on the basin floor. Time structure and isochron maps were made to locate associated updip slump scars.

## INTRODUCTION

During the past five years there has been a boom in 3-D acquisition throughout the Midland Basin in West Texas (Figure 1). The result has been greatly increased exploration and production success in several trends and renewed interest in the basin by many companies. However, the cost of 3-D data acquisition over large areas (>25 km$^2$) is prohibitive for many small petroleum companies, who are traditionally the main operators in this basin. These companies are limited to analysis of small 3-D seismic surveys. This data limitation makes proper geologic analysis difficult, because the explorationist must interpret the effects of regional geologic processes by studying a small area. In some cases, geophysical anomalies are mapped using these small datasets with little analysis of the regional framework. This framework is necessary to explore for hydrocarbons properly, because seismic data commonly yield a nonunique solution. Identical seismic anomalies may be formed by porous carbonate strata in one depositional system and by tight mudstones and claystones in another.

Our experience demonstrates that a thorough understanding of the regional depositional framework is necessary to maximize drilling success in a 3-D play. Therefore, analysis of 3-D data should begin by analysis of 2-D seismic and well log data. These data are used to determine the regional sequence stratigraphic and structural framework within specific trends and to develop potential leads. Then, and only then, are 3-D data acquired over these leads to turn them into drillable prospects. We believe that this approach results in both increased drilling success and reduced expenditure for 3-D data acquisition.

The purpose of this paper is to document the development of this approach as used to delineate porous debris flows in Leonardian (Permian) strata along the northwest shelf of the Midland Basin (Figure 1). Exploration for these objectives requires techniques different from those used in the highly productive shelf facies. The area is especially complicated, because both siliciclastic and carbonate strata are present. Initially, 2-D seismic and wireline data were used to complete a regional sequence stratigraphic study. Approximately 200 line miles (320 line km) of seismic data and over 50 well logs were analyzed in this phase of the study. Proprietary biostratigraphic analyses by Hollingsworth (numerous proprietary reports completed over several years) were also studied and used to properly evaluate regional correlations. Synthetic seismograms were constructed and used to correlate well log, sample log, and biostratigraphic data with the seismic sections. In this way, the relationship of seismic signatures to specific lithostratigraphic units was ascertained. Variations in clastic input relative to specific sequences were evaluated across the area. Updip slump scars were evaluated relative to downdip debris flows and to both siliciclastic and calcarenite turbidites. The nature and extent of reef development in specific sequences were mapped. Finally, the regional extent of structures created by both reverse and normal faulting were evaluated relative to reservoir and trap potential.

Once interpretation of the regional 2-D dataset was completed, a small portion of the study area was studied using 3-D data. The 3-D seismic data were acquired in the vicinity of a well that has produced from a carbonate debris flow. This well (Well No. 1; see Figure 2) allowed us to evaluate the seismic response to porous debris flows in this area. Amplitude and frequency attributes were studied and related to locations of porous carbonate and siliciclastic strata. Analogs developed as a result of this work were used for exploration in other areas.

## REGIONAL SEQUENCE STRATIGRAPHY

Variations in relative sea level caused by variations in tectonic subsidence and eustacy can result in major changes in stratal patterns of carbonate and mixed carbonate-siliciclastic strata (Sarg, 1988, 1989; Kendall et al., 1991; Handford and Loucks, 1994; and others). These changes in stratal patterns have important implications in prediction of reservoir and seal strata. Therefore, we divided Leonardian strata into seven discrete depositional sequences. These sequences are termed Leonard 1 through Leonard 7 sequences in this paper (Figure 2). Local stratigraphic nomenclature was not used in this area due to the difficulty of regionally correlating specific formations as identified by well log tops.

Each sequence was then divided into systems tracts. Systems tracts are linked, contemporaneous sets of depositional systems (Brown and Fisher, 1977). Specific systems tracts developed in the Midland Basin when sea level was at a certain point relative to the shelf edge and rising or falling at a certain rate. These systems tracts developed as a function of the interaction between eustatic changes in sea level and subsidence. Sea level rise can occur by tectonic subsidence (Jervey, 1988). Therefore, relative sea level fall occurred only when the rate of eustatic fall exceeded the rate of rise associated with subsidence. A eustatic fall that is slower than the rate of rise associated with subsidence results in a slow relative rise (Jervey, 1988; Vail, 1987; Posamentier et al., 1988).

Highstand, transgressive, and lowstand systems tracts were identified in Leonardian strata along the northwest shelf. Lowstand systems tracts were subdivided into prograding complex and lowstand fan intervals. These systems tracts differ from those described by

*Seis Strat Services, Inc., 2202 Greencove Lane, Sugar Land, TX, 77479
†Harvey E. Yates Co., P.O. Box 1933, Roswell, NM 88202

Posamentier et al. (1988), Vail (1987), and others for siliciclastic strata. Leonardian systems tracts in the study area most closely resemble those described by Handford and Loucks (1994) for arid, rimmed, mixed carbonate-siliciclastic-evaporite continental margins.

## HIGHSTAND SYSTEMS TRACT

A large portion of the carbonate strata observed in the study area were deposited during highstand time. The highstand systems tract developed during the late stages of eustatic sea level rise and the beginning of eustatic sea level fall. The rate of fall, however, was less than the rate of sea level rise associated with subsidence. Therefore, a relative rise occurred. Relative rise resulted in an increase in accommodation space for carbonate development (Figure 3).

Carbonate sediment originates largely in situ. Most is produced in the photic zone, which comprises the upper 50–100 m of the water column (Schlager, 1981; Sarg, 1988). As such, most carbonate sediment in the Leonard Group was probably formed during sea level highstands when accommodation space existed on the shelf or platform. Most hydrocarbons in the Leonardian sequences are also found in shelf highstand strata. Porosity in these rocks commonly developed or was enhanced when sea level fell and underlying rocks were exposed to fresh meteoric waters.

### Description

Highstand systems tracts in the study area commonly exhibit two types of progradational fill. At most localities, oblique progradational reflections are observed on seismic data (Figures 2 and 4). These progradational reflections are contemporaneous with concordant reflections in shelf strata. A well-defined shelf slope break is present. Well log patterns associated with this progradational style vary greatly; however, thick aggradational or cleaning-upward deposits of carbonate strata are commonly observed (Figure 5).

At other localities, carbonate systems exhibit gentle sigmoidal reflections that merge with concordant reflections on the shelf. Intervals that show sigmoidal progradation can be correlated with intervals in which reflections show apparent onlap onto the slope (as described in Permian strata by Brown and Loucks, 1994) (Figure 6). The shelf-slope break is commonly less abrupt than is observed along margins characterized by oblique progradation. Well logs exhibit aggradational patterns, and analyses of neutron, density, and sonic logs suggest deposition of fine-grained, nonporous strata. Mounded seismic reflections are also observed in some localities at the shelf margin and occur most frequently in areas characterized by sigmoidal progradation.

### Interpretation

Highstand strata that exhibit oblique progradation were probably deposited in "keep-up" conditions (term from Kendall and Schlager, 1981). Keep-up conditions occurred when carbonate accumulation was high enough to keep up with the rate of sea level rise.

The progradational reflections suggest that abundant detritus was delivered to the shelf margins, where it was deposited on the slope by sediment gravity flows. The flows probably range from megabreccia slumps to carbonate oozes. They include debris flows, although porous debris flows are uncommon during sea-level highstands. During highstand time, clasts which range in size from pebbles to boulders are broken off from various portions of a reef and deposited along the foreslope (Cook and Mullins, 1983; Enos and Moore, 1983; James, 1983). Large amounts of mud are formed simultaneously and also deposited on the foreslope. Mud overwhelms production of coarse-grained detritus during highstand time. This mud mixes with the clasts along the foreslope until a point of critical instability is reached and both mud and clasts are redeposited downslope as mud-supported debris flows. Mud deposited from suspension also infiltrates talus cones on the slope and basin floor (Grammar and Ginsberg, 1992; C. Kendall, 1992, personal communication).

Highstand strata which exhibit sigmoidal progradational patterns and/or apparent onlap were probably deposited in "catch-up" conditions. Catch-up carbonate systems develop when the rate of sea level rise exceeds the rate of carbonate production. In contrast to keep-up highstand strata, much of the deposition on the slope and basin floor probably occurred from suspension. This is especially true in highstand strata characterized by apparent onlap. Many catch-up carbonate systems are mud rich, because most deposition commonly occurs below fair-weather wave base (Sarg, 1988). Fine-grained clay and mud were washed off the shelf and deposited on the slope and basin floor.

Shelf-edge mounds observed in strata characterized by sigmoidal progradation may be keep-up sponge-algal reefs formed during overall catch-up carbonate development. James (1983) observed that shelf-edge mounds formed in these conditions because high turbulence at the shelf margin provided optimum conditions for maximum reef growth. Sediment was constantly removed, the water was clear, and nutrients were constantly moved past reef organisms.

## TRANSGRESSIVE SYSTEMS TRACT

### Description

Transgressive systems tracts formed when the rate of sea level rise greatly increased. Sedimentation was commonly not able to keep pace, and the shelf was flooded (Figure 7). This systems tract is thin in many sequences in the study area and is commonly below seismic resolution. Where it is thick enough to be delineated seismically, it shows concordant reflections on both the shelf and the slope (Figures 2, 4, and 6). Three major facies are observed in the transgressive systems tract: mixtures of carbonate and terrigenous mud, isolated backstepping carbonate mounds (Figure 8), and reworked shelf sands that are correlatable over broad areas. Carbonate/terrigenous muds are most characteristic of this systems tract. They commonly exhibit retrogradational well log profiles or profiles that suggest deposition of thick uniform muds (Figure 5).

### Interpretation

Shelf and slope deposition in the transgressive systems tract consisted largely of carbonate muds. These muds developed as terrigenous deposition moved progressively landward and carbonate production on the shelf was reestablished. The backstepping mounds formed where carbonate production was able to keep pace with rising sea level. The initial mounds (lowstand biohermal buildups) developed on the upper slope. These were drowned as water depth increased during transgression and younger mounds were established farther landward.

Regionally correlatable shelf sands in the transgressive systems tract were probably *not* deposited during this time interval. Instead, they were deposited when the shelf was exposed and sand was moved over it by eolian processes. During transgression, these sands were reworked by shelf currents.

## LOWSTAND SYSTEMS TRACT

The lowstand systems tract was divided into fan and prograding complex subdivisions. In contrast to sequence stratigraphic studies of siliciclastic margins (e.g., Vail, 1987), it was not possible to divide the lowstand into basin-floor fan, slope fan and prograding complex subdivisions. It is not clear whether the lowstand fan represents both basin-floor fan and slope fan (terms from Vail, 1987) deposition. The lowstand fan and lowstand prograding complex together are commonly thinner than the highstand systems tract. However, this systems tract was studied in detail, because porous debris flows are more likely to form in lowstand strata than in highstand or transgressive systems tracts (Grammar and Ginsberg, 1992).

### Description

Lowstand fans in Leonardian strata exhibit a prominent reflection that downlaps onto the underlying sequence boundary on the basin floor. This reflection defines the upper boundary of the lowstand fan. In some cases, these fans downlap onto basin floor and onlap the lower slope (Figures 2, 4). Large slump scars developed near the shelf margin. Some slump scars may have developed into submarine canyons (Figures 9 and 10). The sequence boundary is defined by erosion in these canyons. Lowstand fans observed in the study area are preferentially located immediately downdip of them. Lowstand fan facies exhibit aggradational patterns on well logs (Figure 5). Well and sample log data suggest that lowstand fans in the study area commonly are composed of massive sandstones; interbedded sandstones, siltstones, and carbonates; and/or aggradational carbonates.

Most deposition in the lowstand prograding complex occurs along the slope front and the base of the slope (Figures 2, 4, and 6). Sigmoidal to oblique, progradational reflections that onlap the slope along the underlying sequence boundary are observed. In contrast to highstand strata, most *preserved* lowstand shelf strata are confined to the heads of submarine canyons. These strata onlap the sides of the canyons (Figure 9). Well and sample logs suggest that deposition of both sandstone and carbonates occurred in the lowstand prograding complex. Strata range from thick uniform sands, to mixtures of interbedded sandstones and carbonates, to intervals dominated by carbonates. Most sandstones in the study area were deposited during lowstand time.

A productive debris flow in the study area was penetrated by Well no. 1 (Figure 2). The pay interval is located in the prograding wedge of the Leonard 2 sequence. Analysis of drill cuttings suggests that it is composed of finely crystalline limestone with vuggy to cavernous porosity. Sponge spicules and rare oolites are also present. The cavernous and vuggy porosity was probably developed before resedimentation into deep water.

#### Analysis of Lowstand Facies Using 3-D Data

Once initial sequence stratigraphic analysis of the 2-D data was complete, more detailed analysis of the 3-D dataset was undertaken. Since the 3-D data were shot over a producing debris flow located within a lowstand prograding complex, most of our work with the 3-D dataset focused on delineation of sedimentary features in the lowstand systems tract.

Initial work on the 2-D seismic data suggested that a submarine canyon was developed in the 3-D study area. This canyon (Figures 9 and 10) was further delineated and mapped using the 3-D dataset. It trends southeast and cuts into the upper slope (Figure 11). The submarine canyon formed on a well-defined sequence boundary. A time structure map constructed along this sequence boundary indicates that initial deposition of material transported through the submarine canyon probably occurred in the southwestern portion of the area covered by the 3-D survey—to the south and west of the productive well. However, an isochron map of the pay interval (Figure 12) suggests that the depocenter may have shifted during formation of the lowstand systems tract.

Initial deposition probably occurred in the southwestern portion of the 3-D study area, immediately downdip of the submarine canyon. However, an isochron map (Figure 12) shows that the pay interval is thin to nonexistent in this portion of the 3-D study area. This suggests that the bathymetric low that developed downdip of the submarine canyon was probably filled during early lowstand time. Then the depocenter shifted to the vicinity of the productive well. An isochron thick developed immediately to the east of this well (Figure 12).

The porous carbonate debris flow was too small to exhibit a discrete seismic geometry. However, it exhibits prominent amplitude and average weighted frequency anomalies. A horizon amplitude map (Figure 13) was made on the reflection that occurs at the pay interval. A large percentage of the shelf and slope strata in the 3-D study area exhibit relatively low amplitude. In contrast, high amplitudes are observed over most of the basin floor. A pronounced area of low amplitude, however, occurs in the vicinity of the productive well. An elongate, channel-like area of low-amplitude strata extends from the porous debris flow to the mouth of the submarine canyon (Figure 13).

Similar phenomena are observed on a horizon average weighted frequency map (Figure 14). Most shelf and slope strata are characterized by relatively low-frequency data. The basin floor, however, exhibits high-frequency strata which encircle an area of lower-frequency data. This area roughly coincides with the debris flow as delimited by the low-amplitude signal (Figure 14). The frequency anomaly is more subtle than the anomaly observed on the horizon amplitude map. However, both maps clearly show a change in seismic character associated with the porous debris flow.

An isochron map was also made of the overlying transgressive systems tract and the lower portion of the highstand systems tract (Figure 15). This isochron map records fill of the incised valley and associated submarine canyon. Valley and canyon fill occurred within a short time. A time structure map of the overlying sequence boundary (Figure 16) shows no evidence of either feature.

### Interpretation and Implications for Exploration for Porous Debris Flows

Development of the lowstand fan probably occurred during relative sea level fall. Much of the shelf was subaerially exposed, and major erosion occurred along the shelf margin and upper slope (Sarg, 1988; Handford and Loucks, 1994) (Figure 17). Production of carbonate deposits may have been limited to a thin zone along the upper slope located within the photic zone (Sarg, 1988). Carbonate production ceased on the shelf, which was probably characterized by a broad eolian plain (Handford and Loucks, 1994). Wadis transported sand and silt to the shelf margin. This sand and silt was redeposited downdip by sediment gravity flows. In the study area, both carbonate and siliciclastic detritus was probably transported to the basin floor through submarine canyons and slope gullies. These canyons may have formed by slumping along the shelf margin during initial sea level fall. Much of the eolian lowstand deposition on the shelf was probably not preserved in the geologic record. Instead, these strata were extensively reworked during transgression.

Lowstand prograding complexes probably developed during the initial stages of sea level rise. Erosion of the reef front continued during development of the lowstand prograding complex, and sediment was transported downdip by debris flows. However, backfill of slump scars and submarine canyons began at this point, and detritus was deposited along the slope instead of bypassing it. During lowstand progradation, calcareous sands were created by erosion of exposed carbonate strata and siliciclastic material was deposited on the shelf by eolian processes. Both detrital types were deposited near the shelf-slope break and reworked by nearshore processes. These materials were, in turn, redeposited along the slope and the base of the slope by sediment gravity flow processes, resulting in further progradation of the shelf margin.

Both highstand and lowstand systems tracts in the study area exhibit abundant sediment gravity flows. Many of these flows, however, exhibit a muddy matrix and low porosity and permeability. Nonporous debris flows studied by Loucks et al. (1985) and Hobson et al. (1985b) in Permian strata of the Midland Basin exhibit diverse clast lithology and a mud-rich matrix. The matrix commonly has a different lithology and provenance from that of the clasts. In contrast, porous debris flows exhibit clasts of uniform lithology (Hobson et al., 1985a, 1985b; Harris and Wiggins, 1985). The matrix is generally sand-sized and is either of the same lithology as the clasts (Hobson et al., 1985a) or is siliciclastic (Harris and Wiggins, 1985). Porosity in the clasts is generally inherited. Clasts commonly exhibit secondary porosity created by subaerial exposure before redeposition downdip. Porosity is enhanced by sand-sized matrix material (either carbonate, as observed in the study area, or siliciclastic), which results in abundant initial primary porosity.

Porous debris flows, such as the producing flow in the study area, are most likely to form in lowstand strata (Grammar and Ginsburg, 1992; C. Kendall, 1992, personal communication). When sea level fell to a point at or below the shelf margin, abundant carbonate sand as well as gravel and boulders were formed by reef-front erosion. The gravel- to boulder-sized clasts mixed with the carbonate sand and were transported downdip through a submarine canyon to form the porous deposit.

## CONCLUSIONS

The widespread availability of high-quality 3-D data has led to a renewal of interest in the Permian Basin for many companies. These companies use two different strategies in acquiring these 3-D data. The first strategy is to acquire and interpret large regional surveys. Although this strategy allows the most detailed interpretation, it may not be cost-effective for many petroleum companies. We suggest an alternative approach. Before acquisition of 3-D data begins, we suggest that a detailed sequence stratigraphic study be completed using well log and 2-D seismic data.

Identification of lowstand strata, areas of lowstand input, incised valleys, and slump scars along the reef front are key factors in locating porous debris flows in Permian strata. These steps can occur *before* expensive 3-D data are shot. High-quality 2-D seismic data are available over most of the northwest shelf of the Midland Basin. Incised valleys, submarine canyons, and slump scars can be delineated on these data by first locating sequence boundaries, then examining each boundary for evidence of erosional truncation. If the 2-D grid is dense enough, individual canyons can be mapped within a specific sequence. In many areas of the Permian Basin, sufficient well logs and 2-D seismic data exist to evaluate changes in lowstand thickness and composition within specific systems tracts. This involves a two-stage process. First well log signatures are compared to specific seismic reflection patterns using synthetic seismograms. Then well log correlations are made using the seismic correlations as a guide. We observed major changes in lithology over very short distances in our study area. These lithologic changes made correlation of wireline data without seismic control nearly impossible. Once these correlations are complete, changes in thickness and lithologic composition within specific lowstand intervals can be mapped. These maps can be used, along with seismic data that show submarine canyons and slope gullies, to delineate major point sources along the progradational margins. The best chance for locating productive debris flows may occur immediately downdip of these major point sources. Once they are located, 3-D data acquisition can be concentrated in these areas to isolate potentially productive debris flows.

The 3-D data are then used to delineate subtle differences in seismic attributes and to pinpoint areas of maximum productive potential. For example, the debris flow in the study area was too small to exhibit discrete seismic geometry. However, it exhibited a much lower amplitude than surrounding strata along the basin floor. A small channel-like feature composed of low-amplitude strata extends from the debris flow to the submarine canyon developed on the slope. The average weighted frequency map does not show the same magnitude of difference as the amplitude map. However, the debris flow was also characterized by relatively lower frequency than surrounding strata of the basin floor. We believe that this may be the most cost-effective way to explore in much of the Permian Basin.


### *Acknowledgments*

We would like to thank George Yates, President and CEO of Heyco Producing Co. and Bill Ranck, President of Ranck Exploration for allowing us to publish this manuscript and for encouraging us to work on it. Al Reid did some of the initial work on the debris flow reported on in this paper and graciously allowed publication as well. We also give special thanks to Jim Manatt (formerly of PXC Corporation, now with Providence Technologies) and Colin Macmillian Jr. (President) of PXC for permission to publish pieces of their 2-D seismic dataset over the study area. Finally we would like to thank J. F. Sarg for numerous suggestions during the course of this analysis.

## REFERENCES CITED


Brown, A. A., and R. G. Loucks, 1994, Influence of sediment type and depositional processes on stratal patterns in the Permian Basin Lamar Limestone, McKittrick Canyon, Texas, *in* R. G. Loucks and J. F. Sarg, eds., Carbonate Sequence Stratigraphy—Recent Developments and Applications, AAPG Memoir 57, p. 133–156.

Brown, L. F., and W. L. Fisher, 1977, Seismic stratigraphic interpretation of depositional systems: examples from Brazilian rift and pull-apart systems, *in* C. E. Payton, ed., Seismic Stratigraphy—Application to Hydrocarbon Exploration, AAPG Memoir 26, p. 213–248.

Cook, H. E., and H. T. Mullins, 1983, Basin margin environment; *in* P. A. Scholle, D. G. Bebout, and C. H. Moore, eds., Carbonate Depositional Environments, AAPG Memoir 33, p. 540–617.

Enos, P., and C. H. Moore, 1983, Fore-reef slope environment; *in* P. A. Scholle, D. G. Bebout, and C. H. Moore, eds., Carbonate Depositional Environments, AAPG Memoir 33, p. 508–537.

Grammar, G.M., and R.N. Ginsburg, 1992, Highstand versus lowstand deposition on carbonate platform margins: insight from Quaternary foreslopes in the Bahamas: Marine Geology, v. 103, p. 125–136.

Handford, C. R., and R. G. Loucks, 1994, Carbonate depositional sequence and systems tracts—responses of carbonate platforms to relative sea-level changes, *in* R. G. Loucks and J. F. Sarg, eds., Carbonate Sequence Stratigraphy—Recent Developments and Applications, AAPG Memoir 57, p. 13–42.

Harris, P. M., and W. D. Wiggins, 1985, Allochthonous carbonates of the Getaway Limestone, Upper Permian of the Delaware Basin, *in* P. D. Crevello and P. M. Harris, eds., Deep-Water Carbonates—A Core Workshop, SEPM Core Workshop No. 6, p. 174–211.

Hobson, J. P., C. D. Caldwell, and D. F. Toomey, 1985a, Sedimentary facies and biota of early Permian deep-water allochthonous limestone, southwest Reagan County, Texas, *in* P. D. Crevello and P. M. Harris, eds., Deep-Water Carbonates—A Core Workshop, SEPM Core Workshop No. 6, p. 93–139.

Hobson, J. P., C. D. Caldwell, and D. F. Toomey, 1985b, Early Permian deep-water allochthonous limestone facies and reservoir, West Texas: AAPG Bulletin, v. 69, p. 2130–2147.

James, N. P., 1983. Reef environment, *in* P. A. Scholle, D. G. Bebout, and C. H. Moore, eds., Carbonate Depositional Environments, AAPG Memoir 33, p. 346–453.

Jervey, M. T., 1988, Quantitative geological modeling of siliciclastic rock sequences and their seismic expression, *in* C. K. Wilgus, B. S. Hastings, C. G. St. C. Kendall, H. W. Posamentier, C. A. Ross, and J. C. Van Wagoner, eds., Sea-Level Changes: An Integrated Approach, SEPM Special Publication 42, p. 47–70.

Kendall, C. G. St. C., B. E. Bowen, A. Alsharhan, D. K. Cheong, and D. Stoudt, 1991, Eustatic controls on carbonate facies in reservoirs and seals associated with Mesozoic hydrocarbon fields of the Arabian Gulf and the Gulf of Mexico, *in* A. W. Meyers, T. A. Davies, and S. W. Wise, eds., Evolution of Mesozoic and Cenozoic Continental Margins, 1990 SEPM Research Symposium Proceedings, SEPM, Tulsa, p. 215–238.

Kendall, C. G. St. C., and W. Schlager, 1981, Carbonates and relative changes in sea-level: Marine Geology, v. 44, p. 181–212.

Loucks, R. G., A. A. Brown, C. W. Achauer, and D. A. Budd, 1985, Carbonate gravity flow deposition on low-angle slopes of the Wolfcampian northwest shelf of the Delaware Basin, *in* P. D. Crevello and P. M. Harris, eds., Deep-Water Carbonates—A Core Workshop, SEPM Core Workshop No. 6, p. 56–92.

Posamentier, H. W., M. T. Jervey, and P. R. Vail, 1988, Eustatic controls on clastic deposition I—conceptual framework, *in* C. K. Wilgus, B. S. Hastings, C. G. St. C. Kendall, H. W. Posamentier, C. A. Ross, and J. C. Van Wagoner, eds., Sea-Level Changes: An Integrated Approach, SEPM Special Publication 42, p. 109–125.

Sarg, J. F., 1988, Carbonate sequence stratigraphy, *in* C. K. Wilgus, B. S. Hastings, C. G. St. C. Kendall, H. W. Posamentier, C. A. Ross, and J. C. Van Wagoner, eds., Sea-Level Changes: An Integrated Approach, SEPM Special Publication 42, p. 155–181.

Sarg, J. F., 1989, Middle-Late Permian depositional sequences, Permian Basin, West Texas and New Mexico, *in* A. W. Bally, ed., Atlas of Seismic Stratigraphy, v. 3, AAPG Studies in Geology, 27, p. 140–154.

Schlager, W., 1981, The paradox of drowned reefs and carbonate platforms: Geological Society of America Bulletin, v. 92, p. 197–211.

Silver, B. A., and R. G. Todd, 1969, Permian cyclic strata, northern Midland and Delaware basins, West Texas and southeastern New Mexico: AAPG Bulletin, v. 53, p. 2223–2251.

Vail, P. R., 1987, Seismic stratigraphy interpretation procedure, *in* A. W. Bally, ed., Atlas of Seismic Stratigraphy, v. 1, AAPG Studies in Geology, 27, p. 1–10.

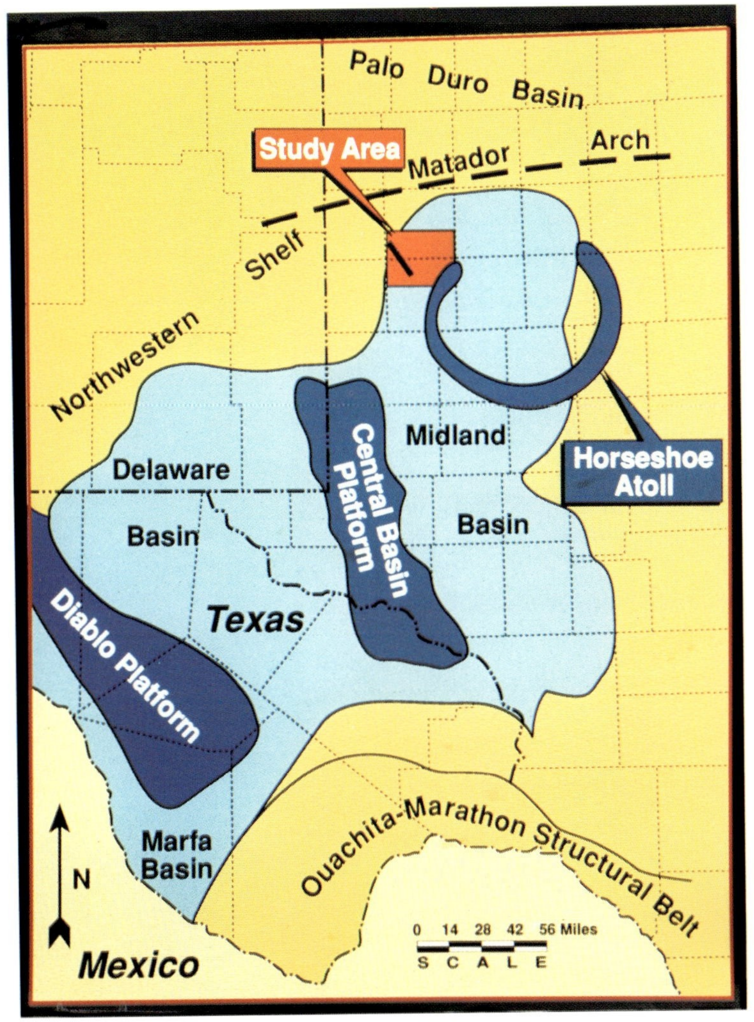


FIGURE 1. Geologic setting and location of the study area. Modified from Silver and Todd, 1969.

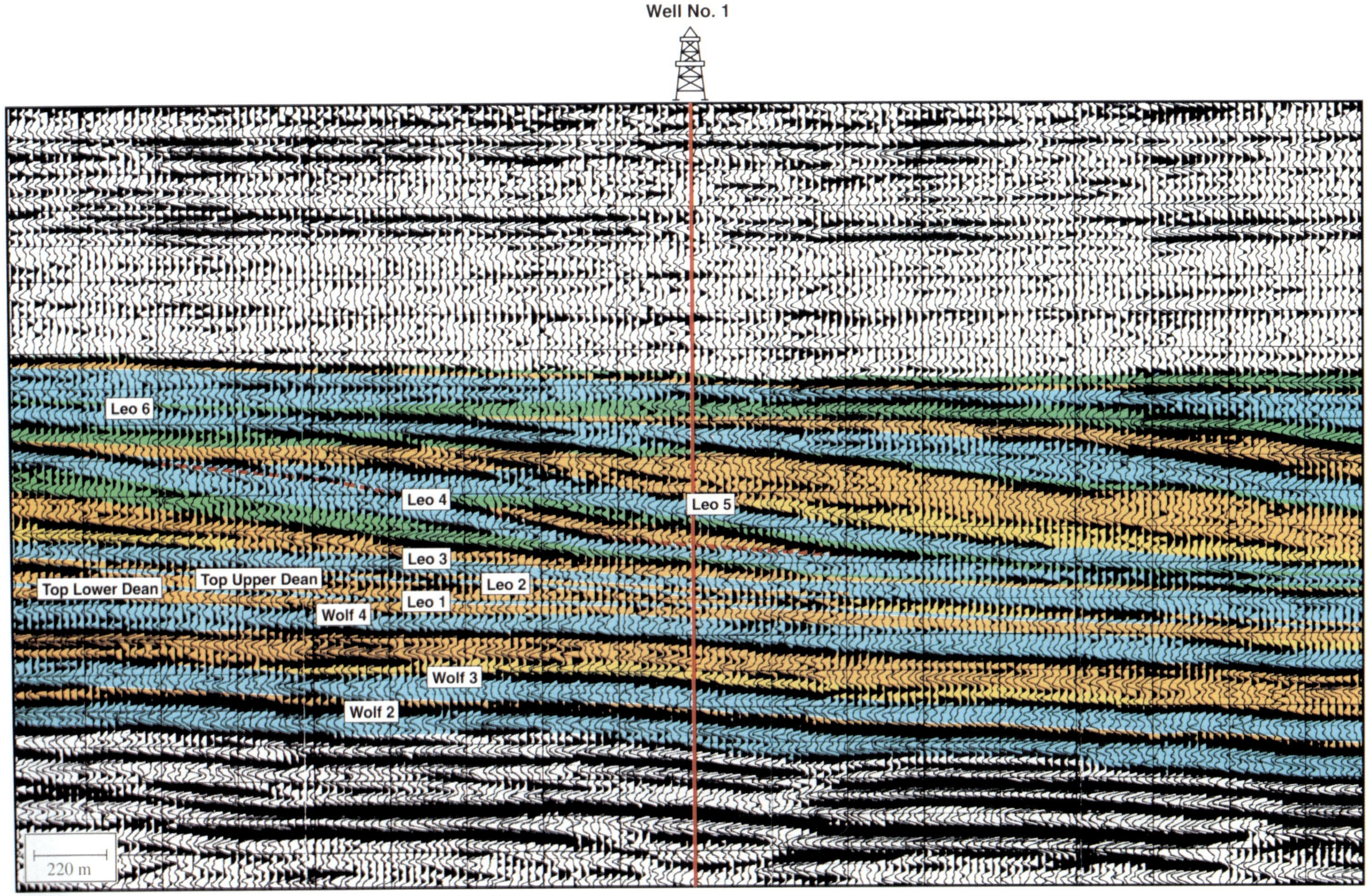


FIGURE 2. Seismic section that shows highstand, transgressive, lowstand prograding complex and lowstand fan systems tracts in multiple sequences of the Leonard and Wolfcamp Groups in the study area, (yellow = lowstand fan, orange = lowstand prograding complex, green = transgressive systems tract, blue = highstand systems tract). Leo 7 (not labeled) is the uppermost sequence shown. The distance between solid horizontal lines is approximately 220 m. The distance between solid vertical lines is 100 ms. Seismic data courtesy of PXC.

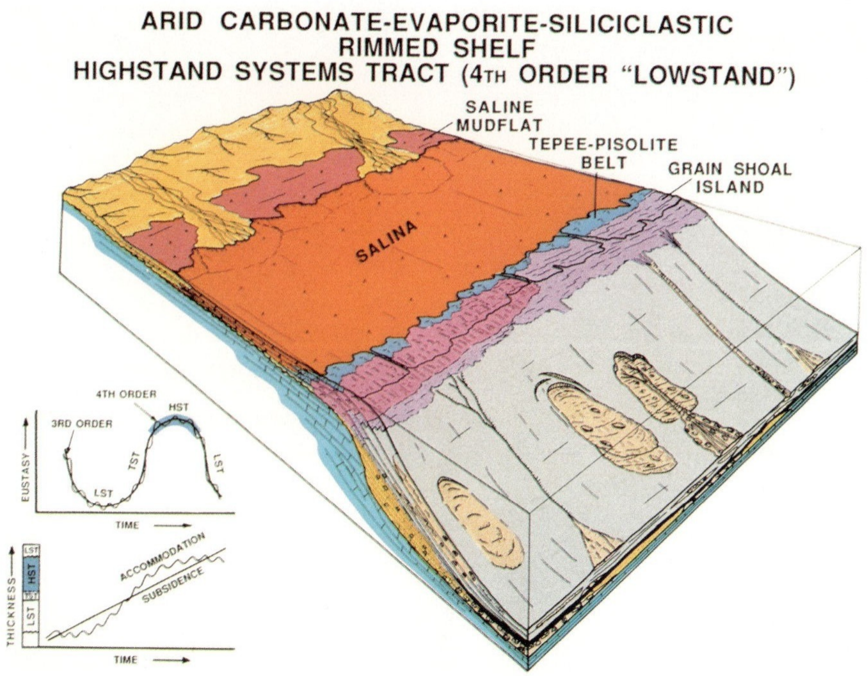


FIGURE 3. Block diagram that shows depositional systems developed during highstand time. From Handford and Loucks, 1994.

FIGURE 4. Highstand transgressive, lowstand prograding complex, and lowstand fan systems tracts in a single sequence. Highstand strata show oblique progradation typical of "keep-up" deposition (yellow = lowstand fan, orange = lowstand prograding complex, green = transgressive systems tract, blue = highstand systems tract). The distance between solid vertical lines is approximately 220 m. The distance between solid horizontal lines is 100 ms. Seismic data courtesy of PXC.

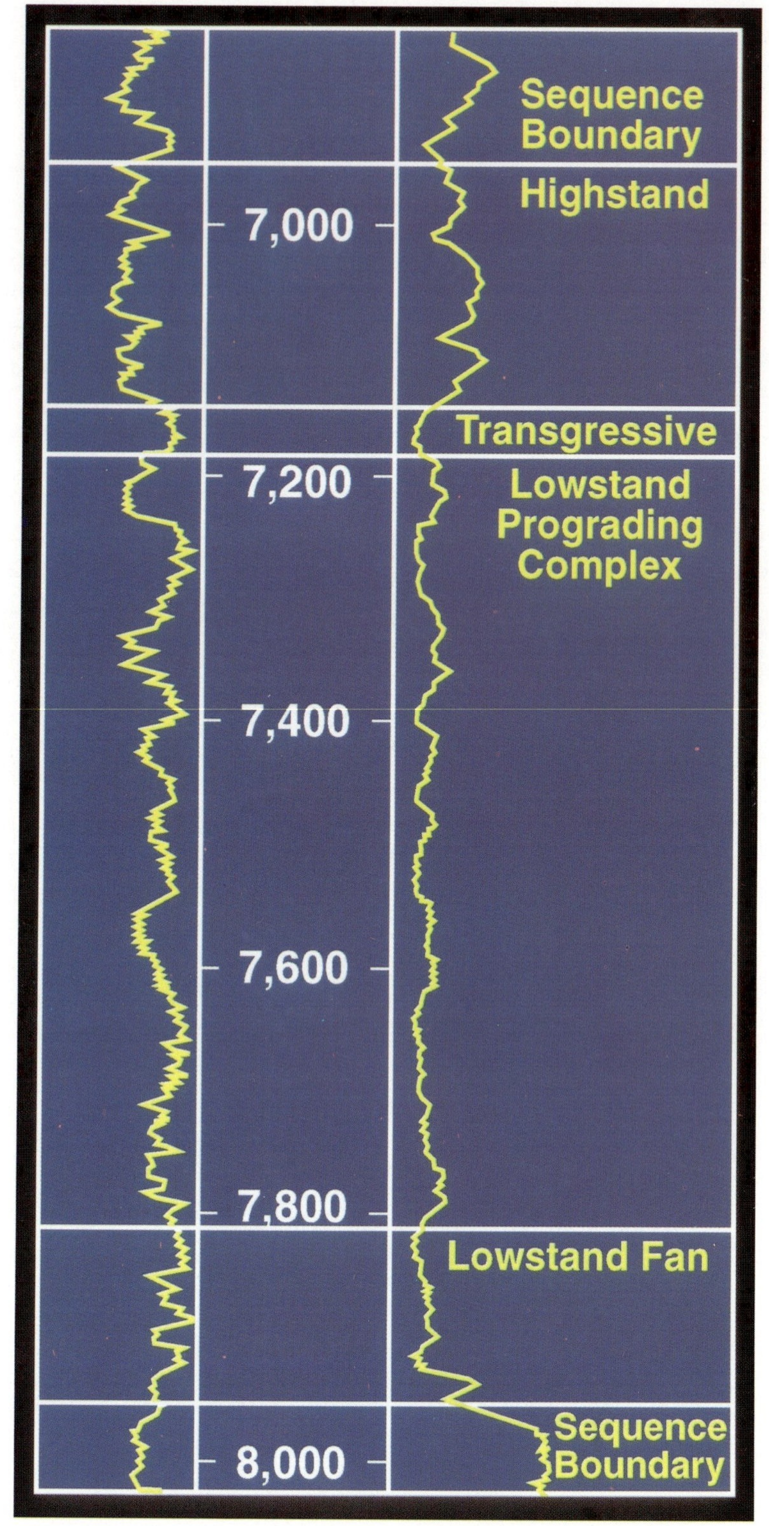


FIGURE 5. Induction and gamma ray well log that shows typical profiles of systems tract in the study area. The log is from "Well No. 1" on Figure 2.

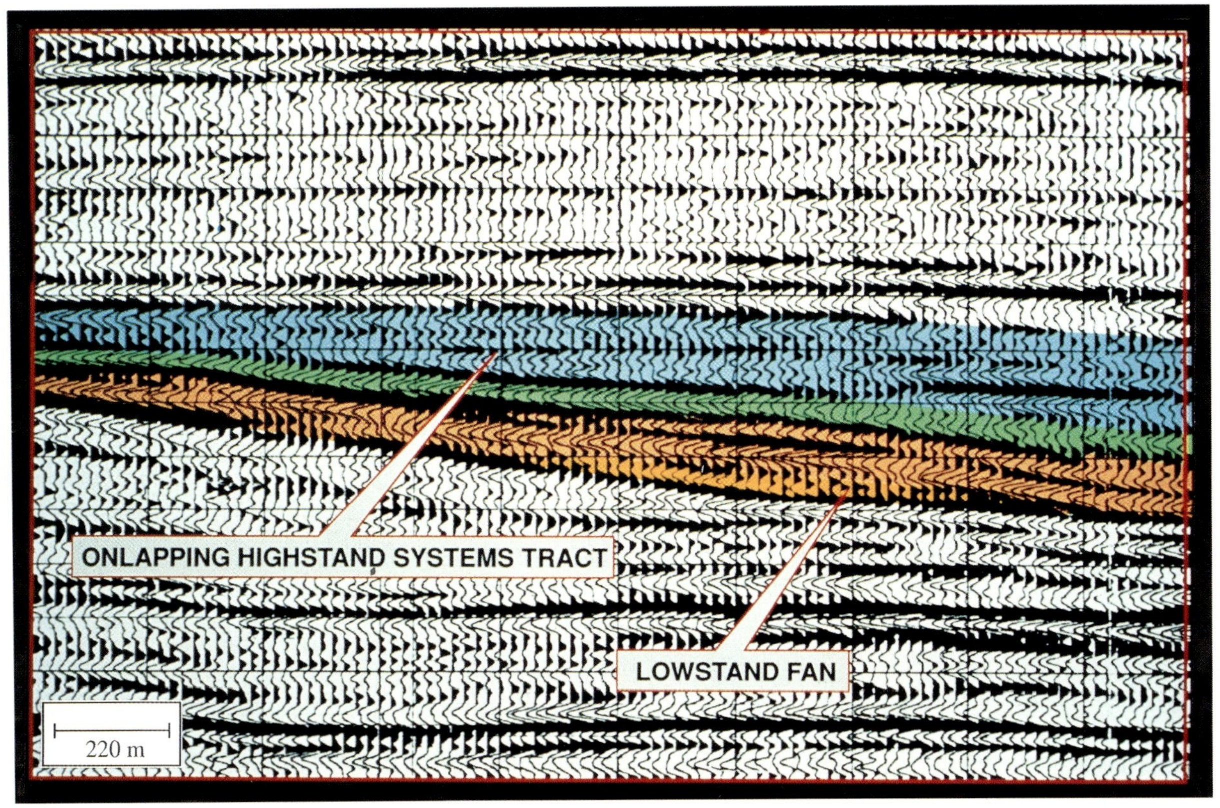


FIGURE 6. Highstand, transgressive, lowstand prograding complex, and lowstand fan systems tracts in a single sequence. Highstand strata show apparent onlap typical of "catch-up" deposition (yellow = lowstand fan, orange = lowstand prograding complex, green = transgressive systems tract, blue = highstand systems tract). The distance between solid vertical lines is approximately 220 m. The distance between solid horizontal lines is 100 ms. Seismic data courtesy of PXC.

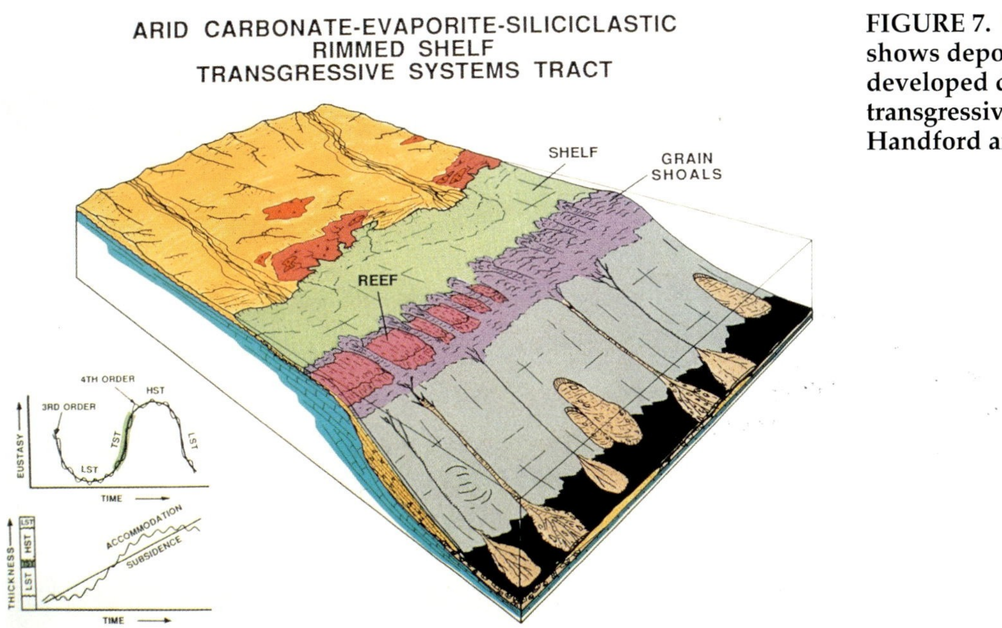


FIGURE 7. Block diagram that shows depositional systems developed during formation of the transgressive systems tract. From Handford and Loucks, 1994.

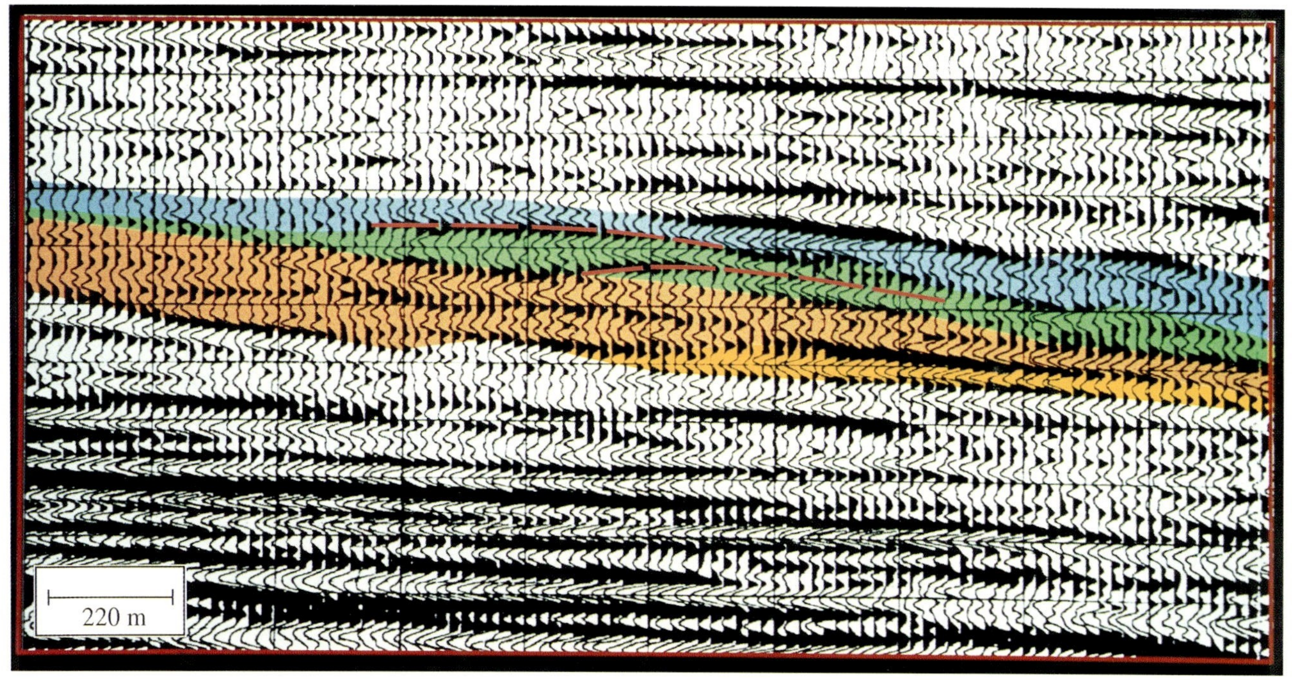


FIGURE 8. Seismic section showing backstepping mounds developed in the transgressive systems tract. The distance between solid vertical lines is approximately 220 m. The distance between solid horizontal lines is 100 ms. Seismic data courtesy of PXC.

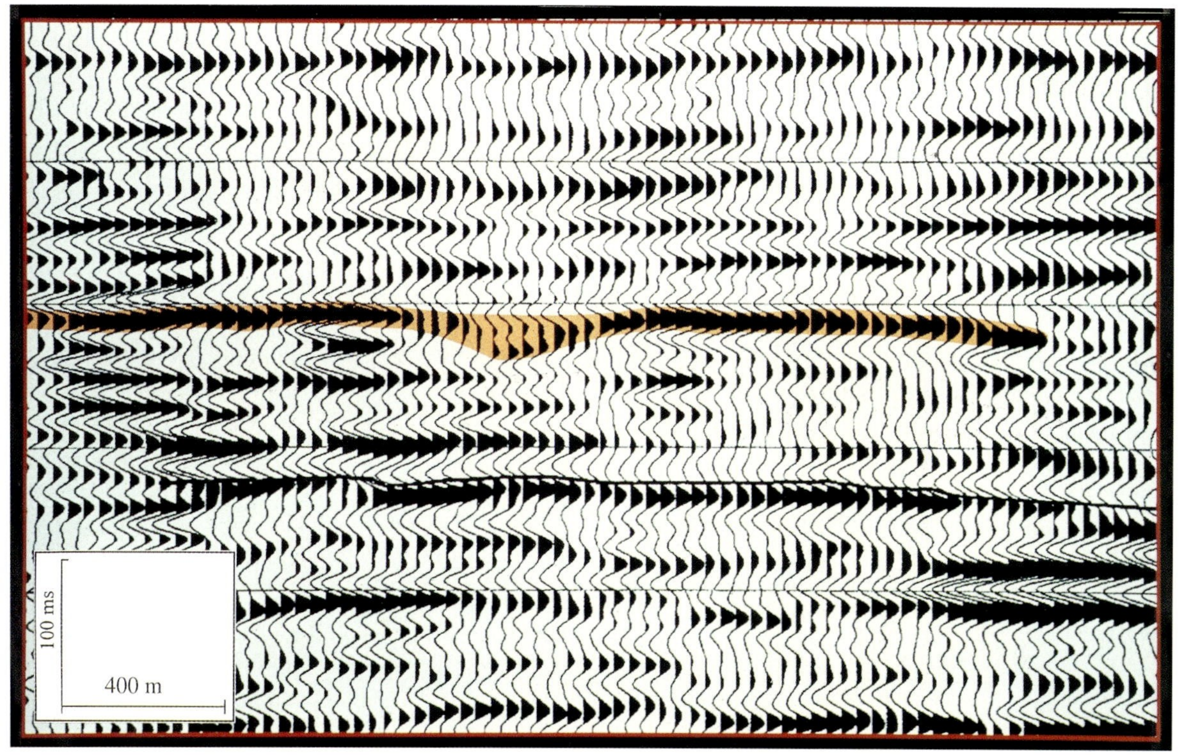


FIGURE 9. Seismic section from 3-D dataset showing a strike view of the head of a submarine canyon developed in the 3-D study area. Seismic data courtesy of Heyco Production Company and Ranck Exploration.

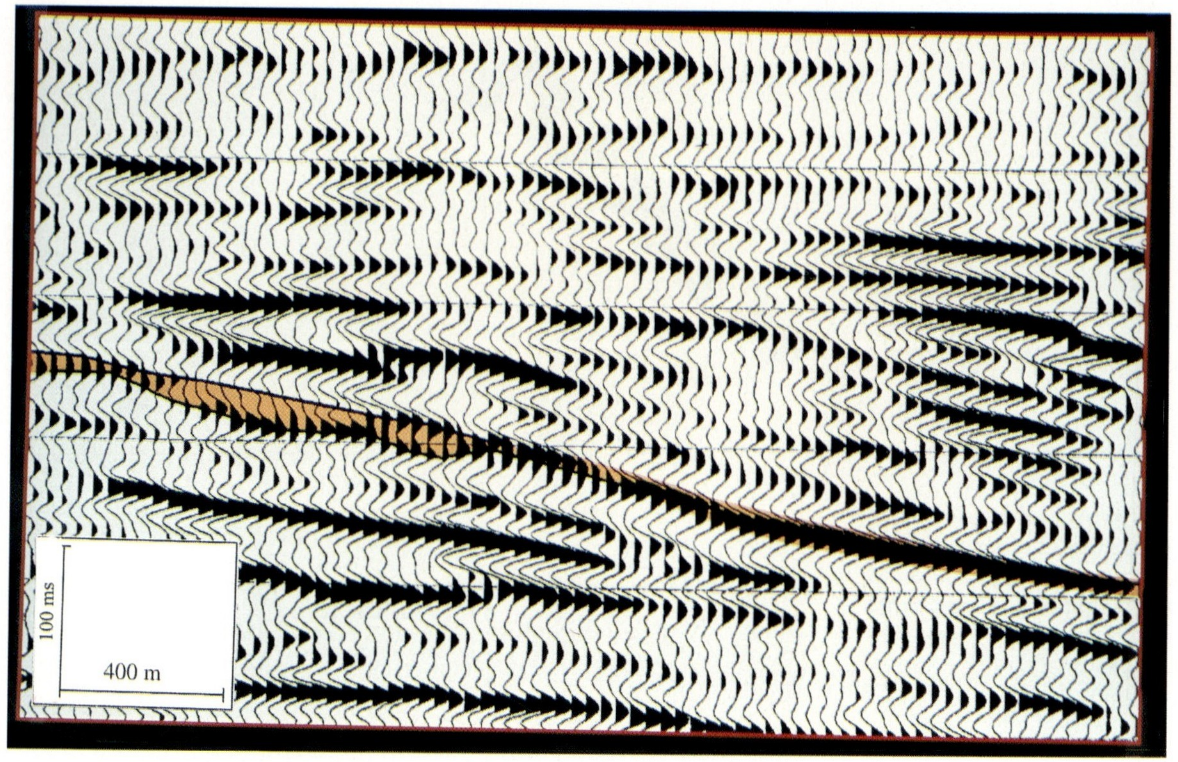


FIGURE 10. Seismic section that shows submarine canyon developed in the 3-D study area. Seismic data courtesy of Heyco Production Company and Ranck Exploration.

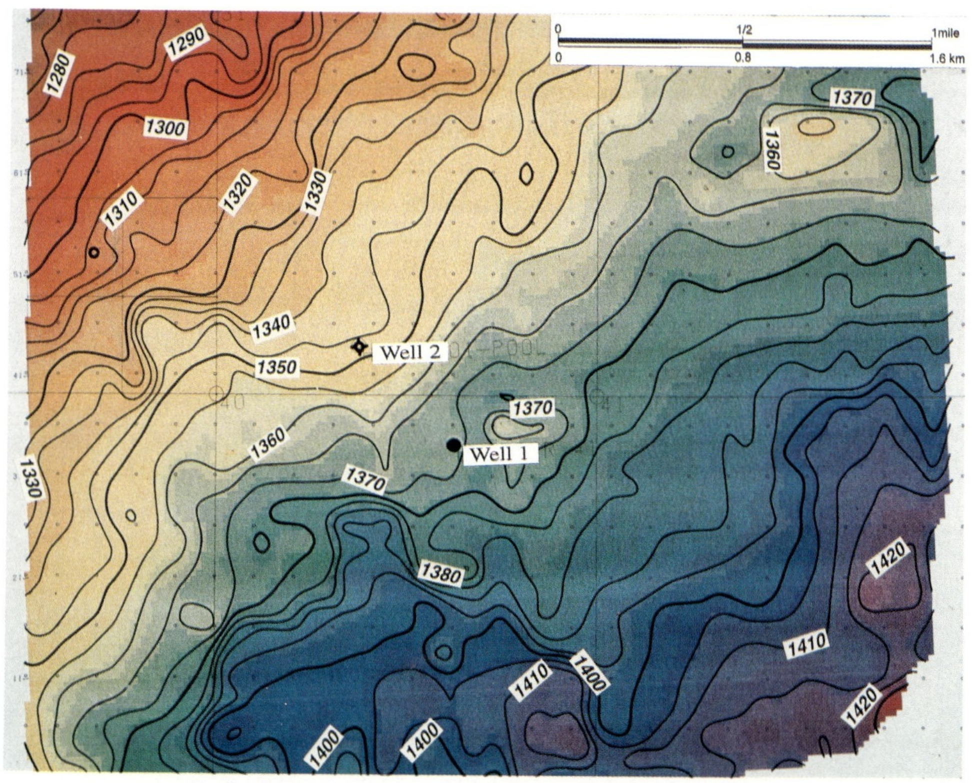


FIGURE 11. Time structure map of the sequence boundary that underlies the pay interval in the 3-D study area. Map published courtesy of Heyco Producing Co. and Ranck Exploration.

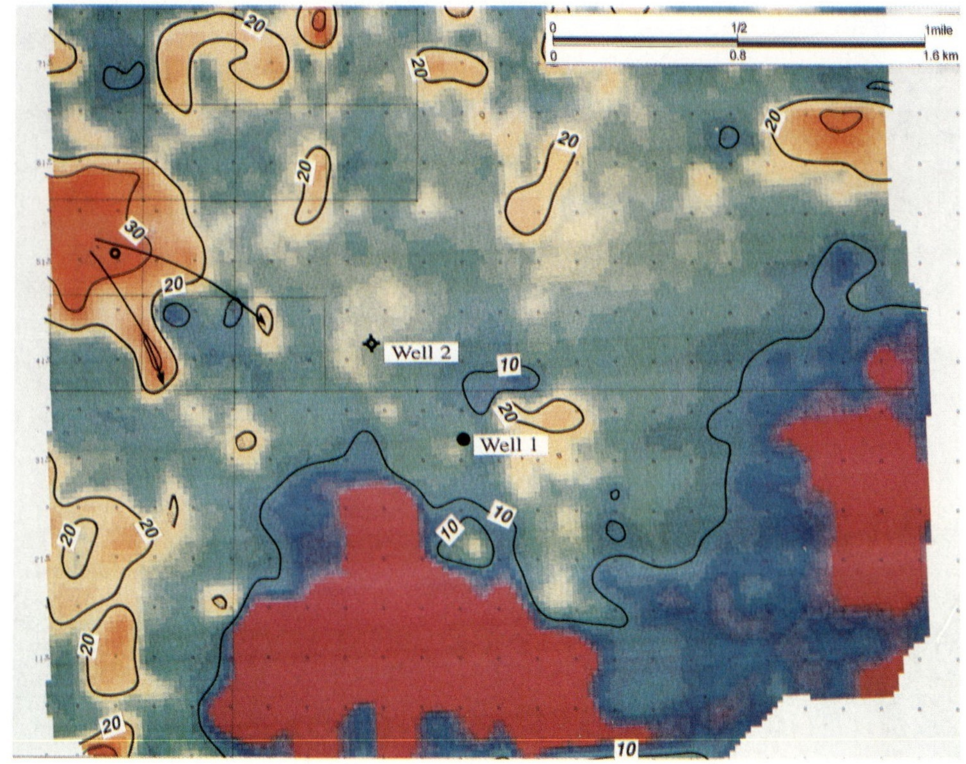


FIGURE 12. Isochron map of the pay interval in the 3-D study area. Map published courtesy of Heyco Producing Co. and Ranck Exploration.

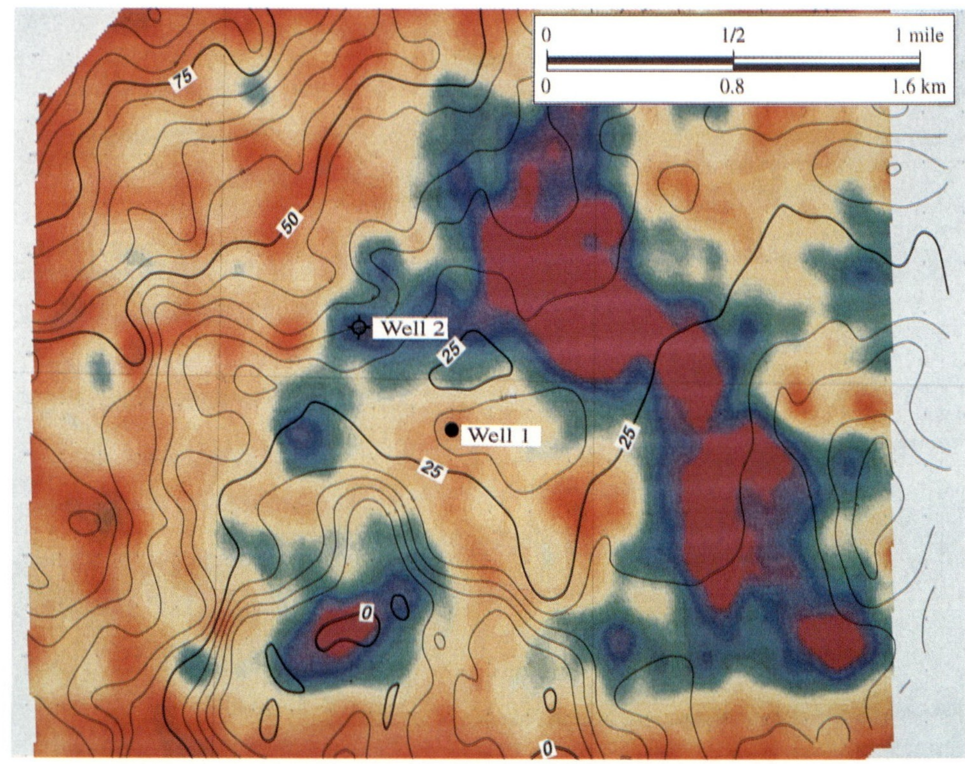


FIGURE 13. Horizon amplitude map of the pay interval. Map published courtesy of Heyco Producing Co. and Ranck Exploration.

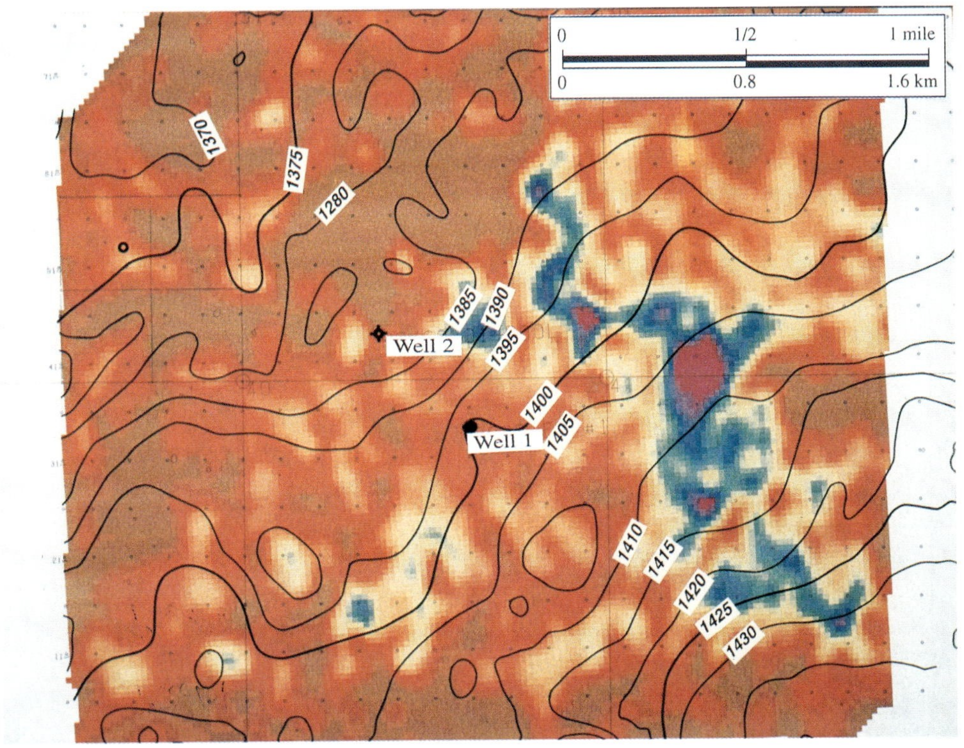


FIGURE 14. Horizon "average weighted frequency" map of the pay interval. Map published courtesy of Heyco Producing Co. and Ranck Exploration.

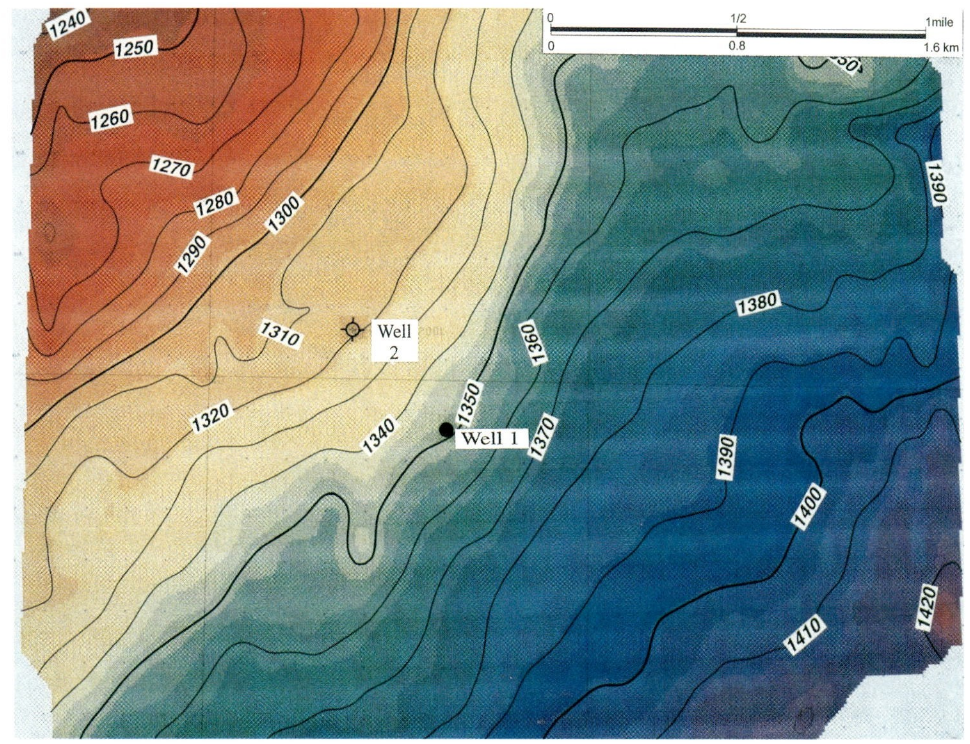


FIGURE 16. Time structure map of the overlying sequence boundary. Map published courtesy of Heyco Producing Co. and Ranck Exploration.

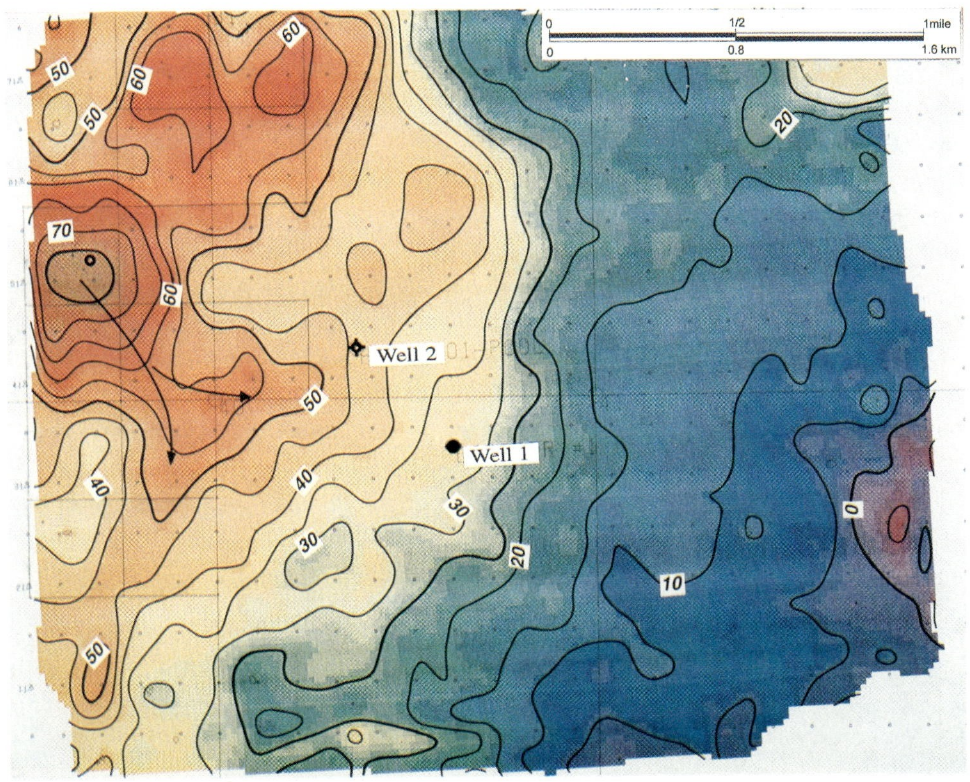


FIGURE 15. Isochron map of transgressive and highstand systems tracts that overlie the pay interval. Map published courtesy of Heyco Producing Co. and Ranck Exploration.

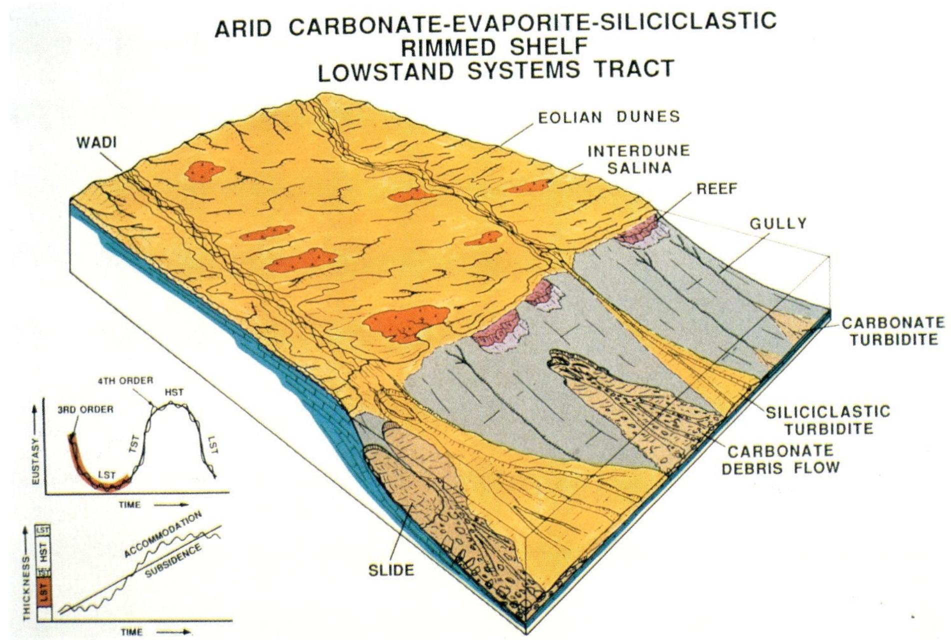


FIGURE 17. Block diagram that shows depositional systems developed during lowstand time. From Handford and Loucks, 1994.

# CHAPTER 18

Arestad, J. F., T. L. Davis, and R. D. Benson, Utilizing 3-D, 3-C seismology for reservoir property characterization at Joffre Field, Alberta, Canada, *in* P. Weimer and T. L. Davis, eds., *AAPG Studies in Geology No. 42* and *SEG Geophysical Developments Series No. 5*, AAPG/SEG, Tulsa, p. 171–178.

# Utilizing 3-D, 3-C Seismology for Reservoir Property Characterization at Joffre Field, Alberta, Canada

**John F. Arestad, Thomas L. Davis, and Robert D. Benson***

## ABSTRACT

A 3-D, three-component (3-D, 3-C) seismic survey was collected over the northeastern limit of Joffre Field, Alberta, for the purpose of detecting porosity within two complex Upper Devonian carbonate reservoirs, Nisku and Leduc. The survey demonstrates the robust nature and power of 3-D, 3-C seismic technology to determine not only porosity, but also permeability. As a result, by integrating results from this new seismic technology with geologic and reservoir production data into a reservoir model, new field extension and field redevelopment targets are identified, and improved reservoir exploitation can be achieved.

## INTRODUCTION

The need for improved hydrocarbon recovery from existing fields requires more efficient reservoir characterization. These characterizations should aim to map the spatial distribution of key reservoir properties, such as porosity, permeability, and water saturation. Accurate determination of these properties is essential for economically viable reservoir development and management. Conventional, compressional seismic surveys (both 2-D and 3-D) have been routinely used for structural mapping of reservoirs and, in certain geological environments, have not effectively characterized these reservoirs. Enormous potential exists for using three-dimensional (3-D), three-component (3-C) seismology as a tool for improved reservoir characterization. By using three-component geophones, ground displacements can be recorded in three orthogonal directions, as opposed to the singular vertical component utilized in conventional seismic recording. The two horizontal components allow the measurement displacements from shear waves that can be used to determine reservoir properties.

The Colorado School of Mines Reservoir Characterization Project (RCP) conducts research into the utilization of 3-D, 3-C reflection seismology for improved reservoir description. These efforts are conducted in a multidisciplinary fashion, involving geological, geophysical, and petroleum engineering researchers. The Joffre Field study emphasizes use of shear waves to discriminate lithologic, porosity, and permeability changes in an area where studies based on compressional seismic data have had limited success at reservoir characterization.

Joffre Field is located in south-central Alberta, between Calgary and Edmonton (Figure 1), and produces oil from Devonian carbonates in the Western Canada Sedimentary Basin. Most of the hydrocarbons produced in Joffre Field are found in the Nisku Formation (often referred to as the D2) of the Upper Devonian (Frasnian) Winterburn Group (Figure 2). The Nisku Formation is a widespread, complex biostromal shelf carbonate and has been completely dolomitized in the Joffre area. The Joffre Field Nisku hydrocarbon trap is diagenetic and has been created by plugging of porosity with anhydrite. This occurs where the Nisku overlie the northwest flank of the Bashaw Complex (aerially large Leduc carbonate buildup). This reservoir was discovered in 1956 and has produced approximately 7 million $m^3$ (43 million stock tank barrels [MMSTB]) of oil, which is approximately 26% of the original oil in place. This 20- to 25-m-thick reservoir is in the lower Nisku and has highly variable vuggy porosity and lithologic content (mainly dolomite and anhydrite). It has two main productive units (an upper grainstone and a lower reef facies), which are separated by a mudstone interval. These two facies contain most of the field reserves. Oil recoveries from the grainstone interval have not been as high as those from the reef facies during waterflooding due to permeability differences. The geologic complexity of this reservoir, its production spacing units, and its hydrocarbon production history suggest that significant additional reserves remain to be recovered. A small, oil-productive Leduc (D3) pinnacle reef (Figure 2) is located within the study area along the edge of the Bashaw Complex below the Nisku reservoir updip limit (Figure 3). This reef has vuggy porosity and has produced several million bbl of oil.

The northeastern edge of the field was selected for the location of the 3-D, 3-C seismic survey collected by the RCP. Widely spaced well control (160 acres) defines the reservoir edge (Figure 3). This 13-$km^2$ (5-$mi^2$) survey was designed to test the imaging potential of 3-D, 3-C seismic data for the Nisku reservoir's porosity and for the oil-productive Leduc pinnacle. The survey was shot in May 1993 and is the first 3-D, 3-C seismic survey to be acquired in western Canada. From this survey, compressional, shear, and converted (shear) wave 3-D datasets were processed and interpreted. Additionally, a dipole sonic log and compressional and shear vertical seismic profiles (VSPs) were collected in the 11-22-39-26 W4 well (11-22), near the center of the survey.

## COMPRESSIONAL 3-D DATA ANALYSES

The Joffre 3-D compressional seismic data are useful for structural mapping and identification of some stratigraphic features. The P-wave data analyzed were migrated, and a small phase rotation was required to match the VSP corridor stack from the 11-22 well. Crossline 32, extracted from the compressional 3-D dataset, shows several key events, including the productive Leduc reef (Figure 4). The upper 1.25 seconds of data are dominantly in Cretaceous clastic rocks and the Upper Devonian carbonate section lies at and below the D1 (Wabamun) event. The location of the Leduc pinnacle reef can be readily identified by the structural drape over its edges, a velocity pull-up below, and amplitude increases of some reflections over it. The Nisku P-wave reflection is generally weak and does not discriminate between porous and tight zones.

A time structure map on the Blairmore coal event (Figure 5) identifies the northeast-oriented edge of the Bashaw Complex, as well as three reefs attached to it. The largest reef, in the center of the survey, is oil productive. The southwesternmost of the two smaller features sitting on the Bashaw edge was drilled (5-23 well) and was tight. Two northeast-oriented linear features in the western side of the survey are fault/fracture zones that locally enhance Nisku reservoir permeability. These features, which had not been previously mapped, are aligned parallel to the present direction of maximum horizontal stress (Bell and Gough, 1979; Zoback, 1992). Wells lying along these features exhibit oil recoveries higher than the field average. Additionally, an S-shaped feature is present on the eastern side of the structure map. A time isochron map from the Second White Speckled shale to the Blairmore coal events (Figure 6) identifies these northeast-oriented fault/fracture zones, which show up as thicker intervals. This suggests that these faults were reactivated during deposition of this sequence. This map also shows that the S-shaped feature is a narrow (~250 m), meandering coal-filled incised channel.

## MULTICOMPONENT 3-D DATA ANALYSES

It is well documented that as shear waves encounter azimuthally anisotropic media, they are split and polarized into fast (S1) and slow (S2) shear waves (Martin and Davis, 1987). The resulting shear wave particle motion will be oriented parallel (S1) and perpendicular (S2) to the direction of maximum horizontal stress (Figure 7). The mechanism for this azimuthal velocity anisotropy is commonly vertical fractures opened by differential horizontal stresses. More recently, it has been recognized that pore structure (elongate pore shapes) can also contribute to this phenomenon (Crampin and Zatsepin, 1995). Analysis of the polarization directions of the split shear waves is critical during processing and interpretation. The shear wave data must be rotated into their natural polarization directions for optimum imaging. Rotation analysis of the Joffre shear wave data yields an average N45°E S1 polarization direction, which is consistent with the present orientation of the maximum horizontal stress field in this area (Zoback, 1992). This trend is also observed in Nisku reservoir production data, indicating higher northeast-southwest directional permeability.

*Colorado School of Mines Dept. of Geophysics, 1500 Illinois St., Golden, CO 80421

The shear wave data volumes were migrated and phase rotated to match the 11-22 VSP corridor stacks, which were verified with synthetic seismograms from the dipole sonic log data. The VSP ties to the P, S1, and S2 surface datasets were used to provide reliable identification of surface data reflection events. This step is crucial for interval work involving more than one data volume, such as velocity (*Vp/Vs;* where *Vp* = compressional wave velocity and *Vs* = shear wave velocity) ratio and anisotropy calculations.

## LEDUC REEFS

A fast (S1) shear wave line (crossline 41) extracted from the 3-D data volume is shown in Figure 8. The producing Leduc reef, present in the middle of this section, is imaged in the 3-D shear wave data. A time slice at 2.772 seconds from the S1 data successfully images the positions of the productive Leduc reef (crossline 40, inline 70), the smaller tight reef (crossline 18, inline 100), and the Bashaw Complex edge (Figure 9). These features can be located on the compressional wave Blairmore coal time structure map (Figure 5) but are not imaged in the compressional data.

Ratios of compressional velocity (*Vp*) to shear velocity (*Vs*) were shown to be extremely powerful for lithologic discrimination by Pickett (1963). Petrophysical determinations of velocity ratios from core measurements and wireline logs in the Nisku and Leduc formations show a strong dependency on porosity, suggesting that *Vp/Vs* ratio mapping should be an effective tool for porosity discrimination in these vuggy dolomites (Sarmiento, 1994; Arestad, 1995). *Vp/Vs* ratios can be computed utilizing interval traveltimes from both the compressional (p) and the shear (s) wave datasets, provided that the P-wave and S-wave reflections originate from identical horizons. Velocity ratios for the 3-D, 3-C seismic data are computed using the methodology discussed by Tatham and McCormack (1991). In order to assess whether velocity ratio mapping can identify Leduc porosity, a *Vp/Vs* map was computed for the Nisku to Cooking Lake interval, which brackets the Ireton and Leduc formations (Figure 10). The map shows higher *Vp/Vs* ratios over the productive portion of the central reef and lower velocity ratios over the tight reef. This map also shows an area of higher *Vp/Vs* values along the Bashaw Complex edge, which may contain porosity. This illustrates how velocity ratio mapping can be used to verify porosity within these reef targets. The P-wave data can identify these reefs but do not discriminate porous from tight reefs. These results could be significant for Leduc exploration, as well as in other carbonate reef or mound plays, such as the Lodgepole Formation (Williston Basin), Chappel Formation (Hardeman Basin), Silurian reefs (Michigan Basin), algal mounds in the Paradox Basin, and many others.

## NISKU RESERVOIR

Shear wave seismic characterization of the Joffre Nisku reservoir incorporated amplitude, velocity ratio, and anisotropy analyses. Examination of the fast (S1) shear wave crossline 41 section (Figure 8) reveals that the Nisku event amplitude is stronger to the west of the Leduc reef where the D2 is productive, which is an effect not observed on the compressional data. Amplitude mapping of the Nisku event on both the fast (S1) and slow (S2) shear wave datasets shows similar features. The higher S1 amplitudes correlate spatially with better producing areas within the field, as depicted on the map showing S1 amplitudes overlain by Nisku cumulative oil production (Figure 11). An exception to this observation is the undrilled broad high-amplitude feature centered at inline 90 and crossline 80. Only the edge of this feature was tested by the 12-26-39-26 W4 well, which had minor Nisku porosity and produced 12 thousand stock tank barrels (MSTB) of oil. The complex nature of this map is consistent with the known rapid lateral and vertical geologic changes within the Nisku reservoir. Amplitude mapping of the Blairmore coal event in the shear wave data also shows the fault/fracture zones seen in the compressional data analysis.

Velocity ratio mapping of the Nisku reservoir interval was conducted using the methodology described previously. The Wabamun (D1) and mid-Ireton reflections were selected for this analysis because they are the strongest continuous events which narrowly bracket the Nisku interval. The resulting *Vp/Vs* ratio map (computed from the S1 and P datasets), overlain by mean Nisku reservoir core porosity contours, is depicted in Figure 12. The *Vp/Vs* ratio near the 11-22 well is approximately 2.0, and the ratio computed from the 11-22 VSP traveltimes is 1.95. Higher velocity ratios are present over the western (oil-productive) part of the survey, and the field edge can be identified by *Vp/Vs* ratio changes. The contours of mean Nisku reservoir core porosities show good agreement with the mapped velocity ratios. Due to the large number of cored wells and the lack of modern wireline logs in the Joffre D2 pool, porosity data used in this study are based on whole core measurements. Mapping of the degree (%) of anhydrite plugging in cores of the lowermost vuggy Nisku reservoir reef facies shows less plugging (i.e., higher porosity) over the higher *Vp/Vs* areas of the survey. Geostatistical analyses exhibit good linear correlation between *Vp/Vs* ratio and mean Nisku core porosities, implying that these velocity ratio maps can be used to map porosity between wells (Cabrera-Garzón, 1995, personal communications). The porosity contours on this map were prepared independently of the velocity ratio analysis. Based on the linear relationship between porosity and *Vp/Vs*, revised porosity contours will follow the velocity ratio map features. The edge of the Nisku reservoir as determined from well control (see Figure 3) is somewhat different from that derived from the *Vp/Vs* map, suggesting that there is potential for field extensions. In order to get a good pressure match in the reservoir simulation, additional reservoir volume had to be added in the area east of the Leduc reef, beyond the limits of the field edge defined by well control, which also supports the field extension potential identified by the velocity ratio mapping.

Although the Nisku reservoir represents about 10% of the interval being measured, the correlation of mapped velocity ratios with reservoir porosities indicates that the higher ratios found in the oil-productive area are related to porosity variations. Petrophysical analysis of wireline log data showed that the dolomite-anhydrite lithologies of this interval vary somewhat across the survey area, but not enough to produce the observed results (Sarmiento, 1994). Pore structure within the vuggy and moldic Nisku reservoir controls the changes of *Vp/Vs* across the survey. Pores with low aspect ratios (i.e., elongate pore shapes) strongly influence this increase in velocity ratios. The porosity of this interval is low (mean porosity of 4–5%) but is the largest contribution to the observed velocity ratio differences. These observations point out the robust nature of velocity ratio measurements for porosity determination in vuggy carbonates.

Local variations of fast (S1) and slow (S2) shear wave traveltimes can be used to map spatial changes of polarization angles, which indicate local stress field or reservoir property variations. Anisotropy is a measure of these changes and can be calculated over an interval using S1 and S2 interval traveltime differences (Lewis et al., 1991). At Joffre, the shear wave data were processed with an N45°E S1 polarization angle; thus, positive anisotropy values indicate northeast-southwest S1 polarization angles and negative values indicate north-west-southeast S1 directions for the interval being analyzed.

Since elongate pore shapes can influence shear wave splitting, changes of anisotropy within a carbonate sequence can indicate the presence of moldic and vuggy porosity, as well as their orientations. The anisotropy map computed over the same interval used in the *Vp/Vs* analysis (Wabamun to mid-Ireton) shows a spatial anisotropy distribution that matches the mean D2 reservoir porosity contours (Figure 13). Due to good correlations between porosity and permeabilities in the Nisku core data, permeability contour maps have the same shape as the porosity contour maps used on this figure. Since the anisotropies are variably negative along the higher-porosity trend at the updip limit of the field, the Nisku pore shapes should be oriented northwest-southeast. Such pore shape orientations might be expected along the diagenetic edge of the D2 reservoir. These variations of pore orientation also influence permeability anisotropy with the reservoir. In order to get a reservoir simulation match, the direction of maximum permeability within the reservoir model had to be modified towards the northwest along the updip field limits, which is consistent with the shear wave anisotropies. This points out the potential of this tool for exploration and reservoir characterization within carbonate rocks.

## INTEGRATED RESERVOIR CHARACTERIZATION

Too often in research efforts, specific findings are published on a stand-alone basis, without the results' being applied to the big picture. The trend towards specialization in this highly technical world often encourages this. The results of applied multidisciplinary research must be integrated together to provide a better understanding of the importance of the individual results and, in the case of reservoir studies, to develop additional recoverable reserves.

The Joffre Nisku reservoir is compartmentalized to the extent that hydrocarbon recoveries are lower than other Nisku fields in this area. With the known 180 million bbl of oil in place within this reservoir, an increase in recovery of 10% is very significant (18 MMSTB). Several key elements were brought together in order to characterize the northeastern end of this field. When these new data are incorporated into a reservoir simulation model, their importance can be evaluated. Results of the analysis of the 3-D, 3-C seismic survey, combined with geologic and reservoir engineering studies, are incorporated into this model.

Significant reserves remain to be recovered from the grainstone facies of this reservoir, and an area of thicker grainstone is found in the center of the 3-D, 3-C seismic survey. The amount of porosity is a very important factor influencing hydrocarbon recoveries in carbonate reservoirs. The *Vp/Vs* mapping has identified areas of better porosity within the field as well as outlined potential field extensions. The compressional seismic volume shows evidence of local northeast-southwest–oriented fault/fracture zones that influence reservoir fluid flow movement. These zones, near good producing wells, must be accounted for in the design of an effective waterflood program. In some cases, these fracture zones may act as conduits for injected water and should be avoided by new production wells. Analysis of the production database indicates a preferential permeability direction oriented northeast to southwest across most of the field. Anisotropy and production data show local changes towards the northwest along the updip field limits. By integrating these individual findings, the resulting reservoir description is much improved and can be used to optimize future production.

As an example, Figure 14 depicts how $Vp/Vs$ mapping can be incorporated with fracture zone maps to identify potential targets within and outside the present limits of Joffre Field. By combining this information with facies distribution and residual hydrocarbon mapping, three potential infill locations were identified with favorable porosity and grainstone thicknesses. One of these locations is near a seismically identified fracture zone. These locations target grainstone facies reserves with horizontal wells. This type of integrated model is a great improvement over past reservoir models and should be utilized to maximize enhanced recovery efforts in this field.

## CONCLUSIONS

The compressional 3-D dataset from the Joffre 3-D, 3-C survey identifies several of the key structural and stratigraphic features of the northeastern updip edge of Joffre Field, such as the Bashaw Complex edge, several Leduc reefs, and northeast-aligned fault/fracture zones. These previously unmapped fracture zones locally enhance the production characteristics of the Nisku dolomite reservoir. The compressional seismic data have been ineffective for detection of the Joffre Nisku reservoir.

Shear wave 3-D data image productive and tight Leduc reefs in the survey area. $Vp/Vs$ mapping of the reef interval shows higher velocity ratios over the productive reef and lower ratios over the tight reefs. These results show that $Vp/Vs$ mapping can be used to identify porosity in carbonate buildups.

Shear wave Nisku reflection amplitudes from the S1 data correlate with reservoir properties. $Vp/Vs$ ratios over the Wabamun to mid-Ireton interval are higher over the productive Nisku reservoir. This is a direct indicator of Nisku reservoir porosity. The velocity ratio mapping has identified both field extension and field redevelopment potential. Anisotropy mapping of the shear wave datasets shows negative areas or northwest-southeast directionality along the updip limit of the Nisku reservoir. This effect is due to elongation of vuggy Nisku pores towards the northwest and influences reservoir permeability anisotropy along the edge of the field. Both $Vp/Vs$ and anisotropy measurements are powerful description tools in this reservoir.

Three-dimensional, three-component seismic technology successfully characterizes the spatial variability of porosity and permeability within the Nisku complex dolomite, which has significant lateral reservoir heterogeneity. Mapping of the spatial distribution of reservoir properties and compartments, when utilized in an integrated fashion with geological and reservoir engineering data, identifies additional recoverable reserves. This technology can contribute towards improved hydrocarbon recovery and exploitation of this and many other types of reservoirs through improved imaging of lateral reservoir heterogeneities.

### *Acknowledgments*

The authors would like to thank the other members of the Joffre research team (Abdul Al-Awar, Karen Bard, Raul Cabrera-Garzón, Bruce Mattocks, Max Rolla, and Victor Sarmiento) for their enthusiastic integrated multidisciplinary research. We would like to express our appreciation for direction, encouragement, and financial support for this research to the RCP–Phase V industry partners.

## REFERENCES CITED

Arestad, J. F., 1995, An integrated multicomponent three–dimensional seismic characterization of Joffre Field, Alberta, Canada: Ph.D. dissertation T-4762, Colorado School of Mines, Golden.

Bell, J. S., and D. I. Gough, 1979, Northeast-southwest compressive stress in Alberta: evidence from oil wells: Earth and Planetary Science Letters, v. 45, p. 475–482.

Crampin, S., and S. V. Zatsepin, 1995, A new understanding of shear-wave splitting: 57th EAEG/EAPG Conference Extended Abstracts, v. 1, p. C041.

Lewis, C., T. L. Davis, and D. Vuillermoz, 1991, Three-dimensional multicomponent imaging of reservoir heterogeneity, Silo Field, Wyoming: Geophysics, v. 56, n. 12, p. 2048–2056.

Martin, M. A., and T. L. Davis, 1987, Shear wave birefringence: a new tool for evaluating fractured reservoirs: The Leading Edge, v. 6, n. 10, p. 22–28.

Pickett, G. R., 1963, Acoustic character logs and their application in formation evaluation, Journal of Petroleum Technology, v. 15, p. 650–677.

Sarmiento, V., 1994, Petrophysical relationships from wireline logs for seismic calibration of the Devonian Nisku and Wabamun formations, Joffre Field, Alberta: M. S. thesis T-4593, Colorado School of Mines, Golden.

Tatham, R. H., and McCormack, M. D., 1991, Multicomponent seismology in petroleum exploration, investigations in geophysics, v. 6: Society of Exploration Geophysicists, Tulsa, Oklahoma.

Watts, N. R., 1987, Carbonate sedimentology and depositional history of the Nisku Formation (within the Western Canadian Sedimentary Basin) in South Central Alberta: CSPG Second International Symposium on the Devonian System, p. 87–152.

Zoback, M. L., 1992, First- and second-order patterns of stress in the lithosphere: The world stress map project, J. Geophys. Res., p. 11,703–11,728.

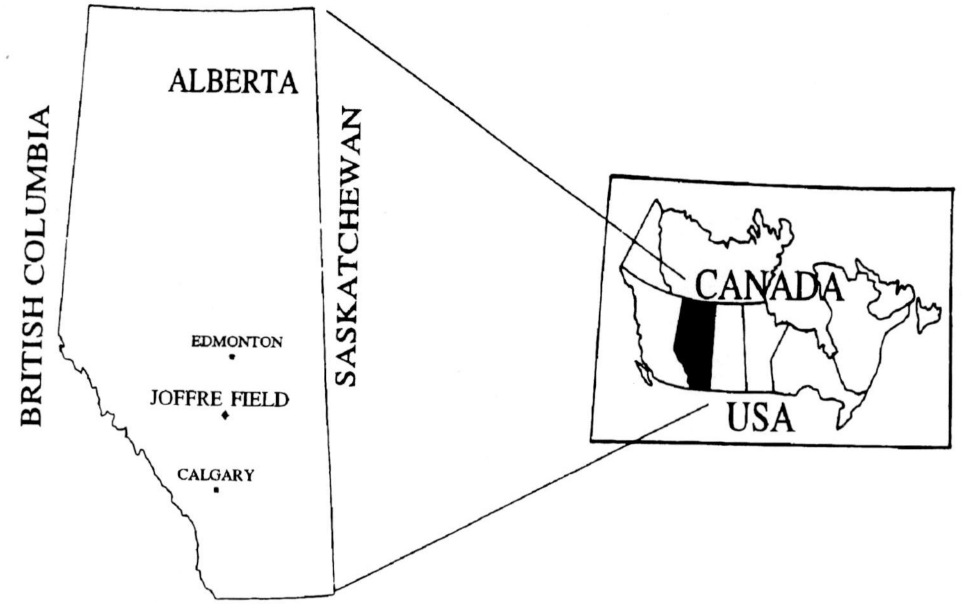


FIGURE 1. Location of Joffre Field (adapted from Sarmiento, 1994).

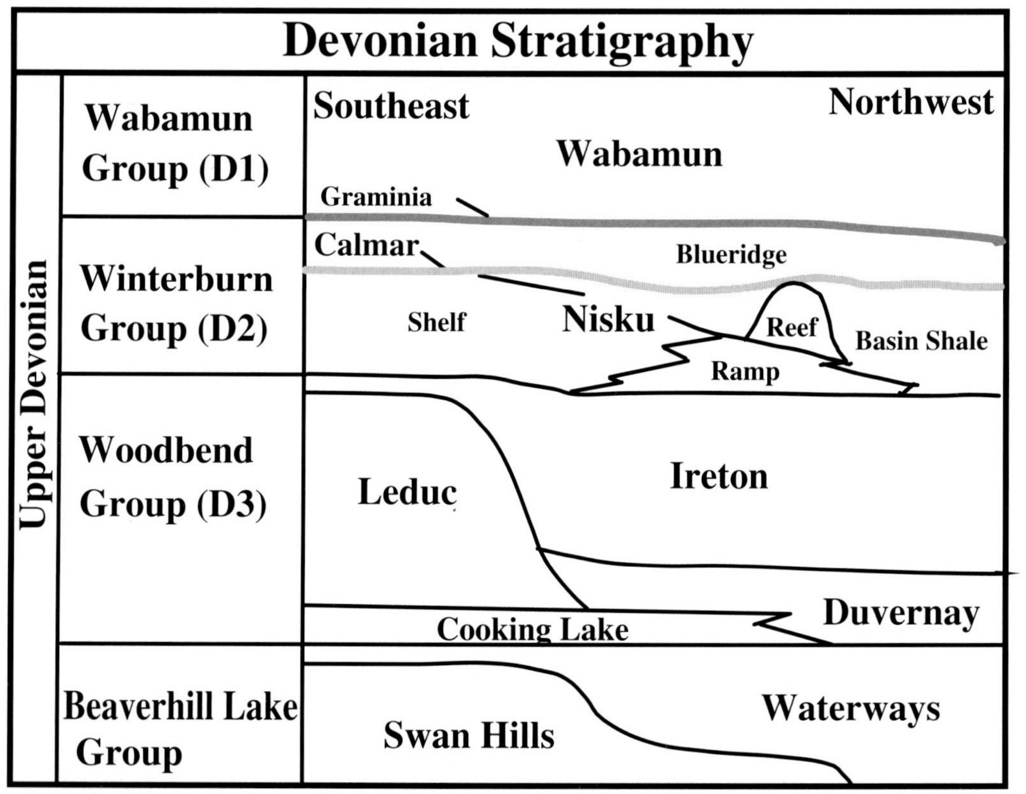


FIGURE 2. Upper Devonian stratigraphic column for the Western Canada Sedimentary Basin (adapted from Watts, 1987). Regional stratigraphic changes are depicted from the Upper Devonian basin edge to the southeast of Joffre towards the basin center to the northwest. Joffre Field lies along the edge of the Leduc carbonate buildup.

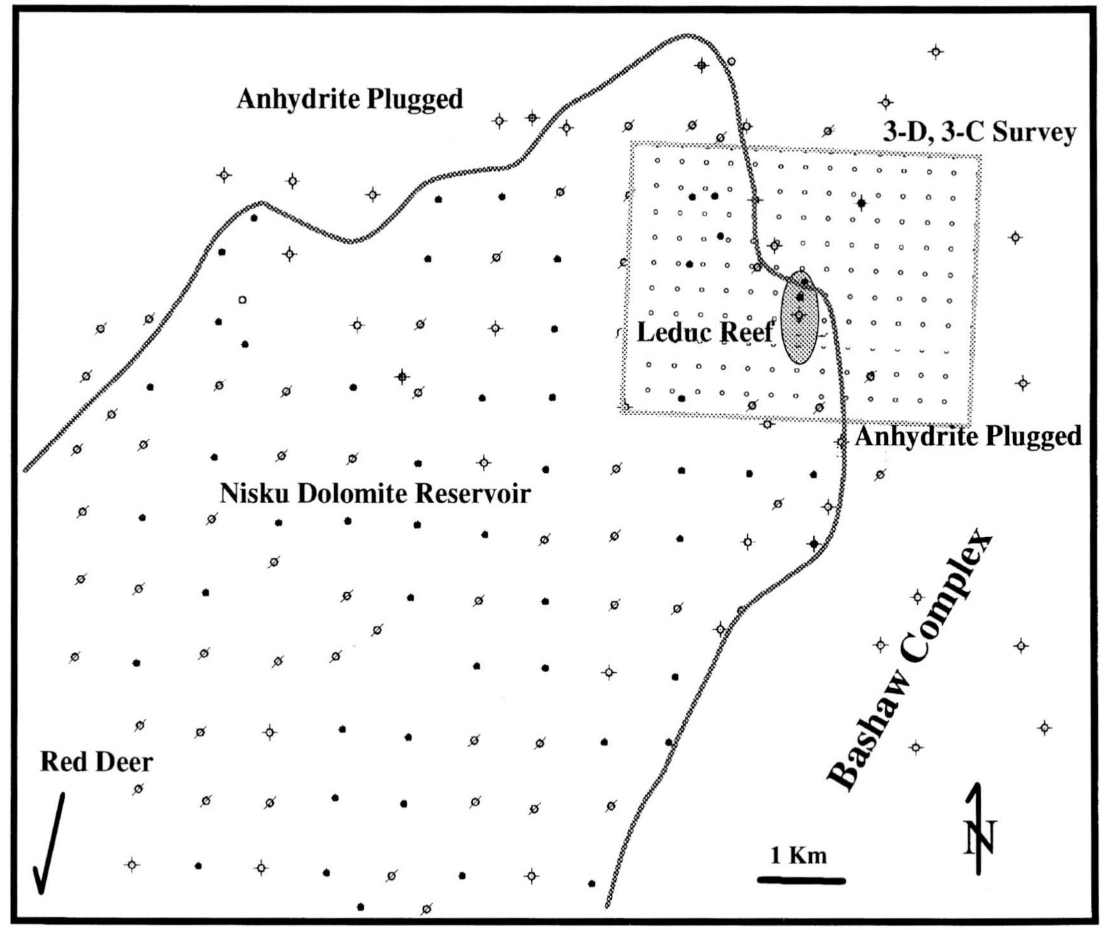


FIGURE 3. Joffre 3-C, 3-D seismic survey location map with well locations. The Nisku (D2) reservoir edge, defined by well control, is indicated by the line between anhydrite and dolomite. Key features include the Bashaw Complex edge and the oil productive Leduc reef.

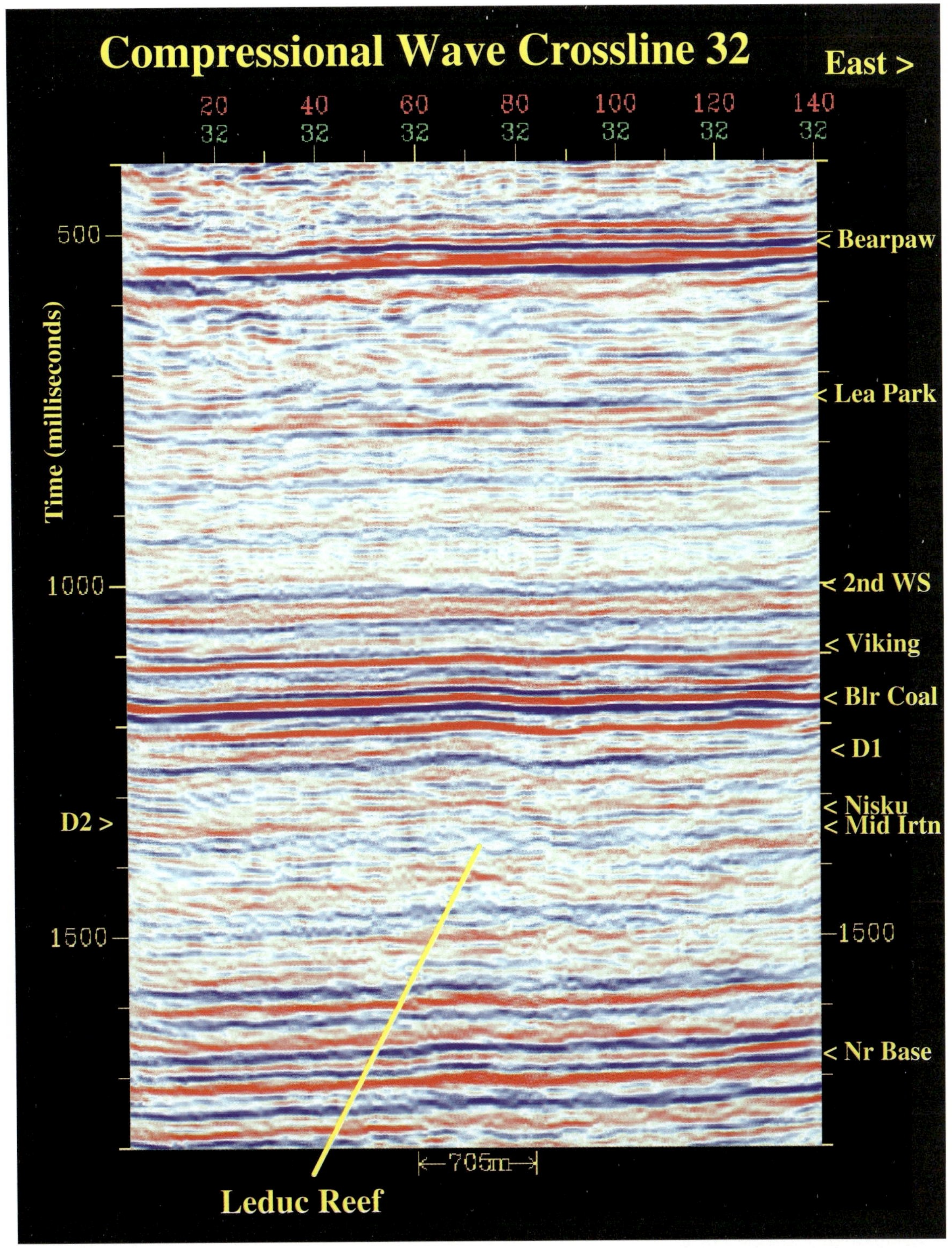


FIGURE 4. P-wave crossline 32 section (west to east), extracted from the Joffre compressional 3-D seismic data volume. Key events are annotated, including the Nisku and Leduc events.

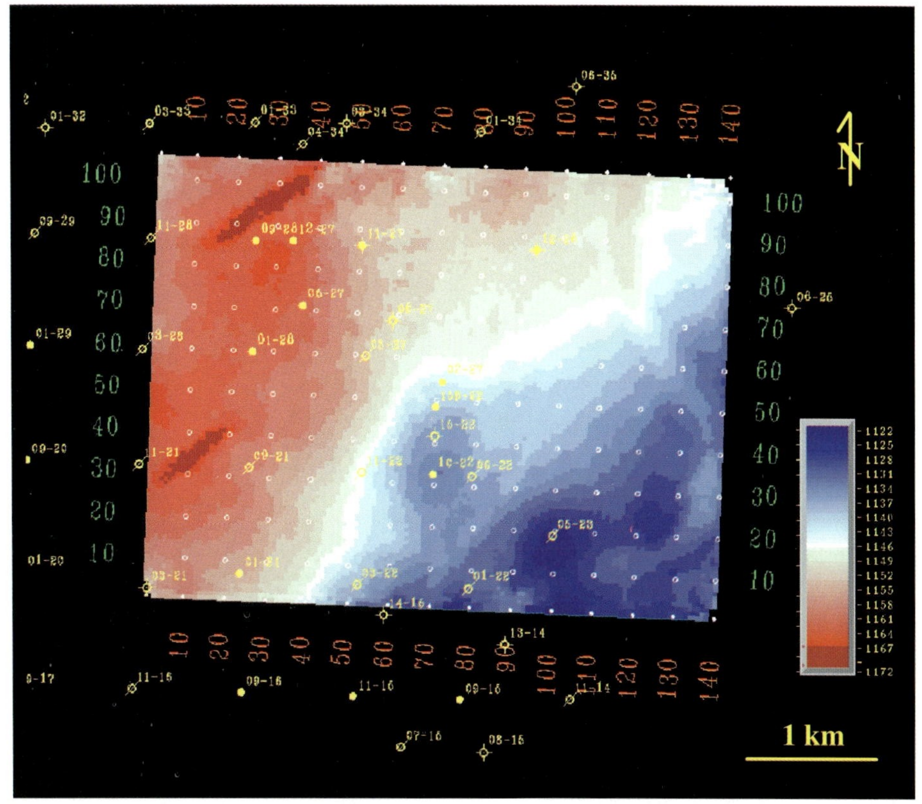


FIGURE 5. Compressional 3-D Blairmore coal time structure map. Shallower times in blue.

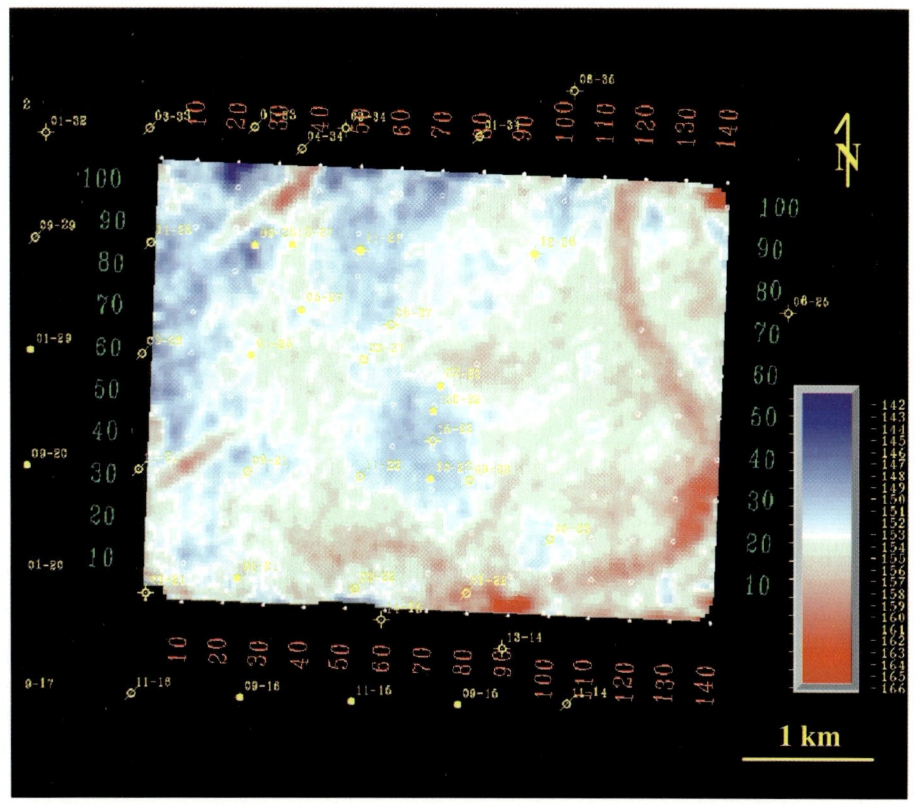


FIGURE 6. Compressional 3-D time interval isochron map from the Second White Speckled shale to the Blairmore coal events. Thicker intervals in red.

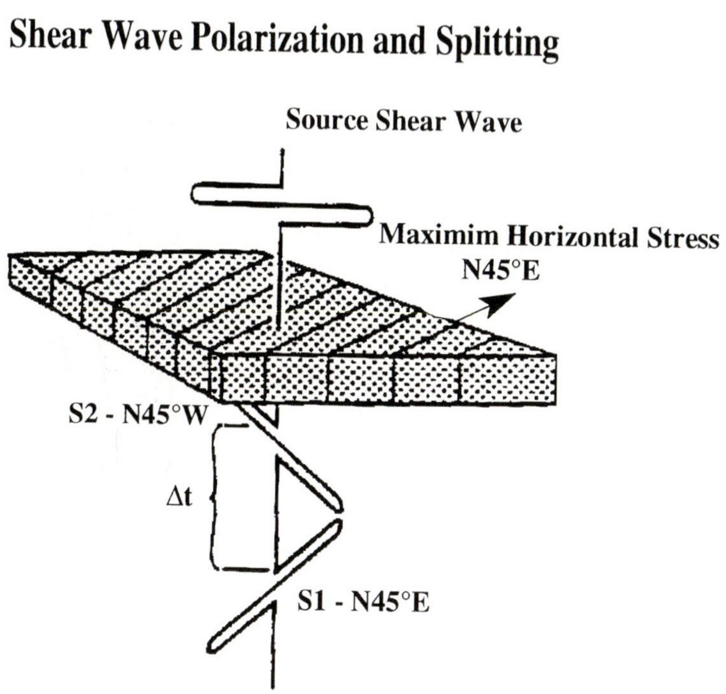


FIGURE 7. Shear wave splitting and polarization schematic. Fast (S1) shear wave is polarized with particle motion parallel to the direction of maximum horizontal stress, which is approximately N45°E in the Joffre Field area.

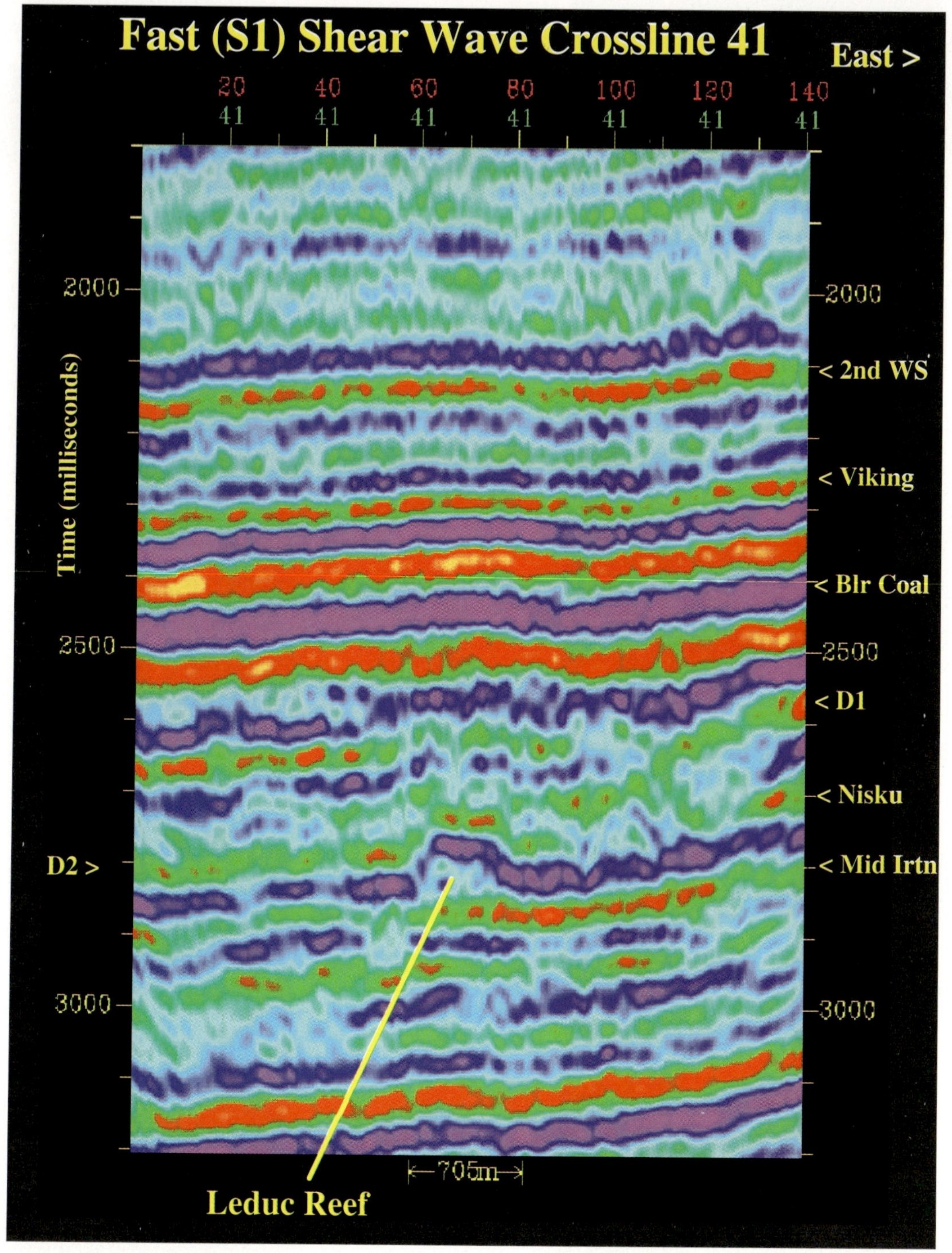


FIGURE 8. Crossline 41 section (west to east), extracted from the Joffre fast (S1) shear wave 3-D data volume with key events annotated. Note that the Leduc reef is well imaged and that the Nisku event has higher amplitudes over the Joffre Nisku reservoir.

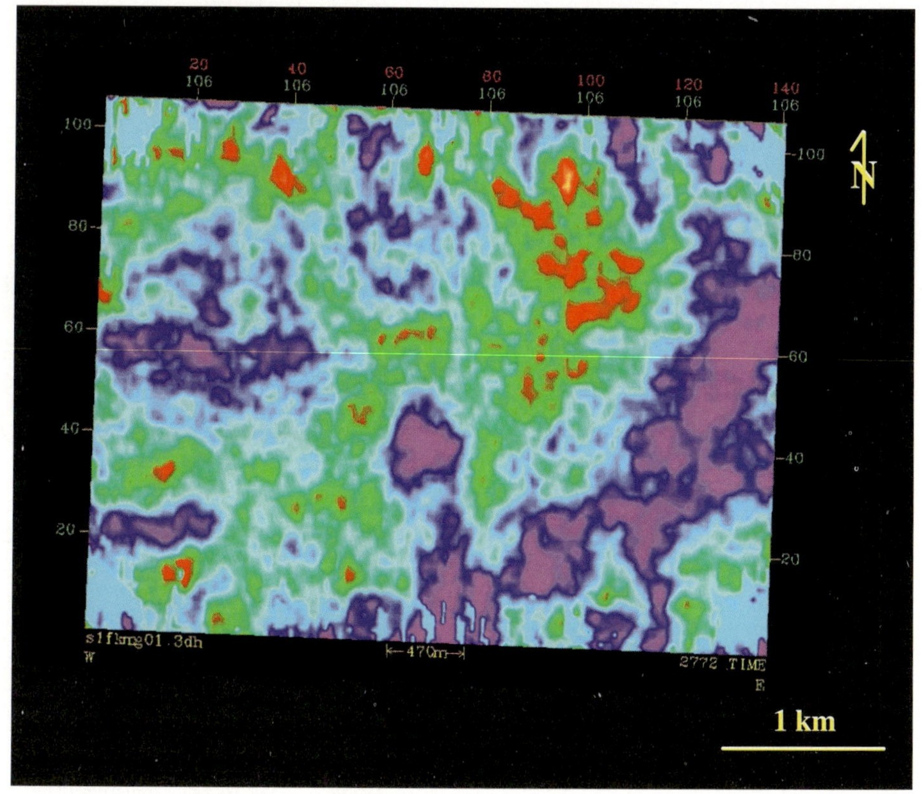


FIGURE 9. Time slice at 2.772 seconds for the fast (S1) shear wave data volume, showing the location of the oil-productive Leduc pinnacle reef and the edge of the Bashaw Complex.

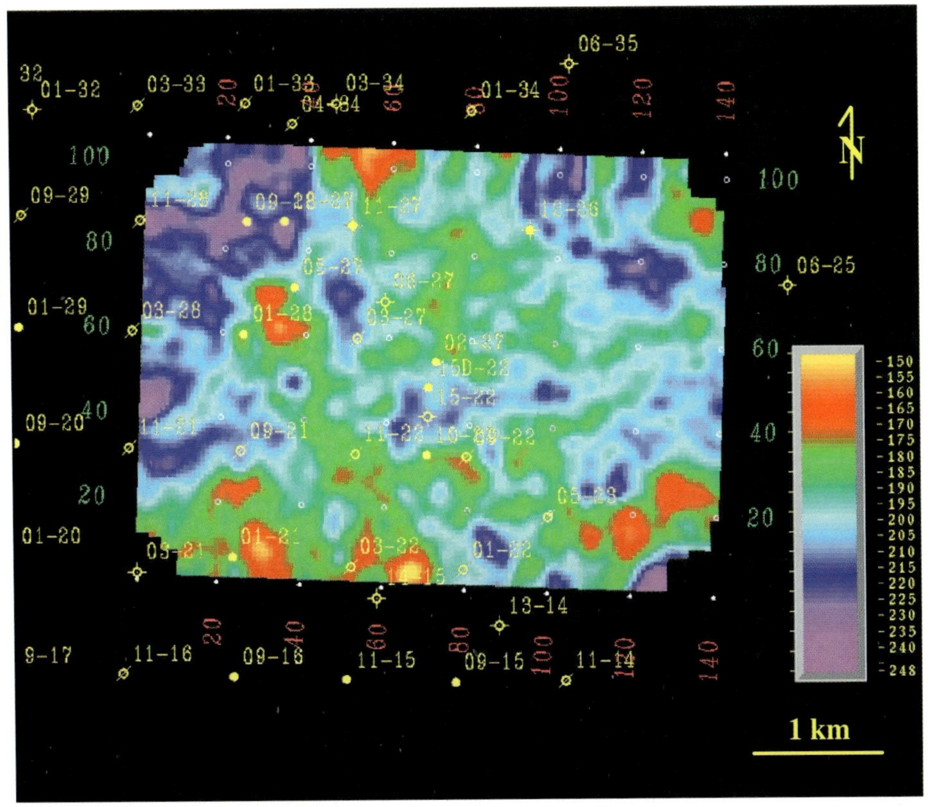


FIGURE 10. Velocity ratio map of the Nisku (D2) to Cooking Lake interval (fast [S1] shear wave data). Data are shown as *Vp*/*Vs* times 100 for display purposes. Blue and magenta indicate higher *Vp*/*Vs* ratios. Note higher *Vp*/*Vs* ratios over productive reef and lower ratios over tight reef.

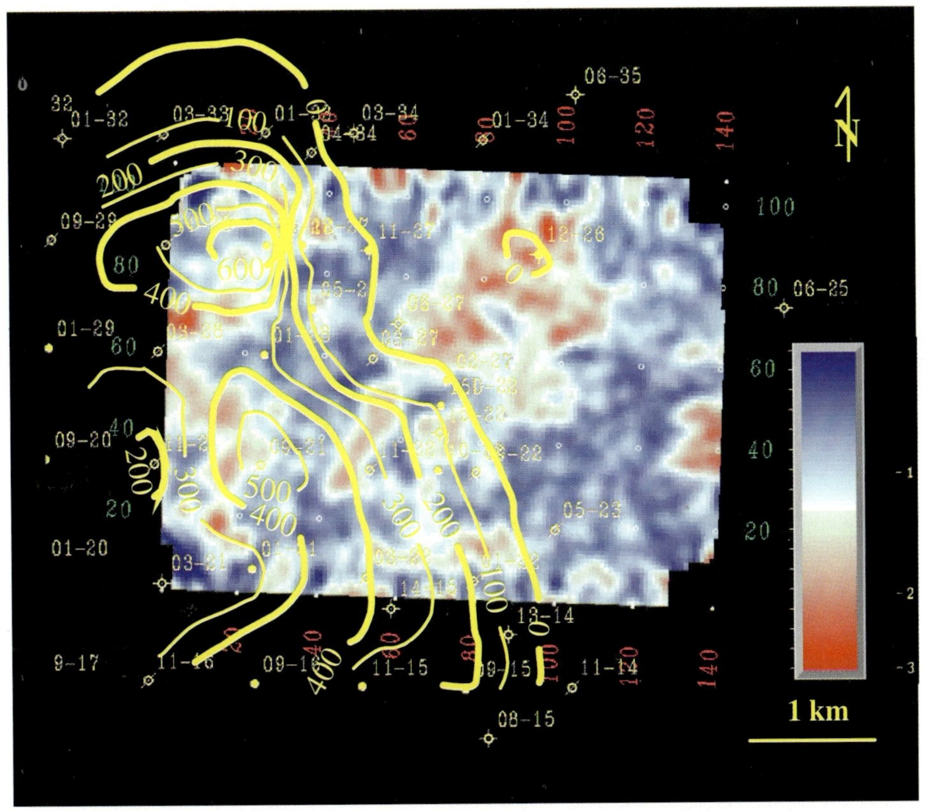


FIGURE 11. Fast (S1) shear wave amplitude map of the Nisku (D2) event, with cumulative Nisku oil-production contours (thousands of barrels). Higher amplitudes in red.

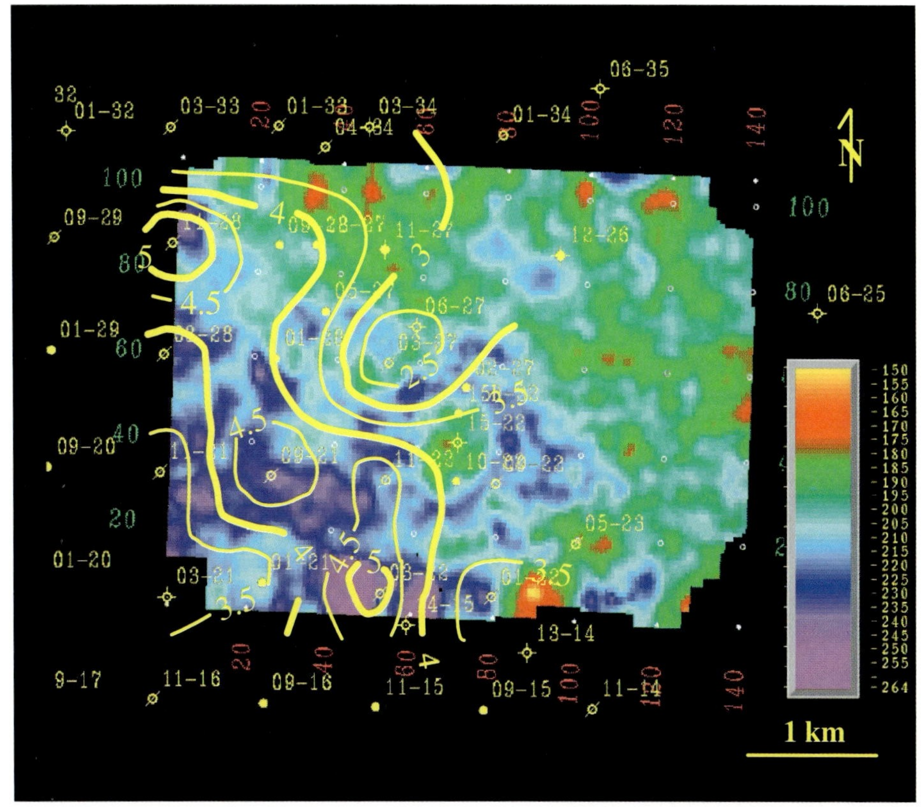


FIGURE 12. Velocity ratio map of the Wabamun (D1) to mid-Ireton interval (fast [S1] shear wave data), with mean Nisku reservoir core porosity contours (%). Data are shown as *Vp*/*Vs* times 100 for display purposes. Blue and magenta colors indicate higher *Vp*/*Vs* ratios. Note higher *Vp*/*Vs* ratios over the Nisku reservoir and areas of higher reservoir porosity.

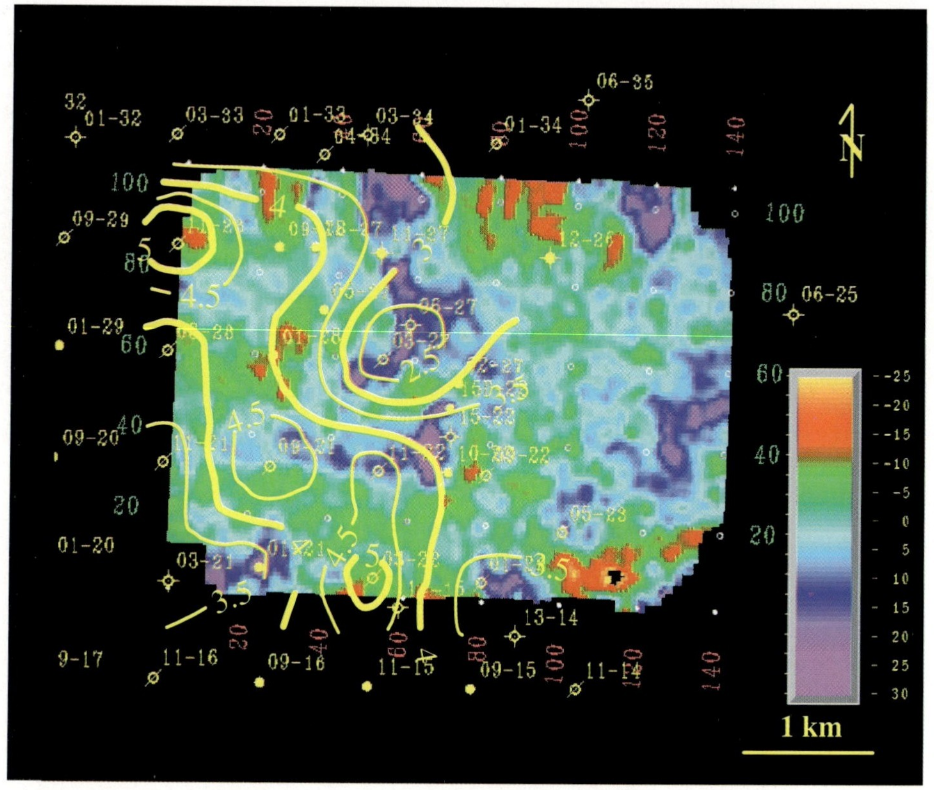


FIGURE 13. Anisotropy (%) map for the Wabamun (D1) to mid-Ireton interval, with mean Nisku reservoir core porosity contours (%). Positive anisotropies are in blue and magenta colors. Note negative anisotropies along the updip Nisku reservoir edge.

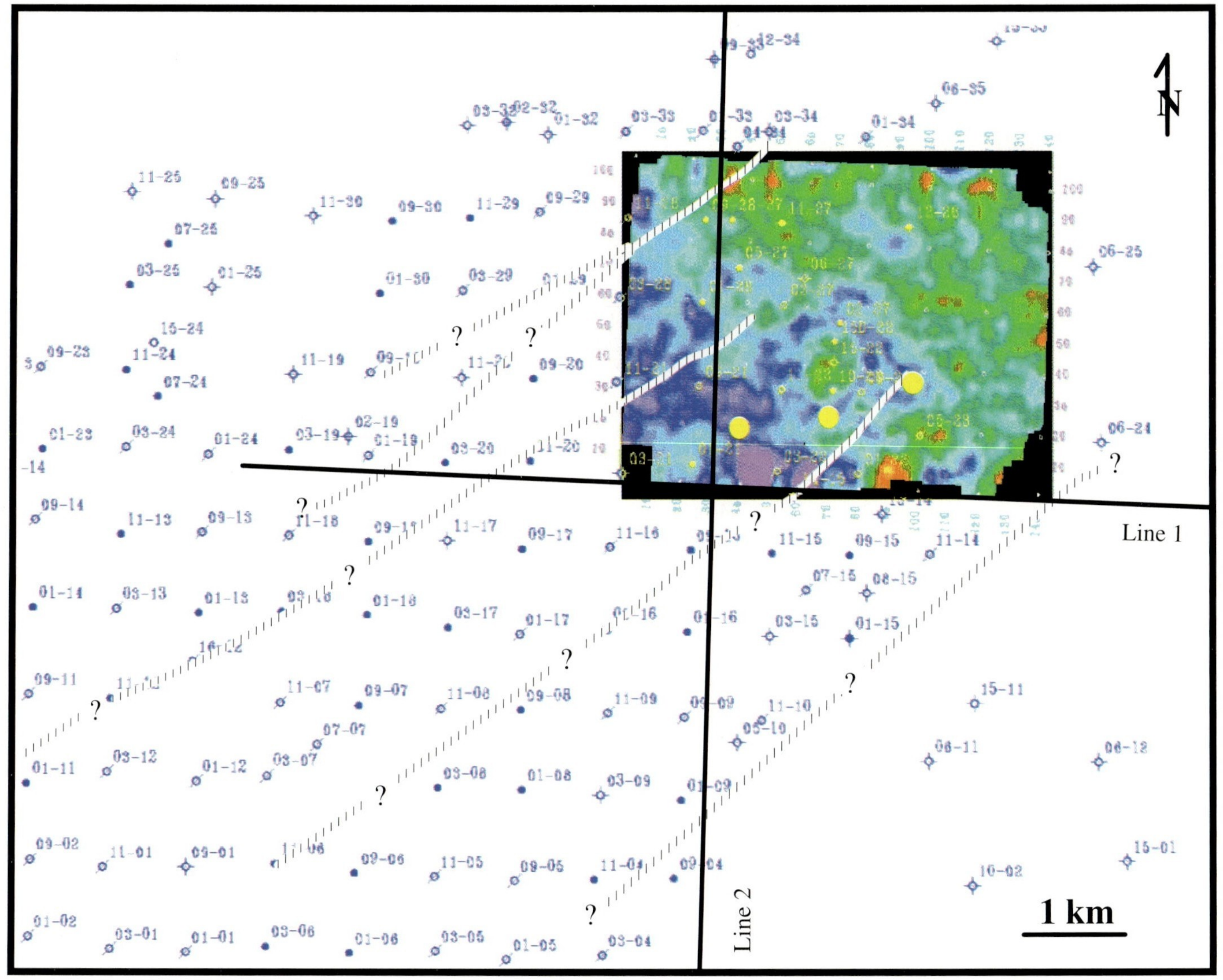


FIGURE 14. Integrated reservoir map incorporating the velocity ratio map (depicting Nisku reservoir porosities) with fault/fracture zones. Blue and magenta colors indicate areas of better reservoir porosity. Yellow dots identify potential drilling targets.

# CHAPTER 19

Uffen, J. D., Swan Hills Unit #1: Adding value with seismic data through reservoir delineation and characterization, *in* P. Weimer and T. L. Davis, eds., *AAPG Studies in Geology No. 42* and *SEG Geophysical Developments Series No. 5*, AAPG/SEG, Tulsa, p. 179–188.

# Swan Hills Unit #1: Adding Value with Seismic Data Through Reservoir Delineation and Characterization

J. Douglas Uffen*

## ABSTRACT

An intensive geological study by the operator of the production Unit suggested that clean, unswept reef edge oil could exist in the reservoir. 3-D seismic data were acquired to delineate the reef edge to mitigate drilling risk associated with tapping into these potential reserves. Environmental concerns resulted in modifications to the original program design. Data quality exceeded expectations. Lithologic model-based seismic inversion was conducted to enhance confidence in the location of the reef edge and to investigate whether porosity predictions were possible. Log balancing, pattern recognition event picking, and isovelocity mapping of the reef and platform section were technological innovations required to address this mandate. Reef edge predictions and porosity predictions appeared trustworthy with resolution limits stretched to ⅙th of a wavelength. The seismic program proved to add value by mitigating drilling risk for two horizontal wells and one vertical well. The drilling results reaffirmed predictions, and unit production increased.

## INTRODUCTION

The town of Swan Hills is located approximately 200 km northwest of Edmonton, Alberta, Canada (Figure 1). The townsite is situated on the southeast side of the Swan Hills Unit #1 production Unit. The Unit is located along the trend of several longstanding producing fields from the Swan Hills Formation of the Beaverhill Lake Group. Home Oil Company Limited of Calgary has been the operator of Swan Hills Unit #1 since its creation and held an 18.32% working interest in the unitized lands. Through a corporate acquisition in September 1995, the Unit is now operated by Anderson Exploration Ltd. The operators of neighboring units along with Home Oil make up the membership of the Swan Hills Synergy Group. The Synergy Group (Figure 1) joined forces to lower operating costs for these fields through cooperative efforts that also included field training and emergency response plans for their respective Unit partners. At the time this paper was written, all companies noted in Figure 1, with the exception of Shell, were members of Swan Hills Unit #1.

The field was discovered in March of 1957, just two months after the discovery of Virginia Hills. The field was unitized in the Beaverhill Lake zone in 1963. The Unit boundary was subsequently revised in 1983 and comprises 44,160 acres (69 mi$^2$ or 176.6 km$^2$) or approximately two townships of land (Figure 1). The unitized lands are estimated conservatively to contain 1413 million bbl oil in place. Cumulative production through October of 1993 was 540 million bbl, with daily oil production rates of 27,000 bbl/day, 37 Mcf/day of solution and breakthrough gas, and 130,000 bbl/day of water. Watercuts vary by geologic zone and geographic area within the reef. The average field watercut at the time of this study was 86%. Waterflooding started as early as 1965, eight years after the initial discovery. Shown in Figure 2 is the geologically defined reef edge of Swan Hills Unit #1. The reef, as shown, is divided into four phases of enhanced recovery. Phases 1 and 2 commenced miscible flooding with an ethane-rich solvent in 1985 and 1988, respectively, while phases 3 and 4 were being waterflooded. The water and gas production from Swan Hills Unit #1 is a result of the miscible flooding and water injection plans. The majority of the water is recycled and re-injected.

## GEOLOGY

The producing zone for the Unit is the Mid-Devonian Swan Hills reef and the reefal platform. Drilling depths vary across the unit from 7700 to 8000 ft (2350 to 2440 m), with a topographic elevation averaging 3500 ft (1067 m). During 1990, Home Oil initiated a computer-based geologic reservoir modeling study consisting of information from 485 wells to understand more fully the reservoir architecture of the field. Of the 485 wells used, 360 were reefal penetrations within the unit. The study took two years to complete. It involved digitizing logs for all the wells, correcting logs of varying vintages, reviewing historical core descriptions, and mapping the data.

The reef and platform are subdivided into different layers or stages and mapped accordingly. The platform consists of six layers, progressively named from bottom to top—the P1 to P6. The reef is subdivided from its base to full build-up into eight layers, namely the B1, B2A, B2B, B3A, B3B, B4, B5, and B6 (Figure 3). Porosity is preferentially developed in the reef, although porosities up to 7% are possible in the platform in some geographic locations. Some reefal layers possess "hot streaks" of high-grade porosity in excess of 18% over local areas. In some local geographic areas, vertical impermeable barriers exist that disappear laterally. This heterogeneity complicates the design and implementation of waterfloods, miscible floods, and well workovers as the prediction of fluid type at varying levels becomes critical. The most homogenous layer, and the one which contains the largest share of the reserves proportionately, is the B2A. The B2A layer is approximately 20 ft (6 m) thick and averages a fairly consistent 8% porosity. Proportionately, the B2A buildup is much thinner than full reefal buildup of all eight layers, which can reach over 225 ft (69 m).

The resultant geologic perspective enhanced the knowledge of the reservoir and identified further opportunities. The geologic study identified potential for "attic" oil that could be structurally trapped and not yet swept, and for untapped oil along the reef edge which would also constitute untapped reserves. The emphasis of this paper documents only the pursuit of possible low watercut oil along the reef edge, because it was for this purpose that 3-D seismic imaging was investigated successfully and deemed to add value.

The northeast side of the reef posed the greatest likelihood for clean reef edge oil, because it was in the structurally highest position. It was in this area that the 3-D seismic program was conducted. Depicted in Figure 4 is a portion of the Swan Hills reef isopach map from the geological study for the northeast side of the reef. The geological study noted two reef edge wells that were producing a few barrels of oil per day with no water, well into their production history. Due to low porosities and poor permeabilities, it is believed that these wells continued to produce clean oil at small volumes, possibly because they were separated from the areas of water injection by impermeable barriers (Figure 4). Despite the wells presence, the existence of clean reef edge oil was not believed to be proven, only indicated as a potential avenue to pursue for added value.

The 12-13-67-10 W5M well, situated at the south end of the 3-D survey (Figure 4), was an injector in the platform and did not produce B2A reserves despite having B2A section present. Due to voidage replacement problems associated with unit platform production east of the reef, the well could not be worked over and converted back to a B2A producer. Situations such as these suggested that new drilling might be required to access these potential reserves, although re-entries of some existing boreholes proximal to the reef edge were still a possibility.

In order to mitigate drilling risk associated with these potential reserves, the prediction of the reef edge became critical. Figure 3 depicts the conceptual geologic scenario. With wells drilled on 160-acre (0.64 km$^2$) spacings, interwell distances averaged 3000 to 3500 ft (914 to 1067 m). Depending upon the position of the reef edge between the wells shown in Figure 4 by the gray solid line, considerable reserves could exist updip from the last reefal well, beyond those anticipated based on the geological study. The reefal well depicted in Figure 3 acts as the last draw-down point for unitized reserves. Promontories of reefal growth could also exist beyond this position, where the reef could have buildups above the B2A layer, which would enhance the potential reserves.

*Home Oil Company Ltd., 324 8th Avenue S.W., Calgary, Alberta, Canada T2P 225. Currently affiliated with Boyd PetroSearch, Suite 700, 540 5th Avenue S.W., Calgary, Alberta, Canada T2P 0M2

## SEISMIC EVALUATION

Seismic technology was investigated to see if it could be employed for added value to predict the reef edge and thereby increase the opportunity for drilling success for untapped and potentially clean reef-edge oil reserves. Historically, this Swan Hills field was discovered through geological efforts and optimized with an engineering and geological approach during the past few decades. Only since the introduction of the common-depth-point-seismic method and digital recording in the early to mid-1960s have seismic data been used extensively to mitigate exploration risk. The only seismic data available to the Unit, and of no real value to Unit partners, were some single-fold data from 1958. More modern geophysical techniques had not been employed. This situation is not uncommon to the petroleum industry, because it has been only in the past few years that geophysical techniques have been applied more frequently and rigorously to characterize reservoirs. Recently, reservoir geophysical applications have become a major growth aspect for the seismic method. Hence, this case history serves to document how geophysics added value in a reservoir development sense to a stalwart producing core property.

Seismic modeling was conducted to determine whether the reef edge was discernible with seismic data. Based upon sparse local sonic well control with estimates of resolution limits, it was anticipated that the reef edge could be delineated by inference from seismic tuning phenomena to a vertical resolution of approximately 50 ft (15 m). Comparing this resolution threshold to the known geology, it is noted that the 50-ft (15-m) resolution limit also corresponded to a B2A layer that would contain a full B2A buildup with 20 ft (6 m) of net pay and 8% porosity (Figure 4). Thus, once a geophysical edge was drawn, one could anticipate from the geological study that economic reservoir would exist. Therefore, the value of the seismic technique was determined to be the delineation of the reef edge to mitigate drilling risk for these potential untapped reserves.

## ACQUISITION DESIGN

As previously noted, the northeast side of the reef was believed to be the least risky area to pursue this concept of clean reef-edge oil, because it was structurally the highest portion of the reef. The survey was designed to be 8.74 mi$^2$ (22.64 km$^2$) in size and is shown areally in white with orange outline in Figure 4. Source lines were laid out in a north-south orientation to permit the greatest use of natural access. Receiver lines were positioned orthogonally in an east-west direction. Source-line and receiver-line spacings were chosen to be 400 m. Shot point and receiver group spacings were selected at 40 m, thereby providing 20 by 20-m bins. The receiving patch approximated 3600 by 4000 m. Approximately 900 groups of phones from 11 receiver lines were live for each shot. The fine spatial sampling selected was preferable not only for picking a reef edge with minimal spatial uncertainty, but also to mitigate spatial aliasing and to permit higher frequencies in the dataset.

For operational convenience, dynamite was the selected sound source, because limited access to the environmentally sensitive areas such as the Swan River ecosystem in the center of the survey required hand-cut source lines. A drill on a Nodwell mount, transported by helicopter, was employed to access this region. Forestry sample plots were avoided altogether. Figure 5 shows the modeled fold diagram for the 3-D volume and compares it to the actual processed version. Fold aberrations were noted in the initial design about the Swan River, but they were appreciably mitigated as several shots were skidded. Comparing the modeled version to the actual data, it is noted that source and receiver line positions were altered in the field to make use of further natural access and to make up for lost fold in this area. This was also true for the southeast and west central portions of the survey, where a lake and a reservoir had to be avoided. Some of the pipelines within the survey paralleled the existing access, which required numerous pushouts where shotholes were offset into the bush to adhere to government regulations associated with pipeline clearance. Due to further environmental concerns with tall timber and high timber assessments, all receiver lines were hand cut. Environmental officers were pleased with the operations despite three shotholes' being drilled too close to the Swan Hills River ecosystem.

A DFS-V recording unit was used during January 1992, one month before the seismic acquisition operations commenced, to conduct source testing. Two sites were selected, one to the north on high ground and the other in the southern portion of the proposed survey, topographically lower by approximately 500 ft (152 m). Tested parameters included charge size, hole depths, and patterned holes versus single holes. A charge size of 2 km buried at a depth of 15 m was selected as optimum. Patterned holes appeared to be slightly better but operationally much more difficult to conduct throughout the survey due to topography, environmental constraints, and government regulations associated with pipelines. Noise tests were also conducted around pumpjacks. It was noted that a region of 60–80 m, or 7–9 traces, would be contaminated differentially with noise due to downhole rod strings. The impact of this was deemed to be negligible. The greater risk for noise was thought to be the 4-26 water plant beside the reservoir on the west central side of the survey. The plant derives the water from an adjacent reservoir and thus sources much of the water for injection in phases 3 and 4. Because it was in a flanking position to the survey and back of the reef edge, the chance that the data would be noisy was a risk worth taking.

## DATA QUALITY

Seismic operations commenced in February and continued into March when the balance of the recording was conducted. A DFS-VII recording unit was contracted to record the seismic data. Temperatures held consistently to daytime highs of –15 to –20 degrees Celsius, which were ideal for seismic operations. Data quality exceeded expectations as the raw data were reviewed (Figure 6). As shown by the amplitude spectrum, frequencies up to 70 Hz were recovered in the zone of interest before deconvolution or any processing of the dataset. After a processing runstream, which included surface-consistent static and surface-consistent deconvolution followed by zero-phase deconvolution processed to a 37-fold stack, the dataset contained and sometimes exceeded 80 Hz in the zone of interest. The dataset was migrated with a generic post-stack finite difference time migration algorithm.

Noisier areas of the dataset had no correlation to pumpjacks in the field or even to the 4-26 water plant. On the last three days of recording, temperatures averaged 3–5° Celsius, and field records degraded from exceptional to acceptable quality. At the time, the most southerly portion of the survey was being acquired. The two most likely reasons for the tolerable degradation in the data in this area were the temperature during recording and the skewing of shots about a lake.

An example of the stacked and migrated dataset is shown in Figure 7. Depicted is crossline 174, oriented in an east-west direction through the middle of the data in the vicinity of the Swan River ecosystem (Figure 4). The high end of the frequency bandwidth was 80 to 85 Hz in the vicinity of the Swan Hills reef, which is situated just below the red line in Figure 7.

## SEISMIC INTERPRETATION

The Swan Hills reef edge was discernible on this 3-D dataset. The dense well control afforded by this production mode application provided plenty of calibration points for subtle amplitude variations. Figure 7 shows a reef edge and two wells, one reefal and the other basinal. The interpretation of the position of the reef edge depended on understanding the tuning of the seismic wavelet. As a result, determining the reef edge was difficult. The red line approximates the spatial positioning of the reef edge between the two wells.

Calibrating to the known well control, various inlines and crosslines were interpreted first on the workstation. Every inline and crossline was interpreted individually and sequentially twice to confirm the interpretation of the reef edge. Figure 8 depicts the resultant interpreted edge and compares it to the geological edge derived from the geological study. The serrated appearance of the seismically defined edge is immediately apparent. In the northern part of the survey, the edge was pulled back behind the geological edge, but in the southern portion, three regions were defined where potentially unswept reserves might reside. In areas where the reef edge had indications of being porous, vertical resolution limits calibrated to well control were indicating a resolvable limit of 50 ft (15 m), as the model had predicted. In areas where the reef edge was believed to be tight, the vertical resolution limit was 35 ft (11 m), which exceeded expectations.

Once the reef edge was positioned, the exercise became one of looking for 5-1/2-in. and 7-in. casing as re-entry candidates for kick-off points for new horizontal wells. At the time, technology did not exist to re-enter 4-1/2-in. casings. All of the well casings in the vicinity of the potential untapped reserves were in fact 4-1/2-in. This forced the commitment to drill a new hole. Upon review of the geophysical interpretations between companies and scrutiny of various reef edges, the commitment was made to drill the 9-23 grassroots horizontal well (Figure 8). A horizontal well approach was preferred because the trajectory could be redesigned for success in the event of failure. First, drilling from within the reef outwards, the operator and the unit partners had a greater chance of success for a producing well if the geophysically defined reef edge were inaccurate. Drilling from a basinal position towards the reef heavily relied upon the accuracy of the edge. A vertical hole would test the concept at one point only. Second, with success, a horizontal well would provide appreciably higher deliverability rates from this carbonate reservoir. The 9-23 well trajectory of the borehole was designed to go due north for 175 m to encounter porous reef, as anticipated by the geological study, and thereby guarantee success. Then the trajectory would turn 37° eastward until it reached the sidewall of the reef to confirm or discredit the seismically defined reef edge.

## LITHOLOGIC MODEL-BASED 3-D INVERSION

As the 9-23 well license application and surface lease preparation were proceeding, seismic inversion work was initiated to investigate whether enhanced confidence in the interpreted reef edge could be acquired. It was also hoped that the removal of the seismic wavelet and enhanced frequency content would provide a uniform vertical resolution threshold that was not geologically dependent. The interpretation problem posed in this application was the resolution of a reef edge below tuning frequencies. On conventional 3-D seismic data, the reef top could be interpreted only by inference through seismic tuning phenomena. The porosity information from the seismic data was definitely not quantifiable, so this became a third objective for the inversion process. The inversion process was highly calibrated with information from the geologic study to address the prediction of the reef edge and potentially reefal porosity.

The inversion process was calibrated with the benefit of the extensive geological study. Sixteen wells within the 3-D survey area had sonic logs, 13 of which were of suitable quality and length to be used in the inversion process. The log dataset consisted of highly detailed data for any given control point, as compared to the seismic data, which consisted of lower vertical resolution but enhanced spatial resolution. To make the well log data more accurate, the log curves were shifted in accordance with the parameters determined in the geological study to compensate for different readings associated with varying vintages of logs, washouts, and tool calibration errors. Log balancing software was developed to facilitate these corrections. The software corrected for baseline shifts in the logs and depth-varying drifts in readings, both resulting from varying vintages of logs.

Once corrected, the logs were then used for a reflectivity color correction of the seismic data, which corrected the seismic response to the reflectivity portrayed by the well log data. This step acted as a normalization of the seismic data to the well log data at the known control points. If it were not done, wavelet estimation errors would result and the inversion product would be contaminated by these discrepancies. In fact, the inversion product, which would be trying to remove the wavelet from the seismic, could remove some of the reflectivity.

The next step was to extract a wavelet from the seismic data. A simultaneous least-squares wavelet extraction was conducted on the dataset at the well control calibration points. The resultant wavelet was quite stable and exhibited a uniform phase of 300–310°. The amplitude spectrum was fairly smooth, indicating that we had a fairly good match to the seismic data.

The inversion method employed was a 3-D lithologic model–based inversion method, which was available at the time in a beta-test release form from Ulterra Geoscience Limited. The software requires an initial well log model and then uses additional structural information for the inversion process. A single structural pick in the area, the Duvernay of the Late Devonian Period, situated roughly 30 ms above the zone of interest, was used to give the basic structural control information within the 3-D volume for the inversion process.

## INVERSION INTERPRETATION

The 3-D inversion dataset was then loaded onto a workstation for interpretation. The top of the reef could now be resolved uniquely, as opposed to by inference through wavelet tuning phenomena (Figure 9). The reef edge was more clearly defined by changes in velocity. The reef top was picked with the aid of the workstation line by line, just as it was done with the migrated data. Variances in the interpreted edges did occur, but only in one small geographic location did the edge move as much as 150 m horizontally. The anticipated potential error in the positioning of the reef edge by the original migrated data was believed to be a breadth of 150 m. What was observed by means of the inversion confirmed that prediction. The inversion-defined reef edge, however, was believed to have an accuracy within 50 m, or 2-1/2 bin cells or traces. Confidence in the reef edge was appreciably heightened.

Not only was the lateral accuracy or resolution enhanced, but the vertical resolution was also improved. The reef top could now be seen directly. Correlating with well control, the vertical resolution limits were increased to 30 ft (9 m) of reef growth independent of lithology and tight versus porous edges. As quickly as we commenced to celebrate the resolution improvement to 1/6th of a wavelength, we realized that our success became our instant demise. A vertical resolution limit of 30 ft (9 m) increased our drilling risk a bit, and the whole purpose of this exercise was to mitigate drilling risk for the potential low watercut oil along the reef edge. Figure 10 illustrates the conundrum. When the interpreter predicted a reef edge with 30 ft (9 m) of buildup, a new well was no longer likely to encounter the porous B2A section. Impermeable B1 section could be all the reservoir section that would be present. Rather than filter back the data, which goes against the grain of any geophysicist who continuously tries to push the envelope of subsurface imaging, something else had to be done.

During interpretation of the inversion data, it was noted that the contact between the platform of the reef and the underlying Watt Mountain shales was also discernible (Figure 9). Instead of picking the base of the platform upon which the reef grew by the same methodology as the reef top, a pattern recognition algorithm was developed. Using the manual pick for the top of the reef, the algorithm picked the base of the platform quite accurately. This innovation was mildly noteworthy, for at the time, to the knowledge of the author, no inversion software would permit the interpretation of events not already picked and supplied to it previously.

Once the reef top and base platform picks were interpreted, it became possible to create an isochronal map for the reef buildup and platform (Figure 11). Keeping in mind that the isochronal thicknesses were small and that error bars of plus or minus 2 ms (the data sample rate) applied to each of the two interpreted events, a gross correlation of reef thickness to reef buildup could be inferred. An isochronal value of 12 ms approximated a section of reefal platform, the B1 and B2A layers. A thickness of 8–9 ms approximated a section comprising platform and B1 layers only. The interpreter was able to crudely predict, despite the errors associated with the measurements, the possibility that the B2A section would be present.

Being able to intepret the top of the reef and the base of the platform provided additional prediction potential. The fine calibration of the seismic data to the well log data provided the opportunity to investigate and stretch the limits of porosity prediction potential. On a bin cell-by-bin cell basis, an average velocity was derived for the section between the base of the platform and the top of the reef. This isovelocity map (Figure 12) depicts the average combined velocity for the reef-to-base platform section. Despite the fact that the colors for the map were identical to those for the inversion profiles, color variations existed between the two displays, due to the vertical averaging of the velocities within a thin isochronal section in the creation of this map. Areas of lower velocity or implied enhanced porosity were indicated by the map. In the northern one-third of the survey area, lower average velocities were indicated, which agreed with the findings of the geological study. Wells in this region portrayed thicker and more consistent or "blocky" porosity. In the middle of the data volume, a higher predominance of fuschia indicated higher-velocity material. Well logs in this region displayed lower-grade and streakier porosity. In an effort to see further detail, a hypersensitive colored display of this map was created (Figure 13). Interpreting this isovelocity map on a bin cell-by-bin cell basis exceeded the accuracy of the measurement in this instance, because of the errors associated with the picking of the reef top and base platform. However, regions of spatially expansive response (100 by 100 m) were thought to be more trustworthy.

As noted before, the drilling of the 9-23 grassroots horizontal well was proceeding, and at the time that this inversion work was completed, the sonic log for the well had been recorded. This provided an excellent opportunity to test the spatial prediction capability of the inversion data. Figure 14 depicts an inversion profile down the exact drill-bit trajectory of the 9-23 well. The surface lease in 1-23 was selected as the point at which to spud the well, which had a 200-m build angle to the horizontal section. The well bore was steered to change its azimuthal heading beginning at crossline 271. The sonic log for the entire 550 m of horizontal section is shown below at the same scale as the inversion profile. The porosity is shaded with a 3% dolomite cutoff. The actual vertical trajectory through the reef is shown by the yellow line on a vertically exaggerated scale, and it is important to note that throughout the entire section, the bit stayed within the B2A layer of reef growth.

Comparing the sonic log to the velocity predictions of the inversion data, one notes a remarkably good correlation. Near the heel of the horizontal well—the point of first penetration of the reservoir by the horizontal well—the inversion profile predicted a predominance of high-velocity material for the reef and platform section, denoting less porosity. Halfway along to the toe of the well, where total depth was reached, lower-velocity material was predicted, indicating potentially better porosity. The sonic log confirmed this prediction with the occurrence of spotty porosity in the heel of the well; porosities varied from 5 to 8% and porosities towards the toe exceeded 18% in some places. Geologically, the reefal material encountered by the bit in the heel constitutes tighter reef channel facies; the high-porosity material was porous reef margin. These empirical observations lead to the correlation of porosity with velocity prediction by the isovelocity map shown in Figure 15. The technology existed at the time this paper was written to directly correlate porosity to velocity and to create an implied porosity map. However, this technology had not been developed by the time of this project, and a quantitative approach was undertaken.

Interpreting the isochronal map concurrently with the isovelocity map for the reef and platform section, the interpreter believed that he had fully recouped the increased drilling risk caused by the enhanced resolution thresholds derived from the inversion dataset. Recalling that modeled vertical resolution was touted to be 50 ft (15 m) with a virtual guarantee from the geological study that 20 ft (6 m) of 8% porosity would exist, and that the migrated seismic data provided resolution limits of 35 ft (11 m) for tight reef edges and 50 ft (15 m) for porous ones, despite having a 30-ft (9-m) resolvable limit with the inversion product, the interpreter was able to predict reefal thickness and average porosity for the combined reefal and platform sections to mitigate the risk associated with enhanced resolutions. In fact, this added prediction ability mitigated drilling risk marginally further beyond that of the original 50-ft (15-m) resolvable limit.

## DRILLING AND PRODUCTION RESULTS

The 9-23 well had 550 m of horizontal section in the B2A reservoir. Geologic predictions indicated that at the end of the well bore 37 ft (11 m) of reef buildup resided underneath the drill bit. The well bore did not extend all the way to the seismically defined reef edge, because differential sticking of the drilling pipe and lost drilling circulation forced an end to drilling. Fear was growing that the well would be lost, while an economic success was already achieved. The drill bit proved the existence of the B2A section 110 m beyond the zero reef edge contour as shown by the geological study. Total depth was called 200 m along this trajectory from the seismically defined reef edge. Initial production rates were 1110 bbl of fluid per day, with a 3% watercut. Average field watercut in the area at the time was 92% with an average field watercut of 86%. The well has already reached payout for Unit partners because the economics were buoyed by horizontal well royalty relief. Payout would have been reached sooner had an additional lateral not been drilled. One year into production, the well is producing at 1180 bbl of fluid per day, with oil rates at 270 bbl per day and a 78% watercut. This portion of the reservoir is highly underpressured. Mail ballots were issued and approved by Unit partners to convert two nearby downdip producers to injectors to provide pressure support. Water production rates are anticipated to stabilize once the injectors pressure up the reservoir.

The second well drilled was 14-34, which was a vertical hole drilled as the last drainage point for the Unit (Figure 13). This well benefited from an interpretation of the migrated stacked data and the inversion data. The well was drilled almost to total depth when loss of circulation occurred. A drill bit and drill string were left in the hole. The reservoir characterization of the isovelocity map was used to guide the whipping of the well into better porosity (Figure 15). A total of 15 ft (5 m) of B1 reefal buildup was encountered, with no B2A present. The carbonate platform section contained porosities exceeding 7%. The well was completed as the best platform producing well in the unit. It is believed that the porosity prediction of the inversion was accurate; however, it should be noted that it is not possible to predict in what layer the porosity resides. Second, it should be noted that the isochron map failed to predict the presence of B2A, but that the extra buildup required for B2A to develop was well within error limits. Initial production rates were 390 bbl per day, with a 15% watercut. One year after initial production, the well was still producing 240 bbl a day, with no water. The initial water recovered during the production test was load fluid which escaped into the reservoir due to lost circulation. Payout will be longer for this well, not because of well performance but due to the lost drill string, the additional cost for the whip, and the fact that this well did not qualify for royalty relief.

The third well drilled was 12-13, another grassroots horizontal well. Operational constraints dictated that only surface leases prepared in advance could be used for the drill rig. This dictated using the 1-23 surface location once again and drilling southeast. The inversion profile guided the trajectory of the well bore and anticipated poorer porosity development than that encountered in 9-23. A sonic log was not acquired in this hole, because cost overruns were being incurred. After an acid wash, initial production was 1140 bbl a day of fluid, with a 5% watercut. The well was shut-in later for two months to address maintenance problems on the semisubmersible pump. It returned to production and is now on pumpjack at 230 barrels per day net oil, with no water. Like 9-23, this well is in an underpressured part of the reservoir and will benefit from the downdip injection in the future. Payout will be longer than anticipated for this well, because two mud motors were lost in this hole. Like 14-34, this well did not qualify for royalty relief.

Initial production rates from the three wells totalled 2640 barrels per day, which constituted a 10% increase in unit production. At the time this paper was written, despite the lower oil production volumes, the three wells still comprised 3% of Unit production. Payout on the last two wells is anticipated, as neither is making any water one full year into their production history. The source of the water for 9-23 is thought to be the two injectors to the west, 10-27 and 12-26, which are injecting water into porous reef margin facies in stratigraphically higher layers that are connected to the B2A reef margin encountered in the toe of the 9-23 well (Figure 15). That quality of reservoir was not encountered in either of the last two wells. Pressuring the reservoir about the two horizontals will undoubtedly help to control water influx. For the 9-23 well, the reservoir quality was so exceptional and so well connected to the injectors, that the water influx occurred quite early.

## DISCUSSION OF RESULTS

The project constituted many firsts for the people and the organization involved. Modern geophysical techniques were employed on this aging core property that had benefited for three decades from a predominant engineering perspective supported by geological work. For the people involved on the team and within the production department, it was the first time they had worked directly with a geophysicist. The success of this project propelled the status of a geophysicist from the role of providing service to the engineering discipline in the production department to that of a bona fide team member. Requests for geophysical input on core properties increased appreciably. For production management, it was a new challenge, and a welcomed one, to understand a new discipline. For the geophysicist involved, the project required time, patience, and perseverance but was very gratifying. For the operator and unit partners, the project added value.

The geophysical discipline was able to add value through reservoir delineation and characterization by mitigating the risk associated with drilling for clean, unswept reef-edge oil. The concept of clean reef-edge oil in an aging reservoir was proven by these drilling results. Innovations in the field of seismic inversion were required to successfully meet the mandate of reducing drilling risk. Internally, the probability of success in drilling for these potential reserves moved from 50% based upon the geological study, to 85% once the seismic work was completed. The log balancing algorithm, the pattern recognition algorithm used to pick the base of the platform, and the isovelocity map were principal technological innovations required to further mitigate drilling risk. The ability to crudely predict porosity exceeded the goals of the study, because it was the intent to use 3-D seismic data to delineate the reef edge. Only through considerable calibration of the data, governed by the geological study, and a desire to test and push resolution limits was it possible to predict porosity. The seismic data netted back its value within the first month of combined production from the three wells.

## CONCLUSIONS

The initial geological concept of clean oil existing in this aging reservoir was proven. The seismic program delineated the reef edge and thereby added value by reducing drilling risk associated with the capturing of these reserves. The positioning of the reef edge also assisted with drilling program design in terms of how best to access these potential reserves. The inversion processing increased interpreter confidence in the reef edge. However, the real value from the inversion was the ability to crudely quantify porosity and to predict the best locations and trajectories for further drilling. An initial production increase to the Unit of 10% was realized, with rates stabilizing at a 3% production increase to the Unit. Enhancement techniques and the continual improvement of engineering technologies should further enhance this contribution.

### *Acknowledgments*

Any project of this magnitude involves several parties. The contribution of the senior Unit partners (Gulf, Mobil, Amoco, and Imperial) is greatly appreciated. The author wishes to thank Home Oil Company Limited of Calgary and the Swan Hills Unit partners for permission to release this material. Judi Dobbin and Roy Benteau of D & S Reservoir Consultants Ltd., the geologists who conducted the geological study, contributed greatly to the calibration of the seismic data for the inversion work and assisted in the drilling program design for the horizontal wells. One of the figures included in this paper came from Judi's mapping and is submitted here with her approval. Kevin Powell, the reservoir engineer, is noted for his contributions during drilling program design and implementation. Dan Botterill, Engineering Coordinator for Swan Hills Unit #1, the accepted economic potential of this project, which opened the door for the geophysical discipline to contribute. Andrew Holder, a formally trained geophysicist and Vice-President of Exploration, gave his technical support during the internal approval process and provided considerable input during the proofing of the text. Kerry Allred, geophysical operations, spent considerable time overseeing operations and dealing with design modifications. Richard Garcia, a graphic artist, contributed through the creation of figures in both hard copy and slide form.

The author also wishes to acknowledge Hampson-Russell Software Limited for its support in the amendments to its inversion programs. Scott Jamieson of Ulterra Geoscience Limited performed the inversion work. The isochronal map is submitted as a figure in this paper with the approval of Ulterra Geoscience Limited.

The author is indebted to Rick Wallace of Ulterra Geoscience Limited for his contributions throughout the inversion process. Rick's progressive thinking, combined with a willing innovative desire to push technology, was a key to the ability to quantify porosity.

Several other contractors should be noted for their involvement and support. Globestar Positioning is recognized for its supervision of cutting operations and for the GPS surveying. Veritas Seismic Data Processors unraveled surveying problems and processed the 3-D seismic data. The portable drill on a Nodwell mount that was contracted from Destiny Drilling Limited enhanced our fold coverage in environmentally sensitive areas. Halliburton Geophysical Limited contributed through the contracted use of their DFS VII recording unit.

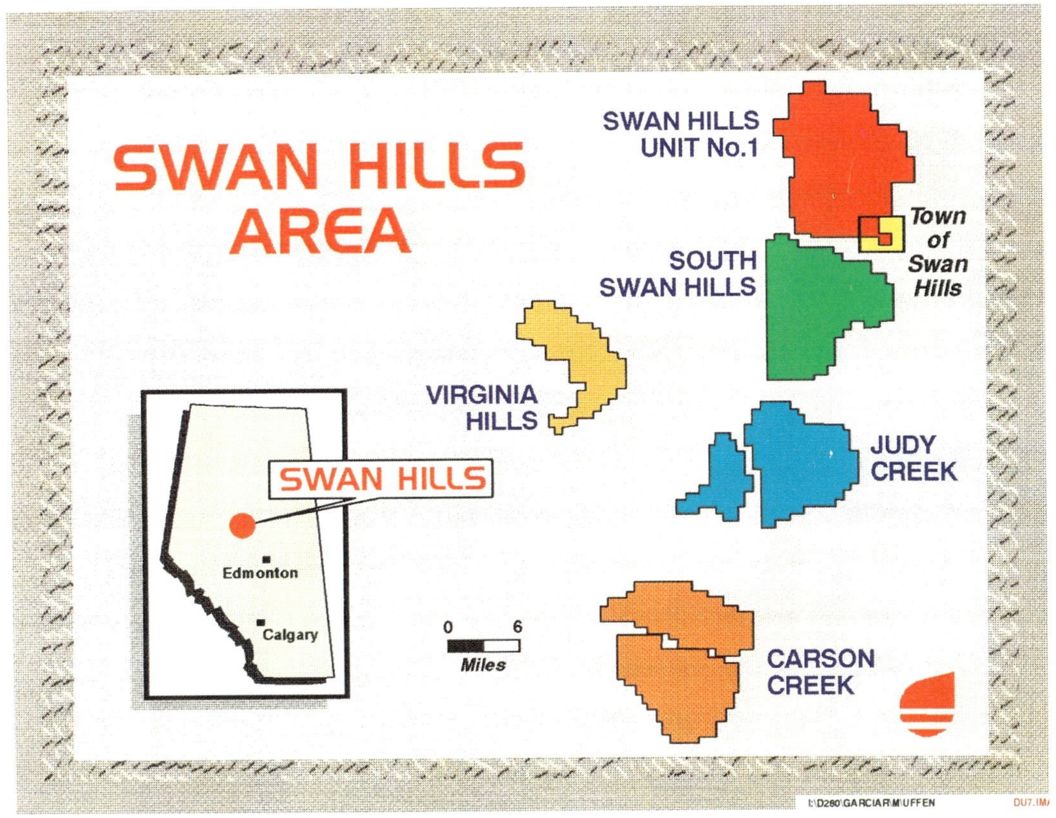


FIGURE 1. Location of Swan Hills in both displays. Members of the Swan Hills Synergy Group and their respective operated units are depicted. Swan Hills Unit #1: Home Oil (operator); South Swan Hills: Amoco (operator); Judy Creek: Imperial (operator); Carson Creek: Mobil (operator); Virginia Hills: Shell (operator); South Swan Hills: Gulf (senior partner).

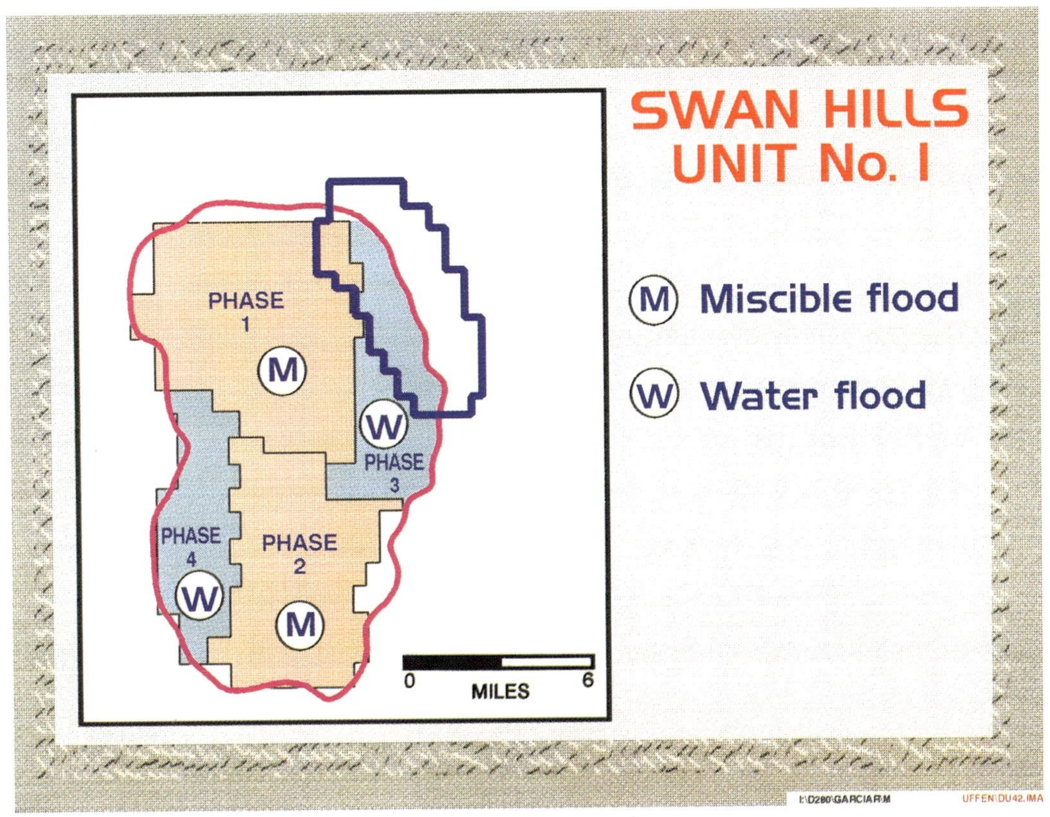


FIGURE 2. The Swan Hills reef outline is depicted, with the four phases of enhanced recovery denoted by the legend. The initial 3-D seismic program outline is shown in blue on the northeast flank of the reef.

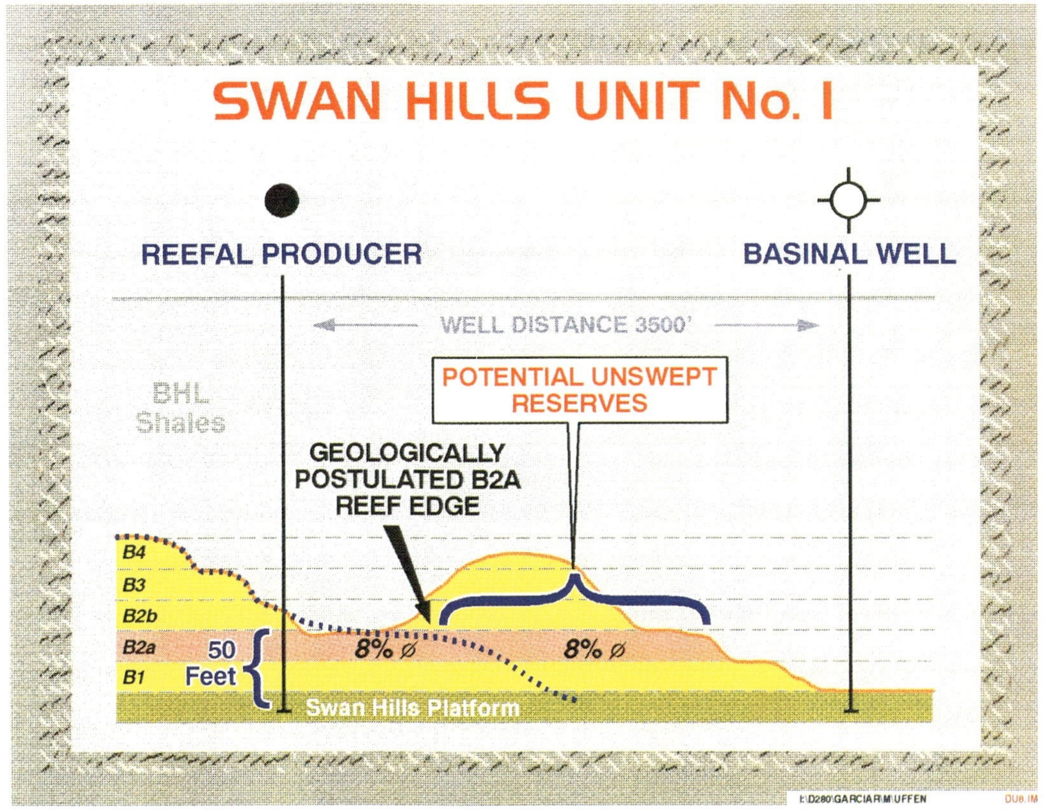


FIGURE 3. Schematic geologic cross section illustrating the position of the reef edge between the two wells shown, which is critical in determining the volume of reserve left unswept by the downdip producing well. Determining the actual position of the reef edge is vital in this reserve prediction exercise.

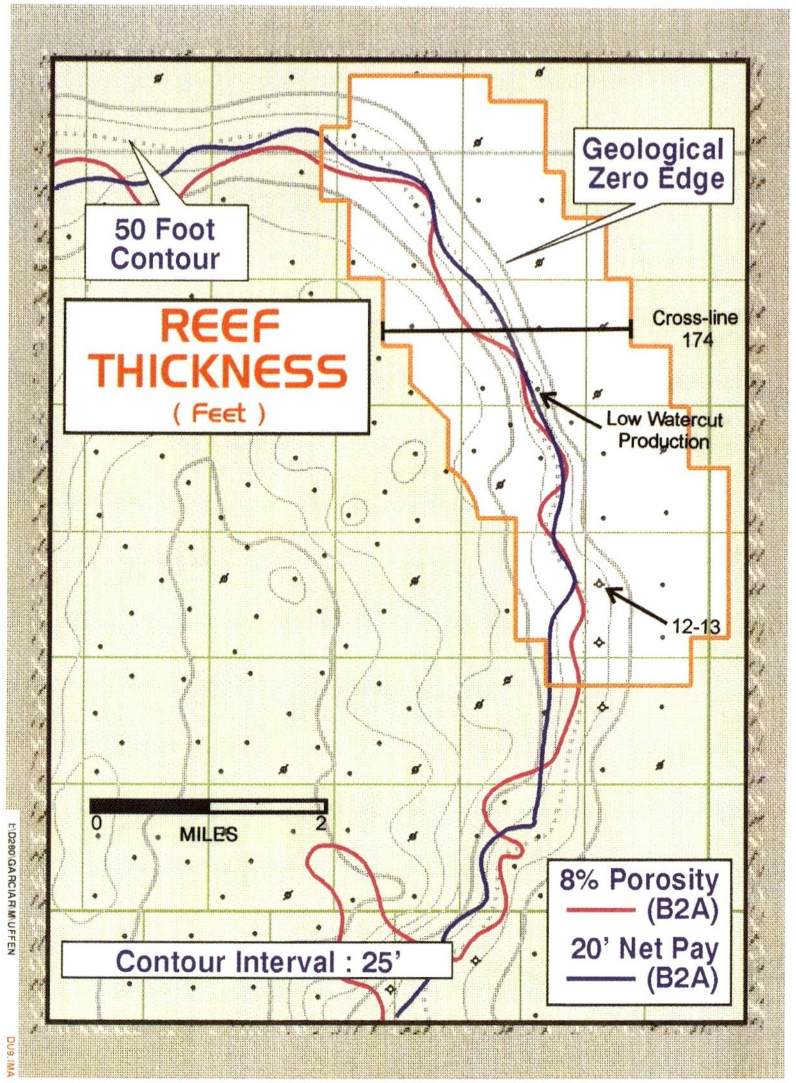


FIGURE 4. Swan Hills reef isopach (courtesy of Judi Dobbin) with 25-ft (7.6-m) contour intervals. The 20-ft (6 m) net pay isopach for the B2A layer, constituting full B2A development, is superimposed as a blue line. The B2A net porosity line of 8% is also shown in pink.

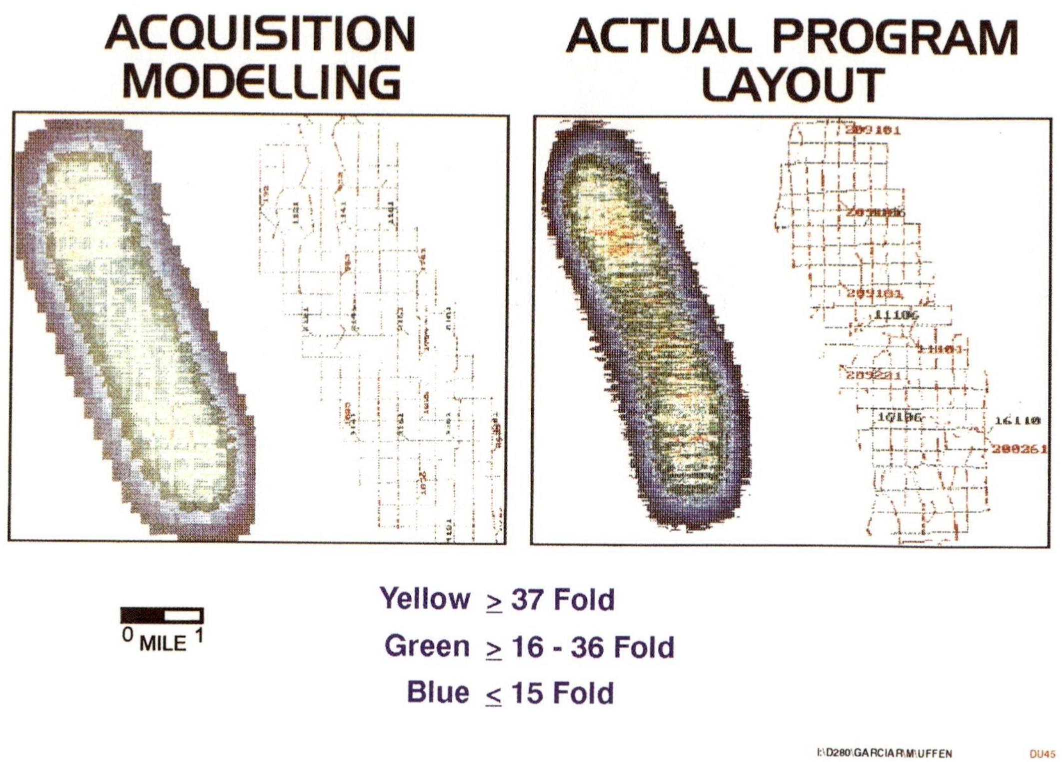


FIGURE 5. The modeled fold diagram and program layout are shown beside the resultant processed version. Alterations in the positioning of source and receiver lines as environmental concerns associated with sensitive areas were addressed are evident.

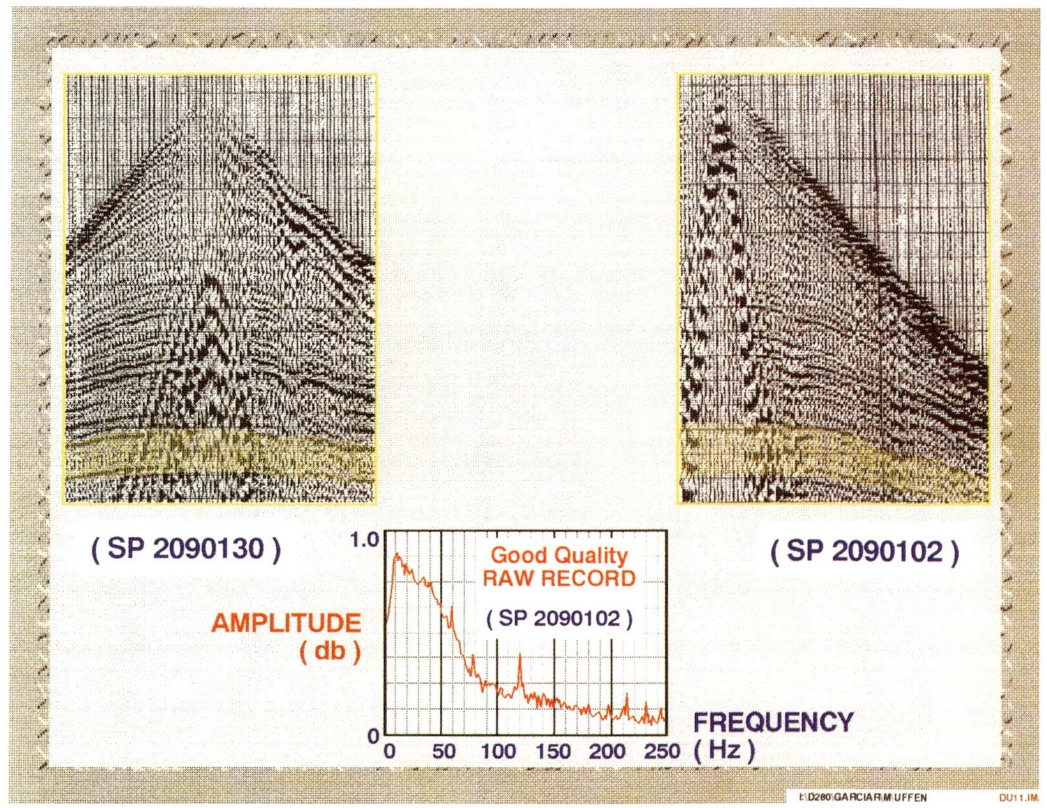


FIGURE 6. Two seismic shot records, with the amplitude spectrum of one of them (shown below). The area of interest for the Swan Hills reef is highlighted in yellow.

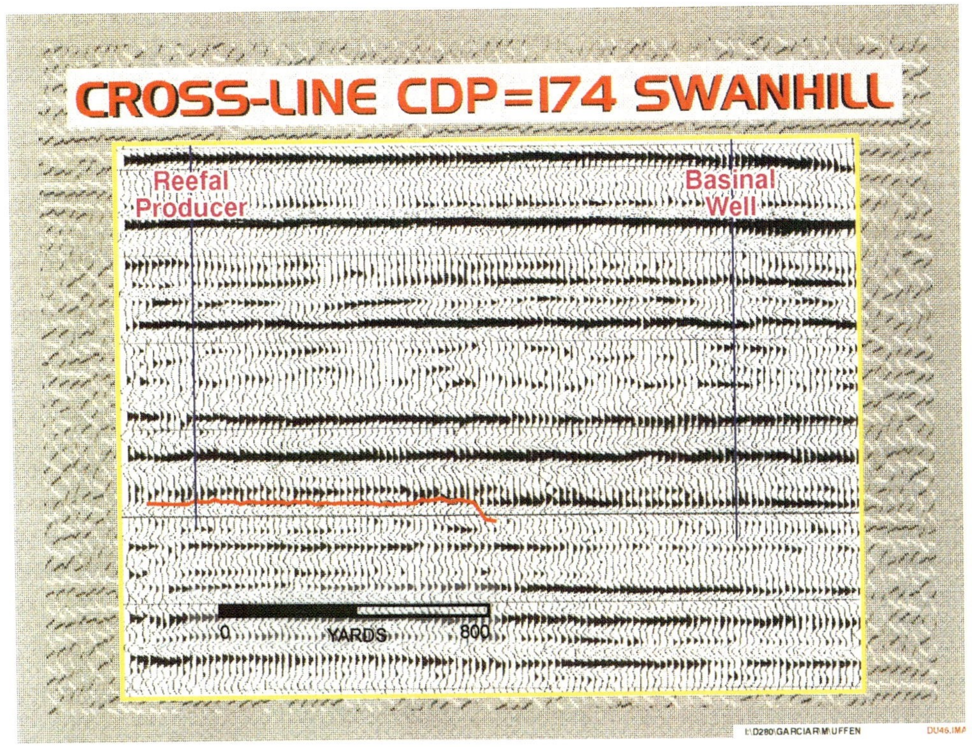


FIGURE 7. Crossline 174, an east-west profile, is taken from the middle of the 3-D volume in the area of the Swan River. Adjacent well control is shown, with the original interpreted reef edge in red.

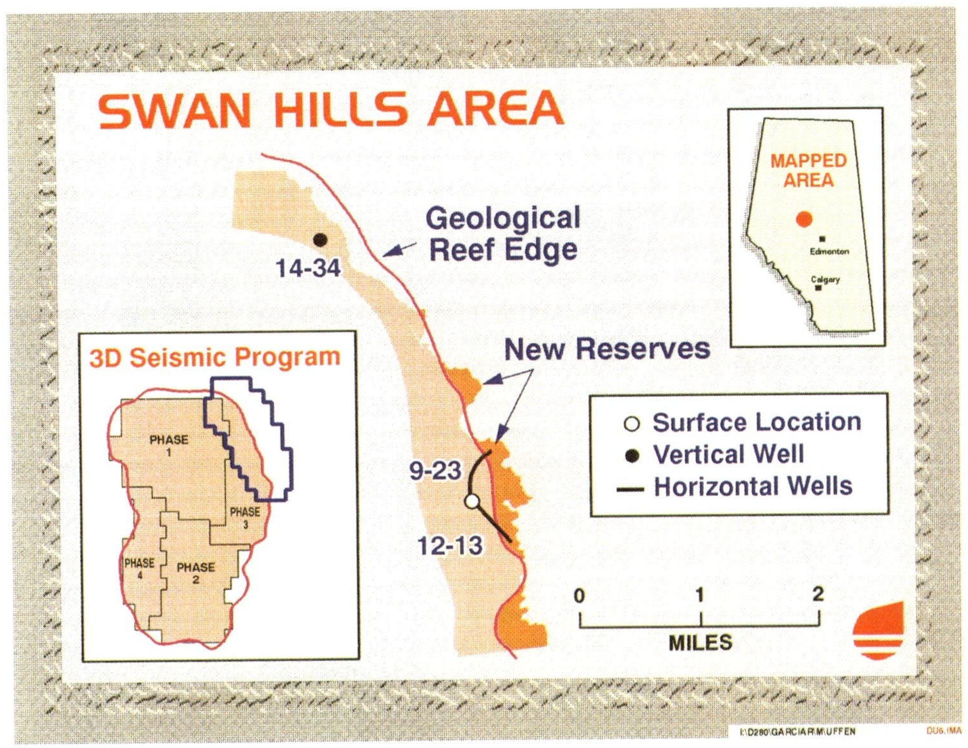


FIGURE 8. Interpreted edge of field from seismic, which has a serrated appearance, is shown with the geological zero edge superimposed. The three wells drilled with the aid of the seismic data are also shown.

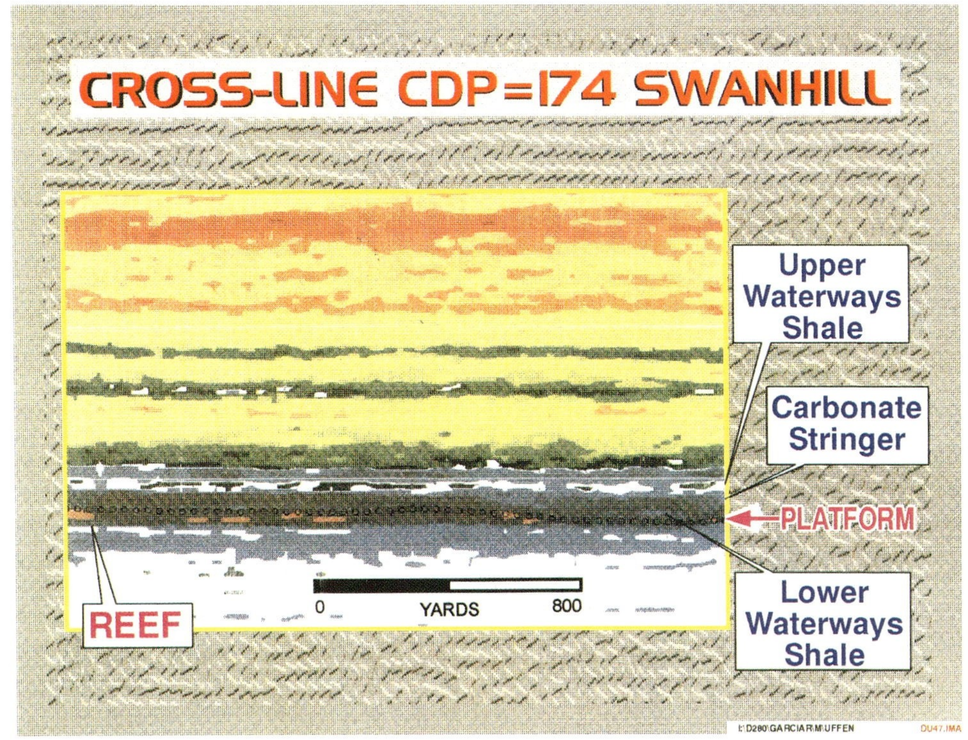


FIGURE 9. Inversion profile for crossline 174 is illustrated, with the interpreted top of the reef pick denoted by the circles and the edge shown by the structural flexure in this picked event. Comparing the inversion profile to the seismic profile for the same crossline (Figure 7), one recognizes improved clarity of the reef edge with the inversion profile, because the top of the reef is shown uniquely.

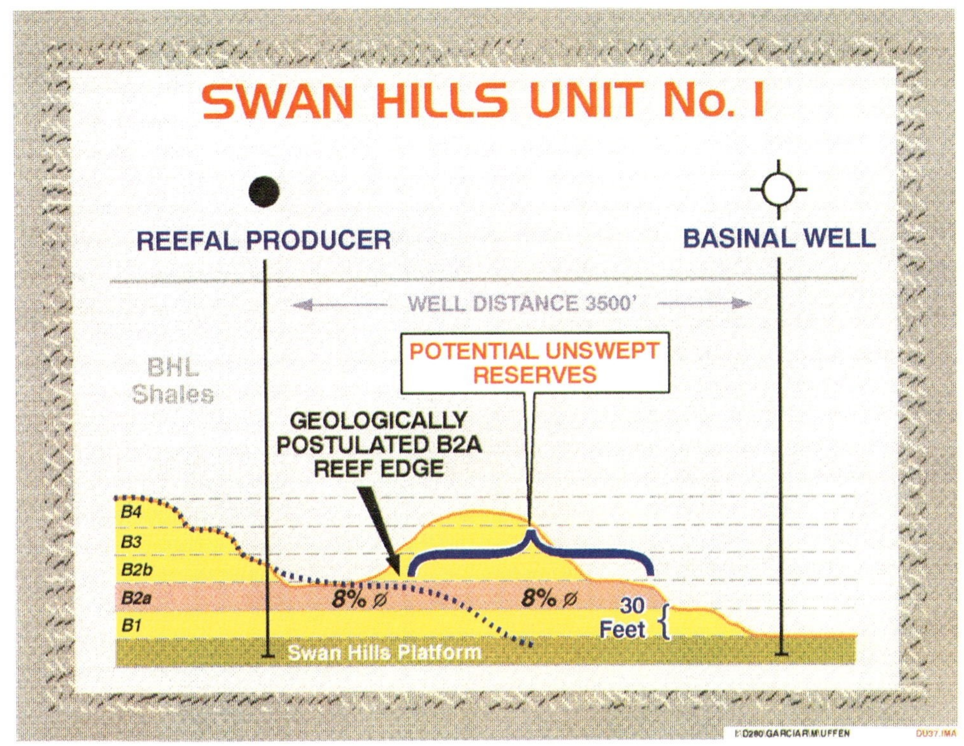


FIGURE 10. New schematic cross section demonstrates the impact of the 30-ft (9-m) vertical resolution limit. At these resolution limits, B2A reserves could no longer be assured.

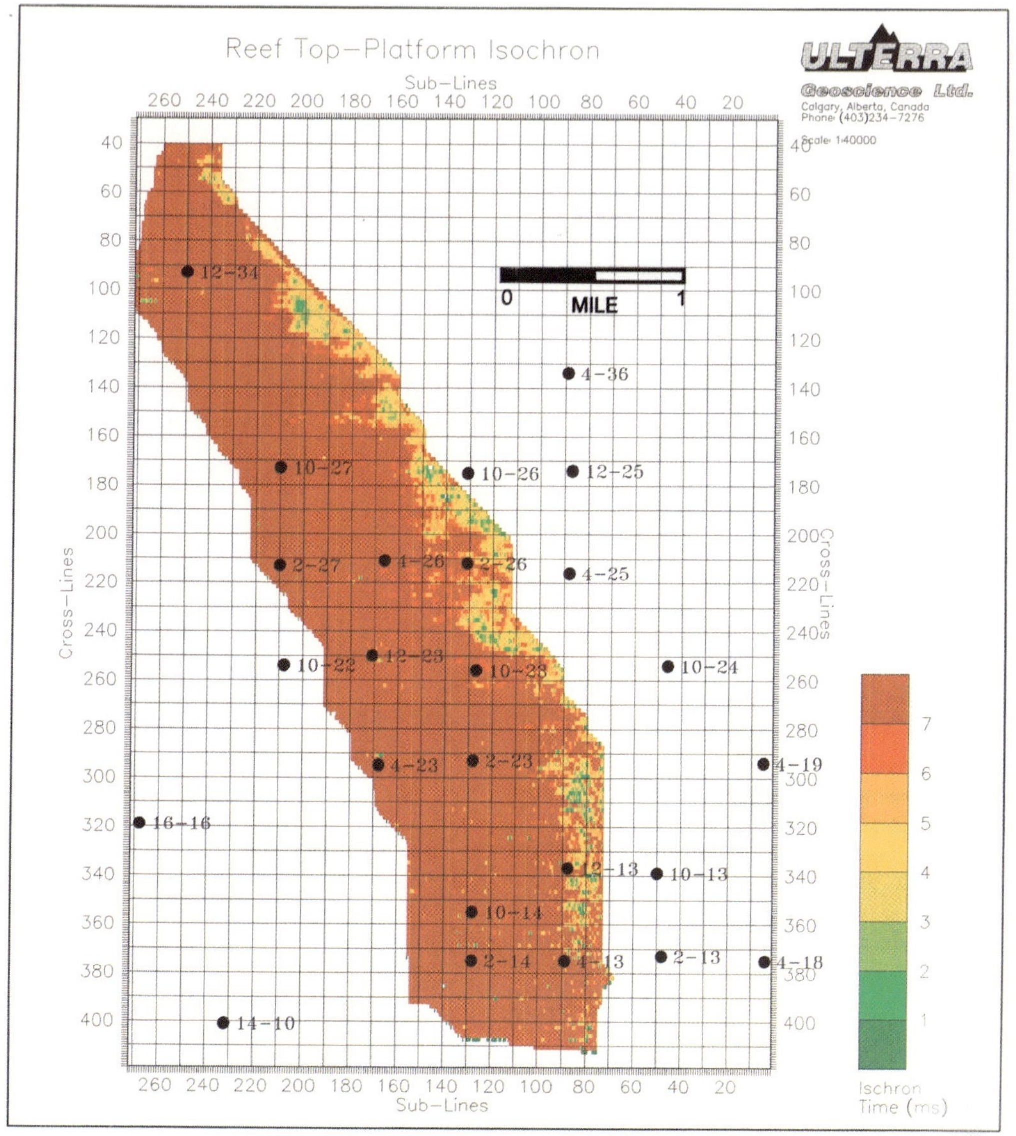

FIGURE 11. Isochronal map (courtesy of Ulterra Geoscience Limited) from the top of the reef to the base of the platform, depicting the platform and reefal thickness within the 3-D data volume. The color contrast between red and yellow approximates the inversion interpretation of the reef edge.

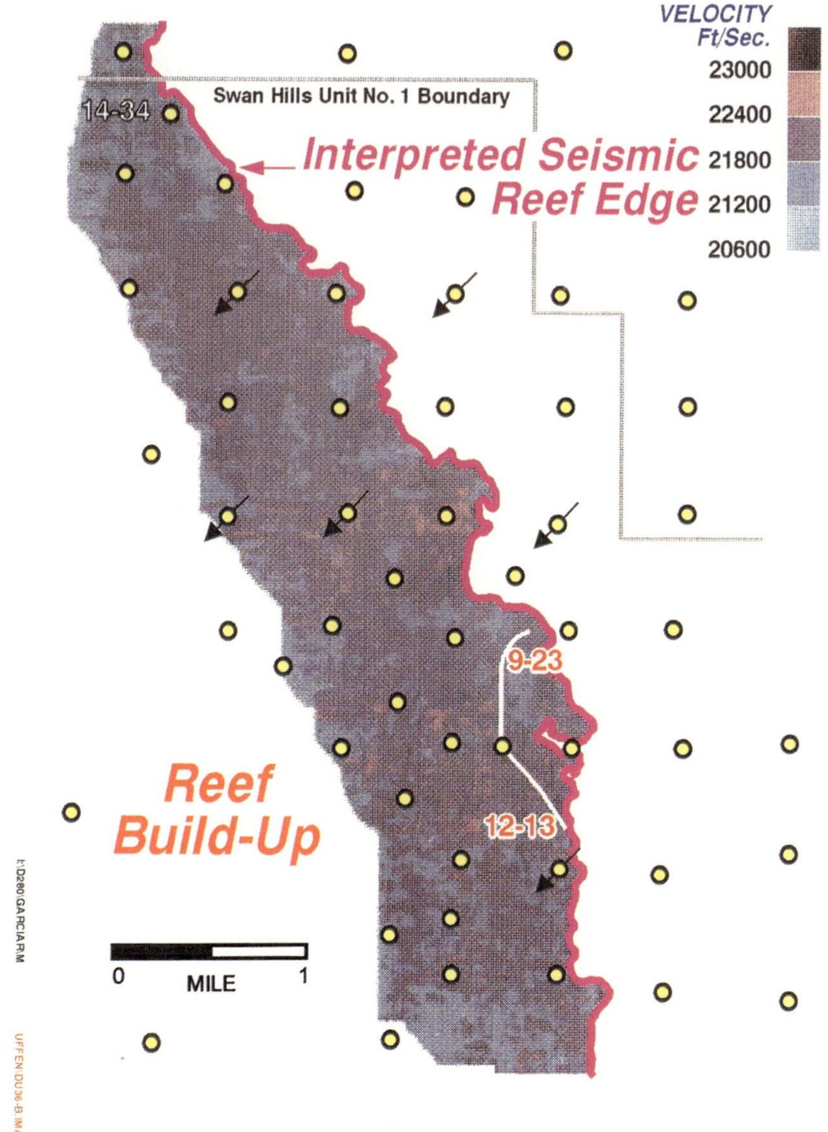

FIGURE 12. Isovelocity map of the reef and platform sections color codes the average velocity on a bin by bin basis throughout the interpreted reefal data volume. A color shift on the map towards lower velocities occurs as the average velocity of the reef is depicted. Comparing this to the inversion profiles, one sees a shift of one gradation of color.

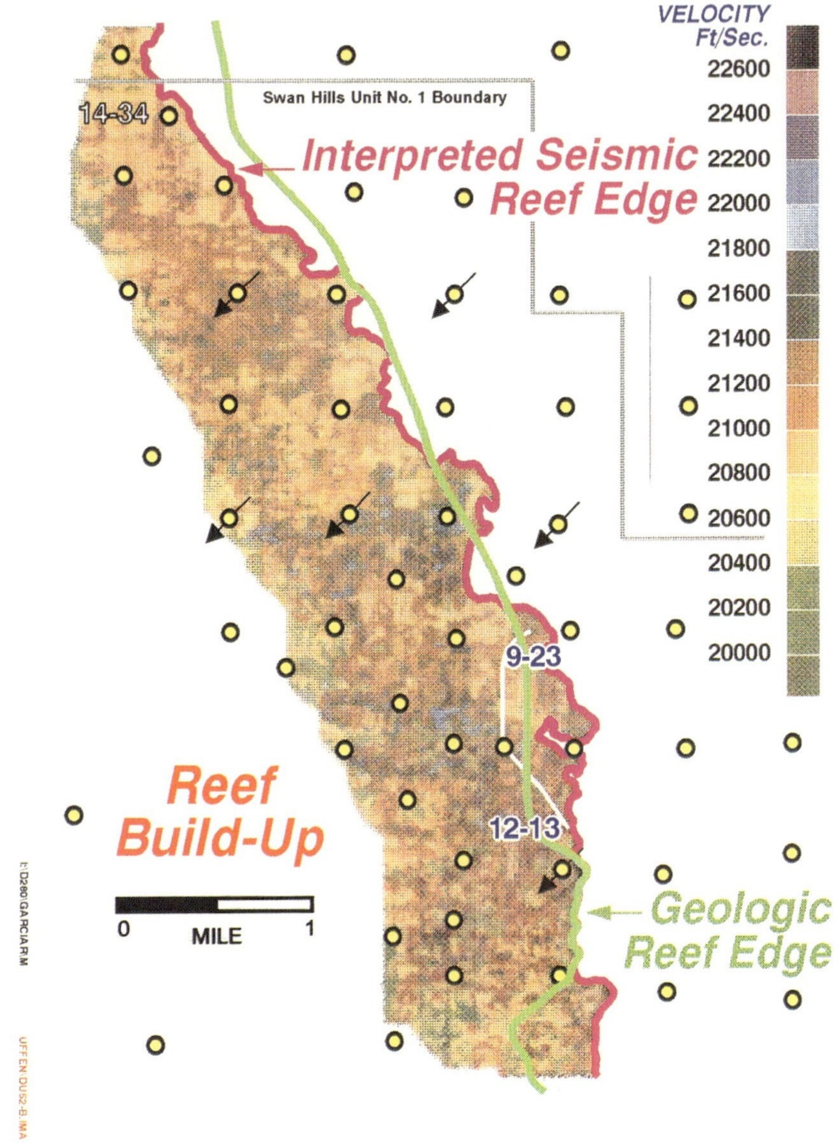

FIGURE 13. Isovelocity map of the reef and platform sections is shown here in a hypersensitive color display. The geological zero reef edge is superimposed.

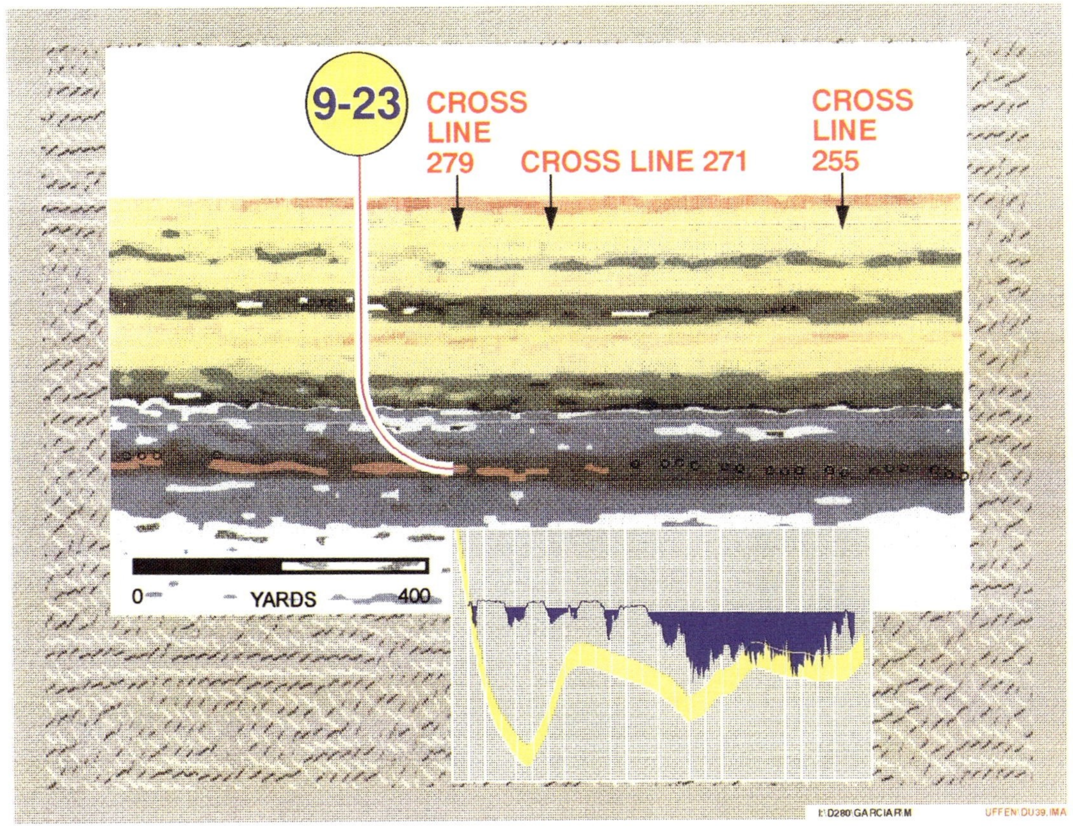


FIGURE 14. This inversion seismic profile, constructed with the aid of a workstation, follows the trajectory of the 9-23 well bore. The well bore is shown by the red line until it intersects the reef, with the total depth drilled indicated by the end of the sonic log. The recorded sonic log is depicted below, with porosity shaded in blue and the vertical trajectory, shown here vertically exaggerated, drawn in yellow.

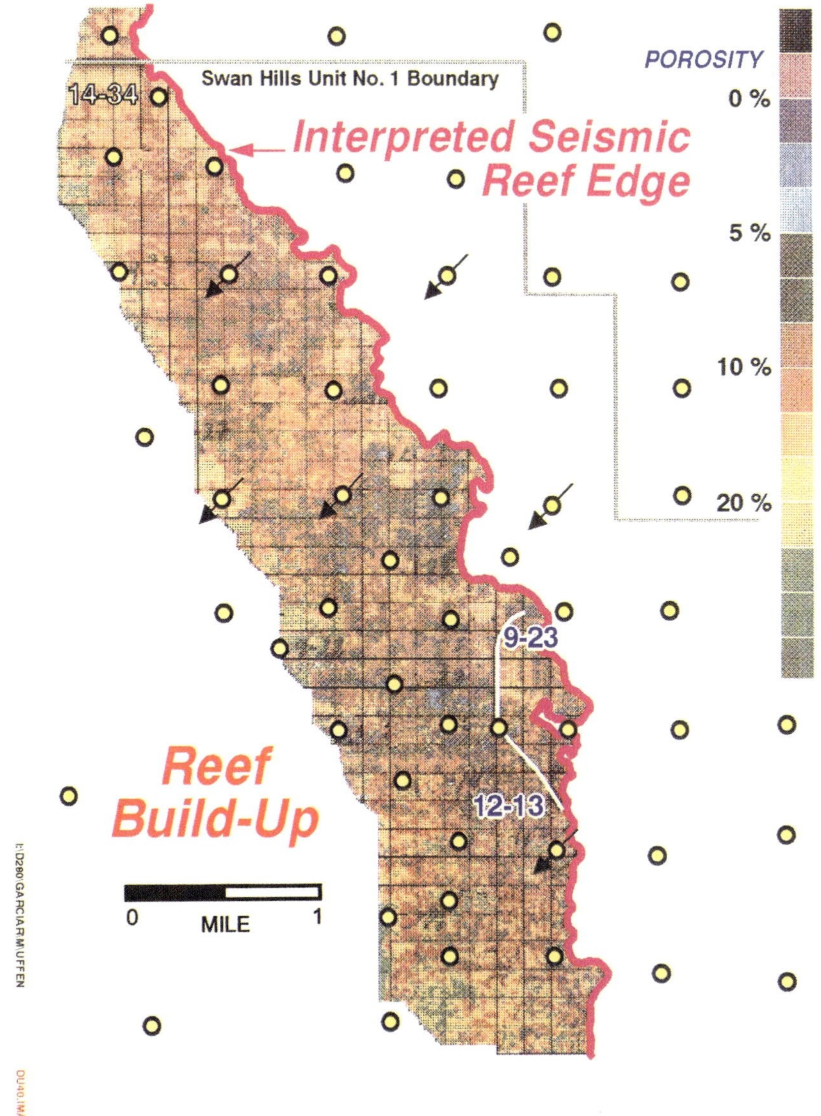


FIGURE 15. Implied porosity map of the reef and platform section, derived by empirical observation by the interpreter. Technology to create this map by direct methods now exists.

# CHAPTER 20

Edwards, D. J., Using 3-D seismic data to investigate tectonic controls on Middle Devonian reef development, north-central Alberta, Canada *in* P. Weimer and T. L. Davis, eds., *AAPG Studies in Geology No. 42* and *SEG Geophysical Developments Series No. 5*, AAPG/SEG, Tulsa, p. 189–196.

# Using 3-D Seismic Data to Investigate Tectonic Controls on Middle Devonian Reef Development, North-Central Alberta, Canada

Darran J. Edwards*

## ABSTRACT

3-D seismic data are used to unravel the tectonic history of productive Middle Devonian (Givetian) patch reef dolomites of the Keg River Formation in the eastern Peace River Arch area of north-central Alberta. Here, subtle reactivations of an orthogonal network of Precambrian normal faults has directly influenced reef initiation and development. A patch reef trend consisting of irregular, low-relief buildups, together with smaller, neighboring bodies, is clearly identified to be rooted on raised Precambrian structures. On-structure amplitude dimming of the normally strong near–Keg River event marks the location of buildups. 3-D mapping allows faults imaged adjacent to Precambrian on-structure dimmed zones to be traced beneath reefs. Such structures constitute important fluid migration conduits, often controlling later dolomitization and hydrocarbon potential of Keg River Formation reefs.

## INTRODUCTION

In the past, 3-D surveys were once awkward and expensive to acquire. For this reason, only about ten 3-D surveys had been shot in Canada before 1984. With the introduction of multiline instruments in 1984, the number of 3-D surveys worldwide increased considerably, and the number acquired during 1988 in Canada exceeded that of the rest of the world (Hawthorne and Webster, 1989). However, compared to those elsewhere, potential anomalies in Alberta tend to be relatively small targets, such as pinnacle reefs, and this is often reflected in the reduced size of surveys in Alberta. As expected from the mature exploration nature of the Western Canada Sedimentary Basin, the density of conventional seismic coverage and well control is considerable in most areas. Clearly, most reefs should already have been found. However, reefs continue to be discovered, owing to factors such as small size, geological complexity, poor 2-D seismic data quality, and changing economics. 3-D seismic imaging provides an excellent exploration tool to inject new life into old fields by targeting such anomalies.

In this paper, the interpretational benefits of a 3-D survey acquired during the winter of 1991–1992 in the eastern Peace River Arch area of north-central Alberta are discussed. The survey was oriented toward development through improved imaging of Precambrian structure and Devonian reefs. From a 3-D perspective, the tectonic framework is described and significant lateral variations in seismic amplitude are investigated. Improved structural and reef definition is essential to facilitate further field development.

*Department of Geology and Geophysics, The University of Calgary, 2500 University Drive N.W., Calgary, Alberta, Canada, T2N 1N4.

## GEOLOGICAL SETTING

Structurally, north-central Alberta is dominated by the Peace River Arch, a large northeast-trending positive feature cored by Precambrian crystalline rocks and created by pre–Middle Devonian uplift. This feature was a principal influence on the sedimentation of Devonian carbonate and clastic rocks in north-central Alberta (Cant, 1988). Since the discovery of oil at Amoco's 2/3-11-96-6W5 test well in 1983, the eastern Peace River Arch has been an active exploration area. In the five years following, an estimated 55 million bbl ($8.7 \times 10^6$ m$^3$) of recoverable oil were discovered, and anywhere from 20 to 30 operating companies have been actively pursuing the Keg River–Senex play. The Keg River Formation of the Elk Point Group (Middle Devonian) is the principal reservoir facies along high-energy carbonate depositional trends consisting of patch reefs. Apart from initial discoveries at the Senex, Panny, Trout, and Kidney Fields (Figure 1), this play is characterized by a large number of oil pools with an anomalously low average pool size of less than $1 \times 10^6$ m$^3$ (Podruski et al., 1988). Recovery factors range from 15 to 50%, owing to the reefal depositional facies and tectonic-diagenetic enhancement of porosity and permeability. In assessing the remaining potential in this play, Podruski et al. (1988) estimated a median expectation value of $33 \times 10^6$ m$^3$ OIP.

During Keg River Formation time, the Senex area was part of a fringing shelf surrounding the Elk Point Basin. Shoreward of the shelf edge, low-relief patch reefs developed, typically on or above prominent basement structures. In some cases the top of the underlying block was probably emergent, because the reefs grew on the windward side of the high (Dunham et al., 1983). The Keg River Formation is typically productive where it is structurally closed across Precambrian highs. Typically, it is draped across these structures as a result of differential compaction of lower Paleozoic sediments. These clastic sediments can also be productive. Recognition of closed Precambrian structure is thus vital to successful Keg River Formation exploration.

## MIDDLE DEVONIAN STRATIGRAPHY

The generalized Devonian stratigraphic nomenclature for the eastern Peace River Arch area is summarized in Figure 2. The overlying or flanking Precambrian granite-gneiss basement rock is a series of siliciclastic sands, conglomerates, and carbonate interbeds. The distribution of diachronous, coarse-grained Granite Wash or basal clastic deposits was strongly influenced by Precambrian paleotopography. They are commonly derived from erosion of that surface and may depositionally pinch out against or drape over basement highs, creating potential reservoir targets. Many of the larger Precambrian boulders show evidence of encrustation with stromatoporoids (Campbell, 1987).

The Keg River Formation (Elk Point Group) consists of carbonate with a variable degree of argillaceous content. Patch reef dolomites of the Upper Keg River Member were originally deposited under higher-energy conditions rooted on or near raised Precambrian structure (Figure 3). Bioclastic gravel-stromatoporoid sequences appear to form the bulk of the deposits associated with these reefal facies. The reservoir is a clean, recrystallized dolomite with good to excellent vuggy porosity and solution-enlarged fractures.

The Keg River Formation is overlain by the Muskeg Formation, an interlayered sequence of salts and anhydrites. The basal anhydrite of this unit seals the Keg River Formation reservoir facies. Although this basal anhydrite has a slightly higher velocity and density than the underlying Keg River Formation, the contact between the two units cannot be mapped on seismic data. However, it is possible to seismically delineate the contact between the basal anhydrite and the overlying salt; this is referred to as the basal anhydrite/Keg River event (Mitchell, 1988), or near–Keg River event (Anderson et al., 1988). Figure 4 shows a characteristic well tie to synthetic data illustrating these relationships.

## EASTERN PEACE RIVER ARCH TECTONISM

A typical 2-D seismic section for the eastern Peace River Arch area (Figure 5) illustrates the characteristic undulatory nature of reflective horizons. The Precambrian surface consists of anomalous structural highs, with relief typically ranging from 10 to 100 m. Generally, such structures have been regionally mapped from available well data as having discrete trends and being areally closed. This pattern is consistent with the idea of a surface fractured by conjugate faults, both parallel and perpendicular to the northeast-southwest orientation of the Peace River Arch.

The first documented episode of reactivated normal faulting affecting the Arch occurred during deposition of the Elk Point Group in the Middle Devonian. Granite Wash detritus was eroded from uplifted blocks and interbedded with Elk Point sediments, indicating episodes of tectonism (Campbell, 1987). It is therefore apparent that paleotopography at this time was controlled by pre–Middle Devonian normal block faulting and subsequent erosion of that surface. There is further evidence, both geological (e.g., Cant, 1988) and geophysical (e.g., Anderson et al., 1988), of renewed faulting since Devonian time, either on new surfaces or as reactivations of former faults.

From Figure 5, direct identification of faulting using 2–D seismic data is seen to be difficult, due to the limited-offset, subvertical nature of these basement displacements. Furthermore, these subtle, extensional faults are thought to have undergone wrench reactivation involving lateral displacement. Extrapolation of focal mechanism evidence by Horner et al. (1994) from adjacent, low-level (M < 4) seismicity suggests that a dextral strike-slip component to the faulting may indeed be present. Seismic events are observed to *smear* horizontally across such discontinuities, with no conspicuous breaks, producing the characteristic undulatory paleotopography. Detection of these structures in the eastern Peace River Arch has significant implications in terms of the proposed development of overlying Keg River Formation patch reefs.

## THE ROLE OF 3-D SEISMIC DATA IN KEG RIVER EXPLORATION

The subtle nature of the faulted, time-structural relief on the Precambrian surface is particularly evident on Figure 5. As mentioned previously, recognition of closed Precambrian structure is vital to successful exploration for this play, and identification of possible tectonism on 2-D seismic profiles can be open to question. The Keg River Formation patch reefs themselves are also limited in amplitude, growing up to a maximum thickness of approximately 50 m over uplifted blocks. Although 2-D data can often successfully locate significant Precambrian paleostructure and overlying Keg River Formation buildups, smaller anomalies frequently lack definition and are not clearly resolved. These could easily be overlooked, particularly since this Keg River–Senex play is characterized by a large number of oil pools with an anomalously low average pool size. Patch reefs are frequently small and irregular, commonly growing over the fault planes that had once provided the loci for initiation of reef growth. Delineation of these anomalies using only cross-sectional interpretations thus becomes less reliable. Detailed mapping of Precambrian fault blocks and Keg River Formation reefs through a plan view would seem to be essential for accurately determining what their morphology is and whether such carbonate buildups are structurally closed across these Precambrian highs.

Since a direct relationship between reef development and fault movement has been established for this area (e.g., Campbell, 1987), it may be possible to predict further Keg River–Senex play pools by extrapolation of proven basement trends. Smaller satellite patch reefs may be located along strike from larger buildups, previously discovered using 2-D seismic data.

3-D seismic imaging would seem to provide an excellent opportunity for examining the size, shape, and trend of these Keg River Formation patch-reef anomalies, together with the fault-bounded highs on which they grow. 3-D data allow considerably improved interpretability over conventional 2-D profiles.

## 3-D SEISMIC SURVEY

A 3-D seismic survey was acquired by Western Geophysical Ltd. in December 1991 and January 1992 across a Keg River–Senex prospect in the eastern Peace River Arch area in north-central Alberta. The exact location of the survey within this active exploration area is not revealed for proprietary reasons. An area of 16 km$^2$ (6.2 mi$^2$), measuring 5.8 km (3.6 mi) by 2.8 km (1.74 mi), was covered with a shotpoint spacing of 60 m (197 ft) and a line spacing of 300 m (984 ft), resulting in an average coverage of 20-fold. The relevant acquisition and processing parameters are summarized in Tables 1 and 2, respectively. A single-hole dynamite source with a 1-kg charge was used. Processing followed a relatively standard set of procedures through to one-pass 3-D migration.

The primary purpose for shooting this survey was field development, considering this area already contains a number of established wells. It was anticipated that a 3-D dataset could contribute to enhancing the recovery from producing pools by better 3-D delineation of these irregular Keg River Formation patch reefs. Such 3-D seismic imaging may also add to reserves by finding supplementary accumulations near existing pools. Any subsequent well programs would obviously benefit from such a survey, helping to alleviate the uncertainty of drilling in this area.

## 3-D INTERPRETATION

Two representative vertical sections from the 3-D grid are presented: (1) an inline section oriented west to east across one of the larger Precambrian highs (Figure 6) and (2) an arbitrary crossline section that parallels the north-northeast–south-southwest crest of a series of such positive structures (Figure 7). The relative positions of these two profiles are shown in the 3-D survey grid of Figure 8.

Synthetic and field data correlate well at the Precambrian, near–Keg River (or basal anhydrite), and Watt Mountain events. These horizons are annotated pink, yellow, and red, respectively (Figures 6 and 7). Like those observed in Figure 5, these units have an undulatory morphology and appear to drape across anomalous Precambrian structures. This is attributed to differential compaction of the overlying Paleozoic section. However, in the near–Keg River–Precambrian interval, carbonates appear to thin significantly from off-structure to on-structure locations, suggesting some form of rejuvenated tectonism.

The Precambrian surface itself has gently undulating relief with no definite offsets. These sections illustrate that all the major structural highs in this field are penetrated by wells, frequently with successful results. These positive structures exhibit a maximum amplitude of 40 ms, corresponding to approximately 100 m. This value may be slightly overestimated since overlying lateral velocity variations, including those due to patch reef growth, above such paleohighs may manifest anomalous structure on the Precambrian surface.

A significant reduction in amplitude on the normally strong Precambrian event is clearly associated with these structural highs. The near–Keg River event shows similar amplitude changes moving

**TABLE 1. Acquisition Parameters for Eastern Peace River Arch 3-D Survey**

Shot by: Western Geophysical
Recording Instrument: I/O system one
Number of channels: varies (approximately 300)
Fold: 20
Sample interval: 2 ms
Record length: 3 s
Format: SEGD demultiplexed

**Sources**
- Shotpoint spacing: 60 m
- Shotline spacing: 300 m
- Source: dynamite (1 × 1 kg)
- Shot depth: 15 m

**Receivers**
- Geophone station spacing: 60 m
- Geophone line spacing: 250 m
- Geophone: LRS-1011 14 Hz 70% damping
- Geophone array: 9 over 20 m

**TABLE 2. Processing Parameters for Eastern Peace River Arch 3-D Survey**

**Prestack**
Preserved amplitude compensation
Instrument and geophone dephasing
Trace editing
Surface-consistent deconvolution
- 200–1100-ms design window
- 320-ms operator length
- 0.1% prewhitening

GLI weathering statics
- datum: 600 m ASL
- replacement velocity: 2200 m/s

Surface-consistent residual statics
Interactive velocity analysis
NMO correction and muting
- distance limit to 1500 m

3-D Invest random noise attenuation
- 30% noise add-back
- 200 ms Δ T

CDP trim statics (non-surface-consistent)
Stack

**Poststack**
One-pass 3-D migration with 100% stacking velocities
Single-gate equalization
- window: 700–1300 ms

Bandpass filter: 10/15–60/70 Hz

laterally across such structures. Such on-structure amplitude dimming frequently indicates the presence of a Keg River Formation patch reef. As demonstrated by modeling (Mitchell, 1988; Anderson et al., 1988), the loss of amplitude can be attributed to interference effects resulting from a thinning of basal anhydrite (Muskeg Formation) over the Keg River buildup and from Granite Wash or basal clastic unit thinning (or even absence) over Precambrian structure. This provides a sharp contrast with the off-structure amplitude character. Representative velocities for these units are indicated in a typical on-structure velocity log, shown in Figure 4. In addition, 3-D physical modeling in this area, as part of an ongoing Ph.D. study at the University of Calgary into basement-carbonate interaction in central Alberta, confirms such a model.

The Precambrian surface time slice (Figure 9) confirms that 3-D seismic data are an invaluable tool when it comes to accurately delineating paleostructure in terms of orientation, shape, and size of basement highs. These features, although subtle on both 2-D data and vertical sections from the 3-D grid, become well defined on horizontal sections by a series of elevated structures aligned along a crest-line oriented northeast to southwest. A number of smaller peripheral basement highs lie adjacent to this main trend. This paleosurface represents only 40 ms of time-structural relief, from 816 ms to 856 ms. A 3-D northwest perspective view of this horizon further exaggerates this faint time-structural relief (Figure 10); closed, Precambrian structural highs can be easily distinguished.

The Precambrian paleorelief is consistent with a conjugate fault pattern (Figure 11), where fault blocks were rounded and individual fault scarp surfaces smoothed through ensuing erosional processes. The main set of paleohighs essentially constitutes a northeast-trending horst structure, compartmentalized into a series of discrete blocks by northwest-southeast–oriented normal faulting. However, even within this 16 km$^2$ area, the size, complexity, and closure nature of these fault blocks vary considerably. These northwest- and northeast-aligned orthogonal pairsets correlate with similar regional trends, identified using potential-field data (Edwards et al., pers. comm.,)

The morphologies of the near–Keg River (basal anhydrite) and Watt Mountain time slices (i.e., the tops of the respective units) on Figures 12 and 13, respectively, closely parallel that of the Precambrian surface, confirming the draped nature of these events due to differential compaction. Similar time-structural relief is observed. Consistent lateral thickness variations above the Keg River Formation rule out the possibility of reactivated tectonism as a possible source for these similarities.

The hot colors, red and yellow, represent time-structural highs, whereas the cool colors, blue and purple, correspond to depressed paleotopography. In most cases, the extent of Keg River Formation patch-reef development conforms to the hot colors shown on Figure 12. Comparison with the Precambrian time slice confirms that the Keg River Formation carbonate buildups are directly rooted above paleohighs on the Precambrian surface. This display indicates the morphology of these buildups as typically elongate and irregular. On comparing this time slice with an inline section that intersects this structure (Figure 6), it becomes apparent that the aberrant tops and margins of these reefs are best interpreted from the perspective of a time slice. The areas designated by these colors also contain the majority of established oil wells for this field. Nevertheless, exploration potential exists through further development of these established reefs, now that their size and shape are accurately known, and via the number of neighboring, smaller buildups that can be distinguished. Certain satellite reefs are associated with unsuccessful drilling where the wells have apparently just missed such buildups. Some dry and abandoned wells are slightly obscured by darker colors (off-structure) on time slice displays.

Comparing Figure 14, the Precambrian–Near–Keg River Formation (Pc-nKR) isochron map, with Figure 9 indicates that the overall Keg River Formation sequence is commonly thinner above Precambrian highs where patch reefs have grown. Close investigation of the Precambrian surface around the largest paleohigh structure (Figure 15) reveals a marked reduction in amplitude underneath full reef buildups. Northwest-oriented fault trends may be followed on either side of this main structure but not directly over the top. Such dimming is a response to the on-structure thinning of the Granite Wash unit. This location also corresponds to the pronounced thinning of the Pc-nKR isochron and the full Keg River Formation patch reef development. Faulting may be projected along trend through the dimmed zone on the Precambrian surface, thereby indirectly implying the presence of structure. Carbonate buildups growing laterally over fault scarps, whose raised side initiated growth, seems to be a common feature of Keg River deposition in this area.

## SUMMARY


This study demonstrates the interpretational value of a 3-D survey versus 2-D seismic data in the eastern Peace River Arch area. Here, Keg River Formation patch reefs (Middle Devonian) were initiated through paleobathymetric differences exerted by Precambrian block faulting and subsequently grew laterally to overlie these periodically active fault planes. Ambiguous 2-D interpretations are overcome by 3-D time slice displays of the tectono-depositional framework. From a time slice perspective, structures can be interpreted directly instead of by inference. The horizontal slice allows an accurate delineation of the main patch reef trend and smaller neighboring buildups, in addition to the complex orthogonal faulting that initiated shoaling. A schematic diagram summarizing local reef-fault block interaction is shown in Figure 16.

Using 3-D data, undulating Precambrian paleostructure is interpreted as orthogonal normal faulting, with fault blocks subsequently smoothed through ensuing erosion, producing Granite Wash detritus. The Precambrian fault map consists of northwest-southeast and northeast-southwest oriented pairsets. Local variation of the size, complexity, and closure nature of discrete fault blocks is observed. Precambrian on-structure amplitude dimming indicates a thinning of the Granite Wash unit. Rooted above such structures, Keg River Formation patch reefs are often seen to be small and irregular with limited vertical relief. As a result of basal anhydrite (Muskeg Formation) thinning, an amplitude reduction on the near–Keg River event suggests the presence of an underlying reef. Using time slice data, extrapolation of preferred fault trends imaged away from reefs indirectly implies the presence of faulting beneath buildups. These structures may represent potential fluid-migration conduits or produce fracture-related secondary porosity within buildups.

Through areally defining basement faulting and reef bodies, 3-D mapping gives the additional control necessary to accurately locate new wells in this established field. 3-D survey costs could easily be recovered through improved well success ratios. There is the potential for the discovery of more reservoirs in the Peace River Arch area using 3-D exploration philosophies oriented toward this basement-reef tectonic control model, and 3-D seismic data should play a significant role in future field appraisal and development planning.


### *Acknowledgments*


I thank International Colin Energy Corp., particularly Doug McLachlan, for making available the eastern Peace River Arch dataset. I am also grateful to Exploration Innovations Ltd. for facilitating workstation access and assistance and to Sigma Explorations Ltd. (Barry Korchinski) for 2-D data availability. At the University of Calgary, Jim Brown and Don Lawton provided useful advice at various stages of this work. Helpful suggestions by Tom Davis (Colorado School of Mines) are also appreciated. This paper was presented, in part, at the 1994 AAPG Convention, Denver, Colorado.


## REFERENCES CITED


Anderson, N. L., R. J. Brown, and R. C. Hinds, 1988, A seismic perspective on the Panny and Trout fields of north-central Alberta, Canadian Journal of Exploration Geophysics, v. 24, p. 154–165.

Brown, D. L. and D. L. Brown, 1987, Wrench-style deformation and paleostructural influence on sedimentation around a cratonic basin, *in* J. A. Peterson, D. M. Kent, S. B. Anderson, R. H. Pilatzke, and M. W. Longman, eds., Williston Basin—Anatomy of a Cratonic Oil Province: Rocky Mountain Association of Geologists, Denver, Colorado, p. 57–70.

Campbell, C. V., 1987, Stratigraphy and facies of the Upper Elk Point Subgroup, northern Alberta, *in* F. F. Krause and O. G. Burrows, eds., Devonian Lithofacies and Reservoir Styles in Alberta: Second International Symposium on the Devonian System, Calgary, Canadian Society of Petroleum Geologists, p. 243–286.

Cant, D. J., 1988, Regional structure and development of the Peace River Arch, Alberta: a Paleozoic failed-rift system?: Bulletin of Canadian Petroleum Geology, v. 36, p. 298–295.

Dunham, J. B., G. A. Crawford, and W. Panasiuk, 1983, Sedimentology of the Slave Point Formation (Devonian) at Slave Field, Lubicon Lake, Alberta, *in* P. M. Harris, ed., Carbonate Buildups—A Core Workshop, n. 4, p. 73–111.

Hawthorne, R., and R. Webster, 1989, 3-D seismic, *in* N. L. Anderson, L. V. Hills, and D. A. Cederwall, eds., The CSEG/CSPG Geophysical Atlas of Western Canadian Hydrocarbon Pools: The Canadian Society of Exploration Geophysicists/The Canadian Society of Petroleum Geologists, First Western printing, Calgary, Alberta, Canada, p. 311–317.

Horner, R. B., J. E. Barclay, and J. M. MacRae, 1994, Earthquakes and hydrocarbon production in the Fort St. John area of northeastern British Columbia: Canadian Journal of Exploration Geophysics, v. 30, p. 39–50.

Mitchell, K. I., 1988, A random walk through a Keg River carbonate shoal using 3-D seismic to examine a typical anomaly in the Kidney/Panny area (abs.): Canadian Journal of Exploration Geophysics, v. 24, p. 89.

Podruski, J. A., J. E. Barclay, A. P. Hamblin, P. J. Lee, K. G. Osadetz, R. M. Procter, and G. C. Taylor, 1988, Conventional oil resources of western Canada: Geological Survey of Canada Paper 87–26, p. 32–36.

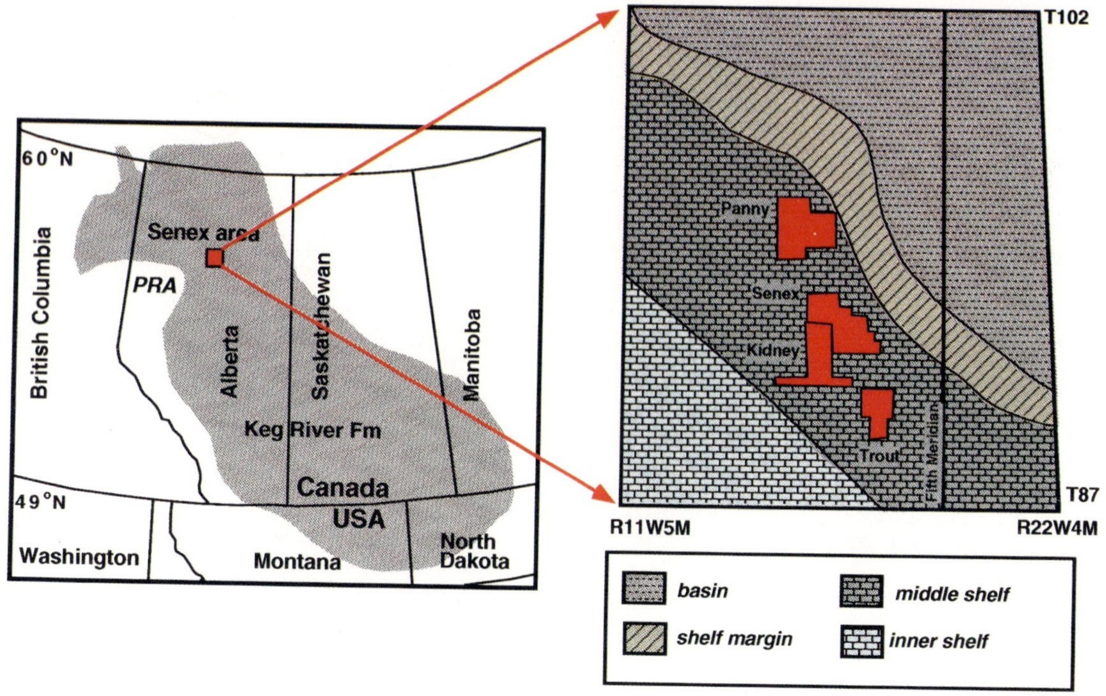


FIGURE 1. Location map of the eastern Peace River Arch (PRA) area, Alberta, showing the outline of the Elk Point Basin, in which Keg River Formation patch reefs were deposited. The main producing reefs and this 3-D seismic survey are situated on a middle carbonate shelf fringing the Arch (after Campbell, 1987). For proprietary reasons, the exact positioning of this survey cannot be shown.

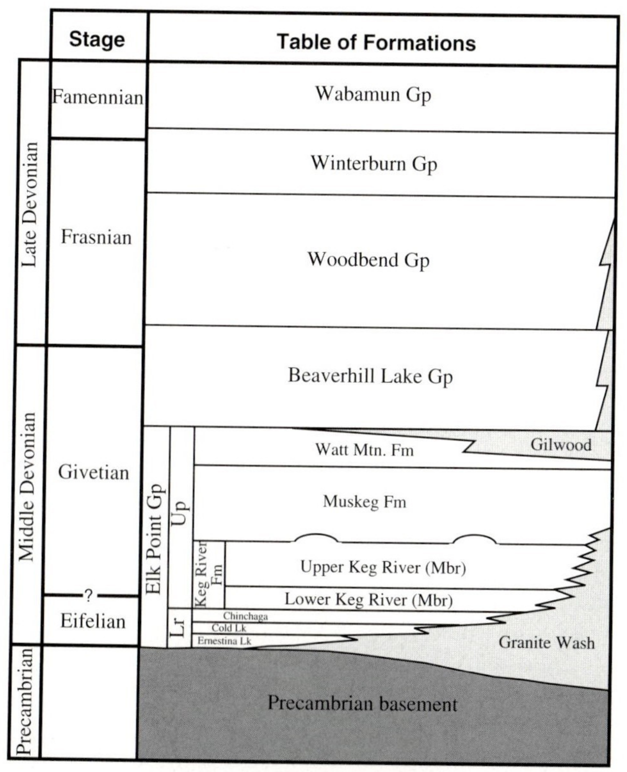


FIGURE 2. Generalized Devonian stratigraphy of the eastern Peace River Arch area. (Modified from Campbell, 1987; Cant, 1988.)

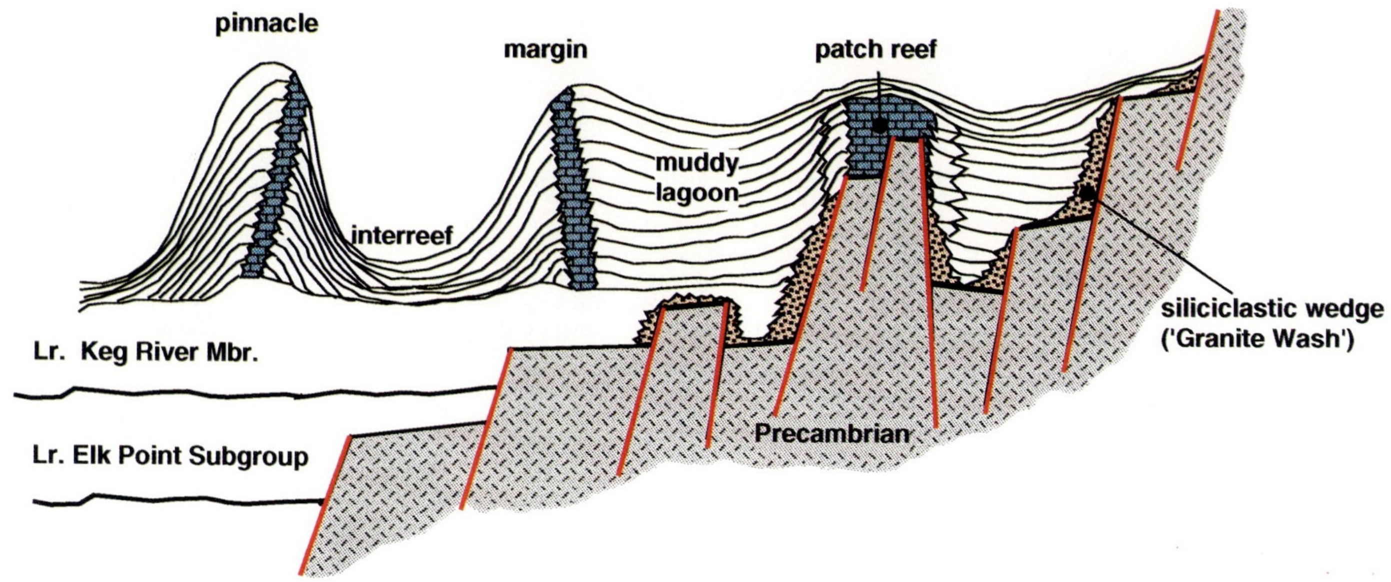


FIGURE 3. Schematic cross section, eastern Peace River Arch area, illustrating the direct relationship between Precambrian structure and Keg River Formation patch reef development (after Campbell, 1987).

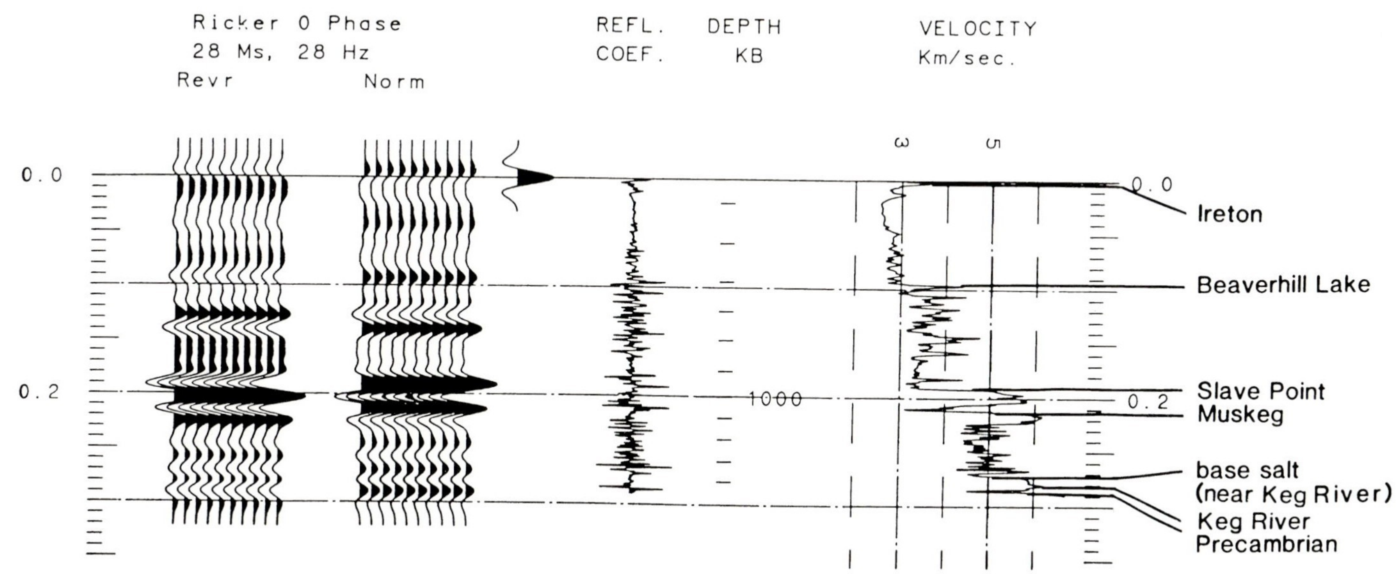


FIGURE 4. Characteristic velocity log and synthetic well tie for the Ireton Formation–to–Precambrian interval in the eastern Peace River Arch area.

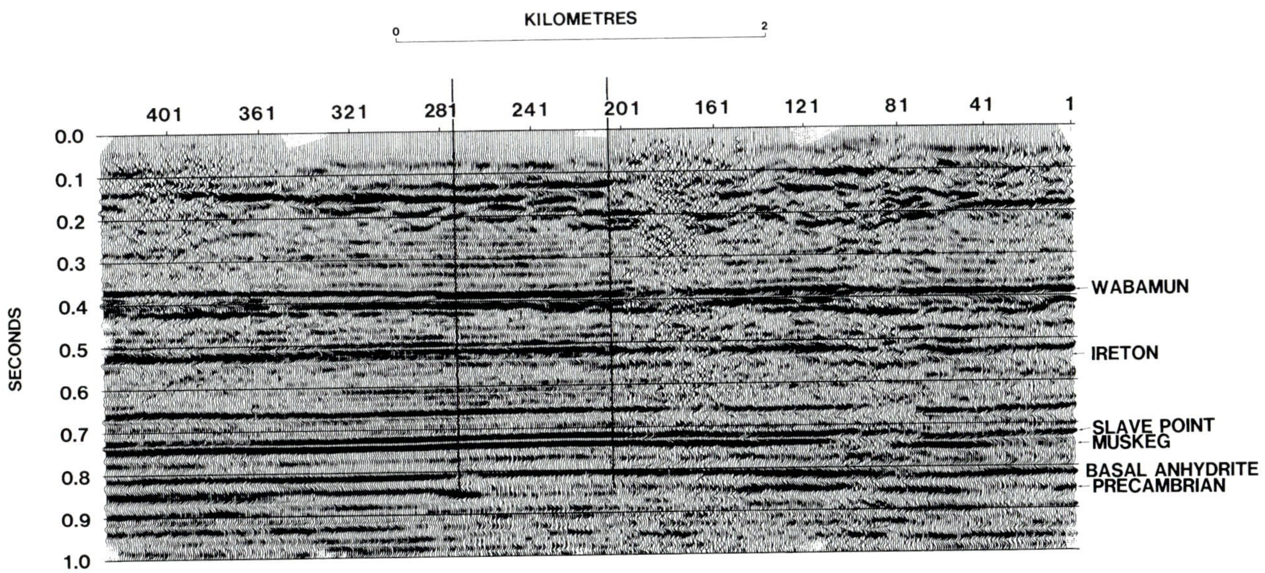


FIGURE 5. 2-D template line, eastern Peace River Arch. Note the characteristic undulatory nature of the Precambrian basement surface and draping of overlying units, both of low time-structural relief (after Anderson et al., 1988).

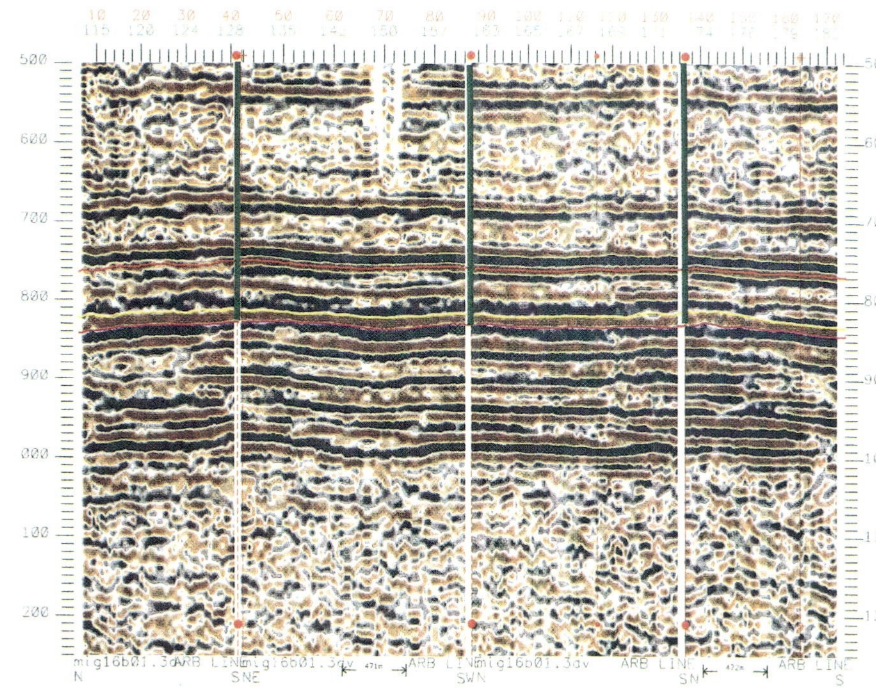

FIGURE 7. A representative crossline, arbitarily oriented northeast to southwest to follow a crestline defining the principal trend of Precambrian highs. This line parallels the maximum development of Precambrian paleostructure and intersects wells producing from Keg River Formation reefs, localized over such structures. The color annotation convention for seismic events is the same as for Figure 6.

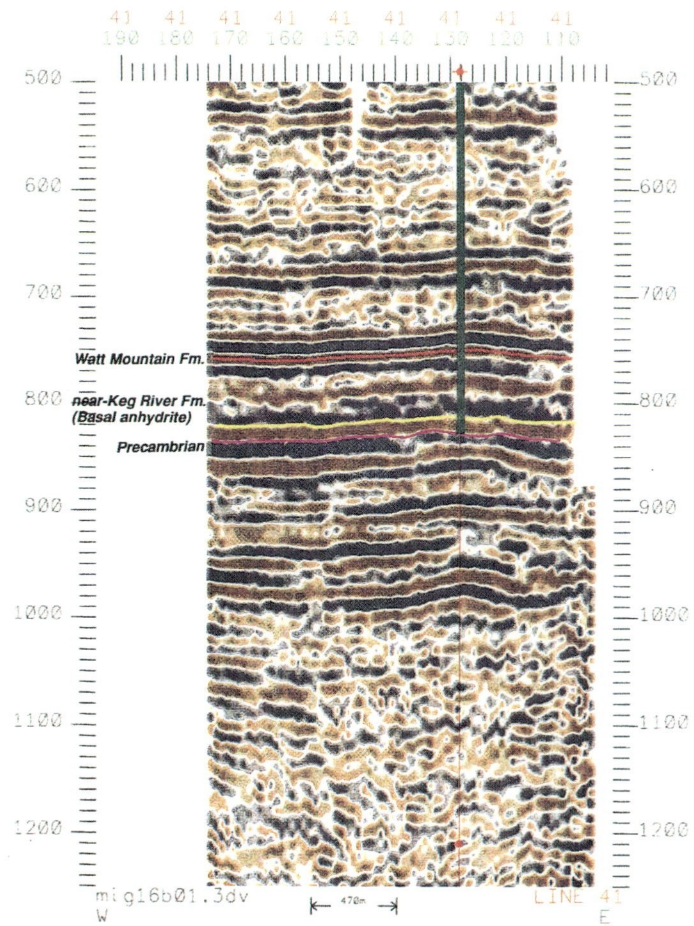


FIGURE 6. A representative inline (41), oriented west to east across one of the larger Precambrian highs and overlying Keg River Formation patch reefs. The Precambrian, near–Keg River Formation (basal anhydrite), and Watt Mountain Formation events are shown in pink, yellow, and red, respectively. The depth of well penetration is indicated by green.

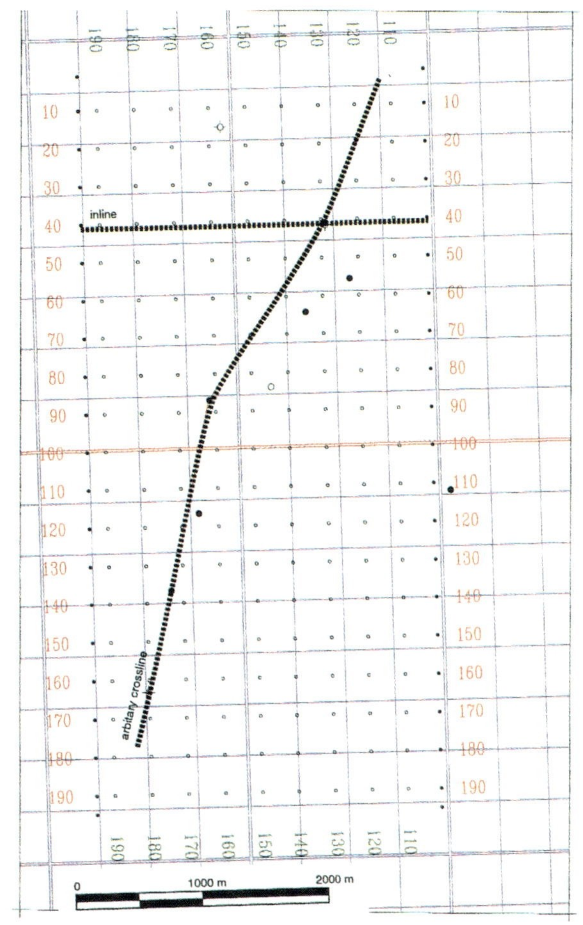


FIGURE 8. 3-D survey grid showing the relative positions of featured vertical seismic lines.

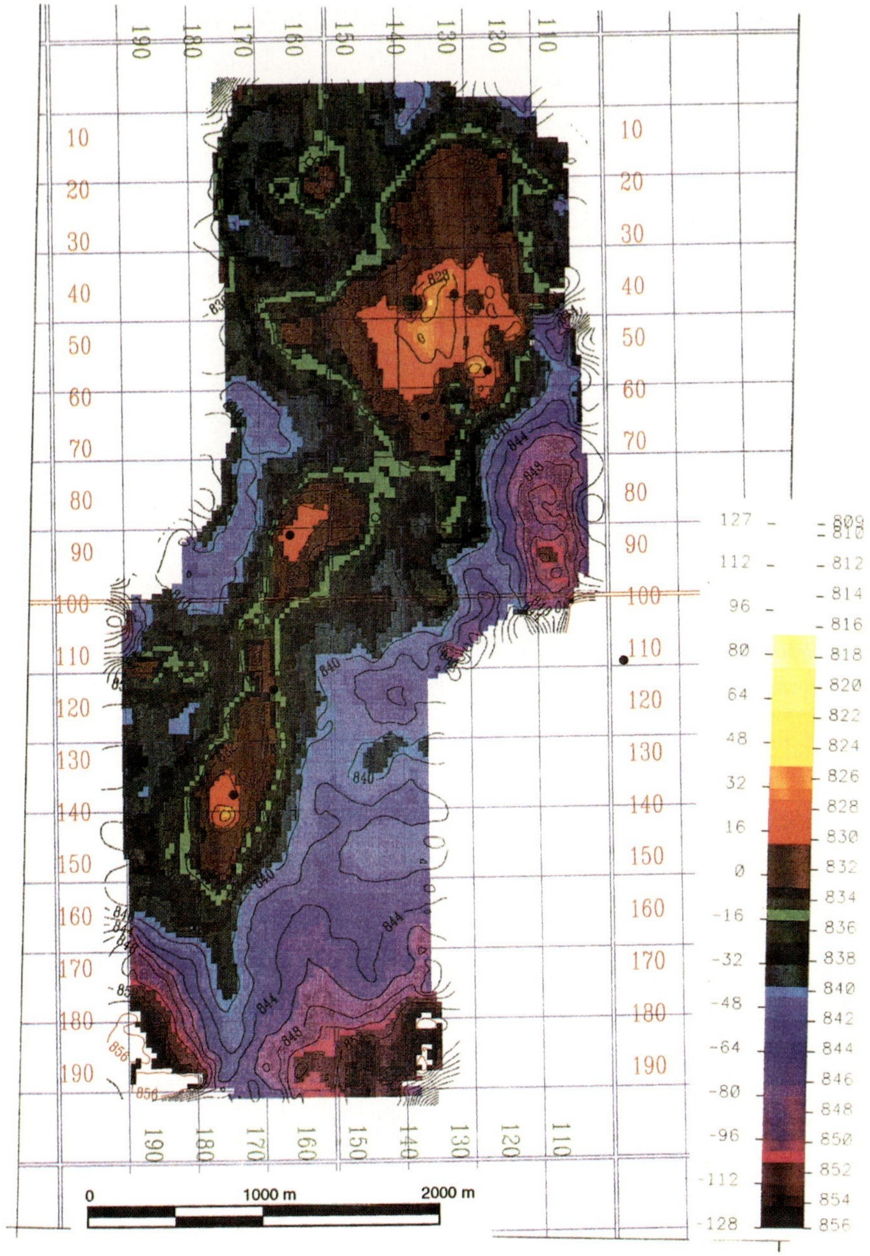

FIGURE 9. Precambrian time slice; the color bar represents time-structural relief in ms.

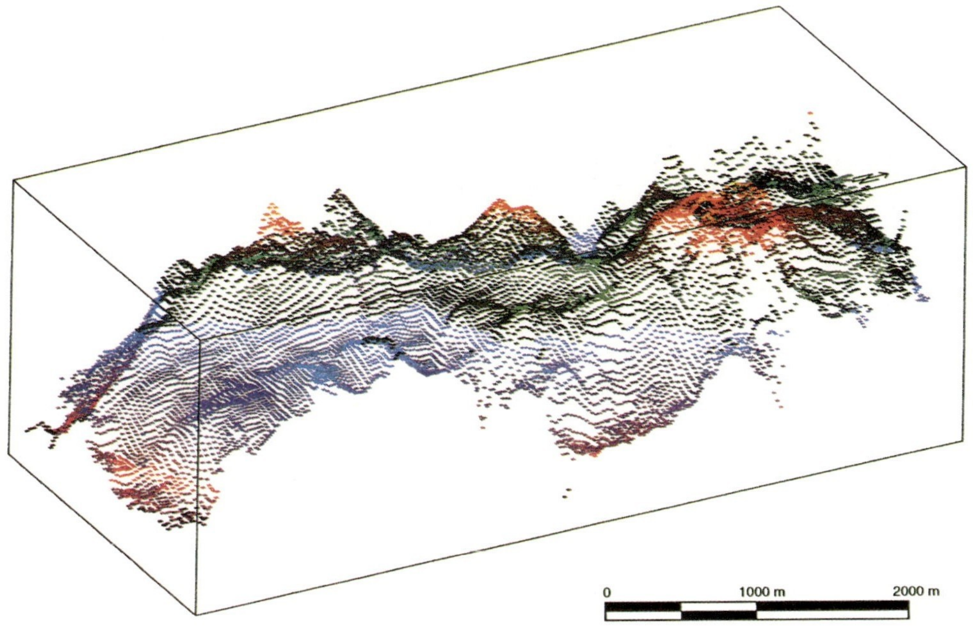

FIGURE 10. 3-D perspective view of the Precambrian paleosurface. The color change from yellow/red to blue/purple is the same as in the Precambrian time slice in Figure 9.

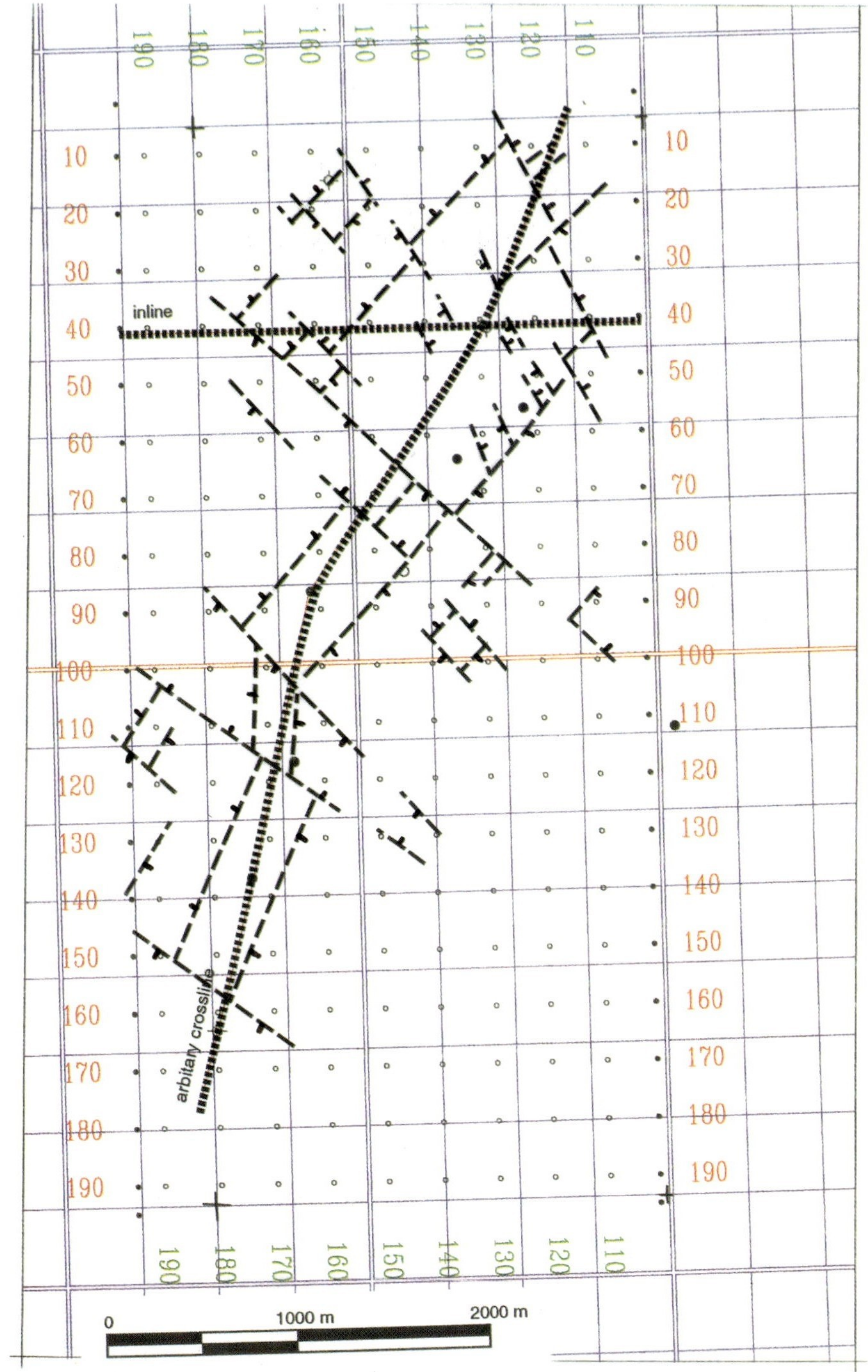


FIGURE 11. Precambrian fault structure map: tectonic interpretation of the Precambrian time slice shown in Figure 9.

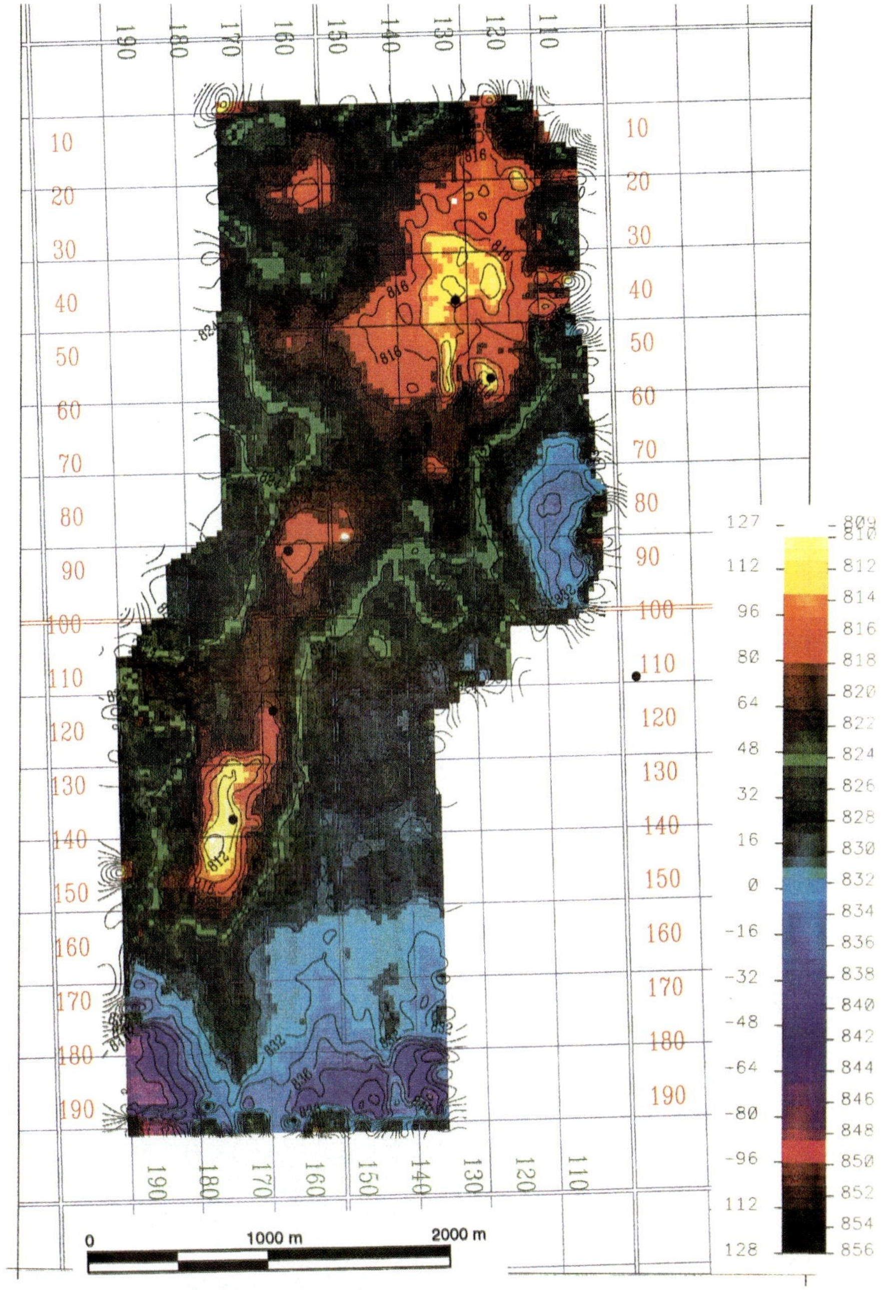


FIGURE 12. Near–Keg River Formation (basal anhydrite) time slice. The color bar represents time-structural relief in ms.

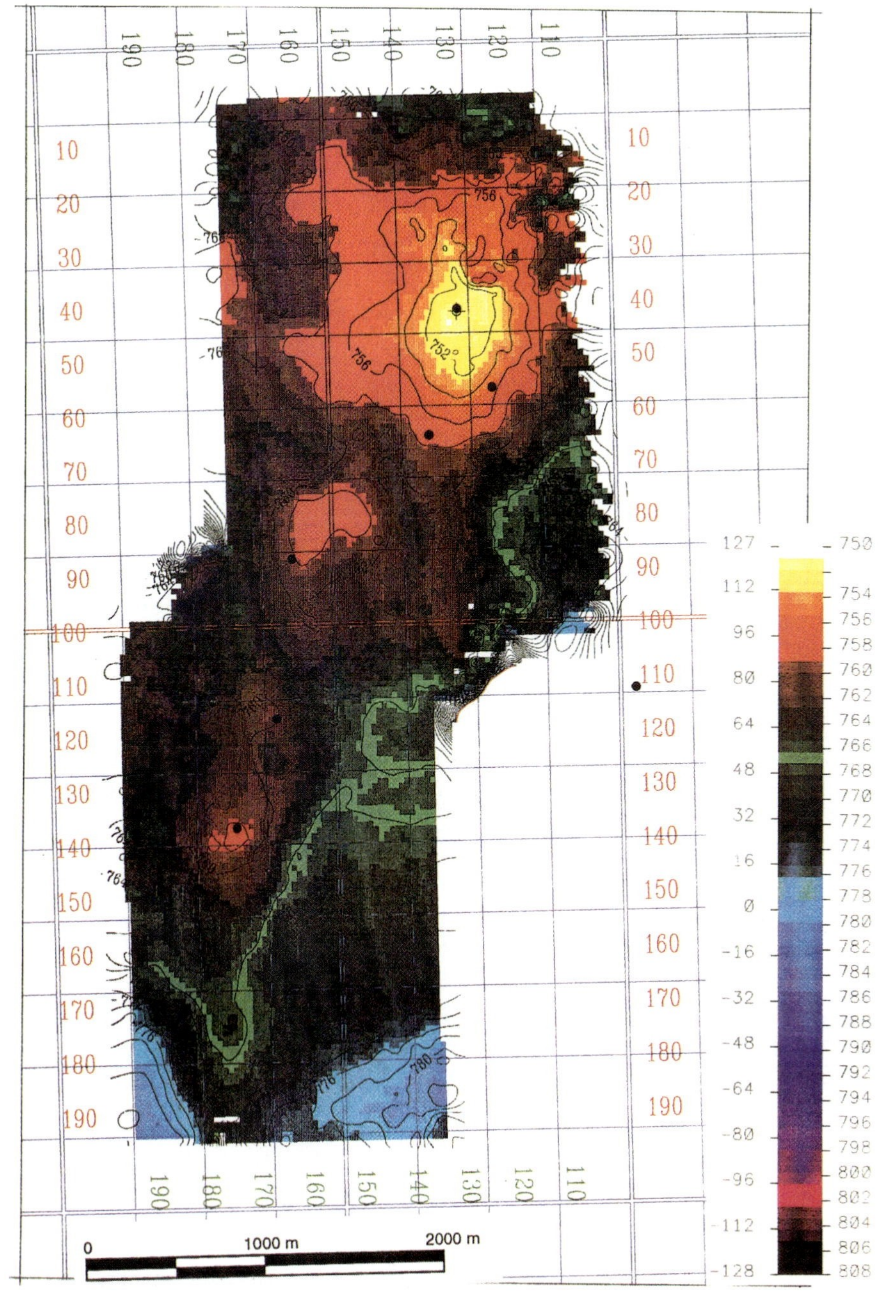


FIGURE 13. Watt Mountain Formation time slice. The color bar represents time-structural relief in ms.

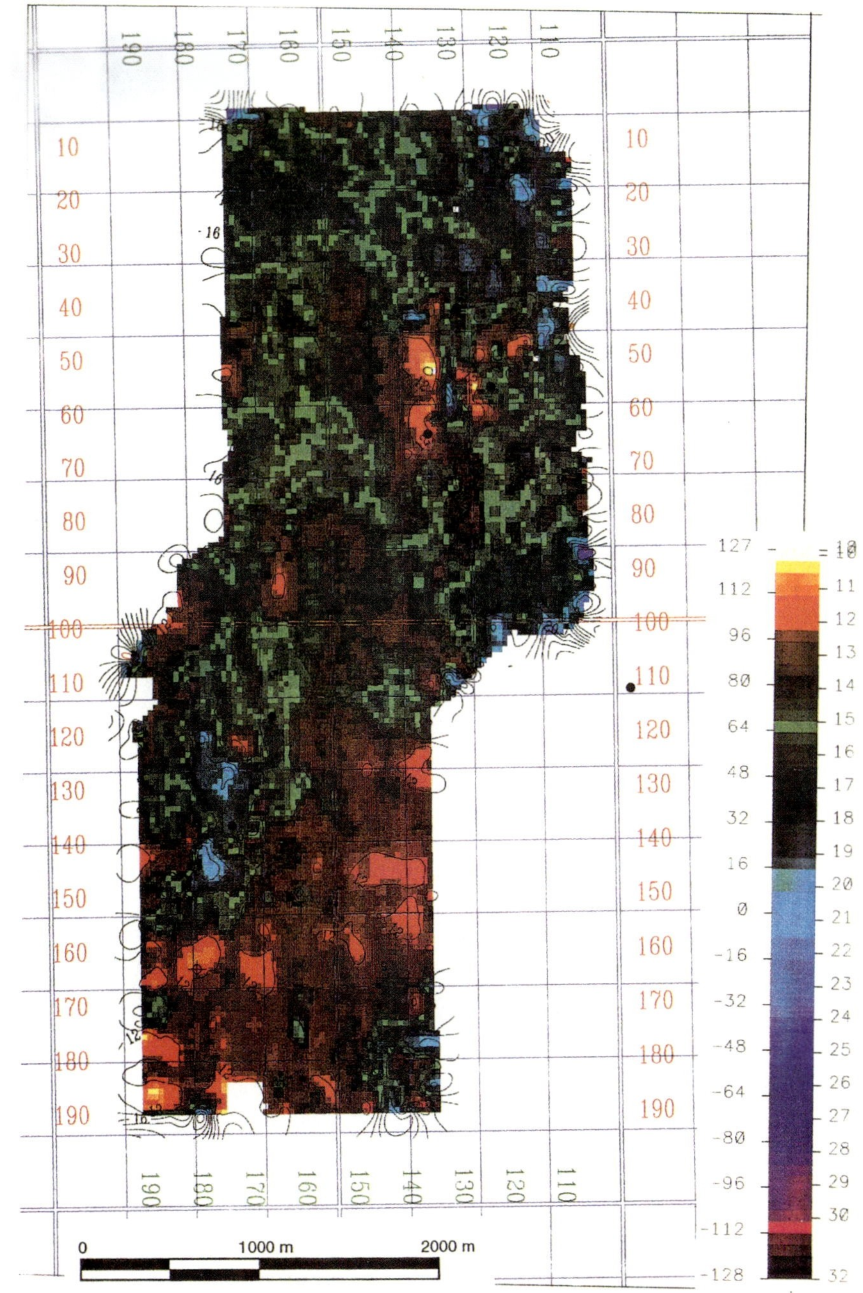


FIGURE 14. Precambrian–near–Keg River Formation isochron. The color bar units are ms.

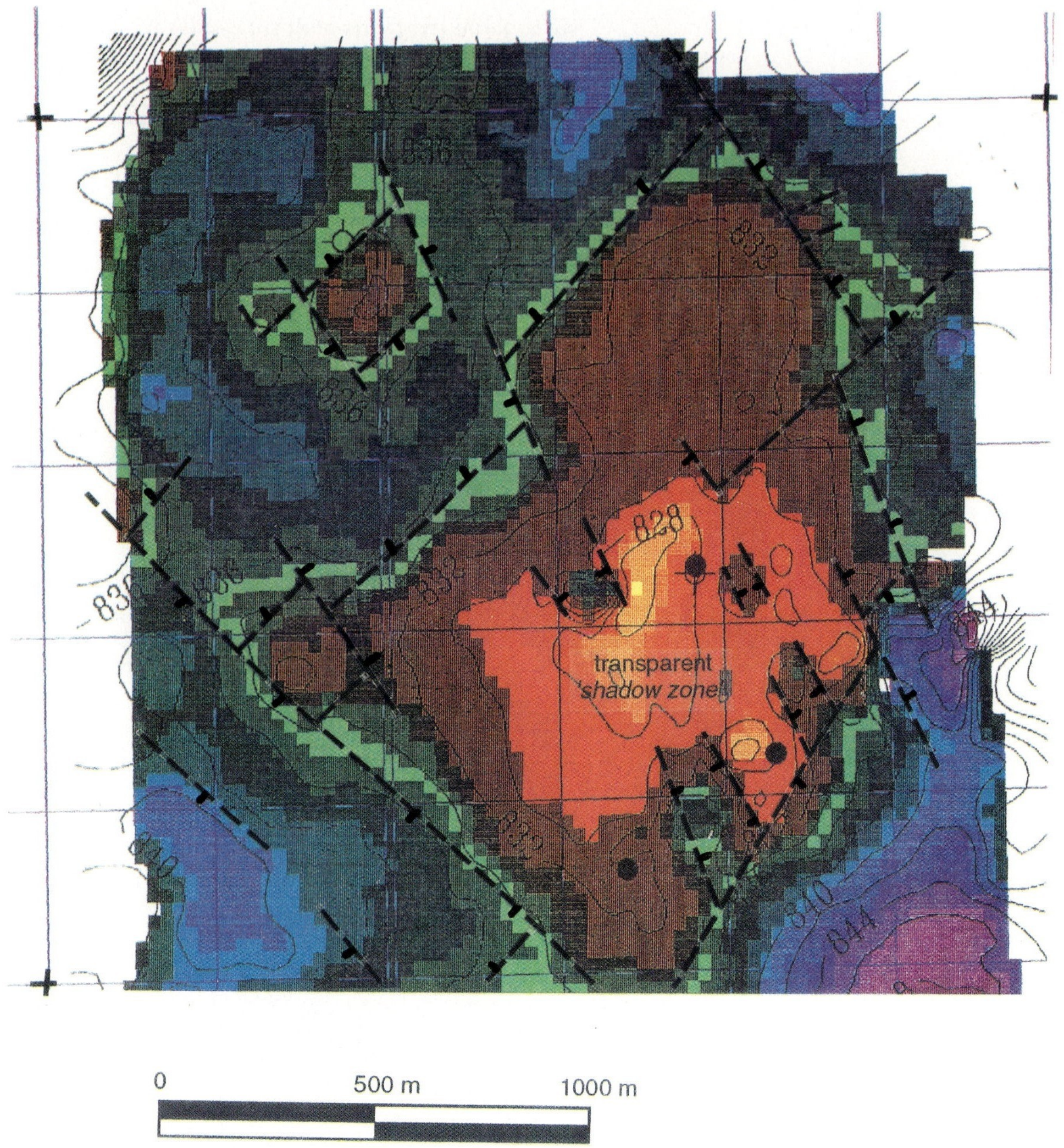


FIGURE 15. Interpreted Precambrian time slice for the northern part of the survey area, covering the largest patch reef–basement paleohigh expanse. The color bar is shown in Figure 9. A lack of apparent structure is a result of an amplitude dimming of the Precambrian event.

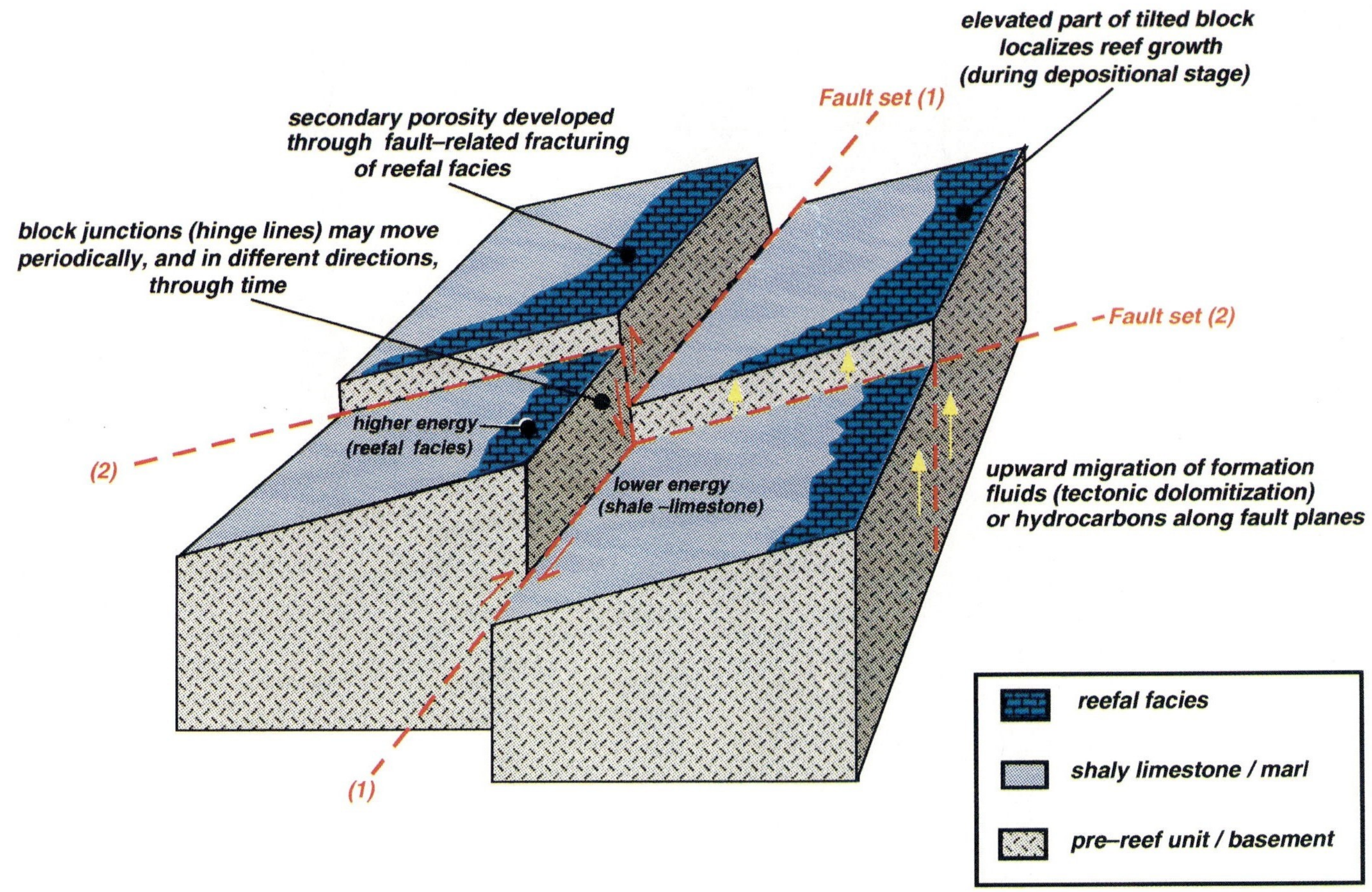


FIGURE 16. Schematic block diagram showing the types of reef–fault block interaction representative of the eastern Peace River Arch area (modified from Brown and Brown, 1987).

# CHAPTER 21

van de Sande, J. M. M., Prediction of reservoir parameters from 3-D seismic data for the Zechstein 2 carbonate play in the northeast Netherlands, *in* P. Weimer and T. L. Davis, eds., *AAPG Studies in Geology No. 42* and *SEG Geophysical Developments Series No. 5*, AAPG/SEG, Tulsa, p. 197–204.

# Prediction of Reservoir Parameters from 3-D Seismic Data for the Zechstein 2 Carbonate Play in the Northeast Netherlands

Jan M. M. van de Sande*

## ABSTRACT

The Zechstein 2 Carbonate reservoir is part of the Basal Zechstein Unit, which developed in the northeast Netherlands mainly as an anhydrite platform at the southern fringe of the Southern Permian Basin. The facies within the Zechstein 2 Carbonate Member reflect platform, slope, and basinal settings. The subsurface data collected in the exploration and development of the Zechstein fields have led to a detailed sedimentological model for the Zechstein 2 Carbonate Member.

Prospects in the Zechstein 2 Carbonate Member are seismically defined as structural highs at the Top Zechstein 2 Anhydrite level. As exploration moves towards the search for more subtle traps, 3-D seismic data become indispensable.

Basal Zechstein isochore maps determined on 3-D seismic volumes display the topography of the Basal Zechstein platform. Using the sedimentological model, these maps are used to estimate gross reservoir thickness. Variations in reservoir parameters, especially porosity, of the Zechstein 2 Carbonate Member are reflected in the seismic response. Through synthetic seismic modeling, 3-D seismic data allow spatial prediction of average reservoir porosity. The Collendoornerveen exploration well demonstrates the successful application of the quantitative interpretation techniques.

## INTRODUCTION

The Nederlandse Aardolie Maatschappij (NAM) has actively pursued the exploration of the Zechstein 2 Carbonate play over the last 40 years. The first successful well, Coevorden-1, was drilled in 1948 and the most recent one, Collendoornerveen-1, in 1990.

In the Zechstein 2 Carbonate play, abundant gas charge into the dip and fault-closed structures is sourced by the underlying coal measures of the Westphalian A and B (Carboniferous). Seals are provided by the overlying Zechstein 2 Anhydrite Member and the Zechstein 2 Halite Member. The main uncertainty in this play is the reservoir quality, especially the porosity and permeability of the carbonates.

The wealth of geological and geochemical data which has been collected in the exploration and development of the Zechstein fields has led to the establishment of detailed depositional and diagenetic models for the Zechstein 2 Carbonate Member (Clark, 1986; Van der Baan, 1990). More recently, interpretation of the extensive 3-D seismic coverage and integration of quantitative prediction techniques, as well as results of detailed fracture studies, have added to the comprehension and provided powerful additional tools in the exploitation of the play.

The experience gained within NAM over the years of successful exploration and development of the Zechstein 2 Carbonate Member has yielded a comprehensive toolbox of predictive techniques that can be employed successfully in the continuing exploration program in the Netherlands and elsewhere within the Zechstein hydrocarbon province (van de Sande et al., 1996).

This paper demonstrates that 3-D seismic data can effectively and successfully be used to estimate gross reservoir thickness and average reservoir porosity. A sedimentological model is used to make indirect estimates of gross reservoir thickness from time isochore measurements. A geophysical synthetic seismic data model is used to estimate the average reservoir porosities from reflection attribute measurements. The porosity prediction presented in this paper is a follow up of the work by Maureau and van Wijhe (1979), who qualitatively determined the porosity of the Zechstein 2 Carbonate Member using 2-D seismic data. Using similar models, predicted porosities are quantified using 3-D seismic data.

The following sections describe the geological setting of the Zechstein 2 Carbonate Member, the sedimentological model for the Zechstein 2 Carbonate Member, the interpretation techniques applied to 3-D seismic data to provide information on reservoir parameters such as gross thickness and average porosity, and the application of these techniques in the Collendoornerveen case history.

## GEOLOGICAL SETTING

This paper concentrates on the lowermost part of the Upper Permian Zechstein group, the Basal Zechstein Unit, which includes the important Zechstein 2 Carbonate reservoir (Figure 1).

The Basal Zechstein Unit in northeast Netherlands includes the entire Zechstein 1 Formation and the lowermost parts of the Zechstein 2 Formation (the Carbonate and Anhydrite Members; see Figure 1). It constitutes the nonmobile rock units formed prior to the deposition of the first thick, ductile Zechstein salt, the Zechstein 2 Halite Member. It overlies Carboniferous strata and the Lower Permian Rotliegend Group and was deposited during the Late Permian transgression which flooded the European Permian basins and led to the establishment of the major, intracontinental Zechstein sea. This inland sea extended some 1600 km, from the United Kingdom in the west, to the Baltic regions in the east (Ziegler, 1990). Northeast Netherlands was situated on the southern fringe of the Southern Permian Basin in the zone of transition from basin center to basin fringe with associated facies. In the hot and arid conditions, an anhydrite and carbonate platform developed, represented by the Basal Zechstein Unit. Platform carbonates are only a minor part of the Basal Zechstein Unit, the bulk being anhydrite. The general morphology of the setting in which the unit was formed is illustrated on the log correlation panel, showing the basin-slope-platform transition. (Figure 2).

## SEDIMENTOLOGICAL MODEL FOR THE ZECHSTEIN 2 CARBONATE MEMBER

The facies of the Zechstein 2 Carbonate Member largely reflect water depth (basin-slope-platform), salinity contrasts, sea level fluctuations, and ensuing variations in carbonate production. The position relative to the prevailing northeasterly Late Permian wind direction was also important. In the absence of any significant tides within the inland Zechstein sea, wind-generated wave action was the most important process determining environmental hydraulic energy contrasts between various areas.

High-energy ooidal grainstones developed along the windward face of the platform margin, where wave action was strongest. Triggered by storms or longshore currents, such deposits were constantly transported and winnowed over the platform. Ultimately they formed wave-resistant bars. Comparatively little sediment was shed down the windward slope of the platform, which is therefore steep; but due to the aggradational activity of accumulating wave-resistant bars, the carbonate rim of the platform is relatively thick. By contrast, in leeward settings, carbonate sediments were more easily shed from the platform down the slope, and this process resulted in rather thin platform facies and comparatively thick slope facies. The slopes on the leeward side of the platform have, therefore, a more gentle dip than those on the windward side. The thicker reservoirs are thus found on the slopes of the leeward sides of the platform and on the fringes of the windward sides (see Figure 3).

As with most carbonates, diagenesis largely controls the reservoir quality of the Zechstein 2 Carbonate Member. Each depositional facies (basin, slope, platform, etc.) has a specific diagenetic overprint,

*XEX/1, Business Unit Exploratie, Nederlandse Aardolie Maatschappij, Schepersmaat 2, 9405 TA Assen, The Netherlands

which results in carbonate units with enhanced or deteriorated reservoir quality (Reijers and Bartok, 1985). In the barriers on the windward sides of the platform, early cementation and repeated leaching of the lime grainstones, followed by dolomitization, led to good reservoir potential. On the open shelf on the leeward sides of the platform, dolomitization followed by dedolomitization of the lime wackestones resulted in a fair reservoir potential. Dolomitization of the lime packstones followed by leaching resulted in fair to good reservoir potential in the slope proximal to the platform. Good porosities can occasionally be found in slope slump deposits. In the basin and in the lagoon, cementation has diminished the porosities, and reservoir potential is generally poor.

## 3-D SEISMIC INTERPRETATION

This section describes the interpretation methodology applied to 3-D seismic data to obtain information on the Zechstein 2 Carbonate reservoir.

### STRUCTURAL INTERPRETATION

Zechstein 2 Carbonate prospects are defined on seismic profiles as structural highs at the Top Zechstein 2 Anhydrite level. This level corresponds to a strong reflection at the base of the thick Zechstein 2 Halite Member over most of the region. However, at the edge of the Zechstein basin, the limit of the thick Zechstein 2 Halite Member is seen to correspond to the thickening of the Basal Zechstein Unit as it passes from the basin into the platform area (Figure 4). In the platform areas, the Top Zechstein 2 Anhydrite reflection becomes less distinguishable and is less prominent than the overlying Top Zechstein 3 Anhydrite reflection.

With all of the large, high-relief structures drilled, Zechstein 2 Carbonate exploration has increasingly focused on low-relief structures, particularly in the platform area, where structural definition was previously difficult but where reservoir potential is high.

### GROSS RESERVOIR THICKNESS PREDICTION

Although 3-D seismic data enable the accurate mapping of the top of the Basal Zechstein 2 Anhydrite Member, it is usually not possible to map the reservoir body, the Zechstein 2 Carbonate Member, since there is usually little acoustic impedance contrast between this and adjacent members (Zechstein 2 Anhydrite Member above and Zechstein 1 Anhydrite Member below). However, with 3-D seismic data it is possible to accurately map both the Top Zechstein 2 Anhydrite and the Base Zechstein reflections, together defining the Basal Zechstein Unit (see Figure 2). On seismic profiles, the transition within this unit from basinal to platform setting is clearly visible as a divergence of the two reflections (Figure 4). Across a distance of a few kilometers, the two-way time thickness of the unit increases from some 20 ms in the basinal areas to around 100 ms within the platform areas. When mapped over the entire area of the 3-D surveys, the time difference between the top and base of the unit can easily be calculated and displayed as a time isochore map (Figure 5). This time isochore map, together with the sedimentological model, can be used to estimate gross reservoir thickness.

In a simple model, the leeward side of a platform is prograding, as in the Schoonebeek field, and thick carbonate sections are found on the slope (Figure 3A). A relationship between the thickness of the carbonate member and that of the Basal Zechstein Unit can be established for a prograding platform edge. As the overall Basal Zechstein Unit thickens, so does the Zechstein 2 Carbonate Member, up to a maximum in the upper slope. Beyond this point the carbonate member thins, although the total Basal Zechstein Unit thickens. An equilibrium is reached on top of the platform, where the two thicknesses remain approximately constant.

The windward side of the platform is likely to be aggradational, as in the Collendoorn/Hardenberg fields, as modeled in Figure 3B. In this model, the thickness of the carbonate member generally increases on the slope with increasing thickness of the overall Basal Zechstein Unit and reaches its maximum on the platform fringe.

The abundant well data in northeast Netherlands has been used to plot the encountered carbonate thickness against the time isochore measured on seismic data at the well penetration locations (Figure 6). A relationship can be seen between the two variables, depending on the type of platform margin present. To effectively use the relationship between the Zechstein 2 Carbonate reservoir thickness and the Basal Zechstein Unit time isochore, the platform margin style (progradational or aggradational) has to be established. This is done on the basis of the isochore map in combination with the sedimentation model.

### RESERVOIR POROSITY PREDICTION USING SYNTHETIC SEISMIC DATA

Although it is not possible to resolve the Zechstein 2 Carbonate Member on seismic data, variations in reservoir parameters (porosity and thickness) are reflected in the seismic response, particularly in the platform areas, where porosities are generally highest. The frequency bandwidth of the 3-D seismic data within the Zechstein time window is generally limited to 10–40 Hz. Therefore, reflections from the top of the Zechstein-2 Carbonate and the thin overlying Zechstein-2 Anhydrite and Zechstein-3 Carbonate and Anhydrite are not fully resolved on seismic images over the platform area, and in fact interference results in a tuned response, viz., a composite peak. Where the carbonate porosities are low, as in the basinal and slope areas (e.g., Hardenberg 4, with an average porosity of 4%), there is little or no acoustic impedance contrast between the Zechstein 2 Carbonate Member, and the underlying Zechstein 1 Anhydrite Member (Figure 7A). However, where porosity in the Zechstein 2 Carbonate Member increases, as occurs on the platform (e.g., Collendoorn-1, with an average porosity of 14%), both density and velocity drop, giving rise to an acoustic impedance contrast that results in an additional seismic reflection directly below the regional Top Zechstein 2 Anhydrite reflection (Figure 7B).

Having determined the gross reservoir thickness from the isochore map, it is of interest to predict the porosities for reasonably thick carbonates. The first attempts to predict reservoir porosity were made using 2-D seismic data (Maureau and van Wijhe, 1979). With 3-D seismic data it is possible to make quantitative studies over an area. In order to study which attribute measurements can contribute to the prediction of reservoir porosity, synthetic seismic data were generated and studied.

Petrophysical parameters for the Zechstein-2 Carbonate reservoir and over- and underlying lithologies have been taken from logs from nearby wells. Linear relationships between porosity and density or sonic transit time for a gas-bearing dolomite reservoir were established. These relations were used to generate blocky acoustic impedance models (in two-way traveltime) (Figure 8A), which were subsequently differentiated and filtered to band-limited (10–40 Hz) reflectivity traces (Figure 8B). Several reflection attributes (e.g., loop–amplitude, loop area, loop thickness) were measured on the composite reflection corresponding to the top reservoir and the reflection corresponding to the base of the Zechstein 2 Carbonate Member. (The word loop is used here to describe a part of a seismic wavelet between two zero crossings.) These measurements were statistically analyzed to arrive at a regression equation for the average porosity of the Zechstein-2 Carbonate reservoir. The loop area measurement on the base reflection was selected to predict the average porosity. The loop area measurement is influenced more by the acoustic impedance difference (related to the average porosity) between the reservoir and the underlying anhydrite layer than by thickness variations of the reservoir (tuning effects).

Initially, only the reservoir thickness (20–80 m) and the average porosity (1–19%) were varied, while the thicknesses of the over- and underlying formations were kept constant. It was found that a minimum average porosity of 4% was needed for an acoustic impedance contrast sufficiently strong to give a measurable seismic response of the Base Zechstein-2 Carbonate. Thickness variations in seven over- and underlying formations were introduced to make the model set as representative as possible for the real geological setting on the platform and upper slope areas.

However, in order to arrive at a practical result it was necessary to make the following assumptions:

- The Zechstein Members have a constant lithology over the study area; the reservoir lithology is dolomite.
- The observed variation in seismic response at the Base Zechstein-2 Carbonate is due to changes in reservoir porosity and not to changes in porefill.
- The dolomite reservoir is gas-filled over the study area.

The modeling results indicate that:

- After statistical analysis of the selected seismic attributes, the loop area (the area of the [negative] loop of the seismic trace at the picked horizon) proved to be the most stable to predict average porosities from reflectivity data.
- After calibration between model and seismic data the following linear relationship was derived: average porosity = 4.3 + loop area.
- Average porosities lower than 4% cannot be reliably predicted.
- The uncertainty of the average porosities prediction is +/− 2% (porosity units).
- The net thickness cannot be reliably predicted from the attribute measurements.

## COLLENDOORNERVEEN CASE HISTORY

The use of 3-D seismic data as described above is demonstrated in this section for the assessment of the Collendoornerveen exploration prospect.

The availability of the Schoonebeek 3-D seismic data (bin grid 25 × 25 m, nominal 12-fold) over the Collendoornerveen area enabled maturation of the Collendoornerveen prospect. A low-relief dip-closed structure was mapped at the Top Zechstein 2 Anhydrite level (Figure 9A). The Collendoornerveen structure is separated by a saddle from another dip-closure tested by the wells Marslanden-1 and Collendoorn-1.

The Basal Zechstein time isochore map shows that the closure is situated on the fringe of the platform with a rather steep slope (Figures 5 and 9B). It was concluded that the prospect is situated on the windward side of the platform, where good reservoir potential can be expected. Using the aggradational platform edge model, a gross reservoir thickness of some 50–70 m was predicted: The edge map shown in Figure 9C, indicates the faults in the structure. Fractures may be present orthogonal on these faults which can enhance well productivity.

During the structural interpretation, a strong reflection was noticed below the Top Zechstein 2 Anhydrite level (Figure 10). This reflection was mapped and tied with the neighboring Marslanden-1 (MRS-1) and Collendoorn-1 (CLD-1) wells, where it was seen to correspond to the base of the Zechstein 2 Carbonate Member. The match of the well synthetic of MRS-1 to the seismic line running north-south through the well is shown in Figure 11.

The highest amplitudes of this reflection lie in a narrow band immediately adjacent to the steep, northerly facing slope (Figure 9D). The marked similarity of the reflection distribution to the established sedimentological model suggested that it was a direct indication of porous, high-energy, platform-fringe deposits, with the reflection amplitude variations possibly being a function of reservoir porosity.

Detailed acoustic impedance modeling of the expected sedimentological variations over the prospect indicated that the best predictive seismic attribute is the loop area of the Base Zechstein 2 Carbonate reflection. A linear relationship was established which related the loop area to the average Zechstein 2 Carbonate porosity, enabling a prediction of reservoir porosity to be made over the extent of the mapped reflection (Figure 12).

As a result, the target location for the Collendoornerveen-1 exploration well (CLDV-1) was moved slightly downdip of the structural culmination in order to encounter a predicted local porosity maximum, estimated to be between 16 and 18%.

The exploration well CLDV-1 was successful. The Zechstein 2 Carbonate Member was found to be gas-bearing and to consist of high-energy, platform-margin oolitic, dolomite grainstones. The average porosity was 16% and the reservoir thickness was 63 m. The good porosity and the presence of open fractures resulted in a gas production rate of about $10^6$ $m^3$/day.

## CONCLUSIONS

A number of quantitative interpretation techniques have been established to aid the exploration and exploitation of the Zechstein 2 Carbonate play in the northeast Netherlands. Using 3-D seismic data, it is possible to produce detailed time isochore maps which, in combination with the sedimentological model, indirectly provide an estimate of gross reservoir thickness for progradational or aggradational platforms. With 3-D seismic data, it is also possible to conduct quantitative prediction studies to map average reservoir porosities over exploration prospects and existing fields.

Reasons for the good outcome of the quantitative interpretation study are a good understanding of the geological model, availability of petrophysical data from nearby wells, good well-to-seismic line matches, and extensive statistical modeling.

*Acknowledgments*

I am indebted to the Nederlandse Aardolie Maatschappij, Shell Internationale Petroleum Maatschappij (SIPM), and Exxon Company International (ECI) for granting permission to publish this paper. I would like to thank my colleagues within Shell who have contributed to the results presented in this paper: Neil Casson, Vincent van Hoegaerden, Henk Rebel, Tom Reijers, and Bob van Wees.

## REFERENCES CITED

Clark, D. N., 1986, The distribution of porosity in Zechstein Carbonate, *in* J. Brooks, J. C. Goff, and B. van Hoorn, eds., Habitat of Palaeozoic Gas in NW Europe: Geological Society London Special Publication 20, p. 121–143.

Maureau, G. T. F. R., and D. H. van Wijhe, 1979, The prediction of porosity in the Permian (Zechstein 2) Carbonate of Eastern Netherlands using seismic data: Geophysics v. 44, p. 1502–1517.

Reijers, T. J. A., and P. Bartok, 1985, Porosity characteristics and evolution in fractured Cretaceous carbonate reservoirs, La Paz field area, Maracaibo basin, Venezuela, *in* P. Roehl, and P. W. Choquette, eds., Carbonate Petroleum Reservoirs: Springer, New York, p. 409–423.

Van der Baan, D., 1990, Zechstein reservoirs in the Netherlands, *in* J. Brooks, ed., Classic Petroleum Provinces: Geological Society London Special Publication 50, p. 379–398.

van de Sande, J. M. M., T. J. A. Reijers, and N. Casson, 1996, Multidisciplinary exploration strategy in the northeast Netherlands Zechstein 2 Carbonate play, guided by 3-D seismic, *in* H. E. Rondeel, D. A. J. Batjes, and W. H. Nieuwenhuis, eds., Geology of Gas and Oil under the Netherlands: The Royal Geological and Mining Society of the Netherlands/Kluwer Academic Publishers, p. 125–142.

Ziegler, P. A., 1990, Geological Atlas of Western and Central Europe: Shell Internationale Petroleum Maatschappij/Geological Society Publishing House, Avon, 239 p., 56 encls.

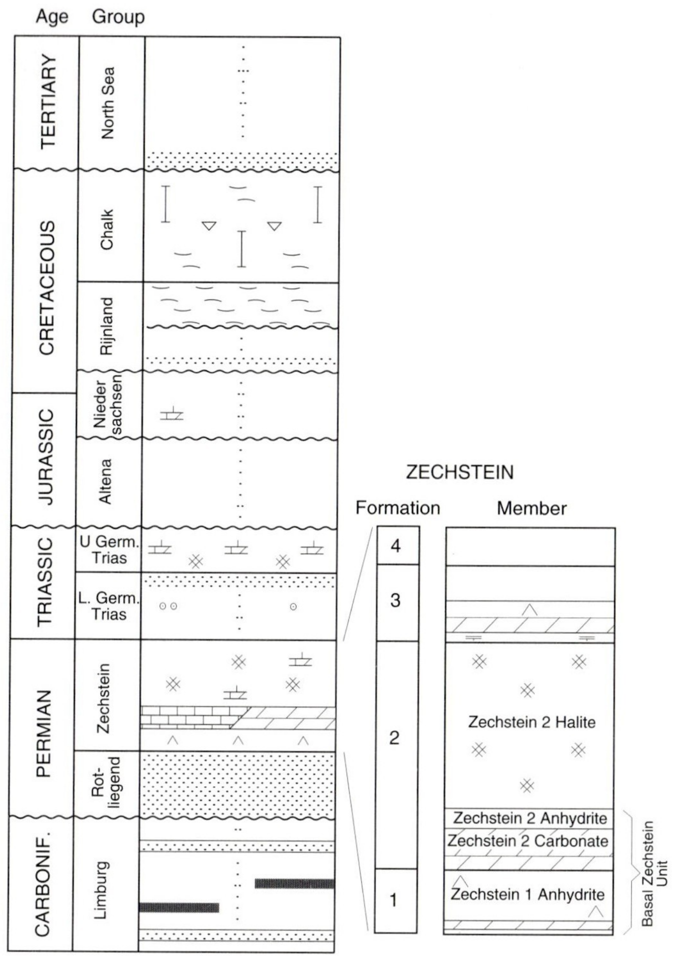


FIGURE 1. Generalized stratigraphy of northeast Netherlands, with details of the Zechstein Group. The Coppershale (approximately 1 m at base Zechstein) is not shown.

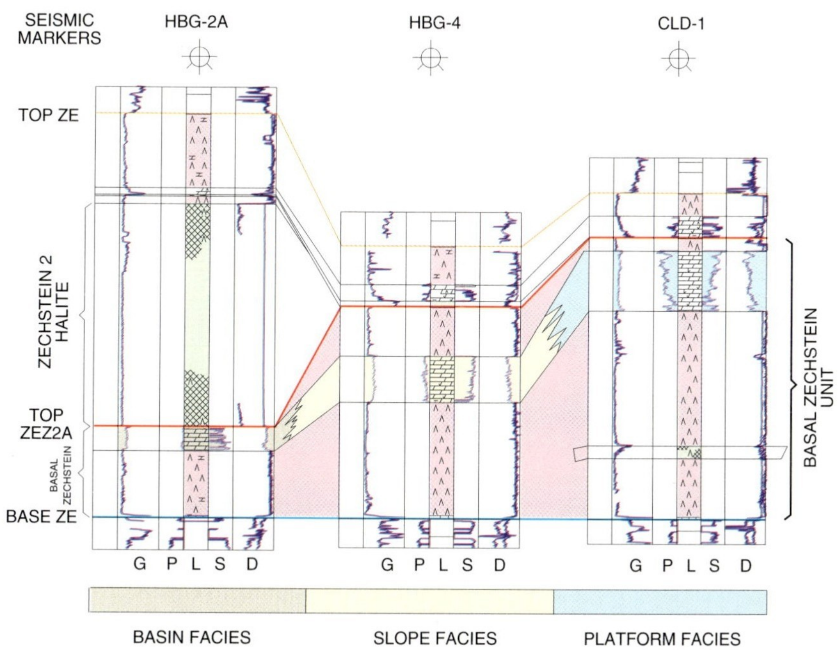


FIGURE 2. Well log correlation panel of the Basal Zechstein Unit using gamma ray (G), porosity (P), lithology (L), gas saturation (S), and sonic and density (D) logs, showing basin, slope, and platform carbonate facies of the Zechstein 2 Carbonate Member (wells: HBG-2A, HBG-4 [Hardenberg], CLD-1 [Collendoorn]).

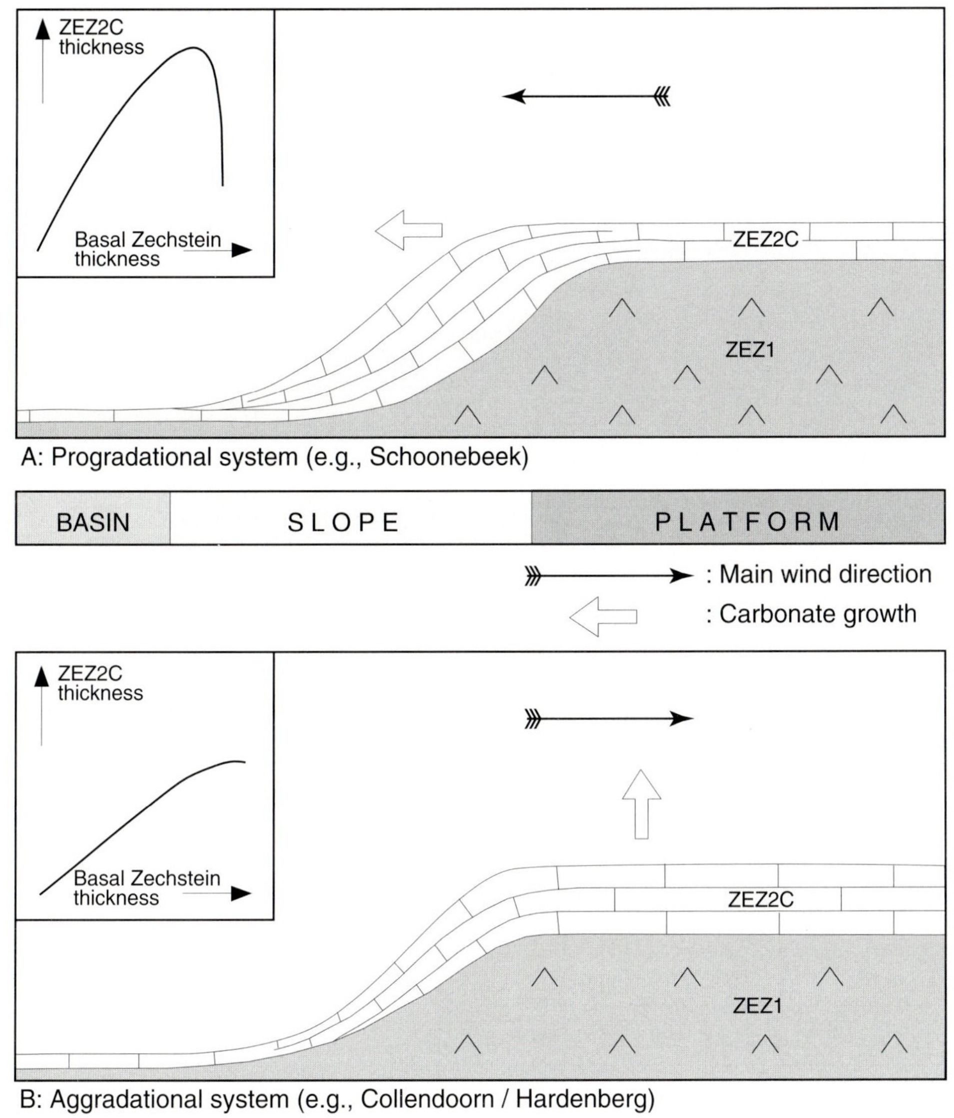


FIGURE 3. Models of the progradational (A) and aggradational (B) platform margin styles. The thickness relationships between the Zechstein 2 Carbonate Member (ZEZ2C) and the Basal Zechstein Unit (ZEZ1) are shown in the insets.

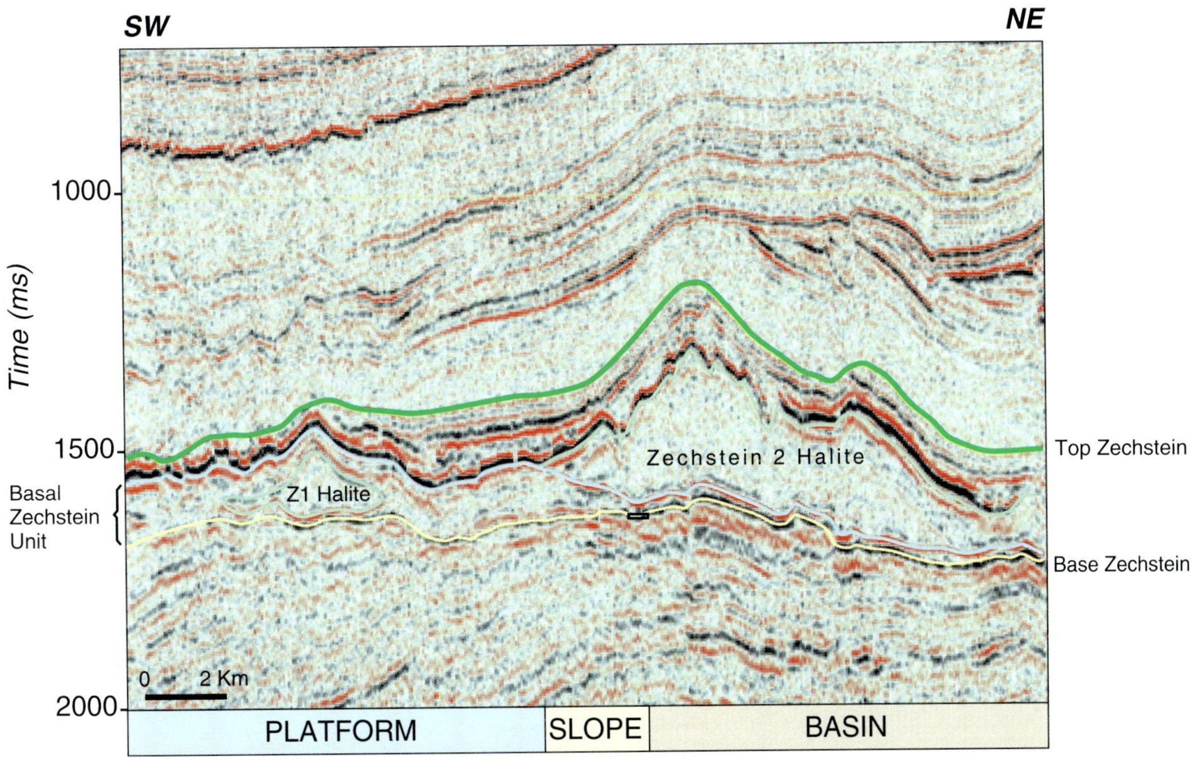


FIGURE 4. Regional seismic line with interpretation of the Zechstein Group, showing the Basal Zechstein Unit in platform, slope, and basinal settings. For location of seismic line see Figure 5.

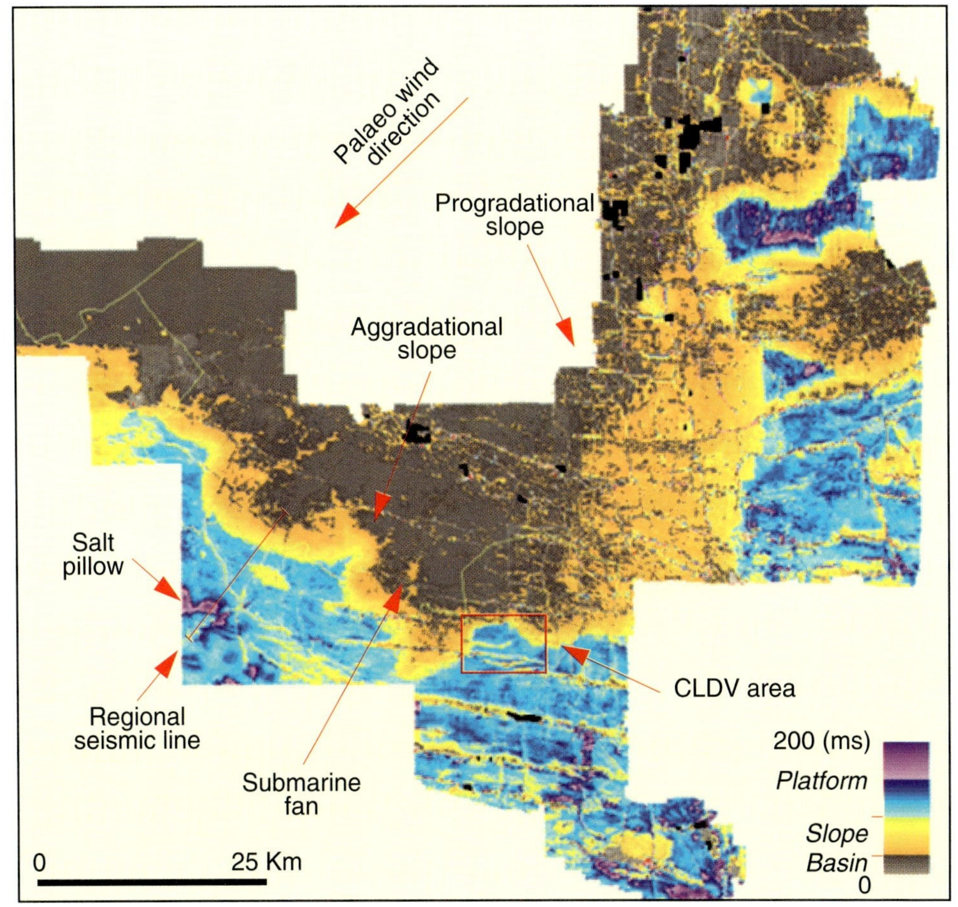


FIGURE 5. Time isochore map of the Basal Zechstein Unit, derived from the 3-D seismic data, showing the areas with platform, slope, and basinal depositional settings. The relatively steep slopes on the northeastern side of the platform and the more gentle slopes on the western side reveal the dominant northeasterly wind direction. Zechstein 1 salt pillows are visible as high isochore values. Submarine fan deposits can be seen in the basin close to the platform edge. The regional seismic line (Figure 4) and the Collendoornerveen (CLDV) area (Figure 9) are indicated.

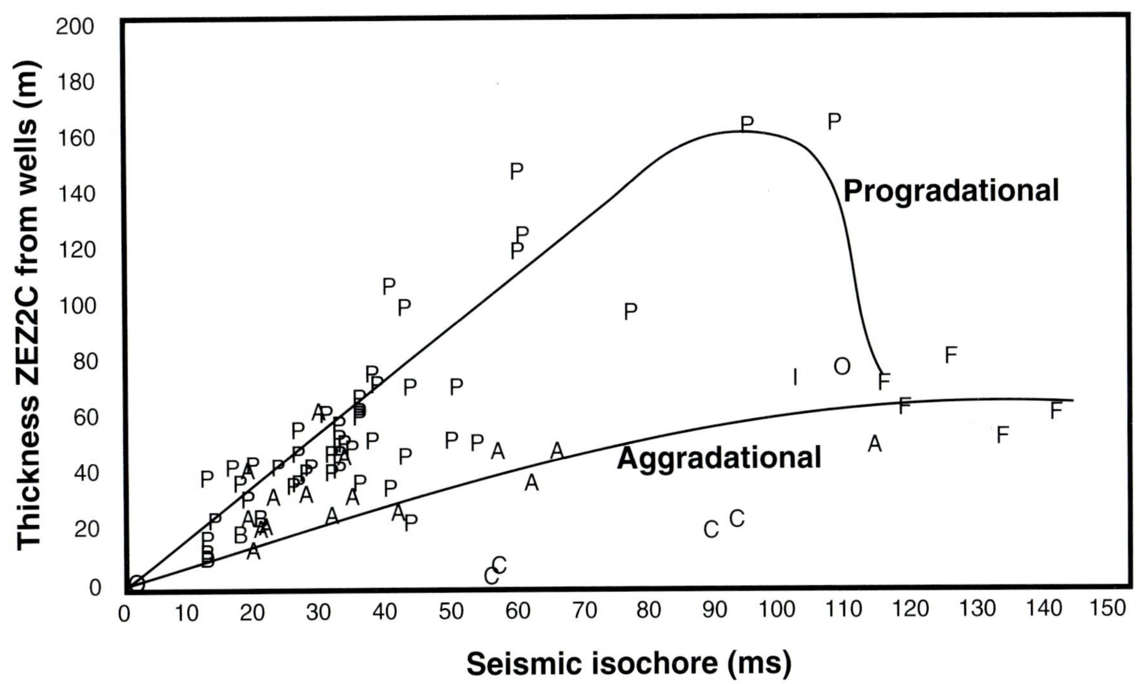


FIGURE 6. Graph showing the relationship between the thickness of the Zechstein 2 Carbonate Member (ZEZ2C) and the time isochore of the Basal Zechstein Unit for wells in northeast Netherlands. The depositional setting (B = Basin, A = Aggradational slope, P = Progradational slope, F = platform Fringe, I = platform Interior, O = Open shelf) of the Zechstein 2 Carbonate Member is determined from the time isochore map of the Basal Zechstein Unit (Figure 5). The two platform margin styles can be discriminated. (C = Cut out by fault)

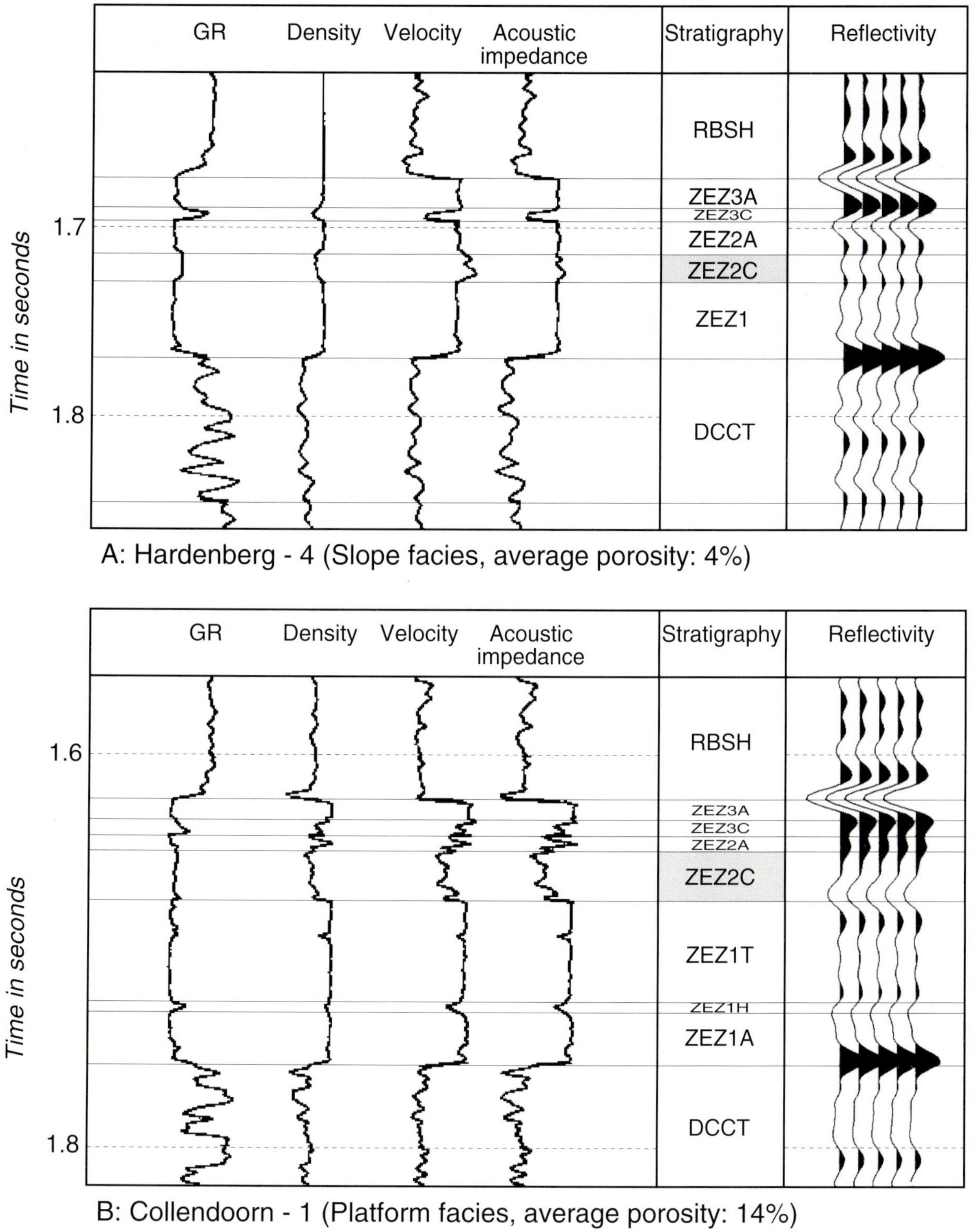


FIGURE 7. Well logs and reflectivity synthetics of the Zechstein Group for slope (A) and platform (B) settings. Density and velocity values are lower and the Base Zechstein 2 Carbonate reflection is stronger if the average porosity of the Zechstein 2 Carbonate Member is higher. GR= gamma-ray; RBSH = Lower Bunbsandstein; DCCT = Tubbergen Sandstone.

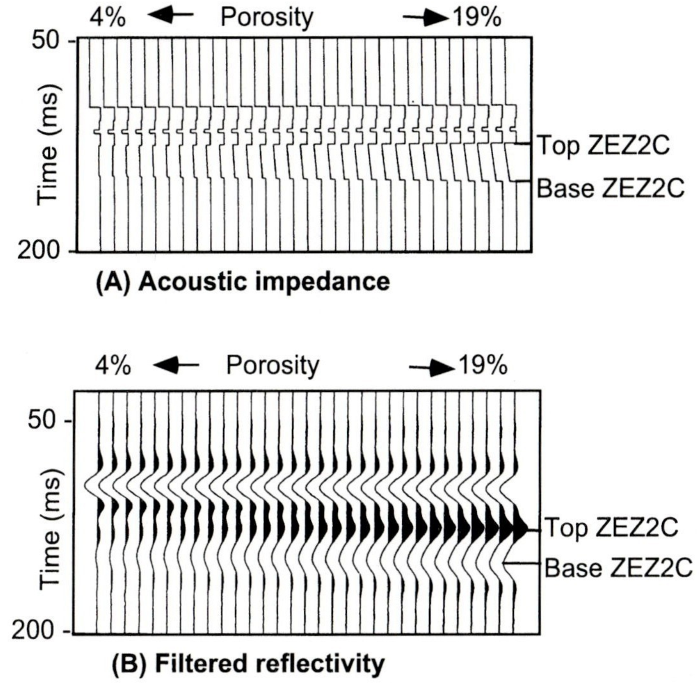


FIGURE 8. Synthetic seismic data of the Basal Zechstein Unit, with linear variation in the average porosity (4–19%) of the Zechstein-2 Carbonate Member. (A) The acoustic impedance traces in two-way traveltime. (B) The filtered reflectivity traces with a frequency bandwidth of 10–40 Hz.

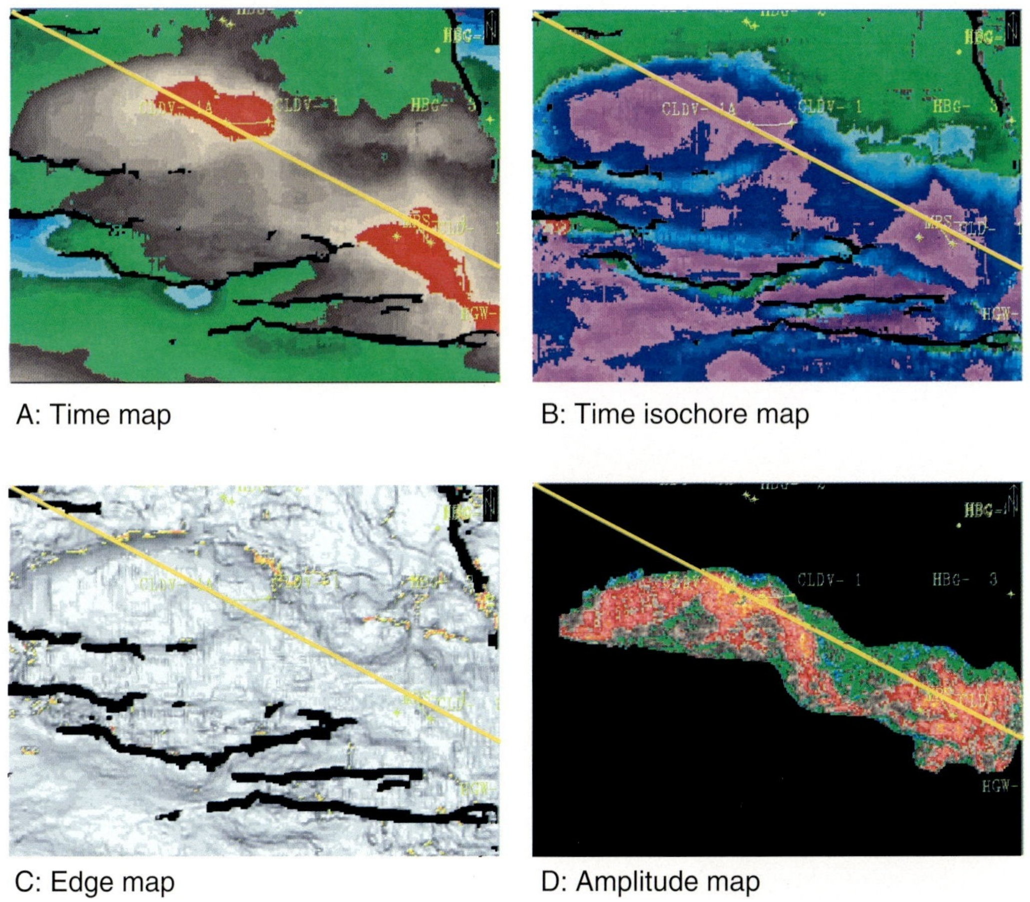


FIGURE 9. Maps showing characteristics of the Collendoorn-nerveen (CLDV) prospect determined on 3-D seismic data. The area mapped measures 8 × 6 km. Its position is shown in Figure 5. The position of the seismic line shown in Figure 10 is indicated. (A) The time map of the Top Zechstein 2 Anhydrite reflection (red = shallow) shows the structural closure (assuming a laterally constant overburden). (B) The time isochore map of the Basal Zechstein Unit (purple = platform) shows the platform edge close to CLDV. (C) The edge map of the Top Zechstein 2 Anhydrite reflection shows the minor faults. (D) The amplitude map of the Base Zechstein 2 Carbonate reflection (red = high) indicates the better porosities at the platform fringe.

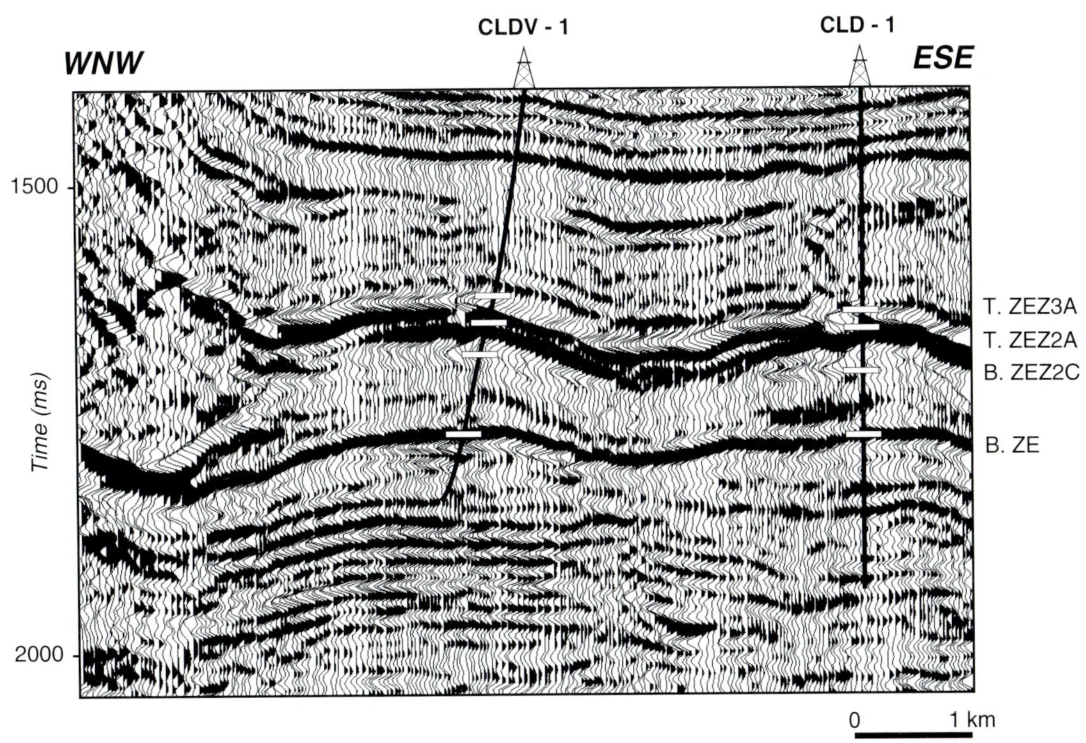


FIGURE 10. Seismic line through the Collendoornerveen structure (see Figure 9), showing the development of the Base Zechstein 2 Carbonate (B.ZEZ2C) reflection. The well trajectories with the seismic reflections mentioned in the margin are indicated for the Collendoorn-1 (CLD-1) and Collendoornerveen-1 (CLDV-1) wells.

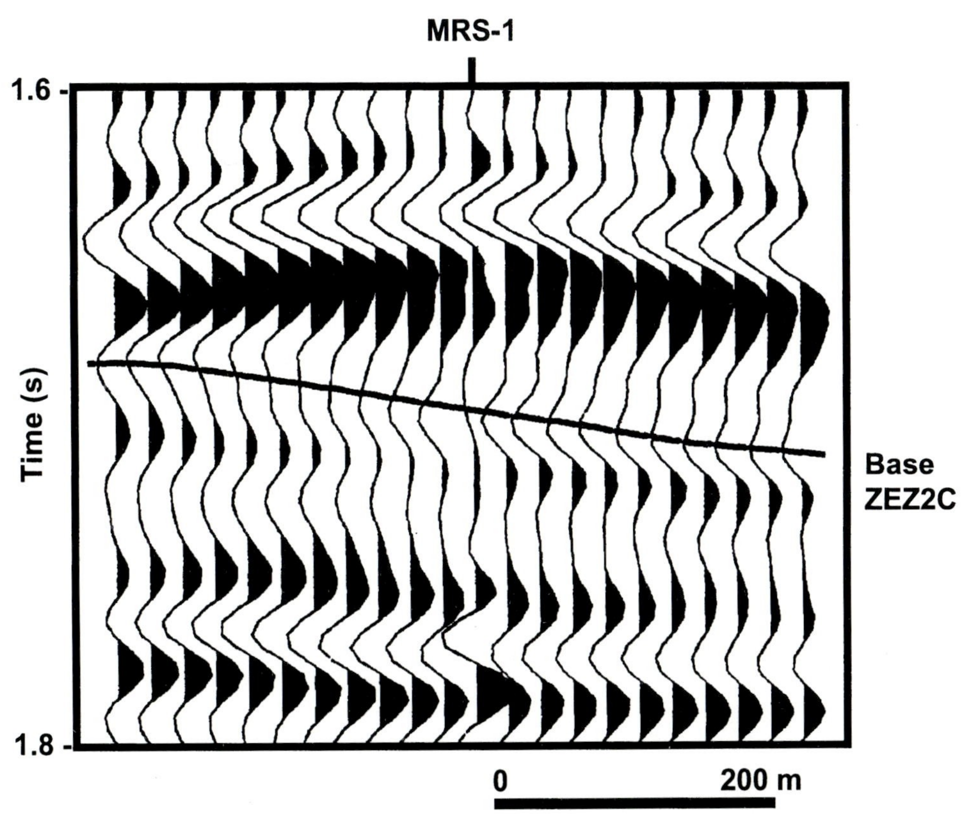


FIGURE 11. Well-to-seismic line match for Marslanden-1 well and north-south seismic line 268 over the Basal Zechstein Unit. The frequency bandwidth is 10–45 Hz.

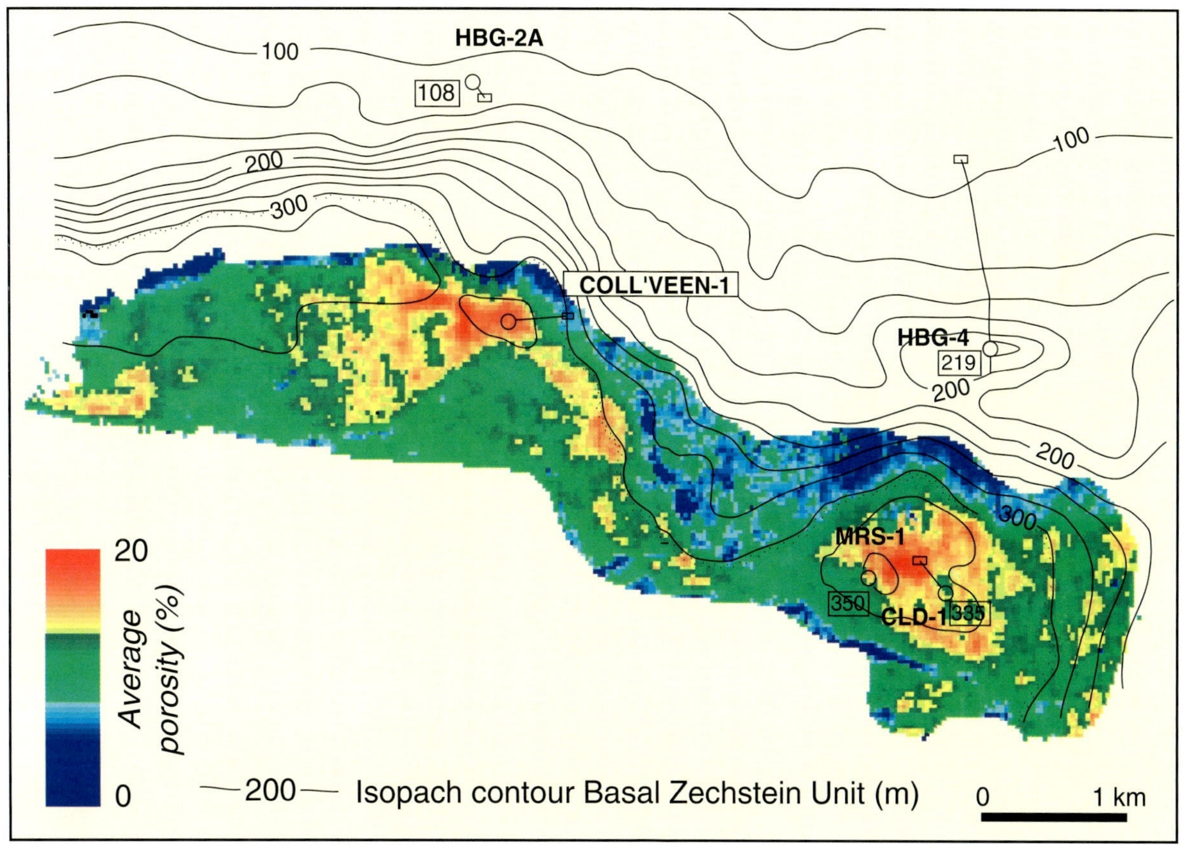


FIGURE 12. Map showing the predicted average porosity of the Zechstein 2 Carbonate Member for the Collendoornerveen prospect. The thickness of the Basal Zechstein Unit is also indicated.

# CHAPTER 22

# Impact of 3-D Seismic Data on Reservoir Characterization and Development, Ghawar Field, Saudi Arabia

Saudi Arabian Oil Company, Impact of 3-D seismic data on reservoir characterization and development, Ghawar Field, Saudi Arabia, in P. Weimer and T. L. Davis, eds., *AAPG Studies in Geology No. 42* and *SEG Geophysical Developments Series No. 5*, AAPG/SEG, Tulsa, p. 205-210.

**Saudi Arabian Oil Company, compiled and edited by David W. Alexander***

## ABSTRACT

The giant Ghawar Field produces oil from a Jurassic carbonate reservoir and is an ideal candidate for mapping and evaluation using 3-D seismic data. 3-D seismic acquisition began over the field in 1992 with the goal of more accurately defining reservoir structure and identifying causes of anomalous reservoir performance. Initial data acquisition occurred in the central 'Uthmaniyah Area of Ghawar. This area was chosen because of a concentration of anomalous reservoir characteristics which have combined to cause a premature advance of injection fluids into the field. Mapping and interpretation of 3-D seismic data reveal much greater structural complexity than was previously recognized, including extensive faulting and fracturing. Seismic attribute data were also used to produce an improved model of reservoir porosity and lithology. Using geostatistical methods, 3-D seismic data were integrated with well data to produce a much improved reservoir model for use in simulation and reservoir management.

## INTRODUCTION

Oil was first discovered at the giant Ghawar Field in 1948, but it was not until 1952 that several subsequent discoveries were recognized as being part of one large oil accumulation. Ghawar Field is located approximately 200 km east and north of Riyadh in the Eastern Province of Saudi Arabia and is the world's largest oil field (Figures 1 and 2). Primary production is from the D member of the Upper Jurassic Arab formation, which is the oldest of four upward-shoaling cycles of carbonate and evaporite. The contrast between the overlying thick and dense anhydrite and the porous carbonates of the Arab-D reservoir provide a strong acoustic impedance contrast and make the reservoir a viable candidate for mapping and evaluation using the 3-D seismic tool (Figure 3). While the porous reservoir is generally highly productive, anomalous reservoir characteristics such as dolomitization, fracturing, and the occurrence of extreme permeability zones complicate reservoir development efforts.

Three-dimensional seismic acquisition was begun in selected areas of Ghawar Field in April of 1992 to more accurately define the reservoir structure and to help identify causes of anomalous reservoir performance characteristics. Since that time, over 3000 km$^2$ of data have been acquired, including the 300 km$^2$ discussed in this paper. Information extracted from these data has become an indispensable tool in the Saudi Arabian Oil Company's reservoir characterization and development efforts in Ghawar and other fields.

*Saudi Arabian Oil Company, Dhahran, Saudi Arabia

## ANOMALOUS RESERVOIR CHARACTERISTICS

The first 3-D data acquired in Ghawar were from the central 'Uthmaniyah Area (Figure 2). This area was chosen for the first survey because of a concentration of anomalous reservoir characteristics, the most critical of which was an anomalous premature advance of injection water into the interior of the field.

Approximately 100 wells had been drilled within the outline of the survey prior to early 1993. Structural mapping based on these wells (Figures 4 and 5) shows a relatively smooth structure with a prominent south-plunging nose and an evenly graduated slope from the line of injector wells to the first row of producers, several hundred feet structurally higher and approximately 2 km to the west. Structure contouring shows no evidence of anomalous patterns which could help explain the unexpected encroachment of injection fluids in the area.

To be able to history-match the reservoir performance observed to date, including the anomalous flood front advance, reservoir simulation engineers were forced to introduce a line of extreme permeabilities of 250 d into a layer of the simulation model (Figure 6).

Thin-bedded superpermeability and total reservoir dolomitization are two additional anomalous reservoir characteristics observed in this part of Ghawar Field. Both of these factors play significant roles that complicate fluid movement in the reservoir. Understanding their distribution and occurrence is a primary goal of interpreters working with the 3-D seismic data (Alexander, 1994).

## RESULTS OF THE 3-D SEISMIC SURVEY

Mapping of the Arab-D using seismic data shows a reservoir with significantly greater structural complexity than was recognized from well control alone (Figures 7 and 8). Faulting, with its associated fracturing, was recognized for the first time with the 3-D seismic interpretation. Some of the faults and fractures appear to be acting as conduits for injection fluids and help explain the premature arrival of these fluids into the interior of the field. The flank of the structure is now recognized as having twice the dip gradient previously mapped, and strong evidence is present that much of its structural character is the result of late-stage shear movement affecting the preexisting Ghawar structure (Alexander, 1994).

Two examples of faulting as seen on the seismic data are shown in Figures 9 and 10. Figure 9 shows a large, high-angle, reverse fault trending North-South near the top of the steep flank of the structure. This fault has a throw of more than 100 ft (30 m), as shown by the offset of the Arab-D and deeper events. Figure 10 is from the northern part of the survey and shows the complexity of some of the faulted areas (Alexander et al., 1994).

The importance of structure, and in particular fractures and faulting, in Arab-D reservoir behavior in Ghawar becomes apparent when we look at a map showing cumulative production/injection values from wells in the 3-D survey area (Figure 11). Not only are large cumulative injection characteristics associated with wells where faulting has occurred on the flanks of the field, but surprisingly low cumulative production totals have occurred in several wells associated with a concentration of southwest-striking faults near what could be called a breach in the south-plunging structural nose. Most of these wells watered out relatively early. It also may be pointed out that many of the best wells, as far as cumulative production is concerned, are those located away from faulting and on or near structural culminations. The correlation between the location of the extreme permeabilities used in the simulation model and the mapped occurrence of these faults should also be noted (Alexander et al., 1994).

Since completion of structure mapping in this area, subsequent and planned wells have been evaluated and located to avoid potential problem areas. It has also been noted that several significant structural highs have yet to be tested within the area, and wells are planned to evaluate several such anomalies.

One goal of this 3-D seismic program is to gain a better understanding of porosity distribution in the reservoir, as well as other reservoir characteristics previously interpreted and mapped using only well data. Figure 12 is a map derived entirely from well data that shows total porosity thickness ($\Phi H$) for the entire Arab-D reservoir. Total porosity of the reservoir shows significant variability and falls by more than 50% from the crest of the structure to the flank injector well locations. Areal distribution of pervasive dolomites can also be interpolated from $\Phi H$ distribution due to the close correlation between high $\Phi H$ and dolomitization. Figure 13 is an equivalent map of $\Phi H$ based on seismic amplitude. The same general trends are evident, but with much more detail derived from the dense seismic coverage.

Using geostatistical methods, seismic data can be reliably integrated with well data to produce a much improved model of the reservoir for use in simulation and reservoir management. This type of porosity mapping has proved to be very useful in Ghawar (Jeffery and Thomsen, 1995).

## DISCUSSION

3-D seismic data have already provided significant benefits in Ghawar Field. With the help of 3-D data, production drilling has been optimized, allowing maintenance of production capacity by identifying optimal places to drill and also areas to avoid. 3-D seismic data have also been useful in optimizing injection drilling by minimizing the uncertainties in flank structural elevations. The data have provided, and will continue to provide detailed understanding of reservoir stratigraphy by more accurately mapping lateral changes in reservoir characteristics, such as porosity and lithology, which have implications for permeability distribution.

We feel that the use of 3-D seismic data and other state-of-the-art tools will help Saudi Aramco improve its effectiveness in managing the hydrocarbon resources of the Kingdom of Saudi Arabia.

## REFERENCES CITED

Alexander, D. W., 1994. Ghawar 3-D seismic: a new look at an old field: Society of Petroleum Engineers 1994 Technical Symposium, Dhahran, Saudi Arabia.

Alexander, D. W., C. W. Anderson, P. Lawrence, and M. A. Thomsen, 1994, Ghawar 3-D seismic: a new look at an old field: Saudi Aramco Journal of Technology, spring, p. 2–9.

Jeffery, R., and M. Thomsen, 1995, Predicting porosity in a Saudi Arabian carbonate reservoir using geologic constraints integrated with 3-D seismic and well data: 1995 AAPG International Conference and Exhibition, Nice, France.

FIGURE 1. Regional setting of the Ghawar Oil Field in northeastern Saudi Arabia.

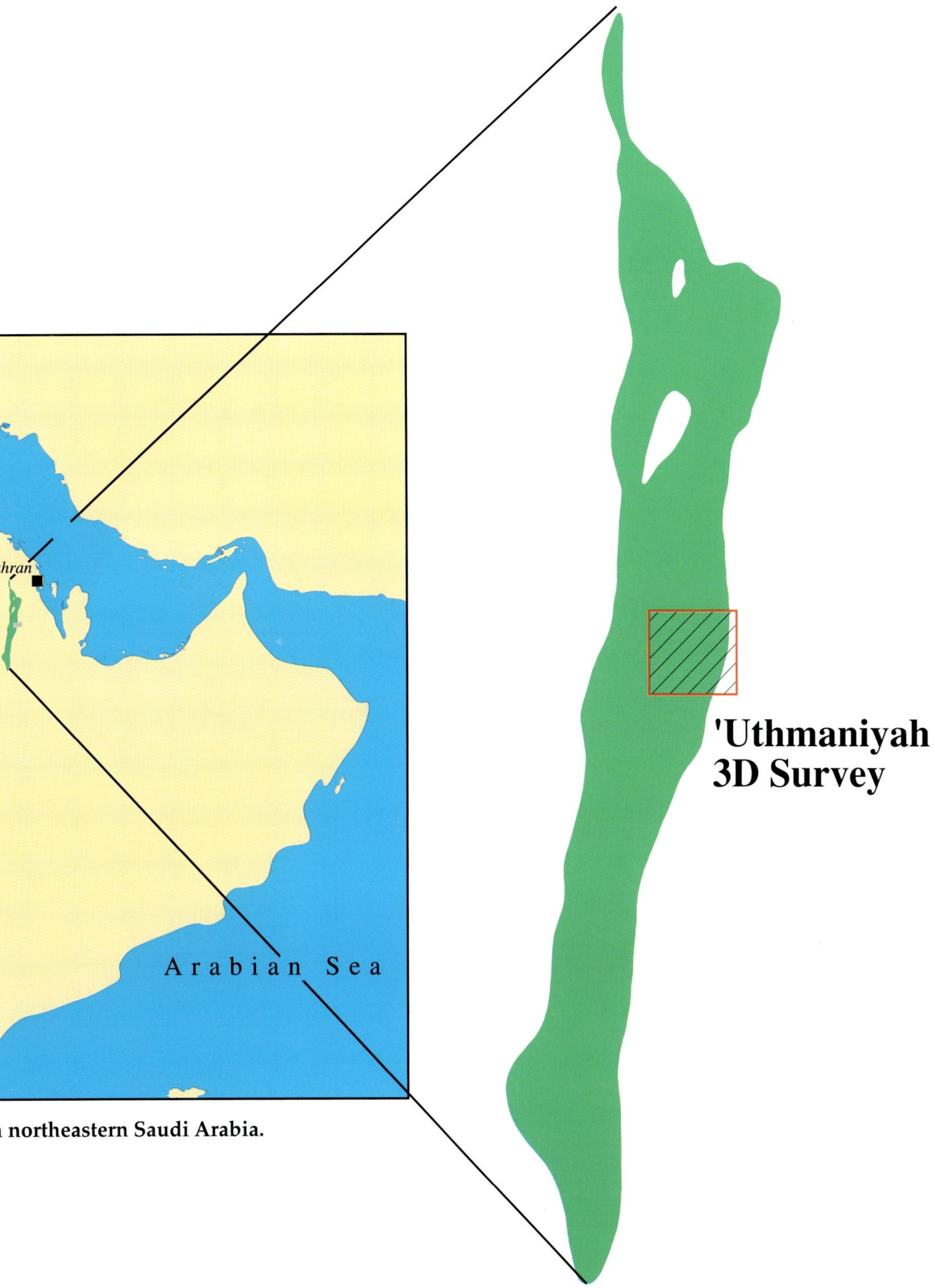


FIGURE 2. Outline of the Ghawar Field and location of the 'Uthmaniyah 3-D seismic survey.

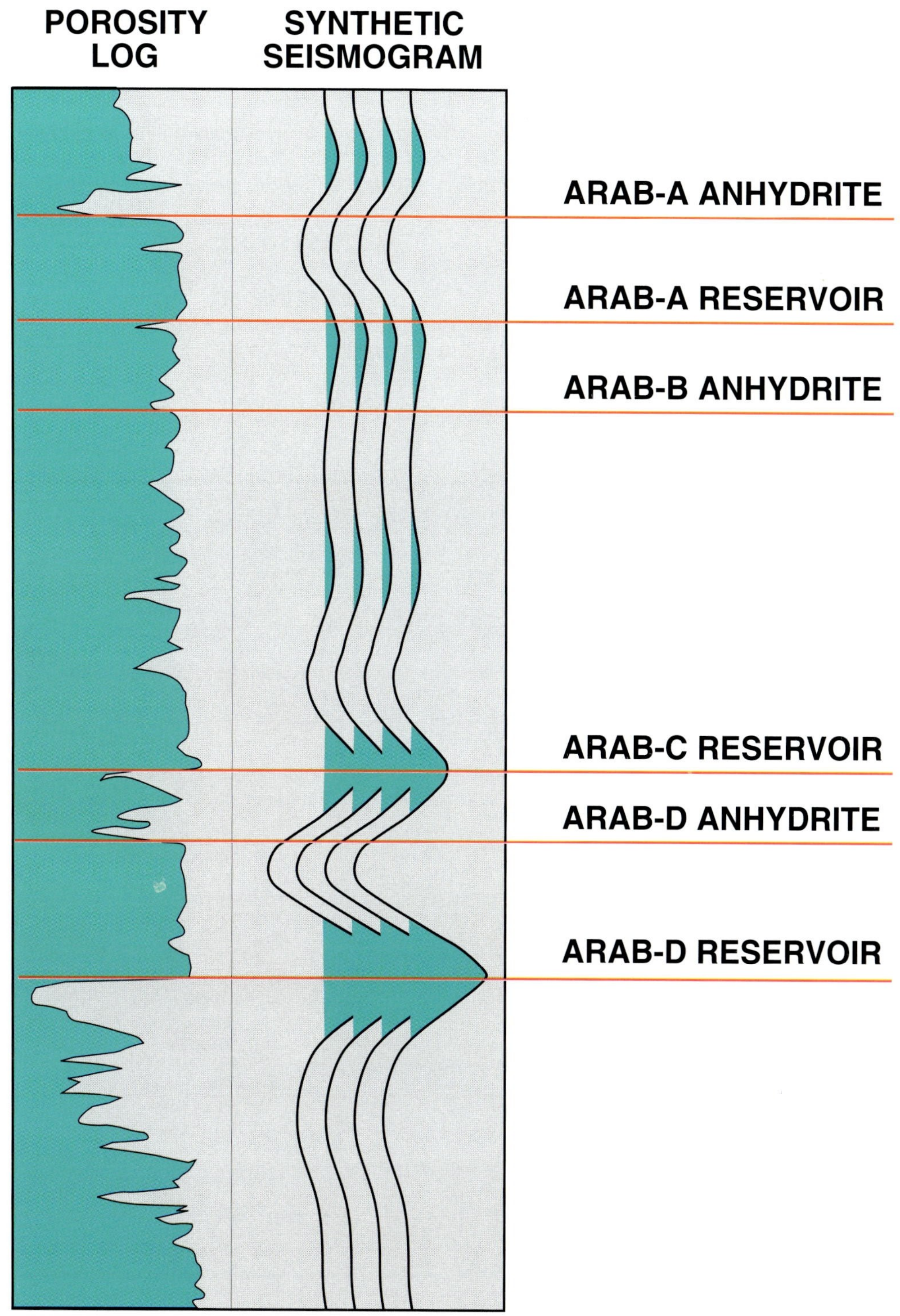


FIGURE 3. The seismic response and porosity log of the Arab-D Formation, Ghawar Field.

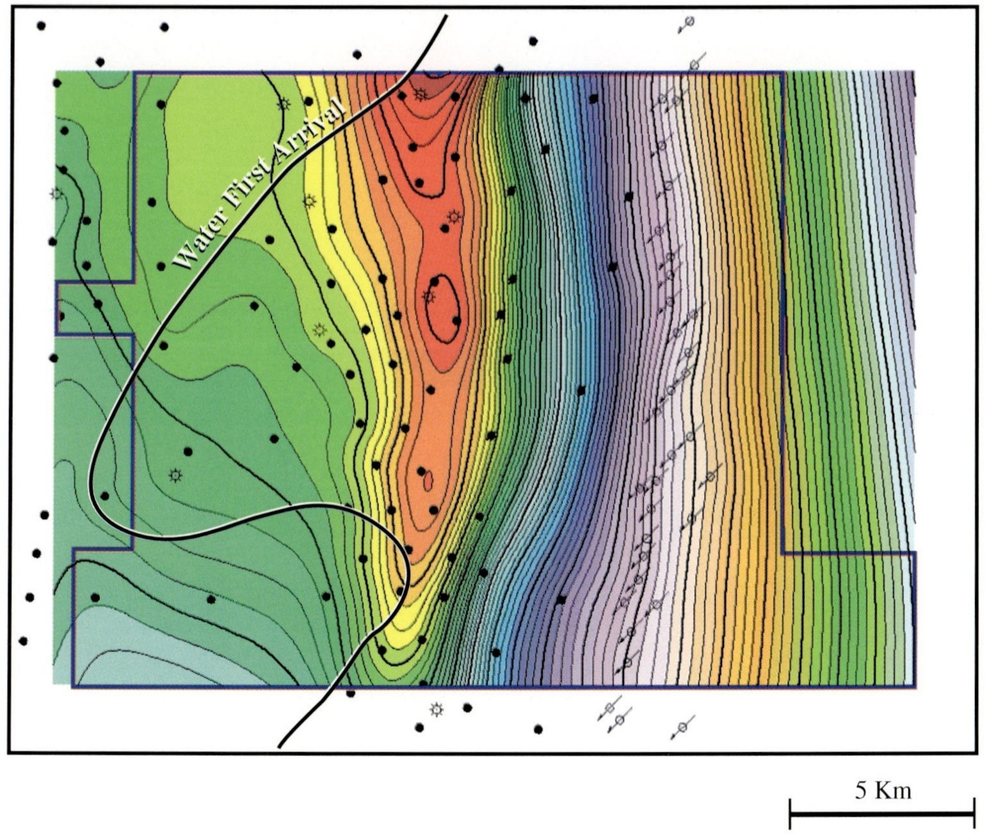


FIGURE 4. Top Arab-D reservoir structure map in the 'Uthmaniyah area, Ghawar Field, based on well log correlations.

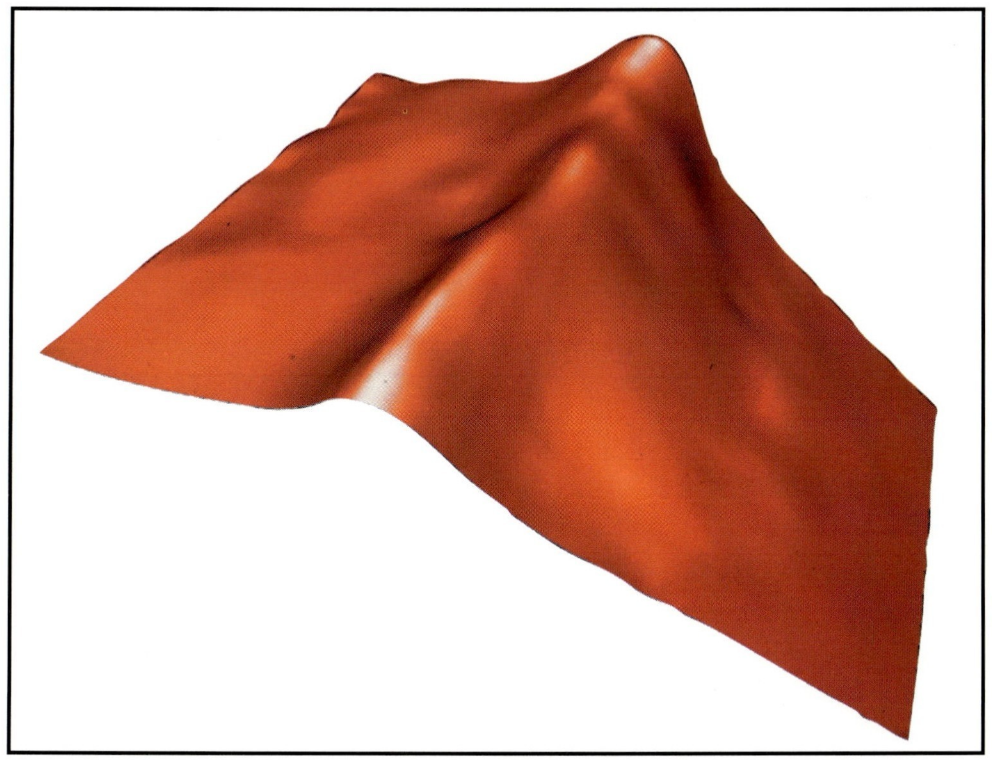

FIGURE 5. 3-D perspective of the top Arab-D reservoir shown in Figure 4.

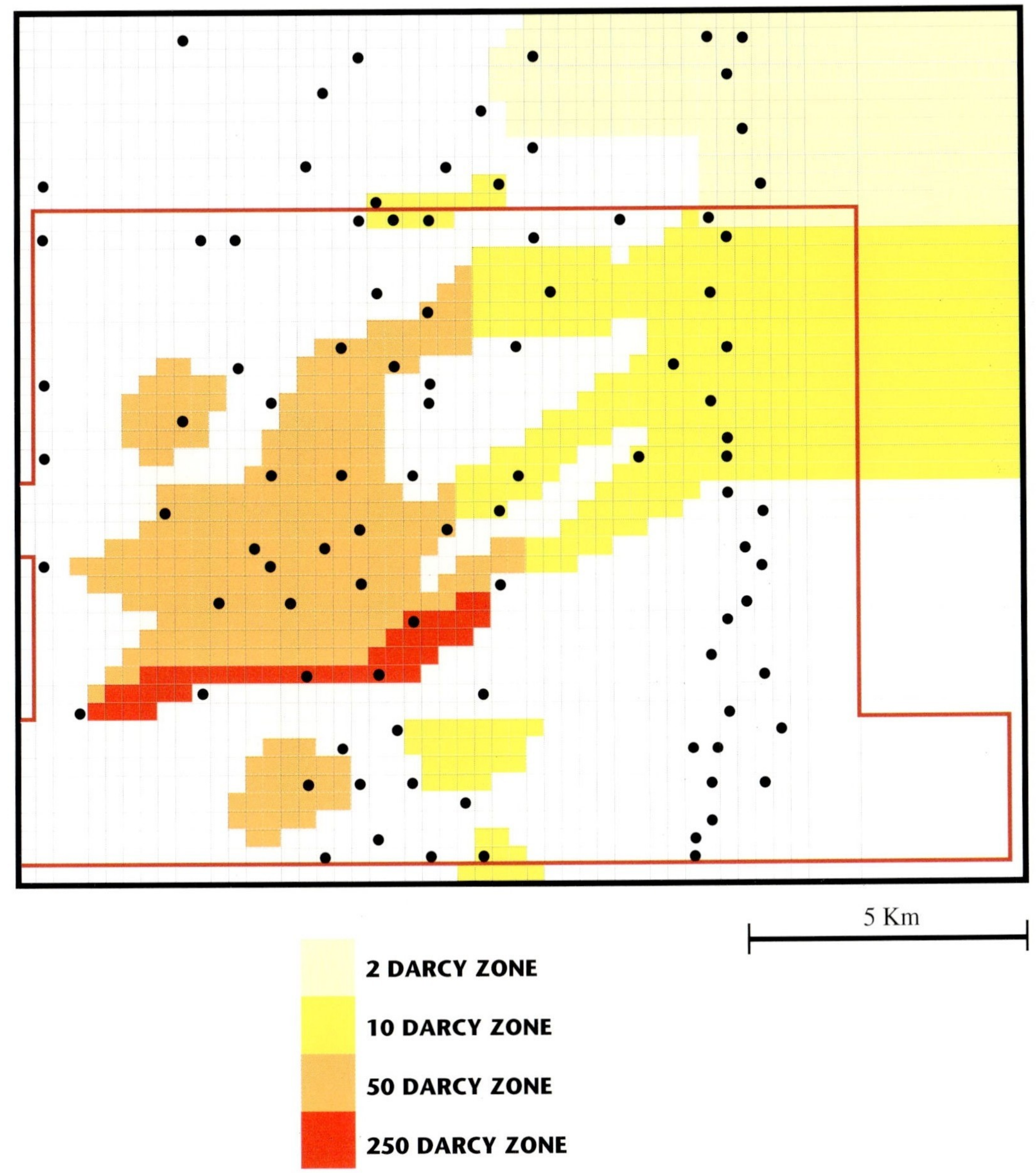


FIGURE 6. Reservoir simulation model of Central 'Uthmaniyah area. Note that to explain the anomalous flood front advance, a narrow band of anomalously high permeabilities (250 d) was introduced in the model.

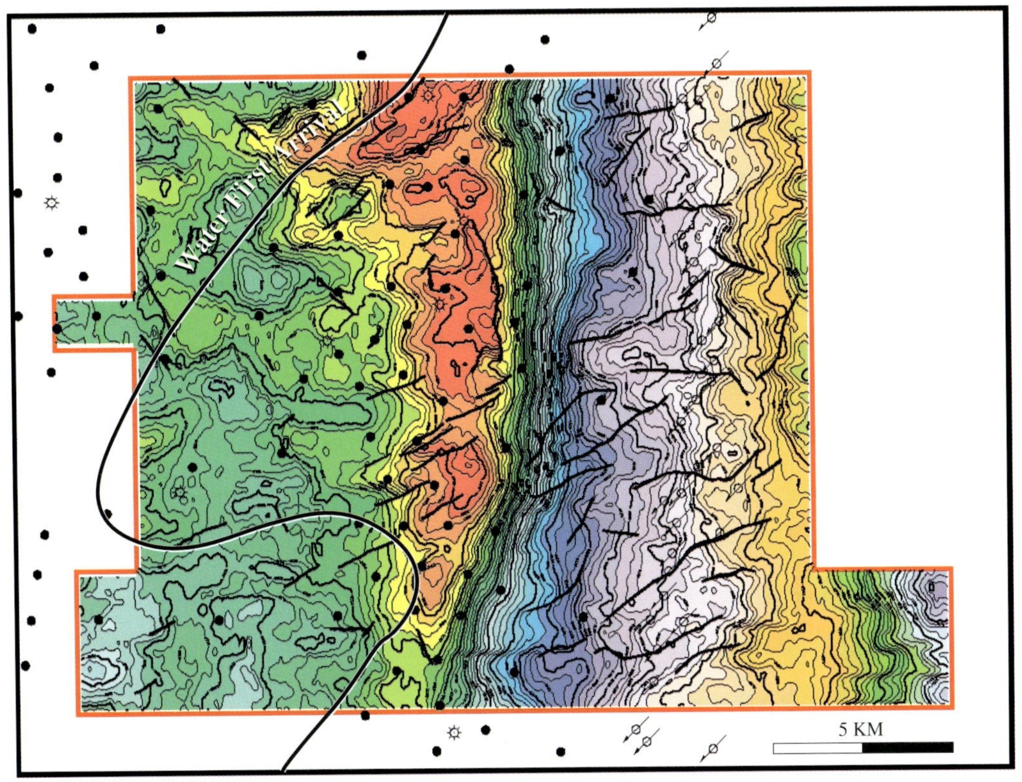


FIGURE 7. Top Arab-D reservoir structure map in the 'Uthmaniyah area, Ghawar Field, based on 3-D seismic data.

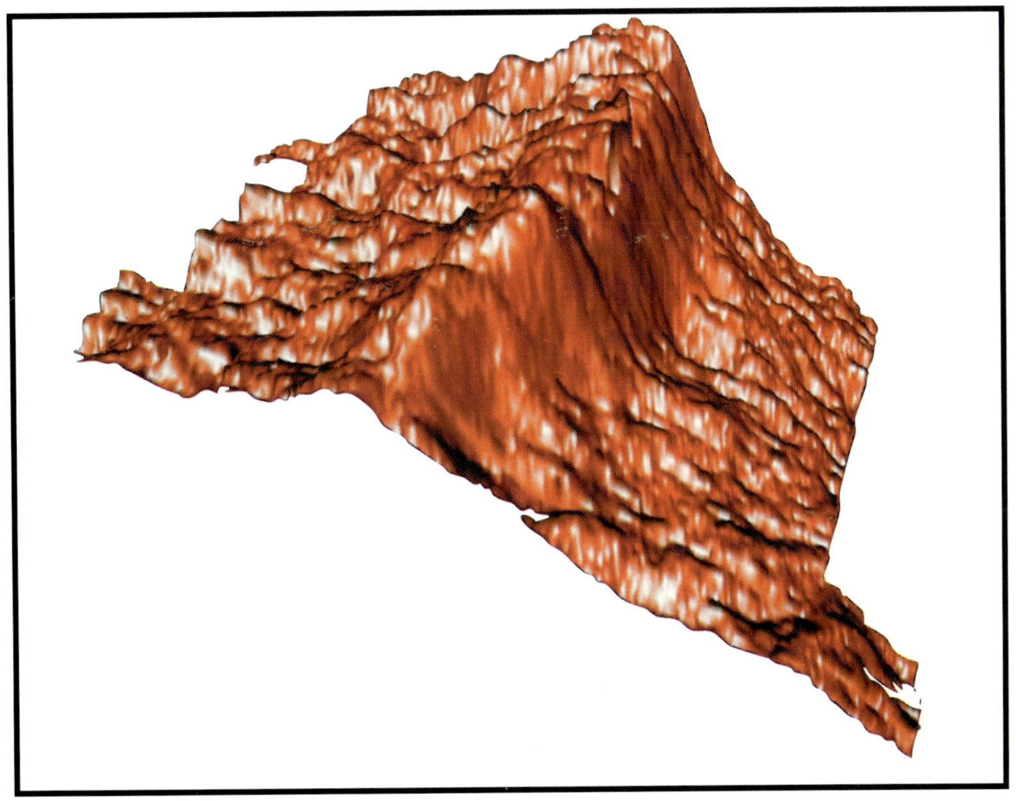

FIGURE 8. 3-D perspective of the top Arab-D reservoir shown in Figure 7.

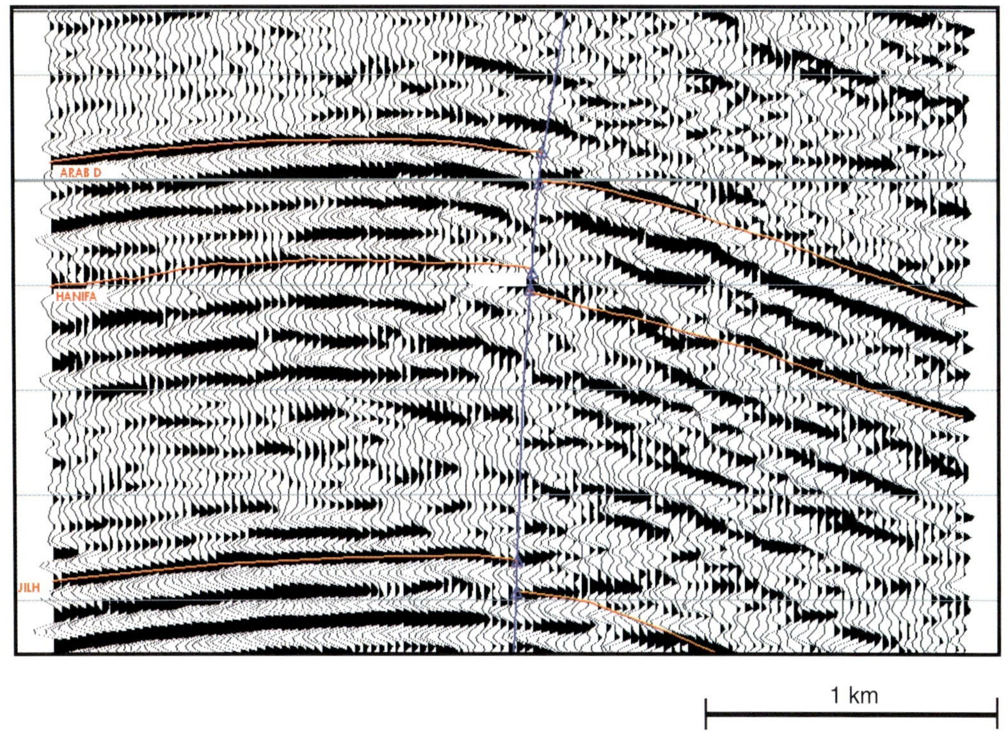


FIGURE 9. Faulting in Central 'Uthmaniyah as seen on seismic example I.

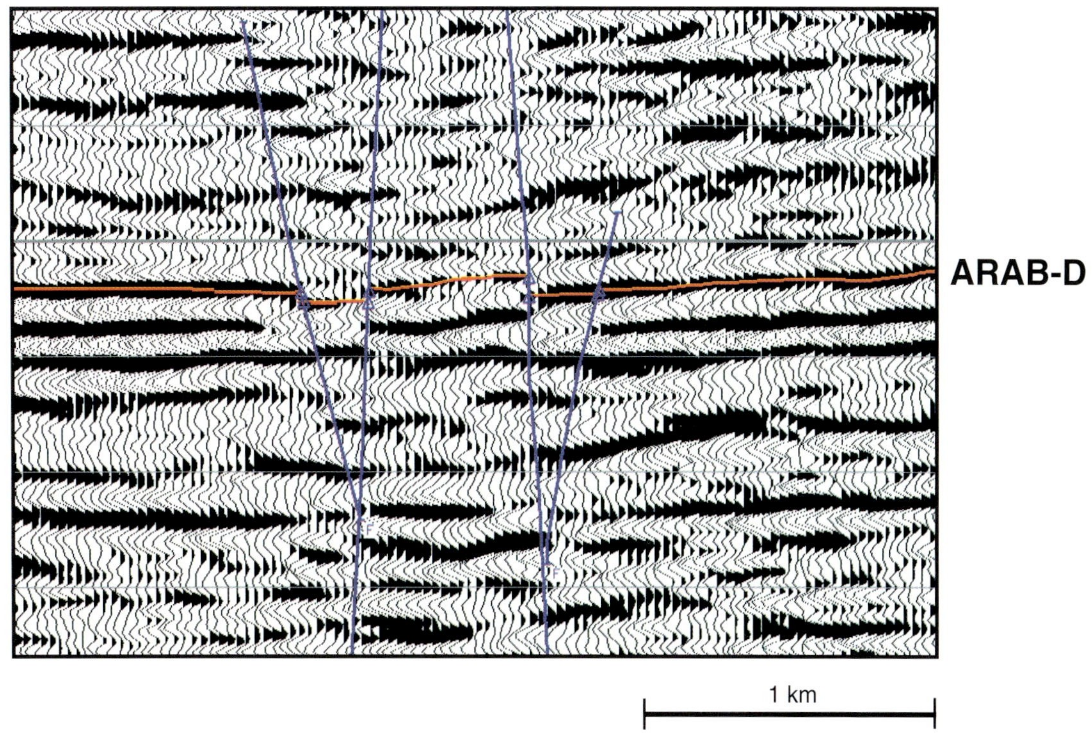


FIGURE 10. Faulting in Central 'Uthmaniyah as seen on seismic example II.

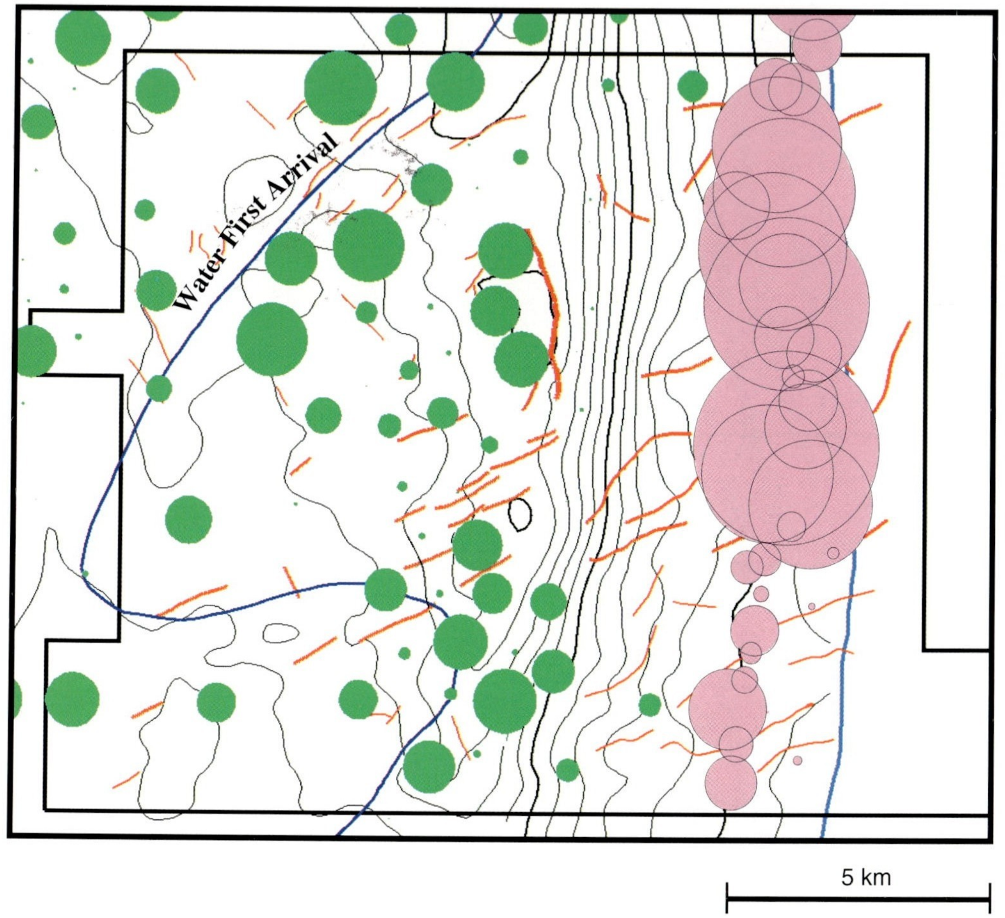


FIGURE 11. Cumulative production/injection in the Arab-D reservoir of the central 'Uthmaniyah area. Also shown are structure of top Arab-D reservoir and faults (see Figure 7).

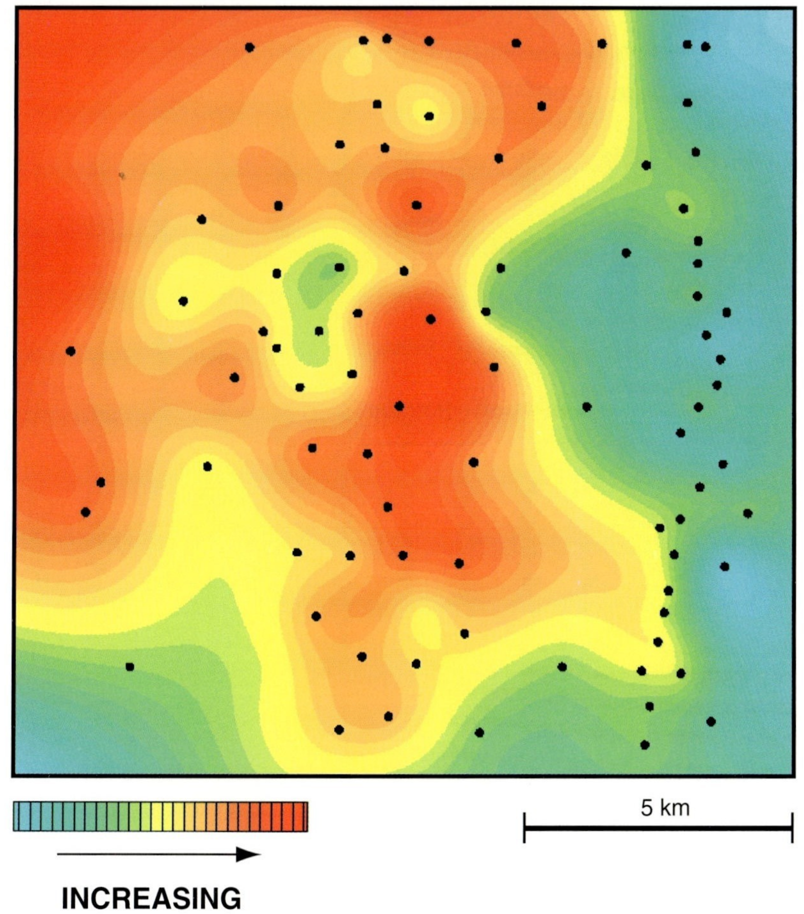


FIGURE 12. Arab-D reservoir ΦH from well data in the central 'Uthmaniyah area.

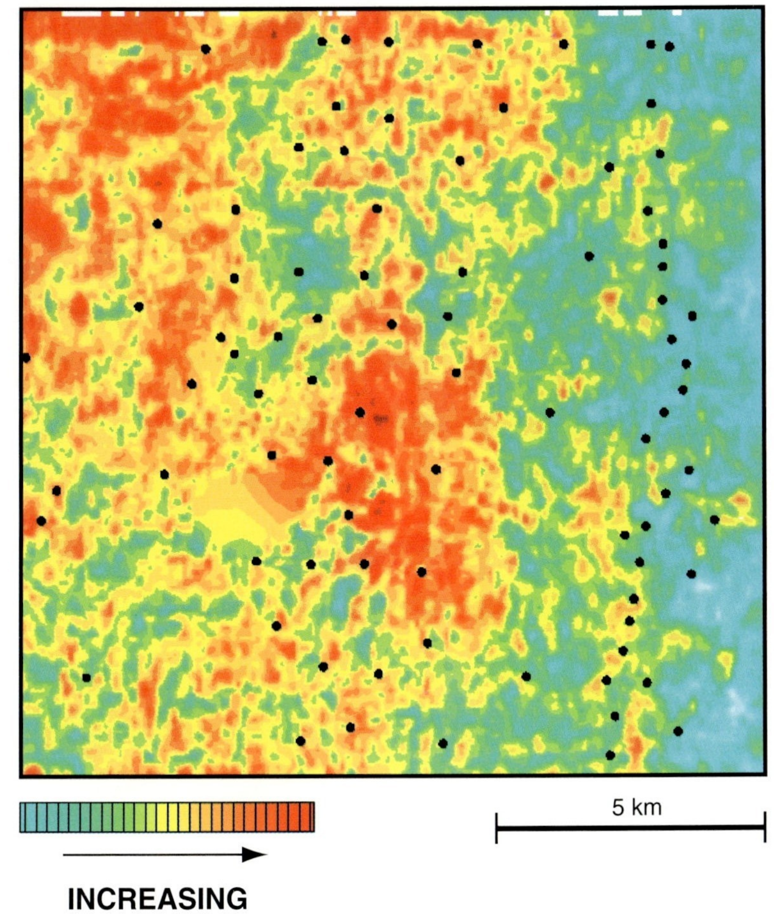


FIGURE 13. Arab-D reservoir ΦH from seismic amplitude and well control, central 'Uthmaniyah area.

Skaloud, D. K., The use of 3-D seismic data as a tool to increase ultimate recovery from the Yibal Shuaiba Reservoir, Oman, *in* P. Weimer and T. L. Davis, eds., *AAPG Studies in Geology No. 42* and *SEG Geophysical Developments Series No. 5*, AAPG/SEG, Tulsa, p. 211–218.

CHAPTER 23

# The Use of 3-D Seismic Data as a Tool to Increase Ultimate Recovery from the Yibal Shuaiba Reservoir, Oman

Dieter K. Skaloud*

## ABSTRACT

The Yibal Field, located in northern Oman, is a mature oil field, with a current total of 436 wells, from which about half of the ultimately recoverable reserves have been produced. Some 100 Shuaiba reservoir production wells have been preselected using structural and amplitude interpretations of a 3-D seismic survey acquired in 1988. The seismic data were integrated with petrophysical, geological, and production models to arrive at optimal well locations and trajectories.

The results of these wells, many of which are horizontal, have been very encouraging in terms of production and recovery improvements and demonstrate the success of the approach. Top Shuaiba seismic attribute data have been shown to have high predictive value for reservoir porosity, hydrocarbon distribution, and production behavior.

New 3-D data were acquired in 1993 to improve the structural model, identify remaining unswept oil, and enable real-time reservoir visualization. Although not yet fully processed and evaluated, interpretations of time-lapse effects in reservoir fluid distribution are presented.

## INTRODUCTION

The Yibal Field, located in the desert of northern Oman, some 300 km SW of the capital, Muscat (Figure 1), is the Sultanate's most important oil-producing structure. Currently, some 32,000 $m^3/d$ of 40° API oil are being produced from the Aptian limestone reservoir of the Shuaiba Formation (Figure 2) at Yibal. The field was discovered in 1962 and has been on production since 1969. It contains 540 million $m^3$ of stock tank oil initially in place. It is developed with an initial nine-spot water-injection pattern, which is being filled in with further horizontal producers. In many cases, well spacings have already been reduced to less than 200 m. To date, a total of 436 wells have been drilled in the field, the large majority of which are producing from the Shuaiba reservoir. There are some 40 horizontal producing sections, including sidetracks from exhausted vertical wells. Recovery factors in some parts of the field are forecast to be as high as 60%.

Besides Shuaiba oil, there is also significant gas production from an Upper Cretaceous Natih reservoir and further oil production from the Lower Cretaceous Khuff Formation.

*Petroleum Development, Oman L.L.C., P.O. Box 81, Muscat 113, Oman
Present address: Shell Nigeria Exploration and Production Company, P.O. Box 10235, London SW19 3ZN, UK

The field's anticlinal structure, most likely salt-induced, has an areal extent of circa 70 $km^2$ at the Shuaiba level, at a depth of around 1400 m below the surface (Figure 3). The maximum column length is 110 m, and the average porosities in the reservoir range from 30 to 40%. There is an abundance of faults, which are known to behave as potential water conduits, leading to early high watercut production in wells drilled too close to them. Water fingers are also known to have developed in higher permeable reservoir units, originating from either these water-conductive faults or the previous practice of water injection into the oil column rather than below the oil-water contact.

Drilling wells in the field itself can be most problematic, owing to the presence of numerous total loss zones in the overburden formations. Losses are in most cases related to the proximity of faults. Gas migration into shallow aquifers through poor cementations or corroded casing has led to an omnipresent risk of subsequent inflow after such losses. Elaborate casing schemes are required to combat the problem, and well-drilling costs are relatively high. Furthermore, the existing oil field infrastructure on the surface often renders it impossible to prepare locations for newly proposed wells, leaving deviated or horizontal well paths as the only solution. This approach has now proven to be a highly cost-effective way of developing reserves. The economic impact increases when existing exhausted vertical well bores are re-entered and horizontally sidetracked.

A 3-D seismic survey of 157 $km^2$ over the Yibal Field was acquired between September and November 1988. The prime objective was to improve the understanding of faulting in the overburden and to assess the risk of mud loss in wells to be drilled. A vibratory energy source was used, and the receiver aperture consisted of four lines of geophones, 200 m apart. The coverage at the reservoir level was 16-fold. The source sweep ranged from 5 to 60 Hz. At the reservoir level, the frequency content was 30–40 Hz.

A second 3-D seismic survey covering the field was acquired in 1993, with the main objective of improving structural resolution at the Shuaiba reservoir level by state-of-the-art acquisition and processing techniques. Multiplicity at the reservoir level was increased to about 36-fold, and the frequency content was improved significantly. The second objective of the 1993 survey was to provide data for possible time-lapse studies. Reprocessing of both the old and the new data is currently underway to compensate for the different acquisition techniques and to allow application of exactly the same seismic parameters when comparing the two datasets.

## SEISMIC INTERPRETATION

Key reflectors in the surveys are the top of the Eocene Rus Formation, two boundaries within the Albian to Cenomanian Natih Formation (Top Natih B and E), the Base Nahr Umr unconformity (which coincides with the top of the Aptian Shuaiba reservoir and the base of the Jilh Formation; Figure 4), and an unconformity near the top of the Permian Khuff reservoir (Figures 2 and 5).

All these horizons are marked by the presence of faults in two dominant directions. One is northwest-southeast, parallel to the strike of the Oman Mountains, and the other, that of the apparently older but re-activated fault, is southwest-northeast. There are also some clear geometrical indications of strike-slip movements along both fault families, such as steeply dipping fault planes and the expression of flower structures on vertical sections and en-echelon fault patterns in plan view (Figures 3 and 5).

The reflection at the top of the Shuaiba reservoir contains large amplitude variations, to the extent of polarity reversal. The time interpretation is therefore not straightforward. Very clear amplitude changes are observed at the edges of the field, coinciding with the 50% oil saturation contour and a marked drop in porosities. Pronounced amplitude dimming notably occurs around most of the field's water-injection wells and near fault zones, which are known to be water-conductive planes (Figures 5 and 6). It can therefore be concluded that the overall reflection amplitude pattern at the top Shuaiba level is related to reservoir properties such as porosity, oil saturation, and pressure.

### SHUAIBA RESERVOIR SEISMIC AMPLITUDES

There is a good well synthetic–to-seismic match, particularly with the 1993 zero-phased data (Figure 4). Most synthetic seismograms also show a reflection coinciding with the oil-water contact (and reduced reservoir porosity) at most of the matched wells, but no horizontal event can be mapped at this boundary. The field had been producing oil for about 20 years prior to the 3-D seismic acquisition. Water was injected initially at the top of the reservoir (full face drive) and later below the oil-water contact (bottom water drive). An even contact may therefore not be expected. Indeed, it is known from the well results that bottom water rise can vary significantly over short distances and that water fingers present in some high-permeability layers within the reservoir will cause tuning effects, rendering a recognition of the present-day contacts impossible.

The generally high porosities of the oil-bearing Shuaiba Formation in Yibal make the reservoir acoustically softer than the overlying shales of the Nahr Umr Formation. Porosities in many central parts of the field do reach up to 40%, while the field average for the top 30 m of the reservoir is 36%. A high oil saturation further enhances the resulting negative acoustic impedance contrast. Where porosity is below a certain value, particularly outside the field boundaries, the Shuaiba Formation is acoustically harder than the overlying seal.

Amplitude versus offset (AVO)–dependent forward modeling illustrates the expected reflectivity at the top of the reservoir and hence the amplitude of the resulting reflection (Figures 7 and 8). A threshold at which polarity reversal from negative to positive amplitude will occur can be expected for a 30% reservoir porosity and 100% oil saturation case with a "realistic" offset of 17° (Figure 7). At the same offset and for the same porosity reservoir, the expected amplitude is negative, however, when the oil is replaced by water, i.e., 0% oil saturation. A polarity reversal would, in the water-filled reservoir, occur for a higher-porosity case, around 32.5% (Figure 8).

It is obvious here that the seismic amplitudes are dominated by the porosity of the reservoir rock. A porosity decrease of as little as 0.5% has the same effect on acoustic impedance contrast as a decrease in oil saturation of 30% (Figure 9).

The porosity distribution in the field, however, is well known thanks to the many well penetrations; furthermore, abrupt significant lateral facies changes do not occur. Consequently, seismic amplitude changes within relatively small areas with well control can be attributed to hydrocarbon saturation variations within approximately the top 30 m of the reservoir. This thickness is taken to be the seismic resolution, based on velocity and frequency content.

To prove this model, a crossplot of well-derived average acoustic impedance values for the top 30 m of the reservoir against the 1988 seismic amplitude values at those wells was made. No good correlation is evident. A somewhat better correlation is found when only those wells drilled within a period of one year before and after the seismic acquisition, i.e., Yibal-220 to -274 (Figure 10), are considered.

Scatter may be expected and can be explained by one or more of following:

- saturation data logged at the time a well was drilled (sometime between 1962 and 1993) will not be the same as at the time the seismic acquisition took place (1988);
- water fingers near the top of and within the reservoir (from early water injection into the oil-bearing reservoir and/or via faults) result in tuning effects which may cause misinterpretations and certainly modify the amplitude response;
- the acoustic impedance of the overlying shales is not accurately known; it is considered to be constant for practical purposes, but any variations will affect the measured top reservoir amplitude;
- there may be amplitude versus offset (AVO), or rather offset distribution, effects that have not been accounted for;
- there is petrophysical evaluation uncertainty (refer to the influence of porosity on acoustic impedance, Figures 7 and 8);
- the seismic data at reservoir level have a relatively low resolution of 30–40 Hz;
- the applied processing sequence cannot be considered optimal for real amplitude preservation;
- the effect of reservoir pressure and temperature on acoustic impedance has not been modeled here; pressure variations might have a considerable effect on the reflection amplitude.

## PRACTICAL APPLICATION OF AMPLITUDE INFORMATION IN FIELD DEVELOPMENT

Despite the above factors, it has proven to be highly practical in Yibal to use Top Shuaiba reservoir lateral amplitude variations as a guide to find areas of unswept oil. The predictive value of the seismic amplitude has been confirmed by many wells. Rather than a simple "pick"-amplitude, data volume–related window measurements such as reflection heterogeneity or similar amplitude summation methods, starting above and extending some way across the reservoir interval, have the advantage of "absorbing" any uncertainty in the actual time pick.

The well spacing has been reduced in some patterns to as little as 170 m in what is generally agreed to be a highly complex fault-compartmentalized reservoir. In the case of horizontal wells, the distance of some parts of the trajectory to neighboring wells is even shorter. A number of horizontal wells have trajectories curved in the horizontal plane, which were to some extent determined by following or linking up the highest-amplitude areas.

The seismic data are also extensively used to identify and describe potential drilling hazards, both at the reservoir level and particularly in the overburden. This allows wells to be planned in a cost-effective manner and has an impact on the technical success rate of Yibal wells (Skaloud et al., 1992).

The power of using amplitude information for selecting new infill well locations for further field development may well be twofold. First, a high seismic amplitude or equivalent window measurement in a pattern with good porosity and pressure control can indicate a significant oil column with high saturations. That, together with supporting petrophysical and production information from nearby wells, is often sufficient to economically justify a new production well with a view to increasing recovery efficiency. Second, most faults and fractures in the field have become water-conductive features under the prevailing dynamic reservoir conditions and should be avoided in new wells. Even small subseismic faults, whether water conductive or not, are often expressed as lineations with subtle amplitude changes. By targeting new wells in the center of high-amplitude areas, however small, the distance to any surrounding fault should be maximized. This normally results not only in finding an unswept oil column but also in delayed high-rate water production.

Detailed history matching of wells in simulation models has been possible only by the incorporation of fault interpretations on the 3-D seismic data. The resulting improved understanding of water movement in the reservoir has increased confidence in high predicted recovery factors and allowed significant upgrades to recoverable reserves. The initiation of an ongoing major infill development project has been a direct result.

## A FEW EXAMPLES—CONCEPTUAL AND OPERATIONAL

### The "Classical" Infill Pattern

An initial nine-spot water injection pattern, in which a quarter pattern and a so-called high-density infill were completed, is represented in Figure 11. Both the Top Shuaiba seismic amplitude map and the production performance of the wells in this pattern are shown in Figure 11. The latter are represented as circles; the size of each circle relates to gross production in mid-1992, and the fractions shown in red and blue represent oil and water, respectively. There is a coincidence and good qualitative correlation of oil producers with a low watercut in areas of relatively high seismic amplitude (in red) and of producers with a high watercut in low seismic amplitude areas (in yellow and blue).

Note, however, the marked difference between wells Y-365 and Y-367, both recent high-density infill wells. These two wells have been drilled on their geometrically correct infill locations. They are both situated in relatively high-amplitude areas of only slightly different magnitude. Well Y-365 found an unswept oil column with oil saturations around 95%. The well produced oil at a net rate of 200 $m^3$/day, with 5% water. Well Y-367, however, encountered an entirely swept reservoir, with a residual oil saturation of only 60%. The initial production was nearly 100% water. These two wells are mentioned again in the section on time-lapse effects, since the difference between the two neighboring and almost contemporaneously (in 1992) drilled reservoir sections is much better expressed on the 1993 seismic data, a possible time-lapse effect.

### The Flank Horizontal with Production Logging Data

Well Yibal-384 is a horizontal well on the southeastern periphery of the field. It was designed to stay as far as possible from mapped faults and to connect areas of high amplitude, indicative of high porosity and/or oil saturations at the top of the reservoir. The well was drilled very close to the top of the reservoir (Figure 12). Logs showed a number of intervals with less than 75% oil saturation. The bulk of the 1000-m-long trajectory, however, had oil saturations above 90%, and the well was completed in the open hole (barefoot). After the well had produced for some 12 months, with increasing volumes of water, a production logging tool (PLT) was run to identify the water influx zones and to allow remedial action. The PLT identified one 300-m interval of predominant water production towards the middle part of the well bore. This interval coincides with the zone of most pronounced change in the seismic trace character (Figure 12).

### The Contingency and the Horizontal Well Drilled from the Other Side

Well Yibal-406 was spudded to become a horizontal infill well. Severe losses while drilling in the Natih Formation resulted in a necessary extra casing string, which in turn did not allow the well to be drilled horizontally as planned. An alternative location, in reach of a technically feasible deviation from the existing well bore, was found and targeted. Figure 13 shows this location; it is in a structurally favorable position in the field with an extraordinarily high-amplitude reflection from the Top Shuaiba. The well, drilled with an S-bend deviation but vertically at the reservoir level, encountered an unswept oil column of some 60 m and oil saturations of around 95%. The well produced oil at an initial net rate of about 400 $m^3$/d, free from any water. A nearby older well, in an area with subdued seismic amplitude, found lower-porosity reservoir and a much shorter column.

The structure and amplitudes of the originally planned horizontal trajectory, however, still looked attractive. Because of the overburden problems, it was decided to start drilling from the opposite side. A low-amplitude area, possibly intersected by subseismic faults (of less than 5 m throw), is seen on the seismic data in the second quarter of the 800-m-long horizontal trajectory. The well logs show low oil saturations over a 200-m interval exactly coinciding with the area of low seismic amplitudes (Figure 13).

## CONSIDERING TIME-LAPSE EFFECTS ON SEISMIC DATA

The (interpreted) seismic data represent an image of the reservoir at the time of acquisition. Appreciating the total amount of fluid moved, the complexity of the movements that have taken place between 1988 and 1993, and the above-mentioned modeling results, it can be assumed that production-related effects can be seen by comparing the two datasets. It was, however, thought to be unwise and impractical to conduct the new survey with the same acquisition and processing parameters as the old one. Instead, considering the fashion in which the older data were successfully used, priority was given to obtaining a dataset of the best possible quality.

The seismic reshoot was economically justified on the basis of a cost-benefit analysis. Acquisition and processing costs for the survey compare favorably with the amount saved for wells that would potentially be suboptimally sited without new 3-D seismic data and hence not contribute to production. Further potential benefit is likely to come from higher initial production rates from wells optimized with the new seismic data.

The new data cover 200 $km^2$ and were acquired between May and October 1993, again using a vibratory technique, but with denser vibrator point spacing compared with that used in 1988 (double zigzag versus single zigzag) and with a 5- to 80-Hz sweep. The resulting nominal stack fold at the Shuaiba reservoir level is 32-fold.

Figure 14A, B shows seismic cross sections through wells Y-365 and Y-367 on the 1988 and the 1993 data, respectively. The different seismic character at the two well locations on the 1993 data can be measured and visualized—for instance, by an amplitude-summation attribute (reflection heterogeneity), measured from 12 ms above to 16 ms below the reflection maximum, displayed as Figure 14C. This attribute is shown in map form as Figure 15. The porosity and saturation profiles of the two wells are shown as Figure 14D, E; the wells are described in more detail in the "Classical Infill Pattern" section of this chapter. Although many differences in both acquisition and processing may actually contribute to the different seismic responses between 1988 and 1993, the correspondence of a seismic attribute trend with well logs and production behavior appears too good to be merely coincidental. It is therefore suggested that a time-lapse effect can be observed here between 1988 and 1993, with the calibrating wells logged in 1992.

The state-of-the-art processing has now been finalized, and both datasets are currently being reprocessed to remove as much as possible any differences that are related to different acquisition and processing parameters.

## CONCLUSIONS

The Yibal Shuaiba is a mature field, currently having some 300 oil-producing wells. Water movements are complex, well spacing is dense, and more than half of the field's reserves have been recovered. In spite of this, the initial production rates of newly drilled wells are at least as high as they were for wells in the past. Extremely high recovery factors, up to 60% in many parts of the reservoir, were forecast, resulting in a significant reserves increase with an ambitious infill drilling program.

An important step that led to this result is the integration of 3-D seismic data with petrophysical, geological, and production models to arrive at optimal well locations and trajectories. Seismic data have a high predictive value for porosity, hydrocarbon distribution, and even production behavior for the light oil, high-porosity carbonate reservoir described here.

Although the data are not yet reprocessed to remove variances related to different acquisition and processing parameters, interpretations of time-lapse effects in reservoir fluid distribution can be made with a repeat seismic survey.

This Yibal case study is an example that shows how important it can be to fully incorporate production seismology in field development activities.

### *Acknowledgments*

Permission from the Ministry of Petroleum and Minerals of Oman, the management of Petroleum Development Oman LLC, and Shell Internationale Petroleum Maatschapij to publish this paper is gratefully acknowledged. The author wishes to thank Calum Macdonald for the seismic modeling work and along with Jos Bonnie and Hans Tetteroo, for help in the preparation of figures.

## REFERENCES CITED

Skaloud, D. K., S. A. Rackley, C. F. M. Bos, and N. Holleman, 1992, Impact of 3-D seismic on well targeting, well design and reservoir modeling in the Yibal Shuaiba Reservoir, Oman: SPE 24488, presented at the 5th Abu Dhabi Petroleum Conference, May 18–20, 1992.

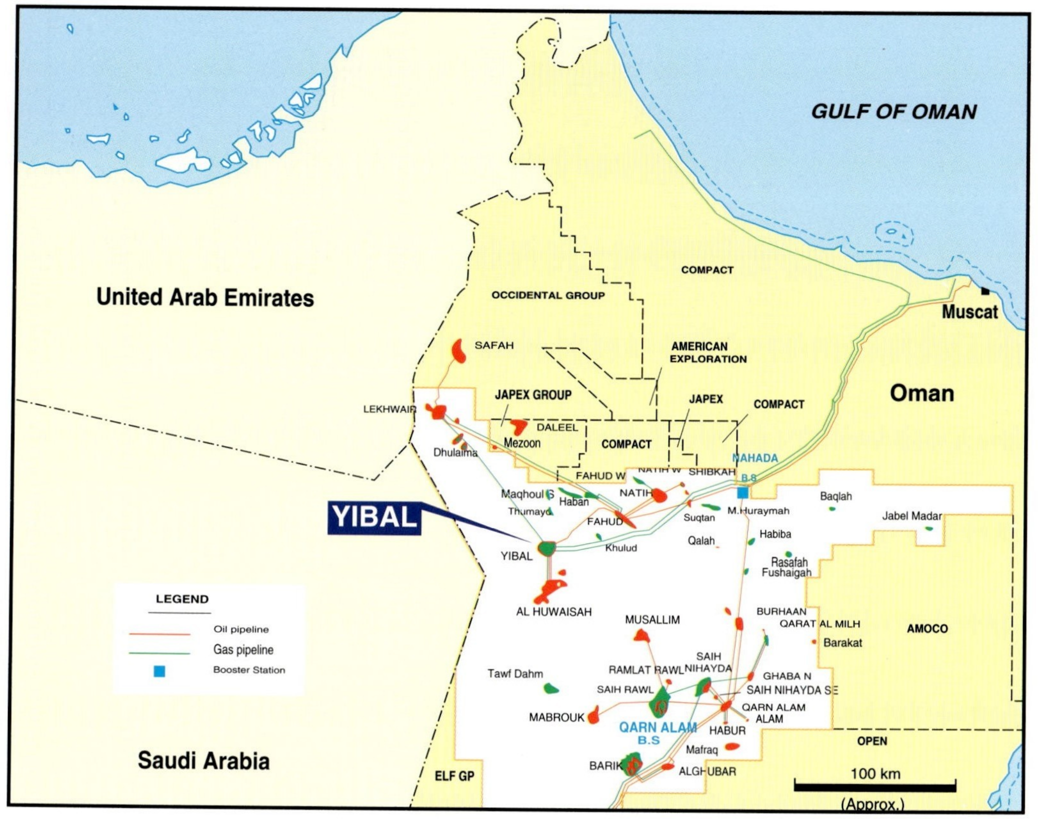


FIGURE 1. Location map of North Oman, highlighting the Yibal Field.

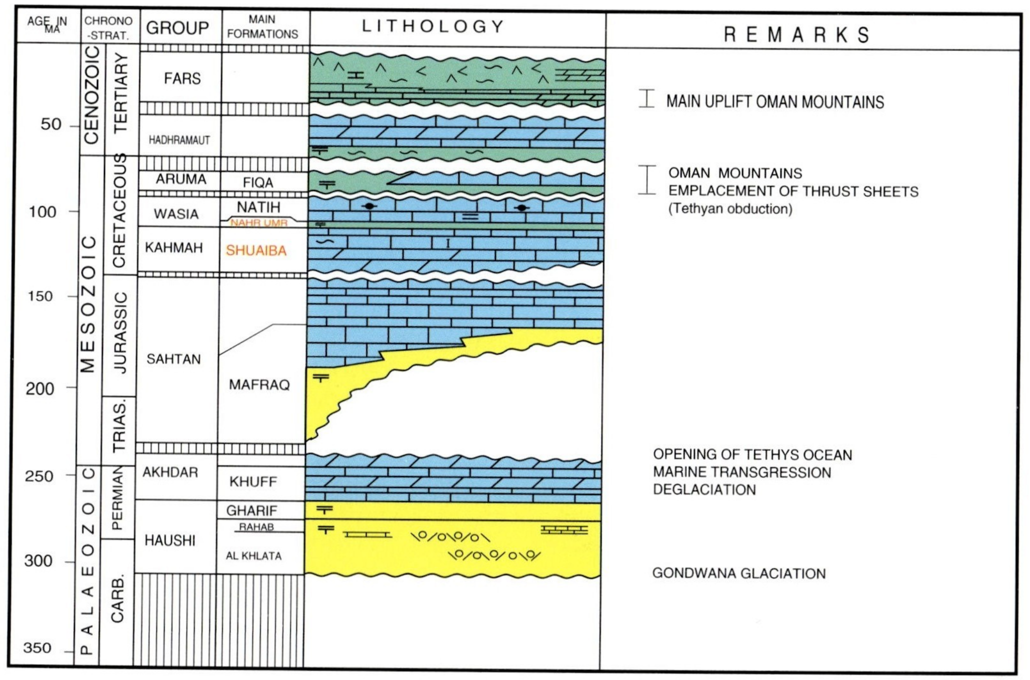


FIGURE 2. Overview of post-Devonian stratigraphic column found in Oman. The formations forming the reservoir-seal pair discussed here are highlighted in red.

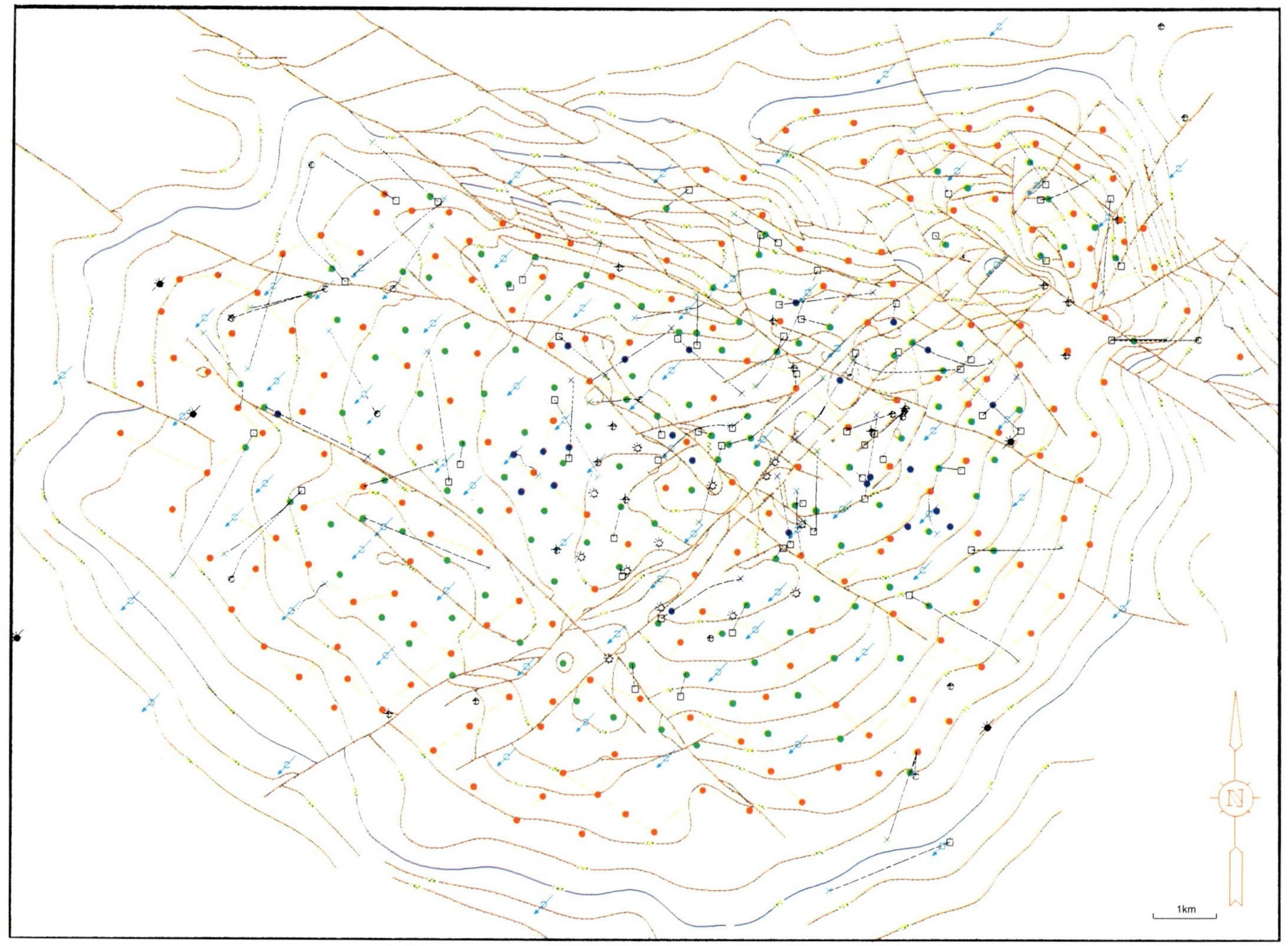


FIGURE 3. Yibal Field, Top Shuaiba Formation structure map, indicating size, shape, and structural style of the accumulation. The blue contour is the original 50% oil saturation level at 1340 m subsea; the contour interval is 10 m.

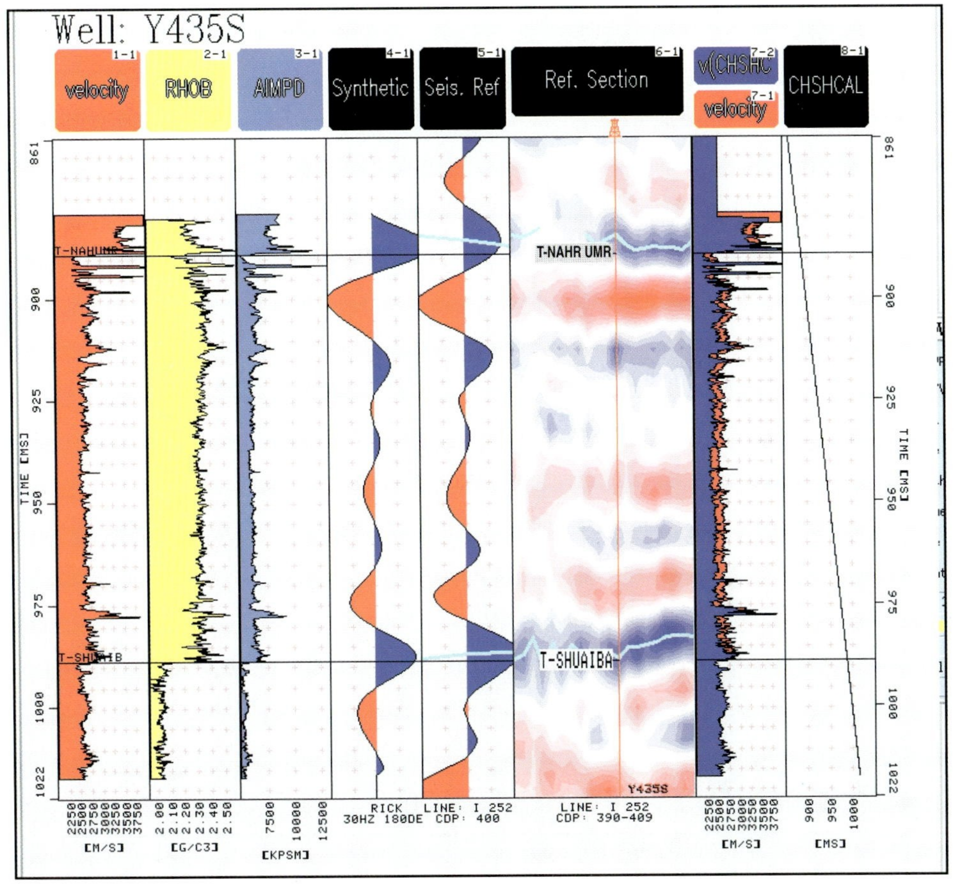


FIGURE 4. Well synthetic display for a typical Yibal well. Tracks 1 through 3 show velocity, density, and impedance logs. In track 4, the well synthetic is displayed and can be compared with a real seismic well trace shown in track 5. A seismic line, centered on the well, is shown in track 6. Note the decrease in acoustic impedance at the Nahr Umr-to-Shuaiba interface, resulting in a positive (SEG-normal) reflection amplitude at top reservoir.

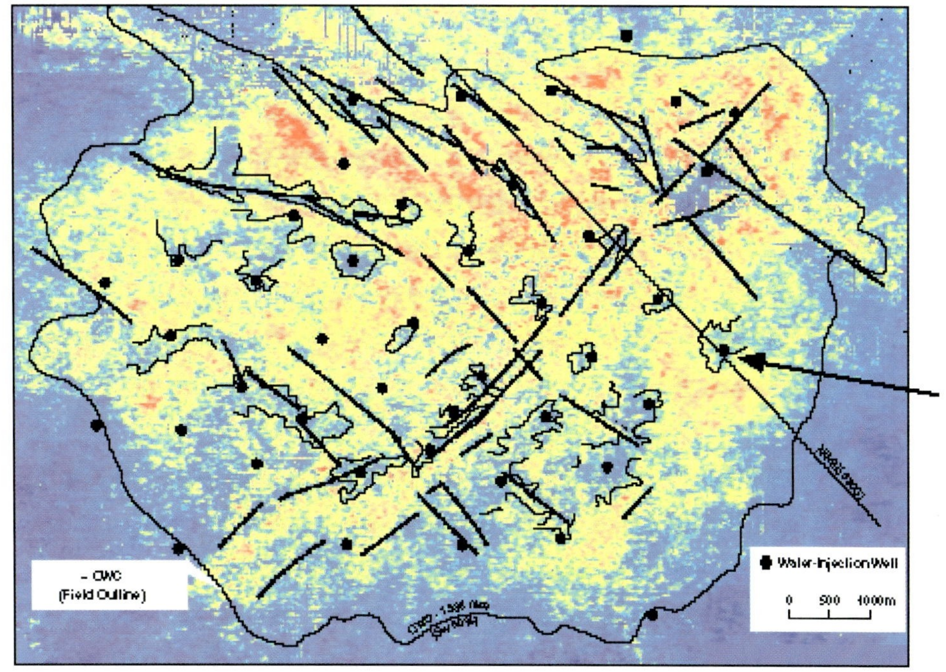


FIGURE 6. Top Shuaiba reservoir amplitude map (color scale from red—high positive amplitude—to yellow and blue—low positive and negative amplitude). The field outline, main faults, and all water-injection wells are indicated. Note the lower amplitudes around most water injectors, where the acoustic impedance contrast was reduced by injection water replacing oil at the top of the reservoir. This practice was discontinued, and water is now injected below the 50% oil saturation level.

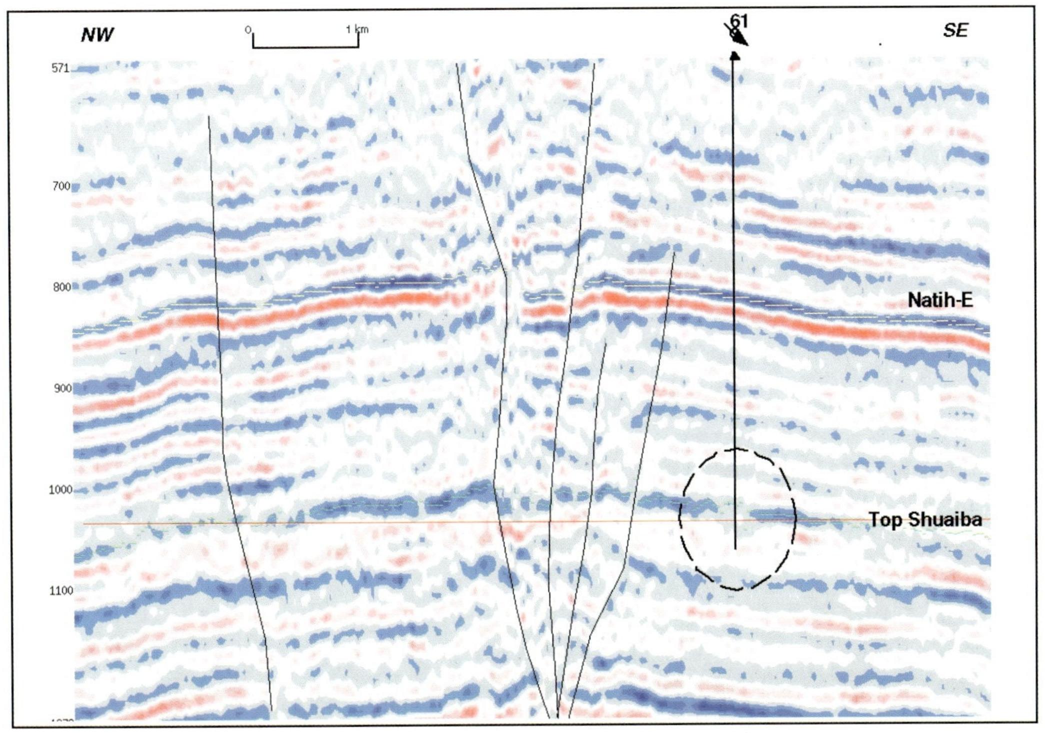


FIGURE 5. A typical northwest-southeast section through the Yibal 3-D data. Note an interpreted polarity reversal at the intersection of the top Shuaiba with the 50% oil saturation level (thin red line at approximately 1025 ms) and the marked amplitude dimming around water-injection well Y-61 (circled). The projection of this section is indicated on the map in Figure 6.

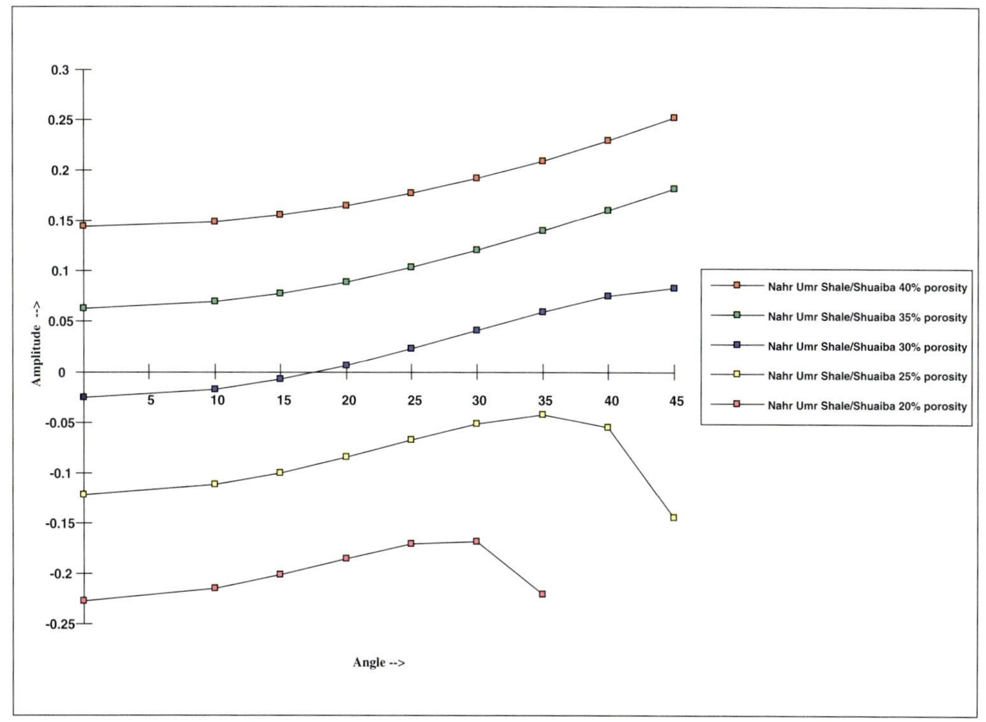


FIGURE 7. Modeling of amplitude versus offset for the Nahr Umr shale to oil-filled (100% $s_o$) Shuaiba reservoir interface for five different Shuaiba reservoir porosities. Note that the reservoir with 30% porosity will have a very different seismic response from the one with 35% porosity.

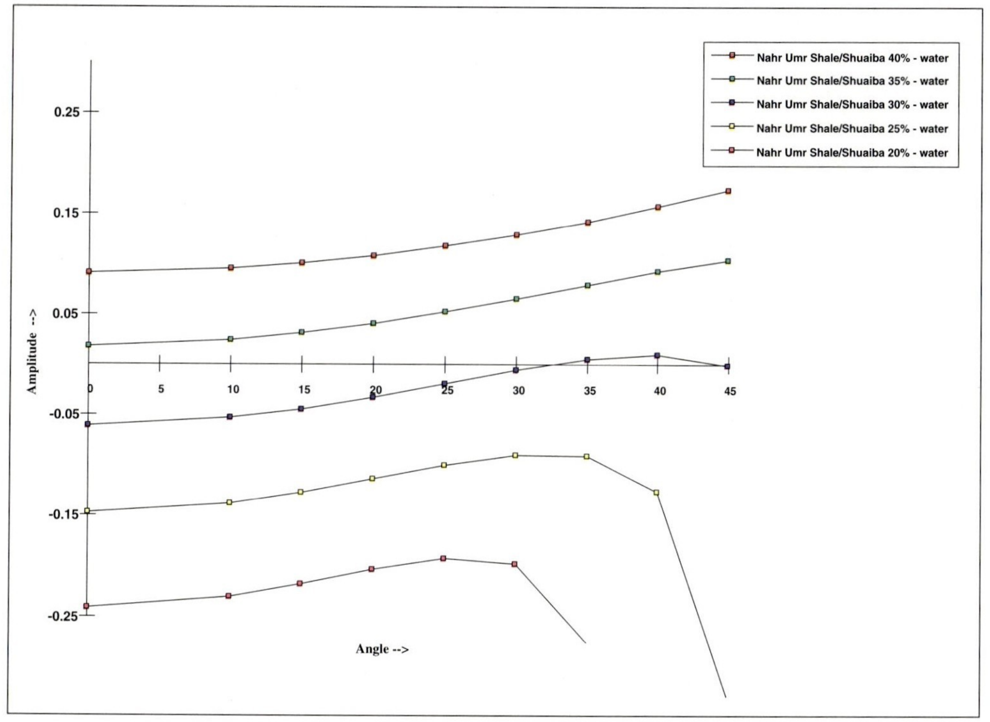


FIGURE 8. Same as Figure 7 but for a water-saturated (0% $s_o$) Shuaiba reservoir. Seismic amplitude at the Nahr Umr shale–to–Shuaiba reservoir interface is shown to be reduced for all porosities and at all offsets (angles). The higher the porosity, however, the more significant the influence of porefill on the reflection amplitude.

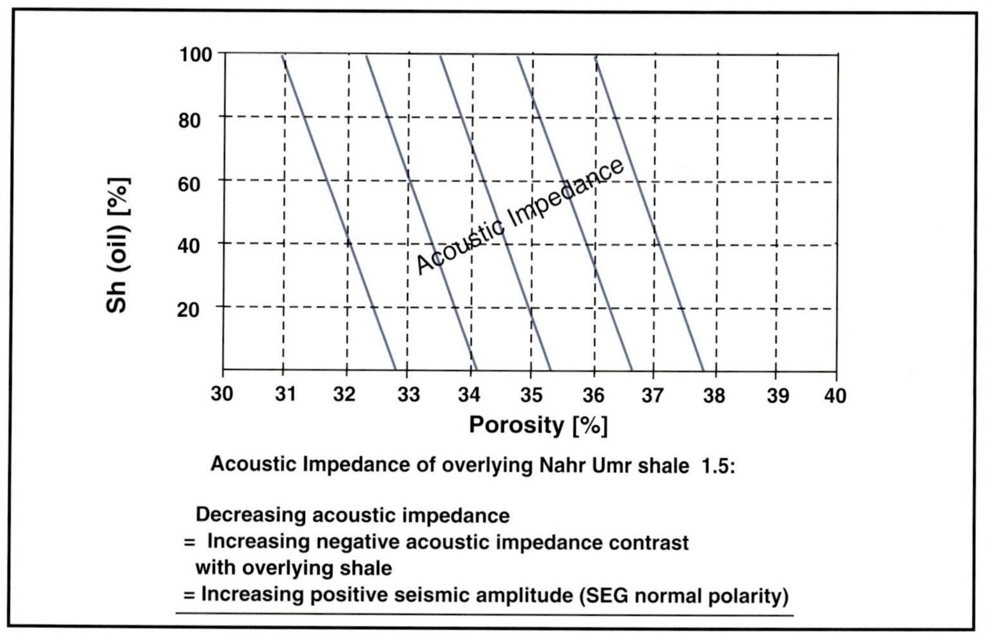


FIGURE 9. Relationship among porosity, oil saturation, and acoustic impedance for the Shuaiba reservoir. This graph visualizes the predominance of porosity over saturation in its effect on seismic response. An increase in porosity of 1% has the same effect as an increase of 60% in oil saturation.

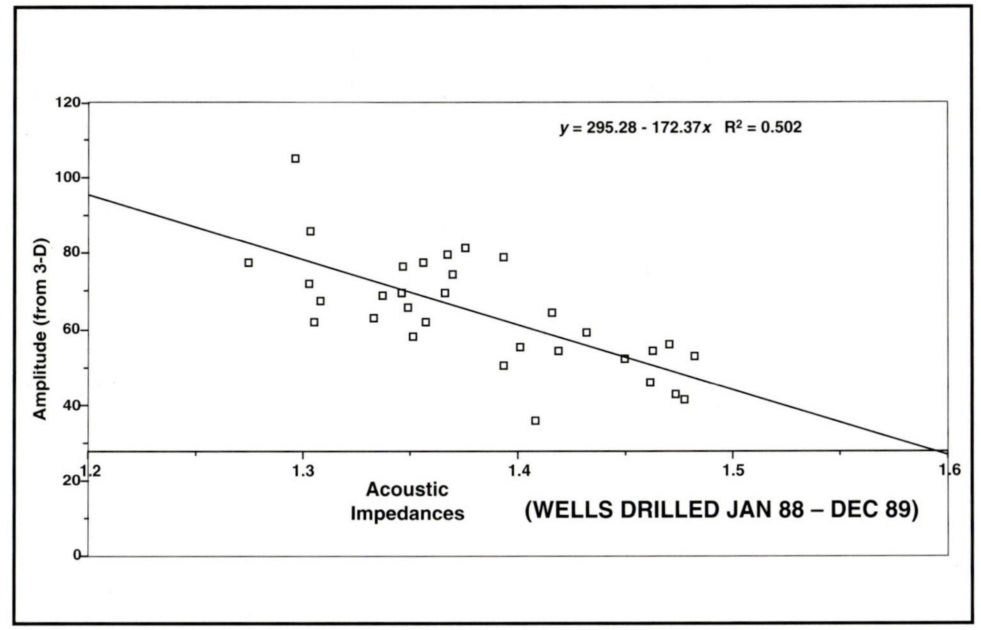


FIGURE 10. Top Shuaiba seismic amplitude (1988 3-D survey) versus acoustic impedance calculated from petrophysical well data. Only wells drilled between one year before and one year after the seismic acquisition (Y-220 to Y-274) are plotted. Only for this period can a reasonable correlation be made between well log data and seismic data, which both rely on a static measuring method in a dynamic reservoir situation.

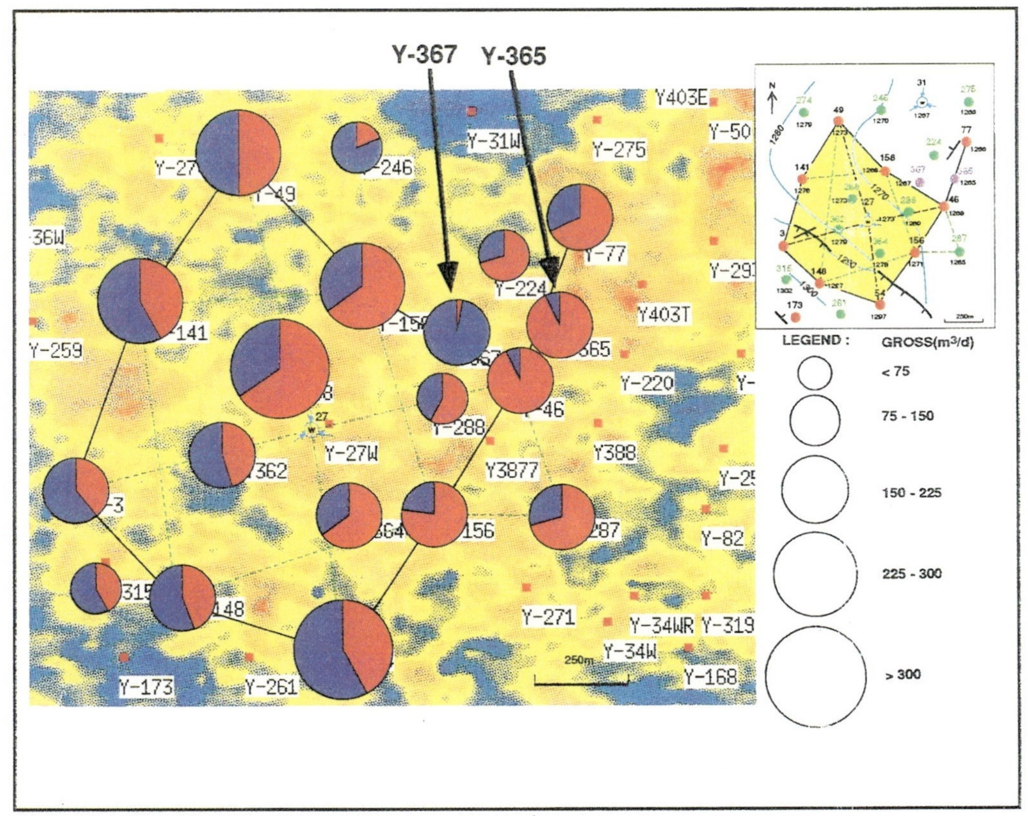


FIGURE 11. Detail of top Shuaiba reservoir amplitude map, showing one representative development pattern; production data are shown as "bubbles," whose size represents gross liquid production rate in m³/d, and whose red-filled part repesents the net oil fraction of this (blue represents the water fraction). It can be seen that the locations of more "dry" oil producers coincide with relatively higher seismic amplitude areas (red and yellow on map).

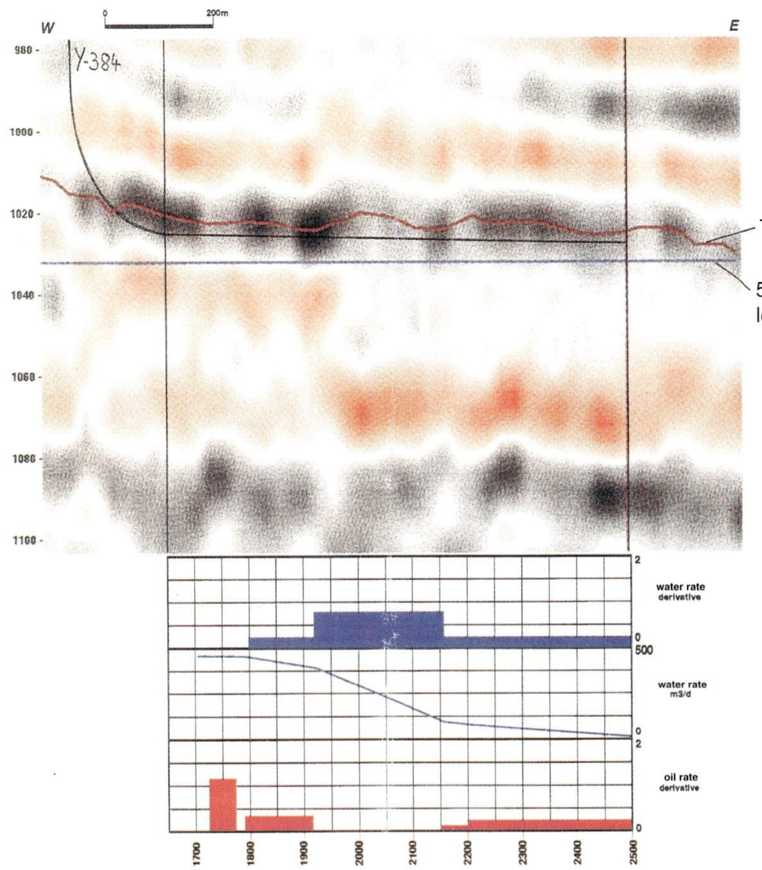


FIGURE 12. Seismic section along the Yibal-384 horizontal well trajectory on the southeastern flank of the field. The top of the reservoir and the approximate level of the 50% oil saturation level are indicated. Below the section, the result of a production logging tool interpretation is shown. Note the coincidence of a low Top Shuaiba reservoir amplitude with irregular traces over the reservoir interval and the section of the well bore in which the most water influx is observed.

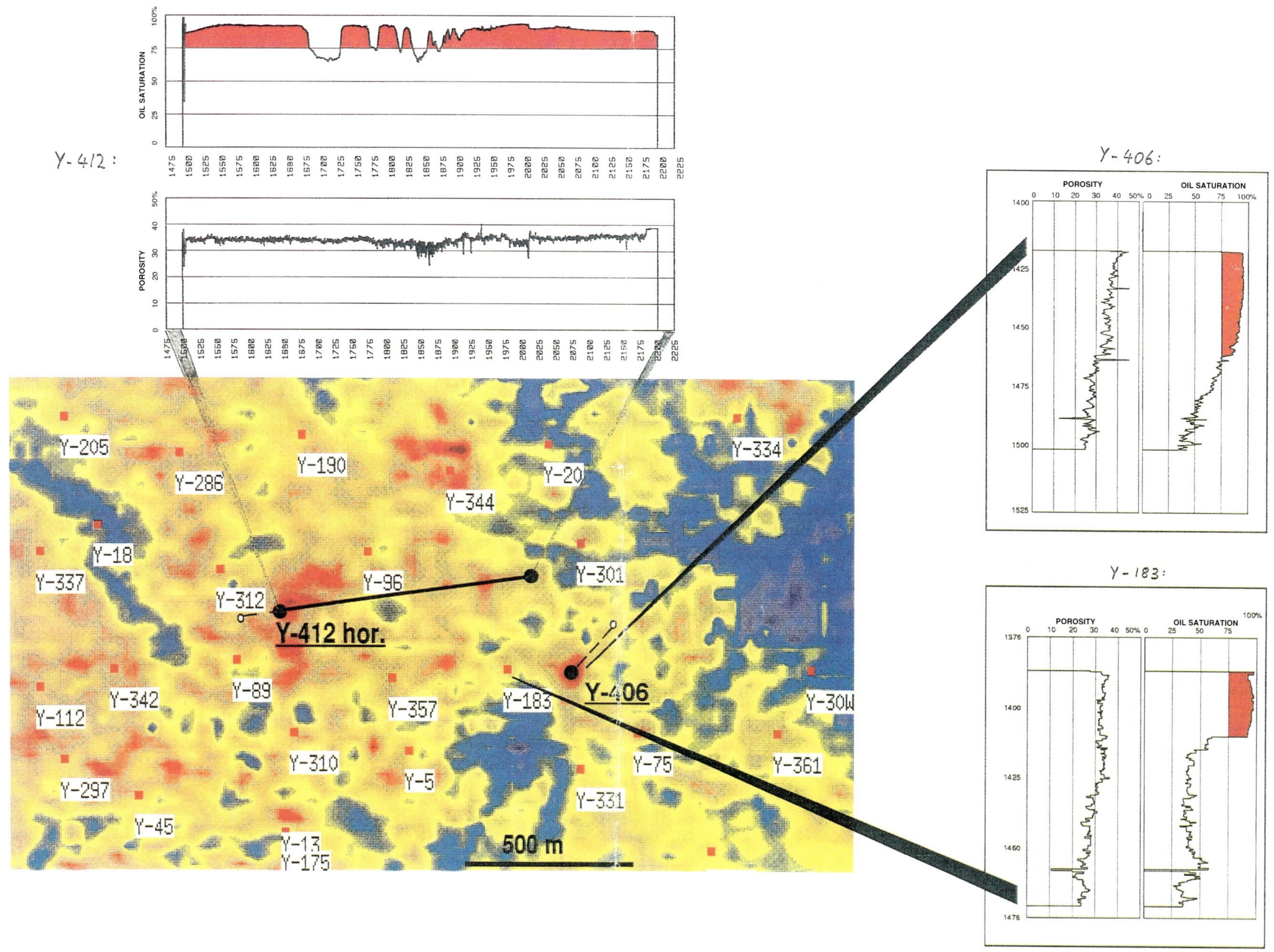


FIGURE 13. Detail of top Shuaiba reservoir amplitude map and three key wells; Y-406 was drilled (with an S-bend deviation) to target a "sweet spot" of the reservoir in a structurally favorable position, indicated by high amplitude (red on map). The well encountered some 60 m of reservoir with oil saturation above 50% (of which 45 m was above 75%). An older well, Y-183, with top reservoir at the same true vertical structural depth and only 200 m away, had found only some 30 m of reservoir with oil saturation above 50% (of which 25 m was above 75%). Note, however, that the porosity at the top of the Y-406 section appears to be extremely high (40%) in comparison with Y-83 (35%); this porosity difference is sufficient to explain the observed amplitude variation on the map between the two points, only further enhanced by a longer and more highly oil-saturated column. Y-412 was drilled as a horizontal well, 3–5 m below the top of the reservoir, with the objective of linking two high-amplitude areas and potential reservoir "sweet spots." The logs show a fairly constant porosity; however, the lower-saturated, disturbed interval in the second quarter of the horizontal hole, between 1675 and 1875 m (ah bdf), precisely matches a relatively low-amplitude area.

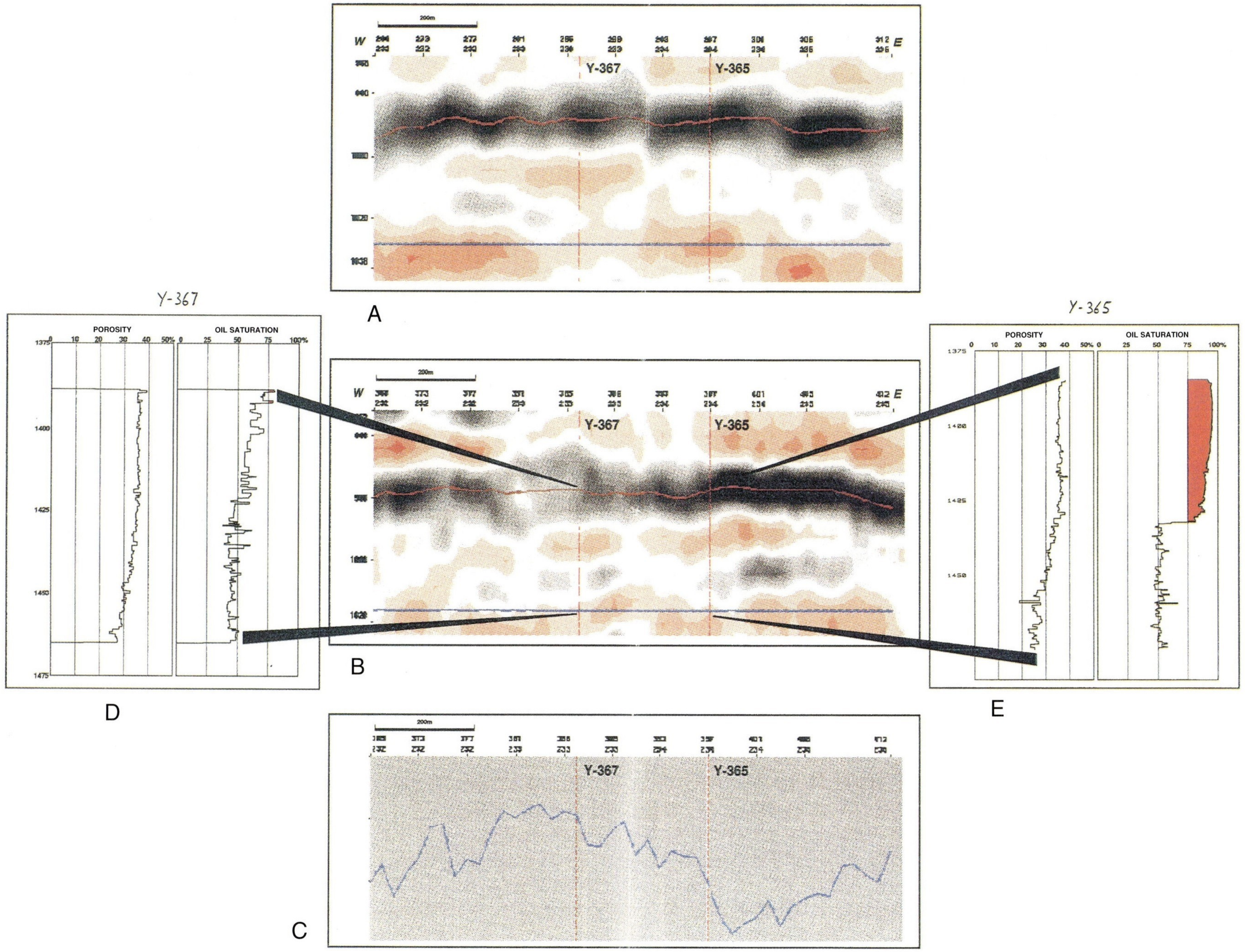


FIGURE 14. (A) Section from the 1988 seismic data through two wells drilled at approximately the same time, in 1991, only 250 m away from each other. These wells are also marked on the amplitude map of Figure 11. (B) Equivalent section from the 1993 data. (C) Display of "reflection heterogeneity" for the 1993 data along this section, a summation of vertical amplitude changes from sample to sample in the seismic volume, here over an interval from 12 ms above to 16 ms below the interpreted top Shuaiba; scale irrelevant. A pronounced difference between (A) and (B) in the appearance of the reflection at the Top Shuaiba level is seen. (D) Porosity and saturation profiles for well Y-367. (E) Porosity and saturation profiles for well Y-365. At the same porosity, oil saturation in well Y-365 is well above 75%, whereas in Y-367 it is 30–40% lower. This difference is matched by the seismic response at the two well locations on the 1993 data. It is not observable on the 1988 data, which may permit the assumption that flushing of the reservoir at the Y-367 location has occurred after 1988 (i.e., a time-lapse effect detectable on seismic).

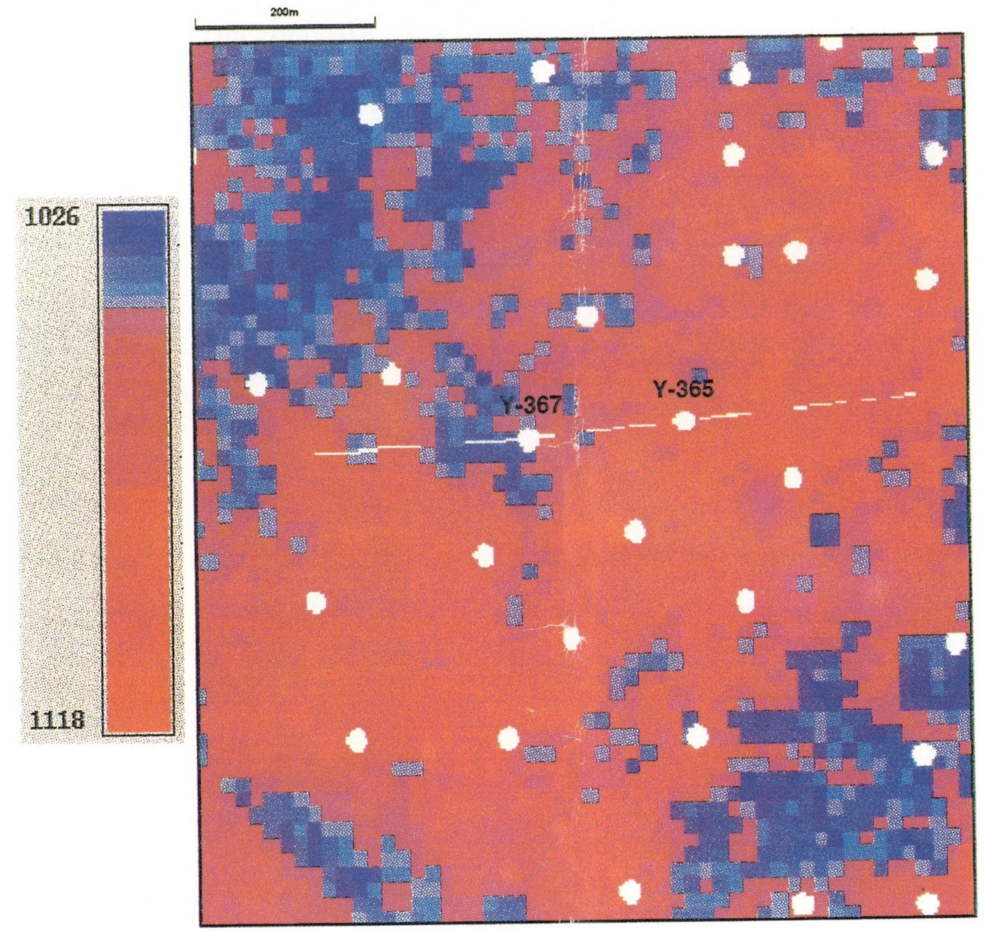


FIGURE 15. The "reflection heterogeneity" described in Figure 14C, from the 1993 seismic data, in map form. This picture suggests some form of breakthrough from a blue area in the west towards Y-367, perhaps along a subseismic fault or fracture zone. (The absolute scale here is relative and meaningless.) Seismic attribute mapping on the 1988 data does not reveal this phenomenon (compare Figure 11).

Ng Tong San, and M. Mohamad, A quantitative analysis of seismic reflection in a gas-bearing carbonate buildup, offshore Malaysia, *in* P. Weimer and T. L. Davis, eds., *AAPG Studies in Geology No. 42* and *SEG Geophysical Developments Series No. 5*, AAPG/SEG, Tulsa, p. 219–224.

## CHAPTER 24

# A Quantitative Analysis of Seismic Reflection in a Gas-Bearing Carbonate Buildup, Offshore Malaysia

Ng Tong San and Marzuki Mohamad*

## ABSTRACT

Miocene carbonate buildups are extensively developed in the Luconia Province offshore Sarawak, Malaysia. Some 200 carbonate buildups have been seismically mapped. In the early exploration days, the carbonate structures were mainly defined using 2-D seismic data.

The advances of 3-D seismic data allow seismic attributes to be used in the development of carbonate gas fields, both for structural mapping and for stratigraphic interpretations. Both vertical and lateral porosity variations in carbonate reservoirs can be calibrated to seismic amplitude and acoustic impedance. This paper outlines a method that highlights the relationship between the thickness of the reservoir and seismic isochore values. These techniques can be used to select the optimum location for development wells.

## INTRODUCTION

The study area is located in offshore Sarawak in Luconia geological Province (Figure 1). The study area, Luconia, is bounded to the east by the Baram Delta, to the south by the Balingian Province, and the west by the Rajang Delta. Carbonate buildups are major exploration targets in the Luconia geological province. Twenty gas fields have been discovered in these features.

Carbonate buildups occur primarily in the middle to late Miocene strata (Epting, 1980). In the south, some of the carbonate buildups were drowned by rapid sea level rise as early as the middle Miocene time, whereas in the central area, some of the carbonate buildups were killed during the late Miocene by influx of siliciclastic sediments from the Baram Delta or the Balingian Province. The carbonate buildups in the northern area are exceptional, because they began to grow as early as the late Oligocene, and some of them, e.g., G10, are still flourishing today (Figure 2) (Ho, 1978).

The source rock for the gas is mainly terrestrial derived, consisting of coal and coaly shale deposited in a lower coastal plain environment during the late Oligocene to early Miocene age.

The 3-D seismic data clearly imaged the carbonate buildup where the top of the carbonate has a prominent high-amplitude reflection (Figure 3). Furthermore, the 3-D seismic data imaged the internal geometry of the carbonate buildups as a series of layered porosity zones.

*Petronas Carigali Sendirian Berhad, P.O. Box 12407, 50776, Kuala Lumpur, Malaysia

## QUANTITATIVE ANALYSIS OF SEISMIC REFLECTION ATTRIBUTES

A simple method using seismic amplitude and seismic two-way isochore to predict reservoir porosity and thickness, respectively, is implemented (Meckel and Nath, 1977). The seismic two-way isochore value is the seismic two-way time between the peak and trough of the seismic reflection which correlates to the porosity zone. The first step is to analyze the 3-D seismic data by calibrating it to a synthetic seismogram. The synthetic seismogram was generated with a zero-phase wavelet of bandpass frequency similar to that of the 3-D seismic data. The direct approach established a linear relationship between porosity and seismic amplitude and between porosity zero thickness and seismic two-way isochore (Ng, et al., 1993). The final product of these exercises was porosity- and porosity zone–thickness maps that were generated from seismic amplitude and seismic two-way isochore maps, respectively.

## CALIBRATION OF SYNTHETIC SEISMOGRAM TO 3-D SEISMIC DATA

The initial procedure in carrying out the quantitative analysis of the seismic reflection attributes is to correlate the well data to the 3-D seismic data by means of the synthetic seismogram. A total of two wells' results were used in this exercise. The good match between the synthetic seismogram and 3-D seismic data suggests a zero-phase wavelet output of the 3-D seismic data is required in this study (Figures 4 and 5). This is important because we assumed a simple geologic model in which the porosity zone is thicker than the "tuning" thickness. Therefore, the boundary at the porosity zone in a carbonate buildup will correspond to a peak or trough of seismic reflection, which has allowed us to calibrate the peak or trough of the seismic reflection wavelet to porosity. This exercise also confirms that at the zone of interest, the 3-D seismic data are free of reflection multiples. The synthetic seismogram also indicates that the internal geometry of the carbonate is characterized by layers of seismic reflections that correlate to the porosity zonation. The less porous zones have a higher acoustic impedance value than the porous zone, which cause the acoustic contrast at their boundaries to generate seismic reflections. This synthetic seismogram matches well with the 3-D seismic data (Figure 5), indicating that the 3-D seismic data could be used to image the porosity layering. The seismic data also suggest that the lateral extent of the porosity layering is quite wide.

## POROSITY-SEISMIC AMPLITUDE

The tops of the layered porosity zones (more porous zones) in the carbonate buildups which correspond to a peak of the seismic reflection, were interpreted through automated horizon picking using a seismic interpretation workstation. The seismic amplitude of this interpreted peak seismic reflection was then extracted from the 3-D seismic database. This process was repeated for the less porous zone, but in this case, the automated horizon picking followed the trough of the seismic reflection. At the well locations, the seismic amplitude corresponding to the top of the porosity zone was calibrated to the wireline log porosity using a simple linear plot (Figure 6). In this study, low seismic amplitudes are related to higher porosity values and high seismic amplitudes are related to lower porosity values. Quantitatively, the less porous zone, (porosity range 22–29%) is related to negative seismic amplitudes; the more porous zone (porosity range 29–36%) is related to positive seismic amplitudes.

Figure 7 is a map generated from the seismic amplitudes that correlate to the less porous zone. The seismic amplitude is negative; hence, the low absolute seismic amplitudes are related to higher porosities, and the high absolute seismic amplitudes are related to lower porosities. The porosity map derived from seismic amplitude indicates lateral variations of porosity within the top zone. This is due to higher porosity values in the reefal area compared to the lagoonal area to the north.

Porosity can also be calibrated to acoustic impedance, and a porosity map derived from impedance is similar to a porosity map derived from seismic amplitude. However, this calculation method is more complicated and is not discussed in this paper.

## ZONE INTERVAL—SEISMIC TWO-WAY ISOCHORE

The thickness of the porosity zone interval was calibrated to the seismic two-way isochore. A plot of porosity zone interval thickness versus seismic two-way isochore suggests a linear relationship (Figure 8). A higher value isochore is related to a thicker porosity zone interval. The porosity zone interval thickness map is derived directly from the seismic two-way isochore map (Figure 9). The thickest porosity zone intervals are located in the north and in the northwestern corner of the area and are associated with tidal channels surrounded by lagoonal facies. Over the main reefal area, the reservoir thickness indicates little variation. In general, the porosity zone interval exceeds the tuning thickness, except in the western area, where the seismic two-way isochore is less than 10 ms.

## DISCUSSION

In this study, simple relationships between porosity and seismic amplitude and between porosity zone interval and seismic two-way isochore were assumed and, in general, were verified by the well data. The bed thickness is thicker than the "tuning" thickness which supports these linear relationships of porosity zone thickness and seismic two-way isochore (Widess, 1973). Furthermore, the gas column in the carbonate buildups is very thick; hence the seismic reflection amplitude inside the reef is the result of porosity variation (Ng and Leong, 1990). The end products of this study are porosity and porosity-thickness maps which indicate detail variation at 25-m intervals—the 3-D seismic data cell size. In contrast, porosity and porosity-thickness maps generated from well data and 2-D seismic data are very smooth and lack detail.

However, the seismic calibration exercise should be repeated when more well data are available. For a more complex relationship among porosity zone interval, seismic amplitude, and seismic two-way isochore, the mapping to the reservoir quality suggested by Stanulonis and Tran (1992) is recommended. A statistical approach of seismic attributes calibration by neural networks (Johnston, 1993) is required when several seismic attributes are used in the analysis.

The new geologic models that come out of this study helped petroleum engineers select better production well locations and optimum perforation intervals.

## CONCLUSION

Seismic data can be optimized and integrated with geological data to reduce the risk involved in developing hydrocarbon resources. Our case study suggests that a simple method of relating porosity to seismic amplitude and zone interval to two-way isochore is effective and efficient and provides a good geological model for reservoir management and hydrocarbon development.

### *Acknowledgments*

We thank the management of PETRONAS for allowing us to publish this paper. Special thanks are also due to Amiruddin Mansor, geophysicist at Petronas Carigali, for his technical contribution.

## REFERENCES CITED

Epting, M., 1980, Sedimentology of Miocene carbonate buildups, central Luconia, offshore Sarawak: Geological Society Malaysia Bulletin 12, p. 17–30.

Ho, Kiam Fui, 1978, Stratigraphic framework for oil exploration in Sarawak: Geological Society Malaysia Bulletin 10, p. 1–13.

Johnston, D. H., 1993, Seismic attribute calibration using neural networks: Expanded Abstracts, SEG 1993, p. 250–253.

Meckel, L. D., Jr., and A. K. Nath, 1977, Geological considerations for stratigraphic modelling and interpretation: AAPG Memoir 26, p. 417–438.

Ng, T. S., and L. S. Leong, 1990, Application of Gassman's model in hydrocarbon bearing reservoir characterisation (abs.): Petroleum Geology Seminar, Kuala Lumpur.

Ng, T. S., I. M. Shuhud, and L. S. Leong, 1993, A quantitative study of the seismic time-amplitude reflection characteristics in an oilfield (abs.): Petroleum Geology Seminar.

Stanulonis, S. F., and H. V. Tran, 1992, Method to determine porosity-thickness directly from 3-D seismic amplitude within the Lisburne Carbonate Pool, Prudhoe Bay: The Leading Edge, v. 11, n. 1, p.14–20.

Widess, M. B., 1973, How thin is a thin bed?: Geophysics, v. 38, p. 1176–1180.

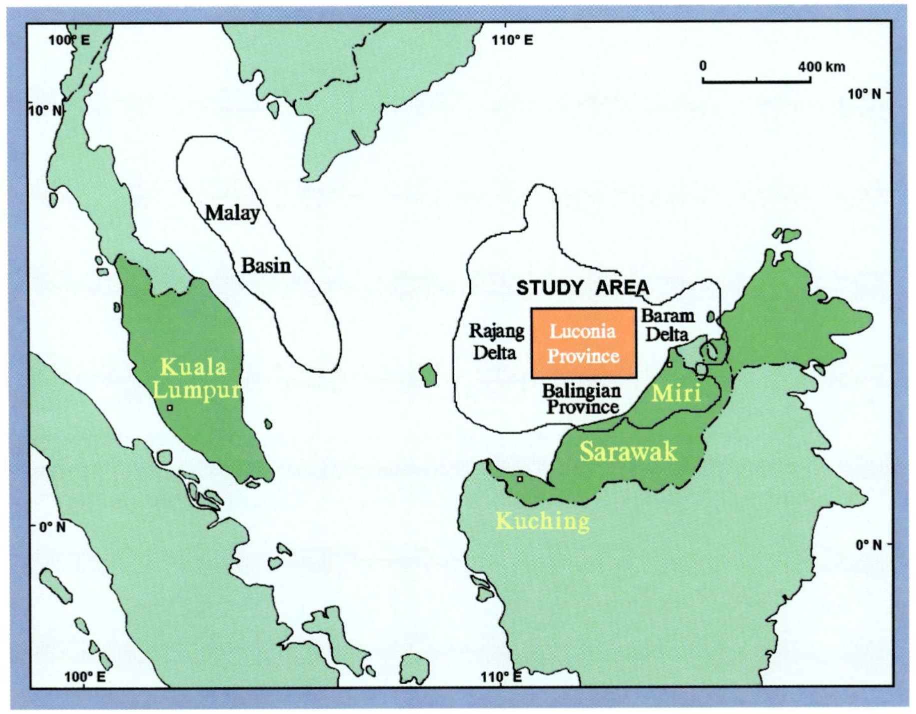


FIGURE 1. Location map showing the study area in the Sarawak basin.

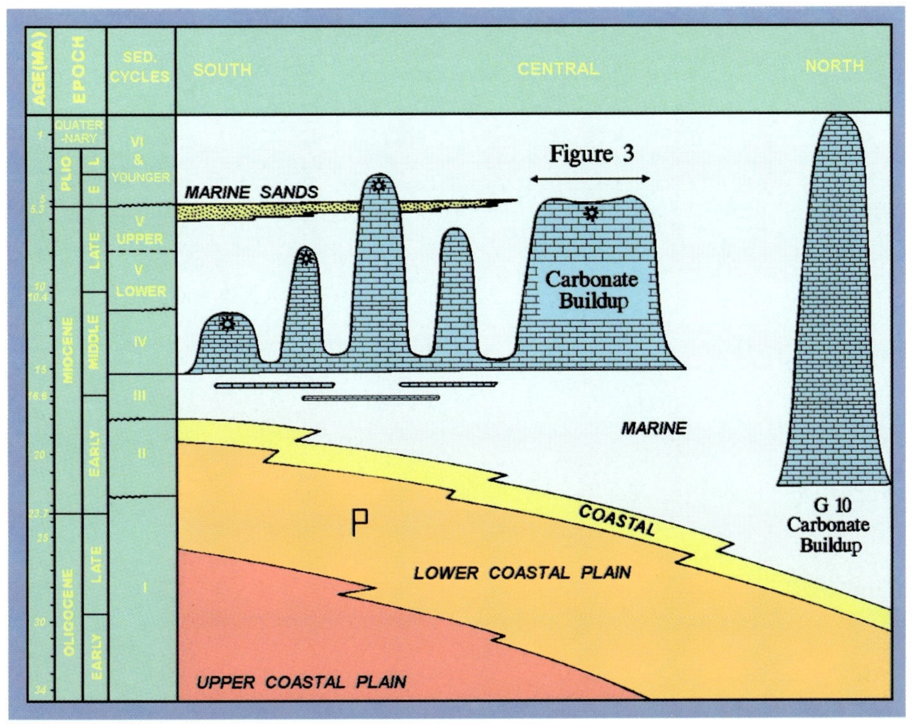


FIGURE 2. Schematic diagram showing the distribution of lithofacies in the Oligocene-Miocene strata, Sarawak Basin. The main growth of the reefal carbonate buildups was during the middle to late Miocene age.

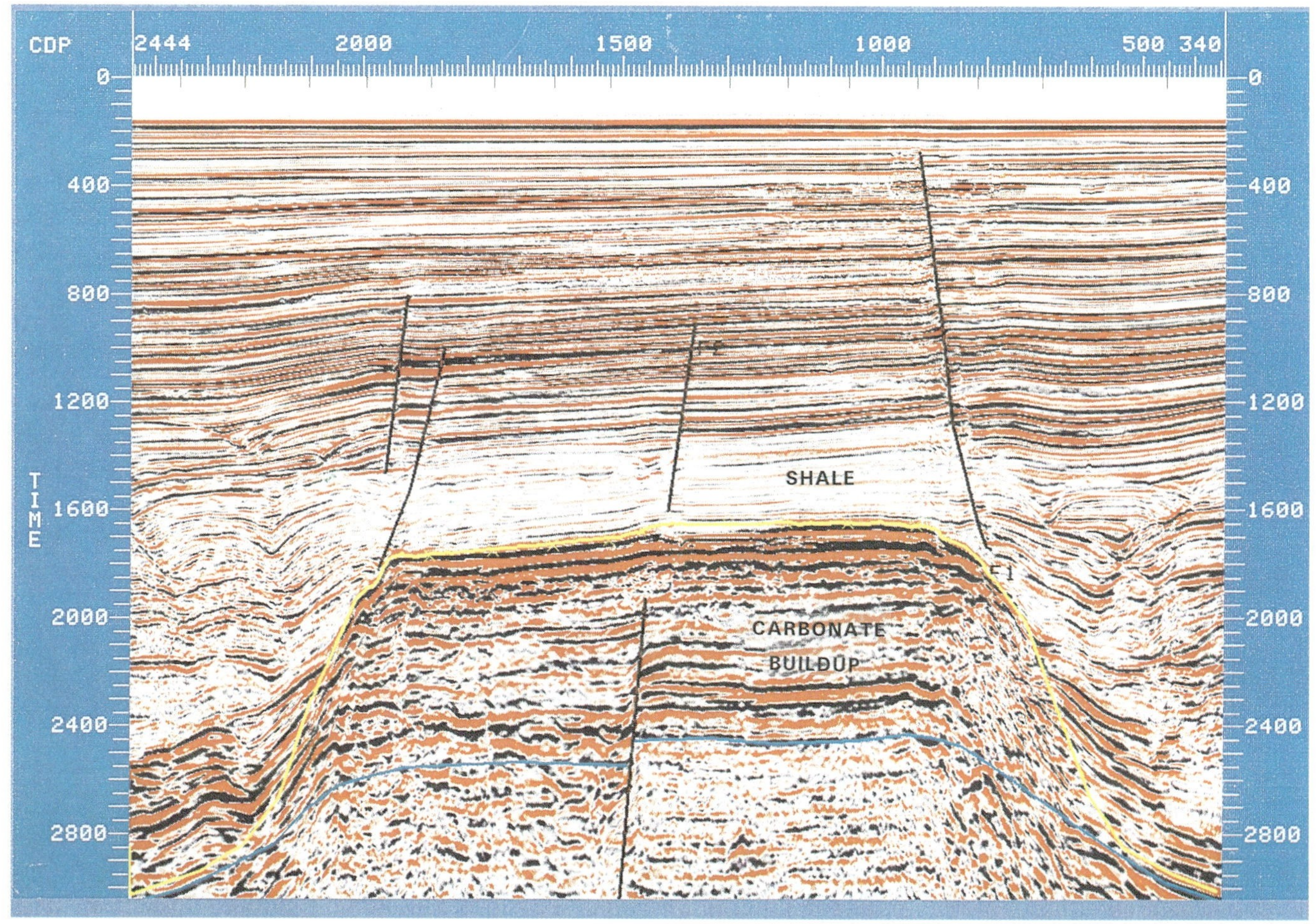


FIGURE 3. Seismic profile showing that the top of the carbonate buildup is associated with a strong seismic reflector at around 1650 ms. The seismic reflector layering within the carbonate buildup is the result of horizontal porosity variation. Overlying the top of the carbonate buildup is a low-amplitude reflector zone that represents the shaly interval.

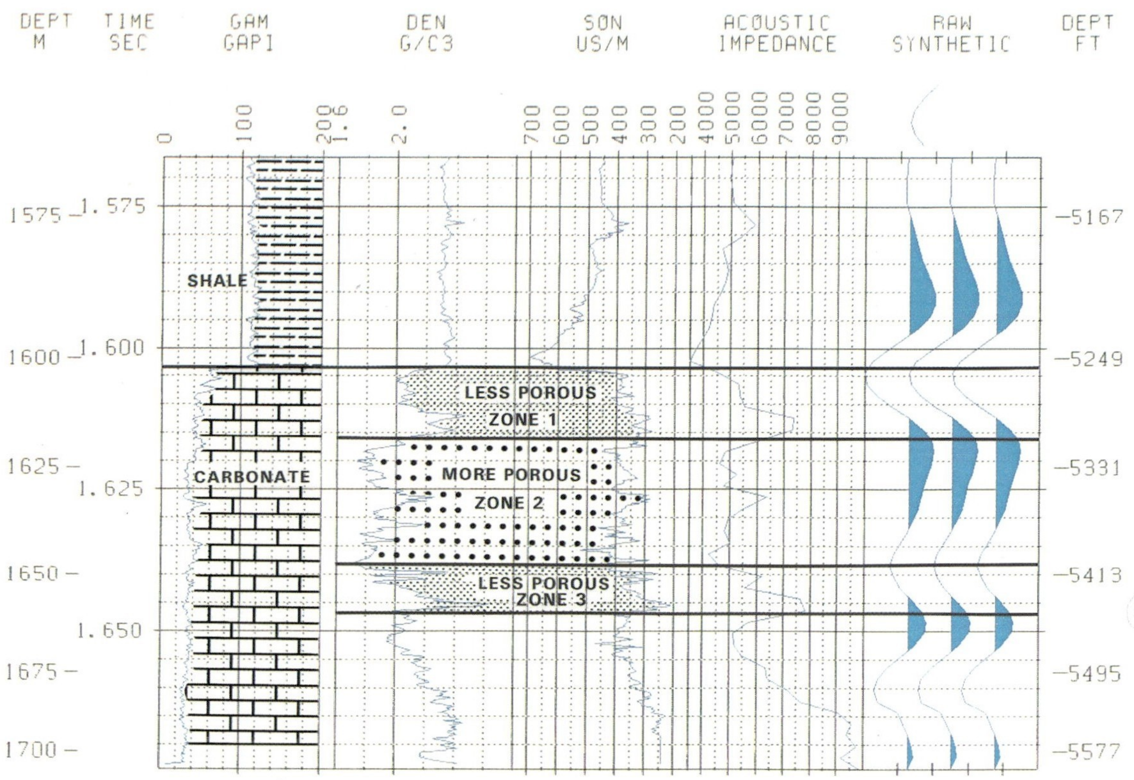


FIGURE 4. The synthetic seismogram is generated using a zero-phase wavelet with a bandpass of 10–40 Hz. The top of the carbonate buildup has a lower gamma ray value and a lower density but higher sonic values than the overlying shaly sequence.

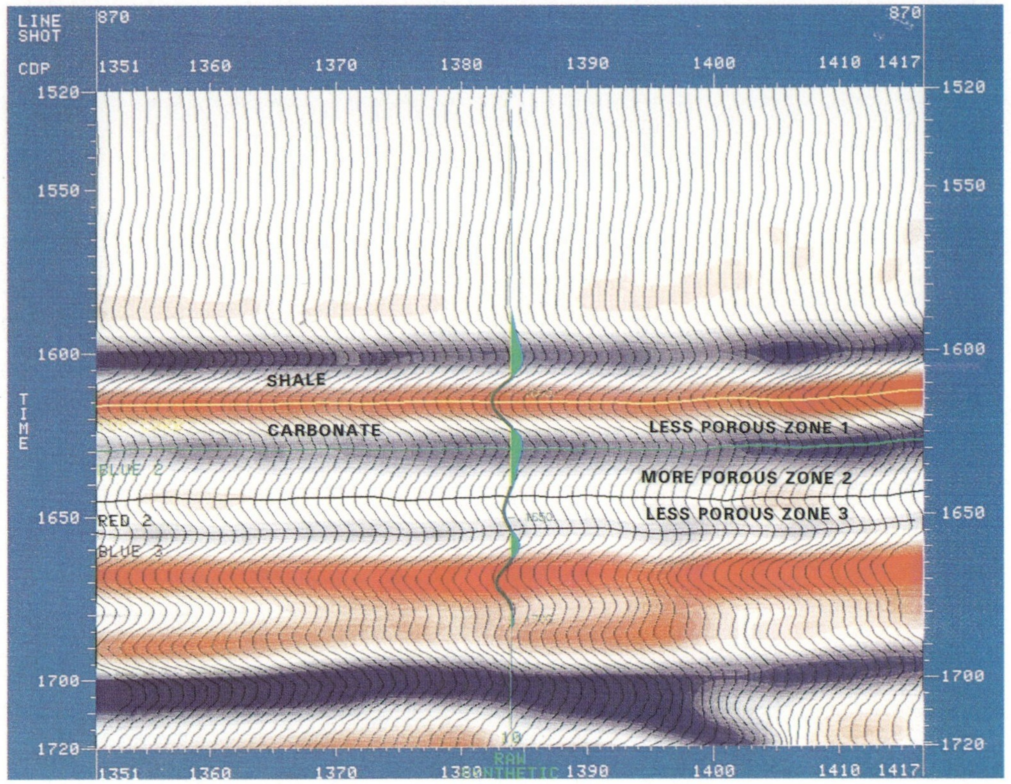


FIGURE 5. A magnified cross section of Figure 3, highlighting the horizontal layering of the carbonate buildup interval structure.

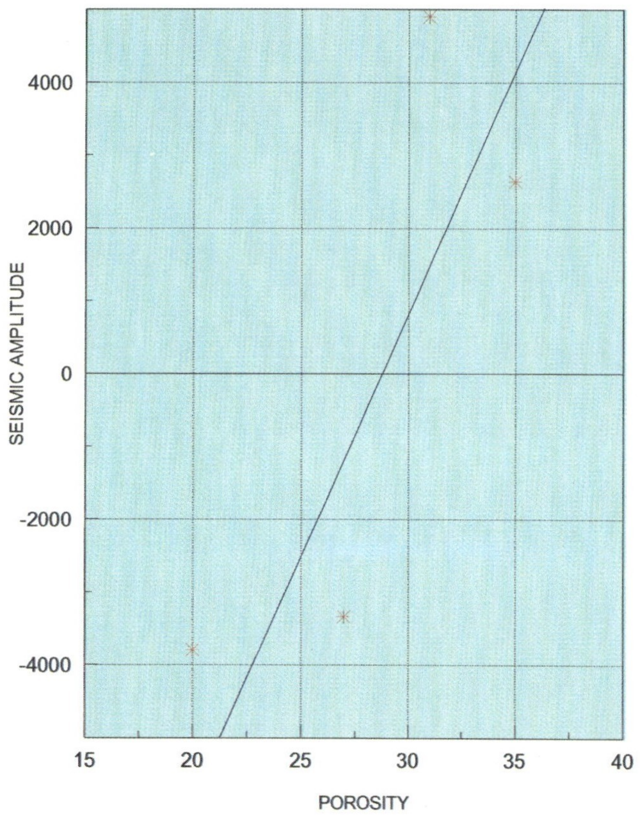


FIGURE 6. A plot of seismic amplitude versus porosity suggests a linear relationship.

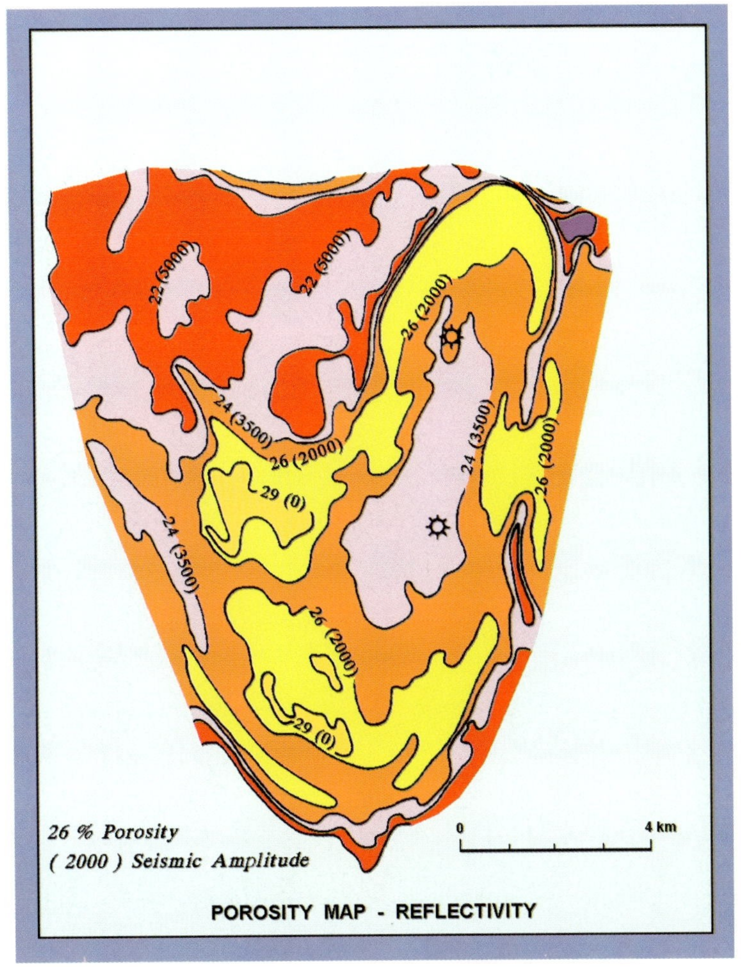


FIGURE 7. Porosity map derived from the seismic amplitude map. Note that the seismic amplitude is negative, which relates to the less porous zone.

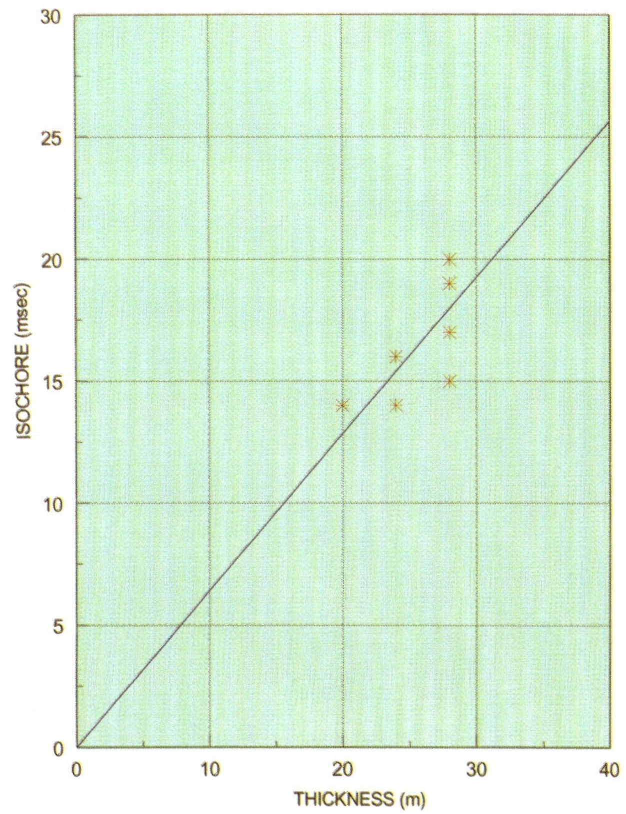


FIGURE 8. A plot of seismic two-way isochore against porosity zone interval suggests a linear relationship.

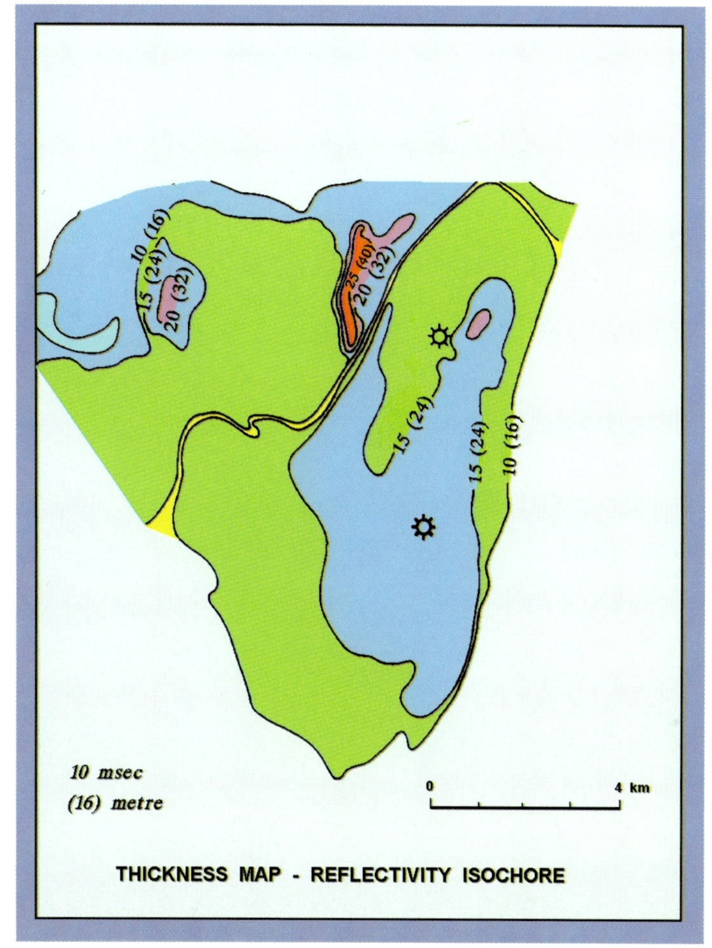


FIGURE 9. Map highlighting the main carbonate buildup that was studied. The porosity zone thickness is extracted from the seismic two-way isochore. The reefal facies has an elongated geometry in the eastern area. The northwestern area consists of lagoonal facies.

# CHAPTER 25

# Polygonal Fault Systems: A New Type of Fault Structure Revealed by 3-D Seismic Data from the North Sea Basin

Cartwright, J. A., Polygonal fault systems: a new type of fault structure revealed by 3-D seismic data from the North Sea Basin, *in* P. Weimer and T. L. Davis, eds., *AAPG Studies in Geology No. 42* and *SEG Geophysical Developments Series No. 5*, AAPG/SEG, Tulsa, p. 225–230.

Joseph A. Cartwright*

## ABSTRACT

This paper describes a new class of fault structure: the polygonal fault system. This complex fault system is composed of almost randomly oriented minor extensional faults organized in a polygonal network. The component faults are typically 500–1000 m long and have throws ranging from 10 to 100 m. They are closely spaced (200–400 m), and fault mapping is possible only with 3-D seismic data.

The polygonal fault system deforms Lower Tertiary shale sequences over a large part of the North Sea Basin and is restricted in extent to low-permeability facies in slope and basin plain regions. The faulted intervals are bounded above and below by completely undeformed sequences. These key observations are taken to suggest a genetic model in which the fault system formed through a compaction-related mechanism.

## INTRODUCTION

3-D seismic data have two main advantages over 2-D data when used for structural interpretation: (1) the potential gains in resolution of structures achieved through better migration and (2) the potential for better correlation of structures through higher spatial sampling density of the subsurface. The aims of this paper are to describe a new type of geological structure, the polygonal fault system, and to show how this discovery was possible only with the enhanced correlation of faults achieved through the denser sampling of 3-D surveys. The essential features of the new structure type are described using a case study interpretation of a 3-D seismic survey from block 21/14 of the North Sea (UK Sector). The expression of the polygonal faults on this dataset is typical of that seen throughout the central region of the North Sea Basin and is used to illustrate the general characteristics of the structure. The paper concludes with a short discussion of the genesis of polygonal faults.

## STRATIGRAPHIC SETTING OF THE POLYGONAL FAULT SYSTEMS

The polygonal fault systems described here affect Paleocene to Miocene clastic sedimentary sequences in the central portion of the North Sea Basin. The lithostratigraphy of the deformed units is dominated by mudstones and silty mudstones that accumulated in prodelta, slope, and basin plain depositional environments (Ziegler, 1982; Lovell, 1990) in a partly enclosed marine basin. Gravity-flow sandstones are common in the lower part of the succession and are widely developed on the northwestern slope of the basin and in the basin axis, where they form reservoirs for a substantial proportion of the discovered petroleum reserves of the North Sea (Bain, 1993). The region over which the polygonal fault system developed is illustrated in Figure 1. This area encompasses most of the commercial production from Tertiary reservoirs.

*Department of Geology, Royal School of Mines, Imperial College of Science, Technology and Medicine, Exhibition Road, London SW7 2BP, UK

## 2-D SEISMIC EXPRESSION OF THE NORTH SEA TERTIARY

A representative regional 2-D seismic section from the central North Sea (Figure 2) is used to illustrate some of the key features of the seismic expression of the Tertiary of the North Sea Basin. The Tertiary is divided into two major seismic-stratigraphic units with markedly contrasting acoustic characters. The Paleocene to early Miocene interval (unit 1) is characterized by a pervasive, discontinuous reflection character, whereas the middle Miocene to Recent interval (unit 2) is highly continuous. Beneath unit 1, the Mesozoic succession is characterized by high lateral continuity of reflections.

The discontinuous reflection character of the Lower Tertiary might be considered to be a depositional response and interpreted as a high-energy, discontinuous-to-chaotic seismic facies. This stratigraphic interpretation is consistent with the widespread development of turbiditic facies in this interval, as proven by well data (Bain, 1993). The external forms of parts of the interval are often marked by mounded reflection geometry (Mitchum, 1985), and this is often interpreted in a seismic-stratigraphic sense as being indicative of submarine fan development. Mounds identified in this way are common exploration targets in the central North Sea (Bain, 1993).

An alternative interpretation of the discontinuous "seismic facies" of unit 1 is shown in Figure 2B. The interpretation shows the Lower Tertiary to be pervasively deformed by many small extensional faults. The fault throws are small, typically a few tens of meters (10–50 ms), and they are closely spaced (200–500 m). These factors combine to make fault interpretation difficult if the frequency content is low, as applies in Figure 2. The systematic alignment of reflection terminations can easily be overlooked, and the appearence is typical of a chaotic or discontinuous seismic facies rather than of a structural disturbance of originally continuous reflections.

An important feature of the extensional faulting observed in the Lower Tertiary is that there appear to be at least two distinct layers (tiers) in which the faults are concentrated. These faulted tiers are related to the gross sequence stratigraphy, in that the tiers are separated by regionally condensed sections (Cartwright, 1994a, 1994b). This structural framework can be seen in Figure 2B, where sets of faults are grouped into layer-bound intervals (tiers) with a distinctive structural style. These are separated by thin intervals cut by only a small proportion of the faults in the tiers above and below.

The small extensional faults affecting the Lower Tertiary are never fully resolved on 2-D data (Figure 2B). The fault interpretation is subjective, particularly in the low-reflection-contrast intervals. Some crosscutting relationships between faults can be inferred, but it is clear that there are additional complexities not revealed by the 2-D migration. Tying dip and strike lines does not resolve this problem and often leads to a suspicion that there are numerous low-angle structures on any given 2-D section, irrespective of the line orientation. Attempts to map these small faults with 2-D datasets with typical North Sea line spacings of more than 500 m are inconclusive because of difficulties in picking faults consistently from section to section and because of spatial aliasing problems.

## 3-D INTERPRETATION OF LOWER TERTIARY FAULTS: CASE STUDY OF BLOCK 21/14

The block 21/14 3-D survey covers a 6 × 10-km area with a line spacing in both directions of 25 m. The survey is located close to the axis of the Tertiary North Sea Basin (Figure 1), in a region where the Tertiary is predominantly composed of mudstone or silty mudstones and where the gross isopach is close to a maximum value for the basin. The geological context is illustrated with reference to a 2-D section tied to well 21/14b-2 (Figure 3). The Permian to Late Jurassic rift-fill sequence is overlain by the Cretaceous post-rift sequence. The Tertiary can be subdivided into a lower faulted interval and an upper undeformed interval (cf. Figure 2). Several tiers can be identified within the faulted interval, but the fault pattern is difficult to interpret in detail.

The fault pattern in the Lower Tertiary is much easier to interpret on the 3-D data (Figure 4). Faults are clearly recognizable throughout the Eocene-to–middle Miocene interval and are easy to interpret in the high-reflectivity parts of the section. The faults are all extensional, with small throw values (10–50 ms, approximately equivalent to 10–50 m). Many of the faults dip in one direction, giving a structural style of rotated domino blocks. Individual faults are generally restricted in vertical extent to only a part of the entire deformed interval. A small number of faults crosscut most of the interval. The pattern of faulting is seen to vary from one stratigraphic level to the next, but the extent of mechanical coupling between the different sets of faults is not clear, since changes often occur within the low-reflectance intervals. This packaging of sets of faults within stratigraphic units represents the tier structure (Cartwright, 1994a, 1994b). A similar structural style is apparent on the orthogonal lines through the 3-D survey (Figure 5).

Although it is much easier to interpret faults on the 3-D data than on the 2-D data, the structure is so complex that it is difficult to make consistent fault interpretations on a section-by-section basis. The approach used for mapping the fault structure is to concentrate on mapping closely spaced horizons and to use these to analyze the 3-D structure. Tests run using autocorrelation routines on the Landmark workstation (ZAP! 3) give good results for fault mapping, but the best fault maps were constructed by picking every line in the grid. The total survey consists of 400 crosslines (east-west) and 240 inlines (north-south) at 25 m spacing. By picking fault breaks on every line, any aliasing problems are automatically eliminated, and the dense sampling creates a unique fault correlation.

An example of one of the time structure maps produced by this high-resolution picking method is shown in Figure 6. The map has not been processed or edited; it represents the raw first-time picks of the horizon on a manual line-by-line interpretation. The faults are identifiable as linear, and often irregular, gaps in the otherwise continuous horizon display. The gaps represent the fault heave, and the irregularity in their trace reflects a combination of genuine fault plane irregularity and interpretational error. Editing would clean up these fault traces, but the map is presented in the unedited form to show the quality of the resolved fault pattern obtainable from first-time picking, provided the spatial sampling is high.

The outstanding feature of the fault pattern seen in Figure 6 is the polygonal geometry of the fault system. The pattern does not resemble typical tectonic fault systems, in that there is no systematic offset of one trend by another and there is a much broader range in fault orientations than would be expected for a tectonic fault set. The closest analog for this type of fault pattern is dessication cracks in mud flats, but these are usually much smaller structures (cm- to m-scale) than the km-scale structures imaged in Figure 6. Mud cracks are vertical tensile fissures, open at the surface, whereas the structures in Figure 6 are normal faults in cross section, with dips ranging from 30 to 70°.

Figure 7 shows three structure maps from successively shallower horizons within the Eocene-to-middle Miocene deformed interval. The fault patterns visible on each of these maps are also polygonal, but there are important differences among these maps. The spacing of faults, the linearity versus curvature of individual fault segments, and the orientations show that each horizon map represents a different faulted tier. Horizon mapping of closely spaced markers thus confirms the interpretation that the Lower Tertiary faulted interval consists of a number of mechanically distinct tiers. This complex, multitiered polygonal geometry is summarized in Figure 8.

The polygonal geometry of the fault system explains many of the anomalous features observed on 2-D data. Any 2-D section through a polygonal network will intersect faults at a range of angles, and few of the intersections are likely to be orthogonal and sharply imaged. This explains why there are often smeared terminations of stratal reflections at fault cutoffs and numerous low-angle fault cuts on any 2-D profile, adding to the already difficult interpretational challenge of consistency in fault picking.

## GENESIS OF THE POLYGONAL FAULT SYSTEM

The occurrence of the layer-bound polygonal fault systems in the deepest water, fine-grained sequences of the slope and basin floor (Figure 1) led Cartwright (1994a, 1994b) to suggest that they formed in response to collapse of shallow-formed overpressure in rapidly deposited units. The multiple-tier structure was interpreted as evidence that the pressure release and fluid escape phenomena were combined responses to episodic compactional disequilibria (Cartwright, 1994a), with the deformation of each individual tier being somehow related to a single cycle of overpressure collapse.

Polygonal fault systems of the scale described here have not previously been recognized in the subsurface. The nearest analog to these structures are the giant dessication polygons formed by isotropic volumetric contraction on playa mud flats in the western United States (Neal et al., 1968), but these are superficial surface structures restricted to a layer that is thin relative to the polygon diameter. The fault polygons in the Lower Tertiary affect layers with thickness comparable to the polygon spacing and persist for hundreds of meters into the subsurface.

All types of polygonal fracture systems share a common mechanical boundary condition: that the stress state in the plane in which the polygons are developed is either isotropic or close to isotropic (Price and Cosgrove, 1990). The polygonal pattern forms because there is no strong preferred orientation(s) adopted by the propagating fault planes. The layer-bound extensional faults in the Lower Tertiary of the North Sea can be regarded as being directly analogous to mud cracks, but they develop to accommodate volumetric contraction resulting from compaction-driven fluid expulsion rather than contraction during periods of surface evaporation and dessication (Cartwright and Lonergan, 1996). The in situ stress conditions for these two cases are similar in the horizontal dimension, but because the compaction dewatering occurs at depth, the vertical effective stress exceeds the horizontal effective stresses, and inclined failure planes result, rather than vertical tensile fissures.

## SUMMARY AND CONCLUSIONS

The description of fault structure in block 21/14, presented above, is an excellent example of how the high spatial sampling density of 3-D seismic data can lead to the recognition of a fundamentally different type of fault system. Without the sampling density of 3-D data, the polygonal pattern could not have been proved. Any uncertainties in picking faults on individual lines are minimized in their impact on the final fault correlation by the combination of a series of closely-spaced adjacent picks. Attempts to map this type of fault pattern with 2-D data are doomed to fail, because of aliasing problems and the uncertainties in fault interpretation on any particular section.

The recognition of the fundamental polygonal nature of the fault system has been one of the critical factors leading to a fuller understanding of their mode of formation. The ability to map closely spaced horizons has confirmed that the faults are organized in tier structures and that these tiers are related to the stratigraphy. These observations from the 3-D interpretation have clarified the relationships between episodic compaction and fault development. The polygonal faulting is now seen to be an intrinsic product of a cyclic process of rapid deposition, pressure buildup, pressure release, dewatering, and volumetric contraction. Overpressured shales are widely developed in many different basin settings, and if this model for the development of polygonal fault systems is generally applicable, it suggests that polygonal fault systems in mudstones should not be an isolated occurrence restricted only to the North Sea Tertiary. It is possible that examples of layer-bound extensional faults from other types of basins that have been interpreted from 2-D data as gravity collapse faults or minor tectonic faults may prove to have a polygonal geometry, and hence an origin, similar to those described here. 3-D data are the ultimate key to the wider recognition of this complex and interpretationally challenging new class of structures.

### *Acknowledgments*

I am very grateful to Fina Exploration Ltd. and particularly to Joe Staffurth and Rod Laver for their help in supporting the research on which this paper is based. I thank Fina and partners in 21/14b (Texaco UK Ltd., Santa Fe Ltd., and Purbeck Oil & Gas Ltd.) for access to and permission to publish the 3-D data, and Geco Prakla Ltd. for permission to include excerpts of their excellent 2-D regional seismic data from the North Sea. Andy Strachan is particularly thanked for his monumental mapping effort between games of cricket.

## REFERENCES CITED

Bain, J. S., 1993, Historical overview of exploration of Tertiary plays in the UK North Sea, *in* J. R. Parker, ed., Petroleum Geology of Northwest Europe: Proceedings of the 4th conference: London, Geological Society of London, p. 5–13.

Cartwright, J. A., 1994a, Episodic basin-wide fluid expulsion from geopressured shale sequences in the North Sea basin: Geology, v. 22, p. 447–450.

Cartwright, J. A., 1994b, Episodic basin-wide hydrofracturing of overpressured Early Cenozoic mudrock sequences in the North Sea Basin: Marine and Petroleum Geology, v. 11, p. 587–607.

Cartwright, J. A., and L. Lonergan, 1996, Volumetric contraction during the compaction of mudrocks: a mechanism for the development of regional-scale polygonal fault systems, Basin Research, v. 8, p. 126–137.

Lovell, J. P. B., 1990, Cenozoic, *in* K. W. Glennie, ed., Introduction to the Petroleum Geology of the North Sea (3d ed.): Oxford, United Kingdom, Blackwell, p. 151–169.

Mitchum, R. M., 1985, Seimic stratigraphic expression of submarine fans, *in* O. R. Berg and D. G. Woolverton, eds., Seismic Stratigraphy 2: AAPG Memoir 39, p. 117–138.

Neal, J. T., A. M. Langer, and P. F. Kerr, 1968, Giant dessication polygons of Great Basin playas: GSA Bulletin, v. 79, p. 69–90.

Price, N. J., and J. W. Cosgrove, 1990, Analysis of Geological Structures: Cambridge, England, Cambridge University Press, 502 p.

Ziegler, P. A., 1982, Geological Atlas of Western and Central Europe: The Hague, Shell Internationale Maatschappiij, B.V., 130 p.

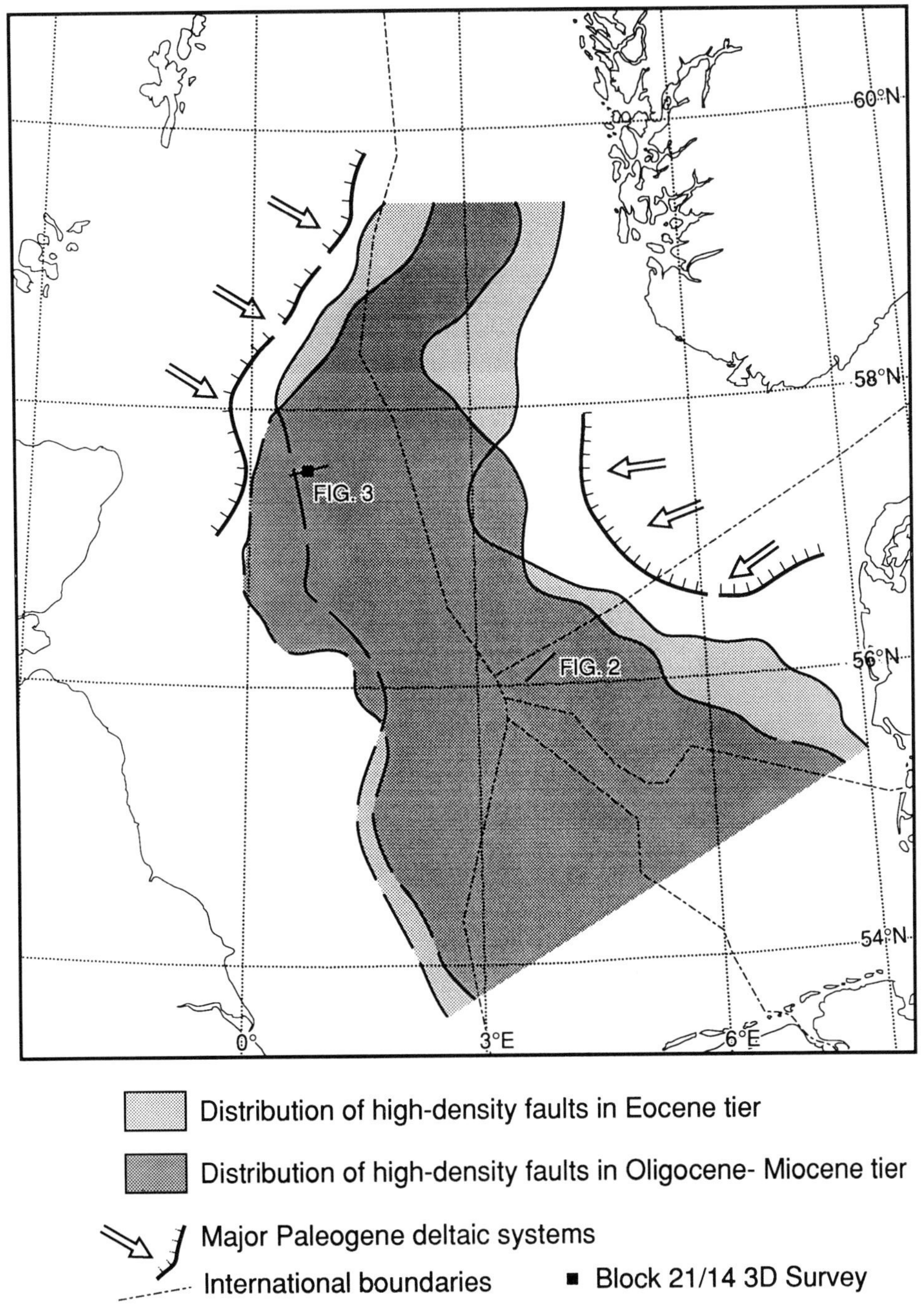


FIGURE 1. Location map of study area in the North Sea Basin. The distribution of sequences in the lower Tertiary affected by small extensional faults is shown in stipple in two tiers. These regions are located in the center of the basin, in the deepest water paleoenvironments of slope and basin floor, where the sediments are mostly mudstones or silty mudstones. This distribution is strong evidence for a lithofacies control on fault genesis. Also shown are the positions of two major deltaic inputs into the basin, in the northwest from the Shetland Platform and in the southeast from the peri-Baltic region. These deltaic depositional systems are not affected by the faulting. The locations of regional seismic sections (Figures 2 and 3) and the 21/14 3-D survey are given on this figure.

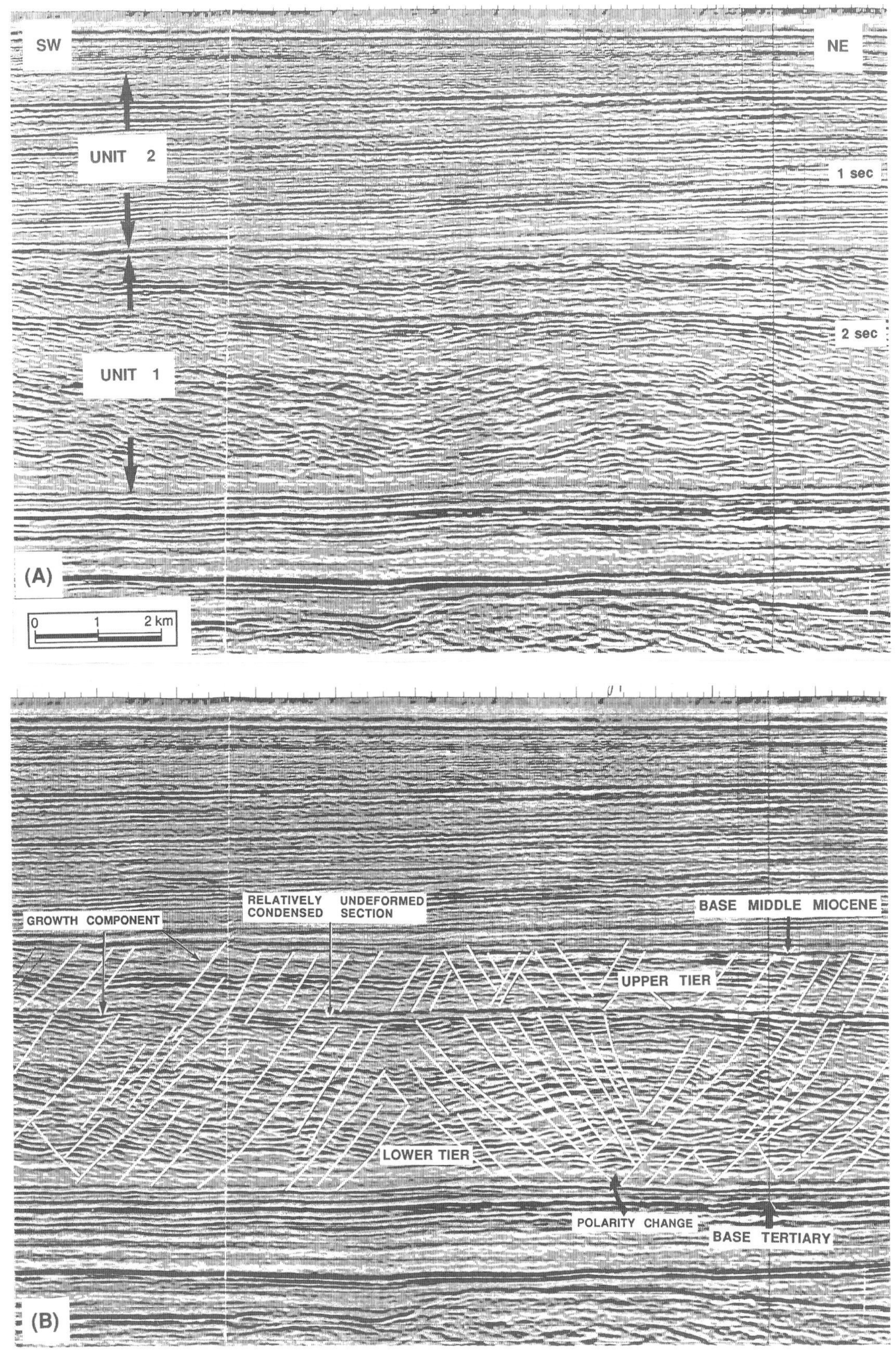


FIGURE 2. 2-D regional seismic section, published courtesy of Geco Prakla from the Danish Sector of the North Sea (location shown in Figure 1). This profile shows the typical seismic expression of the Paleocene-through–middle Miocene interval as a highly discontinuous pattern of reflections, bounded above and below by much higher-continuity reflection intervals (Figure 2A). Seismic-stratigraphic interpretation would classify this as a discontinuous seismic facies, with the wavy pattern possibly suggestive of sedimentary mounding. A contrasting interpretation of this interval as pervasively deformed by many small throw extensional faults arranged in stratigraphically bounded packages (tiers) is shown in Figure 2B. The small throws (10–50 m) mean that the faults can easily be overlooked on low-frequency 2-D sections. The tiers are characterized by sets of faults with a particular style that mostly do not connect to faults in adjacent tiers. Tier boundaries often correspond to stratigraphically condensed intervals. Growth sequences (small divergent wedges) are often seen at the upper boundaries of tiers, suggesting that these small faults were active at the depositional surface and that tiers represent successive deformational episodes.

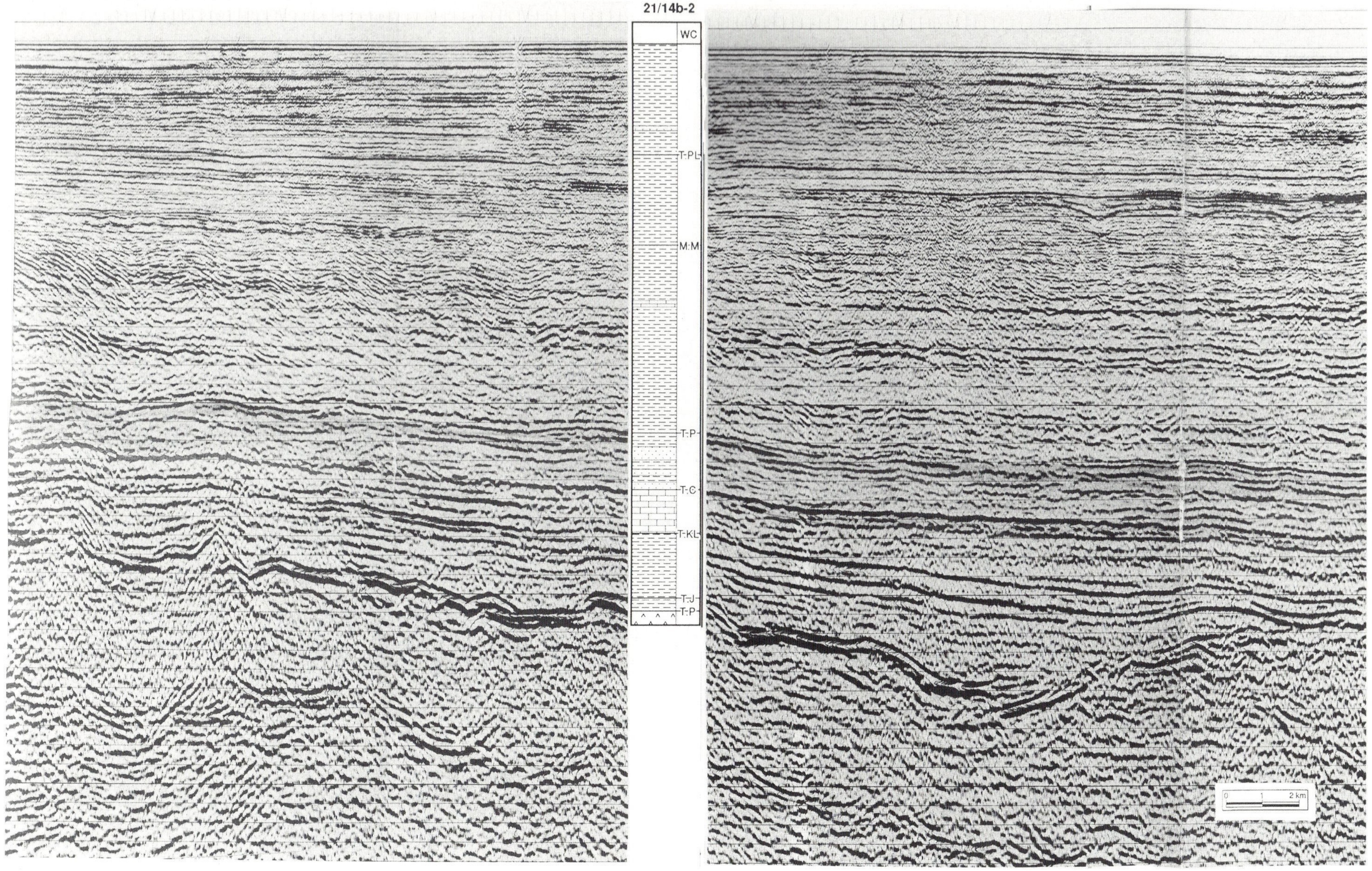


FIGURE 3. Regional 2-D seismic section across the Central Graben in the UK Sector reproduced by courtesy of Geco Prakla. This section shows the geologic context for the 21/14 3-D survey, near the base of the basin slope during the early Cenozoic. The stratigraphy is indicated on the 21/14b-2 well located on this section. The deformed interval is Eocene to middle Miocene age, and consists almost entirely of deep marine mudstones and silty mudstones. Detailed fault geometries within the deformed interval are difficult to interpret on this section, but the faults are organized into sets of rotated blocks facing mostly in an upslope direction. At least two tiers are recognizable on this section (c.f. Figure 2). The underlying units are characterized by high continuity seismic facies. The rift structure deeper in the section can be seen to have had little influence on the stratal geometry of the lower Tertiary interval.

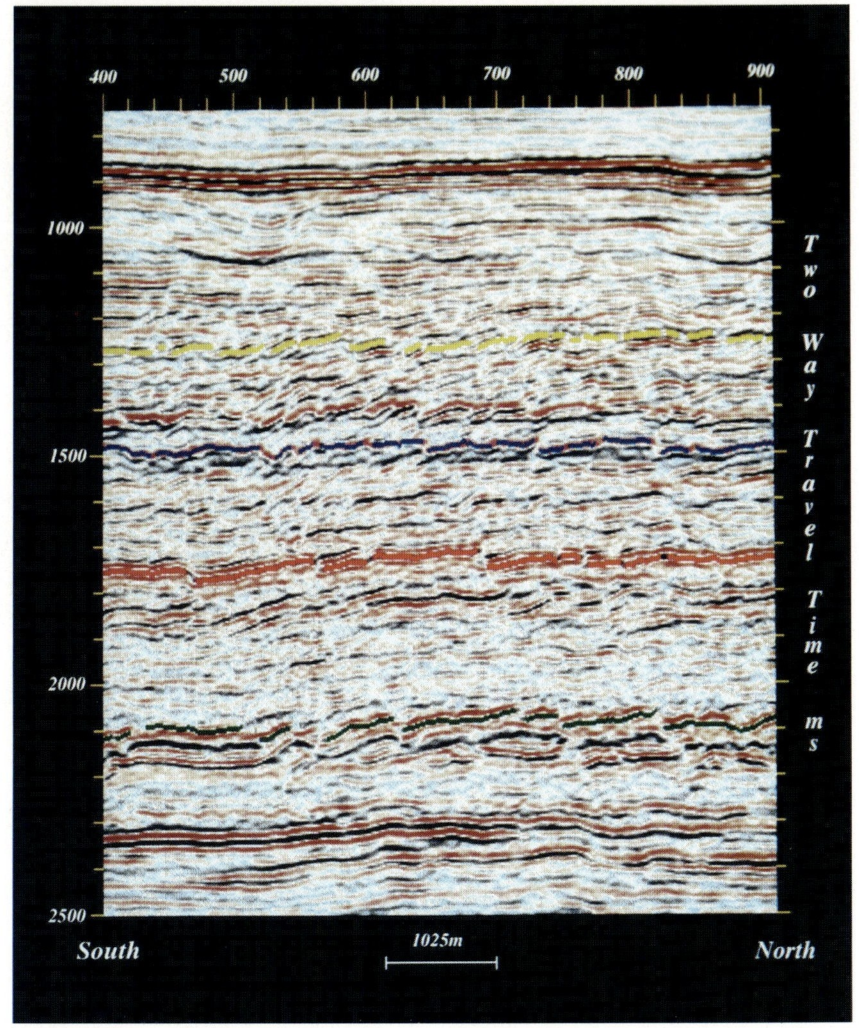


FIGURE 4. An east-west seismic panel from the 21/14 3-D seismic survey, reproduced courtesy of Fina Exploration Ltd. and partners. Vertical exaggeration is about 4×. The four picked horizons are: Green = early Eocene; Red = late Eocene; Blue = middle Oligocene; Yellow = early Miocene. The Eocene-to–middle Miocene interval is pervasively deformed by many small extensional faults. Fault planes are planar to gently listric, and average fault plane dips range from 30 to 70°. The fault throws visible on this section range from 10 to 60 ms (c. 10–60 m). The spacing and arrangement of fault sets change upwards through the faulted interval; it is simple and easy to interpret in the lower interval, particularly in the high-amplitude reflective sequence, but more complex near the top. Note that the faulted interval is enclosed within sequences that are totally undeformed.

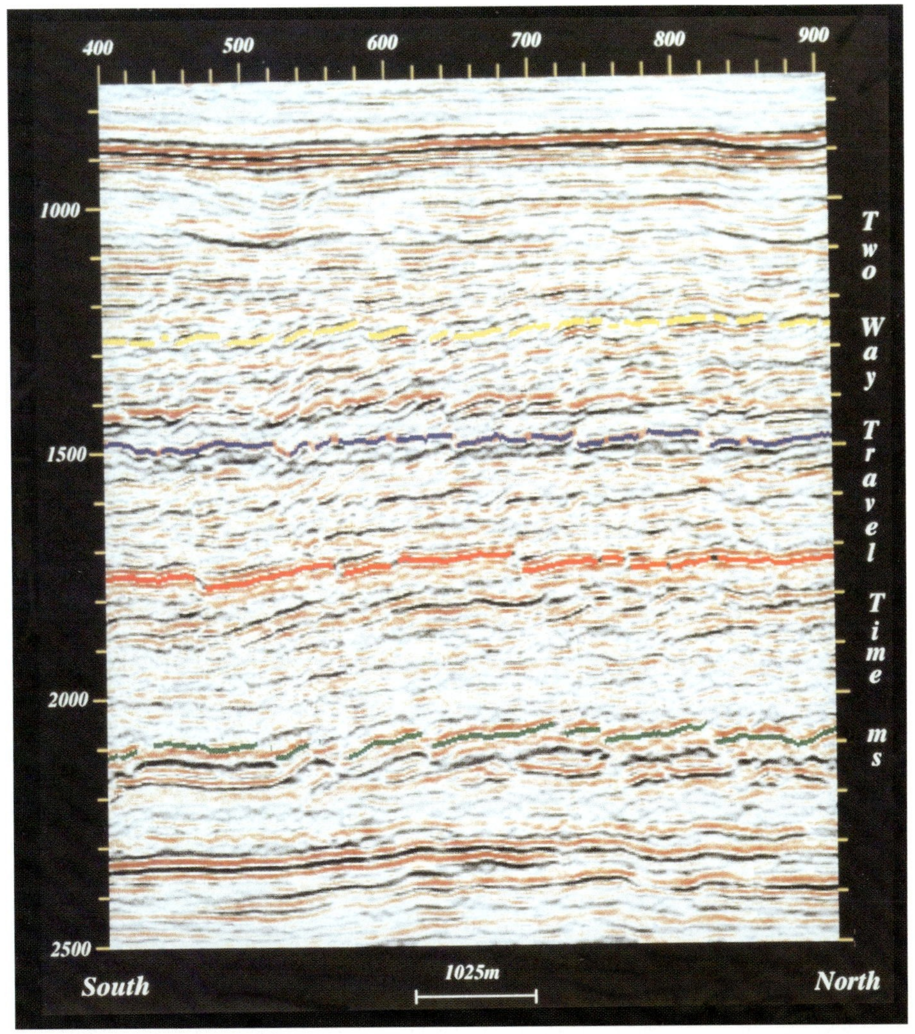


FIGURE 5. A north-south seismic panel from the 21/14 3-D seismic survey, reproduced courtesy of Fina Exploration Ltd. and partners. This section is displayed at the same scale and with the same color code as Figure 4. The structural style is similar to that on the orthogonal section. Fault sets dip predominantly in one direction, and the intensity of the faulting increases in the upper part of the deformed interval. Only a small proportion of the faults can be traced completely through the deformed interval. This section and Figure 4 are representative of all the dip and crosslines in this survey; only minor differences are noted through the data volume.

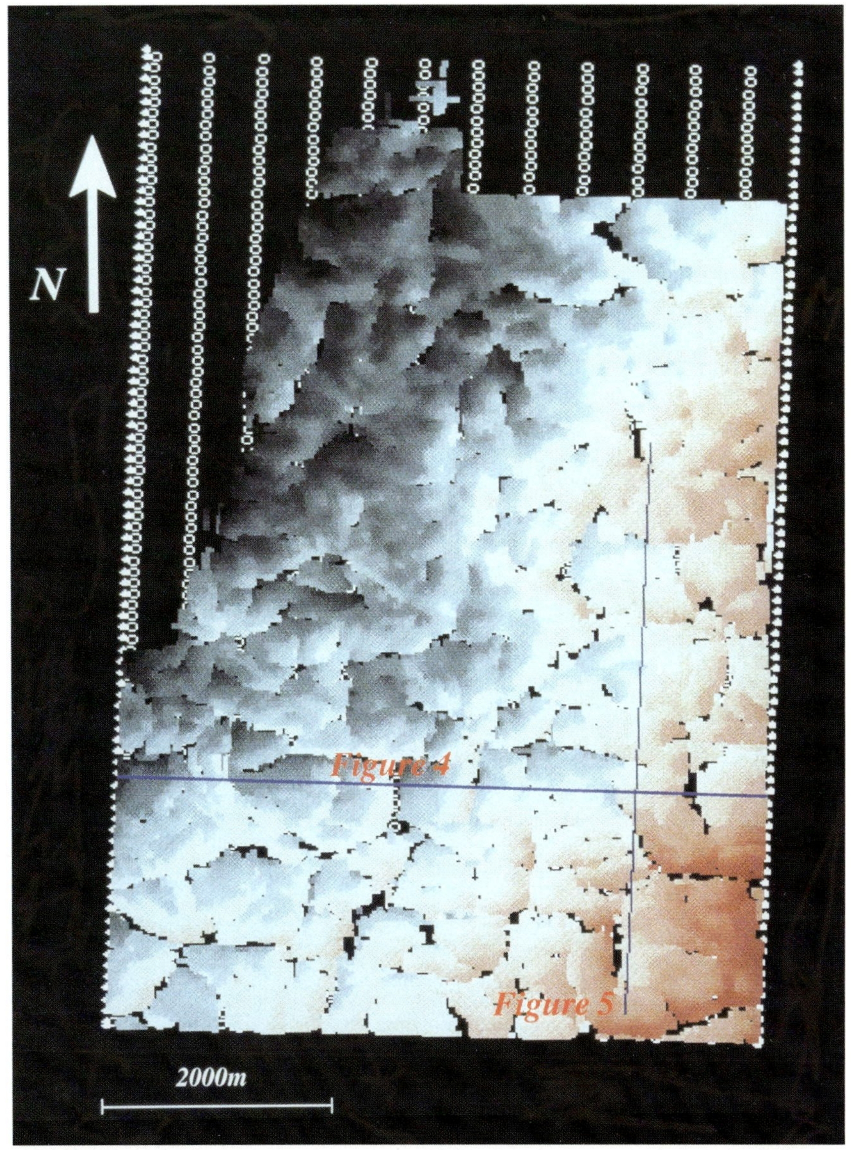


FIGURE 6. Time structure map of the early Eocene horizon (green marker on Figures 4 and 5), based on a line-by-line interpretation of the 21/14 3-D survey. The structural relief from the shallow northwest part of the survey to the deeper southeast region is approximately 50 ms (c. 50 m). The outstanding feature of this map is the randomly oriented fault pattern. Faults are recognizable as linear or curvilinear discontinuities in the horizon map. These faults are arranged in a polygonal network, and numerous three- and four-way fault intersections can be seen. The black zones represent horizon cutouts (fault heave) of the largest faults. The average spacing of the faults is about 450 m in any orientation. This polygonal pattern could not be identified with any grid spacing greater than about 100 m. Any 2-D section through this fault system would intersect a high proportion of the fault planes at a highly oblique angle. This helps explain the difficulty of interpreting these faults on 2-D data.

FIGURE 7. Compilation of three time structure maps from the 21/14 3-D survey. These three maps are of the red (A), blue (B), and yellow (C) markers on Figures 4 and 5. Structural relief (black to red gradations are shallow to deep) on all three horizons is less than 100 m. As with Figure 6, all three maps are characterized by polygonal fault geometry, but there are important differences in spacing, orientation, and degree of curvature of the faults among these maps. The contrast in fault characteristics is sufficient to infer that these four horizons are representative of different tiers. The tiers are not fully decoupled, since some faults evidently repeat on successive maps.

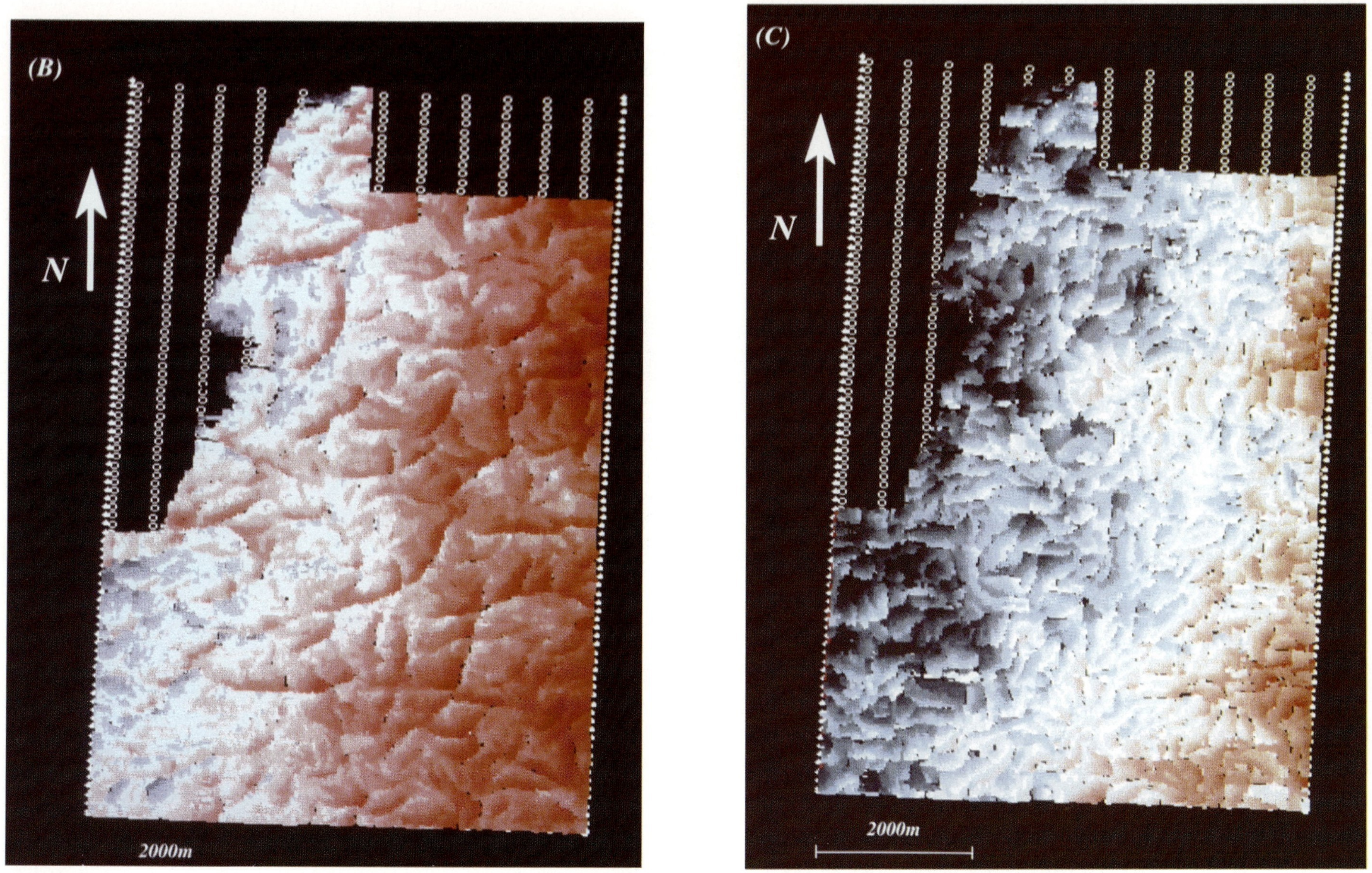


FIGURE 7. (Continued) Compilation of three time structure maps from the 21/14 3-D survey. These three maps are of the red (A), blue (B), and yellow (C) markers on Figures 4 and 5. Structural relief (black to red gradations are shallow to deep) on all three horizons is less than 100 m. As with Figure 6, all three maps are characterized by polygonal fault geometry, but there are important differences in spacing, orientation, and degree of curvature of the faults among these maps. The contrast in fault characteristics is sufficient to infer that these four horizons are representative of different tiers. The tiers are not fully decoupled, since some faults evidently repeat on successive maps.

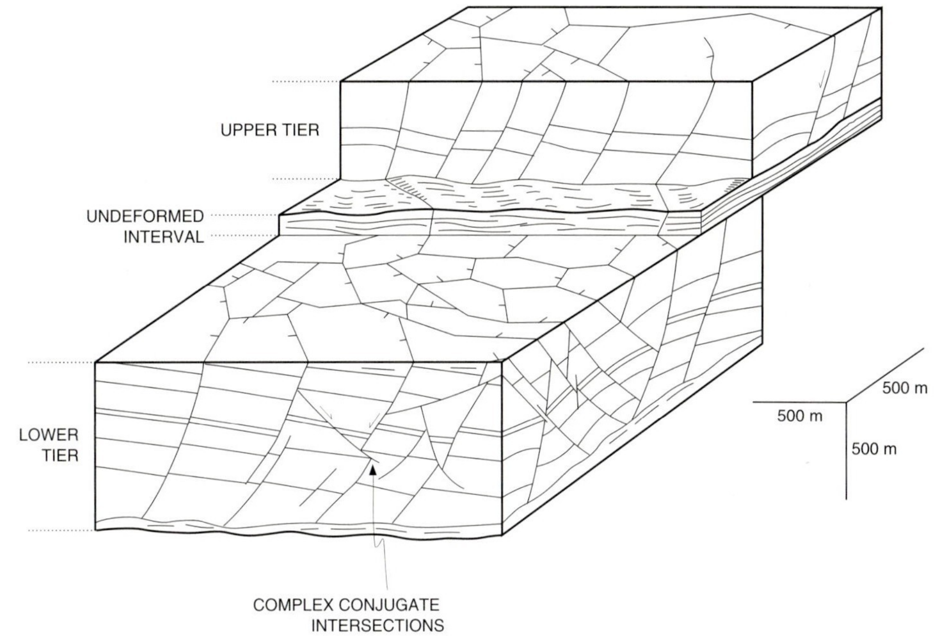


FIGURE 8. Synoptic block diagram showing the structural framework of the multitiered polygonal fault system affecting the lower Tertiary in the North Sea. On any 2-D section through the volume, the faults appear as sets of inclined extensional structures, often dipping uniformly in groups over considerable distances (cf. Figure 3). Because the fault structure consists of groups of skewed prisms, interspersed with cones or pyramids, on any 2-D section there will be a high proportion of oblique fault intersections on the seismic data. This adds to the already complex seismic imaging problem and explains why it is so difficult to interpret these faults on 2-D data.

# CHAPTER 26

Mansfield, C. S., Fault displacement distributions: insights using high-resolution 3-D seismic data, *in* P. Weimer and T. L. Davis, eds., *AAPG Studies in Geology No. 42* and *SEG Geophysical Developments Series No. 5*, AAPG/SEG, Tulsa, p. 231–236.

# Fault Displacement Distributions: Insights Using High-Resolution 3-D Seismic Data

Christopher S. Mansfield*

## ABSTRACT

Detailed mapping of the distribution of vertical displacement (throw) over the planes of a number of growth faults in the Gulf of Mexico has revealed anomalous patterns that are characterized by subhorizontal minima. The subhorizontal arrangement of the anomalies is inconsistent with models of fault growth that require uniform slip distributions in conjunction with radial tip-line propagation at each slip event. An alternative model is presented whereby evolution of a fault is established by linkage between precursor fault segments in the slip direction. Overlap and linkage of fault tips in the slip direction result in relay structures that are subparallel to fault strike and therefore subhorizontal on normal faults. Relay structures orthogonal to the slip direction are likely to have a low preservation potential and will be poorly resolved on cross-sectional seismic profiles. However, their associated low displacements leave a strong signature in the fault displacement fields.

## INTRODUCTION

The increasing use of 3-D seismic data in structural interpretations has revealed that faults in the subsurface rarely exist as single, isolated features but more commonly make up one element of a complex array of faults interconnected along strike. Consequently, where it has been possible to map displacement over the planes of these faults, the distribution has been shown to exhibit a prominent departure from the quasi-elliptical displacement field predicted for the model of an "ideal, isolated blind fault" (Barnett et al., 1987). Elongated subvertical anomalies in the displacement patterns are seen to be associated with the points of overlap and linkage, i.e., with the long axis subparallel to the fault slip vector.

This paper presents evidence that subhorizontal displacement anomalies, i.e., anomalies oriented parallel to fault strike, also exist. Using a high-resolution 3-D seismic dataset discrete measurements of vertical displacement (throw) have been made at very small intervals over the planes of several growth faults from a survey located in the Gulf of Mexico. The high frequency of sampling reveals numerous irregularities in the patterns of displacement that point to local controls of the distribution. A series of tests on the data have shown that recognition of these anomalies is highly sensitive to the interval at which the fault displacements are sampled due to their limited amplitudes and dimensions. The key to a consistent and accurate definition of the anomalies is utilization of the full spatial resolution of the 3-D seismic data, to give a high sampling density of displacement measurements and precise definition of stratal relationships close to the fault planes. The paper concludes with a discussion of possible explanations for the origin of the anomalies and the implications these have for subsurface fluid migration regimes and for development of potential adjacent hydrocarbon traps for which the faults provide closure.

*Department of Geology, Imperial College of Science, Technology and Medicine, Prince Consort Road, London SW7 2BP, UK

## GEOLOGICAL SETTING

The dataset used in this study is part of a high-resolution, speculative 3-D seismic survey provided by Geco-Prakla. Located offshore Louisiana in shallow water (Figure 1), the survey area covers a part of the continental margin that is dominated by a thick, mixed siliciclastic sequence of deltaic and prodeltaic sediments of predominantly Late Cenozoic age, overlying Early Cenozoic slope and basin plain shales and Mesozoic evaporites (Jackson and Galloway, 1984). Gravitational collapse of prograding deltaic sequences across much of the present day shelf region of the Gulf of Mexico has resulted in the development of an extensive series of growth faults that in the region covered by the 3-D survey is transitional between two major trends present along the northern coast (Figure 2).

The faults have gentle listric geometries at depth (Figure 3), becoming subvertical with well-defined upper tips. A thickening of strata from footwall to hangingwall, recording fault growth contemporaneous with sediment deposition, is indicated by the seismic data at most intervals across all of the faults in the survey area, with expansion ratios (downthrown thickness divided by upthrown thickness; Thorsen, 1963) averaging about 1.1 but approaching 2.6–2.7 for some stratigraphic intervals. The largest measured vertical displacement is that on Fault 2, at 160 ms one-way traveltime, which at an average velocity of 3500 m/s for the entire depth range of the fault corresponds to a throw of 560 m.

## DISPLACEMENT MAPPING TECHNIQUE

Fault interpretation and displacement mapping were carried out on a workstation using Landmark seismic interpretation software. A series of strike-normal profiles across each of the faults was selected at 50-m intervals, and the two-way traveltimes of reflections from the footwall and hangingwall cutoffs of selected horizons were then recorded at intervals of approximately 25–50 ms down the fault planes. Local errors in displacement values related to drag folding adjacent to the fault planes were corrected using the technique employed by Chapman and Meneilly (1991). To minimize spatial distortion of the distributions resulting from curvature of the fault planes along strike, the data were then projected onto strike-parallel vertical planes for contouring.

The high frequency content of the dataset and the limited expansion of sequences in the hangingwalls allows precise jump correlations to be made across the faults. Therefore, the accuracy with which the traveltimes of individual reflections can be measured is estimated to be ±1 ms. However, uncertainty in the position of stratal terminations does introduce a minimal error associated with the position in depth of the recorded displacements. This is a function of the frequency content of the seismic data and is estimated in this case to be approximately equivalent to ±12 m.

## DISPLACEMENT DISTRIBUTIONS

The patterns of vertical displacement distribution for Faults 1 and 2, shown in Figure 4, are typical of the distributions recorded on all the mapped faults. The contours indicate a general increase in displacement with depth from an upper tip-line towards the deepest mapped parts of the faults. Locally, however, the highly erratic nature of the contours points to a strong influence from underlying smaller-scale distributions that define systematic departures from a smooth, idealized distribution.

Since many of these irregularities are subtle, resulting from local influences that are limited both in magnitude and dimension, maintaining a consistently high sampling density is absolutely vital for resolving their displacement patterns (Figure 5). Without the availability of 3-D seismic data providing for a sufficient sampling interval, spatial aliasing and subsequent smoothing will result, with a consequent failure to identify anomalous patterns of throw.

Despite the high sampling density and clarity of the seismic data, no systematic pattern in the arrangement of the imaged anomalies is apparent from the plots of displacement distribution. Although the faults exhibit a general increase in displacement with depth, the presence of numerous local minima results in a decrease in displacement with depth over limited areas of the faults, i.e., local negative displacement gradients. The magnitude of these minima is as great as 16 ms, much larger than the estimated measurement precision (±1 ms), so they cannot be artifacts arising from resolution limitations in the methodology. Filtering the data for the effects of differential compaction and changes in interval velocity across the faults is insufficient to eliminate most of the anomalies, leaving a distribution of steeper gradients that represent a real variability in the throw distributions of the faults (Figure 6). The resulting patterns are disordered, reflecting the complexities seen in the original plots of throw distribution, but the most striking feature is the arrangement of some of the regions of negative gradient as subhorizontal semicontinuous bands along the fault planes, conceivably resulting from a lithological or structural controlling mechanism.

## SIGNIFICANCE OF DISPLACEMENT ANOMALIES

The results of displacement mapping presented in this paper have shown the pattern of throw over two growth faults to be dominated by local anomalies, many of which are aligned subhorizontally along the fault planes. This distribution is inconsistent with the subvertical geometry of anomalies associated with points of linkage between neighboring faults interacting along strike. However, linkage between individual fault segments in the dip direction—"dip linkage"—was proposed by Laubscher (1956) as a possible evolutionary mechanism for some of the larger normal faults he had mapped using seismic data in the Eastern Venezuelan Basin. This was based on structural style and the variable expression of drag folding adjacent to the fault planes. In the same way that the linkage of individual segments has been proposed as a component element of growth along the strike of normal faults (e.g., Peacock and Sanderson, 1991), so the evolution of en-echelon segments could be viewed as precursors to the development of a throughgoing fault surface in the slip direction (Figure 7). The overlap of tip-lines in the dip direction produces subhorizontal relay structures that are elongated approximately parallel to the strike of the fault. Since the linkage points correspond to relict fault tips and are associated with points of displacement minima, the subhorizontal nature of the relay structures results in subhorizontal displacement anomalies across the fault plane.

The existence of unlinked or partially linked fault structures in a hydrocarbon province has obvious implications for prospectivity, in terms of subsurface fluid migration regimes and the development of hydrocarbon traps adjacent to the fault planes. The impact, with reference to potential reservoir charging, depends on the timing of fault linkage relative to fluid movements and to the development of favorable structural traps. If a throughgoing fault structure is not completely developed, then the fault as a whole has little potential as a pathway for upward fluid migration. Incomplete linkage may either impede flow into the upper part of a fault structure or stop it altogether. Alternatively, the fluids may leak completely into adjacent permeable strata, depending on the precise structural configuration, the degree of fracturing in the region of linkage, and the favorable juxtaposition of porous lithologies with any area of leakage. Also, where a fault provides closure for a hydrocarbon reservoir, for example, in tilted or folded strata adjacent to the fault plane, the potential of that reservoir is critically dependent on the sealing capacity of the fault. A key component of fault sealing potential is shale smear, particularly in the interbedded sand-shale lithologies typical of such deltaic settings. The amount of smear is believed to be critically dependent on the amount of displacement accommodated locally by the fault. Therefore, the pattern of fault sealing potential will mimic closely that of the displacement distribution, such that even where two faults are completely linked, the shale smear is so incompletely developed as to be ineffective as an impermeable barrier to fluid movements. If the fault is only partially linked, then it is unlikely that any fluids will be retained at all in adjacent reservoirs.

An example of possible unlinked, collinear fault segments interpreted from the seismic data across Fault 2 is shown in Figure 8. Such discontinuities or steps on a fault surface with a sense of offset that is orthogonal or highly oblique to the slip direction will have a low preservation potential during continued slip and therefore are unlikely to be well imaged by seismic data. However, offsets, bends, and breached relay structures along the strike of normal and strike-slip faults create rough fault plane topologies that are known to act as barriers to earthquake rupture distributions (e.g., Sibson, 1985). In a similar manner, dip-linkage sites might also be expected to arrest or impede slip distribution, partitioning slip between component segments and so accentuating the displacement minima at linkage points. Consequently, while seismic data may be unable to resolve the structure (Figure 9), their mechanical influence on the local accommodation of slip may leave a strong signature in the displacement field which can be imaged by detailed mapping of the fault displacement distribution.

## CONCLUSIONS

(1) The distributions of throw across several well-imaged growth faults are seen to be characterized by highly erratic contours and numerous contour closures, which represent many small, but systematic, departures from a uniform, idealized distribution.

(2) The resolution of the true spatial relationship of these small-magnitude anomalies is critically dependent on the spatial sampling density. Maintaining a consistently high sampling density is vital to avoid spatial aliasing of the anomalies.

(3) Mapping the distribution of displacement gradients indicates that many of the regions of negative gradient, precursors to local displacement minima, define subhorizontal discontinuous bands across the fault planes. Filtering the data for the effects of differential compaction and changes in interval velocity across the faults does not eliminate the anomalies.

(4) A model of approximately collinear, offset, linked and unlinked fault segments in the dip direction is proposed, whereby the structure evolves from the slip-direction propagation and coalescence of previously discrete smaller fault segments—"dip-linkage." The continued partitioning of slip between fault segments produces irregular displacement distributions, with minima marking the points of overlap and linkage. Detailed displacement mapping utilizing the full spatial resolution of 3-D seismic data is therefore considered essential for imaging these structures.


### *Acknowledgments*

Thanks are due to Geco-Prakla for kindly providing the seismic dataset upon which this work is based and for their permission to publish the results. I would also like to thank Joe Cartwright for his review and helpful criticism, and Shell International Petroleum Company Limited for providing a research studentship.


## REFERENCES CITED


Barnett, J. A. M., J. Mortimer, J. H. Rippon, J. J. Walsh, and J. Watterson, 1987, Displacement geometry in the volume containing a single normal fault: AAPG Bulletin, v. 71, p. 925–937.

Chapman, T. J., and A. W. Meneilly, 1991, The displacement patterns associated with a reverse-reactivated normal growth fault, *in* A. M. Roberts, G. Yielding and B. Freeman, eds., The Geometry of Normal Faults: Geological Society of London Special Publication, 56, p. 183–191.

Jackson, M. P. A., and W. E. Galloway, 1984, Structural and depositional styles of Gulf Coast Tertiary continental margins: application to hydrocarbon exploration. AAPG, Continuing Education Course Note Series No. 25, 226 p.

Laubscher, H. P., 1956, Structural and seismic deformations along normal faults in the Eastern Venezuelan Basin. Geophysics, v. 21, p. 368–387.

Peacock, D. C. P., and D. J. Sanderson, 1991, Displacements, segment linkage and relay ramps in normal fault zones: Journal of Structural Geology, v. 13, p. 721–733.

Sibson, R. H., 1985, Stopping of earthquake ruptures at dilational fault jogs: Nature, v. 316, p. 248–251.

Thorsen, C. E., 1963, Age of growth faulting in south-east Louisiana: Transactions of the Gulf Coast Association of Geological Societies, v. 13, p. 103–110.

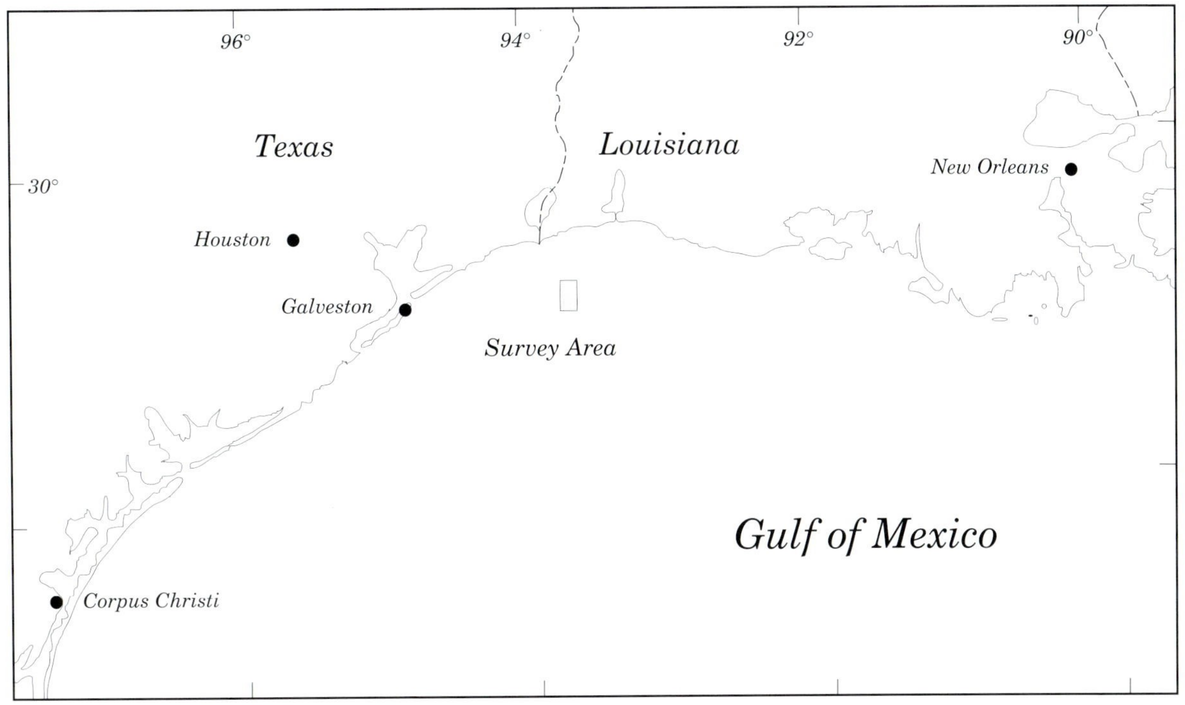


FIGURE 1. Location map showing position of the 3-D seismic survey in the Gulf of Mexico.

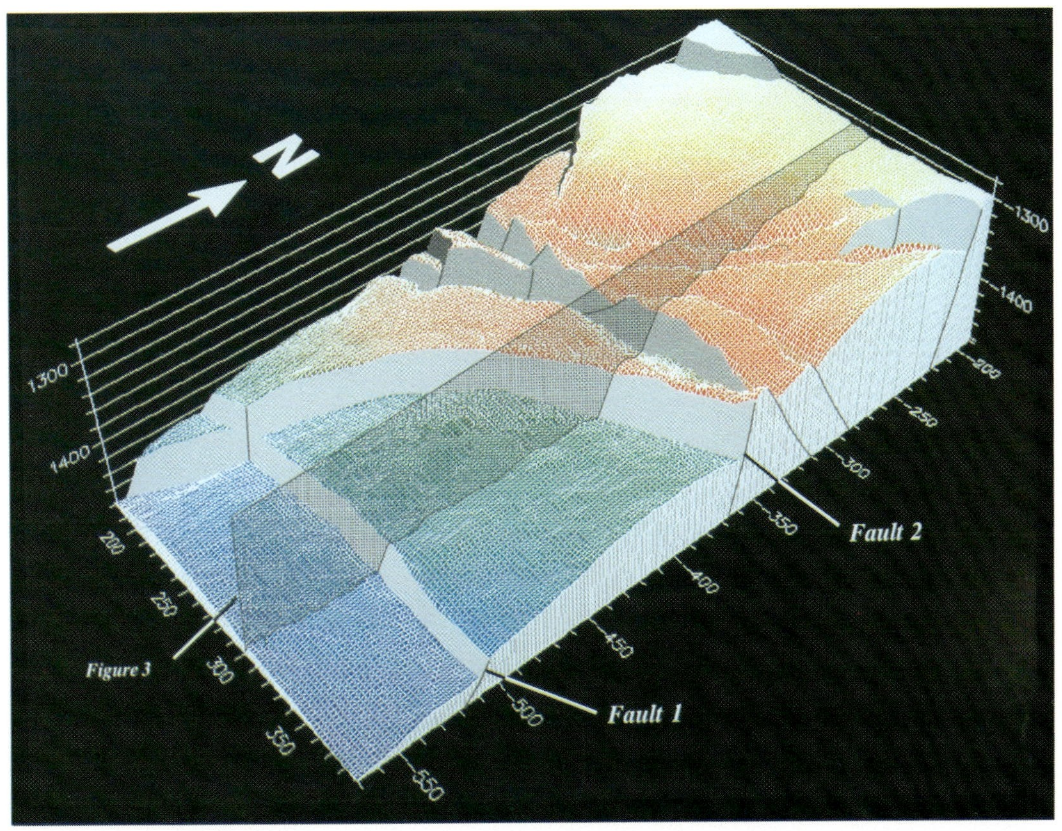


FIGURE 2. 3-D perspective view of an intermediary-level horizon interpreted from the seismic data, colored according to depth in milliseconds (two-way traveltime). The figure shows two prominent fault sets, one striking approximately east-west and the other approximately northeast-southwest, of which the faults labeled 1 and 2 are selected for discussion of their displacement distributions. The indicated vertical plane represents a seismic cross section presented as Figure 3. Survey dimensions are approximately 5 × 10 km.

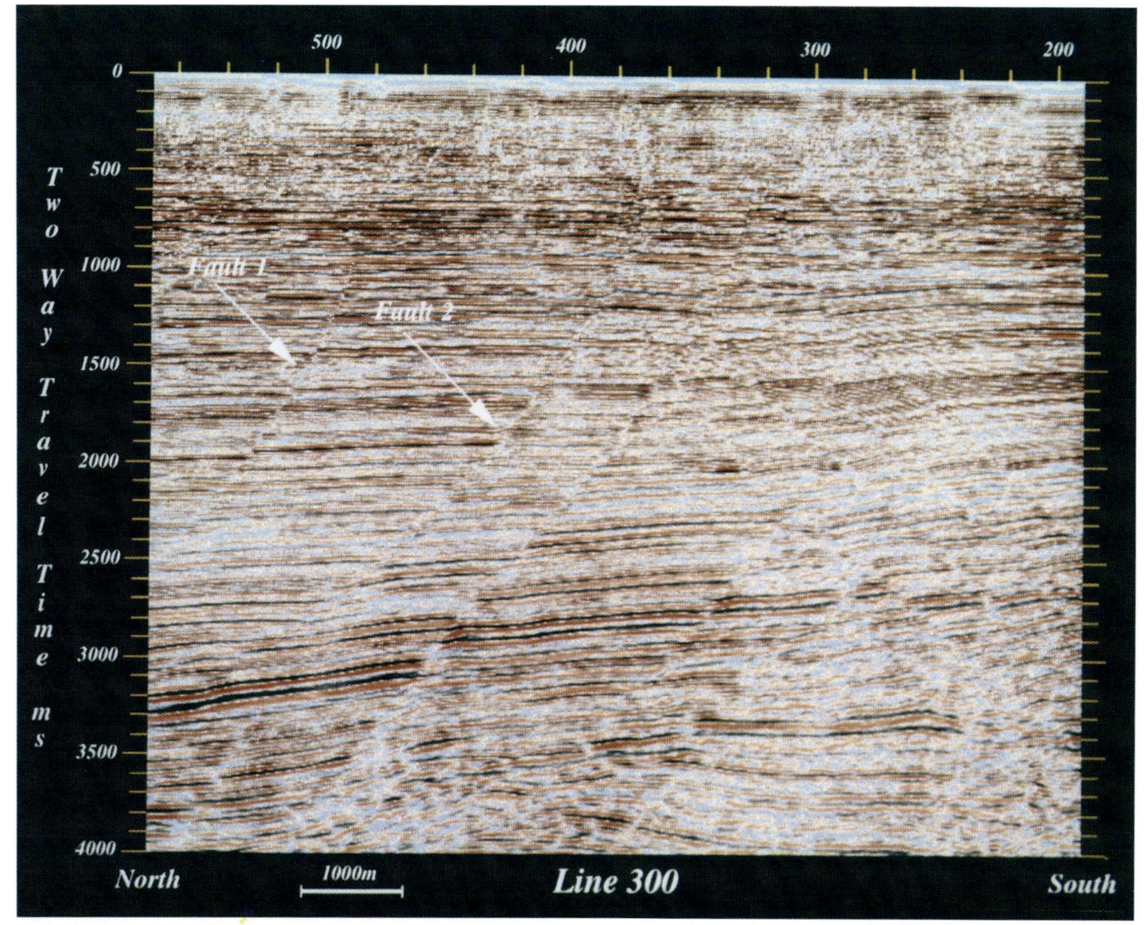


FIGURE 3. Representative north-south seismic cross section showing fault structure over the survey area. The alternating sands, silts, and clays of the deltaic deposits are expressed in the seismic data as a highly reflective sequence, with good lateral continuity. Their gentle basinward regional dip and strong reflection character mean that the seismic data have been well migrated, succeeding in almost perfect positioning of the fault planes, even in the deepest parts of the dataset.

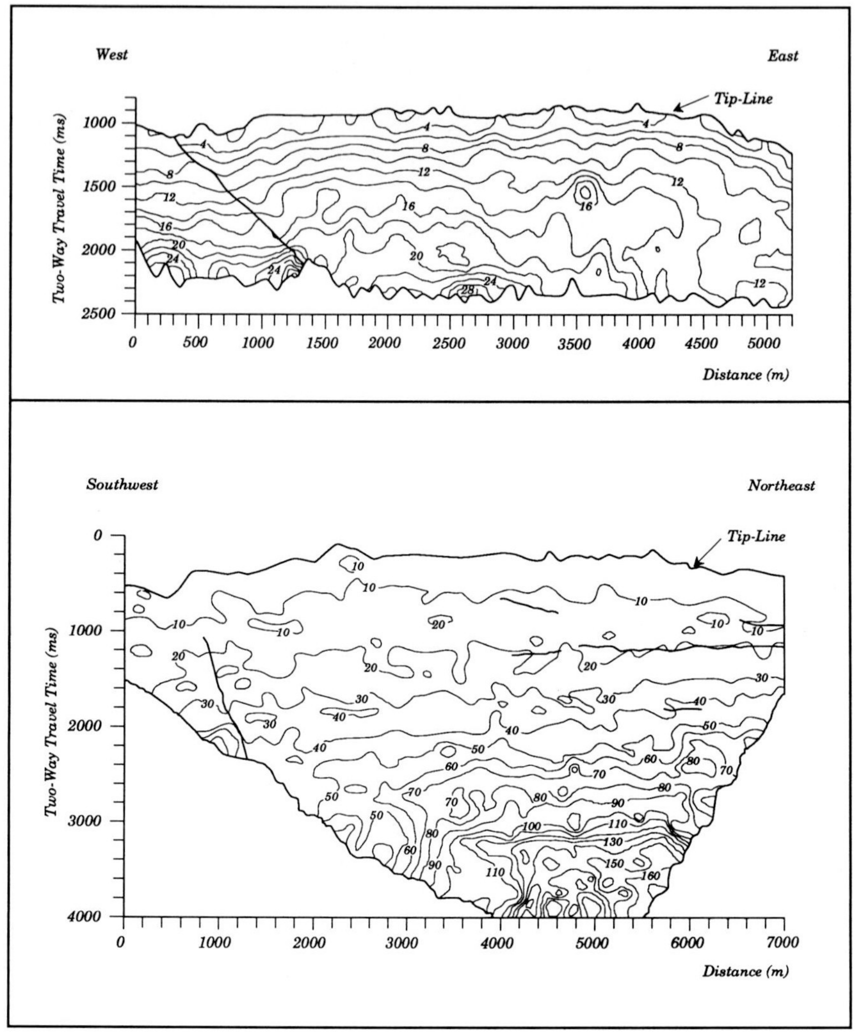


FIGURE 4. Vertical strike projections of the distribution of throw on Faults 1 and 2. The patterns are defined from 1384 and 2319 discrete displacement measurements, respectively, and are contoured in milliseconds (one-way traveltime). The distributions in each case are bounded laterally by the edges of the dataset and vertically by a lower limit of measurement and a well-resolved upper tip-line. Branch lines defining points of intersection with neighboring faults are expressed as bold lines, locally trending oblique to the contours. Note that these maps are not at the same scale.

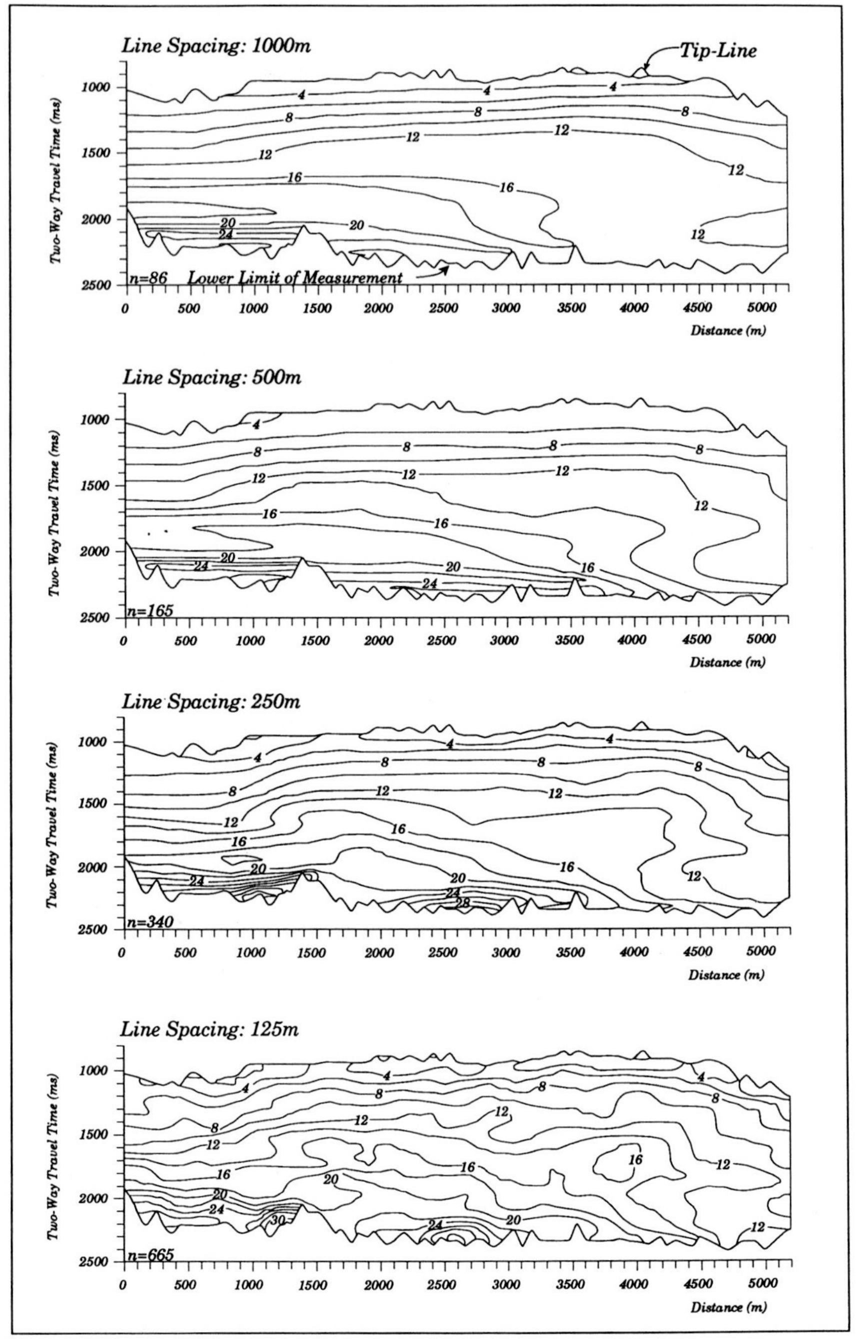


FIGURE 5. A series of vertical displacement distribution maps showing differences in the resolved throw patterns on Fault 1 derived from sample profiles at different separations. Spatial aliasing at the largest separations means that the mapping has failed to resolve any of the local high-frequency throw variations. Consequently, the distributions are defined by smooth, gently arcuate contours with mild local gradients that represent a uniform and systematic increase in displacement with depth. However, with a reduction in the profile spacing, a greater proportion of the small local displacement patterns are resolved and the contours exhibit an increasingly complex and erratic distribution.

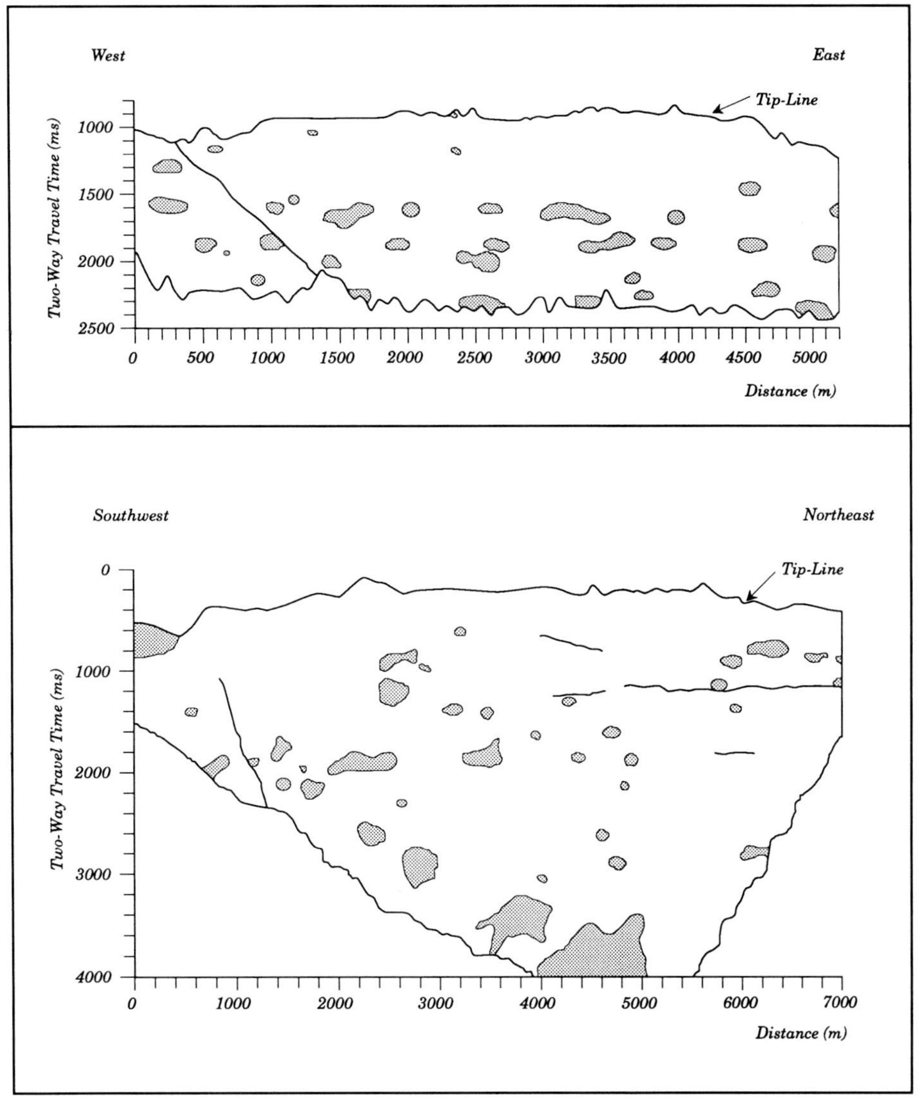


FIGURE 6. Vertical strike projections of displacement gradients on Faults 1 and 2. Regions of negative displacement gradient (a decrease in displacement with depth) are defined by discrete closures with stippled shading, lying in prominent subhorizontal bands along the fault planes. Note that the maps are not at the same scale.

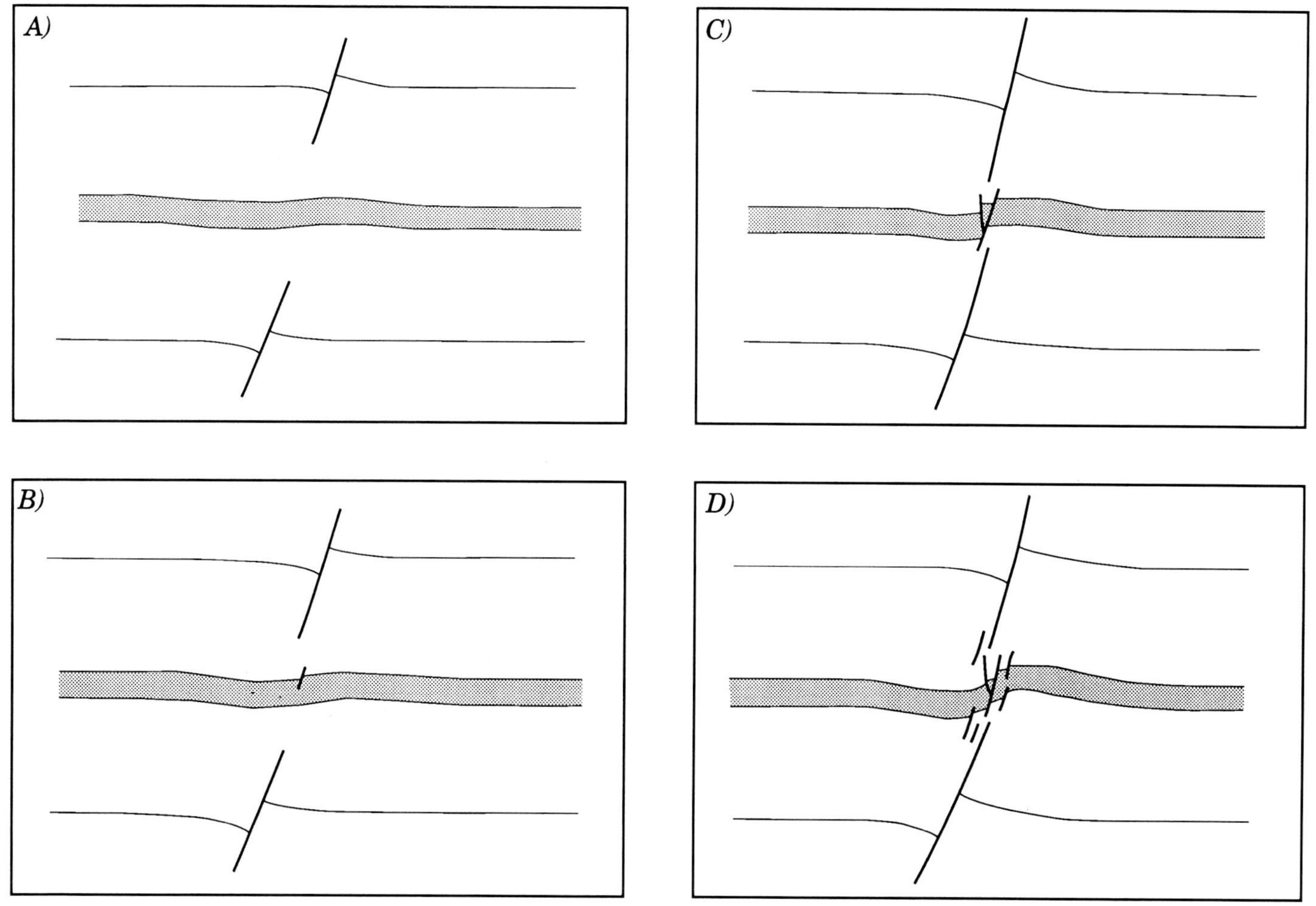


FIGURE 7. Four-stage model depicting the evolution of a throughgoing normal fault plane from the dip linkage of two initially isolated fault segments. (A) Initial extension is accommodated by ductile flexuring and the nucleation of discrete fractures in narrow, inclined deformation bands. Later, brittle deformation is concentrated within these narrow zones, the nucleation of small slip surfaces originating at points of weakness in the material. (B), (C) As extension continues, new embryo faults are nucleated; meanwhile continued shear on existing discontinuities results in radial propagation of their tip-lines at each slip event. (D) The alignment of neighboring faults associated with their initial nucleation positions and collinear propagation trajectories results in eventual overlap of the tip regions, while deformation in the relay structure results in eventual linkage to form increasingly larger slip surfaces.

A B

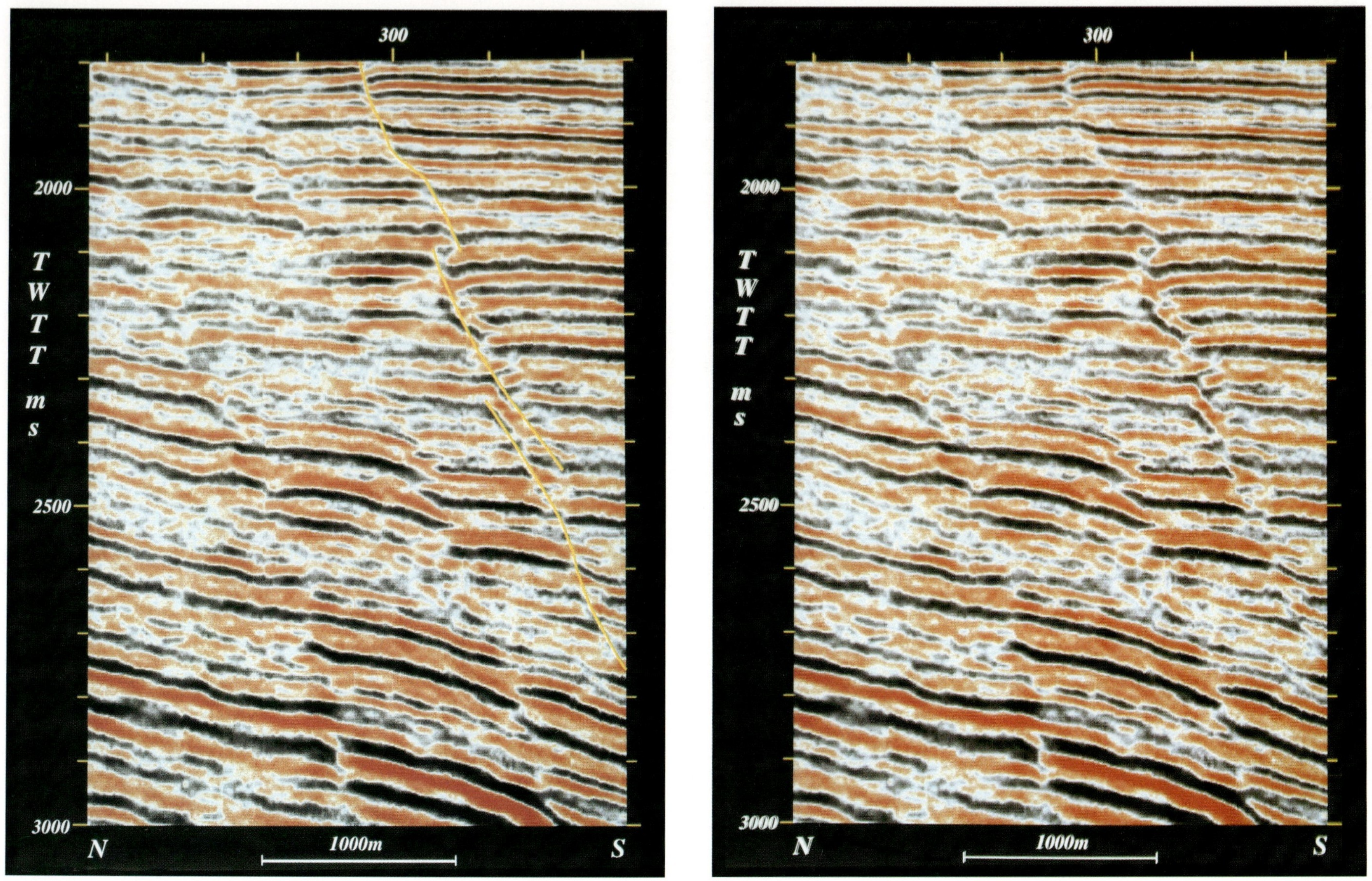


FIGURE 8. Comparative uninterpreted (A) and interpreted (B) seismic cross sections showing proposed configuration for part of Fault 2 as a series of discrete, unlinked fault segments in the slip direction.

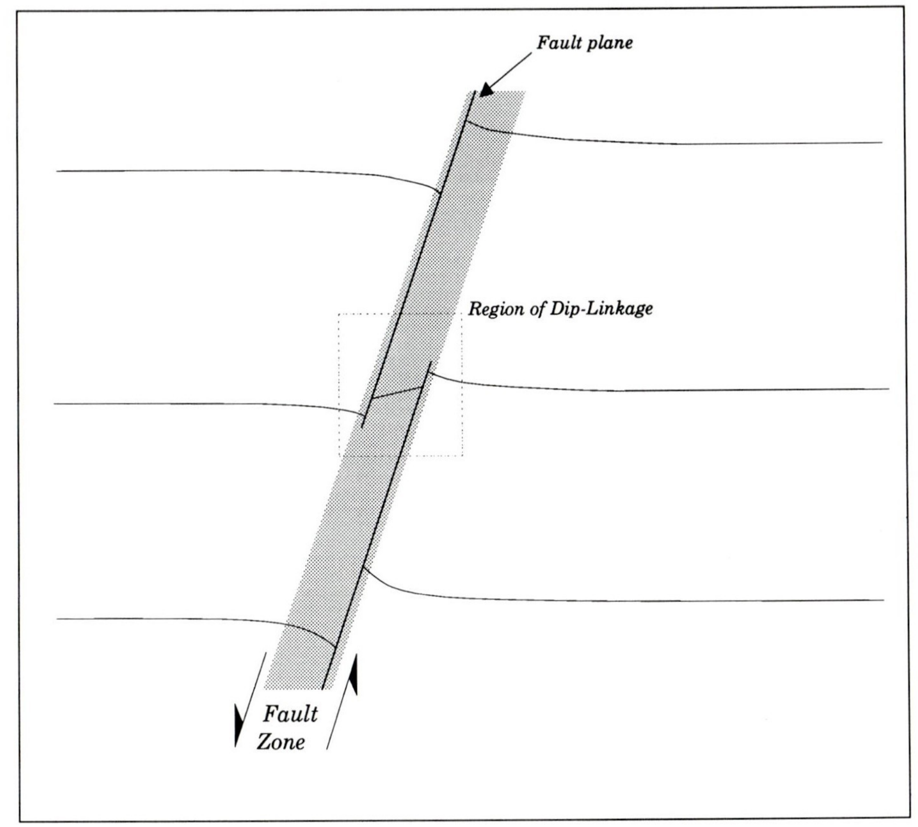


FIGURE 9. Illustration depicting how relay structures in the slip direction may fail to be defined by seismic data. Finite trace spacing and limited frequency content combine to image faults not as single slip surfaces but as zones, the width of which potentially hides the detail of many smaller structures.

Johnson, E. H., The structure of the Grant Canyon and Bacon Flat oil fields, Nevada, as revealed by 3-D seismic data, *in* P. Weimer and T. L. Davis, eds., *AAPG Studies in Geology No. 42* and *SEG Geophysical Developments Series No. 5*, AAPG/SEG, Tulsa, p. 237–246.

## CHAPTER 27

# The Structure of the Grant Canyon and Bacon Flat Oil Fields, Nevada, As Revealed by 3-D Seismic Data

Eric H. Johnson*

## ABSTRACT

The 20-million-bbl Grant Canyon oil field and its satellite feature, the 1-million-bbl Bacon Flat field, are located at the eastern edge of Railroad Valley, in Nye County, Nevada. An 11 $mi^2$ (28 $km^2$) 3-D seismic survey was acquired in 1993 to image the complicated structures in the area.

Interpretation ambiguities were reduced by calibrating the seismic data with known geology from 21 wells drilled prior to the 3-D survey and four subsequent wells. Still, interpretation of this complex area was arduous, hampered by extreme velocity variations in the overlying basin-fill sediments that degraded seismic data quality and skewed the imaged time structures.

Integrated seismic and well data show that the reservoirs of the Grant Canyon and Bacon Flat fields are remnants of detached Devonian carbonate strata that were emplaced over younger Paleozoic strata. Following emplacement, the reservoir rocks were faulted, eroded, and buried under more than 3500 ft (1070 m) of basin-fill sediments as Railroad Valley developed. Beneath the Devonian reservoirs in the oil fields, a flexure in the younger Paleozoic strata plunges northwesterly into the Railroad valley basin. Under the flexure, a flank of the Cretaceous Troy Pluton dips about 30° northwest. The overlying mass of Paleozoic rocks has slipped down the surface of the pluton, probably contributing to much of the faulting and subsequent erosional scouring of the field structures.

To date, the 3-D seismic survey has contributed to the discovery of 206 ft (63 m) of additional oil column in the Bacon Flat field and a residual oil accumulation on the flank of the Grant Canyon field. Other oil traps within the survey area may be masked by velocity gradients in the basin fill or by the misinterpretation of seismic data in areas lacking well control.

## INTRODUCTION

Railroad Valley, in Nye County, eastern Nevada, is a Tertiary rift basin (Grabb, 1994) that formed in the Basin and Range physiographic province of the Great Basin (Figure 1). The valley is flanked by the Pancake Range on the west and the Grant Range on the east. The Bacon Flat and Grant Canyon oil fields are located along the eastern edge of Railroad Valley, 65 miles (105 km) southwest of the town of Ely (Figure 1). The Bacon Flat field was discovered during July 1981 by the Northwest Exploration Company. The #1 Bacon Flat (1 BF) discovery well, located in the center of the southwest Section 17, T7N-R57E (Figure 2), drilled a faulted west-plunging nose observed on poor quality seismic data. During September 1983, the company moved one mile (1.6 km) to the southeast, updip from the 1 BF well, and drilled the #1 Grant Canyon (1 GC) discovery well in the south-west-northwest Section 21 (Figure 2).

The Bacon Flat well established Nevada's first oil production from Devonian carbonate strata, and the Grant Canyon field has demonstrated the remarkable reservoir potential of those rocks. At rates up to 4000 BOPD per well, the Grant Canyon field has produced 20 million bbl of oil from an oil reserve that comprised about 300 acres (Figure 2) and now is nearly depleted.

The development of the Grant Canyon and Bacon Flat fields, detailed by Johnson (1994b), progressed by "hit and miss" drilling; 21 wells were drilled in the area (Figure 2) before a 3-D seismic survey was completed, during 1993. About 100 line miles (160 km) of 2-D seismic data already covered the area, but the structural and stratigraphic complexities could not be deciphered from those data. The 3-D seismic data helped to delineate the field structures and resulted in two new oil wells.

*Equitable Resources Energy Company-Balcron Oil Division, 1601 Lewis Avenue, Billings, MT 59102

## STRATIGRAPHY

The general stratigraphy of Railroad Valley is shown in Figure 3. Some of the formations have been "thinned" extensively by attenuation faulting (Fryxell, 1988, 1991; Lund et al., 1991, 1993; Camilleri, 1992; Walker et al., 1992) and locally may be only 10–20% of their usual thicknesses. Cambrian and Ordovician clastic and carbonate rocks account for about 4000 ft (1220 m) of the basal strata in the area. Those rocks are overlain by a comparable thickness of Silurian and Devonian carbonate strata, beginning with about 400 ft (120 m) of the Silurian Laketown Dolomite. The Devonian strata include 500–1000 ft (150–305 m) of the Sevy Dolomite, 700–1000 ft (215–305 m) of the Simonson Dolomite, and up to 2000 ft (610 m) of the Guilmette Formation, composed of interbedded limestones and dolostones (Kleinhampl and Ziony, 1985).

Generally, the Devonian-Mississippian Pilot Shale is absent or less than 20 ft (6 m) thick throughout the Railroad Valley area. The Lower Mississippian Joana Limestone ranges from 200–700 ft (60–215 m) thick and is about 650 ft (200 m) thick beneath the Grant Canyon field. The Upper Mississippian Chainman Shale is more than 2200 ft (670 m) thick in the center of Railroad Valley, but is only 130–630 ft (40–192 m) thick in wells in the vicinity of the Grant Canyon and Bacon Flat fields. The Chainman Shale is the primary oil source in the Railroad Valley basin (French, 1983, 1989; Poole and Claypool, 1984).

The overlying Pennsylvanian Ely Limestone has been highly eroded throughout the area. The Ely Limestone is about 700 ft (215 m) thick in the Gulf well at the north end of the Bacon Flat field (Figure 2) and probably has been truncated by erosion beneath the Grant Canyon field, as shown in Figure 4. There are no Permian- through Cretaceous-age strata in the area due to erosion or nondeposition.

A Late Cretaceous–age intrusive crops out in the Grant Range east of the Grant Canyon field (Fryxell, 1984). A flank of the intrusive dips about 30° northwest beneath the Grant Canyon and Bacon Flat fields (Figure 4) (Johnson, 1994b). Named the Troy Pluton, it is part of a batholith that underlies the east half of Railroad Valley (Blank, 1993).

The Paleogene strata in the Railroad Valley area include the lacustrine limestones and shales of the Sheep Pass Formation, which rarely exceed 800 ft (245 m) in total thickness, and the ash flow volcanic deposits of the Garrett Ranch Group, which can be up to 3000 ft (915 m) thick. In the vicinity of the Bacon Flat and Grant Canyon fields, Sheep Pass sediments have been found only northeast of the Grant Canyon field, where 820 ft (250 m) were drilled by the 8 GC well (Figure 2). Garrett Ranch deposits have been encountered in a number of wells skirting both oil fields (BTA, 24–17, 2 BF, 5 BF, 6GC, and 8 GC wells) (Figure 2), the thickest being 454 ft (138 m) in the 2 BF well.

Lastly, Neogene and Quaternary fluvial and lacustrine clastic rocks were deposited unconformably in the developing Railroad Valley structural basin. Over 10,000 ft (3000 m) of those sediments, herein called the "basin fill," have accumulated in the center of the valley. The unconformity at the bottom of the basin fill forms the top seal for all of the oil fields in Railroad Valley to date (Figure 1). Generally, whatever formations underlie that unconformity become the oil reservoirs.

## GEOLOGICAL SUMMARY

The Grant Canyon and Bacon Flat oil fields are erosional remnants of detached Devonian strata (Johnson, 1993, 1994b). The structure and extent of the detached rocks are shown in Figure 2. The remnants of Devonian strata rest on faulted blocks of younger Paleozoic strata, as shown in Figure 4. The Grant Canyon and Bacon Flat field structures overlie a northwest-plunging flexure in the Paleozoic strata beneath the fields, as depicted by 3-D seismic data described later. For simplification, some faults have been omitted from the figures.

From 3-D seismic and well data, I interpret that erosion exposed northwest-dipping Paleozoic strata; Pennsylvanian Ely Limestone was exposed in the Bacon Flat area and Mississippian Chainman

Shale and Joana Limestone were exposed in the Grant Canyon area. Then Devonian carbonate strata detached from the Grant Range to the east and slid westward over the Paleozoic strata. This slide mass was emplaced before the end of Paleogene time (Johnson, 1994b) due to any of several mechanisms, including the tectonic denudation of the Grant Range (Camilleri, 1992; Fryxell, 1984, 1988, 1991; Lund and Beard, 1992; Lund et al., 1991, 1993), uplift of the Troy Pluton (Barker and Peterson, 1991; Blank, 1993), or rift offset development (Grabb, 1994).

Following emplacement of the detached slide mass in the Bacon Flat–Grant Canyon area, normal faults broke the underlying Paleozoic strata into horst and graben blocks, probably as a result of slippage of the entire mass of Paleozoic strata down the northwest-dipping surface of the Troy Pluton. Due to the faulting and subsequent erosion, the Grant Canyon and Bacon Flat fields were sculpted into hills with over 1000 ft (305 m) of relief (Figures 2 and 4). Erosion removed the detached Devonian strata between the oil fields, isolating the fields and exposing the underlying Ely Limestone (Figures 2 and 4).

Early in Railroad Valley's evolution, Paleogene Sheep Pass lacustrine sediments and Garrett Ranch volcanic strata were deposited in the topographic lows of an extensional terrain. During the Neogene and Quaternary, the early valley topography was inundated by basin-fill lacustrine and fluvial clays, sands, and conglomerates. Those deposits buried and sealed the existing Bacon Flat and Grant Canyon field structures. About 5000 ft (1520 m) of basin-fill sediments overlie the Bacon Flat field, and 3500 ft (1070 m) cover the Grant Canyon field (Figure 4).

Two horizons in the basin fill, labeled BF-2 and BF-1 in Figure 4, are distinct stratigraphic markers on resistivity logs. The uniform thickness of this basin-fill interval indicates that structural activity in the area diminished after burial of the Grant Canyon and Bacon Flat field structures (Figure 4). Maps of the BF-2 and BF-1 horizons by McCutcheon and Zogg (1994, their Figures 9 and 10) show some indication of the underlying flexure, but not the faulted oil-field structures.

Beginning in late Miocene time, hydrocarbons were generated from the Chainman Shale after burial in the Railroad Valley basin subjected the formation to maturation temperatures (French, 1983, 1989; Duey, 1983; Poole and Claypool, 1984; Barker and Peterson, 1991; İnan and Davis, 1994; Peterson, 1994). Oil migrated into the Bacon Flat and Grant Canyon field structures, and it is likely that oil is still generating and migrating in the area (Barker and Peterson, 1991; Hulen et al., 1994).

The detached carbonate reservoirs in the fields are exceptionally porous and permeable; lost circulation material pumped into the 7 GC well in the Grant Canyon field later was recovered from the 10 GC well bore, 550 ft (170 m) away (Johnson, 1994b). Read and Zogg (1988) attributed the outstanding reservoir qualities to fracturing and brecciation by tectonics and karsting during subaerial exposure. Aided by strong, natural water drives, the reservoirs easily flowed oil at rates of thousands of barrels per day per well (Johnson, 1994b).

An abnormally high geothermal gradient in the immediate area of the oil fields (Duey, 1983; Peterson, 1994, citing Duey) has contributed to reservoir temperatures of 115°C in the Grant Canyon field and 121°C in the Bacon Flat field and indicates the presence of geothermal activity beneath the fields. As described by Hulen et al. (1991, 1994), French and Hulen (1993), and Goff et al. (1994), geothermal fluids could have leached the carbonate rocks and precipitated silica and clay in the tops of the structures, enhancing both the reservoirs and seals of the oil fields. Oil accumulation may have been augmented by the movement of large volumes of water in an upwelling geothermal plume.

## BASIN-FILL VELOCITIES

Lateral changes in density and velocity in the basin fill are produced by the interfingering of fluvial and lacustrine deposits, ranging from volcanic clays to conglomerates containing eroded carbonate boulders. Vreeland and Berrong (1979) reported basin-fill velocities ranging from 5000 to 16,000 ft/s (1520 to 4880 m/s). A 500-ft (150-m)-thick interval in the 8 GC well measures about 17,000 ft/s (5180 m/s). Lateral variations of such magnitude in the basin fill modify the appearance of deeper structural features on seismic data. They also introduce static shifts in seismic data that degrade data quality during stacking and migration (Johnson, 1994a).

In the Bacon Flat–Grant Canyon area, average basin-fill velocities from the surface to depths of about 5000 ft (1520 m) range between 9000 and 11,000 ft/s (2740 and 3350 m/s). Generally, the basin-fill velocities increase from northwest to southeast. This is illustrated by comparing Figure 5, a time structure map on the top of the Paleozoic strata in the area, to the structure map of the oil fields on the same mapping surface (Figure 2). Figure 5 exhibits a prevalent tilt to the northwest that diminishes the closure of the time structures to the southeast compared to the actual closure of the field structures.

The substantial discordance possible between real and apparent structure over a short distance is evident when comparing well velocity surveys (where seismic traveltimes are measured from known depths) acquired in the 3 GC, 8 GC, and 10 GC wells. Figure 6 shows that a two-way traveltime of 0.840 s corresponds to a depth of 3650 ft (1115 m) in the 3 GC well and an average velocity of 8700 ft/s (2650 m/s). In the 8 GC well, 2600 ft (790 m) to the east, the same time corresponds to a depth of 4650 ft (1420 m), which is 1000 ft (305 m) deeper and equates to an average velocity of 11,000 ft/s (3350 m/s). Because of this velocity gradient, a surface with 1000 ft (305 m) of dip between the two wells would appear to be flat on seismic data.

The parallelism of the three velocity curves below 2000 ft (610 m) (Figure 6) indicates that the upper 2000 ft (610 m) of the basin fill are responsible for the velocity differences. In fact, the largest variations generally occur within the first 1000 ft (305 m) of sediments. Unfortunately, those shallow basin-fill velocities are measured only by well velocity surveys, which are seldom acquired. Sonic logs record velocities only below the surface pipe in a well, which is set to 500 ft (150 m) or deeper. This exacerbates the problem of estimating seismic velocities for time-to-depth conversion of seismic data. Some of the structural aberrations introduced by velocity gradients in the Grant Canyon–Bacon Flat area are pointed out in the following discussions of the 3-D seismic sections (Figures 7–10).

## 3-D SEISMIC SURVEY

The Grant Canyon 3-D seismic survey was acquired and processed during 1993 by a Railroad Valley AMI partnership headed by Apache Corporation. More than 700 recording channels were utilized to obtain 30-fold data every 60 × 60 ft (18 × 18 m). A 60-ft (18-m) bin size was needed to image steep dips without aliasing, aid in lateral velocity analyses, and resolve the complex structures targeted (Zwart, 1994). An 11-mi$^2$ (28 km$^2$) layout (Figures 11–14) was centered over the Bacon Flat and Grant Canyon fields, with sufficient perimeter to image the flanks of the fields, dipping up to 45°, although not everywhere at full fold. Generally, surfaces dipping more than 45° were not imaged, due to the limited size of the survey.

Vibroseis with a 10–80 Hz sweep provided the energy source, except in a patch of wet ground over Grant Canyon field where dynamite was used in mini-hole patterns. The vibroseis and dynamite data were comparable in signal strength and were carefully phase matched before being merged during processing.

## 3-D SEISMIC SECTIONS

Three vertical cross sections through the 3-D seismic volume transect the Grant Canyon and Bacon Flat oil fields. These seismic sections are reproduced in Figures 7, 9, and 10. They are located on the structure map (Figure 2) and the seismic time slices (Figures 11–14). The seismic sections were constructed to pass through wells to optimize the tie between known geology and the seismic data. Another seismic section, shown in Figure 8, clarifies the stratigraphy of the Bacon Flat field.

The zero time seismic datum is +4800 ft (+1460 m) msl and time values are two-way traveltimes below that datum. The horizontal and vertical scales of the seismic sections are roughly 1:1. Time-to-depth conversions can be approximated using an average velocity of 10,000 ft/s (3050 m/s) for the basin-fill sediments.

### SEISMIC SECTION A-A′

Seismic section A-A′ (Figure 7) is a dip section traversing from northwest to southeast through the Bacon Flat and Grant Canyon fields (Figures 2, 11–14). Its traverse is the same as geologic cross section A-A′ (Figure 4), so a direct comparison can be made. To the northwest, the Paleozoic strata are imaged beneath the onlapping Oligocene volcanic strata in the vicinity of the BTA well, which penetrated 72 ft (22 m) of volcanic rocks.

At the Bacon Flat field, the top of the Devonian slide mass is at –560 ft (–171 m) in the 1 BF discovery well and at –214 ft (–65 m) in the 23-17A well (Figures 4 and 7). The wells are about 450 ft (135 m) apart, requiring an average dip of 38° between the wells to account for the 346-ft (105-m) difference in depth. Figure 7 illustrates that an erosional escarpment defines the west flank of the Bacon Flat field. At about 1.150 s, the escarpment is incised by a terrace. This can be seen even more distinctly in Figure 8, a seismic segment that traverses the field perpendicular to the escarpment (as located in Figure 2). The pinchout of the Bacon Flat Devonian reservoir to the east is somewhat interpretive in Figure 7 but is unequivocal in Figure 8.

Finding Pennsylvanian strata between, and structurally low to, the Devonian reservoir rocks of both fields produced a variety of geologic cross sections as reviewed by Johnson (1994b). Figures 2, 4, and 7 illustrate that the Bacon Flat and Grant Canyon fields are separated by down-to-basin faulting and erosion that has removed a portion of the Devonian slide mass, thus isolating the oil reservoirs of the fields.

There is a significant velocity difference in the basin fill between the 23-17A and 6 GC wells. The elevation of the Paleozoic rocks changes 651 ft (198 m) between the wells (Figure 4). This should equate to a two-way traveltime difference of more than 100 ms, yet a difference of only about 40 ms is observed on the seismic data (Figure 7).

Also note that the volcanic top in the 6 GC well appears to be structurally high to the Devonian top in the 23-17A well, but it is actually 96 ft (29 m) lower.

A lateral velocity difference in the basin fill also occurs over the Grant Canyon field, where the 9 GC well encountered the Devonian reservoir at +1212 ft (+370 m), and the 10 GC well penetrated the reservoir at +1057 ft (+322 m), 155 ft (47 m) structurally lower (Figure 4). In Figure 7, the 10 GC location appears to be structurally high to the 9 GC well, no matter whether the top of the reservoir is picked as shown (based on sonic data from the wells) or the high-amplitude reflection at about 0.750 s is chosen. Note that the high-amplitude reflection would be the logical pick in the absence of well data.

Based on the 3-D data, the 10 GC location was drilled because it appeared to be structurally high to both the 9 GC well and the 4 GC well, which penetrated the reservoir at +741 ft (+226 m). In fact, the 10 GC well is the second highest well in the Grant Canyon field and is a "dry hole" only because the oil column at that location already had been drained by the 9 GC well (Figures 2 and 4). But what if the 3-D survey had been acquired before the 9 GC well was drilled? Since the 9 GC location appears to be downdip from the structural apex on the 3-D data, the 9 GC well might not have been drilled and 155 ft (47 m) of additional oil column would have been missed.

About 2500 ft (760 m) southeast of the 10 GC well, the 2 GC well encountered the detached Devonian strata at –11 ft (–3.4 m) and penetrated a total of 1587 ft (484 m) of Paleozoic rocks above the pluton (Figures 4 and 7). At the THORN well, only 2600 ft (790 m) east of the 2 GC well, the Paleozoic strata have pinched out. Again, a velocity gradient in the basin fill between the 2 GC and THORN wells masks the true structure. The base of the basin fill in the wells differs by only about 30 ms (Figure 7), which should equate to about 150 ft (45 m). The actual structural difference is 485 ft (148 m) (Figure 4).

## SEISMIC SECTION B-B′

Seismic section B-B′ (Figure 9) traverses the Grant Canyon field from southwest to northeast (Figures 2, 11–14) through the flexure that plunges beneath the Grant Canyon and Bacon Flat fields. Basin-fill velocity variations along this traverse are relatively minor. The Grant Canyon discovery well (1 GC) is respectably positioned on the broad flexure of Paleozoic rocks that it was targeting. The well encountered a prolific Devonian carbonate reservoir at +409 ft (+125 m) and penetrated about 150 ft (45 m) of oil column. Who could have imagined that the apex of the oil field, only 1200 ft (370 m) to the north (Figure 2), would be over 800 ft (240 m) structurally higher?

Just southeast of the apex, Figure 9 images the field structure at the 10 GC well location, where Devonian rocks were encountered at +1057 ft (+322 m). To the northeast of the 10 GC well, the Devonian reservoir was penetrated at +741 ft (+226 m) in the 4 GC well, but the reservoir rocks are absent in the 8 GC well, which encountered the Chainman Shale at –1292 ft (–394 m), beneath Sheep Pass and volcanic strata. The 4 GC and 8 GC wells are only 1800 ft (550 m) apart, but the elevation of the field structure drops about 2000 ft (610 m) in that distance, an average dip of 48°. This steeply dipping flank of the Grant Canyon field is poorly imaged by the 3-D seismic data; Figure 9 would probably be misinterpreted without well control.

Above the southwest flank of the Grant Canyon structure, the 43-20 well was drilled to test high-amplitude reflections between 0.800 and 0.900 s that appear to be structurally high to the 1 GC well and truncated by an unconformity. Unfortunately, the reflections were caused by conglomerate beds in the basin fill, and the underlying Devonian strata were penetrated 488 ft (149 m) structurally low to the 1 GC well.

Another attempt to test similar-looking reflections to the east of the 1 GC well was successful. The 22-21 well (Figure 2) found an estimated 100,000 bbl of oil in Devonian rocks, structurally high to the 1 GC well but far below the remaining oil column in the Grant Canyon field. Apparently, the oil was trapped by a small fault, shown in Figure 2, that prevented drainage by updip wells.

## SEISMIC SECTION C-C′

Seismic cross section C-C′ (Figure 10) traverses the Bacon Flat field from south to north, intersecting three field wells and two flank wells (Figures 2, 11–14). Like Figure 9, Figure 10 images the northwest-plunging flexure of Paleozoic strata upon which the Grant Canyon and Bacon Flat reservoir mounds are perched. A more extensive mass of detached Devonian strata was present in the area prior to flexuring and faulting, which would have fractured the carbonate strata and accelerated erosion.

The basin-fill velocities are much slower to the north of the Bacon Flat field than to the south. The effect of this velocity gradient is illustrated by comparing the seismic times and depths of the Pennsylvanian Ely Limestone in the Gulf well and the 2 BF well. In Figure 10, the time difference between the wells is about 150 ms, so it would appear that the Ely Limestone in the Gulf well is about 750 ft (230 m) structurally low to the 2 BF well. In fact, the Ely Limestone in the Gulf well, at –1562 ft (–476 m), is 50 ft (15 m) structurally high to the 2 BF well.

Five dry holes surrounded the 1 BF discovery well before the 24-17, 23-17, and 23-17A wells were drilled (Figures 2 and 10). During January 1992, the 24-17 well was drilled 800 ft (245 m) southeast of the 1 BF well. Beneath 35 ft (11 m) of volcanic rocks, the 24-17 well penetrated the detached Devonian rocks 73 ft (22 m) structurally low to the 1 BF well. The oil column at that location already was depleted and the well was plugged. During August 1992, the 23-17 well was drilled 900 ft (275 m) north of the 1 BF well. The 23-17 well encountered the oil reservoir 140 ft (43 m) structurally higher than the 1 BF discovery well, at an elevation of –420 ft (–128 m). The well flowed 5400 BOPD and no water (Johnson, 1993).

After the success of the 23-17 well, but prior to the completion of the 3-D seismic survey, the Bacon Flat reservoir was remapped using 2-D seismic data and dip meter data from wells. The apex of the field as shown by Johnson (1993) was interpreted to be midway between the 23-17 well and the BTA well, just west of the 1 BF well and the fault shown in Figure 2. That interpretation would have resulted in another dry hole.

The 3-D seismic data indicated that the apex of the Bacon Flat reservoir was situated midway between the 23-17 and 24-17 wells (Figures 2 and 10). There, the 23-17A well was drilled during December 1993, and encountered 206 ft (63 m) of additional oil column, structurally high to the 23-17 well. The 23-17A well should recover over 300,000 bbl of oil of new reserves, and the total production from the Bacon Flat field should approach 1 million bbl of oil.

## TIME SLICES

Figures 11 through 14 are four time slices of the 3-D seismic data. The time slices are located in Figures 7, 9, and 10 by arrows at the left edges of the figures. Beginning with a slice through the top of the Grant Canyon reservoir, the time slices intersect the Paleozoic strata at progressively deeper levels, with the last slice intersecting the top of the Bacon Flat reservoir. The seismic peaks are blue and the troughs are red, with the brightest colors depicting the highest amplitudes of the data. White areas represent low amplitudes, such as zero crossings between peaks and troughs, or a lack of signal strength.

The time slices show the locations of the seismic cross sections (Figures 7, 9, and 10), which can be referenced to see where the slices intersect specific seismic reflections. Of course, the time slices are imaging the same velocity-induced structural aberrations recognized during the analysis of the seismic sections. Although seismic time is constant across a slice, structural depths vary, and the apparent strikes of the seismic events depicted on the slices may differ from the true strikes.

## TIME SLICE AT 0.760 S

Figure 11 is a time slice of the 3-D data at 0.760 s two-way traveltime, 50 ms, or about 300 ft (90 m), below the apex of the Grant Canyon reservoir mound shown in Figures 7 and 9. A circular blue seismic event that appears to be continuous on the time slice joins the 1 GC, 22-21, and 10 GC wells. This circular event appears to indicate a time closure southeast of the structural apex of the Grant Canyon field, as located by the 9 GC well. Actually, the circular blue event does not portray the topography of the field structure. Figure 9 shows that the seismic peak at 0.760 s that equates to the blue event is 70 ms above the Devonian reservoir in the 1 GC well and 50 ms below the top of the reservoir in the 10 GC well.

In the east corner of the survey area, a band of northeast-trending reflections defines the intersection of the time slice with the flank of the Troy Pluton (Figure 7). The first indications of a northwest-trending flexure are a depression in the basin fill to the southwest of the oil fields and a structural low developing to the northeast.

## TIME SLICE AT 0.840 S

Figure 12 is a time slice at 0.840 s two-way traveltime through the middle of the Grant Canyon reservoir mass (Figures 7 and 9), about 800 ft (250 m) below the apex of the Grant Canyon field and about 500 ft (150 m) lower than Figure 11. The high-amplitude reflections from the carbonate strata exhibit a rough pattern of closure in the field area, centered around the 4 GC well. The time slice does not show a simple concentric pattern, as might be anticipated from the smooth dome-shaped field structure mapped in Figure 2, because the seismic reflections within the reservoir mass represent strata that are quite faulted and unconformable with the erosional surface of the reservoir (Figures 7 and 9).

The intersection of the time slice with the Troy Pluton is located by a dashed line trending southwest through the THORN well (Figure 12), which penetrated the Pluton at about 0.840 s (Figure 7). The truncation of the high-amplitude reflections along the dashed line locates the onlap of the Paleozoic strata onto the pluton. Trending northwest from the Grant Canyon field, an arcuate pattern indicates drape of the basin-fill sediments over the underlying flexure in the Paleozoic strata.

## TIME SLICE AT 0.980 S

Figure 13 is a time slice at 0.980 s two-way traveltime. The slice is beneath the detached Devonian rocks at the Grant Canyon field and images the underlying Paleozoic section (Figures 7 and 9). The plunging flexure in the Paleozoic strata beneath the Grant Canyon field is manifested by the northwest-directed nose exhibited by the seismic reflections in the field area. The nose is skewed slightly south due to a velocity gradient in the basin-fill sediments and is modified by a velocity "pull-up" caused by the overlying mound of Devonian carbonate strata (Figures 7 and 9). The projection of the flexure to the northwest of the Grant Canyon field is reflected by trends in the overlying basin-fill sediments. To the southeast of the field, the intersection of the time slice with the pluton is indicated by a dashed line.

## TIME SLICE AT 1.080 S

Figure 14 is a time slice at 1.080 s two-way traveltime and clips the apex of the Bacon Flat field structure. A nose of high-amplitude reflections projects from the Grant Canyon field area to the Bacon Flat field, depicting the Paleozoic strata beneath the Grant Canyon field and the volcanic strata between the fields (Figure 7). The structural separation between the fields in the vicinity of the 6 GC well (Figure 2) is not clear because high-amplitude reflections from the volcanic strata are merging with the Paleozoic reflections. As located by a dashed line, the time slice intersects the flank of the Troy Pluton at the same location as the 2 GC well (Figure 7).

Figure 15 is an enlargement of the Bacon Flat field area as shown by the box in Figure 14. A small time closure is evident immediately south of the 23-17A well, which penetrated the structural apex of the field (Figures 2 and 10). This time closure has been skewed to the south by a velocity gradient alluded to during the discussion of seismic section C-C′ (Figure 10). The time slice has imaged the apex of the field distinctly, but a well located in the center of the time closure could be over 200 ft (60 m) structurally low to the true apex of the Bacon Flat field as mapped in Figure 2.

## CONCLUSIONS

The 3-D seismic survey of the Bacon Flat–Grant Canyon field area imaged the small, complex field structures reasonably well (Figures 7–10). Shortcomings in the quality and resolution of the 3-D data are primarily attributed to the lateral velocity variations in the basin-fill sediments and to the structural and stratigraphic complexities of the area. The steepest flanks of the structures were poorly imaged because most reflections from their surfaces emerged beyond the perimeter of the survey area.

The interpretation of the 3-D seismic data was facilitated by integrating data from 21 wells already drilled in the survey area. In fact, both well data and seismic data were needed to resolve the complicated structural relationships encountered in some of the wells. Integrating the well data also disclosed velocity gradients and stratigraphic ambiguities that could not have been discerned from the 3-D seismic data alone. Therefore, the structural and stratigraphic interpretation of the 3-D data may not be reliable in areas lacking well control.

Utilizing the 3-D seismic data, four additional wells were drilled in the Grant Canyon and Bacon Flat fields. Two new oil wells were completed: the 23-17A well in the apex of the Bacon Flat field (Figures 2 and 10) and the 22-21 well on the southeast flank of the Grant Canyon field (Figure 2). Those two wells will produce about 400,000 bbl of oil that would not have been found without interpretations incorporating the 3-D seismic data. A third well located using the 3-D data, the 10 GC well, turned out to be the second highest well drilled in the Grant Canyon field (Figures 2, 4, 7, and 9). The 10 GC well failed only because an 800-ft (245-m) oil column originally at that location had already been depleted.

Lastly, the 43-20 well was located on the southwest flank of the Grant Canyon field (Figures 2 and 9). Clearly a wildcat well in an area this complex, the 43-20 well evaluated a structural and stratigraphic anomaly imaged by the 3-D seismic data (Figure 9). Conglomerate beds had produced reflections that were thought to represent Devonian strata, but this misinterpretation could not have been discerned prior to drilling the well.

The total cost to acquire and process the 11-mi$^2$ (28-km$^2$) 3-D seismic survey was about $750,000. This expense has been justified by the improved delineation of the field structures and the new oil reserves encountered. The 3-D survey would have been even more cost-effective had it been completed earlier during the development of the Grant Canyon and Bacon Flat oil fields.

### *Acknowledgments*

I thank Makoil, the new operator of a Railroad Valley AMI partnership, Equitable Resources Energy Company, and the other partners for permission to publish the 3-D seismic data. The assistance of Bob Trotter and Randy Zickuhr in drafting and reproducing the figures was greatly appreciated. Reviews by Paul Weimer and Bob Hardage improved the continuity of the paper.

## REFERENCES CITED

Barker, C. E., and J. A., Peterson, 1991, Burial history of the Mississippian Chainman Shale and Eocene Sheep Pass Formation, Railroad and White River valleys, eastern Nevada, *in* D. M. H. Flanigan, M. Hansen, and T. E. Flanigan, eds., Geology of the White River Valley, the Grant Range, Eastern Railroad Valley and Western Egan Range, Nevada: Nevada Petroleum Society, 1991 Fieldtrip Guidebook, p. 37–46.

Blank, H. R., 1993, Basement structure in the Railroad Valley–Grant Range region, east-central Nevada, from interpretation of potential field anomalies, *in* R. W. Scott, P. S. Detra, and B. R. Berger, eds., Advances Related to United States and International Mineral Resources—Developing Frameworks and Exploration Technologies: U.S. Geological Survey, Bulletin 2039, p. 25–33.

Camilleri, P. A., 1992, Extensional geometry of a part of the northwestern flank of the northern Grant Range, Nevada—inferences on its evolution: The Mountain Geologist, Rocky Mountain Association of Geologists, v. 29, p. 75–84.

Duey, H. D., 1983, Oil generation and entrapment in Railroad Valley, Nye County, Nevada, in The Role of Heat in the Development of Energy and Mineral Resources in the Northern Basin and Range Province: Geothermal Resources Council, Special Report 13, p. 199–205.

French, D. E., 1983, Origin of oil in Railroad Valley, Nye County, Nevada: Wyoming Geological Association, Earth Science Bulletin, v. 16, p. 9–21.

French, D. E., 1989, Hydrocarbon potential of Mississippian Chainman Shale, Railroad Valley, Nevada (abs.): AAPG Bulletin, v. 73, p. 356–357.

French, D. E., and J. B. Hulen, 1993, Origin and nature of the hydrocarbon seal in Railroad Valley, Nye County, Nevada (abs.): AAPG Bulletin, v. 77, p. 1448.

Fryxell, J. E., 1984, Structural development of the west-central Grant Range, Nye County, Nevada: Ph.D. dissertation, University of North Carolina—Chapel Hill, 139 p.

Fryxell, J. E., 1988, Geologic map and descriptions of stratigraphy and structure of the west-central Grant Range, Nye County, Nevada: Geological Society of America, Map and Chart Series MCH064, 16 p.

Fryxell, J. E., 1991, Tertiary tectonic denudation of an igneous and metamorphic complex, west-central Grant Range, Nye County, Nevada, in G. L. Raines, R. E. Lisle, R. W. Schafer, and W. H. Wilkinson, eds., Geology and Ore Deposits of the Great Basin, v. I: Nevada Geological Society, p. 87–92.

Goff, F., J. B. Hulen, A. I. Adams, P. E. Trujillo, D. Counce, and W. C. Evans, 1994, Geothermal characteristics of some oil field waters in the Great Basin, Nevada, in R. A. Schalla and E. H. Johnson, eds., Oil Fields of the Great Basin: Nevada Petroleum Society, Special Publication, p. 93–106.

Grabb, R. F., 1994, Extensional tectonics and petroleum accumulations in the Great Basin, *in* R. A. Schalla and E. H. Johnson, eds., Oil Fields of the Great Basin: Nevada Petroleum Society, Special Publication, p. 41–55.

Hulen, J. B., L. C. Bortz, and S. R. Bereskin, 1991, Geothermal processes in evolution of the Grant Canyon oil reservoir, Railroad Valley, Nye County, Nevada, *in* D. M. H. Flanigan, M. Hansen, and T. E. Flanigan, eds., Geology of the White River Valley, the Grant Range, Eastern Railroad Valley and Western Egan Range, Nevada: Nevada Petroleum Society, 1991 Fieldtrip Guidebook, p. 47–54.

Hulen, J. B., F. Goff, J. R. Ross, L. C. Bortz, and S. R. Bereskin, 1994, Geology and geothermal origin of Grant Canyon and Bacon Flat oil fields, Railroad Valley, Nevada: AAPG Bulletin, v. 78, p. 596–623.

İnan, S., and A. Davis, 1994, The history of oil generation in Pine and Railroad valleys, eastern Nevada, *in* R. A. Schalla and E. H. Johnson, eds., Oil Fields of the Great Basin: Nevada Petroleum Society, Special Publication, p. 57–84.

Johnson, E. H., 1993, A look at Bacon Flat, Grant Canyon oil fields of Railroad Valley, Nevada: Oil and Gas Journal (May 17), v. 91, n. 20, p. 64–68.

Johnson, E. H., 1994a, Using seismic, gravity and magnetic data to explore for oil in the Great Basin, *in* R. A. Schalla and E. H. Johnson, eds., Oil Fields of the Great Basin: Nevada Petroleum Society, Special Publication, p. 115–120.

Johnson, E. H., 1994b, Geologic and seismic analysis of the Bacon Flat–Grant Canyon area, Nye County, Nevada, *in* R. A. Schalla and E. H. Johnson, eds., Oil Fields of the Great Basin: Nevada Petroleum Society, Special Publication, p. 227–240.

Kleinhampl, F. J., and J. I. Ziony, 1985, Geology of northern Nye County, Nevada: Nevada Bureau of Mines and Geology, Bulletin 99A, 172 p.

Lund, K., and L. S. Beard, 1992, Extensional geometry in the northern Grant Range, east-central Nevada—implications for oil deposits in Railroad Valley, *in* C. H. Thorman, ed., Application of Structural Geology to Mineral and Energy Resources of the Central and Western United States: U.S. Geological Survey, Bulletin 2012, p. 11–19.

Lund, K., L. S. Beard, and W. J. Perry, Jr., 1991, Structure of the northern Grant Range and Railroad Valley, Nye County, Nevada—implications for oil occurrences, *in* D. M. H. Flanigan, M. Hansen, and T. E. Flanigan, eds., Geology of the White River Valley, the Grant Range, Eastern Railroad Valley and Western Egan Range, Nevada: Nevada Petroleum Society, 1991 Fieldtrip Guidebook, p. 1–6.

Lund, K., L. S. Beard, and W. J. Perry, Jr., 1993, Relation between extensional geometry of the northern Grant Range and oil occurrences in Railroad Valley, east-central Nevada: AAPG Bulletin, v. 77, p. 945–962.

McCutcheon, T. J., and W. D. Zogg, 1994, Structural geology of the Grant Canyon–Bacon Flat area, Nye County, Nevada—implications for hydrocarbon exploration in the Great Basin, *in* R. A. Schalla and E. H. Johnson, eds., Oil Fields of the Great Basin: Nevada Petroleum Society, Special Publication, p. 201–221.

Peterson, J. A., 1994, Regional geology and paleotectonic development of the Railroad Valley area, *in* R. A. Schalla and E. H. Johnson, eds., Oil Fields

of the Great Basin: Nevada Petroleum Society, Special Publication, p. 15–40.
Poole, F. G., and G. E. Claypool, 1984, Petroleum source-rock potential and crude-oil correlation in the Great Basin, *in* J. Woodward, F. F. Meissner, and J. L. Clayton, eds., Hydrocarbon Source Rocks of the Greater Rocky Mountain Region: Rocky Mountain Association of Geologists, p. 179–230.
Read, D. L., and W. D. Zogg, 1988, Description and origin of the Devonian dolomite oil reservoir, Grant Canyon field, Nye County, Nevada, *in* S. M. Goolsby and M. W. Longman, eds., Occurrence and Petrophysical Properties of Carbonate Reservoirs in the Rocky Mountain Region: Rocky Mountain Association of Geologists, p. 229–240.
Vreeland, J. H., and B. H. Berrong, 1979, Seismic exploration in Railroad Valley, Nevada, in G. W. Newman and H. D. Goode, eds., Basin and Range Symposium and Great Basin Field Conference: Rocky Mountain Association of Geologists and Utah Geological Association, Guidebook, p. 557–569.
Walker, C. T., R. D. Francis, J. G. Dennis, and W. W. Lumsden, 1992, Cenozoic attenuation detachment faulting—a possible control on oil and gas accumulations in east-central Nevada: AAPG Bulletin, v. 76, p. 1665–1686.
Zwart, D. W., 1994, 3-D finds more drill sites for Apache in its prolific Grant Canyon field (interview): Hart's Oil and Gas World, v. 86, n. 3, p. 62–64.

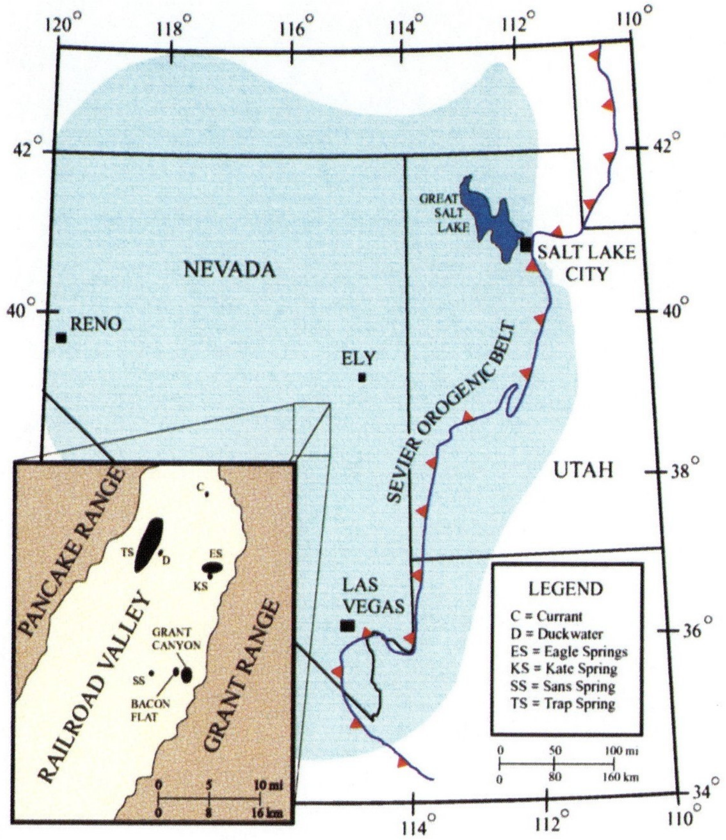


FIGURE 1. Location map showing the Great Basin of Nevada and Utah (shaded area). The enlargement shows Railroad Valley in eastern Nevada. The oil fields in Railroad Valley are located and named in the legend.

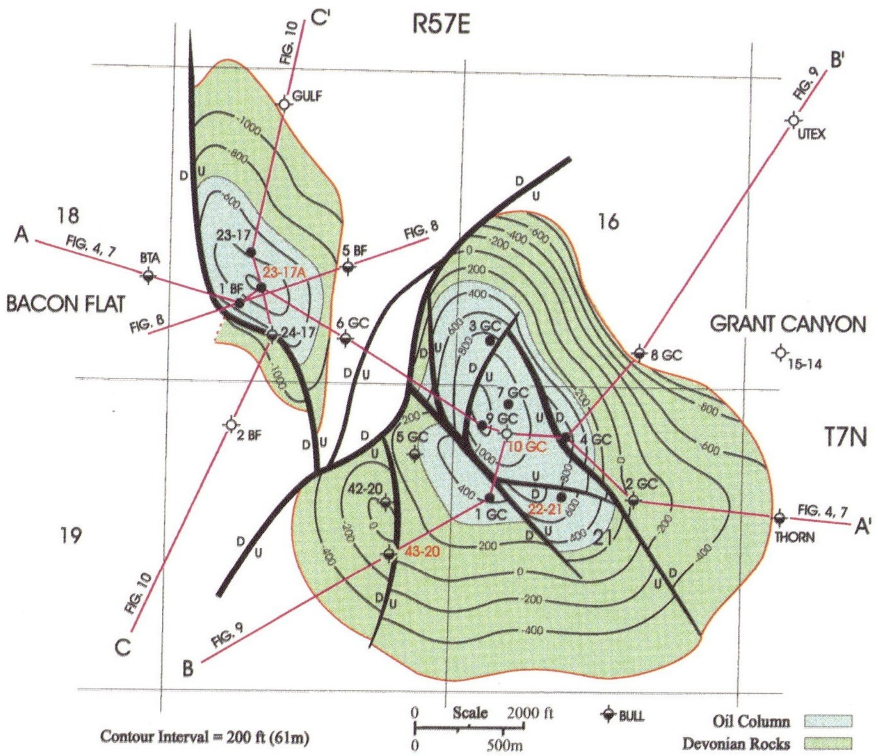


FIGURE 2. Structure map on the detached Devonian strata of the Grant Canyon and Bacon Flat oil fields, showing the extent of the reservoir mounds and the areas covered by the original oil accumulations. The traverse shared by geologic cross section A-A′ (Figure 4) and seismic cross section A-A′ (Figure 7) is shown. Also located are the traverses of seismic sections B-B′ and C-C′ (Figures 9 and 10) and a seismic segment through the Bacon Flat field (Figure 8). Four of the 25 wells shown on the map postdate the 3-D seismic survey and are labeled in red.

| Ma | PERIOD | | SYMBOL | RAILROAD VALLEY NEVADA | |
|---|---|---|---|---|---|
| 0 | QUATERNARY | | Tbf | BASIN FILL CLASTICS | |
| | TERTIARY | NEOGENE | | | |
| | | PALEOGENE | Tgr | GARRETT RANCH GROUP | R |
| | | | Tsp | SHEEP PASS FORMATION | S R |
| 67 | CRETACEOUS | | Ki | TROY INTRUSIVE | |
| 140 | JURASSIC | | | | |
| 200 | TRIASSIC | | | | |
| 250 | PERMIAN | | | | |
| 290 | PENNSYLVANIAN | | IP | ELY LIMESTONE | R |
| 330 | MISSISSIPPIAN | | M | CHAINMAN SHALE | S |
| | | | | JOANA LIMESTONE | R |
| 365 | DEVONIAN | | D | PILOT SHALE | |
| | | | | GUILMETTE FORMATION | R |
| | | | | SIMONSON DOLOMITE | R |
| | | | | SEVY DOLOMITE | |
| 405 | SILURIAN | | S | LAKETOWN DOLOMITE | |
| 425 | ORDOVICIAN | | O | CARBONATES AND CLASTICS | |
| | CAMBRIAN | | Є | | |
| 570 | | | | | |

FIGURE 3. Generalized stratigraphy of Railroad Valley, Nevada, with established hydrocarbon source rocks (S) and reservoir rocks (R) indicated.

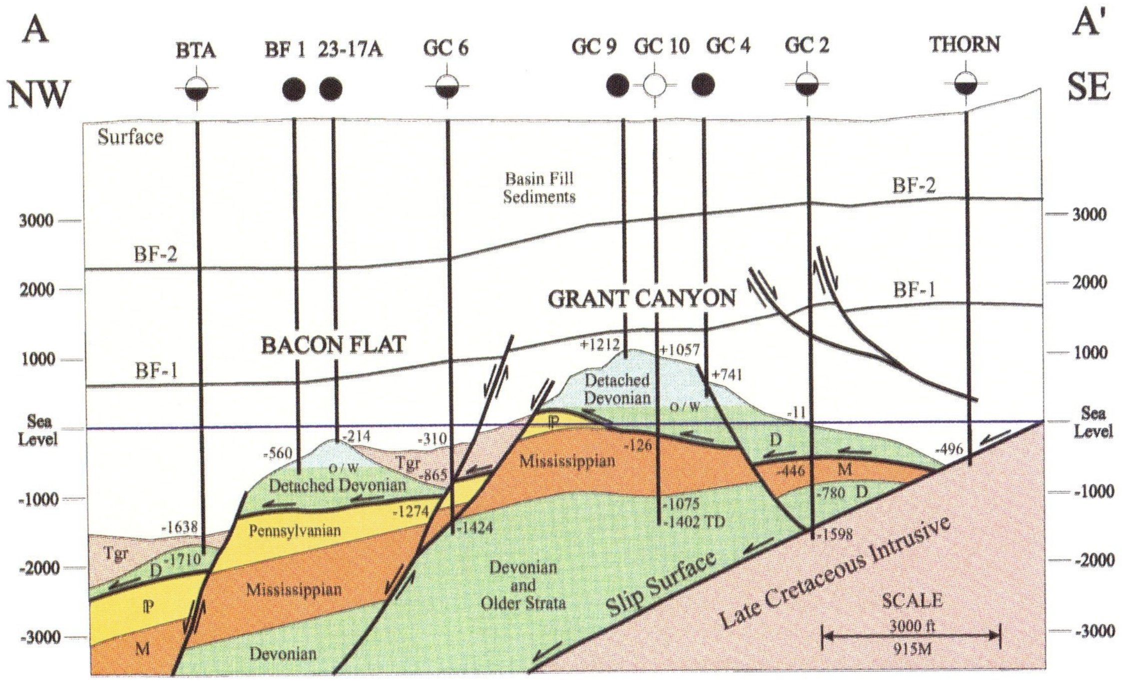


FIGURE 4. Geologic cross section A-A′ through the Bacon Flat and Grant Canyon oil fields as located in Figure 2, showing the detached Devonian reservoir mounds (with original oil columns) resting unconformably on younger Paleozoic strata. Pennsylvanian and Mississippian strata were truncated by erosion before emplacement of the detached Devonian strata. Movement of the entire mass of Paleozoic strata down the "slip surface" of the underlying Troy Pluton contributed to faulting of the strata. The uniform basin-fill interval between the BF-2 and BF-1 resistivity horizons indicates systematic burial of the field structures with minimal structural reactivation. Figure 4 may be compared to seismic section A-A′ (Figure 7), which shares the same traverse. The 23-17A and 10 GC wells postdate the 3-D seismic survey.

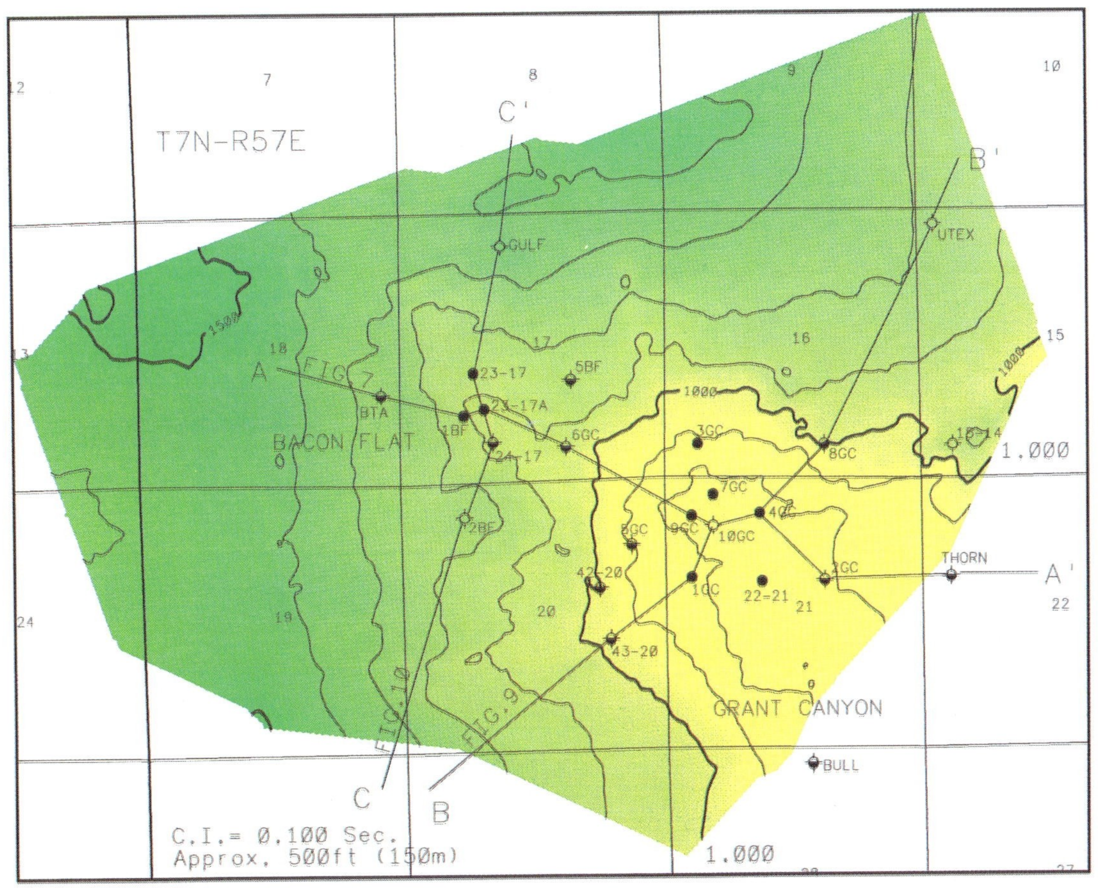


FIGURE 5. Seismic time map on the top of the Paleozoic strata. Compare Figure 5 to Figure 2. Due to a velocity gradient in the basin fill, the time map exhibits a northwest tilt that has diminished time closure and skewed the time structures to the southeast relative to true structure. Time contours are 100 ms, or about 500 ft (150 m). For scale, the section grid is 1 × 1 mi (1.6 × 1.6 km).

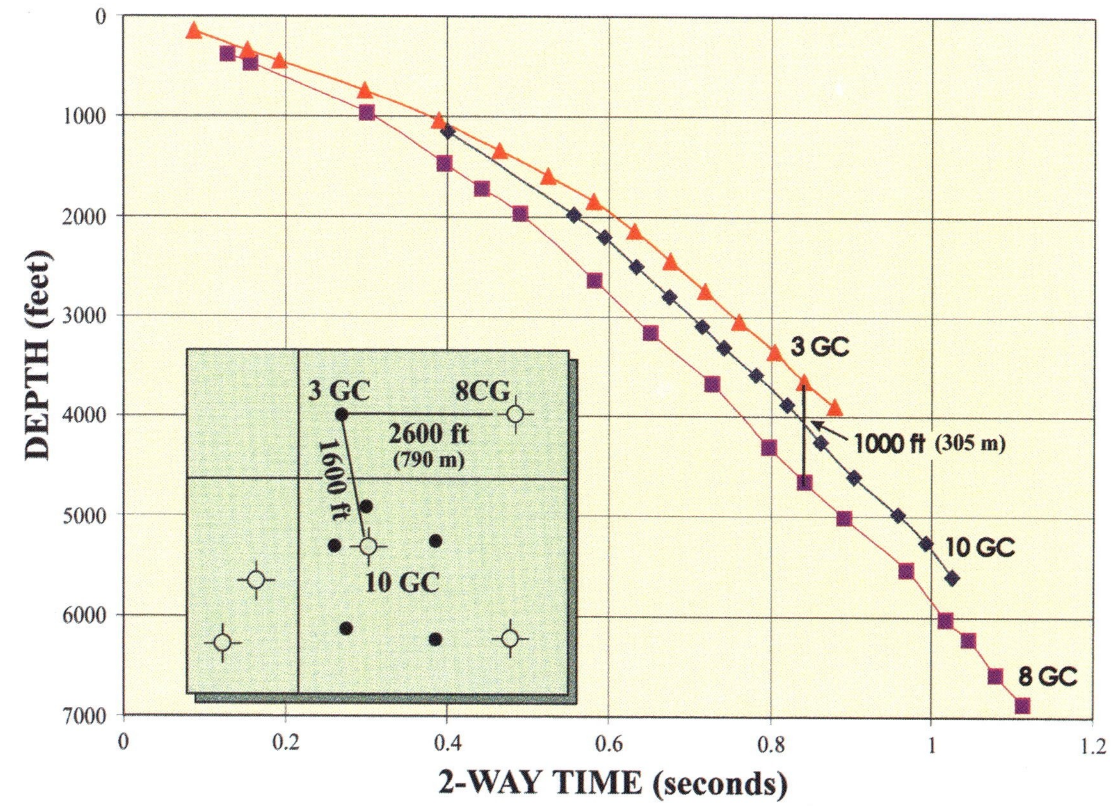


FIGURE 6. Graph of well velocity survey data for the 3 GC, 8 GC, and 10 GC wells. Due to varying velocities in the basin fill over short distances, any 2-way traveltime equates to substantially different depths in the wells and vice versa. Mostly, the upper 2000 ft (610 m) of basin-fill sediments produced the velocity differences between the wells.

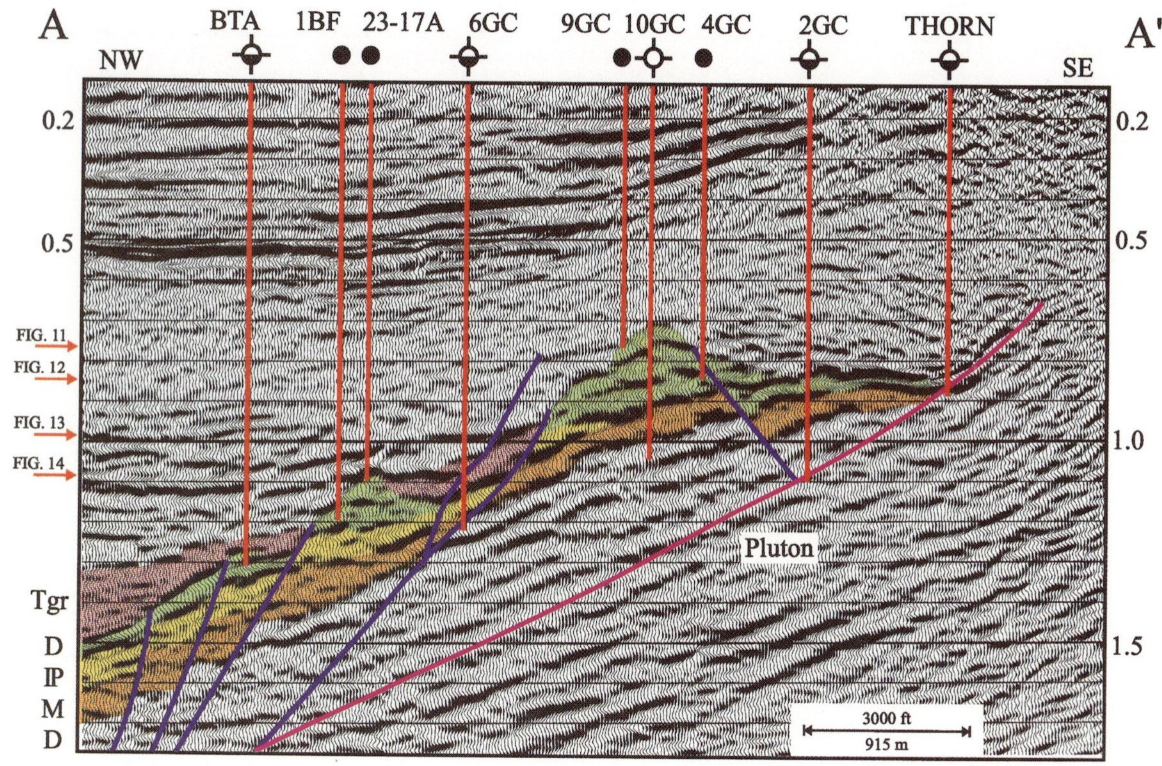


FIGURE 7. Seismic section A-A′ through the Bacon Flat and Grant Canyon fields as located in Figures 2 and 11–14, showing the detached Devonian strata overlying younger Paleozoic strata. A velocity gradient in the basin fill has tilted the Paleozoic strata to the northwest, diminishing the southeast time closure of the field structures compared to Figure 4. Without well data, interpretation of the seismic data would be perplexing. The 23-17A and 10 GC wells postdate the 3-D seismic survey.

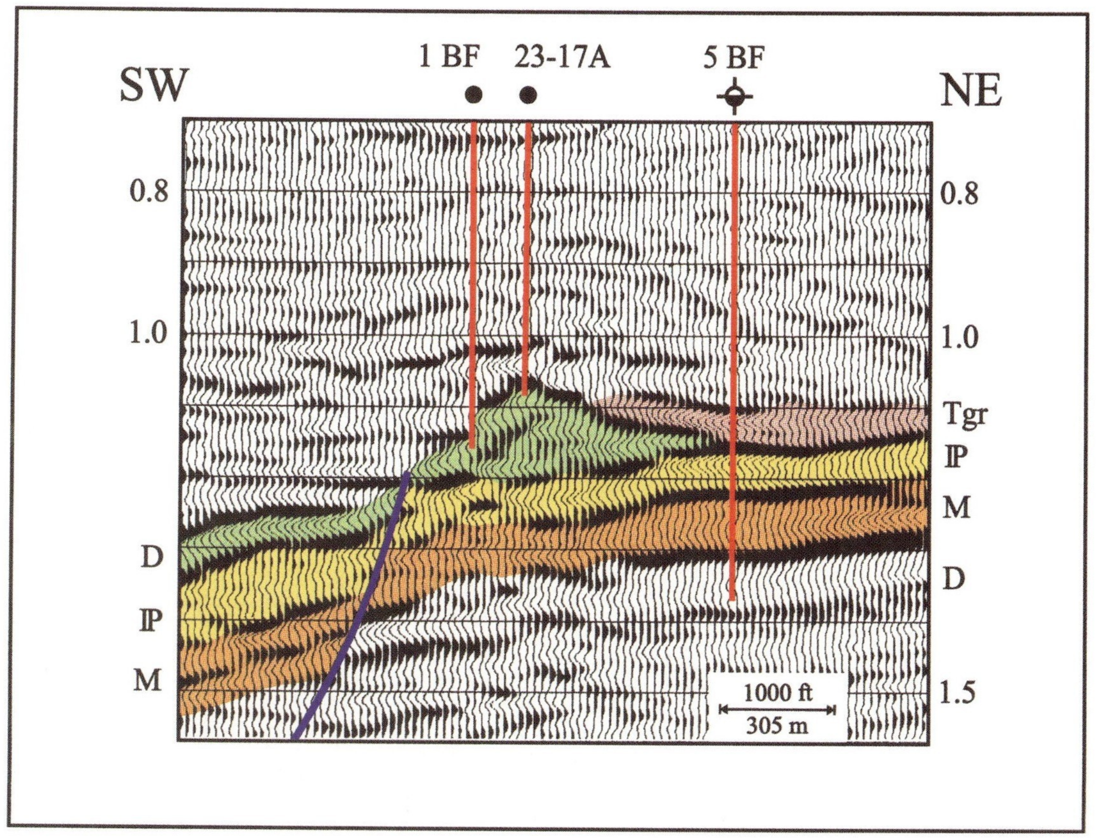


FIGURE 8. Seismic segment through the Bacon Flat field as located in Figure 2. Note the terraced escarpment forming the southwest flank of the field and the distinctly imaged pinchout of the reservoir mound to the northeast. Pennsylvanian and older substrata, penetrated in the 5 BF well, extend beneath the detached Devonian strata of the Bacon Flat field. The 23-17A well postdates the 3-D seismic survey.

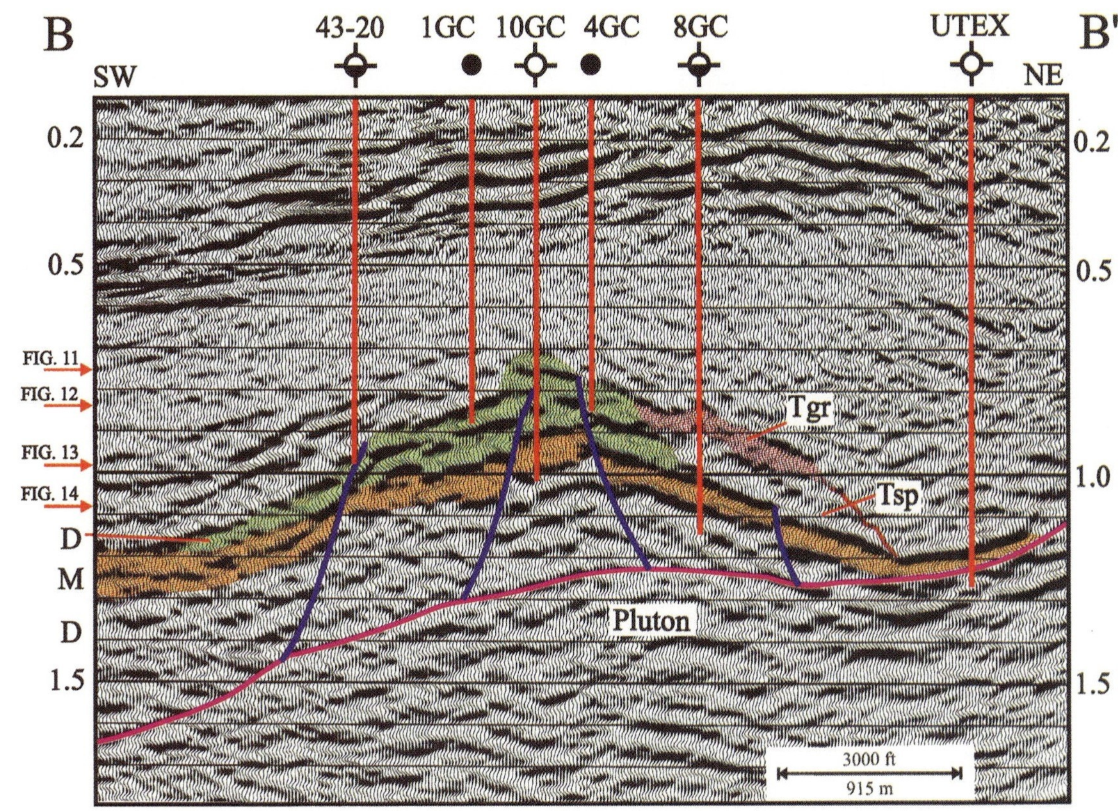


FIGURE 9. Seismic section B-B′ through the Grant Canyon field as located in Figures 2 and 11–14. The detached Devonian strata overlie a flexure formed by Mississippian and older strata. The sub-parallelism of the upper basin fill and the Paleozoic strata forming the southwest flank of the flexure indicate that much of the southwest dip of the Grant Canyon structure is due to a late stage of basin development. Without well control, the structural and stratigraphic changes between the 1 GC, 10 GC, and 8 GC wells would be ambiguous. The 43-20 and 10 GC wells postdate the 3-D seismic survey.

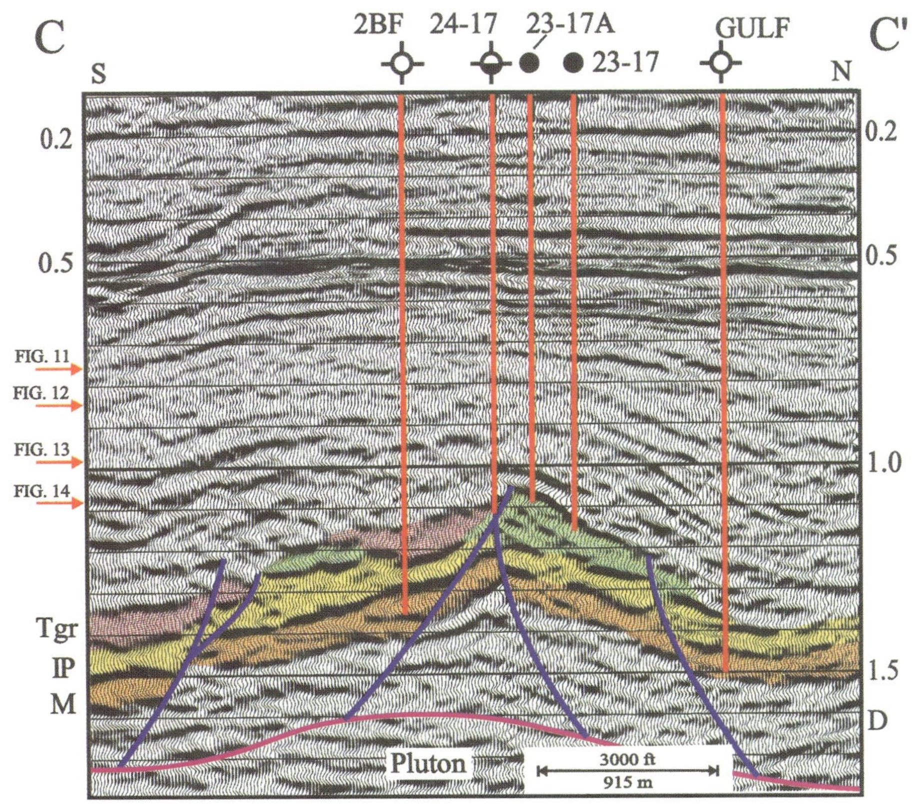


FIGURE 10. Seismic section C-C′ through the Bacon Flat field as located in Figures 2 and 11–14. The detached Devonian reservoir rocks (green) overlie Pennsylvanian and older strata that were subsequently flexed and faulted. Faster velocities in the basin fill south of the Bacon Flat field and slower velocities to the north cause a velocity gradient that has tilted the seismic image of the Paleozoic strata. The Pennsylvanian strata, imaged about 150 ms later in the GULF well compared to the 2 BF well, would appear to be about 750 ft (230 m) structurally lower. Actually, the top of the Pennsylvanian strata is 50 ft (15 m) structurally higher in the Gulf well than in the 2 BF well. The 23-17A well postdates the 3-D seismic survey.

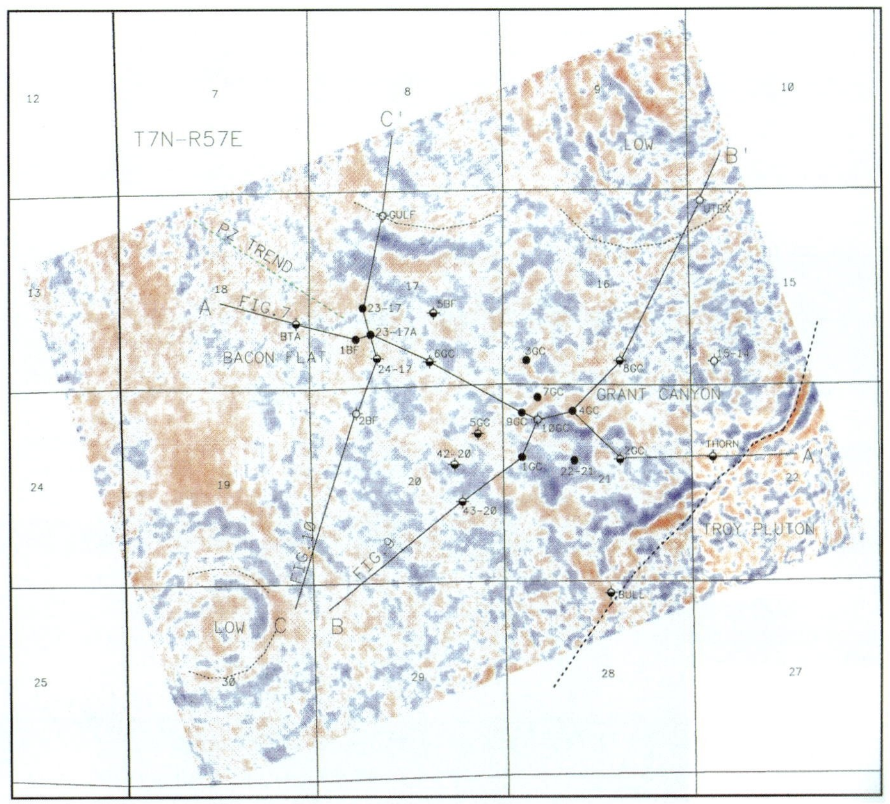


FIGURE 11. Time slice at 0.760 s through the apex of the Grant Canyon field as located in Figures 7, 9, and 10. The seismic peaks are blue and the troughs are red, with the brightest colors depicting the highest amplitudes and white indicating little or no amplitude. The traverses of the seismic sections (Figures 7, 9, and 10) are shown. The circular blue event joining the 1 GC, 22-21, and 10 GC wells in the Grant Canyon field suggests a time closure to the southeast of the structural apex (located by the 9 GC well). Actually, the reflections corresponding to this circular event do not depict the time structure of the Devonian reservoir in either the 1 GC well or the 10 GC well (Figure 9). A depression in the basin fill southwest of the oil fields and a low developing to the northeast of the fields are labeled "LOW." "PZ TREND" locates the trend of the underlying flexure in the Paleozoic strata. The intersection of the time slice with the Troy Pluton is shown by a dashed line. For scale, the section grid is 1 × 1 mi (1.6 × 1.6 km).

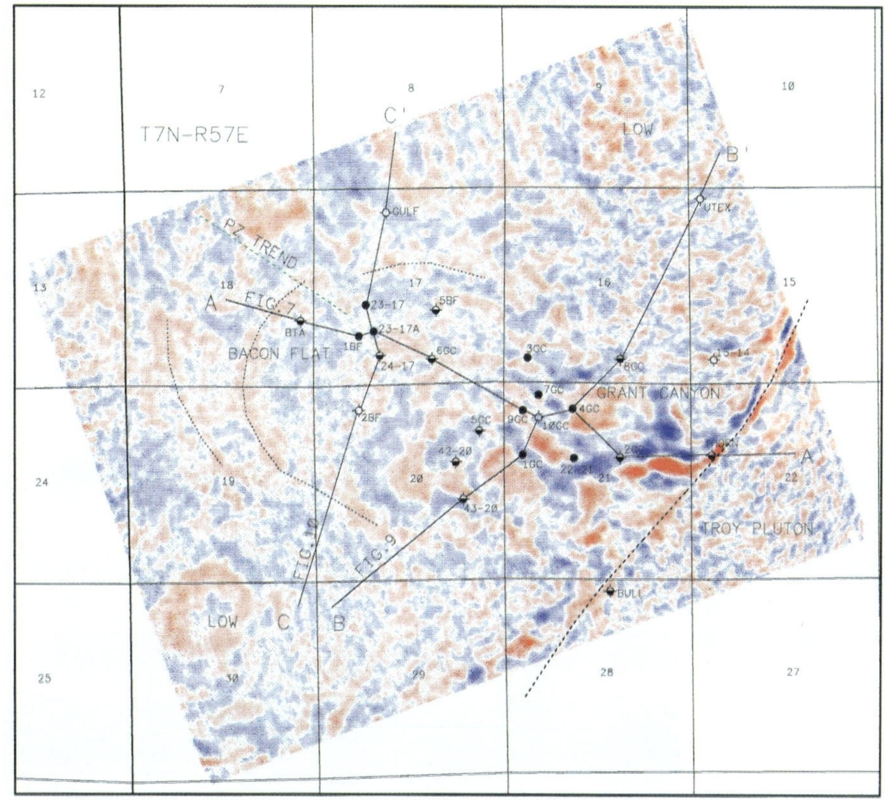


FIGURE 12. Time slice at 0.840 s as located in Figures 7, 9, and 10. The seismic peaks are blue and the troughs are red, with the brightest colors depicting the highest amplitudes and white indicating little or no amplitude. The traverses of the seismic sections (Figures 7, 9, and 10) are shown. The detached carbonate strata in the Grant Canyon field area are delineated by high-amplitude reflections. Those reflections are truncated to the southeast of the field, where the time slice intersects the Troy Pluton, located by a dashed line passing through the THORN well. "PZ TREND" locates the trend of the underlying flexure in the Paleozoic strata. An arcuate pattern in the basin fill above the flexure is highlighted by dashed lines. The structural lows in the basin fill to the southwest and northeast of the oil fields are labeled. For scale, the section grid is 1 × 1 mi (1.6 × 1.6 km).

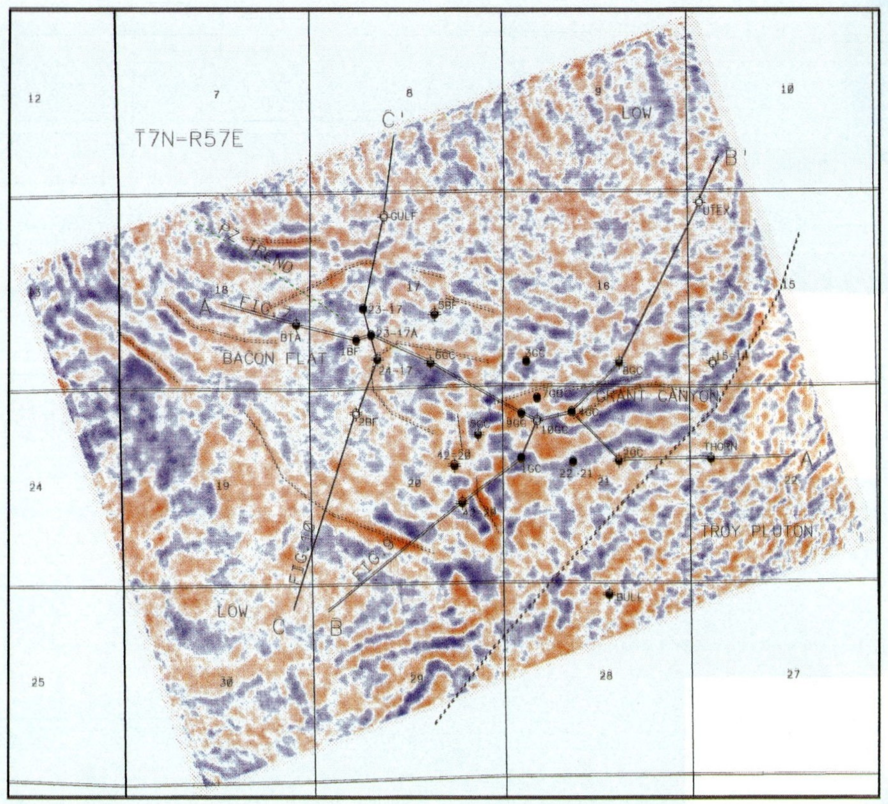


FIGURE 13. Time slice at 0.980 s as located in Figures 7, 9, and 10. The seismic peaks are blue and the troughs are red, with the brightest colors depicting the highest amplitudes and white indicating little or no amplitude. The traverses of the seismic sections (Figures 7, 9, and 10) are shown. "PZ TREND" locates the trend of the plunging flexure in the Paleozoic strata under the basin fill. The time slice has intersected the flexure beneath the Grant Canyon field, resulting in a distinct nose-shaped pattern of high-amplitude reflections. The reflections are truncated to the southeast of the field where the time slice intersects the Troy Pluton (dashed line). The lows to the southwest and northeast of the oil fields are labeled. For scale, the section grid is 1 × 1 mi (1.6 × 1.6 km).

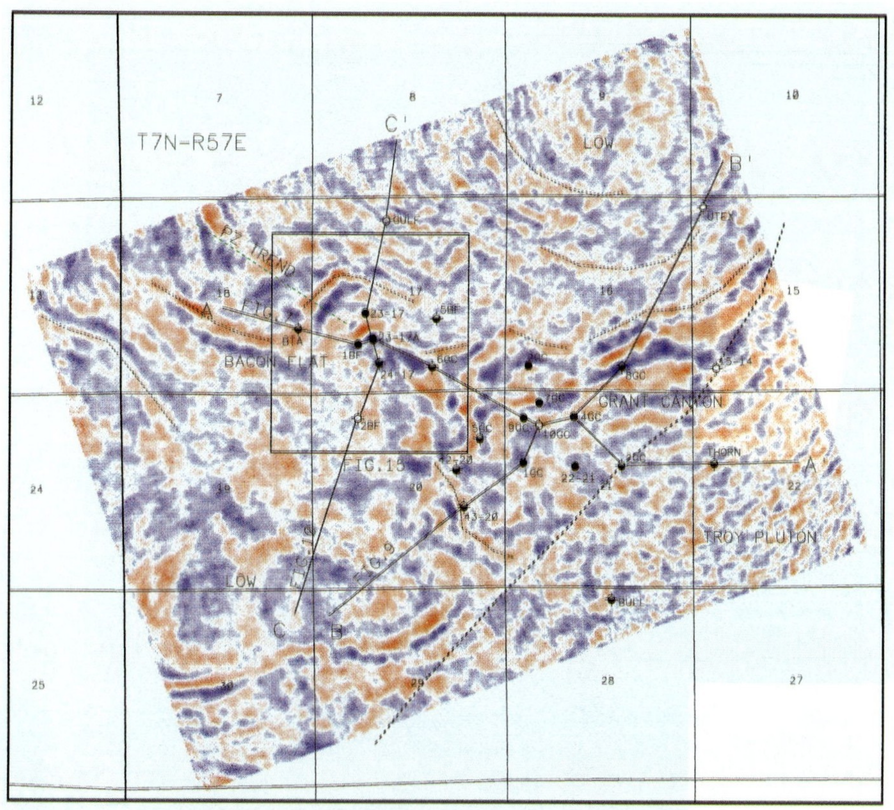


FIGURE 14. Time slice at 1.080 s as located in Figures 7, 9, and 10. The seismic peaks are blue and the troughs are red, with the brightest colors depicting the highest amplitudes and white indicating little or no amplitude. The traverses of the seismic sections (Figures 7, 9, and 10) are shown. "PZ TREND" locates the trend of the plunging flexure of Paleozoic strata beneath the basin fill. The time slice intersects the flexure and also the volcanic strata between the fields (Figure 7) and exhibits a distinct nose of high-amplitude reflections that project from the Grant Canyon field area to the Bacon Flat field. Separation of the Bacon Flat and Grant Canyon field structures is obscured by a velocity gradient in the basin fill and reflections from the volcanic strata. As shown by a dashed line, the time slice intersects the Troy Pluton at the same location as the 2 GC well (Figure 7). The area of Figure 15 is shown by the box. For scale, the section grid is 1 × 1 mi (1.6 × 1.6 km).

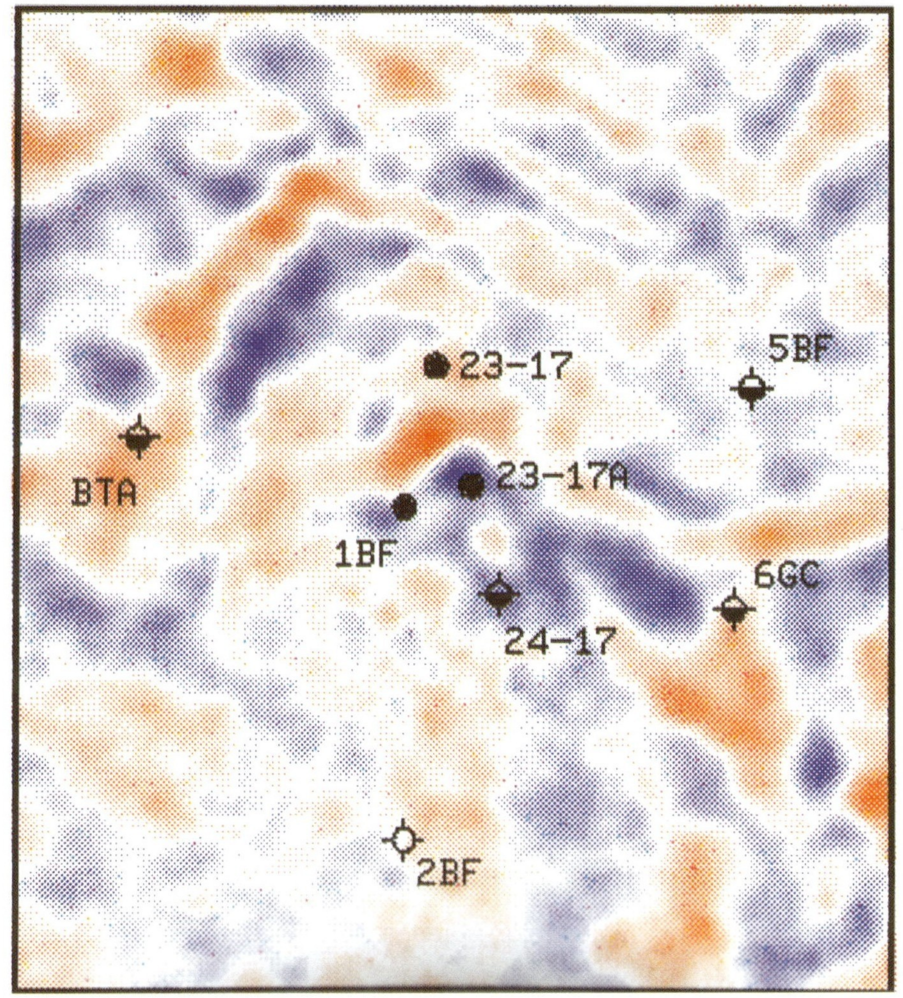


FIGURE 15. Enlargement of the boxed area in Figure 14, showing the Bacon Flat field. Due to a velocity gradient in the basin fill, the apparent apex of the field is skewed to the south of the known apex as located by the 23-17A well (Figures 2, 7, 8, and 10). Closure on the southeast flank of the field is diminished by the velocity gradient and masked by high-amplitude reflections in the volcanic strata in the vicinity of the 6 GC well (Figure 7). For scale, the 24-17 and 6 GC wells are 1500 ft (460 m) apart.

Aminzadeh, F., 3-D salt and overthrust seismic models, *in* P. Weimer and T. L. Davis, eds., *AAPG Studies in Geology No. 42* and *SEG Geophysical Developments Series No. 5*, AAPG/SEG, Tulsa, p. 247–256.

## CHAPTER 28

# 3-D Salt and Overthrust Seismic Models

Fred Aminzadeh*


## ABSTRACT

We have designed two 3-D geologic models and simulated realistic 3-D surveys based on those models. This has been a collaborative effort of the Society of Exploration Geophysicists (SEG), the European Association of Exploration Geophysicists, and the United States Department of Energy National Laboratories. The project is usually referred to as the SEG/EAEG Modeling, or SEM, project. The two salt and overthrust 3-D models and the respective synthetic datasets will be made available to both the oil industry and academia. This SEM project has been a sizable undertaking, with computational requirements of several million dollars. By the end of 1995, about 800 gigabytes of data had been generated, completing Phase A and Phase B of the project. "Classic datasets" containing subsets of calculations conducted for Phases A and B have been defined. Any other subsets of data could be extracted from the whole dataset using a procedure devised at Lawrence Livermore National Laboratory (LLNL). Additional calculations of about three terabytes, Phase C, will be carried out as the project is completed around mid-1997. These datasets have already been used by many for testing and validating processing algorithms and aiding in survey design. Their use as standard 3-D datasets is expected to increase for many years.


## INTRODUCTION

Three-dimensional model data are useful for testing 3-D processing algorithms (e.g., migration, velocity analysis, multiple suppression), understanding wave propagation in a complex 3-D medium, choosing proper acquisition parameters, and testing various data compression and data transmission techniques. They can also be used for training as well as benchmarking different hardware platforms.

Examples of models that have gained widespread popularity for various types of testing and benchmarking are the 3-D French model (French, 1974) and the 2-D Marmousi model (Lailly and Versteeg, 1990). In recent years, advances in computer technology have enabled bigger problems to be modeled. This capability, combined with the ever-increasing complexity of the structures that are being explored and developed, has necessitated construction of more complicated 3-D models.

After a number of surveys of the SEG membership and discussions during various workshops, we have made final decisions on different aspects of the 3-D models. Some of the issues that had to be decided were:

- structural/stratigraphic nature of the model,
- model representation,
- acquisition parameters,
- numerical techniques for generation of seismograms,
- model type (acoustic vs. elastic), and
- computer resources requirements and contributors.

Given the constraints of available computer resources and the immediate needs of the industry, the following compromises had to be made:

(1) Structural problems are the main focus of this modeling effort, although the overthrust model includes some stratigraphic features (channel sand).
(2) Geological models are represented using University of Nancy's GOCAD Project software.
(3) Acquisition parameters are decided in different phases of the project based on the available resources.
(4) Marine acquisition geometry is used for both models to avoid land topography problems.
(5) The finite difference method is used for calculations.
(6) The calculations are based on an acoustic medium.
(7) Computer resources of the U.S. National Laboratories are used to construct the model responses.

Two acoustic models were chosen to carry out the simulation: a salt dome model with features similar to those of the U.S. Gulf of Mexico and an overthrust model that was a 3-D extension of the earlier 2-D Marmousi model. Working groups were formed to finalize separate details of the project, such as (1) construction of the Overthrust Model, (2) construction of the Salt Dome Model, and (3) selection and porting of the finite differences codes to different platforms and execution of the numerical computations.

Other working groups have been formed to help finalize the survey parameters and to develop a scheme for storing and distributing large volumes of the numerical calculations. For other background information and more details about the evolution of this project, refer to Aminzadeh et al. (1994a, 1994b, 1995, 1996) and Kunz et al. (1995).

## SALT DOME MODEL

The design of the salt model was accomplished through the participation of 27 representatives (experienced in salt tectonics, seismic modeling, and seismic imaging) from 23 organizations representing oil companies, seismic contractors, and academia. It was agreed by these representatives to base the salt model on a typical Gulf Coast salt structure. Special care was taken to ensure both its geological reasonability and its adequacy as a testing mechanism for seismic imaging algorithms (particularly subsalt).

Figure 1 shows the 3-D perspective of the salt model with the salt sill, different faults, sand bodies, and lenses. The salt model exhibits the following major features: (1) a northwestwardly plunging stock; (2) a secondary reactivation crest, located southward of the stock; (3) a broad, low-relief eastern flank; (4) a faulted southern flank with minor toe thrust; and (5) a rounded overhang on the west flank. The northern flank is bounded by a counterregional fault, with attendant downthrown rollover structure in the outboard section. Two down-to-the-basin faults occur beneath the salt sill; one offsets the salt base. The salt edge crosscuts stratigraphy at a low angle and varies in shape around the sill perimeter. The eastern flank exhibits a mildly downturned edge. The southern flank exhibits an upturned edge; the western flank has a rounded edge; and the northeastern edge of the salt contains a salt weld. The overall model dimensions are 13.5 × 13.5 × 4.2 km in the *x*, *y*, and *z* directions, respectively. The maximum sill length is approximately 9.5 km, and the maximum sill width is approximately 6.6 km. The salt crest is at approximately 325 m.

The model contains five sands, some of which are gas charged. At least one sand will contain both a gas/oil and an oil/water contact. The two shallowest sands truncate against the sill base. One of the intermediate sands is stratigraphically isolated, and the other conforms to the subsalt anticline. The model contains a shale sheath that is modeled to be geopressured. The seafloor (bathymetry) map exhibits a counterregional fault scarp and a bathymetric rise associated with the sill crest. The map also exhibits a shelf break at the southeast end of the model.

This model is designed in part to be able to test various imaging algorithms in different geologic settings: salt flank, salt overhang, and subsalt. The velocity model is generated using a 20-m grid with 15-Hz center frequency for the wavelet. The velocities surrounding the salt body are typical of Gulf of Mexico sediments and are described by compaction gradients based on $(K, V_0)$ curves ($V_0$ is the initial velocity, and $K$ is the velocity gradient that varies spatially) and a geopressure surface. The following expression is used to define the 3-D velocity field:

$$V(x,y,z) = V_0 + z\,K(x,y) - GP, \qquad (1)$$

where GP is the deceleration term for velocities below the geopressure surface. The water and salt velocities and GP are chosen to be 1500, 4481, and 46 respectively. Figure 2 shows another 3-D display of the model with two perpendicular 2-D velocity profiles. Three parallel cross sections from the three dimensional velocity grid of Figure 2 are depicted in Figure 3. Color velocity scales to the right of Figure 3 are in meters per second, with blue indicating the high salt velocities. Green color indicates relatively lower sediment velocities of the overpressure zones.

We make a constant-density assumption due to the limitations of the finited difference code. The model also contains five point diffractors and a flat horizon above the model basement, which may be

*Unocal Corporation, 14141, SW FWY, Sugar Land, TX 77487

used as a means of evaluating the quality of seismic imaging algorithms. Since the constant density assumption and the simple velocity gradient structure given by Equation 1 result in no reflectivity in the sediment surrounding the salt, we use the "spike" method to create reflections from geologic boundaries. This method increases the velocity of the function for each successive layer as shown in Figure 4. As shown in Figure 5, the synthetic seismogram of the AA′ cross section of Figure 3 exhibits proper reflections from geologic boundaries using the spike velocity profile of Figure 4.

## OVERTHRUST MODEL

A complex thrusted sedimentary succession has been constructed on top of a structurally decoupled extensional and rift basement block. The thrust unit has been built from a collection of 2-D balanced cross sections. Figure 6 shows a 3-D perspective of the overthrust model. The basement (not shown) underneath the overthrust structure includes an extensional fault block system, with a sediment cover on the basement blocks. Above this, there is an erosional truncation, and there is a structural uncoupling between basal and uppermost units. A petroleum trap has been introduced between the two reverse faults in the heavily folded thrust. Also observe the sand channel included in the west side of the model. Figure 7 shows the original 2-D cross sections that were used in constructing the 3-D overthrust model. The model represents two converging thrusts plus an additional blind thrust which dies out laterally. The faults in the overthrust and the basement have been validated by the Fault Analysis Group of the University of Liverpool. The model is overlain by a flat sedimentary layer underneath the seabed. Figure 8 shows various cross sections of the overthrust model. The model has dimensions of 20 × 20 × 3.5 km, giving a zone of maximum seismic coverage of 10 × 10 × 3.5 km. In total, it has 17 layers. The model presents varying degrees of complexity, with a central thrust-faulted anticline and an external monocline and flat zone. The top of the overthrust is eroded and covered by a surface layer to represent recent sediments. Channels and crevasse splay lenses are present in some layers. The layer thickness is preserved, and the percentage shortening at any point across the thrust structure is approximately constant. Fault planes have been checked by 3-D visualization to ensure smoothness and regularity. Figure 9 shows the process by which fault rifts were built by the fault analysis group of the University of Liverpool.

The basement of the overthrust model is composed of alternating layers of carbonate and siliciclastic sedimentary rocks which drape tilted, extensional fault blocks. There is a petroleum trap in the tilted strata draping the middle block. The upper boundary of this extensional unit is an erosional truncation. This layer is in turn overlain by a sedimentary succession in which an initial salt layer is overlain by predominantly marine sedimentary stratigraphy composed of two major carbonate successions and siliciclastic and argillaceous sediments. The salt layer acts as decollement surface for two thrust planes which deform the upper sediment column. The lower of the two thrusts converges laterally with the upper plane. The layers in the area bounded by the faults are folded into an anticline with a fold axis that varies laterally to create a closed petroleum trap in the sandstone layer. The folds contain reverse limbs at the point of minimum structural depth. There is a bulge in the footwall of the thrust sequences that dies out laterally. The thrust layers are capped by a thin sedimentary veneer and a sea layer. Several constructive criticisms on the model have been provided by various companies, the major recommendations being:

- The model should be an onshore rather than an offshore structure, with substantial topography and a complex weathering zone.
- The model should include even more 3-D properties.
- The velocity distribution should be constructed from an existing database.
- Some features (imbrication of thin beds, lenses, pinchouts, etc.) should be included to address the problem of seismic resolution.

While these points are all valid, the choices made were mostly constrained by the available computational resources and the limitations of the code chosen based on various other considerations.

## COMPUTATIONS AND THE 3-D FINITE DIFFERENCE CODE

We concluded early on that 3-D structural problems should be the primary focus of this effort. Because extensive wave propagation computations through an elastic medium in a survey-size setting is still beyond reach of today's technologies, a full-sized seismic survey will therefore be limited to acoustic computations only.

Several 3-D finite difference (FD) codes have been provided by various organizations for acoustic wave propagation computations in the 3-D model. The Computation Working Group conducted extensive evaluations of these candidate codes. The FD 3-D wave propagation code provided by the Institut Francais du Petrole (IFP), with a second-order scheme in time and a tenth-order scheme in space, was chosen for the entire simulation. The time steps are 1 ms, while the space steps ($x$, $y$, $z$) are 25 m. The parallel version of this code is implemented on different supercomputers: Paragon, Meiko CS-2, Cray Research c-916, Connection Machine CM-5, and Fujitsu VP 2400. The entire calculation is divided into segments to be carried out on various machines located in different laboratories. For this reason, extensive work was done to ensure that different platforms would yield the same results.

These calculations are done under the Department of Energy (DOE) Gas and Oil National Information Infrastructure (GONII) project. Four national laboratories—Lawrence Livermore, Los Alamos, Oak Ridge, and Sandia—are contributing to these efforts. Aside from computation, resources and technologies available at the national laboratories, such as networking, high-speed I/O, mass data storage, collaborative tools, distributed computing, visualization, and computer security, are being used to increase the utility of this project to the oil industry.

## SALT CALCULATION

The Phase A acquisition program for the salt model contains two perpendicular lines which span the model with a shot spacing of 80 m. Each of these lines is a broad-line swath across the model. Figure 10 shows the orientation of these lines and the receiver grid pattern for shot line 1. Each line has 138 shots. In addition, two downhole shots were calculated at depth locations of 900 and 2300 m at a vertical well located at the intersection of the two lines. All Phase A calculations (total of 278 shots) have been completed. The data have been stored on the on-line storage system at LLNL.

The Phase B acquisition program for the salt model contains 35 shot lines which span the model with a shot spacing of 320 m. All of the lines are broad-line swaths across the model. In Figure 11, the large square with the coarse grids displays the acquisition geometry. In total, Phase B represents a coarse source spacing 3-D survey across the entire model. The two wells in the salt model are included whenever the wells are within the computational grid. All Phase B calculations have been completed, and the data now reside on the on-line storage system at LLNL.

Phase C plans for the salt model are expected to increase number of total shots by a factor of four. The acquisition for Phase C consists of infilling Phase B shots within a selected area of the model so the resulting shots have an 80-m shot spacing inline and a 160-m shot spacing crossline (Figure 11). The selected area contains 24 shot lines, each containing 25 shots from phase B. To infill these shots to an 80-m spacing inline will require an additional 1728 shots. To reduce the crossline shot spacing to 160 m will require an additional 25 shot lines. Each of these new shot lines is offset 40 m in the inline direction from the Phase B lines to simulate a typical marine acquisition. Phase C will use 97 shots from Phase A, 600 shots from Phase B, and an additional 4103 shots that still need to be acquired. Figure 11 shows the two crossing shot lines from Phase A, in yellow and green, and the coarse shots from Phase B. The Phase C infill shot lines are displayed in yellow, and the new shot lines for Phase C are shown in purple. To be consistent with shots previously acquired, the new shots for Phase C will be acquired using the 279 × 279 receiver geometry used in Phases A and B. As can be seen in Figure 11, Phase C provides dense 3-D shot coverage over the stem, crest, and toe rollers of the salt.

## OVERTHRUST CALCULATION

Overthrust calculations were also done in two phases, A and B. Plans for additional calculations are underway. Figure 12 shows the acquisition geometry for Phases A and B and two components of Phase C (C1 and C2).

Phase A (539 shots) was done along the shot line on Figure 12. Figures 13 and 14 show a sample 3-D shot over the overthrust model and the respective time slices of the wave propagation through the model. Phase A results allow comparisons between 2-D and 3-D processing for dip and strike lines. In addition, data corresponding to different well acquisitions are recorded to study well-to-surface seismic ties in a 3-D complex area. Phase A consists mainly of two lines :

- a dip line, 15.55 km long, with a shot interval of 50 m, and
- a strike line, covering 10 km, with an intershot spacing of 50 m.

In addition, data corresponding to various well acquisitions are recorded to study well-to-surface seismic tie in a 3-D complex area.

- A set of 52 wells is located along the dip line to record normal-incidence VSPs. The first well is located at X = 700 m; the wells are spaced 300 m apart. The receivers are located from top to bottom (4000 m) in each well with a vertical spacing of 50 m.

The objective is to help the interpretation of the surface seismic events.
- A large 2-D walk-away (VSP) is recorded during the computation of the dip line shots.

Phase B (1767 shots) corresponds to a coarse source spacing 3-D acquisition over an area of 15 × 10.2 km (Figure 12). These data cover most of the structures of the model, in particular the anticline thrust limited by the two main faults. This phase consists of a set of shot lines in the dip direction, with a line spacing equal to 600 m.

Phase C (1620 shots) contains two parts. Phase C1 is a dense 3-D survey over a small portion of the model with a shot spacing of 100 m in both the $x$ and $y$ directions. Phase C1 covers the intersection of the two main faults of the Overthrust Model. Phase C2 completes the Phase B acquisition pattern and leads to a more realistic simulated land 3-D survey with 100 m between shots and 600 m between shot lines over the whole area of 15 × 10.2 km surveyed in Phase A and Phase B of the acquisition program. See Figure 12 for details.

## LLNL'S FAST STORAGE FACILITY FOR THE SEM DATASET

Lawrence Livermore National Laboratory (LLNL) has developed tools for accessing more than one terabyte of modeling data. This SEM dataset (SEMD) resides at LLNL's FAST storage facility. Because of the massive size of the modeling project data, tools for easily extracting useful subsets of data are crucial. Four distinct methods for doing these extractions are being pursued: classic datasets, web-based forms, an X-Windows–based tool, and an SEG-P1 parser.

The first three methods will be accessed via the World-Wide Web (WWW). The easiest option for the user will be to download directly from a set of predefined "classic" datasets extracted from the SEMD. These classic datasets, which are interesting subsets of data culled from the project data, are defined next.

For users wanting to build their own subsets of the SEMD, a Website for selecting the model, lines, and seismic traces of interest is provided. Once the selection has been made, the requested data will automatically be extracted from the FAST storage system and sent to the requester in the desired format on the Internet. Interactive searching, browsing, and downloading of the project data can be accomplished using the SEMD home page, http://WildCat.ocf.llnl.gov/SSD/ on the World-Wide Web.

A more sophisticated X-based graphical interface to SEMD extraction will also be provided. This tool will allow the user to set up a virtual seismic survey on either the salt or the overthrust model data. The user will be able to control all aspects of the shot and receiver layout. The result of this operation will be a reasonable sized collection of traces, which will be automatically extracted from the appropriate shot files in the storage facility.

The final method for extracting data from the modeling project data will be a program which uses the shot and receiver information from a user-provided SEG-P1 file to drive the data extraction process. The SEG-P1 file is an industry standard file which contains seismic survey details. This method allows investigators to use their favorite survey tools to define the shot and receiver locations. The same low-level data extraction tools used by the X-based graphical interface described earlier will be employed by this method for creating the subsets. Figure 15 shows a menu for selecting the desired subsets of SEMD.

## CLASSIC DATASETS

To provide industry and academia with manageable, consistent, and useful subsets of the SEM dataset, several classic datasets (CDS) have been defined. These datasets will be made available upon request. The following are the CDS for the salt model. Similar CDSs are being defined for the overthrust model. In the near future, access to these datasets will be provided through the Internet.

### PHASE A CLASSIC DATASET 0–CDS A0

This is composed of a raw shot gather at grid node 384,441 with a square of 279 × 279 receivers. In addition, it includes a simulation of 3-D plane wave source (zero-offset data). The sample interval is 8 ms; the recording time is 5 s. The total dataset size is approximately 193 mb. This location is at the intersection of Phase 1, Line 1 and Phase 2, Line 2 at the crest of the salt.

### PHASE A CLASSIC DATASET 1–CDS A1

This is Line 1 (the yellow line on Figure 10) with 138 shots simulating a six-streamer marine acquisition with 65 groups per streamer. The group interval is 40 m, near offset is 160 m, and far offset is 2760 m. The sample interval is 8 ms and the recording time 5 s. The dataset size is approximately 133 mb. CDS A2 is the same as CDS A1, except it contains the blue line on Figure 10, perpendicular to line 1, as described earlier.

### PHASE B CLASSIC DATASET 1–CDS B1

This dataset includes selected shots from Phase 2. It contains 156 shots of streamer marine acquisition with 65 groups per streamer. The group interval is 40 m, the near offset is 160 m, and the far offset is 2760 m. The sample interval is 8 ms and the recording time 5 s. The dataset is approximately 150 mb.

Other classic datasets for specific applications (e.g., velocity analysis) are under development.

## SOME PRELIMINARY APPLICATIONS

The following important preliminary experimentation has already been done with these models.

### PSEUDO-VSP GENERATION

An overthrust model shot gather has been used to test a new idea for generating a pseudo-VSP section by Alai et al. (1995). This tool has proven to be useful in understanding the nature (multiple or primary) and the origination point of different events on a seismic trace. Figure 16A shows a dip section of the velocity volume of the SEM model where the vertical line depicts the location of the gather, or "well." Figure 16B shows a slice through the migrated volume that images the sand channel. Using the velocity profile of Figure 16C and a slice of the shot record of Figure 16E, the pseudo-VSP section shown in Figure 16D is generated. This section helps establish the relationship between events in the original gather and migrated section.

### CAN WE IMAGE BELOW SALT?

O'Brien and Gray (1996) uses the AA' 2-D velocity profile shown in Figure 3 and the corresponding "exploding reflector" zero-offset section of Figure 5 to answer this question. They conclude that the SEM data provide the necessary challenge for improving the imaging software. In addition, other processing issues, such as multiple elimination techniques, can be tested using these SEM models. For example, Figure 17 shows O'Brian and Gray's final imaging result after elimination of surface multiples and limiting the range of the data to 0–8000 ft (2438 m).

### GULF OF MEXICO SUBSALT SEISMIC IMAGING PROJECT

The Gulf of Mexico Subsalt Seismic Imaging Project was initiated under the Cooperative Research and Development Agreement (CRADA) LA94C1051, between the DOE and several oil companies. Under this project, an exploding reflector modeling experiment has been performed using the SEMD and the data have been migrated using the salt model velocity grid of SEM. Figures 18 and 19 show some of the migration results. Figure 18 shows the match between the original model surfaces and the migrated events. In this figure, we see a thin sand lens below salt and only the top portion of a second sand layer, also below salt. In Figure 19 we see the event imprint of two thin lenses on a depth section. These results are encouraging in that we see that it is possible to image events below salt in the SEM.

## CONCLUSIONS

The SEM project goal has been to design two 3-D models, salt and overthrust, and simulate realistic 3-D surveys over these complicated structures through numerical calculations. The datasets generated through this project in four national DOE laboratories will be made available to industry and academia. Many groups have already begun using and testing the datasets for various applications. In addition, a new umbrella Advanced Computational Technology Initiative (ACTI) project has been initiated to enhance the value of the SEM effort by developing and testing new modeling, processing, and analysis tools using the results of SEM as its basis. See House et al. (1996) for more detail. The strong support and involvement of industry at different stages of the SEM project has been a key to its success up to this point. We expect increased utilization of these datasets over the next several years, especially as Phase C data become available.

*Acknowledgments*

This has been a truly collaborative effort with contributions from the oil industry, academia, and national laboratories. I would like to specially thank Tim Kunz of Amoco, Kay Wyatt of Phillips Petroleum, Fabien Bosquet of Gocad, Laurence Niccoliti and Pierre Duclos of IFP, Norm Burkhard of Lawrence Livermore National Laboratory, and Fabio Rocca of Politecnico di Milano who have made significant contributions during the course of this project.

## REFERENCES CITED


Alai, R., W. E. A. Rietveld, C. P. A. Wapenar, and A. J. Berkhout, 1995, From seismic surface measurements to pseudo VSP data: a new tool in 3-D seismic interpretation: Proceedings of SEG-Rio Conference, August.

Aminzadeh, F., F. Rocca, K. Wyatt, P. Lailly, and N. Burkhard, N. , 1994a, Progress report from the SEG/EAEG 3-D Modeling Committee: The Leading Edge of Geophysics, v. 13, n. 2, p. 105–109.

Aminzadeh, F., N. Burkhard, L. Nicoletis, F. Rocca, and K. Wyatt, 1994b, SEG/EAEG modeling committee project: 2nd report: The Leading Edge of Geophysics, v. 13, n. 9, p. 137–145.

Aminzadeh, F., N. Burkhard, T. Kunz, L. Nicoletis, and F. Rocca, 1995, 3-D modeling project: 3rd report, The Leading Edge of Geophysics, v. 14, n. 2, p. 125–128.

Aminzadeh, F. N. Burkhard, J. Long, T. Kunz, and P. Duclos, 1996, Three dimensional SEG/EAEG models, an update: The Leading Edge of Geophysics, v. 15, n. 2, p. 131–134.

French, W. S., 1974, Two-dimensional and three-dimensional migration of model experiment reflection profiles: Geophysics, v. 39, p. 265–277.

House, L., M. Fehler, F. Aminzadeh, J. Barhen, and S. Larsen, 1996, A national laboratory-industry collaboration to use SEG/EAEG model data sets,: The Leading Edge of Geophysics, v. 15, n. 2, p. 135–136.

Kunz, T., F. Bosquet, and F. Aminzadeh, 1995, Constructing and visualizing the SEG/EAEG salt models: Proceedings of Archie Conference on Visualization Technology to Find More Oil and Gas, May 14–18, 1995, The Woodlands, TX.

Lailly, P., and R. Versteeg, 1990, Marmousi model, Proceedings of the EAEG Research Workshop.

O'Brien, M. J., and S. H. Gray, 1996, Can we image beneath salt?: The Leading Edge of Geophysics, v. 15, n. 1, p. 17–22.

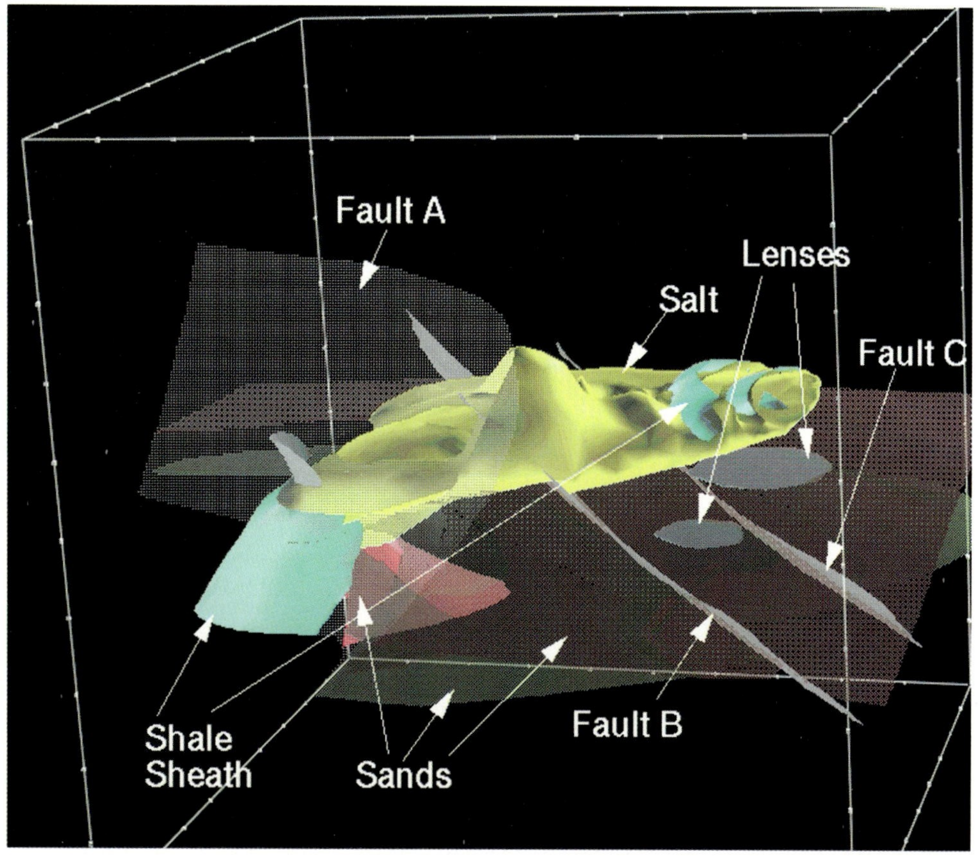


FIGURE 1. The 3-D perspective of the salt model.

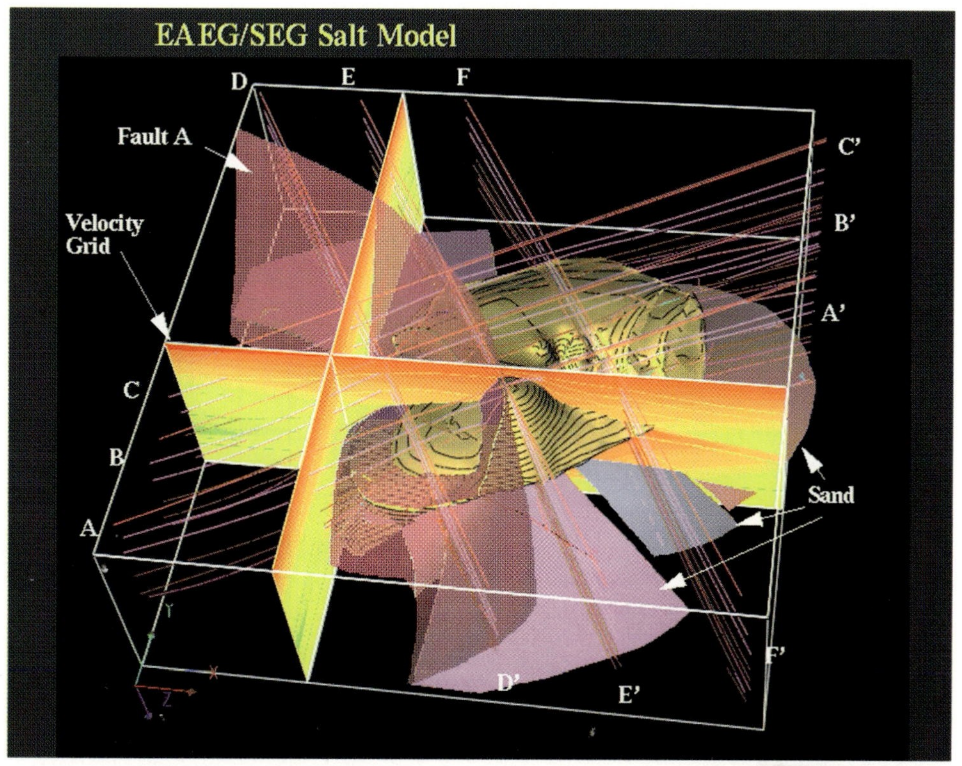


FIGURE 2. The salt model with two velocity profiles.

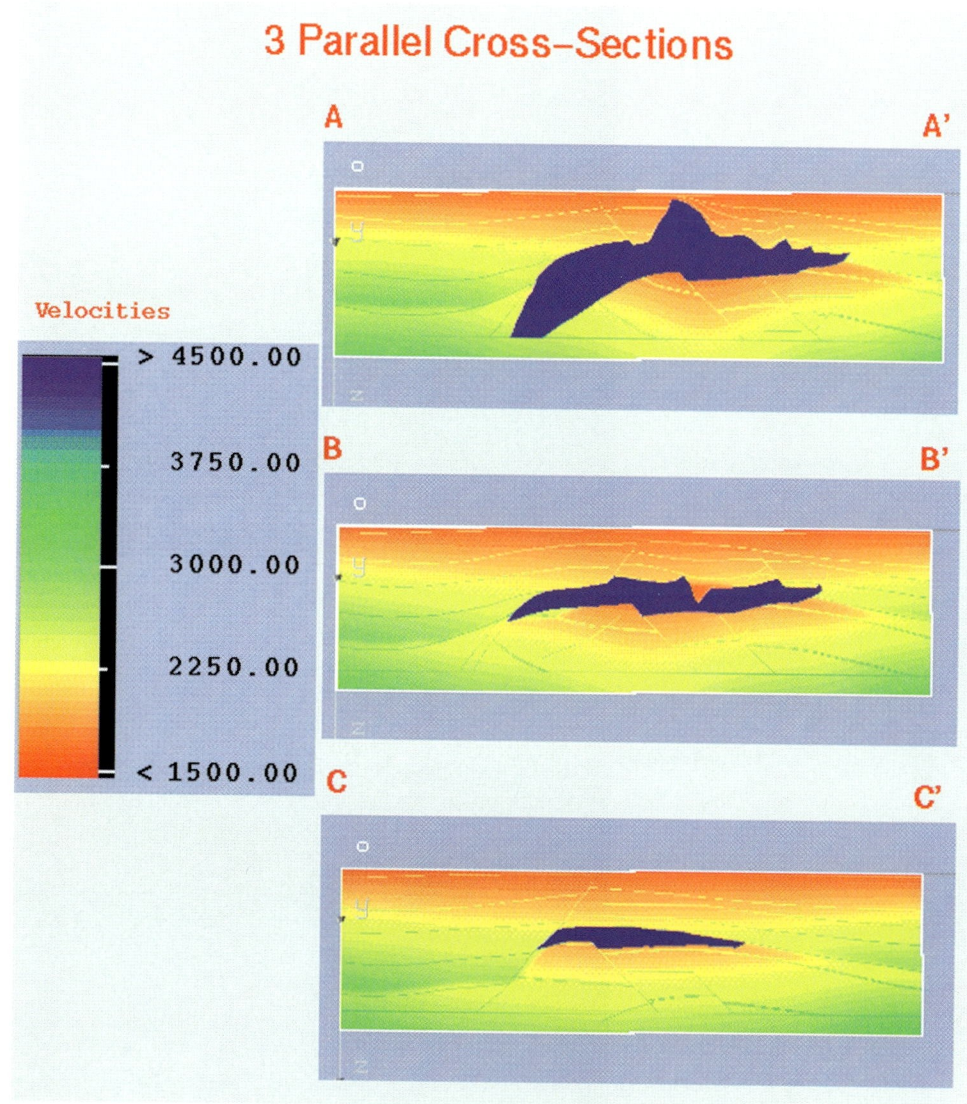


FIGURE 3. Velocity profiles on three parallel cross sections.

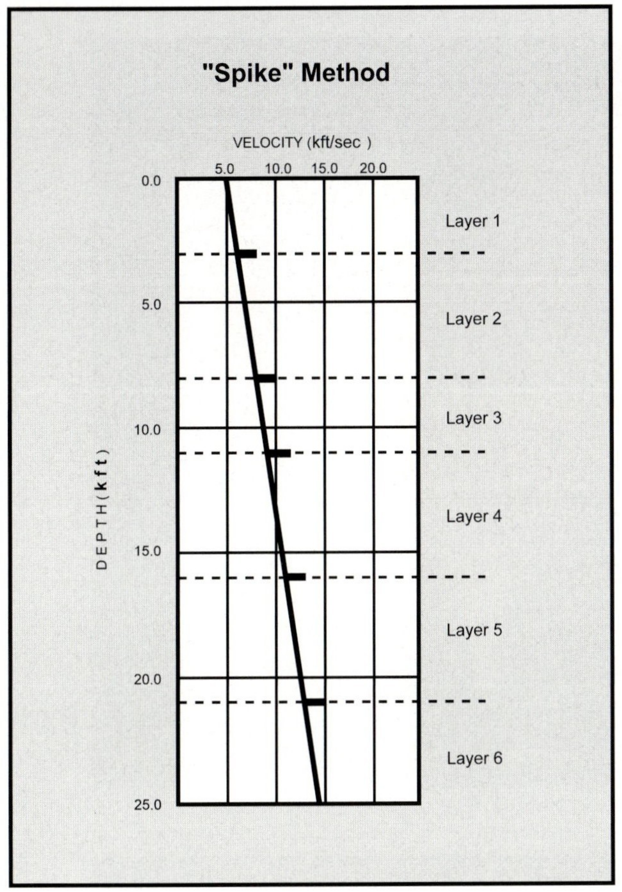


FIGURE 4. Velocity profile using the "spike" method.

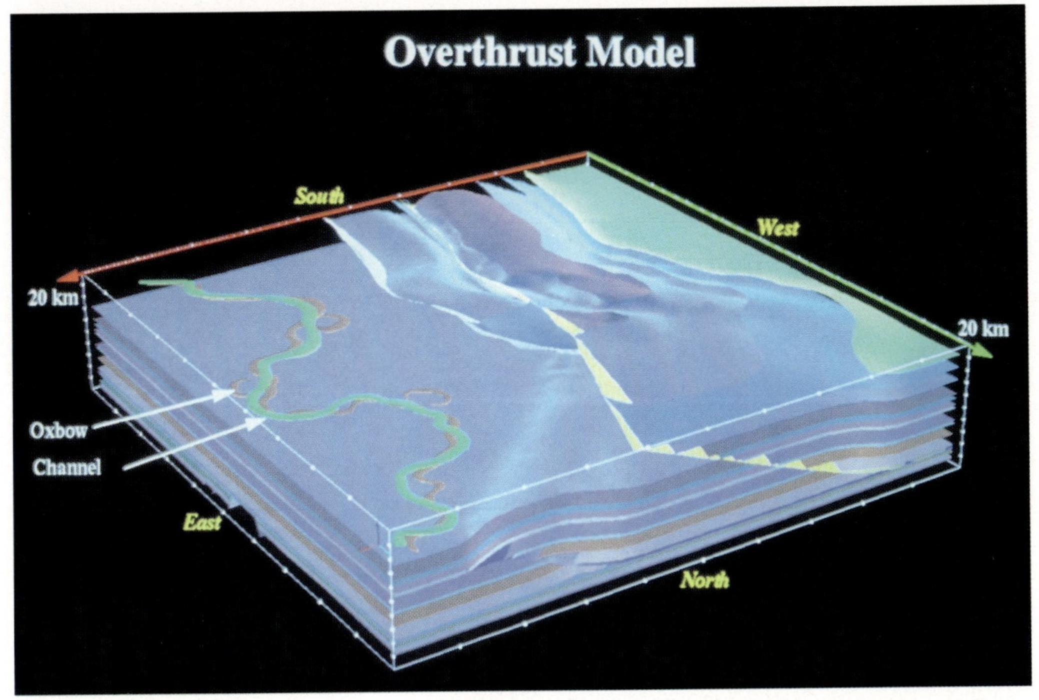


FIGURE 6. The 3-D perspective of the overthrust model. Figure reproduced from 35 mm slide.

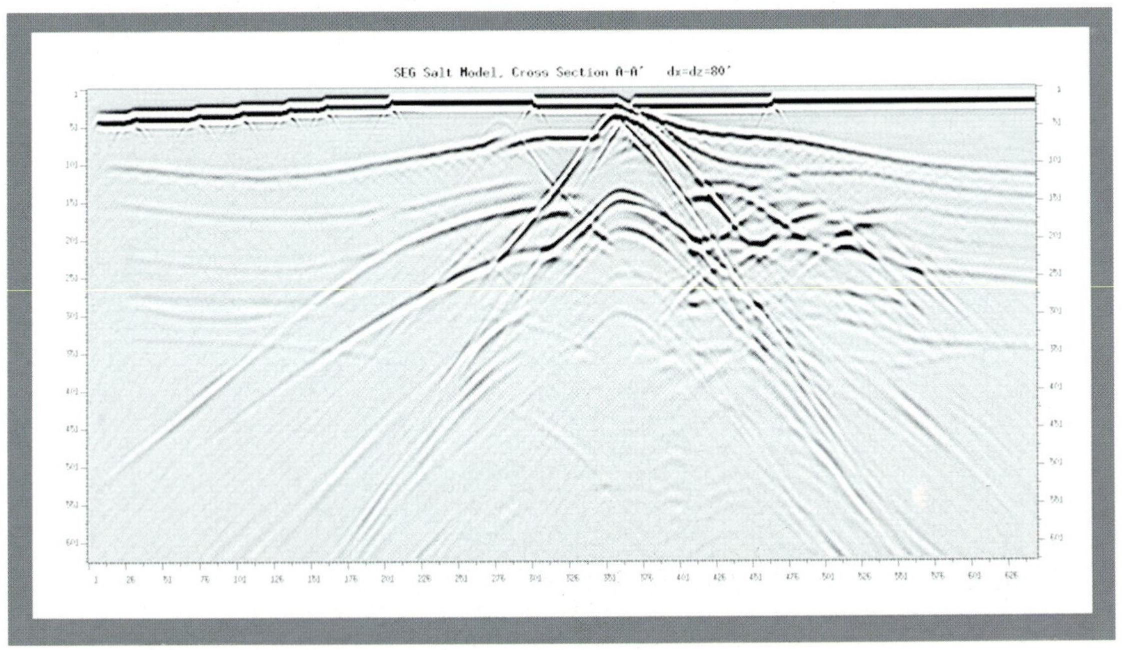

FIGURE 5. 2-D synthetic seismogram of the AA' cross section.

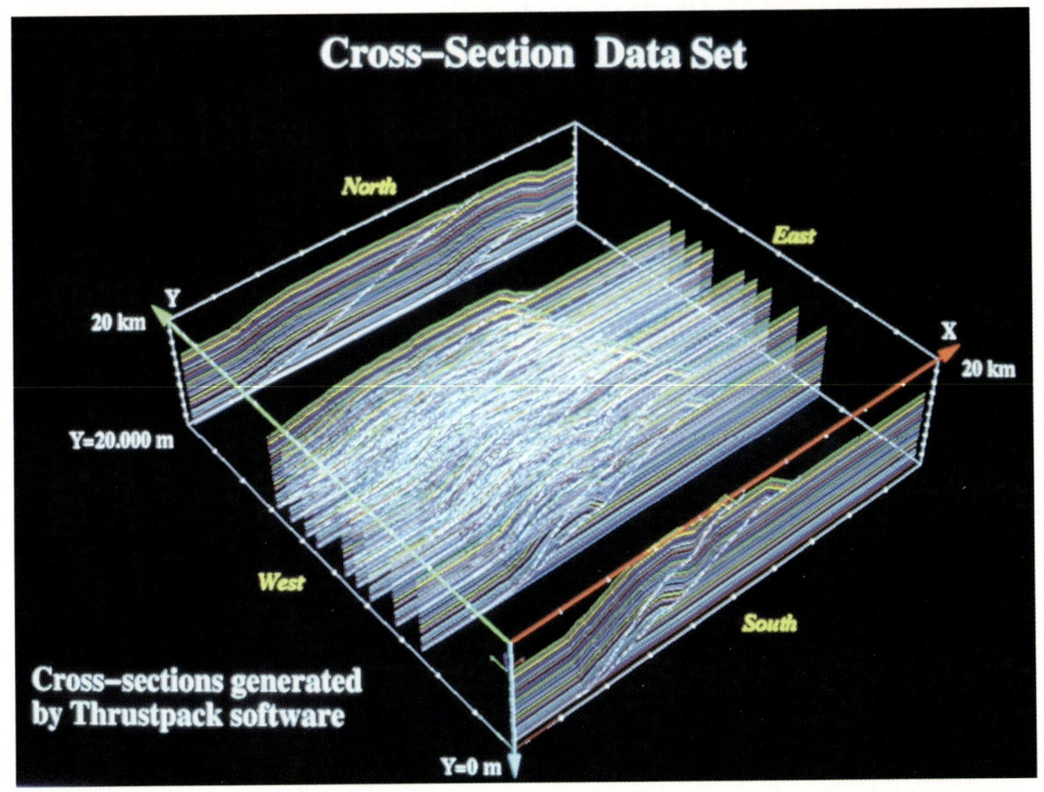


FIGURE 7. Cross section dataset used to generate the 3-D overthrust model. Figure reproduced from 35 mm slide.

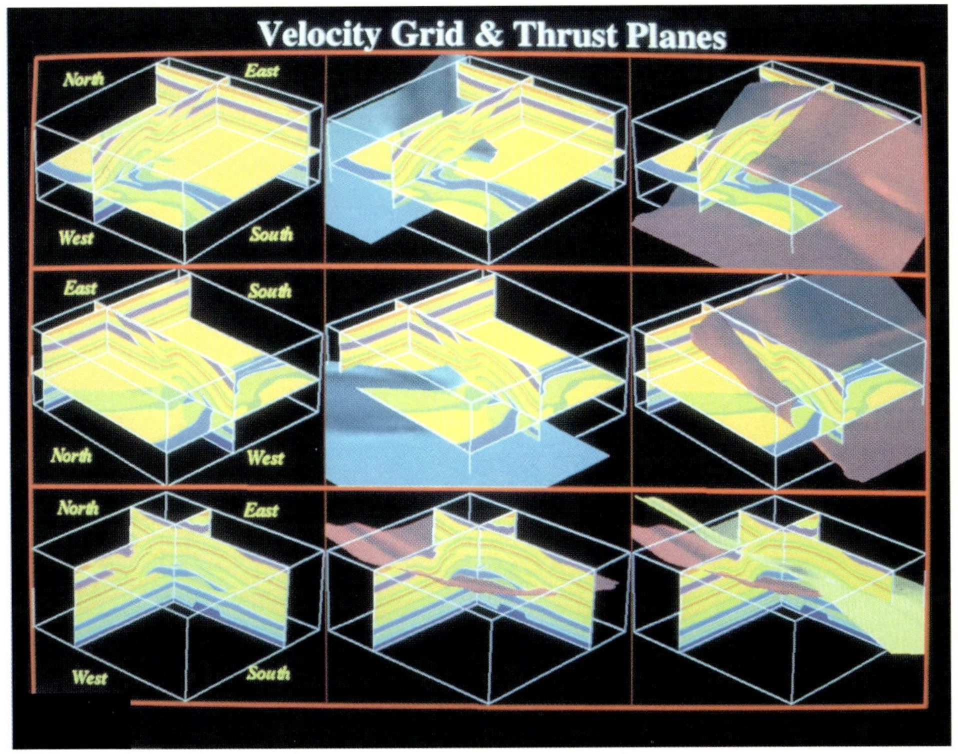


FIGURE 8. 2-D slices through the overthrust model, showing the respective positions of the thrust planes. Figure reproduced from 35 mm slide.

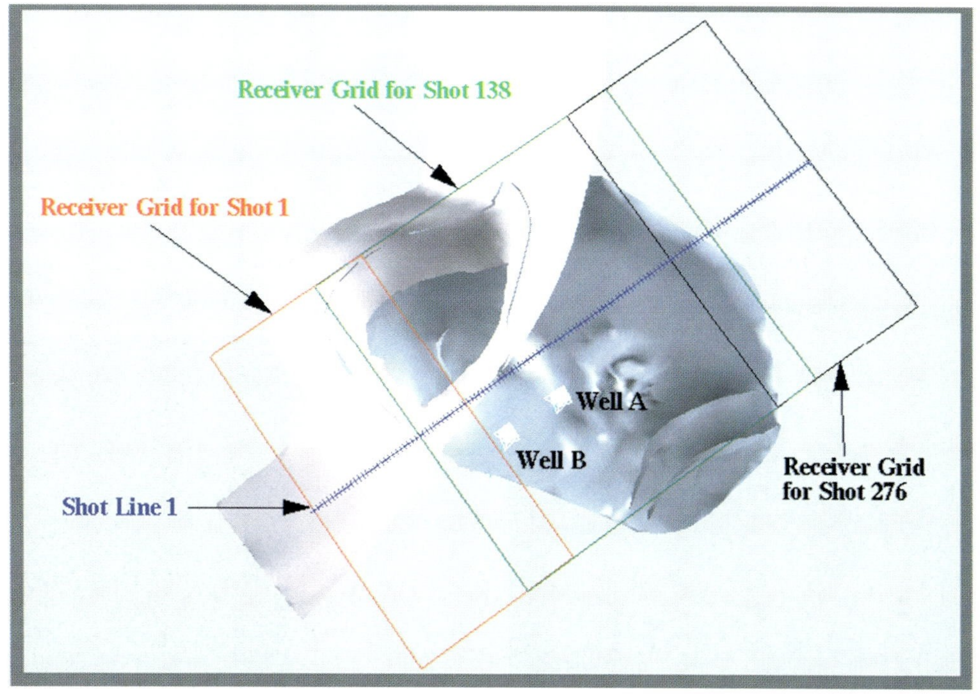


FIGURE 10. Orientation of two lines for phase A calculations.

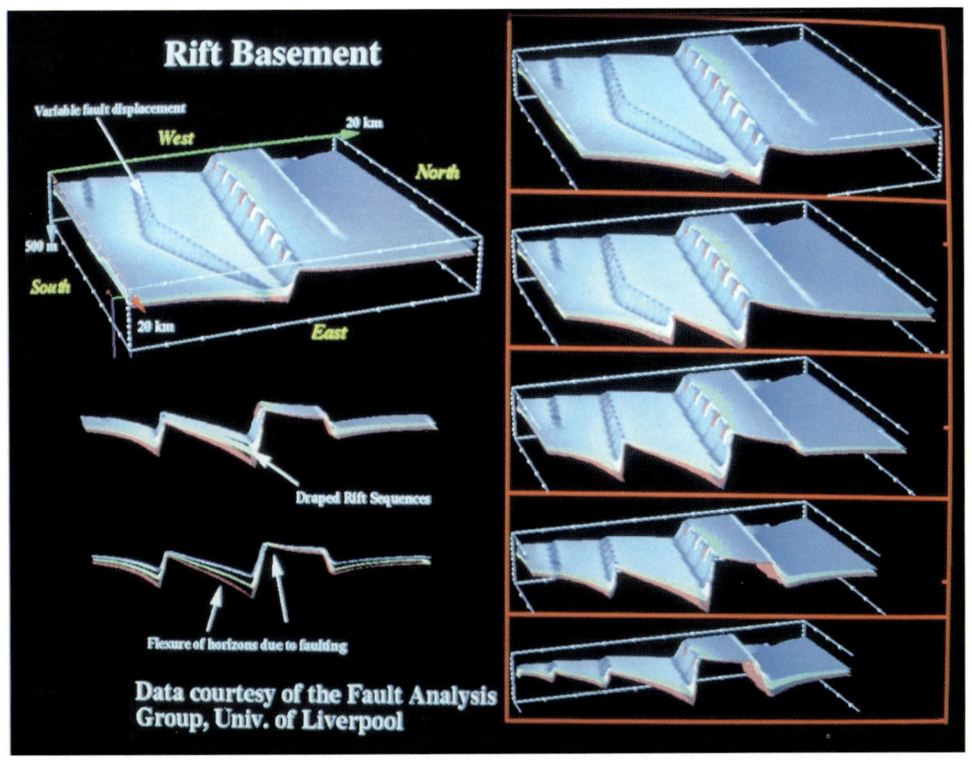


FIGURE 9. Building of the fault planes. (Courtesy of fault analysis group, University of Liverpool.) Figure reproduced from 35 mm slide.

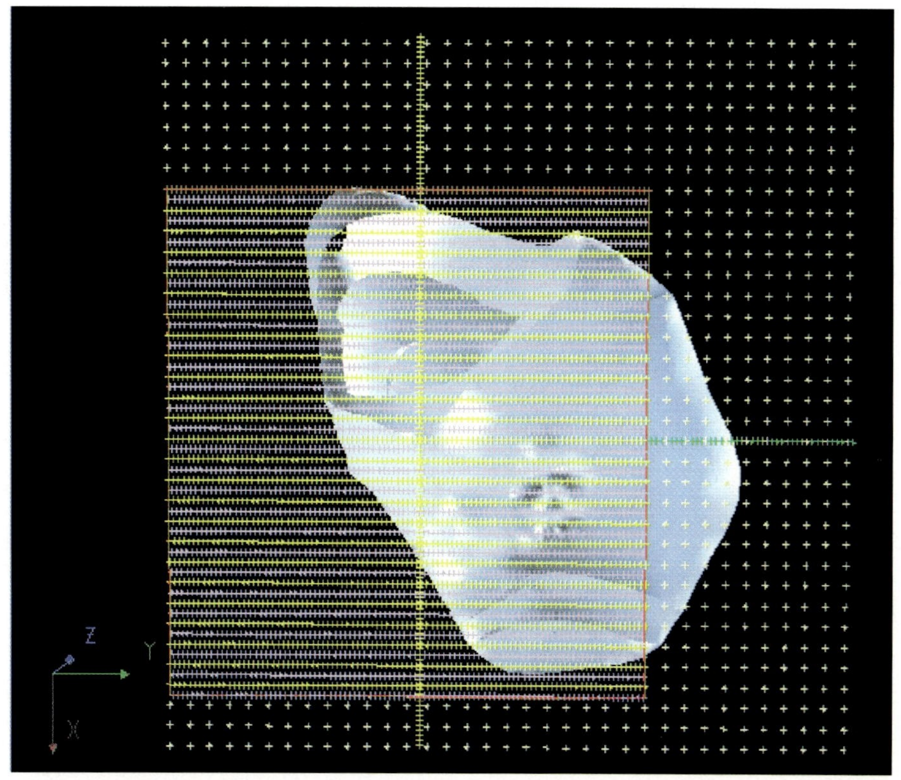


FIGURE 11. Phase B and Phase C acquisition with an overlay of phase A lines.

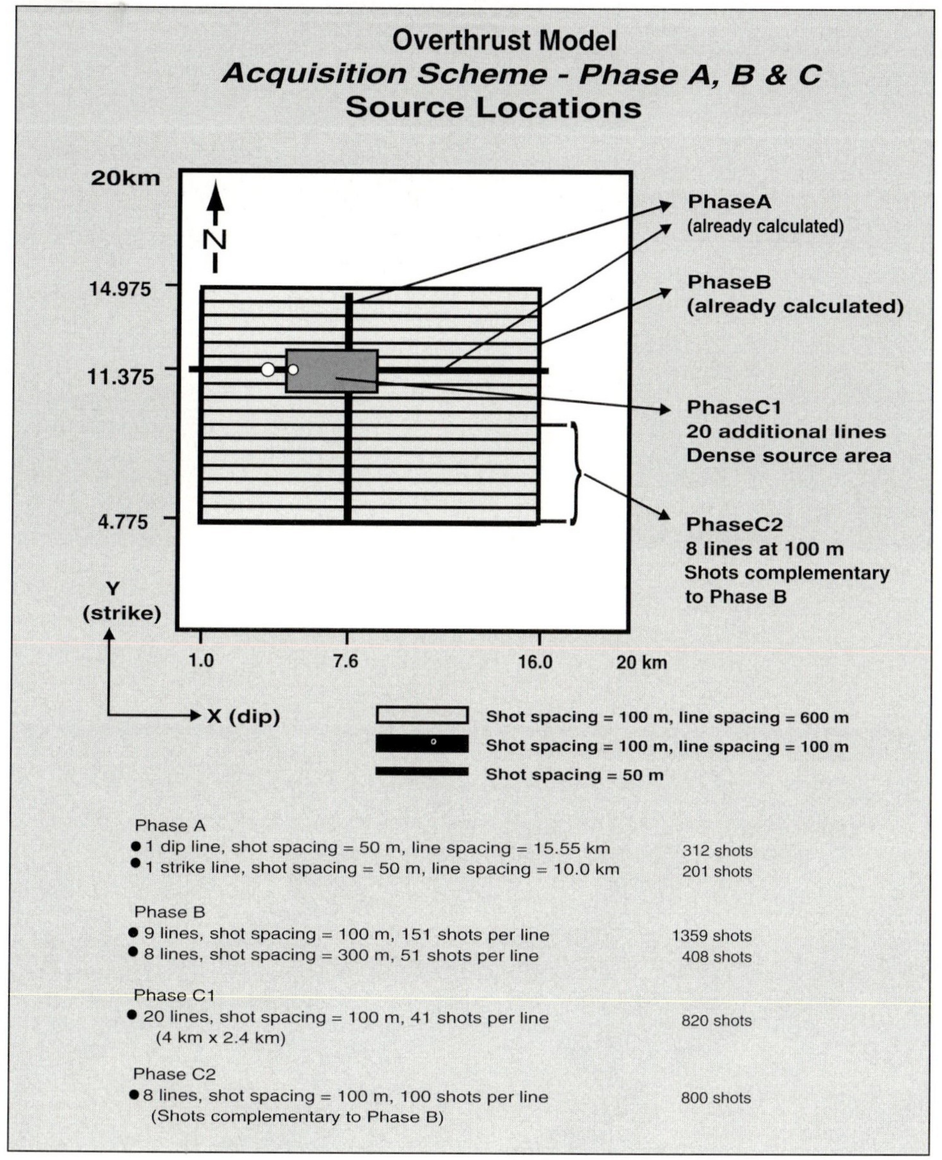


FIGURE 12. Overthrust acquisition for phases A, B, and C.

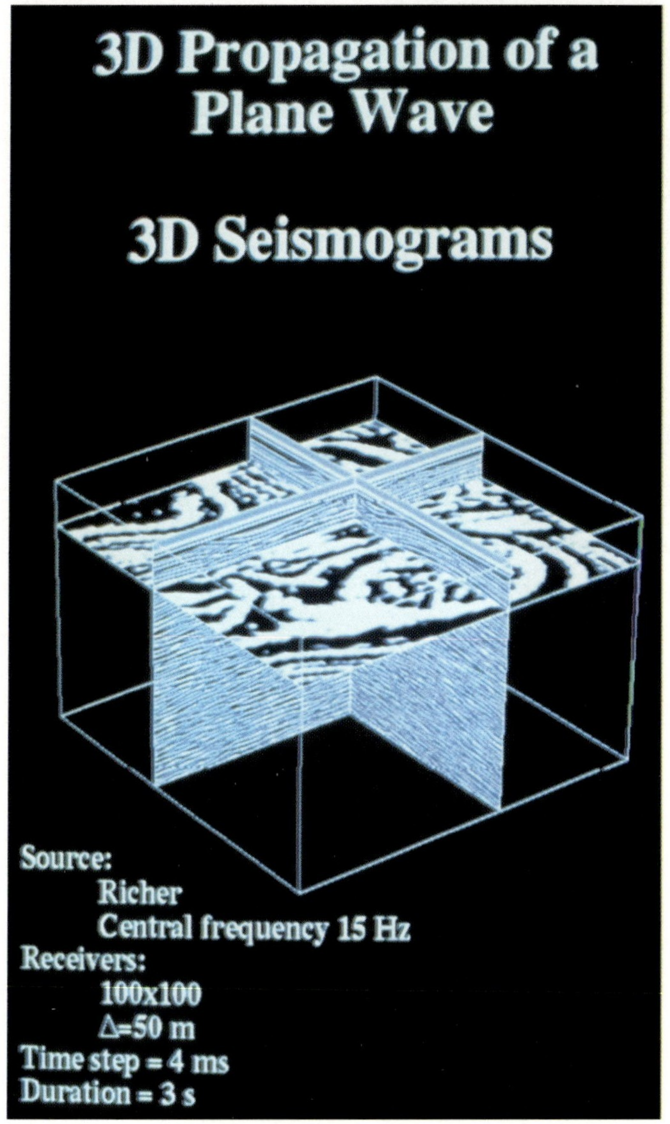


FIGURE 13. Sample 3-D shot for overthrust model. Figure reproduced from 35 mm slide.

FIGURE 14. Time slices of wave propagation through the overthrust model. Figure reproduced from 35 mm slide.

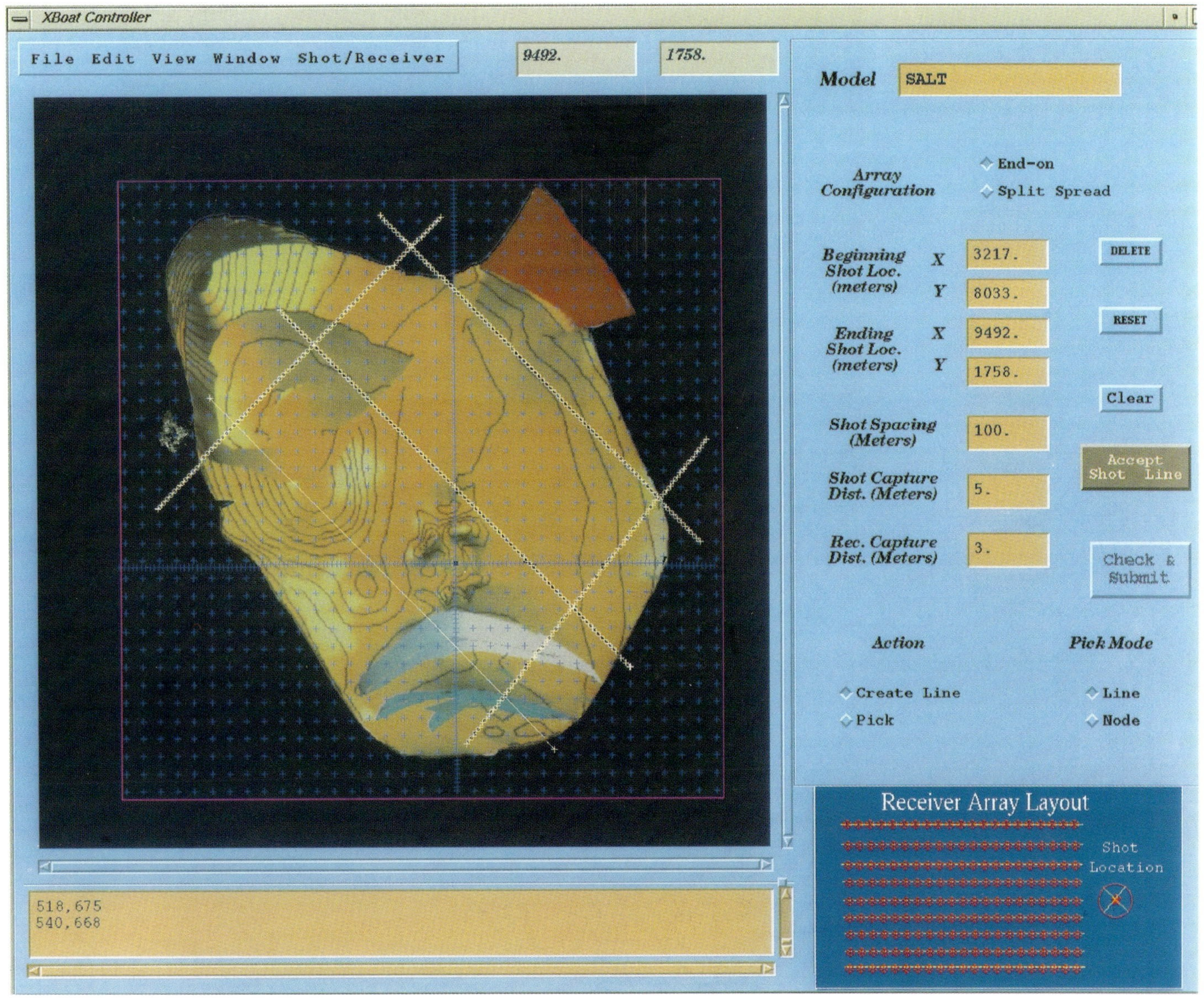


FIGURE 15. A display of the menu-driven program to retrieve desired subsets of the SEMD.

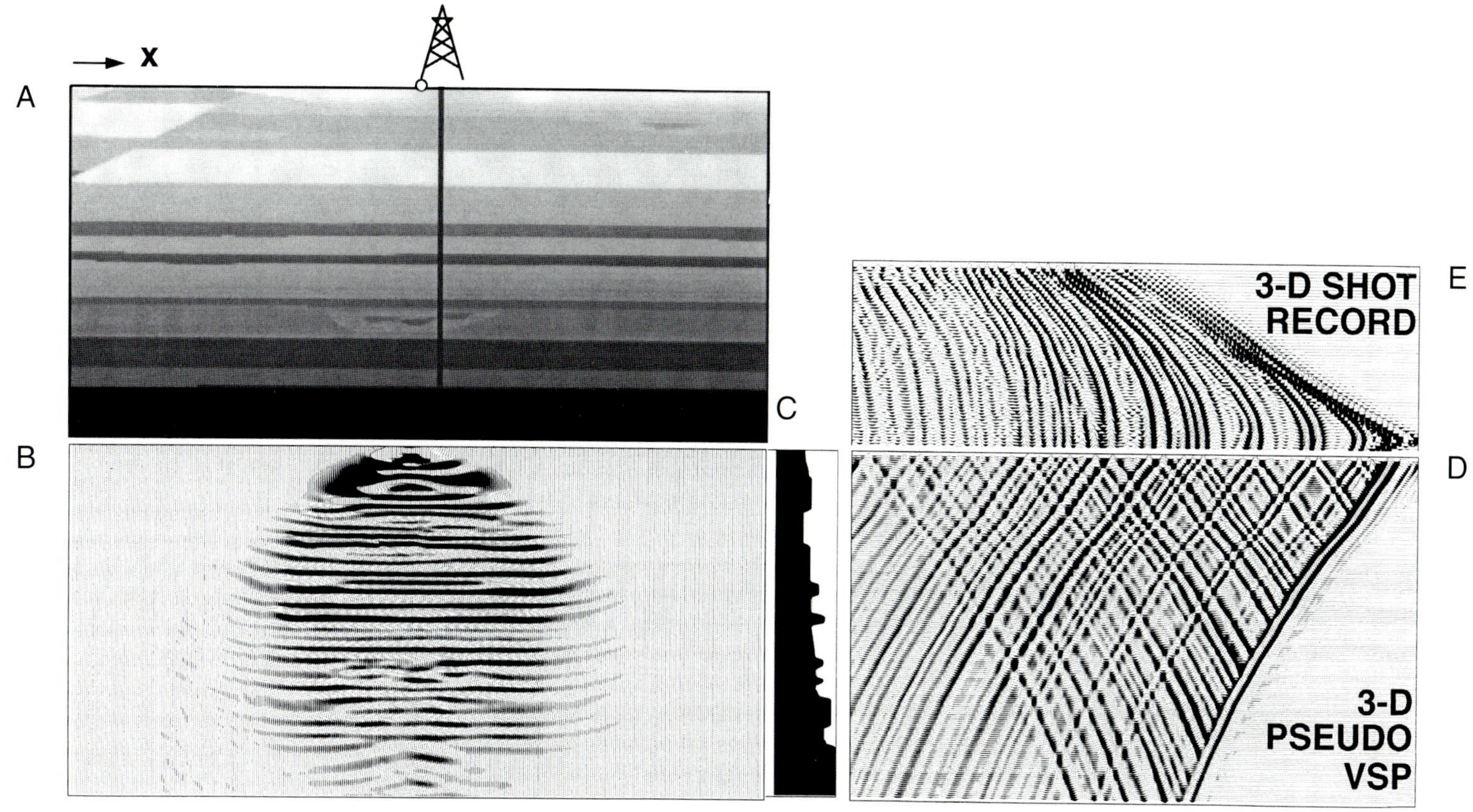


FIGURE 16. Generation of pseudo VSP sections: (A) A 2-D cross section of the overthrust model. (B) 3-D migrated shot record. (C) velocity profile at well location. (D) 3-D pseudo-VSP. (E) 3-D shot record. (From Alai et al., 1995).

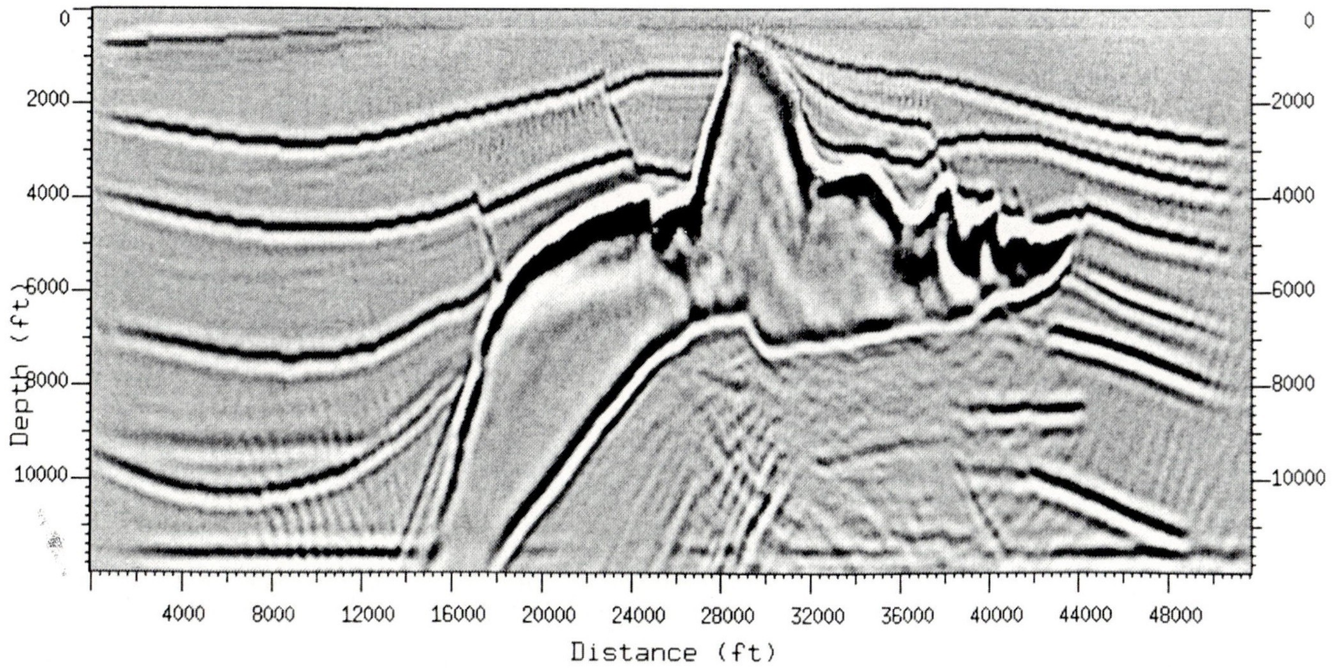


FIGURE 17. Imaging below the salt body after elimination of multiples (from O'Brien and Gray, 1996).

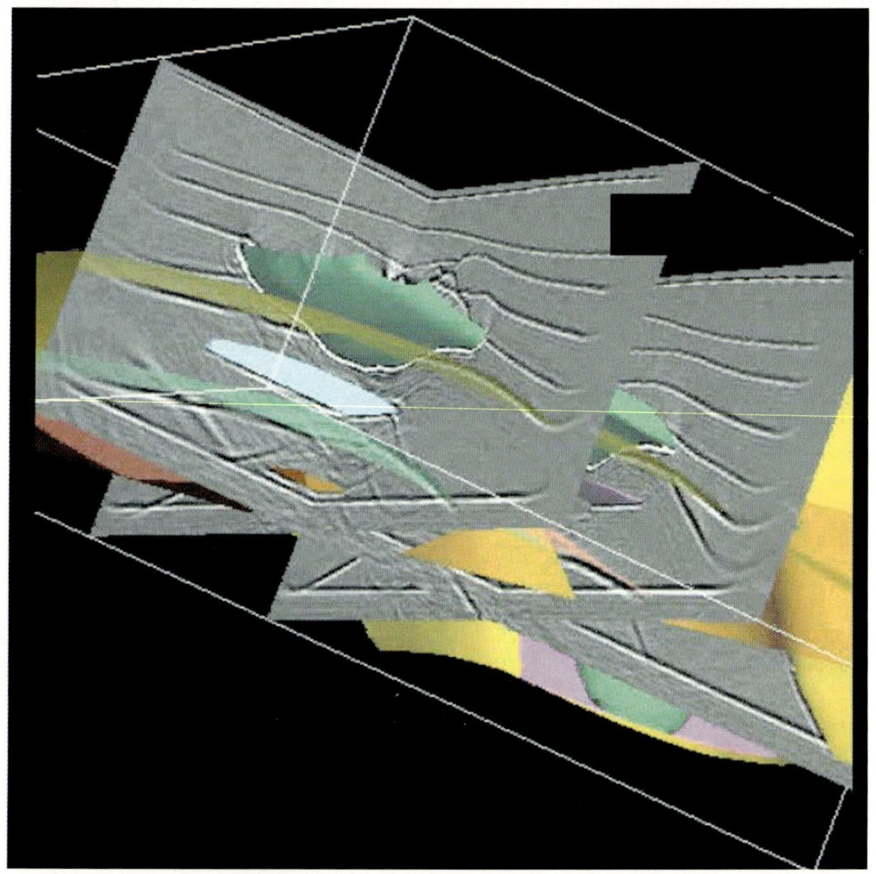

FIGURE 18. 3-D migrated volume overlaid on the 3-D geologic model. (Courtesy of Gulf of Mexico Subsalt Seismic Imaging Project, CRADA LA94C1051.)

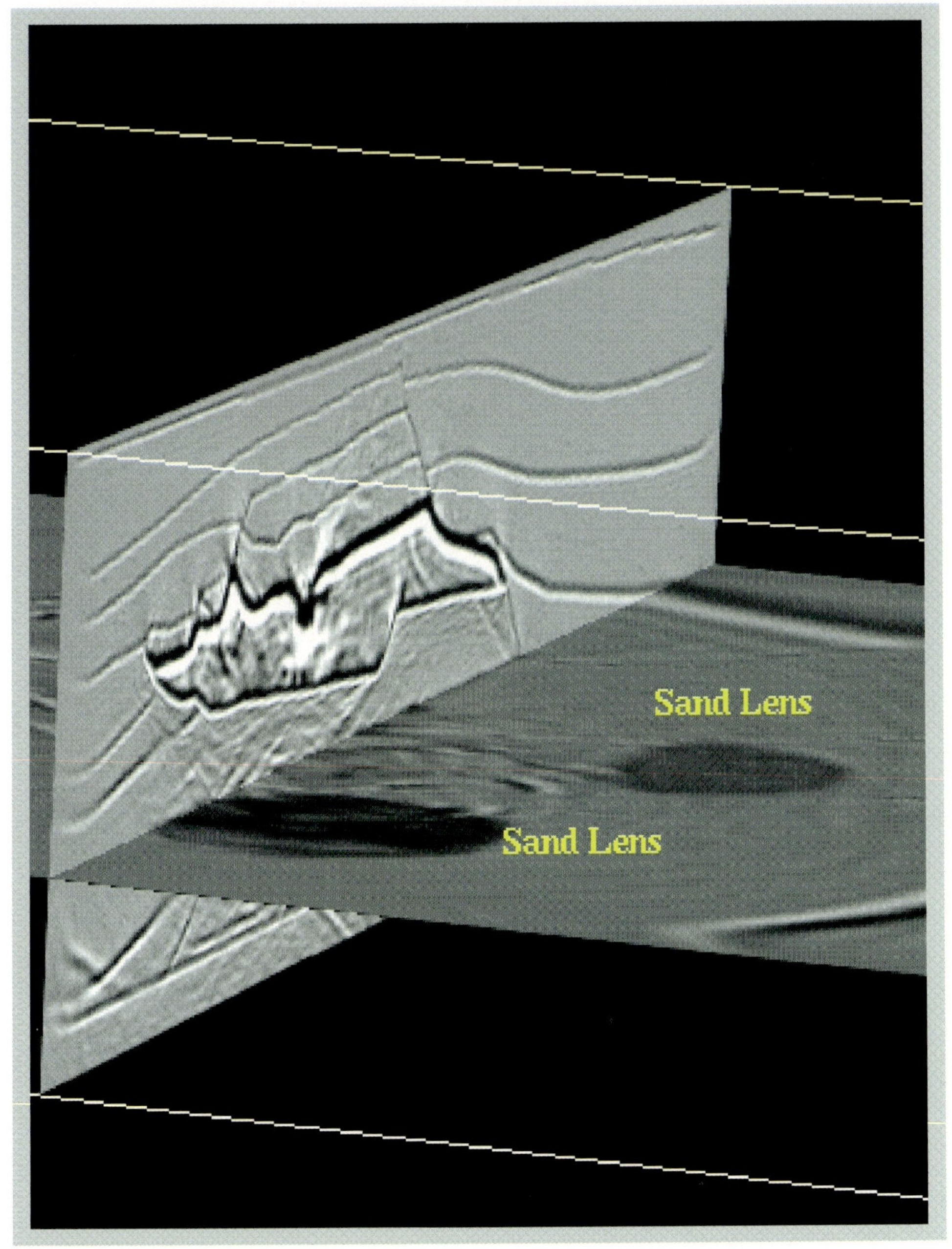


FIGURE 19. Imaged thin sand lens below salt. (Courtesy of Gulf of Mexico Subsalt Seismic Imaging Project, CRADA LA94C1051.)

# CHAPTER 29

House, J. R., T. M. Boyd, and F. P. Haeni, Haddam Meadows, Connecticut: a case study for the acquisition, processing, and relevance of 3-D seismic data as applied to the remediation of DNAPL contamination, *in* P. Weimer and T. L. Davis, eds., *AAPG Studies in Geology No. 42* and *SEG Geophysical Developments Series No. 5*, AAPG/SEG, Tulsa, p. 257–266.

# Haddam Meadows, Connecticut: A Case Study for the Acquisition, Processing, and Relevance of 3-D Seismic Data as Applied to the Remediation of DNAPL Contamination

Jason R. House*, Thomas M. Boyd*, and F. Peter Haeni†

## ABSTRACT

Haddam Meadows State Park in Haddam, Connecticut, serves as a test site for the Branch of Geophysical Support of the USGS in Connecticut. At this site, various geophysical techniques are tested for detection of structures that could influence dense non-aqueous phase liquids (DNAPL) migration. DNAPLs sink through the water column until they reach an impermeable boundary. The geology of Haddam Meadows consists of Ordovician metamorphic bedrock, approximately 43 m deep, overlain by glaciofluvial and alluvial deposits. If no aquitards exist above bedrock, any DNAPLs present at this location should collect in topographic lows in the bedrock. A 3-D seismic dataset was collected at the site to determine the feasibility of using this methodology to constrain bedrock topography. Data were recorded from an evenly spaced shot grid along five parallel receiver lines. Although acquisition was rapid (two days), the offset, azimuth, and full-fold ranges of the dataset are restricted by the nonoptimized shot-receiver pattern, limiting our ability to examine bedrock structure. Data are reduced and interpreted using conventional oil-exploration methods. Two prominent reflections are observed. Both of these tie to a vertical seismic profile (VSP) dataset collected in the same area. The upper reflector corresponds to the refusal depth of the well from which the VSP was acquired. This reflector can be interpreted as either the top of fractured bedrock or the top of an overlying till layer. On the basis of previous refraction studies, we prefer the former interpretation. The lower reflector shows less topographic relief and is approximately 61 m deep. It is distinguished by larger amplitudes and is spatially more continuous. We interpret this reflection as originating from the top of competent bedrock. Amplitude anomalies along each reflection are examined. We interpret the observed amplitude anomalies as being indicative of mineralogic changes within the bedrock, which could be related to remineralization of open fractures or to the presence of pegmatites. Compared to drilling shallow test wells or acquiring less costly geophysical datasets, 3-D seismic surveys at this site have limited economic viability. Investigations that might be deemed suitable for the application of 3-D near-surface seismic include those that (1) contain aquicludes of sufficient thickness, lateral extent, and velocity contrast to be resolved by seismic methods, (2) have targets that are relatively deep (>15 m), (3) display significant spatial variability in target depth, target properties, or target continuity, and (4) require the use of noninvasive methods.

*Department of Geophysics, Colorado School of Mines, Golden CO 80401
†U.S. Geological Survey, 450 Main Street, Room 525, Hartford CT 06103

## INTRODUCTION

Over the past 15 years, the development of 2-D, seismic reflection methodologies for delineating shallow (less than 100 m) structure has made significant strides. Much of this development has taken the form of direct technology transfer from the oil industry. For example, during the 1980s, with the microcomputer revolution, seismic reflection methodologies used in the near-surface evolved from those requiring few recording channels and minimal processing (e.g., the optimum offset method of Hunter et al., 1984) to the now widespread practice of recording many (48 or more) channels and the use of the common midpoint (CMP) method. Numerous examples now exist in the literature, primarily via the efforts of the Kansas Geological Survey and the University of Kansas, of the effectiveness of the CMP method for constraining near-surface structure applicable to engineering problems (e.g., Steeples et al., 1986; Miller et al., 1989; Steeples and Miller, 1990; Miller et al., 1995).

Although technology transfer has provided engineering geophysicists with the basic tools by which near-surface structure and stratigraphy can be imaged, this has not eliminated the need for continued research aimed specifically at near-surface applications of reflection seismology. For example, features associated with near-surface observations often do not have direct analogs with studies focusing on deeper structure (e.g., Steeples et al., 1995a). Also, the relatively high frequency of the propagated wavefield (in some cases as high as 500 Hz) and the shallow depth of the targets involved can alter the emphasis of the relevant physics applicable to the problem (e.g., Black et al., 1994).

While the engineering community has embraced 2-D seismic reflection methodologies, the petroleum industry has rapidly expanded its exploration and reservoir development program through the use of 3-D reflection surveys. This report documents an example of using 3-D seismic reflection technologies developed in the oil industry for a near-surface engineering problem. Although the results described here are encouraging, we acknowledge that the potential uses for 3-D seismic data in defining near-surface structure are limited. This limitation is primarily associated with the cost of data acquisition and processing of 3-D seismic data compared with the cost of drilling shallow test wells or acquiring less costly geophysical datasets. 3-D seismic surveys will be economically feasible in areas that have specific targets previously identified by other methods. Such areas might be candidates for 3-D seismic methods because the nature of the contamination at the site requires the use of noninvasive exploration methods to help guide remediation or because a detailed knowledge of lateral variations in stratigraphy, aquiclude variability, or bedrock topography is needed.

Determining geologic structure such as bedrock topography is relevant to the mitigation of a common class of groundwater contaminants known as dense nonaqueous phase liquids (DNAPLs). Several common examples of DNAPLs include trichloroethylene (TCE), methylene chloride, perchloroethene (PCE), chloroform, and 1,1,1-trichloroethane. These chemicals are commonly used as solvents and cleaning fluids and constitute one of the most common types of groundwater contamination. Hence, methodologies that constrain structures that could trap DNAPL contaminants would help to expedite remediation of a number of sites. DNAPLs have specific gravities greater than 1, are only slightly soluble in water, and are generally less viscous than water (Preslo and Stoner, 1991). Thus, they tend to sink through the water column and collect on impermeable layers in the subsurface. DNAPL flow is controlled primarily by gravity, and not by the regional hydrologic flow regime. Therefore, methods that can be used to map the lateral extents of aquitards and bedrock topography are required for identifying possible sites of DNAPL collection.

In 1993, Bison Inc., the USGS, Geometrics Inc., the Kansas Geological Survey, and Bay Geophysical Inc. collaborated in acquiring a 3-D, vertical-component, seismic reflection dataset at Haddam Meadows State Park, in Haddam, Connecticut. The intent of this survey was to determine the applicability of near-surface 3-D seismic surveys in delineating bedrock topography that could control the migration of contaminants such as DNAPLs. Although not contaminated, the Haddam Meadows site was chosen because of its near-surface water table, its relatively simple geology, the presence of a sharp bedrock interface at relatively shallow depths, a well-documented borehole, and its proximity to several 2-D refraction profiles.

The survey, as described below, was specifically designed to minimize costs related to data acquisition. Unlike the petroleum industry, the environmental industry is cost-driven rather than profit-driven. Seismic methodologies must compete with other relatively inexpensive geophysical techniques, such as ground-penetrating radar, electrical, gravity, electromagnetic, or magnetic methods, and often with more tangible borehole and/or cone penetrometer studies. As a result, the survey design chosen for this 3-D experiment will appear unusual to those familiar with 3-D exploration surveys. Like all survey design choices, the survey presented in this investigation suffers from a number of shortcomings. These can be attributed to the compromises that were made to minimize acquisition costs.

## SITE OVERVIEW

The Haddam Meadows State Park is located on the eastern bank of the Connecticut River in the northeastern corner of the Haddam Quadrangle. The site of the 3-D seismic reflection survey is located within the northern confines of the park, just southwest of the Connecticut River (Figure 1). The regional geology of the area consists of the core members of the Ivoryton synform of Ordovician age (Lundgren, 1979). Based on the depth of borehole refusal, bedrock is believed to be approximately 43 m deep at the site and consists of Ordovician age muscovite bearing rocks of the Collins Hill Formation. The predominant rock type is muscovite-biotite schist, intermixed with muscovitic quartz-feldspar gneisses and calcite bearing silicate gneisses. The rocks containing significant amounts of muscovite are characterized as highly friable, and the entire formation weathers rapidly upon exposure. The bedrock is known to be mineralogically heterogeneous and is locally intruded by pegmatites of Devonian to Permian age (Lundgren, 1979).

Surficial geology at Haddam Meadows consists primarily of glacial and alluvial deposits. Glacial till was deposited atop bedrock following Wisconsinan glaciation. From borehole cuttings, this till layer is believed to be between 1.5 and 3 m thick and composed of poorly sorted sediments ranging in size from clay-sized particles to boulders. Overlying the glacial till are glacial outwash deposits, generally referred to as stratified drift, that are in turn overlain by fine sands and silts deposited by the Connecticut River during recent times. These materials have been well described in glacial valleys to the west of the Connecticut River (Haeni, 1978). The stratified drift materials were deposited by glacial melt water. Unlike the glacial till, the action of the melt water produced well-sorted sand- to gravel-sized sediments. A stylized cross section of the site geology is shown in Figure 2.

Thicknesses of these units are estimated from cuttings recovered from a borehole located in the northeastern corner of the survey area (HM1 in Figure 3). The borehole reached refusal at a depth of 43 m. Although no cuttings were recovered from this depth, it is believed that this depth represents the top of bedrock, based on unpublished refraction profiles shot in the area. The borehole also indicates that the depth to water table in the area is approximately 3 m. This relatively shallow depth is not unexpected given the site's proximity to the Connecticut River.

## SURVEY DESIGN AND DATA ACQUISITION

### SURFACE SEISMIC DATA

Unlike most 3-D land reflection surveys that are acquired by moving a swath of receivers as source location varies (see Stone [1994] for a thorough description of 3-D survey designs), receiver locations for this survey were held fixed while shot locations varied. This acquisition geometry is referred to in Stone (1994) as a full-range survey. Among the benefits of this type of acquisition scheme is the relative speed at which small 3-D datasets can be acquired. The entire data volume described in this study, some 30,000 traces collected within a 91 × 110-m area, was acquired in two days. In addition, because source-receiver azimuths contributing to a single bin are derived from all source locations, full-range surveys have the potential of delivering a good azimuthal distribution of offsets within each bin. As will be shown, however, this potential is not fully realized in this survey because of the limited offset range over which useful reflections are recorded.

Figure 3 shows a schematic of the acquisition geometry used in this study. Five lines of 24 receivers each were laid out parallel to one another in a roughly north-south direction. A regularly spaced shot grid of 247 shots was superimposed on the receiver lines. Each shot was recorded simultaneously on all 120 channels using a Bison seismograph system. Receiver lines were spaced 15.2 m apart with an inline station interval of 3 m. (Please note that the acquisition geometry was implemented in the field using English units. All metric conversions are given to the nearest 0.1 m.) Source lines were spaced 7.6 m apart with an inline source interval of 6.1 m. The source consisted of a 12-lb sledgehammer striking downward on a steel plate. At least eight hammer blows were stacked per shot point. This number, however, varied for each shot point, requiring us to balance shot ensembles prior to performing stacking or true amplitude recovery. Receivers consisted of single 30-Hz vertical component geophones. Data were sampled at 0.25 ms, and a 400-Hz anti-alias filter was applied before sampling. Elevations were measured to the nearest 0.03 m at each source location. Elevation differences across the survey amounted to 0.88 m. Between source locations, elevations are linearly interpolated.

Although the survey design allowed rapid data acquisition, it does have several significant shortcomings. These include the following:

(1) CMP bins are suboptimal. The acquisition geometry forces bins to be rectangular with dimensions of 1.5 m in the inline (north-south) direction and 3.81 m in the crossline (east-west) direction. These bins, particularly in the crossline direction, are relatively large for near-surface surveys. In 2-D seismic reflection studies of the near-surface, a CMP spacing of 1 m is typical (e.g., Steeples and Miller, 1990).

(2) Fold coverage varies significantly over the survey area. In addition to creating elongate bins, the fold coverage varies from a nominal fold of 72 at the center of the survey to single-fold at the survey edges. The peak fold at the survey center is, however, deceiving. After inside muting to remove ground roll and top muting to remove refracted arrivals and normal moveout (NMO) stretched reflected arrivals, the actual fold at the survey center is reduced to about 14. Within 21 m of the survey's center, usable fold decays to 6. As expected, many of the offsets collected provided little useful information. Thus, recording every shot on all 120 channels, other than being expedient, provided little additional information.

(3) One of the potential benefits of acquiring a 3-D data volume is the ability to use this volume to examine azimuthal variability in seismic velocity that might be related to subsurface anisotropy. In near-surface applications such as this, reflection anisotropy from a bedrock reflector may be indicative of preferential fracturing that could prevent the bedrock surface from acting as an effective aquitard. To detect such variability it is necessary to ensure that each CMP bin has a relatively uniform contribution of offsets at a variety of azimuths. The azimuth of the CMP bin contribution is defined as the angle from north to the line connecting the source to the receiver.

Figure 4 shows the azimuth contributions for reflected arrivals observed at all offsets for nine macrobins. The macrobins are defined by dividing the rectangular survey grid into thirds, in both inline and crossline directions. Due to survey symmetry, only five of the nine macrobins are detailed in Figure 4. Based on reciprocity, azimuths greater than 180° have been mapped to azimuths less than 180° by interchanging source and receiver positions. The east and west edges of the survey, subregions 1, 2, 3, 7, 8, and 9, show a preponderance of north-south–directed path contributions. Thus, there is little hope of extracting azimuthally varying velocity information from these regions. Subregions 4, 5, and 6, however, show a relatively even azimuth distribution, suggesting that azimuthal variability could be examined in these regions. There are, however, additional complications.

Figure 5 shows the distribution of offset versus bin location for two azimuth-limited inline sections. Offsets are limited to 20 and 61 m, corresponding to the offset range obtained from the observed CMP gathers after application of inside and outside mutes. Figures 5A and B show the offset contributions for CMP bins along inline 15, restricted to azimuths between 45 and 135° (east-west) and between 135 and 215° (north-south), respectively. The same is shown for inline 16 in Figures 5C and D. Note that for both lines, the offset contributions for azimuths between 135 and 215° is well distributed. When looking at adjacent inlines restricted to azimuths between 45 and 135°, however, the distribution of offsets is limited to different ranges for adjacent inline sections. For inline 15, offsets are limited to between 30 and 46 m. For inline 16, offsets are limited to between 38 and 55 m. On azimuth limited stacks, this range limiting of offsets produces strong north-south amplitude banding, even through macrobins 4, 5, and 6. Thus, the range-limited nature of the offset contributions within azimuth windows reduces our ability to examine the 3-D volume of azimuth-dependent properties.

A better offset-versus-azimuth distribution could be obtained using this basic acquisition scheme if the receiver grid is rotated out of line of the shot grid. For the current acquisition scheme, receivers and sources are laid out in parallel lines. Although this results in an easily implemented acquisition geometry, the resulting dataset suffers from the problems described above. By rotating the receiver lines so that they are no longer parallel to the source lines, a more uniform azimuth-versus-offset distribution could be obtained for any receiver-grid rotation angle other than 0, 45, or 90°.

(4) For most near-surface surveys, migration does not greatly alter the final stacks (Black et al., 1994). For this survey, however, migration could produce significant changes. Using the steepest apparent dip from the shallowest reflection imaged (reflection A, described later), 0.39 ms/m; the average two-way traveltime to this reflection,

75 ms; and a stacking velocity of 1250 m/s, we estimate that migration will produce a horizontal shift in this reflection of as much as 11 m (7 CMPs) and a time shift of as much as 2.2 ms (nine samples). Unfortunately, a full 3-D migration is difficult to perform on this dataset. Of primary concern to us are migration edge effects. The current 3-D survey yields only 60 crossline bins and 21 inline bins. Although, as will be shown later, 2-D migrations in the inline direction yield stable results, crossline migrations are unstable. We believe that this results from the paucity of inline bins available for migration. Because of these complications, the results shown in this report are unmigrated.

To enhance the crossline migration results, more inline bins must be incorporated in future 3-D survey designs. If the full-range survey is used, as in the current study, rotating the receiver grid with respect to the shot grid, as described above, would allow the implementation of a smaller binning interval in the inline direction. With the current acquisition scheme, CMPs fall within the exact center of each bin. By rotating the receiver lines so that they are no longer parallel to the source lines, CMPs scatter throughout the subsurface, allowing potentially smaller bins. This acquisition scheme could, therefore, address concern (2) and increase the likelihood of successfully applying 3-D migration.

(5) The relatively large station spacing, 3 m, aliases ground roll, making it difficult to remove this form of noise using F-K filters, trace mixing, or median filters. We therefore rely on large bottom mutes to remove this source of noise. The application of bottom mutes significantly decreases fold.

### VERTICAL SEISMIC PROFILES

In addition to the surface seismic observations, a VSP was collected in a single well, HM1, whose location is shown in Figure 3. The VSP was collected by lowering a hydrophone into the well and recording successive blows from a sledgehammer source. Two different VSPs were collected: a walkaway VSP in which the receiver was fixed at a depth of 21 m and the source was moved away from the hole in 1.5-m increments from an offset of 0 to 55 m; and a constant-offset VSP in which the source was fixed at an offset of 20 m and the receiver pulled up the borehole in 1.5 m increments from a maximum depth of 39.4 m. Observations were recorded on an EG&G Geometrics 2401 24-channel seismograph at a 0.1-ms sample interval. The data were subsequently resampled at 0.25 ms to match the surface seismic observations. We present results from the constant-offset VSP only.

## DATA PROCESSING

### SURFACE SEISMIC DATA

An example of a raw shot record is shown in Figure 6A. Note that for this figure offsets do not increase uniformly from left to right. Thus, arrivals appear to be less coherent than they actually are and reflections do not show hyperbolic moveout. Also note that the amplitudes of all traces shown in Figures 6A and B have been trace normalized. Several features are apparent in this raw shot record.

First, reflected arrivals are apparent between 60 and 150 ms. The reflections, however, tend to be ringy in character. Closer inspection indicates that this sequence of reflections is composed of two prominent reflections repeated at approximately 20-ms intervals. The stronger of these occurs at a zero-offset time of about 75 ms and is labeled "B" in Figure 6. The weaker occurs at a zero-offset time of about 60 ms and is labeled "A" in Figure 6. We attribute the later arrivals to short-leg multiples generated from the top of the water table. As shown below, the two-way traveltime through the unsaturated alluvium is about 20 ms, and the impedance contrast at the water-table interface is large.

Second, the near-offset traces are dominated by spatially aliased ground roll. Stacked sections produced without attempting to attenuate the ground roll show poor reflection quality. Although, as described previously, nominal fold is as high as 72, after applying an NMO stretch mute, the actual fold can be as much as a factor of five less. Given this low fold, relying on the statistics of a stack to remove high-amplitude noise is ineffective. We attempted various methods of ground roll suppression, including F-K filtering of the raw shot records, F-K filtering of time-shifted shot records, median filtering, and trace mixing. None of these methodologies were successful at producing significant attenuation of the ground roll without introducing significant filter artifacts. We therefore choose to remove the ground roll by bottom muting each shot record. A single bottom-mute was designed and applied to all shot records.

Figure 6B shows the first 100 ms from another shot record. Only inline receivers are shown, and the record has been bottom-muted to remove ground roll. The prominent reflections labeled "A" and "B" in Figure 6A are apparent at offsets greater than 15 m. If we asume, for the moment, the reflectors generating these arrivals are flat, then the observed moveout corresponds to root-mean-square (RMS) velocities of 1185 m/s for reflection A and 1370 m/s for reflection B. Initial estimates of the depths at which these reflectors are located, based on their RMS moveout velocities and their intercept times, are 43 and 55 m, respectively, consistent with reflections at or near the expected depth to the top of bedrock.

In addition to inside-muting the shot records to eliminate ground roll, we top-mute and outside-mute the NMO-corrected CMP gathers. The top mute eliminates direct and refracted arrivals, while the outside mute eliminates post-critically reflected arrivals. The nominal offset used for this outside mute is 60 m. This offset limit is based on a model calculation of the amplitude-versus-offset (AVO) characteristics of the dataset. Figure 7A shows the amplitude-versus-offset characteristics computed from a model assuming a reflector depth of 43 m, a P-wave velocity of 1520 m/s above the reflector and 2250 m/s below, and a p-wave to s-wave velocity ratio of 2. Figure 7B shows a representative 2-D shot record with no amplitude scaling. The critical distance computed from the model is approximately 76 m, consistent with that observed for reflection A in Figure 7B. Before this critical distance is reached, the model predicts—and the data substantiate this prediction—reflection amplitudes decreasing out to offsets of approximately 60 m. Although the choice of a 60-m outside mute is conservative in this case, it was chosen to prevent the stacking of smaller-amplitude arrivals with potentially low signal-to-noise ratios.

A prominent refracted arrival is observed as the first arrival at offsets greater than 12 m (Figure 6B). The moveout velocity of this arrival is approximately 1520 m/s. At offsets less than 12 m, the direct arrival is less well developed due to interference with the airwave. It appears, however, that the moveout velocity of the direct arrival is approximately 300 m/s. (Note that on Figure 6B it is difficult to observe the direct arrival. The high-amplitude, high-frequency event that also has a moveout velocity of 300 m/s evident on Figure 6B is the airwave. By scaling up the amplitudes on the plot, it is possible to detect a direct arrival that arrives just before the airwave.) Using these velocities and the intercept time of the refracted arrival (22 ms), the depth to the refractor is approximately 3 m. The velocities of the overburden and the refractor and the depth to the refractor are consistent with the refracted arrival's being generated from water table.

Based on this interpretation, field statics are computed by manually picking first-break arrival times on all shot records and selecting those offset ranges corresponding to refracted arrivals. Assuming a constant velocity of 300 m/s for the overburden and a datum level 1 m above the highest elevation, a very-near-surface (less than 4.5 m) velocity model is computed from the refracted arrival times using the delay-time methodology (e.g., Lawton, 1989). Surface-consistent source and receiver statics are then computed through this model assuming normal incidence and a replacement velocity of 1520 m/s. The maximum field static computed is 22.8 ms. Because the depth to top of water table is relatively constant (3.5 ± 1 m), the variation in the field static is ± 1.5 ms across the survey.

Figures 8A and B depict a 2-D inline section from our 3-D volume processed without and with the application of refraction statics. A zero-phase 75- to 200-Hz Ormsby filter has been applied to these and all of the other sections shown in Figure 8. A bulk shift of 22.8 ms has been applied to the brute stacked section (Figure 8A) for comparison purposes. The application of refraction statics produces few discernible differences in the final stacked sections. There is a marginal increase in the continuity of reflection A. Reflection B, which is already continuous across the data volume, shows less short-wavelength variation upon application of the statics.

Stacking velocities for the 3-D volume are determined with the use of semblance plots, NMO-corrected CMP gathers, and stacked sections. For the velocity analysis, traces are sorted into CMP macrobins. Each macrobin is constructed by combining five CMP bins in the crossline direction and three CMP bins in the inline direction. Figure 9 shows a representative semblance plot for a CMP macrobin located near the center of the survey. In constructing this semblance plot we employed a 35% stretch mute. Other than the stretch mute, no effort is made to remove the direct or the refracted arrivals. For the final stacks, the direct and refracted arrivals are top-muted on the NMO-corrected CMP gathers.

Three prominent reflectionlike features are apparent in Figure 9, one each at intercept times of 48 (C), 75 (A), and 90 (B) ms. The reflectionlike feature labeled "C" results from stacking refracted arrivals. The two other semblance highs correspond to reflections A and B, identified on Figure 6. The range of acceptable stacking velocities for reflection A is between 1125 m/s and 1300 m/s. For reflection B, the range of acceptable stacking velocities is between 1275 m/s and 1520 m/s. From these stacking velocities and intercept times, the implied interval velocities are between 1125 m/s and 1300 m/s above reflection A and between 1850 m/s and 2430 m/s between reflections A and B. The range of interval velocities above reflection A is consistent with water-saturated alluvium. The range of interval velocities estimated between reflections A and B is consistent with fractured metamorphic bedrock or water-saturated till.

In addition to the field statics, we also apply residual statics to the NMO-corrected CMP gathers. Surface-consistent residual statics are computed via the Gauss-Siedel methodology (Wiggins et al., 1976). Residual statics are limited in size to 8 ms, and the time-gate used for cross-correlation is 100 ms long and centered on reflection B. The largest computed static is 2 ms. Figure 8C shows an extracted 2-D line from the 3-D volume after the application of residual statics. The coherency of the lower reflection is improved, particularly on the right side of the stack. Residual statics also help define the nature of reflection A. Before application of the residual statics (Figure 8B), the upper reflection appears discontinuous. After application of the residual statics (Figure 8C), the upper reflection is more continuous, becoming shallower to the left of the stack, with clearly developed crosscutting relationships displayed.

Finally, we apply a spiking deconvolution operator to the stacked data volume. There are several arguments against the use of deconvolution for near-surface datasets. First, the successful application of deconvolution is dependent on several basic assumptions. One of these is that the subsurface reflectivity can be represented by an uncorrelated, random sequence (e.g., Yilmaz, 1987). One could argue that in near-surface applications the reflectivity sequence is not random, since we often see only one or two reflections on the shot records (e.g., Steeples et al., 1995b). Although it is true that near-surface surveys generally do contain fewer reflections than exploration surveys, this is not the salient point. We suspect that the underlying reflectivity process, on which the randomness requirement is based, is just as random in the near-surface as it is for deeper investigations, albeit with a smaller standard deviation. Although few examples exist, the synthetic reflectivity sequences generated from density logs described by Gochioco (1992) are consistent with this supposition. From our perspective, the most important limitation on the application of deconvolution to near-surface surveys is signal-to-noise (S/N) ratio. Near-surface surveys typically appear to have lower S/N ratios than exploration surveys. The deconvolution process enhances the bandwidth of the dataset at the expense of lowering its S/N ratio. This trade-off may not be justified for many near-surface surveys, given their already low S/N ratio.

Despite this potential pitfall, we believe that deconvolution offers tangible improvements to the final stack. Figure 8D shows a 2-D line extracted from the 3-D volume after the application of a spiking deconvolution operator with a length of 12 ms, and 0.1% spectral prewhitening. As expected, the overall S/N is decreased by the application of deconvolution. This primarily results in a decrease in reflection coherency across the data volume. The wavelet, however, is compressed by about 50%, thus improving vertical resolution. Because we need to identify relatively small changes in bedrock topography, the enhanced vertical resolution is deemed more significant than the loss in S/N. We also note that the short-leg multiples observed on the raw shot records (Figure 6A) are apparent in our stacked sections, both before and after the application of deconvolution (Figures 8C and D). Because these multiple arrivals occur later than our deepest reflection of interest, reflection B, no effort is made to remove them via deconvolution.

As described previously, the appearance of this data volume could be significantly altered with the application of post-stack migration. Because of the difficulties experienced in applying a 3-D migration operator to the data volume, our interpretations are based on the unmigrated data volume as shown in Figure 8D. We believe that the problems associated with applying 3-D migration operators to this data volume are a consequence of the small number of crossline bins (21) available for migration. To illustrate the potential of applying migration to data volumes such as this, we applied a 2-D phase-shift migration (Yilmaz, 1987) to one inline section extracted from the 3-D volume (Figure 8E). The coherency of both reflections A and B is significantly improved, and the steeply dipping reflection, A, has moved updip by an amount consistent with the calculations described previously.

### VERTICAL SEISMIC PROFILE

In addition to the surface seismic observations, the VSP dataset is also processed and interpreted. Figure 10 shows the raw offset VSP (Figure 10A) and representative processing steps (Figures 10B–D) used to generate the vertically extracted trace (Figure 10E). The processing steps illustrated in Figure 10 include (1) phase-rotating the VSP observations by 90° to convert pressure observations to displacement and applying a bottom mute (Figure 10B); (2) removing all of the downgoing energy via 2-D median and F-K filtering after aligning the VSP on first breaks (Figure 10C), (3) converting the remaining upgoing energy to two-way traveltime (Figure 10D), and (4) stacking the final section to produce a vertically extracted trace and bandpass-filtering the result to match the frequency characteristics of the surface seismic sections. Note that the vertically extracted trace shown in Figure 10E has been bulk-shifted upward by 21 ms. This shift mimics the bulk shift applied to the surface seismic observations via the field statics.

As expected, the offset VSP is dominated by downgoing energy (Figure 10B). Before removal of this energy, upward-traveling energy is not apparent. After the removal of the downgoing wavefield, several upward-traveling reflections are apparent. The most prominent of these is labeled “A” in Figure 10C. We note that this reflection intersects the borehole near its depth of termination, 43 m. This reflection is, therefore, apparently associated with the structure responsible for borehole refusal at a depth of 43 m, believed to be the top of bedrock. Several less well-developed reflections are visible below this. Because the borehole does not penetrate more deeply, we cannot assign depths to these reflectors or, for that matter, rule out that they are not multiple reflections.

## INTERPRETATION

### SURFACE SEISMIC-TO-VSP TIE

Figure 11 shows the vertically extracted VSP trace inserted into the 2-D inline stacked section closest to the borehole location. There are three sets of reflections observed on the VSP that tie to the surface seismic observations. The first occurs at approximately 75 ms, the second at 90 ms, and the third at 110 ms. The 75-ms reflection, labeled “A” in Figures 6, 7, 8, and 10, is spatially continuous across the data volume and is 43 m deep at the borehole location. Reflection A appears to be associated with the structure responsible for borehole refusal. Topographic relief on reflection A is quite large. The time of this event on the stacked sections varies from 50 to 100 ms.

The prominent reflections observed on the VSP around 90 ms tie to reflection B, identified on the surface seismic data on Figures 6, 7, and 8. From the 3-D data volume, this reflection is spatially continuous, has locally large relative amplitude, and shows less temporal variability and thus less overall topography than reflection A. Finally, the group of reflections observed on the VSP below reflection B at a two-way traveltime of about 110 ms ties to spatially discontinuous reflections observed on the stacked sections that we have previously identified as multiples.

As a check on the depth conversions presented below, the depth to reflector A can be computed from the interval velocities derived from the surface seismic observations via stacking velocities through Dix's equation. At the borehole, reflector A is estimated to be 43 m deep. This estimate is in excellent agreement with the borehole refusal depth and the depth of intersection of reflector A as observed on the VSP. Because the borehole does not penetrate below 43 m, depths computed for the lower reflections from stacking velocities cannot be independently confirmed. Based on these stacking velocities, however, reflector B varies in depth from 45 to 76 m. This is compared to a depth variation of 30 to 67 m for reflector A.

The structure and amplitude character of reflection A are shown in Figure 12. The amplitudes shown represent the spatial variability of the RMS amplitudes of the reflection defined by mapping the amplitude maximum of the reflection throughout the 3-D data volume and computing RMS amplitudes within a 10-ms window about this peak.

The structure of reflector A is defined by an almost planar interface dipping toward the north west. The maximum dip associated with the interface is approximately 25°. Reflection A cross-cuts deeper reflections (Figure 8C), suggesting that it represents an erosional surface. Its association with more shallow reflections is less well defined, given our inability to map reflections above 50 ms. If it represented an erosional surface related to the Connecticut River, we would expect it to dip toward the river to the northeast. The strike of reflector A, however, runs almost perpendicular to the present course of the Connecticut River. One possible explanation is that reflector A defines the southern bank of a paleochannel that was a tributary of the Connecticut River.

The RMS amplitude anomalies associated with reflection A are relatively small, about a factor of 20. A band of large amplitudes is observed through the middle and northern portions of the survey running north-northeast. Although this band of amplitude anomalies subparallels the local strike of reflection A, it does move downdip toward the northeast, as can be seen in the perspective view included in Figure 12, suggesting that these anomalies are not caused by reflector geometry.

The structure and amplitude anomalies associated with reflection B are also presented in Figure 12. Overall relief on reflector B is approximately 24 m and cannot be as simply described as that observed on reflector A. Identifiable features include (1) a general dip toward the northwest, mirroring that observed for reflector A, (2) a 5-m-high ridge running from south to north through the southwestern corner of the survey, and (3) several localized depressions, two of which, located along the southern portion of the survey, appear to be closed. The larger of these two closed depressions has a relative depth with respect to the surrounding reflector of approximately 1.5 m and a volume of about 790 $m^3$.

Amplitude anomalies associated with reflector B show considerable variation. RMS amplitude deviations are in excess of 400. Amplitude highs form three prominent bands, two trending northeast and one trending northwest. These amplitude highs are not consistently associated with topographic anomalies.

Based on its continuity and large amplitude relative to the other reflections, we interpret reflection B as being generated from the top of unfractured bedrock. There are three possible explanations for the observed amplitude anomalies. (1) Amplitude highs could be associated with local pegmatite intrusions. We would, however, expect pegmatites to be more resistant to weathering than the surrounding bedrock. Because the observed amplitude highs are not associated with highs in reflector topography, this explanation seems unlikely. (2) Amplitude highs could be associated with fracture zones. We note that the distribution of amplitude highs is reminiscent of a conjugate fracture system. Open fractures, however, seem unlikely. If they were open, we would expect amplitude lows and topographic lows to be associated with them. If the fracture zones had been remineralized, however, they could be associated with amplitude highs. Furthermore, the mineralization may not show significantly more resistance to weathering than the surrounding bedrock, thus explaining the lack of correlation between the amplitude anomalies and topography. (3) The amplitude highs could be associated with spatial variability in the mineralogy of the bedrock.

Two possible interpretations exist for reflector A, the upper reflection. Reflector A could represent the top of a glacial till layer, or it could represent the top of fractured bedrock. Both explanations are consistent with the range of interval velocities observed between reflections A and B and with the crosscutting nature of reflection A. We, however, prefer the latter interpretation. This preference is based primarily on previous refraction results. The spatial correlation between the amplitude highs observed on reflector A with those observed on reflector B is suggestive of vertical continuity of the mineralogically heterogeneous bedrock, consistent with this interpretation. We note, however, that the zone of anomaly correlation occurs in a spatially confined region where the topographic difference between reflector A and reflector B is between 8 and 9 m. This difference is approximately one half the wavelength of the reflected wavefield, suggesting that the amplitude anomalies observed on the upper reflector, reflector A, result from wavelet tuning.

If reflector A represents a fractured bedrock surface that is impermeable to DNAPL flow, then our survey suggests that DNAPLs would migrate northwest through the survey region with possible zones of pooling in the southeast and west-central portions of the survey. Given the friable nature of a fractured bedrock surface, it is likely that if DNAPLs were released in this region they would penetrate the fractured interface and begin to collect on the unweathered interface imaged as reflector B. If this were the case, we would expect DNAPL flow to still be dominated by northwestward migration. Pooling would again occur in the southeastern portion of the survey area. Given an estimated volume of 790 $m^3$ for the depression mapped in bedrock in the southeastern portion of the survey, assuming a porosity in the 10–20% range, and assuming every pore is connected and filled, the total volume of DNAPL that could collect in this depression is between 79 and 160 $m^3$.

In this discussion, we have explicitly assumed that, when introduced into the system, DNAPLs penetrate unimpeded to bedrock. This assumption is probably incorrect. There is a strong possibility that there are spatially coherent aquitards in the form of clay layers and stringers in the stratified drift layer located above bedrock. The current dataset is, however, not capable of delineating these potentially important features. Other datasets, such as ground-penetrating radar, which in this region can be used to image to depths of 18–25 m, should be collected in conjunction with the seismic observations to begin to further our understanding of aquitard structure from the surface down to depths greater than 60 m.

## CONCLUSIONS

A 3-D seismic dataset constrained bedrock topography at a site within Haddam Meadow State Park in Haddam, Connecticut. The goal of this project was to test the feasibility of 3-D seismic methodologies of constraining near-surface structure relevant to the remediation of sites contaminated with DNAPLs. With cost one of the primary limitations to wider application of seismic imaging to engineering projects, this test project was designed to minimize acquisition costs. The entire 3-D data volume was recorded by shooting through a fixed 120-channel spread. Although data acquisition was rapid (2 days), the offset, azimuth, full-fold range, and minimum bin size achieved are restricted thereby limiting our ability to examine bedrock structure using all of the potential of 3-D seismic imaging.

Even with these limitations, two prominent reflections were observed and mapped through the subsurface. Both of these reflections tie to an offset VSP acquired in the same area. The upper reflection varies in depth from 30 to 67 m over the survey area and corresponds to the refusal depth of the well from which the VSP was acquired. This reflection is interpreted as the top of fractured bedrock, based on previous refraction studies. The lower reflection shows less topographic relief than the upper reflection, is approximately 61 m deep, and is characterized by larger amplitudes and spatially more continuous reflections than the upper reflector. We interpret this reflection as originating from the top of competent bedrock. Both the upper and lower reflections exhibit zones of high reflectivity that trend northwest and northeast, forming a pattern reminiscent of a conjugate fault system. Because these features are amplitude highs rather than amplitude lows and because they do not correlate to reflector topography, we interpret them as being indicative of mineralogic changes within the bedrock that could be related to remineralization of open fractures.

3-D seismic technology, although capable of constraining structures such as bedrock topography that are relevant to engineering investigations, must compete with inexpensive geophysical techniques such as ground-penetrating radar and electrical methods and with direct drilling and cone penetrometer studies. Thus, its eventual applicability to 3-D seismic methodologies is potentially limited to site-specific investigations. Such investigations might be deemed suitable for a 3-D seismic investigation if (1) the targets, in this case aquitards, are of sufficient thickness, lateral extent, and velocity contrast to be resolved by seismic methods; (2) the targets are relatively deep (>15 m); (3) it is believed that significant spatial variability in target depth, target properties, or target continuity may exist; and (4) noninvasive methods must be employed.

## REFERENCES CITED


Black, R. A., D. W. Steeples, and R. D. Miller, 1994, Migration of shallow seismic data: Geophysics, v. 59, p. 402–410.

Gochioco, L. M., 1992, Modeling studies of interference reflections in thin-layered media bounded by coal seams: Geophysics, v. 57, p. 1209-1216.

Haeni, F. P., 1978, Computer modeling of ground-water availability in the Pootatuck River Valley, Newtown, Connecticut: U.S. Geological Survey, Water Resources Investigations, 77–78 p.

Hunter, J. A., S. E. Pullan, R. A. Burns, R. M. Gagne, and R. S. Good, 1984, Shallow seismic reflection mapping of the overburden–bedrock interface with the engineering seismograph—some simple techniques: Geophysics, v. 49, p. 1381–1385.

Lawton, D. C., 1989, Computation of refraction static corrections using first-break traveltime differences: Geophysics, v. 54, p. 1289–1296.

Lundgren, L., Jr., 1979, The bedrock geology of the Haddam Quadrangle with map: Quadrangle Report 37, State Geological and Natural History Survey of Connecticut, Department of Environmental Protection.

Miller, R. D., N. L. Anderson, H. R. Feldman, and E. K. Franseen, 1995, Vertical resolution of a seismic survey in stratigraphic sequences less than 100 m deep in southeastern Kansas: Geophysics, v. 60, p. 423–430.

Miller, R. D., D. W. Steeples, and M. Brannan, 1989, Mapping a bedrock surface under dry alluvium with shallow seismic reflections: Geophysics, v. 54, p. 1528-1534.

Preslo, L. M., and D. W. Stoner, 1991, The overall philosophy and purpose of site investigations, *in* ed. D. M. Nielsen, Practical Handbook of Ground-Water Monitoring: Chelsea, MI, Lewis Publishers, Inc., p. 69–95.

Steeples, D. W., R. W. Knapp, and C. D. McElwee, 1986, Seismic reflection investigations of sinkholes beneath Interstate Highway 70 in Kansas: Geophysics, v. 51, p. 295–301.

Steeples, D. W., and R. D. Miller, 1990, Seismic reflection methods applied to engineering, environmental, and groundwater problems, *in* S. H. Ward, ed., Geotechnical and Environmental Geophysics: Society of Exploration Geophysicists, Tulsa, p. 1–30.

Steeples, D. W., B. Macy, C. M. Schmeissner, and R. Miller, 1995a, Contrasting near-surface and classical seismology: The Leading Edge, vol. 14, April, p. 271–272.

Steeples, D. W., C. M. Schmeissner, and B. K. Macy, 1995b, The evolution of shallow exploration methods: The Journal of Environmental and Engineering Geophysics, v. 0, p. 15–25.

Stone, D. G., 1994, Designing Seismic Surveys in Two and Three Dimensions, Geophysical References Series: v. 5, Society of Exploration Geophysicists, Tulsa, 244 p.

Wiggins, R. A., K. L. Larner, and R. D. Wisecup, 1976, Residual statics analysis as a general linear inverse problem: Geophysics, v. 41, p. 922–938.

Yilmaz, O., 1987, Seismic data processing, *in* S. M. Doherty, ed., Investigations in Geophysics, n. 2, Society of Exploration Geophysicists, 526 pp.

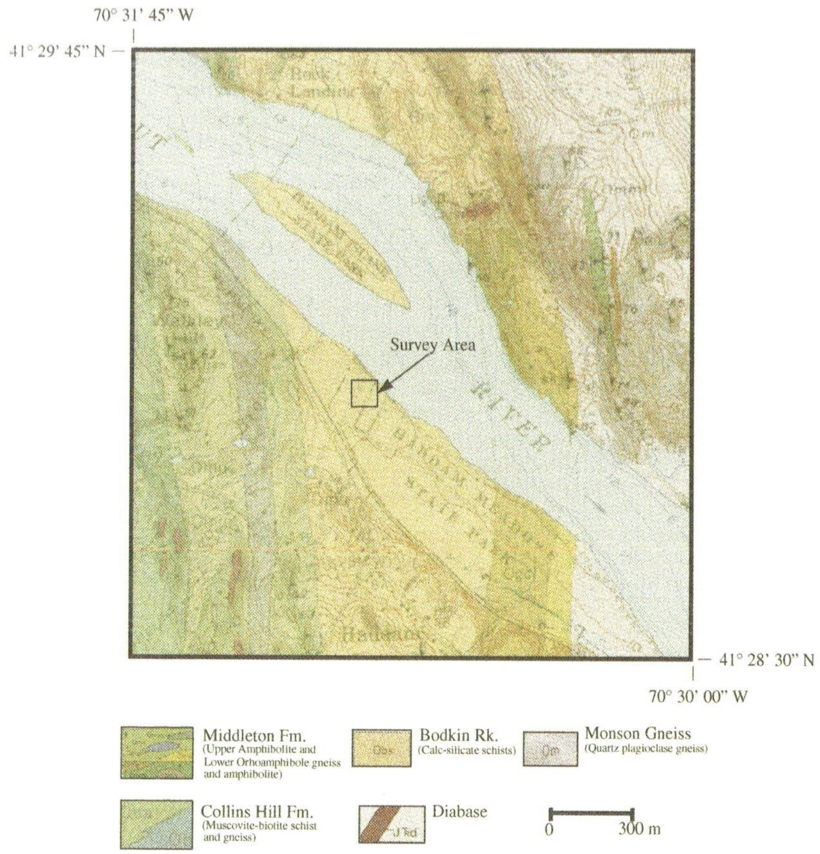


FIGURE 1. Haddam Meadows survey site superimposed upon bedrock geological map of the Haddam Quadrangle (after Lundgren, 1979).

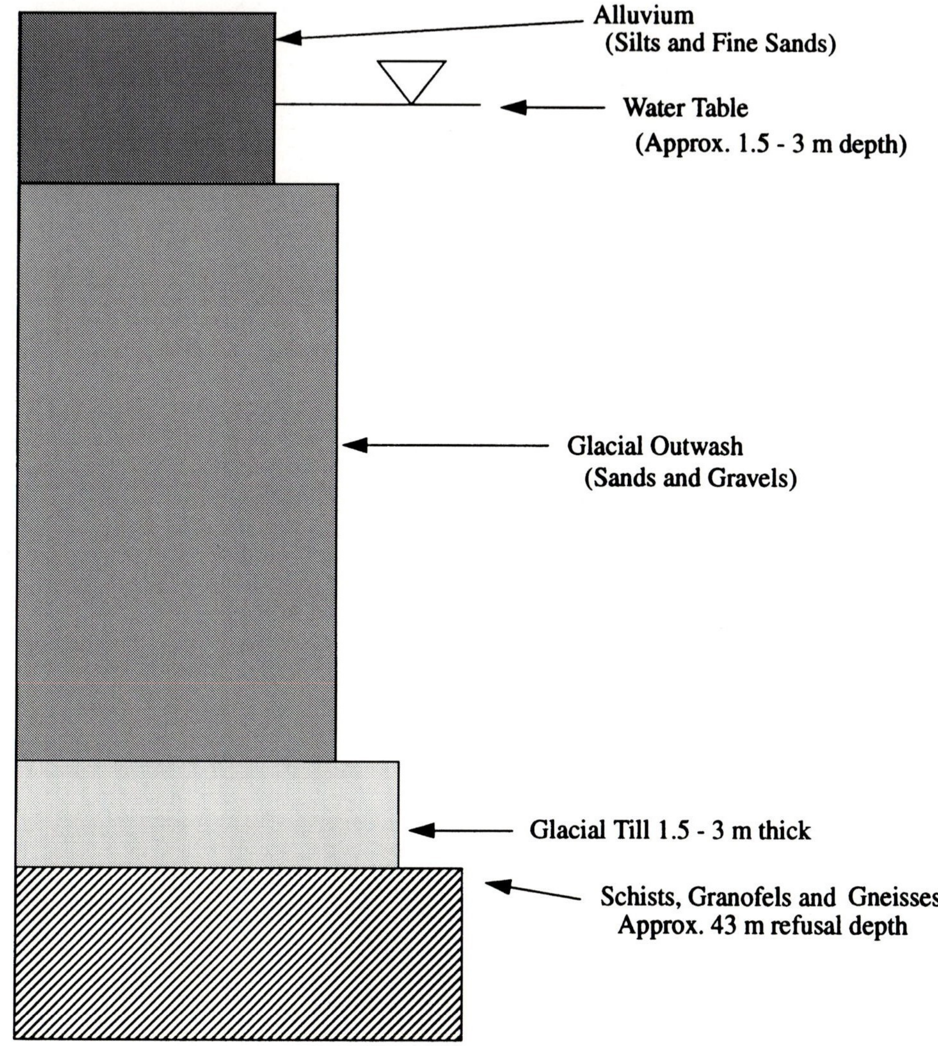


FIGURE 2. Stylized geologic cross section based on well cuttings obtained from well HM1, located in the northeast section of the 3-D survey area. The well penetrated to a depth of 43 m before hitting refusal. No cuttings were retrieved from this depth. Based on previous, unpublished refraction profiles collected by the USGS, however, it is believed that refusal depth corresponds to the top of bedrock.

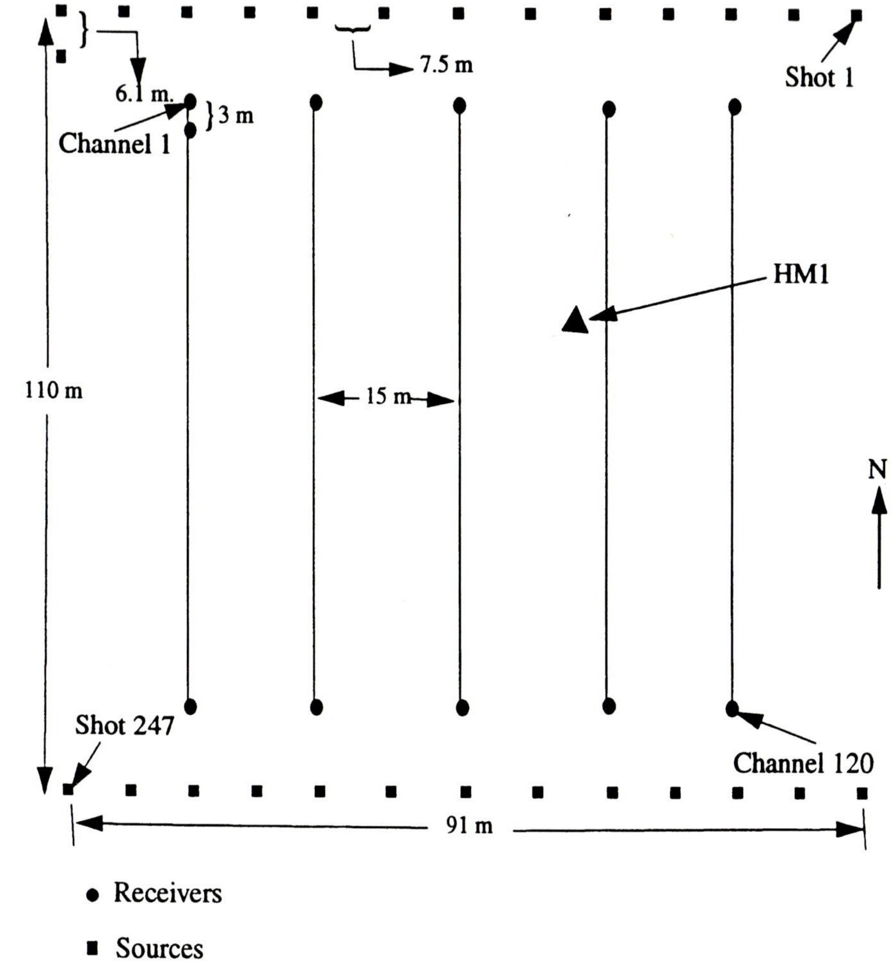


FIGURE 3. 3-D survey design used at Haddam Meadows. The receiver grid consists of five lines of 24 receivers each. Receiver spacing is 3 m. Line spacing is 15 m. The receiver grid remained fixed as shoot locations varied across the survey. Inline shot spacing is 6.1 m. Crossline shot spacing is 7.5 m. This survey design provides for rapid acquisition; however, the bin size, fold distribution, and azimuthal dependence of offsets are limited. HM1 indicates the location of the borehole in the survey area.

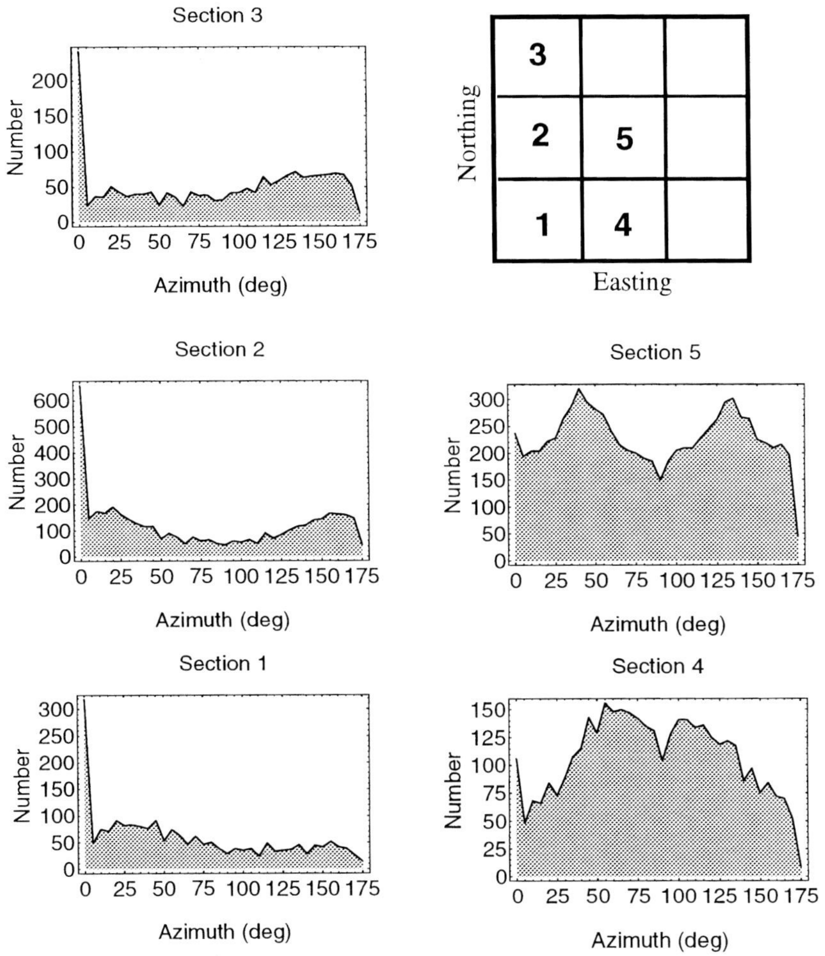


FIGURE 4. Azimuth dependence of reflected ray paths contributing to nine CMP macrobins. Azimuth is defined as the angle from north through east to the straight line connecting each source with each receiver. The CMP macrobins constructed for this figure are defined by group CMP bin points into three macrobins in both the crossline and inline directions. The top left subfigure shows the location of each macrobin with respect to the survey area. The remaining subfigures show the number of raypaths within 5 azimuth increments contributing to each macrobin. Azimuths are plotted between 0° and 180°. Based on source/receiver location reciprocity, azimuths of greater than 180° are mapped between 0° and 180°. The azimuthal distribution of reflected arrivals is shown for only five of the nine macrobins because survey symmetry and source/receiver reciprocity forces the other macrobin locations to be identical to those plotted.

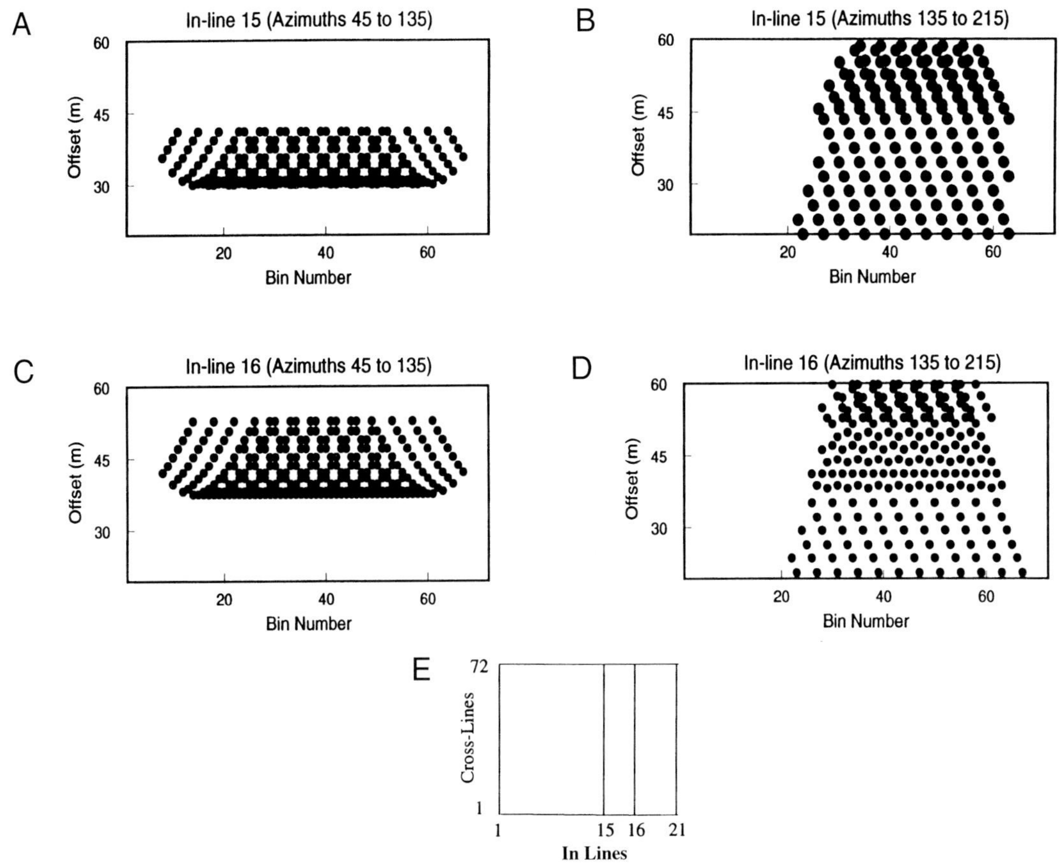


FIGURE 5. Offset versus crossline bin number for inlines 15 and 16 within two azimuth ranges. (A) Inline 15 sorted to azimuths 45–135°. (B) Inline 15 sorted to azimuths 135–215°. (C) Inline 16 sorted to azimuth contributions 45–135°. (D) Inline 16 sorted to azimuth contributions 135–215°. (E) Relative location of inlines 15 and 16 with respect to the survey area. Increasing inline numbers are in the direction of increasing easting value. Increasing crossline numbers are in the direction of increasing northing value.

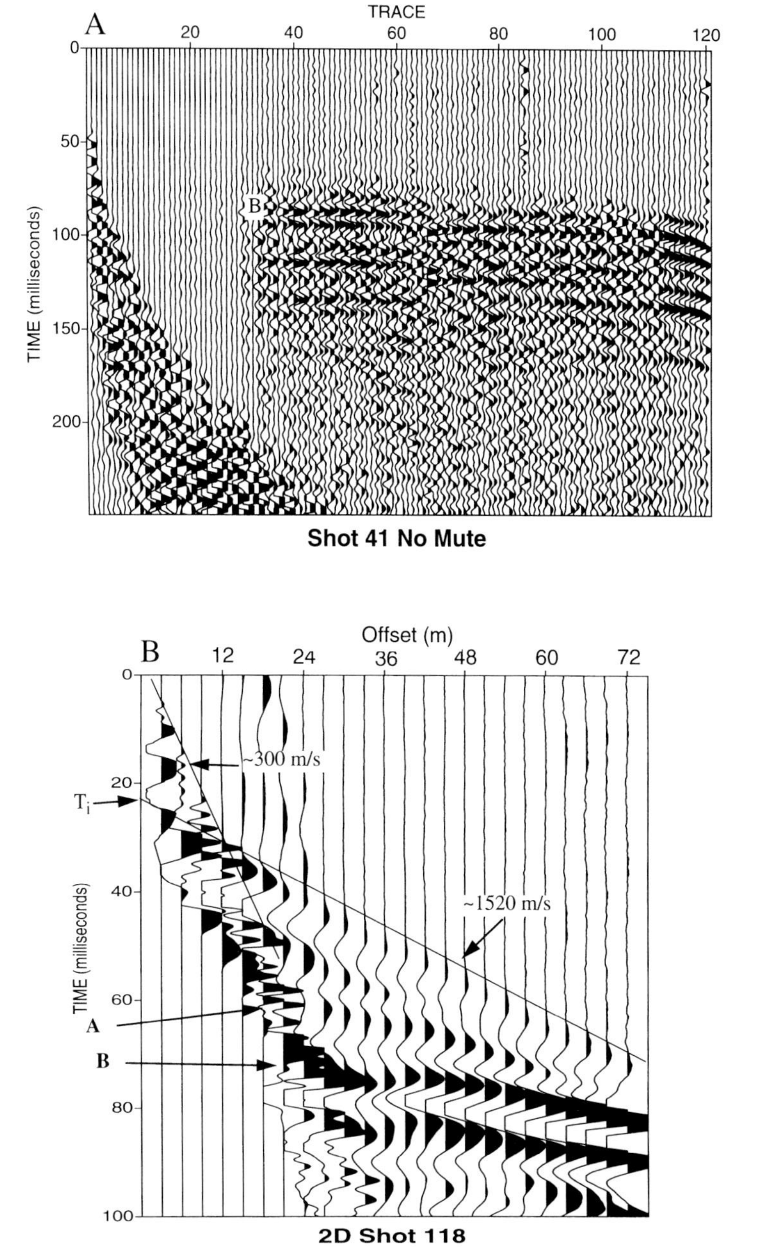


FIGURE 6. Raw (A) and interpreted (B) shot records from the 3-D data volume. The amplitudes associated with trace shown in each record have been equalized. In addition, a bottom mute has been applied to the shot record shown in (B) to remove ground roll. Ground motion recorded at all 120 receiver locations is shown in (A). Because this array is a 2-D geometry, offsets do not increase uniformly from left to right. Ground motion recorded on a 2-D subset of the array is shown in (B). We have interpreted two reflections, labeled "A" and "B," from these records. In addition, the interpretation of the first arrivals is consistent with a near-surface model consisting of unsaturated alluvium overlying saturated alluvium. The estimated depth to the top of the water table, 3 m, is consistent with that observed in borehole HM1.

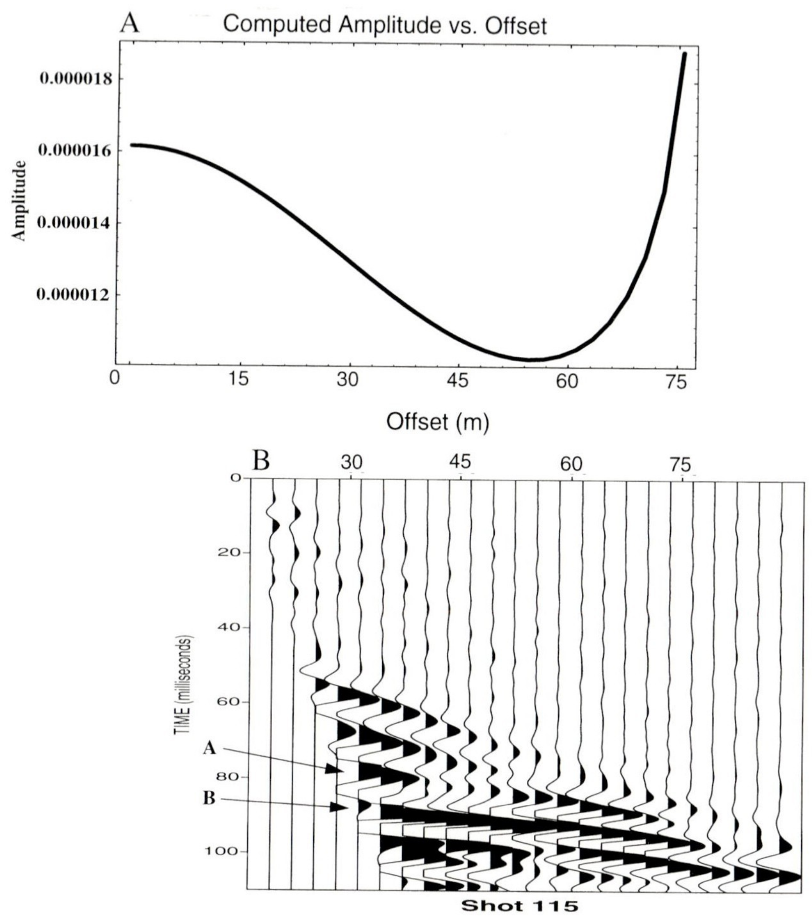


FIGURE 7. Calculated reflection amplitude versus offset (A) compared to a 2-D shot record (B). For this figure, trace amplitude has not been normalized. On the shot record, reflections A and B, identified in Figure 6, are indicated. Amplitude versus offset is computed assuming elastic wave propagation, a reflector depth of 43 m, a P-wave velocity of 1520 m/s above the reflector, and 2250 m/s below, and a Vp/Vs ratio of 2.

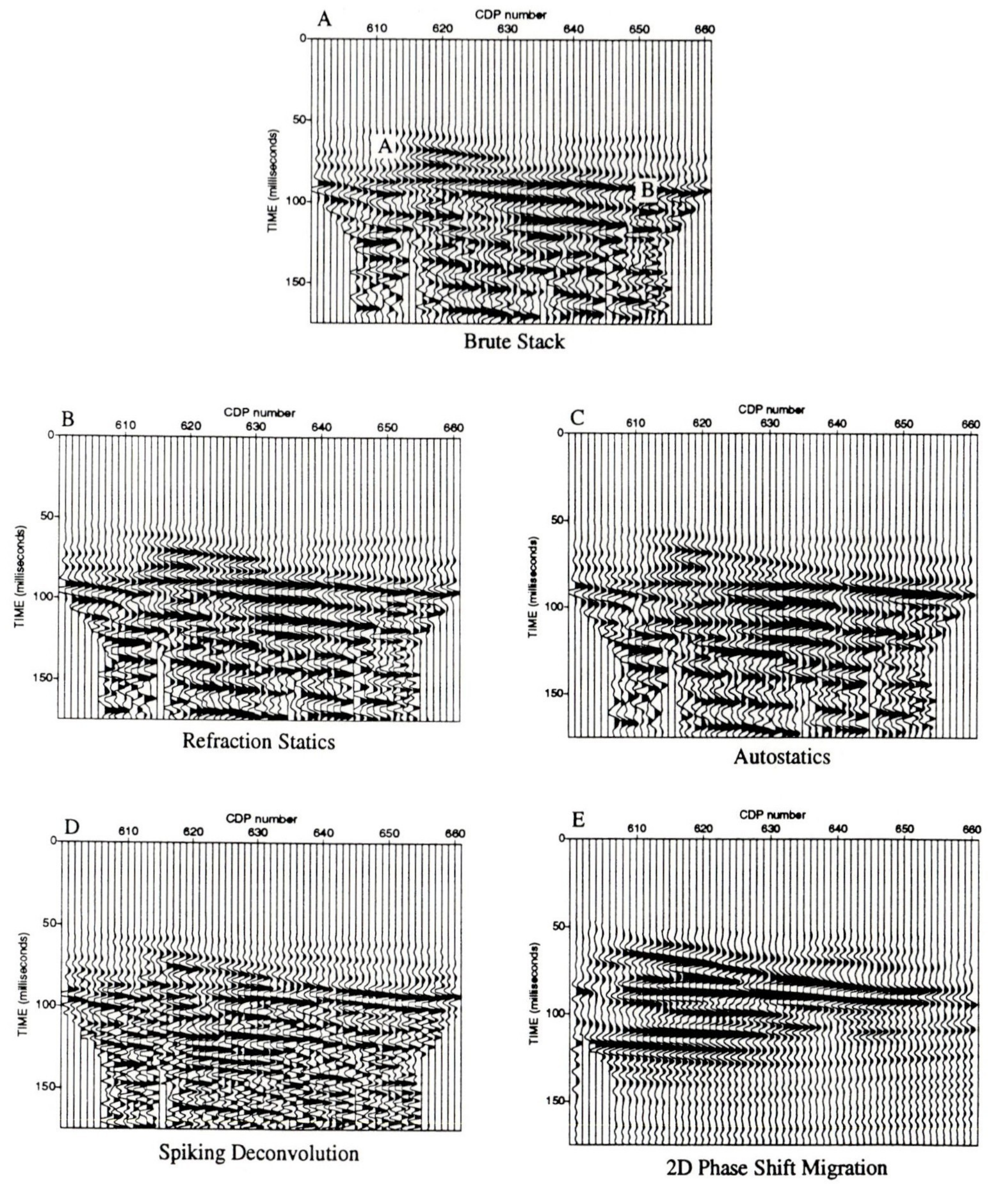


FIGURE 8. 2-D sections (inline 11) extracted from the processed 3-D volume. (A) Brute stack. (B) Stack after application of refraction statics. (C) Stack after application of refraction and residual statics (autostatics). (D) Stack after application of statics (refraction and auto) and deconvolution. (E) Stack after application of statics, deconvolution, and 2-D phase-shift migration. Because, given the limited number of CMP bins in the inline direction, 3-D migration of this dataset does not produce meaningful results, (D) represents a final stacked section.

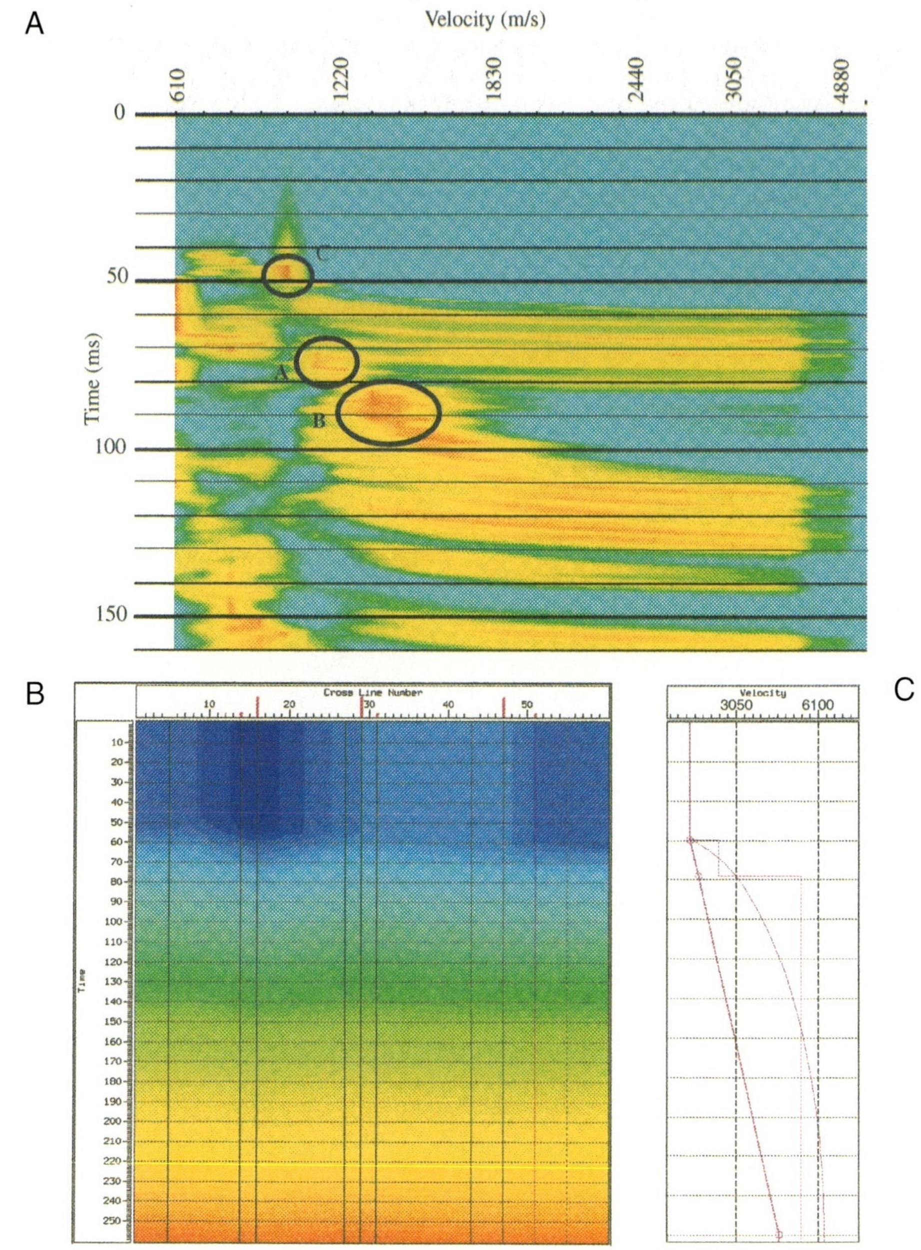


FIGURE 9. Representative semblance plot (A), crossline variation in stacking velocities (B), and velocity-versus-depth profile (C). Semblance and velocity highs are colored red; semblance and velocity lows are colored blue. In computing the semblance, a 35% stretch mute is applied. We interpret three semblance highs labeled "A," "B," and "C." In subfigure (C), stacking, average, and interval velocity variations in time are graphed. Below the semblance high labeled "B," interval velocities are fixed at 5500 m/s.

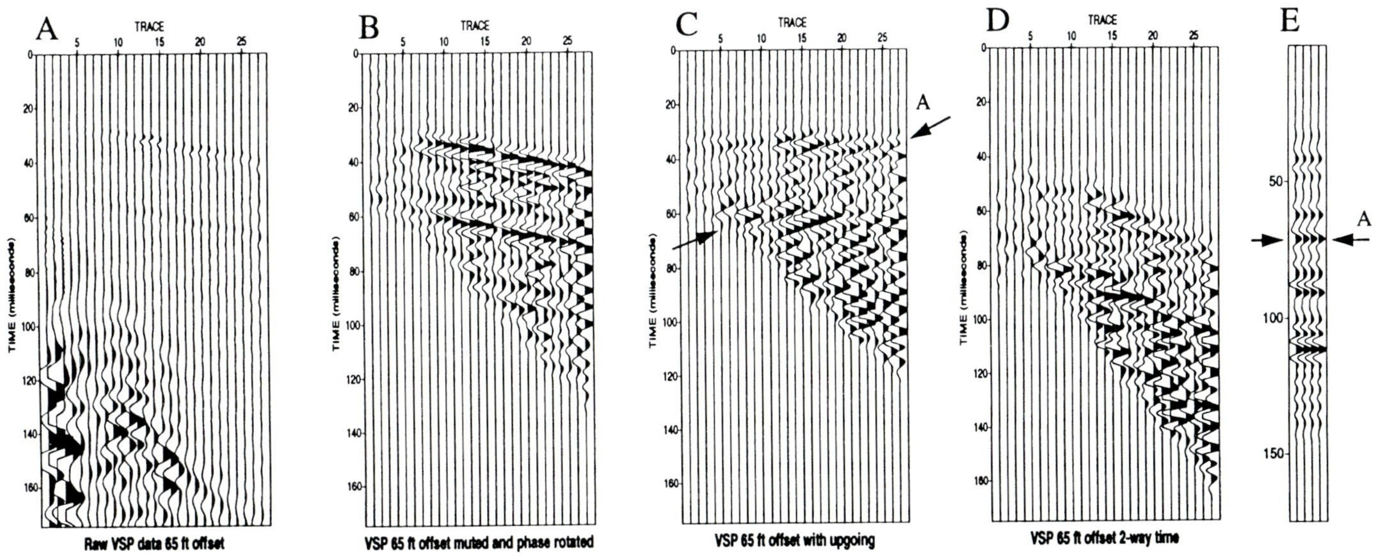


FIGURE 10. Constant-offset, VSP gathers collected from HM1 (A through D) and the vertically extracted trace generated from these (E). Subfigure A is the raw offset VSP, B is the same VSP bottom-muted and phase rotated, C is the extracted upgoing wavefield computed from B, and E is the extracted upgoing wavefield moved out to two-way traveltime.

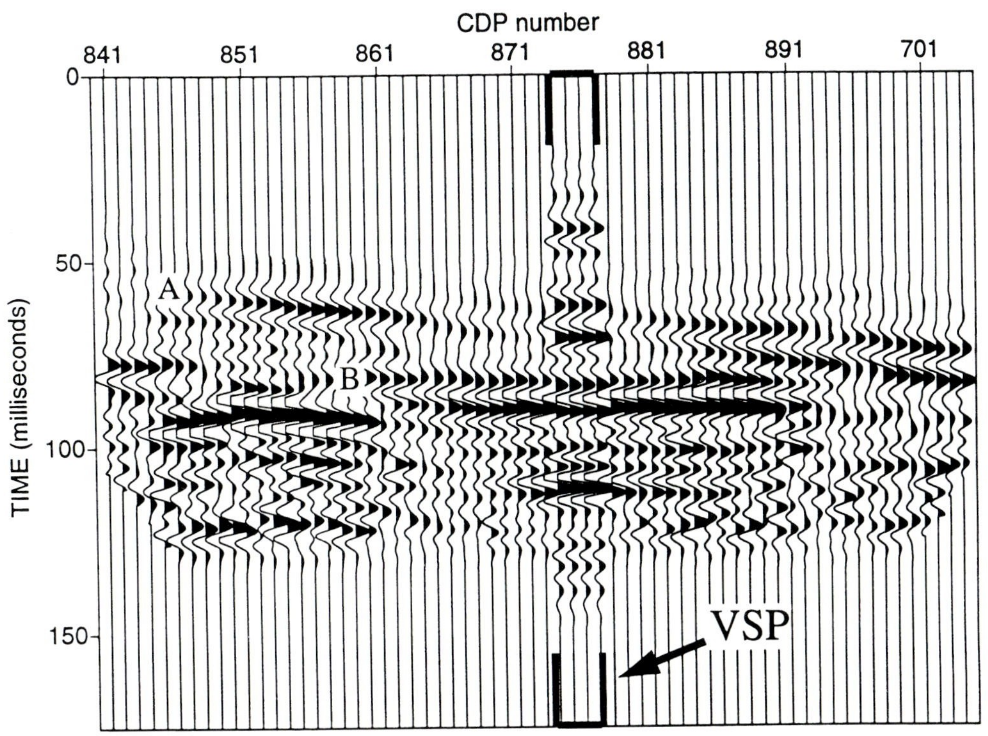


FIGURE 11. VSP tie with sections derived via the 3-D surface seismic experiment. Bracketed traces represent the vertically extracted trace derived in Figure 10 placed within a 2-D seismic section. The 2-D seismic section chosen is that closest to HM1. The vertically extracted trace has been placed in this section midway between HM1 and the VSP source location.

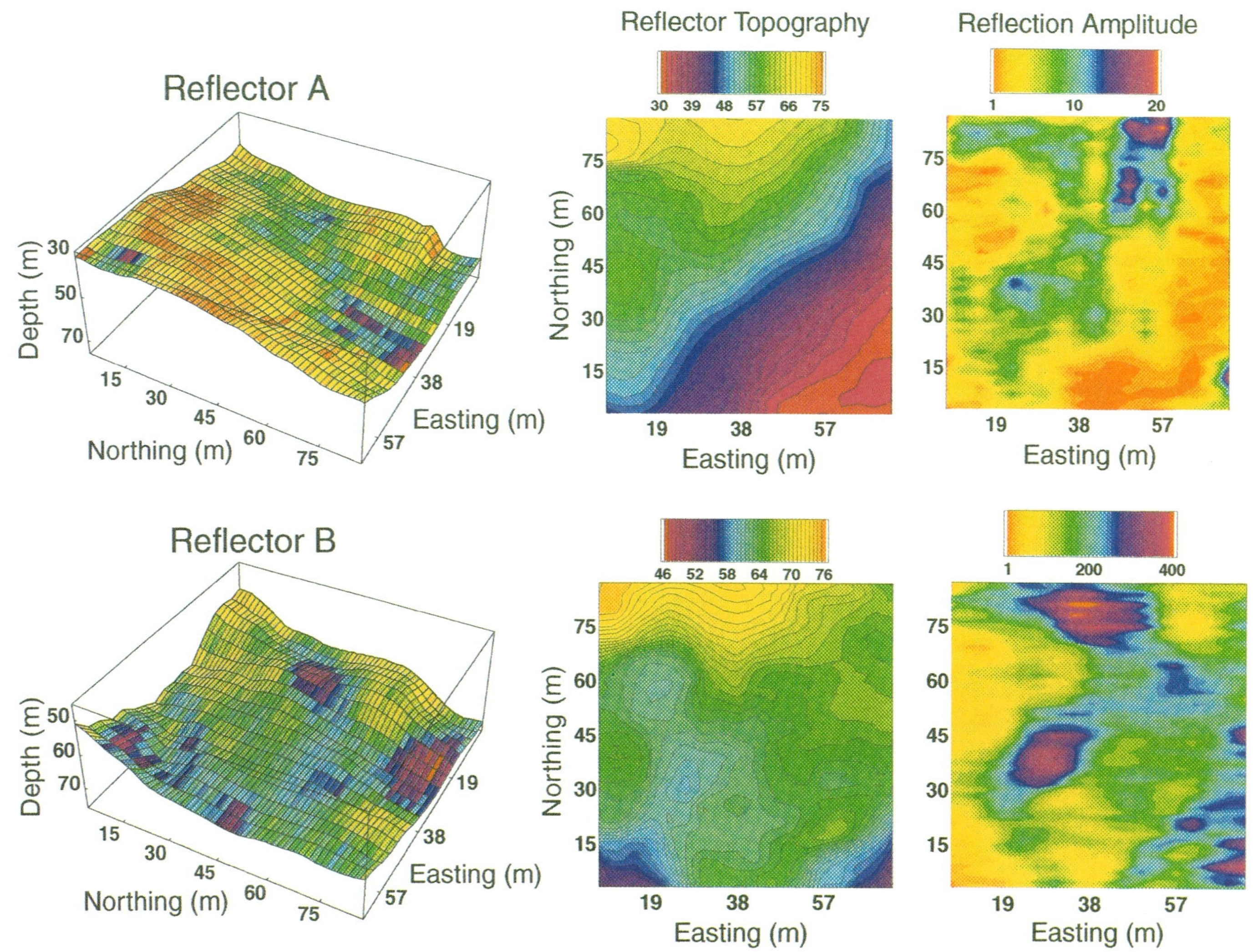


FIGURE 12. Perspective view (A) and contour plot (B) of reflector topography and density plot of reflection amplitude for reflectors A (top) and B (bottom). The coloring shown on the perspective views is based on reflection amplitude. The viewpoint for subfigure A is from the northeast.

# Index

## N



## O



## P



## R



## S

## T



## U



## V



## W



## Y



## Z